Undergraduate Lecture Notes in Physics

Undergraduate Lecture Notes in Physics (ULNP) publishes authoritative texts covering topics throughout pure and applied physics. Each title in the series is suitable as a basis for undergraduate instruction, typically containing practice problems, worked examples, chapter summaries, and suggestions for further reading.

ULNP titles must provide at least one of the following:

- An exceptionally clear and concise treatment of a standard undergraduate subject.
- A solid undergraduate-level introduction to a graduate, advanced, or non-standard subject.
- A novel perspective or an unusual approach to teaching a subject.

ULNP especially encourages new, original, and idiosyncratic approaches to physics teaching at the undergraduate level.

The purpose of ULNP is to provide intriguing, absorbing books that will continue to be the reader's preferred reference throughout their academic career.

Series editors

Neil Ashby
University of Colorado, Boulder, CO, USA

William Brantley
Department of Physics, Furman University, Greenville, SC, USA

Matthew Deady
Physics Program, Bard College, Annandale-on-Hudson, NY, USA

Michael Fowler
Department of Physics, University of Virginia, Charlottesville, VA, USA

Morten Hjorth-Jensen
Department of Physics, University of Oslo, Oslo, Norway

Michael Inglis
SUNY Suffolk County Community College, Long Island, NY, USA

Heinz Klose
Humboldt University, Oldenburg, Niedersachsen, Germany

Helmy Sherif
Department of Physics, University of Alberta, Edmonton, AB, Canada

More information about this series at http://www.springer.com/series/8917

Félix Salazar Bloise · Rafael Medina Ferro
Ana Bayón Rojo · Francisco Gascón Latasa

Solved Problems in Electromagnetics

Félix Salazar Bloise
Departamento de Física Aplicada a los
 Recursos Naturales (FARN)
E.T.S.I. Minas y Energía. Universidad
 Politécnica de Madrid (UPM)
Madrid
Spain

Rafael Medina Ferro
Departamento de Física Aplicada a los
 Recursos Naturales (FARN)
E.T.S.I. Minas y Energía. Universidad
 Politécnica de Madrid (UPM)
Madrid
Spain

Ana Bayón Rojo
Departamento de Física Aplicada a los
 Recursos Naturales (FARN)
E.T.S.I. Minas y Energía. Universidad
 Politécnica de Madrid (UPM)
Madrid
Spain

Francisco Gascón Latasa
Departamento de Física Aplicada II
E.T.S. Arquitectura. Universidad
 de Sevilla (US)
Seville
Spain

ISSN 2192-4791   ISSN 2192-4805 (electronic)
Undergraduate Lecture Notes in Physics
ISBN 978-3-662-48366-4   ISBN 978-3-662-48368-8 (eBook)
DOI 10.1007/978-3-662-48368-8

Library of Congress Control Number: 2016951890

© Springer-Verlag Berlin Heidelberg 2017
This work is subject to copyright. All rights are reserved by the Publisher, whether the whole or part of the material is concerned, specifically the rights of translation, reprinting, reuse of illustrations, recitation, broadcasting, reproduction on microfilms or in any other physical way, and transmission or information storage and retrieval, electronic adaptation, computer software, or by similar or dissimilar methodology now known or hereafter developed.
The use of general descriptive names, registered names, trademarks, service marks, etc. in this publication does not imply, even in the absence of a specific statement, that such names are exempt from the relevant protective laws and regulations and therefore free for general use.
The publisher, the authors and the editors are safe to assume that the advice and information in this book are believed to be true and accurate at the date of publication. Neither the publisher nor the authors or the editors give a warranty, express or implied, with respect to the material contained herein or for any errors or omissions that may have been made.

Printed on acid-free paper

This Springer imprint is published by Springer Nature
The registered company is Springer-Verlag GmbH Germany
The registered company address is: Heidelberger Platz 3, 14197 Berlin, Germany

To our parents

Preface

This work has been developed by the authors after 30 years of teaching several courses of electricity and magnetism. The book contains more than three hundred solved problems, the majority of which have been proposed by the Department of Applied Physics to Natural Resources in official exams of Advance Physics, Physics II and Electromagnetism and Waves in the Mining and Energy School (ETSIME) of the Polytechnic University of Madrid (UPM).

The book has been written for both beginners and advanced students in this subject. However, it may be useful for physicists and engineers, and also for people that work with related topics and need the electromagnetic theory for understanding other disciplines.

The objective of this book is to expose the fundamental concepts of electromagnetism through problems. Starting with this idea each chapter is divided into two parts. The first one contains a brief theoretical introduction where the most important concepts and formulae employed in the chapter are usually presented without demonstrating them. The second one is devoted to the exercises labeled as problems A, B and C, respectively, depending on its difficulty and sometimes thematically. Problems of type A are thought for beginners in electricity and magnetism or for lectures on General Physics, where definitions and concepts about this subject appear for first time. Problems of type B are a bit tougher and can be worked by students who have some basic knowledge in calculus and electromagnetics. For closing each chapter, problems of type C are introduced. These kinds of exercises, even though they are not very difficult, have some conceptual or/and mathematical complications which make them more adequate for advanced lectures. According with the academic level of the student, this chapter distribution gives to the reader the possibility of using the book in a flexible way.

Our experience showed us that the most simplest things may be very difficult for the student at the beginning of learning a subject, if the explanations of the ideas involved are not clear. Sometimes, the supposition by part of the writer that one idea or concept is obvious may lead to waste reader's time. For this reason we have tried to explain the problems in-depth with an emphasis on physical concepts rather than on the mathematical developments.

The book is structured in 14 chapters. It begins with an introductory chapter devoted to the basic mathematical theorems and formulae that are needed for further developments. The next two chapters deal with the electric field in different situations, namely in vacuum and when matter is present. An important topic when studying fields and circuits are the currents; this is the subject of Chap. 4. In the same way as commented for the electric field, Chaps. 5 and 6 study the origin of the magnetic field and the phenomenon of the magnetization. Until this part of the book the techniques for solving the electric and magnetic fields generated under specific circumstances are based on direct calculations. In Chap. 7 other more complicated methods for obtaining these fields are studied. Chapter 8 works out the important topic of the electromagnetic induction. The different causes of producing electromotive force are explained in detail. The understanding of this phenomenon encompasses the knowledge of former chapters. For this reason it is not recommended to be studied without studying previously the fundamentals of the electric and magnetic fields. Chapter 9 refers to energetic aspects of the electromagnetic field and a didactic investigation of the Maxwell equations is left to Chap. 10. The solution of partial differential equations may be very difficult. However, the viewpoint adopted in this chapter is more conceptual than mathematical. In fact, for systems of high symmetry it is possible to find a solution in a simpler way without solving the system of differential equations. In our opinion, numerous questions can be answered using a simple mathematical apparatus, without losing rigor and clarity. Due to the importance of the plasmas, cosmic rays, and machines as cyclotron and betatron, among others, we have included the study of the movement of charged particles in electromagnetic field. This is the subject of Chap. 11. One of the most important consequences of the Maxwell equations is the unification of the electricity and the magnetism and also the light. In this regard in Chap. 12 a general view of the electromagnetic waves is given and in Chap. 13, the phenomena of reflexion and refraction are treated. In this context, an interesting approach to the propagation of electromagnetic waves throughout anisotropic media is dealt with in the last part, Chap. 14.

We would like to express our sincere gratitude, wholeheartedly, to our colleague and friend Faustino Fernández López, for many years of his teaching, experience and for help wherever we needed.

We are also in indebted to Prof. Aristide Dogariu and his scientific team at CREOL of the University of Central Florida. Their valuable discussions in the group meetings provided us very important information which allowed to introduce some ideas and explanations in some parts of this book.

A very special place in this book is for our colleague Dr. James Germann, who reviewed all this work, word by word, and also corrected all equations carefully. We would like to thank him for his invaluable work, dedication, and professionalism. Without his help this book would not have seen the light. We wish him all the best for his future and expect the best in his work with two-photon microscopy and corneal imaging. We hope his work in this book can be appreciated in his scientific career.

Thanks are also due to the Editorial Springer, especially Dr. Claus Ascheron for providing help in the preparation of this book. His generosity and patience with us during the writing is largely appreciated.

Finally, we would like to give special thanks to all our students, without whom we would have never written this book.

Madrid, Spain	Félix Salazar Bloise
Madrid, Spain	Rafael Medina Ferro
Madrid, Spain	Ana Bayón Rojo
Seville, Spain	Francisco Gascón Latasa

Contents

1 A Mathematical Introduction 1
  1.1   Coordinate Systems and Transformations 1
  1.2   Differential Length, Area and Volume 4
  1.3   Scalar and Vector Fields 5
  1.4   Concept and Definition of Regions, Curves and Surfaces.. 6
  1.5   Line, Surface and Volume Integrals. Circulation and Flux 9
  1.6   Gradient 11
  1.7   Curl 13
  1.8   Divergence 14
  1.9   Stokes's Theorem and Divergence Theorem 15
  1.10 Normal Vector to a Surface 16
      1.10.1 Vectorial or Parametric 16
      1.10.2 Explicit 17
      1.10.3 Implicit 17
  1.11 Further Developments 19
  1.12 Classification of Vector Fields 21
  1.13 Obtaining the Scalar Potential 23
      1.13.1 First Method 23
      1.13.2 Second Method 24
      1.13.3 Third Method* 25
  1.14 Vectorial Field from the Vector Potential 26
      1.14.1 First Method 26
      1.14.2 Second Method 28
  Solved Problems 29
  Problems A 29
  Problems B 51

2 Static Electric Field in Vacuum 67
  2.1   Electric Charge 67
  2.2   Coulomb's Law 68
  2.3   Electric Field 70

	2.4	Electrostatic Potential	71
	2.5	Flux of Electric Field. Gauss' Law	73
	2.6	Electrostatic Equations	74
	2.7	Electric Dipole	74
	2.8	Conductors and Insulators	75
	2.9	Biot–Savart-like Law in Electrostatics	76
		Solved Problems	77
		Problems A	77
		Problems B	81
		Problem C	106
3	**Static Electric Field in Dielectrics**		121
	3.1	Polarization	121
	3.2	Polarization Charges	123
	3.3	The D Field	123
	3.4	The Constitutive Equation	125
	3.5	Boundary Conditions	127
	3.6	Coefficients of Potential and Capacitance	128
	3.7	Capacitors	129
		Solved Problems	132
		Problems A	132
		Problems B	139
		Problems C	156
4	**Electric Current**		165
	4.1	Current Density. The Current	165
	4.2	The Equation of Continuity	167
	4.3	Direct Current	167
	4.4	Ohm's Law	168
	4.5	Ohm's Law in a Conducting Straight Wire. Resistance	169
	4.6	Power Supplied by Electric Field. Joule's Law	171
	4.7	Direct Current Generators	172
	4.8	Direct Current Motors	174
	4.9	Ohm's Law in Circuits	175
	4.10	Direct Current Networks	177
		4.10.1 Kirchhoff's Circuit Laws	177
		4.10.2 Mesh Analysis	179
	4.11	Passive Network Equivalence	179
		4.11.1 Resistances in Series, Parallel, Triangle and Star Associations	180
		4.11.2 Resistance in a Conductor with Any Shape	182
	4.12	Thévenin's and Norton's Theorems	184
		Solved Problems	185
		Problems A	185
		Problems B	194
		Problems C	206

5 Magnetostatics ... 223
- 5.1 Differential Equation of the Magnetostatic Field ... 223
- 5.2 Integral Form of the Equations ... 225
- 5.3 Vector Potential ... 227
- 5.4 The Biot–Savart Law ... 229
- 5.5 Forces on Currents ... 230
- 5.6 Magnetic Dipole ... 231
- 5.7 Off-Axis Magnetic Field for Axisymmetric Systems ... 231
- Solved Problems ... 234
- Problems A ... 234
- Problems B ... 266
- Problems C ... 291

6 Static Magnetic Field in Presence of Matter ... 313
- 6.1 Magnetization ... 313
- 6.2 Magnetic Current Densities ... 314
- 6.3 The Magnetic Field **H** ... 315
- 6.4 The Ampère Law of the Magnetic Field **H** ... 317
- 6.5 Basic Kinds of Magnetic Materials ... 318
- 6.6 Description of the Magnetization Curve ... 319
- 6.7 Magnetic Circuits and Electromagnets ... 323
- 6.8 Operating Straight Line and Operating Point ... 326
- 6.9 The Permanent Magnet ... 328
- 6.10 The Demagnetizing Field ... 330
- Solved Problems ... 332
- Problems A ... 332
- Problems B ... 350
- Problems C ... 376

7 Methods for Solving Electrostatic and Magnetostatic Problems ... 419
- 7.1 The Laplace Equation ... 419
- 7.2 The Method of Separation of Variables ... 422
- 7.3 Green's Function Method ... 425
- 7.4 Method of Images ... 428
- 7.5 Application of Complex Analysis to Electromagnetism ... 431
  - 7.5.1 Transforming Boundary Conditions ... 432
  - 7.5.2 Conformal Mapping ... 433
  - 7.5.3 Some Conformal Transformations ... 434
  - 7.5.4 Complex Potential ... 436
- 7.6 Numerical Techniques ... 441
  - 7.6.1 The Finite Difference Method ... 441
  - 7.6.2 Other Important Techniques ... 445
- Solved Problems ... 448
- Problems C ... 448

8 Electromagnetic Induction ... 511
- 8.1 Electromotive Force ... 511
- 8.2 Faraday's law ... 512
- 8.3 Motional Electromotive Force ... 515
- 8.4 The General Law of Electromagnetic Induction ... 518
- 8.5 Self-inductance and Mutual Inductance ... 519
  - 8.5.1 Self-inductance ... 519
  - 8.5.2 Mutual Inductance ... 520
- 8.6 Voltage Between Two Points ... 523
- Solved Problems ... 524
- Problems A ... 524
- Problems B ... 533
- Problems C ... 550

9 Energy of the Electromagnetic Field ... 567
- 9.1 The Electrostatic Energy of Charges ... 567
- 9.2 The Energy of a Capacitor ... 568
- 9.3 The Electrostatic Energy of Distributed Charges ... 569
- 9.4 Relationship Between Force and Electrostatic Energy ... 569
- 9.5 Magnetostatic Energy of Quasi-stationary Currents ... 570
- 9.6 Generalization ... 571
- 9.7 Magnetic Energy in a Hysteresis Loop ... 572
- Solved Problems ... 573
- Problems A ... 573
- Problems B ... 578
- Problems C ... 584

10 Maxwell's Equations ... 599
- 10.1 Generalization of Ampère's Law ... 599
- 10.2 Maxwell's Equations for a Point ... 600
- 10.3 Maxwell's Equations for a Domain ... 600
- 10.4 Scalar Potential ... 601
- 10.5 Surface of Discontinuity ... 602
- Solved Problems ... 604
- Problems A ... 604
- Problems B ... 614
- Problems C ... 618

11 Motion of Charged Particles in Electromagnetic Fields ... 627
- 11.1 Lorentz Force ... 627
- 11.2 Trajectory of a Charge in a Homogeneous Electric Field ... 628
- 11.3 Trajectory of a Charge in a Homogeneous Magnetic Field ... 629
- 11.4 Hall Effect ... 631

	11.5	Trajectory of a Charge in Simultaneous, Homogeneous and Constant, Magnetic and Electric Fields	632
	11.6	The Mass Spectrometer	633
	11.7	The Cyclotron	634
	11.8	The Betatron	635
	11.9	Relativistic Correction	636
	11.10	A Relativistic Particle in an Electromagnetic Field	637
	11.11	Charge in a Homogeneous Electric Field	637
	11.12	Charge in a Homogeneous Magnetic Field	640
	Solved Problems		641
	Problems A		641
	Problems B		652
	Problems C		659
12	**Electromagnetic Waves**		**667**
	12.1	Electromagnetic Wave Propagation: Wave Equation	667
	12.2	Plane and Spherical Waves	668
	12.3	Harmonic Plane Waves in Unbounded Dielectrics	670
	12.4	Polarization	672
	12.5	Intensity and Poynting Vector	674
	12.6	Introduction to Fourier Analysis	676
	Solved Problems		677
	Problems A		677
	Problems B		685
	Problems C		699
13	**Reflection and Refraction**		**715**
	13.1	Laws of Reflection and Refraction	715
	13.2	The Fresnel Coefficients	716
	13.3	Reflected and Transmitted Energy	718
	Solved Problems		719
	Problems A		719
	Problems B		730
	Problems C		740
14	**Wave Propagation in Anisotropic Media**		**749**
	14.1	Concept of Anisotropy	749
	14.2	Susceptibility and Permittivity Tensors Definition	751
	14.3	Maxwell's Equations in an Anisotropic Linear Medium Free of Charges and Currents	753
	14.4	Electromagnetic Waves in Uniaxial Dielectrics	755
	14.5	Propagation of the Energy	759
	14.6	Geometrical Interpretation	761
	14.7	Electromagnetic Waves in Biaxial Crystals	762

	14.8	Crystal Classification	764
	14.9	Retarders	766
		Solved Problems	767
		Problems A	767
		Problems B	771
		Problems C	778

Appendix A: Matlab Programs 787

Appendix B: Electric and Magnetic Properties of Several Materials ... 791

Bibliography 795

Index 799

About the Authors

Prof. Dr. Félix Salazar Bloise studied Physics at the Complutense University of Madrid (1987) and received his Ph.D. degree from the Polytechnic University of Madrid (UPM) in 1992. He was awarded the Ph.D. Prize in the academic year of 1993–94, and in 2004 he won the best textbook of the year award from the General Foundation of the University (FGU). In 1995 he was appointed Associate Professor of Applied Physics at the School of Mines of Madrid (ETSIM). During the periods of 1994–95 and 1997 he was a visiting assistant researcher at the Universität des Saarlandes. Since 2006, he currently is a visiting Professor at Lehrstuhl für Messsystem-und Sensortechnik (Technische Universität München-TUM) and in 2012–16 he has also been a visiting research scientist at Center for Research and Education in Optics and Lasers (CREOL) of the University of Central Florida. He teaches at UPM General Physics, Electrodynamics, and Solid State Physics, and lectures one international course at TUM. His scientific research is focused on the analysis of physical properties of systems by using random electromagnetic fields (speckle) and ultrasound techniques. He is co-author of scientific and technical papers and of several books, and holds some patents. He is a Fellow of the Deutsche Gesselschaft für angewandte Optik (DGaO), Arbeitskreis der Hochschullehrer für Messtechnik (AHMT), and the Optical Society of America (OSA).

Prof. Dr. Rafael Medina Ferro born in Madrid (Spain) in 1961, has been teaching in Higher Technical School of Mining and Energy Engineering, in the Polytechnic University in Madrid since 1986. He was graduated in 1985 and got his Ph.D. in 1989 in Mining Engineering from the Polytechnic University in Madrid. His teaching includes topics in mathematics, physics and mechanical vibrations. He is author and co-author of five books for university students. One of these books won the award of the best textbook from the Polytechnic University in Madrid in 2004. His research has been developed mainly in mechanical vibrations and wave propagation in materials, although he also worked in a European Project about electrostatic effects in powder mixtures. His later research involved impact-echo studies and blasting effects in buildings.

Prof. Dr. Ana Bayón Rojo received her Master of Mining Engineering in 1985 from the Polytechnic University of Madrid (UPM). After working in industry she returned to the University (UPM) where she obtained her Ph.D. in 1992. Her research was recognized by the Ph.D. Prize in the academic year 1992–93. In 1992–93 she was a visiting Assistant Professor at the Institut Français du Pétrol (IFP) in Paris. In 1995 she was appointed Associate Professor of Applied Physics at the School of Mines of Madrid (ETSIM). Her teaching field includes General Physics, Electrodynamics, and Mechanical vibrations, in addition to doctoral courses about non-destructive measurement techniques. In 2004 she was awarded the best textbook of the year for university students by the General Foundation of the University (FGU). Her scientific research is focused on vibration analysis by using optical and acoustical methods. She is co-author of scientific articles, book chapters, and of three books. She is a Fellow of the Acoustical Society of America (JASA).

Prof. Dr. Francisco Gascón Latasa received his Master of Science at the University of Zaragoza in 1958. In the following four years he studied nuclear fusion, obtaining his Ph.D. in 1962. In this year he received a grant from the French Government (ASTEF) for researching nuclear fusion at the C. F. E. Fontenay-aux-Roses (France). In 1963 he was a visiting assistant researcher at the L'École Nationale Supérieure d'Électrotechnique (ENSEEIHT). In 1964, after pausing briefly at the University of Barcelona, Dr. Latasa was elected Associate Professor at the ETUI Industrial of Vitoria. In 1966 he was visiting Professor at the Clarendon Laboratory (Oxford) for research in superconductivity. In 1967 he began work at the ETS Arquitectura in Sevilla until 1978, when he got a full professorship in Applied Physics at the ETSI Minas of Madrid. In 1995 he returned to the University of Sevilla and in 2009 he assumed Emeritus status. He has lectured in many different fields such as Theoretical Physics, Mechanics, Thermodynamics, General Physics, Electrodynamics, and Physics of Continuum Media. He is author of the book *Fundamental of Thermotechnics* (1976) and in 2004 the book *Problems of Electricity and Magnetism* was awarded the best work of the year for students by the General Foundation of the Polytechnic University of Madrid (FGU). His main research fields are holographic and speckle interferometry and the vibration analysis, where he has published many papers and book chapters. He is a Fellow of the Acoustical Society of America (JASA).

Chapter 1
A Mathematical Introduction

Abstract This chapter does not really deal with electromagnetism, but it is needed as the mathematical foundation for the vector treatment of this subject. It is the purpose of this chapter to give a brief exposition of basic mathematical elements to provide an introduction to the field theory which is required for a treatment of electromagnetism.

1.1 Coordinate Systems and Transformations

A point or vector can be represented in any curvilinear coordinate system. A coordinate system is *orthogonal* if the coordinates are mutually perpendicular. The three best-known orthogonal coordinate systems are the Cartesian, the cylindrical and the spherical.

A Cartesian (or rectangular) coordinate system specifies the position of any point P in three-dimensional space by three Cartesian coordinates (x, y, z), its signed distances from three mutually perpendicular planes. A vector **OP** in Cartesian coordinates can be written as (P_x, P_y, P_z) or $P_x\mathbf{u}_x + P_y\mathbf{u}_y + P_z\mathbf{u}_z$, with $\mathbf{u}_x, \mathbf{u}_y, \mathbf{u}_z$ defined as the unit vectors along the x-, y- and z- directions (Fig. 1.1).

In a cylindrical coordinate system, the three coordinates (ρ, ϕ, z) of a point P are defined as (Fig. 1.2):

- The radial distance ρ is the Euclidean distance from the Z-axis to the point P, that is, the radius of the cylinder passing through P.
- The azimuthal angle ϕ is the angle between the X-axis and the line from the origin to the projection of P on the XY-plane.
- The height z is the Euclidean distance from the XY-plane to the point P.

A vector **OP** in cylindrical coordinates can be written as (P_ρ, P_ϕ, P_z) or $P_\rho\mathbf{u}_\rho + P_\phi\mathbf{u}_\phi + P_z\mathbf{u}_z$, with $\mathbf{u}_\rho, \mathbf{u}_\phi, \mathbf{u}_z$ defined as the unit vectors along the ρ-, ϕ- and z- directions.

The relationship between the variables (x, y, z) of the Cartesian coordinate system and those of the cylindrical system (ρ, ϕ, z) are:

$$x = \rho\cos\phi, \quad y = \rho\sin\phi, \quad z = z, \tag{1.1}$$

Fig. 1.1 Point P and unit vectors in the Cartesian coordinate system

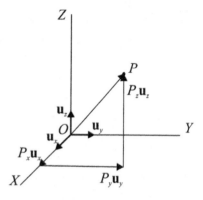

Fig. 1.2 Point P and unit vectors in the cylindrical coordinate system

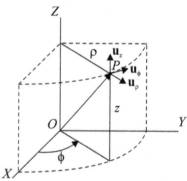

or

$$\rho = \sqrt{x^2 + y^2}, \quad \phi = \tan^{-1}\frac{y}{x}, \quad z = z. \tag{1.2}$$

The relationship between $(\mathbf{u}_\rho, \mathbf{u}_\phi, \mathbf{u}_z)$ and $(\mathbf{u}_x, \mathbf{u}_y, \mathbf{u}_z)$ are

$$\begin{aligned} \mathbf{u}_\rho &= \cos\phi\mathbf{u}_x + \sin\phi\mathbf{u}_y \\ \mathbf{u}_\phi &= -\sin\phi\mathbf{u}_x + \cos\phi\mathbf{u}_y \\ \mathbf{u}_z &= \mathbf{u}_z \end{aligned} \tag{1.3}$$

or

$$\begin{aligned} \mathbf{u}_x &= \cos\phi\mathbf{u}_\rho - \sin\phi\mathbf{u}_\phi \\ \mathbf{u}_y &= \sin\phi\mathbf{u}_\rho + \cos\phi\mathbf{u}_\phi \\ \mathbf{u}_z &= \mathbf{u}_z \end{aligned} \tag{1.4}$$

In a spherical coordinate system, a point P is defined by three coordinates (r, θ, ϕ) (Fig. 1.3):

1.1 Coordinate Systems and Transformations

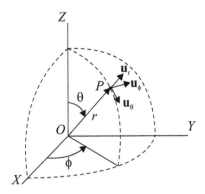

Fig. 1.3 Point P and unit vectors in the spherical coordinate system

- The radius or radial distance r is the Euclidean distance from the origin O to P, that is, the radius of a sphere centered at the origin and passing through P.
- The inclination (or polar angle or colatitude) θ is the angle between the Z-axis and the position vector of P.
- The azimuth (or azimuthal angle) ϕ is the angle between the X-axis and the line from the origin to the projection of P on the XY-plane (the same one in cylindrical coordinates).

A vector **OP** in spherical coordinates can be written as (P_r, P_θ, P_ϕ) or $P_r \mathbf{u}_r + P_\theta \mathbf{u}_\theta + P_\phi \mathbf{u}_\phi$, with $\mathbf{u}_r, \mathbf{u}_\theta, \mathbf{u}_\phi$ defined as the unit vectors along the r-, θ- and ϕ- directions.

The relationship between the variables (x, y, z) of the Cartesian coordinate system and those of the spherical system (r, θ, ϕ) are:

$$x = r \sin\theta \cos\phi, \quad y = \rho \sin\theta \sin\phi, \quad z = r \cos\theta, \tag{1.5}$$

or

$$r = \sqrt{x^2 + y^2 + z^2}, \quad \theta = \tan^{-1} \frac{\sqrt{x^2 + y^2}}{z}, \quad \phi = \tan^{-1} \frac{y}{x}. \tag{1.6}$$

The relationship between $(\mathbf{u}_r, \mathbf{u}_\theta, \mathbf{u}_\phi)$ and $(\mathbf{u}_x, \mathbf{u}_y, \mathbf{u}_z)$ are

$$\begin{aligned}
\mathbf{u}_r &= \sin\theta \cos\phi \mathbf{u}_x + \sin\theta \sin\phi \mathbf{u}_y + \cos\theta \mathbf{u}_z \\
\mathbf{u}_\theta &= \cos\theta \cos\phi \mathbf{u}_x + \cos\theta \sin\phi \mathbf{u}_y - \sin\theta \mathbf{u}_z \\
\mathbf{u}_\phi &= -\sin\phi \mathbf{u}_x + \cos\phi \mathbf{u}_y
\end{aligned} \tag{1.7}$$

or

$$\begin{aligned}
\mathbf{u}_x &= \sin\theta \cos\phi \mathbf{u}_r + \cos\theta \cos\phi \mathbf{u}_\theta - \sin\phi \mathbf{u}_\phi \\
\mathbf{u}_y &= \sin\theta \sin\phi \mathbf{u}_r + \cos\theta \sin\phi \mathbf{u}_\theta + \cos\phi \mathbf{u}_\phi \\
\mathbf{u}_z &= \cos\theta \mathbf{u}_r - \sin\theta \mathbf{u}_\theta
\end{aligned} \tag{1.8}$$

1.2 Differential Length, Area and Volume

Differential elements in length, area and volume are used in integration to solve problems involving paths, surfaces and volumes. The line element or differential displacement $d\mathbf{l}$ (or $d\mathbf{r}$) is given, in Cartesian coordinates, by

$$d\mathbf{l} = dx\mathbf{u}_x + dy\mathbf{u}_y + dz\mathbf{u}_z, \tag{1.9}$$

in cylindrical coordinates by

$$d\mathbf{l} = d\rho\mathbf{u}_\rho + \rho d\phi\mathbf{u}_\phi + dz\mathbf{u}_z, \tag{1.10}$$

and in spherical coordinates by

$$d\mathbf{l} = dr\mathbf{u}_r + rd\theta\mathbf{u}_\theta + r\sin\theta d\phi\mathbf{u}_\phi. \tag{1.11}$$

The surface element or differential normal area is given, in Cartesian coordinates, by

$$d\mathbf{S} = dydz\mathbf{u}_x + dxdz\mathbf{u}_y + dxdy\mathbf{u}_z, \tag{1.12}$$

in cylindrical coordinates by

$$d\mathbf{S} = \rho d\phi dz\mathbf{u}_\rho + d\rho dz\mathbf{u}_\phi + \rho d\phi d\rho\mathbf{u}_z, \tag{1.13}$$

and in spherical coordinates by

$$d\mathbf{S} = r^2 \sin\theta d\theta d\phi\mathbf{u}_r + r\sin\theta drd\phi\mathbf{u}_\theta + rdrd\theta\mathbf{u}_\phi. \tag{1.14}$$

The volume element or differential volume is given, in Cartesian coordinates, by

$$dV = dxdydz, \tag{1.15}$$

in cylindrical coordinates by

$$dV = \rho d\rho d\phi dz, \tag{1.16}$$

and in spherical coordinates by

$$dV = r^2 \sin\theta drd\theta d\phi. \tag{1.17}$$

1.3 Scalar and Vector Fields

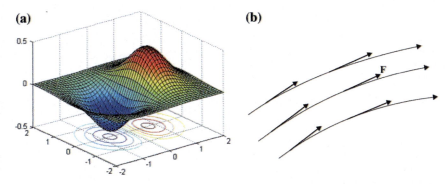

Fig. 1.4 a Scalar field $\varphi = x \cdot e^{-x^2-y^2}$ and its isolines. **b** Vector lines of a vector field

1.3 Scalar and Vector Fields

A scalar is a quantity which is completely characterized by its magnitude. A vector is a quantity which is completely characterized by its magnitude and direction.[1] A **field** is a function that specifies a particular quantity everywhere in a region. Therefore, a **scalar field** associates a scalar value to every point in a space and a **vector field** associates a vector to every point in a space.

In $\Re^n$, a scalar field is a function $\varphi : A \subset \Re^n \to \Re$. If a point **r** is considered, a scalar field can be defined by a function $\varphi = \varphi(\mathbf{r})$. If a Cartesian coordinate system is used, $\mathbf{r} = \mathbf{r}(x, y, z)$ and $\varphi = \varphi(x, y, z)$.

A scalar field φ can be represented by its *level surfaces* or *equipotential surfaces*. Points at which the function φ has the same value, [$\varphi(x, y, z) = c$ with $c \in \Re$ a constant] are said to define a level surface of the function. Such surfaces are usually named by the use of the prefixes *iso* or *equi*. If φ is a function of two variables, $\varphi(x, y) = c$ is a curve along which the function has a constant value and is called a *contour line* or *isoline*. Figure 1.4a is an example of a scalar field and its isolines.

In $\Re^n$, a vector field is a function $\mathbf{F} : A \subset \Re^n \to \Re^n$. A vector field **F** can be represented by its *vector lines*. If we start at a given point of a vector field and consider the vector of the field at that point to be the tangent to a curve passing through the point, the field will determine a set of curves which will at every point have the vector of the field as tangent (Fig. 1.4b). So if **F** is a vector field, a vector line is a curve $\Gamma(t)$ defined by

$$\Gamma'(t) = \mathbf{F}(\Gamma(t)), \tag{1.18}$$

where t is a parameter. If rectangular coordinates are used, $\mathbf{F} = \mathbf{F}(\mathbf{r}) = F_x(x, y, z)\mathbf{u}_x + F_y(x, y, z)\mathbf{u}_y + F_z(x, y, z)\mathbf{u}_z$, and vector lines are determined by

---

[1] Direction and sense.

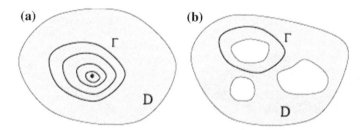

Fig. 1.5 Regions. **a** Simply connected. By deforming the curve Γ we reach a point without leaving the region. **b** Multiply connected. If we deform Γ we cannot obtain a point

$$\frac{dx}{F_x} = \frac{dy}{F_y} = \frac{dz}{F_z}. \quad (1.19)$$

Vector lines are oriented with the same direction of vector field.

1.4 Concept and Definition of Regions, Curves and Surfaces

In this section we try to show some basic concepts that are important for understanding calculation we will develop throughout the next chapters. As we will see, the majority of the integrals used in electromagnetic theory extend over surfaces and curves, and it useful to comment on them to some extent.

A set D is said to be simply connected if any closed curve $\Gamma \subset D$ may be shrunk to a point inside this region. Roughly speaking such a region does not have holes (Fig. 1.5a). In the same way we have a multiply connected region when we cannot reduce a closed curve belonging D to a point without touch its boundaries (Fig. 1.5b). For instance, a sphere is a simply connected set in $\Re^3$, but an infinite cylinder or a torus are not. A region D is said to be star-shaped if there exist a point P in D such that all points inside of this domain can joined with P by means of a straight line (Fig. 1.6).

Let us suppose a vectorial function of scalar variation $\Gamma(t)$ that transforms the points of an interval $I[a, b] \in \Re$ into $\Re^3$. Then the idea is that when the parameter t varies over I, $\Gamma(t)$ draws a curve in the space $\Re^3$.[2] The mathematical form of a curve is the following:

$$\Gamma(t) = x(t)\mathbf{u}_x + y(t)\mathbf{u}_y + z(t)\mathbf{u}_z. \quad (1.20)$$

Two simple curves are represented in Fig. 1.7. Both are of special interest for us, as we will see in Chap. 5. The first one (a) is an open curve and corresponds to a

---

[2]In general a curve is a continuous application $\Gamma(t) : I \subset \Re \rightarrow \Re^n$, but for our practical use in this book we will restrict to the cases of $\Re^3$ and $\Re^2$.

1.4 Concept and Definition of Regions, Curves and Surfaces

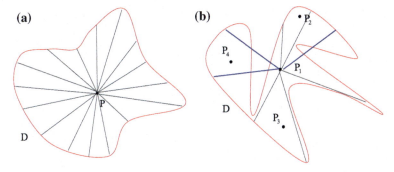

Fig. 1.6 Regions. **a** Star-shaped region. Point P can connect any point inside the domain through a straight line. **b** In this case we cannot find a point P_i, belonging to D, which joins every interior point with a straight line without cutting some part of its boundary

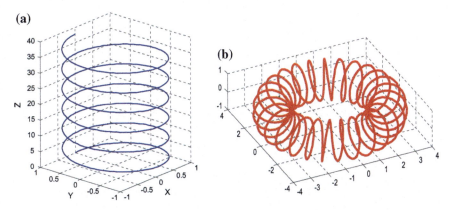

Fig. 1.7 a Helix. This curve may be expressed as $\Gamma(t) = a\cos(t)\mathbf{u}_x + a\sin(t)\mathbf{u}_y + \frac{b}{2\pi}t\mathbf{u}_z$. **b** This curve is represented by $\Gamma(t) = (a + b\sin(\omega t))\cos(t)\mathbf{u}_x + (a + b\sin(\omega t))\sin(t)\mathbf{u}_y + \cos(\omega t)\mathbf{u}_z$

helix. It is the typical form of a finite solenoid. The second one (b) is a closed curve because its beginning coincides with the end. This geometry is described as a toroidal solenoid.

Depending on the problem, an adequate change of parameter may be chosen to obtain the same curve. In this regard we must understand a curve as an equivalence class of equivalent parametric representations. Basically we can distinguish four classes of curves (Fig. 1.8): (a) simple open curves; (b) simple closed curves; (c) not simple open curves, and (d) not simple closed curves. Simple curves, contrary to not simple curves, does not intersects themselves anywhere.

A surface is the image of a continuous transformation of a two dimensional region $D \in \Re^2$ into a subset G of the space $\Re^3$.[3] This transformation must have some properties as the continuity of the partial derivatives, and must admit an inverse

---

[3] There are other definitions of a surface. See, for example, [21].

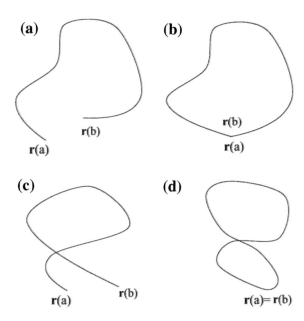

Fig. 1.8 Different kind of curves. **a** Simple open. **b** Simple closed. **c** Not simple open. **d** Not simple closed

transformation between the subset G and $D \in \Re^2$. This last characteristic prevents the surface from cutting itself and guarantees that the geometry of the surface does not depend of the parametrization chosen. Moreover, the transformation must be regular, i.e., the differential $dS : D(\Re^2) \to G(\Re^3)$ exists, which means that we can define a tangent plane for every point belonging to G.

Conceptually, a surface is a geometrical place of the space ($\Re^3$) of two degrees of freedom.[4] For its mathematical representation we have three different forms:

(a) Vectorial or parametric. In this representation each point $P(x, y, z)$ is expressed as a function of two parameters

$$\mathbf{S}(u, v) = X(u, v)\mathbf{u}_x + Y(u, v)\mathbf{u}_y + Z(u, v)\mathbf{u}_z. \qquad (1.21)$$

(b) Explicit form. By this representation one of the variables may be set as a function of the other two. For example, if we have x, y and z, we can express z as a function of x, y as $z = g(x, y)$.

(c) Implicit representation. By this form of surface we find a relationship among variables of the form $S(x, y, z) = 0$.

An important concept when studying surfaces is related with its orientation. We will use this concept also when calculating surface integrals (see next section), then we are going to define it.

If the surface is smooth enough we could locate a vector perpendicular to each point on the surface. In other words, we could find a tangent plane at every point

---

[4] Actually, the same idea may be extended for more dimensions (hypersurfaces), but for the scope of this book such a case is of no interest.

whose normal associate vector coincides with the outward normal at the same point of the surface. However, we could define a similar vector at the same point but inward to the surface. For this reason when dealing with two-sided[5] surfaces it is necessary to say which orientation we have chosen for the calculations. Moreover, in cases for which a surface integral is related with a linear integral (see Stokes theorem), the orientation of the open surface is related with the sense chosen for the vector $d\mathbf{l}$, i.e., with the chosen direction for travelling the closed curve delimiting the surface. For defining correctly such a concept we say that a surface is orientable if there exist local mapping for every region such that the Jacobian of the transformation from one local coordinate system to another is positive.[6]

1.5 Line, Surface and Volume Integrals. Circulation and Flux

We may consider three kinds of integrals: line, surface and volume according to the nature of the differential appearing in the integral. The integrand may be either a vector or a scalar. Certain combinations of integrands and differentials give rise to interesting integrals.

An open line has a beginning and an end. A closed line is one which it is possible to begin at any point, traverse the entire curve in a given sense, and return to the starting point. Therefore, it may be considered to be the boundary of an open surface. So a closed path defines an open surface whereas a closed surface defines a volume.

If $\mathbf{F}$ is a vector field and $d\mathbf{r}$ (or $d\mathbf{l}$) is the line element, a vector representing the differential length of a small element of a defined curve Γ, the line integral of $\mathbf{F}$ along Γ between two points A and B is the integral of the tangencial component of $\mathbf{F}$ along curve Γ,

$$W_F(\Gamma) = \int_{A\ \Gamma}^{B} \mathbf{F} \cdot d\mathbf{r} = \int_{A\ \Gamma}^{B} (F_x(x,y,z)\mathbf{u}_x + F_y(x,y,z)\mathbf{u}_y + F_z(x,y,z)\mathbf{u}_z)(dx,dy,dz) =$$
$$= \int_{A}^{B} (F_x dx + F_y dy + F_z dz). \qquad (1.22)$$

This expression may be written in another equivalent form

$$W_F(\Gamma) = \int_{A\ \Gamma}^{B} \mathbf{F} \cdot d\mathbf{r} = \int_{A}^{B} F \cos\theta dr, \qquad (1.23)$$

---

[5]Not all surfaces have this property. For example, the well known Möbius strip is a non-orientable surface. In fact, when moving along one of its sides, the normal vector reverses its sense at the same point but on the another side.

[6]In differential geometry the open region to be transformed together with a local coordinate system is called a *chart*, and the set of charts covering the entire surface is said an *atlas*.

Fig. 1.9 Circulation of a vector field

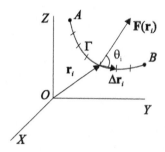

where θ is the angle between **F** and $d\mathbf{r}$ at each point along the curve. This line integral is called the **circulation**[7] of **F** along Γ. Since $\mathbf{F} \cdot d\mathbf{r}$ is a scalar it is clear that the circulation is a scalar. Figure 1.9 shows a visualization of such a line integral, as the limit of the infinite series

$$W_F(\Gamma) = \lim_{N \to \infty} \sum_{i=1}^{N} \mathbf{F}(\mathbf{r}_i) \cdot \Delta \mathbf{r}_i = \lim_{N \to \infty} \sum_{i=1}^{N} F(\mathbf{r}_i) \cdot \Delta r_i \cos \theta_i . \quad (1.24)$$

When the curve Γ is closed, a change is made in the symbol for the circulation over the closed path

$$W_F(\Gamma) = \oint_\Gamma \mathbf{F} \cdot d\mathbf{r} . \quad (1.25)$$

In the language of mathematics (1.22) is known as a differential form of first degree. Differential forms of this kind are expressed as a linear combination of the differentials of the variables, i.e.,

$$dF = F_x(x, y, z)dx + F_y(x, y, z)dy + F_z(x, y, z)dz, \quad (1.26)$$

where the coefficients F_x, F_y, and F_z are real valued functions of x, y and z. An important characteristic of (1.26) is that dF does usually not represent the differential of a function. As a consequence the integral (1.22) or (1.25) will depend not only on the endpoints A and B, but also on the curve Γ chosen for performing the calculation, then the integral around a closed path may or not may be zero. The class of vector fields for which such an integral around any closed curve is zero is of considerable importance in physics, and we will briefly discuss them in Sect. 1.12.

If **F** is a vector field, continuous in a region containing the smooth surface S the surface integral of **F** through S is

$$\Phi_F = \int_S \mathbf{F} \cdot d\mathbf{S} = \int_S \mathbf{F} \cdot \mathbf{n} \, dS = \int_S F \cos \theta \, dS , \quad (1.27)$$

---

[7]Sometimes the expression "circulation" is used only if Γ is a closed line.

1.5 Line, Surface and Volume Integrals. Circulation and Flux

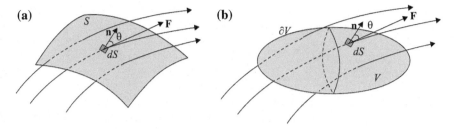

Fig. 1.10 Flux of a vector field through: **a** an open surface; **b** a closed surface

which is called the **flux** of **F** through S. Vector **n** is the unit normal (or normal vector) to S at any point of S and $d\mathbf{S}$ is the surface element, $d\mathbf{S} = dS\mathbf{n}$, as defined in Sect. 1.2. Surface S is arbitrary orientated (Fig. 1.10a). For a closed surface (defining a volume V),

$$\Phi_F = \oint_{\partial V} \mathbf{F} \cdot d\mathbf{S} = \oint_{\partial V} \mathbf{F} \cdot \mathbf{n} \, dS, \tag{1.28}$$

where ∂V is the surface defining the volume V and **n** is the outward unit normal to ∂V (Fig. 1.10b).

If φ is a scalar field and **F** is a vector field, defined in a region V, the two volume integrals in which we are interested are

$$J = \int_V \varphi \, dV, \quad \mathbf{K} = \int_V \mathbf{F} \, dV. \tag{1.29}$$

J is a scalar and **K** is a vector. In **K** there is one integral for each component of **F**. Since dV is as defined in Sect. 1.2, the volume integral can be written using a triple integral,

$$\begin{aligned} J &= \iiint_V \varphi(x, y, z) \, dx \, dy \, dz = \iiint_V \varphi(\rho, \phi, z) \rho \, d\rho \, d\phi \, dz \\ &= \iiint_V \varphi(r, \theta, \phi) r^2 \sin\theta \, dr \, d\theta \, d\phi, \end{aligned} \tag{1.30}$$

in Cartesian, cylindrical and spherical coordinates, respectively.

1.6 Gradient

It is convenient to introduce the idea of the *directional derivative* of a scalar function φ of several variables. The directional derivative of φ at point **r** in the direction of **u** is denoted by $\mathbf{D_u}\varphi(\mathbf{r})$, where **r** is the point where the directional derivative is evaluated and **u** is the unit vector along the considered direction. It may be defined as

Fig. 1.11 Isolines and gradient of the scalar field $\varphi = x \cdot e^{-x^2-y^2}$

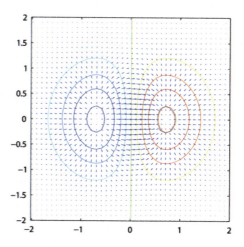

$$D_{\mathbf{u}}\varphi(\mathbf{r}) = \lim_{\Delta r \to 0} \frac{\varphi(\mathbf{r} + \Delta r \mathbf{u}) - \varphi(\mathbf{r})}{\Delta r}. \quad (1.31)$$

Note that partial derivatives $\frac{\partial \varphi}{\partial x}, \frac{\partial \varphi}{\partial y}, \frac{\partial \varphi}{\partial z}$ are the directional derivatives in the directions $\mathbf{u}_x, \mathbf{u}_y, \mathbf{u}_z$, respectively.

It is also convenient to introduce the *del operator*, written ∇, which is the vector differential operator. In Cartesian coordinates,

$$\nabla = \frac{\partial}{\partial x}\mathbf{u}_x + \frac{\partial}{\partial y}\mathbf{u}_y + \frac{\partial}{\partial z}\mathbf{u}_z. \quad (1.32)$$

In cylindrical coordinates,

$$\nabla = \frac{\partial}{\partial \rho}\mathbf{u}_\rho + \frac{1}{\rho}\frac{\partial}{\partial \phi}\mathbf{u}_\phi + \frac{\partial}{\partial z}\mathbf{u}_z \quad (1.33)$$

In spherical coordinates,

$$\nabla = \frac{\partial}{\partial r}\mathbf{u}_r + \frac{1}{r}\frac{\partial}{\partial \theta}\mathbf{u}_\theta + \frac{1}{r \sin \theta}\frac{\partial}{\partial \phi}\mathbf{u}_\phi \quad (1.34)$$

The **gradient** of a scalar field φ at a point is a vector $\nabla \varphi$ (or gradφ) that represents both the magnitude and the direction of the maximum directional derivative at the point. Figure 1.11 shows the example of Fig. 1.4a with its gradients.[8]

---

[8] In the appendix, a MATLAB program to calculate surfaces, isolines and gradients is included.

1.6 Gradient

Using the del operator, the value of $\nabla \varphi$ in the Cartesian coordinate system is

$$\nabla \varphi = \frac{\partial \varphi}{\partial x}\mathbf{u}_x + \frac{\partial \varphi}{\partial y}\mathbf{u}_y + \frac{\partial \varphi}{\partial z}\mathbf{u}_z, \tag{1.35}$$

in cylindrical coordinates

$$\nabla \varphi = \frac{\partial \varphi}{\partial \rho}\mathbf{u}_\rho + \frac{1}{\rho}\frac{\partial \varphi}{\partial \phi}\mathbf{u}_\phi + \frac{\partial \varphi}{\partial z}\mathbf{u}_z, \tag{1.36}$$

and in spherical coordinates

$$\nabla \varphi = \frac{\partial \varphi}{\partial r}\mathbf{u}_r + \frac{1}{r}\frac{\partial \varphi}{\partial \theta}\mathbf{u}_\theta + \frac{1}{r \sin \theta}\frac{\partial \varphi}{\partial \phi}\mathbf{u}_\phi \tag{1.37}$$

We can also take note of the following fundamental properties of the gradient of a scalar field φ:

- The gradient $\nabla \varphi$ of a scalar field φ is a vector field.
- The magnitude of $\nabla \varphi$ equals the maximum rate of change in φ per unit distance.
- $\nabla \varphi$ points in the direction of the maximum rate of change in φ.
- $\nabla \varphi$ at any point is perpendicular to the level surface that passes through that point.
- The directional derivative can be calculated as

$$\mathbf{D}_\mathbf{u}\varphi(\mathbf{r}) = \nabla \varphi(\mathbf{r}) \cdot \mathbf{u}, \tag{1.38}$$

the projection of the gradient in the direction of a unit vector $\mathbf{u}$.

- The differential of a scalar field $\varphi(\mathbf{r})$, $d\varphi = \frac{\partial \varphi}{\partial x}dx + \frac{\partial \varphi}{\partial y}dy + \frac{\partial \varphi}{\partial z}dz$, can be calculated as

$$d\varphi(\mathbf{r}) = \nabla \varphi(\mathbf{r}) \cdot d\mathbf{r}. \tag{1.39}$$

1.7 Curl

The **curl** of a vector field $\mathbf{F}$ at a point $\mathbf{r}$ is an axial (or rotational) vector $\nabla \times \mathbf{F}$ (or curl $\mathbf{F}$)[9] whose magnitude is the maximum circulation of $\mathbf{F}$ per unit area as the area tends to zero and whose direction is the normal direction of the area when the area is oriented so as to make the circulation maximum,

$$\text{curl } \mathbf{F}(\mathbf{r}) = \nabla \times \mathbf{F}(\mathbf{r}) = \left(\lim_{S \to 0} \frac{1}{S} \oint_{\partial S} \mathbf{F} \cdot d\mathbf{l}\right)_{\max} \mathbf{n}, \tag{1.40}$$

---

[9]Because of its rotational nature, rot $\mathbf{F}$ is also used.

where **n** is the outward unit normal to surface S (surrounding **r**) which are bounded by the curve ∂S in which **r** is located and is determined using the right-hand rule.

Note that the curl ($\nabla \times \mathbf{F}$) of a vector field **F** is a vector field which are related with closed circulation of a vector around a point. The direction of the rotational of a vector field at each point P is perpendicular to the plane crossing P for which the circulation is a maximum.

Using the del operator, the value of curl $\mathbf{F} = \nabla \times \mathbf{F}$ in the Cartesian coordinate system is

$$\nabla \times \mathbf{F} = \begin{vmatrix} \mathbf{u}_x & \mathbf{u}_y & \mathbf{u}_z \\ \frac{\partial}{\partial x} & \frac{\partial}{\partial y} & \frac{\partial}{\partial z} \\ F_x & F_y & F_z \end{vmatrix} = \left(\frac{\partial F_z}{\partial y} - \frac{\partial F_y}{\partial z} \right) \mathbf{u}_x + \left(\frac{\partial F_x}{\partial z} - \frac{\partial F_z}{\partial x} \right) \mathbf{u}_y + \left(\frac{\partial F_y}{\partial x} - \frac{\partial F_x}{\partial y} \right) \mathbf{u}_z , \tag{1.41}$$

in cylindrical coordinates

$$\nabla \times \mathbf{F} = \begin{vmatrix} \mathbf{u}_\rho & \mathbf{u}_\phi & \mathbf{u}_z \\ \frac{\partial}{\partial \rho} & \frac{1}{\rho}\frac{\partial}{\partial \phi} & \frac{\partial}{\partial z} \\ F_\rho & F_\phi & F_z \end{vmatrix} = \left(\frac{1}{\rho} \frac{\partial F_z}{\partial \phi} - \frac{\partial F_\phi}{\partial z} \right) \mathbf{u}_\rho + \left(\frac{\partial F_\rho}{\partial z} - \frac{\partial F_z}{\partial \rho} \right) \mathbf{u}_\phi + \frac{1}{\rho} \left(\frac{\partial}{\partial \rho}(\rho F_\phi) - \frac{\partial F_\rho}{\partial \phi} \right) \mathbf{u}_z , \tag{1.42}$$

and in spherical coordinates

$$\nabla \times \mathbf{F} = \begin{vmatrix} \mathbf{u}_r & \mathbf{u}_\theta & \mathbf{u}_\phi \\ \frac{\partial}{\partial r} & \frac{1}{r}\frac{\partial}{\partial \theta} & \frac{1}{r\sin\theta}\frac{\partial}{\partial \phi} \\ F_r & F_\theta & F_\phi \end{vmatrix} = \tag{1.43}$$

$$= \frac{1}{r \sin \theta} \left[\frac{\partial}{\partial \theta}(F_\phi \sin \theta) - \frac{\partial F_\theta}{\partial \phi} \right] \mathbf{u}_r + \left[\frac{1}{r \sin \theta} \frac{\partial F_r}{\partial \phi} - \frac{1}{r} \frac{\partial}{\partial r}(r F_\phi) \right] \mathbf{u}_\theta + \frac{1}{r} \left[\frac{\partial}{\partial r}(r F_\theta) - \frac{\partial F_r}{\partial \theta} \right] \mathbf{u}_\phi .$$

1.8 Divergence

The **divergence** of a vector field **F** at a point **r** is an scalar $\nabla \cdot \mathbf{F}$ (or div **F**) whose value is the outward flux per unit volume as the volume shrinks about P. Hence,

$$\text{div } \mathbf{F}(\mathbf{r}) = \nabla \cdot \mathbf{F}(\mathbf{r}) = \lim_{V \to 0} \frac{1}{V} \oint_{\partial V} \mathbf{F} \cdot d\mathbf{S} , \tag{1.44}$$

where the volume V is enclosed by the closed surface ∂V in which **r** is located.

Note that the divergence ($\nabla \cdot \mathbf{F}$) of a vector field **F** is a scalar field which is related to flux of a vector through a closed surface surrounding each point per unit volume (or source density). Physically, we may regard the divergence of the vector field **F** at a given point as a measure of how much the field diverges or emanates from that

1.8 Divergence

point. It is positive at a source point in the field, and negative at a sink point, or zero where there is neither sink nor source.

Using the del operator, the value of $\text{div}\,\mathbf{F} = \nabla \cdot \mathbf{F}$ in the Cartesian coordinate system is

$$\nabla \cdot \mathbf{F} = \frac{\partial F_x}{\partial x} + \frac{\partial F_y}{\partial y} + \frac{\partial F_z}{\partial z}, \tag{1.45}$$

in cylindrical coordinates

$$\nabla \cdot \mathbf{F} = \frac{1}{\rho}\frac{\partial}{\partial \rho}(\rho F_\rho) + \frac{1}{\rho}\frac{\partial F_\phi}{\partial \phi} + \frac{\partial F_z}{\partial z}, \tag{1.46}$$

and in spherical coordinates

$$\nabla \cdot \mathbf{F} = \frac{1}{r^2}\frac{\partial}{\partial r}(r^2 F_r) + \frac{1}{r\sin\theta}\frac{\partial}{\partial \theta}(F_\theta \sin\theta) + \frac{1}{r\sin\theta}\frac{\partial F_\phi}{\partial \phi}. \tag{1.47}$$

1.9 Stokes's Theorem and Divergence Theorem

Stokes's theorem: Let S be an orientable surface in $\Re^3$ whose boundary is a simple closed curve ∂S, and let $\mathbf{F}(x, y, z) = F_x(x, y, z)\mathbf{u}_x + F_y(x, y, z)\mathbf{u}_y + F_z(x, y, z)\mathbf{u}_z$ be a smooth vector field defined on some subset of $\Re^3$ that contains S. Then

$$\oint_{\partial S} \mathbf{F} \cdot d\mathbf{l} = \int_S (\nabla \times \mathbf{F}) \cdot d\mathbf{S}. \tag{1.48}$$

The direction of $d\mathbf{l}$ and $d\mathbf{S}$ involved in Stokes's theorem is shown in Fig. 1.12 and is determined using the right-hand rule. Stokes's theorem states that the circulation of a vector field $\mathbf{F}$ around a closed path ∂S is equal to the flux of the curl of $\mathbf{F}$ over the open surface S bounded by ∂S provided that $\mathbf{F}$ and $\nabla \times \mathbf{F}$ are continuous on S.

Divergence theorem (or Ostrogradski–Gauss theorem): Let V be a volume in $\Re^3$ whose boundary is a closed surface ∂V, and let $\mathbf{F}(x, y, z) = F_x(x, y, z)\mathbf{u}_x + F_y(x, y, z)\mathbf{u}_y + F_z(x, y, z)\mathbf{u}_z$ be a vector field defined on some subset of $\Re^3$ that contains ∂V. Then

Fig. 1.12 Sense of $d\mathbf{l}$ and $d\mathbf{S}$ involved in Stokes's theorem

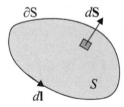

$$\oint_{\partial V} \mathbf{F} \cdot d\mathbf{S} = \int_V \nabla \cdot \mathbf{F}\, dV. \tag{1.49}$$

The divergence theorem states that the total outward flux of a vector field $\mathbf{F}$ through a closed surface ∂V is the same as the volume integral of the divergence of $\mathbf{F}$ over the volume V bounded by ∂V.

Note that whereas Stokes's theorem relates a line integral (circulation) to a surface integral, the divergence theorem relates a surface integral (flux) to a volume integral.

1.10 Normal Vector to a Surface

When calculating some surface integrals, as the flux of a vectorial field across a surface, the problem of obtaining the normal vector $\mathbf{n}$ to a surface arises. The expression for this vector depends on the representation chosen for the surface, i.e., parametric, implicit and explicit.

1.10.1 Vectorial or Parametric

In this case, if we suppose we have u and v as parameters for representing the surface S, the expression of the surface may be written as follows

$$\mathbf{S}(u,v) = X(u,v)\mathbf{u}_x + Y(u,v)\mathbf{u}_y + Z(u,v)\mathbf{u}_z, \tag{1.50}$$

and the normal vector

$$\mathbf{n} = \pm \frac{\dfrac{\partial \mathbf{S}(u,v)}{\partial u} \times \dfrac{\partial \mathbf{S}(u,v)}{\partial v}}{\left| \dfrac{\partial \mathbf{S}(u,v)}{\partial u} \times \dfrac{\partial \mathbf{S}(u,v)}{\partial v} \right|}. \tag{1.51}$$

Note that, in general, the result will depend on u and v. Once we have the normal the element of the differential surface is

$$dS = \left| \frac{\partial \mathbf{S}(u,v)}{\partial u} \times \frac{\partial \mathbf{S}(u,v)}{\partial v} \right| du\, dv, \tag{1.52}$$

and the final formula for the flux

$$\int_S \mathbf{F} \cdot d\mathbf{S} = \int_S \mathbf{F}(\mathbf{S}(u,v)) \cdot \mathbf{n}\, dS = \int_S \mathbf{F}(\mathbf{S}(u,v)) \cdot \frac{\partial \mathbf{S}(u,v)}{\partial u} \times \frac{\partial \mathbf{S}(u,v)}{\partial v}\, du\, dv. \tag{1.53}$$

1.10.2 Explicit

Such a representation is of the form $z = g(x, y)$, but we can put it in vectorial form. In fact, making the substitution $x = u$ and $y = v$, the function $z = g(x, y)$ may be written as $z = g(u, v)$, obtaining for the surface the following expression

$$\mathbf{S}(u, v) = u\mathbf{u}_x + v\mathbf{u}_y + g(u, v)\mathbf{u}_z, \tag{1.54}$$

then for calculating **n** we must first to determine the partial derivatives (1.51)

$$\frac{\partial \mathbf{S}(u, v)}{\partial u} = \mathbf{u}_x + \frac{\partial g(u, v)}{\partial u}\mathbf{u}_z, \tag{1.55}$$

and

$$\frac{\partial \mathbf{S}(u, v)}{\partial v} = \mathbf{u}_y + \frac{\partial g(u, v)}{\partial v}\mathbf{u}_z. \tag{1.56}$$

As a result we have for the vectorial product (1.51)

$$\mathbf{n} = \pm \frac{-\dfrac{\partial g(u, v)}{\partial u}\mathbf{u}_x - \dfrac{\partial g(u, v)}{\partial v}\mathbf{u}_y + \mathbf{u}_z}{\sqrt{\left(\dfrac{\partial g(u, v)}{\partial u}\right)^2 + \left(\dfrac{\partial g(u, v)}{\partial v}\right)^2 + 1}}, \tag{1.57}$$

and

$$dS = \left|\frac{\partial \mathbf{S}(u, v)}{\partial u} \times \frac{\partial \mathbf{S}(u, v)}{\partial v}\right| du\, dv = \sqrt{1 + \left(\frac{\partial g(u, v)}{\partial u}\right)^2 + \left(\frac{\partial g(u, v)}{\partial v}\right)^2}\, du\, dv. \tag{1.58}$$

The expression for the flux in explicit form is

$$\int\!\!\int_S \mathbf{F} \cdot d\mathbf{S} = \int\!\!\int_S \mathbf{F} \cdot \mathbf{n}\, dS$$
$$= \int\!\!\int_S \mathbf{F}(u, v, g(u, v)) \cdot \left(-\frac{\partial g(u, v)}{\partial u}\mathbf{u}_x - \frac{\partial g(u, v)}{\partial v}\mathbf{u}_y + \mathbf{u}_z\right) du\, dv. \tag{1.59}$$

1.10.3 Implicit

As we have seen, when a surface is written in implicit form we obtain a function $S(x, y, z) = 0$. If consider $S(x, y, z)$ as a level surface, we know that the gradient is perpendicular to it, then the normal unitary vector may be found through this mathematical operator,

$$\mathbf{n} = \pm \frac{\nabla S(x,y,z)}{|\nabla S(x,y,z)|} = \pm \frac{\frac{\partial S(x,y,z)}{\partial x}\mathbf{u}_x + \frac{\partial S(x,y,z)}{\partial y}\mathbf{u}_y + \frac{\partial S(x,y,z)}{\partial z}\mathbf{u}_z}{\sqrt{\left(\frac{\partial S(x,y,z)}{\partial x}\right)^2 + \left(\frac{\partial S(x,y,z)}{\partial y}\right)^2 + \left(\frac{\partial S(x,y,z)}{\partial z}\right)^2}}. \quad (1.60)$$

For calculating the flux across the surface S we must project such a surface over some of the three coordinate planes OXY, OYZ or OXZ. If the plane chosen is the OXY we can express $z = f(x, y)$, and for the other projections we can define $x = h(y, z)$ for the plane OYZ and $y = q(x, z)$ for OXZ. In principle the choice of projection has no influence in the procedure, however sometimes we cannot find some of the aforementioned functional relations. For instance, let us suppose we would like to do the calculations by $z = f(x, y)$. In this case we must find the function $S(x, y, f(x, y))$, but it is not possible because the equation $S(x, y, z) = 0$ is not one-valued. Another problem could be that we cannot arrange $z = f(x, y)$ from $S(x, y, z) = 0$. In these cases we can try other projections.

To obtain the flux we start we the scalar product $\mathbf{F} \cdot \mathbf{n}$, were $\mathbf{F} = F_x(x, y, z)\mathbf{u}_x + F_y(x, y, z)\mathbf{u}_y + F_z(x, y, z)\mathbf{u}_z$ which leads to

$$\mathbf{F} \cdot \mathbf{n} = \frac{F_x \frac{\partial S(x,y,z)}{\partial x} + F_y \frac{\partial S(x,y,z)}{\partial y} + F_z \frac{\partial S(x,y,z)}{\partial z}}{\sqrt{\left(\frac{\partial S(x,y,z)}{\partial x}\right)^2 + \left(\frac{\partial S(x,y,z)}{\partial y}\right)^2 + \left(\frac{\partial S(x,y,z)}{\partial z}\right)^2}}. \quad (1.61)$$

We know the relation between the variables, then if we substitute $z = f(x, y)$ in (1.61) we have a function only depending on x and y, i.e., $(\mathbf{F} \cdot \mathbf{n})_{z=f(x,y)} = G(x, y)$ and it follows that dS has the same form as (1.58) that we have seen in Sect. 1.10.2. By combining all in the definition of (1.59) may be obtained

$$\iint_S \mathbf{F} \cdot d\mathbf{S} = \iint_S \mathbf{F} \cdot \mathbf{n}\, dS = \iint_S G(x,y) \sqrt{1 + \left(\frac{\partial f(x,y)}{\partial x}\right)^2 + \left(\frac{\partial f(x,y)}{\partial y}\right)^2}\, dx\, dy. \quad (1.62)$$

This result is sometimes also written as

$$\iint_S \mathbf{F} \cdot \mathbf{n}\, dS = \iint_S \left[\frac{\mathbf{F} \cdot \mathbf{n}}{\cos \gamma}\right]_{z=f(x,y)} dx\, dy, \quad (1.63)$$

γ being the angle between the normal to the surface S and the positive OZ axis, thus

1.10 Normal Vector to a Surface

$$\cos \gamma = \pm \frac{1}{\sqrt{1 + \left(\dfrac{\partial \mathbf{f}(x,y)}{\partial x}\right)^2 + \left(\dfrac{\partial \mathbf{f}(x,y)}{\partial y}\right)^2}}.$$

Alternatively we can use

$$\int\int_S \mathbf{F} \cdot d\mathbf{S} = \int\int_S [F_x(x,y,z)\cos\alpha + F_y(x,y,z)\cos\beta + F_z(x,y,z)\cos\gamma]\, dS, \quad (1.64)$$

where

$$dS \cos \alpha = \pm\, dy\, dz, \qquad (1.65)$$

$$dS \cos \beta = \pm\, dx\, dz, \qquad (1.66)$$

and

$$dS \cos \gamma = \pm\, dx\, dy. \qquad (1.67)$$

Introduction of all these formulae into (1.64) leads to

$$\int\int_S \mathbf{F} \cdot d\mathbf{S} = \int\int_{S_{OYZ}} F_x(h(y,z), y, z)\, dy\, dz + \int\int_{S_{OXZ}} F_y(x, q(x,z), z)\, dx\, dz +$$
$$+ \int\int_{S_{OXY}} F_z(x, y, f(x,y))\, dx\, dy \qquad (1.68)$$

in which S_{OYZ}, S_{OXZ}, and S_{OXY} are the projections of the surface S over the planes OYZ, OXZ, and OXY, respectively. The principal difference between both procedures is that by (1.63) we need only to perform one double integration, but using (1.68) three integrations are required (one for each projection).

1.11 Further Developments

Several operations including gradient, curl or divergence of appropriate kinds of fields may be done.

The curl of the gradient of any scalar field is zero:

$$\operatorname{curl} \operatorname{grad} \varphi = \nabla \times \nabla \varphi = \begin{vmatrix} \mathbf{u}_x & \mathbf{u}_y & \mathbf{u}_z \\ \dfrac{\partial}{\partial x} & \dfrac{\partial}{\partial y} & \dfrac{\partial}{\partial z} \\ \dfrac{\partial \varphi}{\partial x} & \dfrac{\partial \varphi}{\partial y} & \dfrac{\partial \varphi}{\partial z} \end{vmatrix} = 0. \qquad (1.69)$$

The divergence of any curl is also zero:

$$\text{div curl } \mathbf{F} = \nabla \cdot (\nabla \times \mathbf{F}) = \frac{\partial}{\partial x}\left(\frac{\partial F_z}{\partial y} - \frac{\partial F_y}{\partial z}\right) + \frac{\partial}{\partial y}\left(\frac{\partial F_x}{\partial z} - \frac{\partial F_z}{\partial x}\right) + \frac{\partial}{\partial z}\left(\frac{\partial F_y}{\partial x} - \frac{\partial F_x}{\partial y}\right) = 0. \tag{1.70}$$

The divergence of the gradient of any scalar field φ is a scalar field $\Delta\varphi$ (or $\nabla^2\varphi$) that is called the **laplacian** of the scalar field. Hence,

$$\text{div grad } \varphi = \Delta\varphi = \nabla^2\varphi = \nabla \cdot \nabla\varphi. \tag{1.71}$$

Using the del operator, the value of $\Delta\varphi$ in the Cartesian coordinate system is

$$\Delta\varphi = \frac{\partial^2\varphi}{\partial x^2} + \frac{\partial^2\varphi}{\partial y^2} + \frac{\partial^2\varphi}{\partial z^2}, \tag{1.72}$$

in cylindrical coordinates

$$\Delta\varphi = \frac{1}{\rho}\frac{\partial}{\partial\rho}\left(\rho\frac{\partial\varphi}{\partial\rho}\right) + \frac{1}{\rho^2}\frac{\partial^2\varphi}{\partial\phi^2} + \frac{\partial^2\varphi}{\partial z^2}, \tag{1.73}$$

and in spherical coordinates

$$\Delta\varphi = \frac{1}{r^2}\frac{\partial}{\partial r}\left(r^2\frac{\partial\varphi}{\partial r}\right) + \frac{1}{r^2\sin\theta}\frac{\partial}{\partial\theta}\left(\sin\theta\frac{\partial\varphi}{\partial\theta}\right) + \frac{1}{r^2\sin^2\theta}\frac{\partial^2\varphi}{\partial\phi^2}. \tag{1.74}$$

The curl of the curl of a vector field results

$$\text{curl curl } \mathbf{F} = \text{grad div } \mathbf{F} - \Delta\mathbf{F}, \tag{1.75}$$

or

$$\nabla \times (\nabla \times \mathbf{F}) = \nabla(\nabla \cdot \mathbf{F}) - \nabla^2\mathbf{F}, \tag{1.76}$$

where the Laplacian of a vector field is defined as the Laplacian of each component,

$$\Delta\mathbf{F} = \nabla^2\mathbf{F} = \Delta F_x\mathbf{u}_x + \Delta F_y\mathbf{u}_y + \Delta F_z\mathbf{u}_z. \tag{1.77}$$

The most interesting extension of the divergence theorem and of Stokes's theorem is **Green's theorem**, which is

$$\int_V (\psi\Delta\varphi - \varphi\Delta\psi)\,dV = \oint_{\partial V} (\psi\nabla\varphi - \varphi\nabla\psi) \cdot d\mathbf{S}. \tag{1.78}$$

1.11 Further Developments

If $\psi = 1$,

$$\int_V \Delta\varphi \, dV = \oint_{\partial V} \nabla\varphi \cdot d\mathbf{S}, \tag{1.79}$$

the volume integral of the Laplacian of φ over V is the same as the total outward flux of gradient of φ through the closed surface ∂V.

1.12 Classification of Vector Fields

As it was previously commented (see Sect. 1.5), some vector fields have special behavior with respect to the integral (1.22). For this reason, we devoted this paragraph for explaining the most important results referred to them.

Let us suppose a vector field $\mathbf{F}(x, y, z)$ defined in a *simple connected region* D. If the line integral of that field around any closed curve is zero, i.e.,

$$\oint_{\partial S} \mathbf{F} \cdot d\mathbf{l} = \oint_{\partial S} d\mathbf{F} = 0. \tag{1.80}$$

we say that this field is *irrotational* or *conservative*. If we look at Stokes's theorem (1.48), this result may be enounced in an equivalent form by means of the curl of the field. In fact, as (1.80) holds, then $\nabla \times \mathbf{F} = 0$, and the following system of equations apply

$$\frac{\partial F_z}{\partial y} = \frac{\partial F_y}{\partial z}, \quad \frac{\partial F_x}{\partial z} = \frac{\partial F_z}{\partial x}, \quad \frac{\partial F_y}{\partial x} = \frac{\partial F_x}{\partial y}, \tag{1.81}$$

hence, a vector field whose curl is zero *at every point of a simple connected domain* is conservative in that region. When condition (1.80) is satisfied we say that the differential form $d\mathbf{F}$ is *exact*. As a result, the irrotational field $\mathbf{F}$ can always be expressed in terms of another scalar field $V(x, y, z)$, since $\nabla \times (\nabla V) = 0$, that is,

$$\nabla \times \mathbf{F} = 0 \Rightarrow \int_S (\nabla \cdot \mathbf{F}) \cdot d\mathbf{S} = \oint_{\partial S} \mathbf{F} \cdot d\mathbf{l} = 0 \quad \text{and} \quad \mathbf{F} = -\nabla V. \tag{1.82}$$

For this reason $\mathbf{F}$ may be called a *potential field* and V the *scalar potential* of $\mathbf{F}$.

As a corollary of this result we have that the curvilinear integral of such a field between two points A and B in D does not depend on the trajectory chosen for performing the calculation, but only of the endpoints, i.e.,

$$\int_A^B \mathbf{F} \cdot d\mathbf{l} = -\int_A^B \nabla V \cdot d\mathbf{l} = V(B) - V(A). \tag{1.83}$$

Even though the conditions given for potential fields are well posed, some attention should be given for non-simply connected domains. In this case, conditions (1.81)

Fig. 1.13 Multiply connected domain D. The line integral around Γ_1 is not zero. However, the same calculation along Γ_2 is zero. If curve is opened and does not pass throughout the singularities the result will depend only of the endpoints

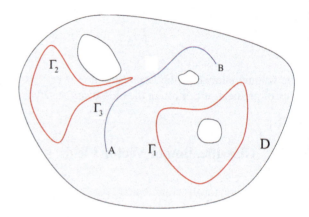

do not guarantee (1.80). In effect, if (1.80) is fulfilled but the closed curve Γ encloses holes (curve Γ_1, Fig. 1.13), the linear integral is not zero, i.e.,

$$\oint_{\partial S} \mathbf{F} \cdot d\mathbf{l} = \oint_{\partial S} d\mathbf{F} \neq 0. \tag{1.84}$$

This result shows that for *multiply connected regions* (see Fig. 1.13) the necessary condition $\nabla \times \mathbf{F} = 0$ to have a potential so that $\mathbf{F} = -\nabla V$ is not sufficient. When it happens we say that the differential form $d\mathbf{F}$ is *closed*. Therefore, the condition to be closed is only (1.81) and we can say that all exact $1-form$ are closed, but the converse is false unless the region G is simply connected. Nevertheless, a comment about it must given.

Observe that in this last case, even the differential $1 - dF$ form is closed (see Fig. 1.13), two possibilities may arise. In fact, let us choose another closed curve as Γ_2 belonging to the multiply connected domain. As Γ_2 does not surrounds any hole, its line integral will be zero as (1.80). It is equivalent to say that, if we do not touch or enclose holes, and $\nabla \times \mathbf{F} = 0$ simultaneously in this subregion of D, we can also find a scalar function $V(\mathbf{r})$ so that $\mathbf{F} = -\nabla V$. Hence, a closed integral must be zero (curve Γ_2 in Fig. 1.13) and an open integral from A to B (curve Γ_3) will only depend on the endpoints A and B.

The aforementioned definitions were focused on the rotational of a vector field. In the same way we can also define other kinds of fields if we look at divergence. So, the vector fields that have zero divergence ($\nabla \cdot \mathbf{F} = 0$) are called *solenoidal* fields. Such fields have neither source nor sinks of flux (see Chap. 5). From the divergence theorem, we get

$$\oint_{\partial V} \mathbf{F} \cdot d\mathbf{S} = \int_V \nabla \cdot \mathbf{F} \, dV = 0. \tag{1.85}$$

A solenoidal field $\mathbf{F}$ can always be expressed in terms of another vector field $\mathbf{A}$ through the curl, since $\nabla \cdot (\nabla \times \mathbf{A}) = 0$, that is,

1.12 Classification of Vector Fields

$$\nabla \cdot \mathbf{F} = 0 \Rightarrow \int_V \nabla \cdot \mathbf{F}\, dV = \oint_{\partial V} \mathbf{F} \cdot d\mathbf{S} = 0 \text{ and } \mathbf{F} = \nabla \times \mathbf{A}. \quad (1.86)$$

$\mathbf{A}$ is called the *vector potential* of $\mathbf{F}$.

A vector field $\mathbf{F}$ is uniquely prescribed within a region by its divergence and its curl. If we let

$$\nabla \cdot \mathbf{F} = \rho, \quad \nabla \times \mathbf{F} = \mathbf{j}, \quad (1.87)$$

ρ can be regarded as a point source density of $\mathbf{F}$ and $\mathbf{j}$ its circulation density. Any vector $\mathbf{F}$ satisfying (1.87) with ρ and $\mathbf{j}$ vanishing at infinity can be written as

$$\mathbf{F} = -\nabla V + \nabla \times \mathbf{A}, \quad (1.88)$$

where V is the *scalar potential* and $\mathbf{A}$ the *vector potential* of $\mathbf{F}$. This is called **Helmholtz's theorem**. This theorem shows that any vector field $\mathbf{F}$ verifying (1.87) can be written as the sum of two vector fields: one irrotational $(-\nabla V)$, the other solenoidal $(\nabla \times \mathbf{A})$.

1.13 Obtaining the Scalar Potential

As we have seen in the preceding section, when $\nabla \times \mathbf{F} = 0$ at every point of a simply connected open set it is possible to find a scalar function $V(\mathbf{r})$ so that $\mathbf{F} = -\nabla V(\mathbf{r})$. As a result, the curvilinear integral of the field $\mathbf{F}$ around any simple closed curve is zero. The aim of this paragraph is to present a procedure for determining the scalar function $V(\mathbf{r})$ from the vectorial field $\mathbf{F}$.

1.13.1 First Method

Let us consider a vectorial function $\mathbf{F}(\mathbf{r})$ in $\Re^3$ which may be obtained from a scalar potential $V(\mathbf{r})$

$$\mathbf{F}(\mathbf{r}) = F_x(x, y, z)\mathbf{u}_x + F_y(x, y, z)\mathbf{u}_y + F_z(x, y, z)\mathbf{u}_z = -\nabla V(\mathbf{r}). \quad (1.89)$$

By identifying them by term of that equality it may be written

$$F_x(x, y, z) = -\frac{\partial V}{\partial x}, \quad (1.90)$$

$$F_y(x, y, z) = -\frac{\partial V}{\partial y}, \quad (1.91)$$

and
$$F_z(x, y, z) = -\frac{\partial V}{\partial z}. \tag{1.92}$$

By choosing, for example the first one, we can integrate in the variable x considering the other variables are constant, i.e.,

$$V = -\int F_x(x, y, z)dx + C(y, z) = -G^x(x, y, z) + C(y, z), \tag{1.93}$$

where $C(y, z)$ is an unknown constant that depends on the rest of variables y and z, and $G^x(x, y, z)$ is the integral of $F_x(x, y, z)$[10] with respect to x. Now, if we introduce this last result into (1.91), we have

$$\frac{\partial V}{\partial y} = \frac{\partial}{\partial y}\{-G^x(x, y, z)\} + \frac{\partial C(y, z)}{\partial y} = -G^x_y(x, y, z) + C_y(y, z) = F_y(x, y, z), \tag{1.94}$$

where the superindex x of $G^x_y(x, y, z)$ represents the integral with respect to the coordinate x and the subindex y the partial derivative respect to y. Integrating again (1.94) but with respect to y,

$$C(y, z) = \int F_y(x, y, z)dy + \int G^x_y(x, y, z)dy + D(z), \tag{1.95}$$

and therefore

$$\begin{aligned} V &= -G^x(x, y, z) + \int F_y(x, y, z)dy + \int G^x_y(x, y, z)dy + D(z) \\ &= -\int F_x(x, y, z)dx + \int F_y(x, y, z)dy + D(z). \end{aligned} \tag{1.96}$$

In (1.96) the constant D is only a function of z. By proceeding in the same way, that is, using (1.92), we have

$$-\frac{\partial V}{\partial z} = F_z(x, y, z). \tag{1.97}$$

1.13.2 Second Method

As $\mathbf{F} = -\nabla V(\mathbf{r})$ for these type of fields, the integral

$$\int_A^B \mathbf{F} \cdot d\mathbf{l} = -\int_A^B \nabla V \cdot d\mathbf{l} \tag{1.98}$$

---

[10] Observe that, in this case for F, this subindex x (or y and z) does not represent the derivative.

1.13 Obtaining the Scalar Potential

Fig. 1.14 This path is formed by three segments parallel to the coordinate axis X, Y, and Z

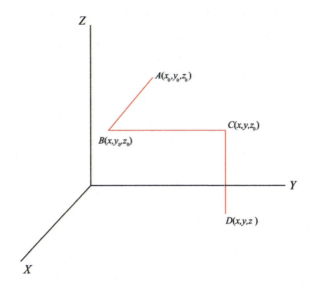

must be true for any curve Γ whose endpoints are A and B. The idea for finding the potential function V is to choose an easily integrable curve and solve (1.98) directly. In fact, let us use the curve shown in Fig. 1.14. This path is composed by three finite line segments, each of them parallel to one of the coordinate axes. The first part of the curve begins at point A, whose coordinates are (x_0, y_0, z_0), and finishes at $B(x, y_0, z_0)$. As this segment is parallel to OX, the projections over OY and OZ are constants between A and B. Following the same idea for the straight lines BC and CD, we can write

$$V(x, y, z) = -\int_{x_0}^{x} F_x(x, y_0, z_0)dx - \int_{y_0}^{y} F_y(x, y, z_0)dy - \int_{z_0}^{z} F_z(x, y, z)dz + R, \tag{1.99}$$

R being a constant.

1.13.3 Third Method*

As we have previously seen mentioned in Sect. 1.4, some kinds of regions have stellar form. In such a domain there is at least one point that may be connected with any point inside of this region. When working with stellar-shaped domains with respect to the origin of coordinates O, the potential $V(x, y, z)$ corresponding to a vectorial field $\mathbf{F}$ can be obtained by means of the following mathematical relation

$$V(x, y, z) = -\int_{0}^{1} \mathbf{F}(\mathbf{P}_1) \cdot \mathbf{r}d\lambda + C, \tag{1.100}$$

Fig. 1.15 A point goes over the line Γ when parameter t varies from 0 to 1

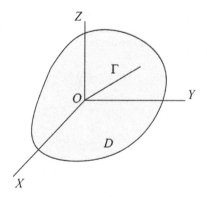

where $\mathbf{r} = x\mathbf{u}_x + y\mathbf{u}_y + z\mathbf{u}_z$ is the vector joining the origin of the reference frame with the point (x, y, z), and $P_1 = (\lambda x, \lambda y, \lambda z)$ for $0 \leq \lambda \leq 1$. Observe that the effect of varying parameter λ is to travel the curve Γ from the beginning to its end (Fig. 1.15).

1.14 Vectorial Field from the Vector Potential

As we have seen in Sect. 1.12[11] a special case when studying vector fields occurs with solenoidal fields. For such fields holds $\nabla \cdot \mathbf{F} = 0$, and then we can express $\mathbf{F}$ as the curl of a vector field, i.e., $\mathbf{F} = \nabla \times \mathbf{A}$. In the same manner studied before for the scalar potential, sometimes we know the vector field $\mathbf{F}$, and we are interested in obtaining the vector potential $\mathbf{A}$ which gives $\mathbf{F}$ through the rotational. The solution to this problem may be found through two different ways.

1.14.1 First Method

By the first procedure we start with the definition of rotational, and we equate each of its components to the field $\mathbf{F}(\mathbf{r}) = F_x(x, y, z)\mathbf{u}_x + F_y(x, y, z)\mathbf{u}_y + F_z(x, y, z)\mathbf{u}_z$, then we can write

$$\left(\frac{\partial A_z}{\partial y} - \frac{\partial A_y}{\partial z} \right) = F_x(x, y, z), \tag{1.101}$$

$$\left(\frac{\partial A_x}{\partial z} - \frac{\partial A_z}{\partial x} \right) = F_y(x, y, z), \tag{1.102}$$

---

[11] We will see again this topic when we study the magnetic vector potential $\mathbf{A}$ in Chap. 5. There, the vector field here denoted by $\mathbf{F}$ will be the magnetic field $\mathbf{B}$.

1.14 Vectorial Field from the Vector Potential

and

$$\left(\frac{\partial A_y}{\partial x} - \frac{\partial A_x}{\partial y}\right) = F_z(x, y, z). \tag{1.103}$$

In this system of equations the A_x, A_y and A_z are the unknowns, and F_x, F_y and F_z the data. As we can suppose, the above complicated system will not have a solution in a simple way. Therefore, it would be necessary to have some compatibility equations in order to give some restrictions to that system, and make it solvable. One easy restriction we can imagine in the properties of the nabla mathematical operator. As we know that $\nabla \cdot (\nabla \times \mathbf{F}) = 0$, we can add to the (1.101), (1.102) and (1.103) the following condition

$$\left(\frac{\partial F_x}{\partial x} + \frac{\partial F_y}{\partial y} + \frac{\partial F_z}{\partial z}\right) = 0. \tag{1.104}$$

For finding $\mathbf{A}$ we can start by choosing a component zero, and with this condition we can try to adjust the other components by playing with the constants that appear in the integrations, as we demonstrate below.

Let us set, for instance, $A_x = 0$. If that holds, the second and third equations of the system reduce to

$$-\frac{\partial A_z}{\partial x} = F_y(x, y, z), \tag{1.105}$$

and

$$\frac{\partial A_y}{\partial x} = F_z(x, y, z). \tag{1.106}$$

Integrating both equations we have

$$A_z = -\int F_y(x, y, z)dx + C(y, z), \tag{1.107}$$

and

$$A_y = \int F_z(x, y, z)dx + D(y, z). \tag{1.108}$$

If we would like to calculate one of the infinite possibilities for $\mathbf{A}$ without more specifications we can put $C(y, z) = 0$ or $D(y, z) = 0$ with the aim to simplify the equations. Let us seek a solution with $C(y, z) = 0$. For this case and employing (1.101) yields

$$\left(\frac{\partial A_z}{\partial y} - \frac{\partial A_y}{\partial z}\right) = -\frac{\partial}{\partial y}\int F_y(x, y, z)dx - \frac{\partial}{\partial z}\int F_z(x, y, z)dx - \frac{\partial D(y, z)}{\partial z} = F_x(x, y, z). \tag{1.109}$$

Supposing there exist continuous partial derivatives for the components of $\mathbf{F}$ in the domain where this function is defined, we can introduce $\frac{\partial}{\partial y}$ and $\frac{\partial}{\partial z}$ under the symbol of integration, that is

$$-\int \frac{\partial F_y(x,y,z)}{\partial y}dx - \int \frac{\partial F_z(x,y,z)}{\partial z}dx - \frac{\partial D(y,z)}{\partial z} = F_x(x,y,z), \quad (1.110)$$

or what is the same

$$-\int \left(\frac{\partial F_y(x,y,z)}{\partial y} + \frac{\partial F_z(x,y,z)}{\partial z}\right)dx - \frac{\partial D(y,z)}{\partial z} = F_x(x,y,z). \quad (1.111)$$

But this expression may be reduced if we take into consideration the restriction (1.104). Owing to that we can substitute $\left(\frac{\partial F_y}{\partial y} + \frac{\partial F_z}{\partial z}\right)$ by $-\frac{\partial F_x}{\partial x}$, thus

$$\int \frac{\partial F_x}{\partial x}dx + \frac{\partial D(y,z)}{\partial z} = F_x(x,y,z) \Rightarrow F_x(x,y,z) - \frac{\partial D(y,z)}{\partial z} = F_x(x,y,z), \quad (1.112)$$

then

$$D(y,z) = H(y), \quad (1.113)$$

that is, this constant of integration (with respect to x) could depend on y. Bringing this last result to (1.108), we know all components of the vector potential.

This method has two variants. The first one consists on the same calculation made but instead of doing the first integrations as (1.107) and (1.108), we could extend such integrals from one initial point $P_0(x_0, y_0, z_0)$ to another generic end point as follows

$$A_z = -\int_{x_0}^{x} F_y(x,y,z)dx + C(y,z), \quad (1.114)$$

and the same for A_y. The only difference is that by this procedure we obtain another form for the constant $D(y,z) = H(y)$, but the basic idea is not affected in any way.

The second variation we can use refers to (1.109). If we do not suppose that we can introduce the partial derivatives into the corresponding integrals we have had for $D(y,z)$

$$D(y,z) = \int \left\{-\frac{\partial}{\partial y}\int F_y(x,y,z)dx - \frac{\partial}{\partial z}\int F_z(x,y,z)dx\right\}dz - \int F_x(x,y,z)dz, \quad (1.115)$$

and now, introducing this expression into (1.108) the value of the component A_y of the vector potential is obtained.

1.14.2 Second Method

In a similar way we saw for the scalar potential in Sect. 1.13.3 another possibility for calculating the vector potential, if the domain is star-shaped, is the following

1.14 Vectorial Field from the Vector Potential

$$A(x, y, z) = \int_0^1 (\mathbf{F}(\mathbf{P}_1) \times \mathbf{r}) \lambda d\lambda, \qquad (1.116)$$

where $\mathbf{P}_1 = (\lambda x, \lambda y, \lambda z)$, λ being a parameter for which $0 \leq \lambda \leq 1$.

Solved Problems

Problems A

1.1 Obtain the expression of the gradient of a scalar function $V(x, y, z)$ in cylindrical and spherical coordinates.

Solution

As we have seen the relation between the differential of a scalar function and the gradient is

$$dV(x, y, z) = \nabla V(x, y, z) d\mathbf{l}. \qquad (1.117)$$

In cartesian coordinates the element $d\mathbf{l}$ is (dx, dy, dz), but if we wish to calculate the expression of $\nabla V(x, y, z)$ in other coordinates we must first change the form of the differential element. In our case we would like to obtain the gradient in cylindrical coordinates, then we put

$$d\mathbf{l} = (d\rho, \rho d\phi, dz) = d\rho \mathbf{u}_\rho + \rho d\phi \mathbf{u}_\phi + dz \mathbf{u}_z. \qquad (1.118)$$

Introducing (1.118) into (1.117) we have

$$dV(\rho, \phi, z) = ((\nabla V)_\rho, (\nabla V)_\phi, (\nabla V)_z)(d\rho, \rho d\phi, dz) = (\nabla V)_\rho d\rho + (\nabla V)_\phi \rho d\phi + (\nabla V)_z dz, \qquad (1.119)$$

where $(\nabla V)_i$ represents each of the components of the gradient along ρ, ϕ, and z, respectively.

On the other hand, the differential of any scalar function is by definition

$$dV(\rho, \phi, z) = \frac{\partial V}{\partial \rho} d\rho + \frac{\partial V}{\partial \phi} d\phi + \frac{\partial V}{\partial z} dz. \qquad (1.120)$$

Therefore, by identifying term by term of (1.119) and (1.120), we can write

$$(\nabla V)_\rho = \frac{\partial V}{\partial \rho}, \qquad (1.121)$$

$$(\nabla V)_\phi \rho = \frac{\partial V}{\partial \phi}, \qquad (1.122)$$

and

$$(\nabla V)_z = \frac{\partial V}{\partial z}, \tag{1.123}$$

hence, the gradient of V in cylindrical coordinates is

$$\nabla V(\rho, \phi, z) = \left(\frac{\partial V}{\partial \rho}, \frac{1}{\rho} \frac{\partial V}{\partial \phi}, \frac{\partial V}{\partial z} \right). \tag{1.124}$$

For calculating a similar expression in spherical coordinates, we will follow the same procedure. Let us write the differential of $V(r, \phi, \theta)$ as function of the gradient, i.e.,

$$dV(r, \phi, \theta) = ((\nabla V)_r, (\nabla V)_\phi, (\nabla V)_\theta)(dr, r \sin\theta \, d\phi, r d\theta), \tag{1.125}$$

and developing the scalar product

$$dV(r, \phi, \theta) = (\nabla V)_r dr + (\nabla V)_\phi r \sin\theta \, d\phi + (\nabla V)_\theta r d\theta. \tag{1.126}$$

The differential with respect to r, ϕ and θ is

$$dV(\rho, \phi, z) = \frac{\partial V}{\partial r} dr + \frac{\partial V}{\partial \phi} d\phi + \frac{\partial V}{\partial \theta} d\theta. \tag{1.127}$$

Now, identifying the corresponding parts of both equalities (1.126) and (1.127), we have

$$(\nabla V)_r = \frac{\partial V}{\partial r}, \tag{1.128}$$

$$(\nabla V)_\phi r \sin\theta = \frac{\partial V}{\partial \phi}, \tag{1.129}$$

and

$$(\nabla V)_\theta r = \frac{\partial V}{\partial \theta}, \tag{1.130}$$

therefore, the gradient in spherical coordinates has the form

$$\nabla V(r, \phi, \theta) = \left(\frac{\partial V}{\partial r}, \frac{1}{r \sin\theta} \frac{\partial V}{\partial \phi}, \frac{1}{r} \frac{\partial V}{\partial \theta} \right). \tag{1.131}$$

1.2 Starting from the differential equation of the field lines, calculate the curve Γ that corresponds to the following vector field

$$\mathbf{F}(x, y, z) = -y\mathbf{u}_x + x\mathbf{u}_y + P\mathbf{u}_z,$$

P being a constant.

Solution

For obtaining lines of the field in cartesian coordinates we use (1.19). If the components of the field are F_x, F_y, and F_z, the following equality holds

$$\frac{dx}{F_x} = \frac{dy}{F_y} = \frac{dz}{F_z}. \tag{1.132}$$

Setting $F_x = -y$, $F_y = x$, and $F_z = P$ in that equation, we have

$$\frac{dx}{-y} = \frac{dy}{x} = \frac{dz}{P}. \tag{1.133}$$

This equation may be separated in two parts,

$$\frac{dx}{-y} = \frac{dy}{x}, \tag{1.134}$$

and

$$\frac{dy}{x} = \frac{dz}{P}. \tag{1.135}$$

The first one can be integrated directly, i.e.,

$$xdx = -ydy \Rightarrow \int xdx = -\int ydy \Rightarrow \frac{1}{2}x^2 = -\frac{1}{2}y^2 + C \tag{1.136}$$

where C is a constant, due to the indefinite integration. Grouping the variables x and y we can write

$$xdx = -ydy \Rightarrow \int xdx = -\int ydy \Rightarrow \frac{1}{2}x^2 + \frac{1}{2}y^2 = C \Rightarrow x^2 + y^2 = 2C = G, \tag{1.137}$$

G being a constant, which is the equation of a circle.

1.3 Calculate the gradient of the following scalar field

$$V(x, y, z) = x + 2y + 3z,$$

at the points of coordinates $P(1, 2, 3)$ and $Q(4, -2, 9)$.

Solution

By directly using the definition of gradient we apply the nabla operator

$$\nabla = \frac{\partial}{\partial x}\mathbf{u}_x + \frac{\partial}{\partial y}\mathbf{u}_y + \frac{\partial}{\partial z}\mathbf{u}_z \tag{1.138}$$

to the scalar function $V(x, y, z)$, i.e.,

$$\nabla V(x, y, z) = \frac{\partial V(x, y, z)}{\partial x}\mathbf{u}_x + \frac{\partial V(x, y, z)}{\partial y}\mathbf{u}_y + \frac{\partial V(x, y, z)}{\partial z}\mathbf{u}_z, \quad (1.139)$$

obtaining

$$\nabla V(x, y, z) = 1\mathbf{u}_x + 2\mathbf{u}_y + 3\mathbf{u}_z. \quad (1.140)$$

If we introduce the points $P(1, 2, 3)$ and $Q(4, -2, 9)$ into (1.140) the result does not vary. Actually, the expression of the gradient for this function does not depend on the variables x, y or/and z, and therefore its value is the same everywhere. The reason for this result may be found in the geometry of the surface $V(x, y, z)$ studied. In fact, considering V as a level surface it represents planes parallel to each other in the space. As we have explained in the theory (see also Problems 1.6 and 1.7), the gradient geometrically represents a vector perpendicular to the *level surface* at each point. In the case of this problem the surface is a plane, which has the same normal vector everywhere, and therefore the result is the same independently of the point chosen for calculating the gradient. As we will see in other problems, usually this is not the case for any scalar function.

1.4 Calculate the flux of the vector field

$$\mathbf{F} = x^2 \mathbf{u}_x + xy \mathbf{u}_y + z \mathbf{u}_z$$

across the plane $2x + 3y + z = 6$ that corresponds to the first octant of the $OXYZ$ system.

Solution

In order to calculate the flux we must first obtain the unitary perpendicular vector to the surface $S(x, y, z) \equiv 2x + 3y + z - 6 = 0$ at any point. With this aim we use (1.60) which leads to

$$\mathbf{n} = \pm \frac{\nabla S(x, y, z)}{|\nabla S(x, y, z)|} = \frac{2\mathbf{u}_x + 3\mathbf{u}_y + \mathbf{u}_z}{\sqrt{4+9+1}} = \frac{2\mathbf{u}_x + 3\mathbf{u}_y + \mathbf{u}_z}{\sqrt{14}}. \quad (1.141)$$

where the sign $(+)$ represents the outward normal. Following Sect. 1.10, introducing it into (1.27) we have

$$\int\int_S \mathbf{F} \cdot d\mathbf{S} = \int\int_S \mathbf{F} \cdot \mathbf{n} \, dS = \frac{1}{\sqrt{14}} \int\int_S (x^2 \mathbf{u}_x + xy\mathbf{u}_y + z\mathbf{u}_z) \cdot (2\mathbf{u}_x + 3\mathbf{u}_y + \mathbf{u}_z) dS =$$
$$= \frac{1}{\sqrt{14}} \int\int_S (2x^2 + 3xy + z) \, dS. \quad (1.142)$$

As we can see we have three variables, but the element dS. To solve this difficulty we choose, from the equation of the surface in implicit form, for instance, $z = f(x, y)$ giving $z = f(x, y) = 6 - 2x - 3y$. Now, substituting this value of z in the above integral and employing (1.141) we obtain (Sect. 1.10.3)

$$\frac{1}{\sqrt{14}} \int \int_S (2x^2 + 3xy + f(x,y)) \sqrt{1 + \left(\frac{\partial f(x,y)}{\partial x}\right)^2 + \left(\frac{\partial f(x,y)}{\partial y}\right)^2} \, dx \, dy. \tag{1.143}$$

The partial derivatives are

$$\frac{\partial f(x,y)}{\partial x} = -2, \tag{1.144}$$

and

$$\frac{\partial f(x,y)}{\partial y} = -3, \tag{1.145}$$

thus the root square

$$\sqrt{1 + \left(\frac{\partial f(x,y)}{\partial x}\right)^2 + \left(\frac{\partial f(x,y)}{\partial y}\right)^2} = \sqrt{1 + 4 + 9} = \sqrt{14}. \tag{1.146}$$

Writing the value of $f(x,y)$

$$\int \int_{S_{OXY}} (2x^2 + 3xy + (6 - 2x - 3y)) \, dx \, dy. \tag{1.147}$$

As we can see this last expression depends only of the variables x and y, therefore we can easily perform the calculation. The domain of integration is surface of the triangle formed by the straight line $6 - 2x - 3y = 0$ and the axis OX and OY. Grouping the variables may be obtained

$$\int \int_{S_{OXY}} (2x^2 - 2x - 3y + 3xy + 6) \, dx \, dy. \tag{1.148}$$

The limits of integration are determined by using the equation $y = 2 - \frac{2}{3}x$. So, $0 \leq y \leq 2 - \frac{2}{3}x$ and $0 \leq x \leq 3$. Substitution of these limits in the integrals

$$\int_0^3 \int_0^{2-\frac{2}{3}x} (2x^2 - 2x - 3y + 3xy + 6) \, dx \, dy = \int_0^3 \left(2x^2 y - 2xy - \frac{3}{2}y^2 + 3x\frac{y^2}{2} + 6y\right)_0^{2-\frac{2}{3}x} dx =$$

$$= \int_0^3 \left(-\frac{2}{3}x^3 + \frac{2}{3}x^2 + 2x + 6\right) dx = \left(-\frac{1}{6}x^4 + \frac{2}{9}x^3 + x^2 + 6x\right)_0^3 = 19.5. \tag{1.149}$$

1.5 A square thin plate of side $L = 20$ cm of a material placed with one of its corners at the origin, has a non-homogeneous density which is given by the following function in two variables

$$\rho(x, y) = 7800 + 10000(x^2 + y^2), \tag{1.150}$$

Fig. 1.16 Three dimensional representation of the scalar function $\rho(x, y) = 7800 + 10000(x^2 + y^2)$ in the domain $[0, 0.2] \times [0, 0.2]$ m². Observe the curves of constant density (level curves) on the plane OXY

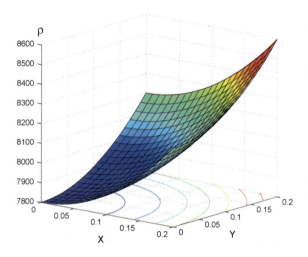

in kg m^{-3}. Calculate: (a) The curves of the same density. (b) The gradient at the points of coordinates $(4, 6)$ cm and $(10, 10)$ cm. Give an interpretation of the result obtained.

Solution

(a) Before beginning to calculate both questions it is important to understand the problem from a geometrical viewpoint.

The physical problem, that is, the density of the plate, is a two dimensional problem, and then the function giving the density has two variables. However, geometrically, we need three dimensions (axis) for representing the scalar function $\rho(x, y)$. Actually, this is a general characteristic of all scalar functions of two variables. If we draw (1.150) we have Fig. 1.16 which corresponds to a paraboloid.

The curves equal density must verify the following equation

$$\rho(x, y) = 7800 + 10000(x^2 + y^2) = C, \tag{1.151}$$

where C is a constant. This equation may be written in an equivalent form by leaving the variables alone in a member

$$x^2 + y^2 = \frac{C - 7800}{10000}. \tag{1.152}$$

As the right side of (1.152) is constant, that equation represent a family of circumferences centered at the origin of coordinates. Mathematically it corresponds to the intersection of the function $\rho(x, y)$ with a plane parallel to the OXY plane, whose distance to the origin of coordinates varies as the constant C changes. Therefore, what we see in Fig. 1.16 is the projection of this intersection on the plane OXY.

(b) To calculate the gradient of a function in cartesian coordinates (1.35) may be used, but considering only two variables, i.e.,

Fig. 1.17 Level curves of the function $\rho(x, y)$

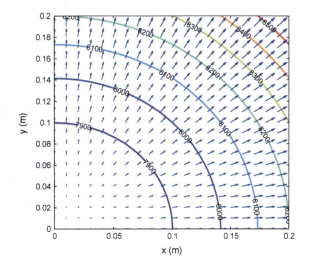

$$\nabla \rho(x, y) = \frac{\partial \rho}{\partial x}\mathbf{u}_x + \frac{\partial \rho}{\partial y}\mathbf{u}_y. \qquad (1.153)$$

Applying that equation we have

$$\nabla \rho(x, y) = 20000x\mathbf{u}_x + 20000y\mathbf{u}_y. \qquad (1.154)$$

By introducing the two points of coordinates (4, 6) cm and (10, 10) cm, we obtain

$$\nabla \rho(0.04, 0.06) = 800\mathbf{u}_x + 1200\mathbf{u}_y. \qquad (1.155)$$

and

$$\nabla \rho(0.1, 0.1) = 2000(\mathbf{u}_x + \mathbf{u}_y), \qquad (1.156)$$

respectively.

Figure 1.17 depicts the gradient of that function for many points. It is interesting to note that the gradient of a function is *perpendicular to the curve of level* of the corresponding problem studied, then the gradient (and the curve of level) is located always on the physical object (plate of material in this case) which, mathematically appears in the domain of the scalar function (plane XY), and *not on the surface that corresponds to the graphic of the scalar function in three dimensions!* (see the note at the end of the next problem).

1.6 The temperature in celsius degrees inside of a metallic sphere of radius $R = 10$ cm is given by the following function

$$T(x, y, z) = 1000(x^2 + y^2 + z^2). \qquad (1.157)$$

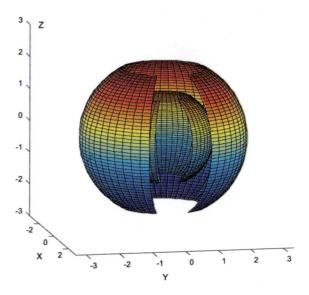

Fig. 1.18 Family of spheres of constant radius. Each surface corresponds to a different constant C in (1.158)

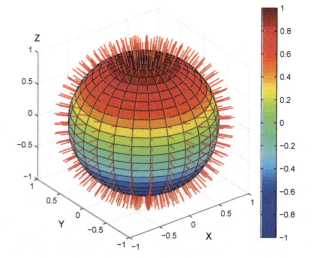

Fig. 1.19 Picture of the gradient of the function $T(x, y, z) = 1000(x^2 + y^2 + z^2) = C$. Note that the vectors are perpendicular to the surface level, which in this case has three dimensions (sphere). This surface is placed in the domain of $T(x, y, z)$, and then it is not possible to represent (visualize) such scalar function $T(x, y, z)$ (we would need four dimensions)

Obtain: (a) The surfaces of constant temperature. (b) The gradient at any interior point of the sphere. (c) A picture of the gradient. Comment on this result.

Solution

(a) As we have seen in the preceding problem, when representing a scalar function of two variables it is necessary to consider one additional dimension (variable) to draw the graphic of this function. The present exercise corresponds to the temperature of a sphere which physically is a three dimensional problem, then the function in

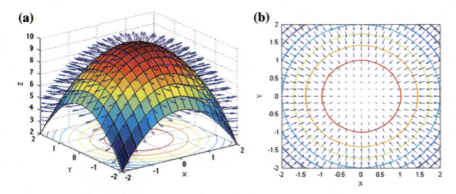

Fig. 1.20 a Vectors perpendicular to a surface $z = f(x, y)$. These normal vectors do not represent the gradient of $f(x, y)$ (error!). **b** Level curves of $z = f(x, y)$ and the corresponding gradient. Observe that the gradient of a function has the same dimension as the number of independent variables (two in this case) and that it must be placed in the domain of the function and not on the three-dimensional graphic $z = f(x, y)$ (**a**)

cartesian coordinates has three variables.[12] If we want to represent in a graphic the value temperature (1.157) at every point inside of the metallic sphere we immediately see that it is impossible. In fact, as the temperature T has three variables, we should have one additional dimension (axis) for drawing the graphic which means we need a four-dimensional space. In general, for all cases of scalar functions of three or more variables, the construction of a graphic is impossible. However, for functions of three variables we can represent the family of surfaces of constant temperature, because them are located in the domain of the corresponding function, which has three dimensions.

To calculate the surfaces inside of the sphere of the same temperature we equal (1.157) to a constant C

$$T(x, y, z) = 1000(x^2 + y^2 + z^2) = C, \qquad (1.158)$$

that is

$$x^2 + y^2 + z^2 = \frac{C}{1000}. \qquad (1.159)$$

This resulting equation corresponds to a family of spheres of different radius, centered at the origin $(0, 0, 0)$ (Fig. 1.18).
(b) By using (1.35), we have for the gradient

---

[12] Be careful when examining the physical dimensions of a problem and the number of variables. Many problems, from the viewpoint of their mathematical representation (variables) may be reduced to a fewer number of variables, even though physically they are posed in space or in a plane. For example, in this problem if we introduce spherical coordinates the temperature becomes $T(r, \phi, \theta) = 1000r$, which is a function of only one variable. However, the physical problem continues being three-dimensional.

$$\nabla T(x, y, z) = 2000x\mathbf{u}_x + 2000y\mathbf{u}_y + 2000z\mathbf{u}_z. \tag{1.160}$$

(c) Picture of the gradient is in Fig. 1.19.

Note: A common error of the students when representing the gradient of a function of two variables is to draw it perpendicular to the function self (graphic in three dimensions-Fig. 1.20a) instead to the level curves in two dimensions (Fig. 1.20b).[13] Conceptually, they confuse the surface $z = f(x, y)$ with the surface level $F(x, y, z) = C$, C being a constant (as, for instance Fig. 1.18), of other three dimensional problems (because both are surfaces). Observe that in this last case the resulting surface $F = C$ is located in the domain of $F(x, y, z)$, then it does not matter with the three dimensional representation of $z = f(x, y)$ of a two dimensional problem.

1.7 Let us suppose a fluid whose particles move with constant velocity (Fig. 1.21). Calculate the rotational of the field velocities. Interpret the result.

Solution

Employing (1.41) to a vector $\mathbf{v}$ constant, we have

$$\nabla \times \mathbf{v} = 0 \tag{1.161}$$

Due to the rotational of the velocity at any point of the fluid is zero, the movement is said to be irrotational or potential. As a result the curvilinear integral (circulation) of the velocity over a closed curve must be zero. In fact, by applying Stokes's theorem we have

$$\oint_\Gamma \mathbf{v} d\mathbf{l} = \int\int_S \nabla \times \mathbf{v} = 0. \tag{1.162}$$

This result means that, in case of a movement potential it is not possible to find closed lines of current.[14] When (1.162) applies we can find a scalar function $\varphi(\mathbf{r})$ so that its gradient gives the velocity, i.e.,

$$\mathbf{v} = \nabla \varphi(\mathbf{r}). \tag{1.163}$$

Observe the similitude between this equation and the relation for the electrostatic field $\mathbf{E} = \nabla V(\mathbf{r})$ (see Sect. 2.4).[15]

---

[13] The same applies for any scalar function of n independent variables, i.e., the gradient has n components located in the domain of the function, and the representation of the hypersurface needs $(n + 1)$ dimensions.

[14] This result is valid if the domain where the fluid moves is a simply connected region. If the region is multiply connected the circulation over a closed curve could be not zero (see Sect. 1.12).

[15] Actually, we could have posed one electromagnetic problem substituting the velocity of the fluid by the electric field $\mathbf{E}$, and we would obtain the same result and interpretation. However, the examples with fluids and other mechanical pictures are easier to understand by the students.

Fig. 1.21 Field of constant velocities. Though the velocity has three components, this picture depicts the velocity on a plane parallel to the vectors for commodity

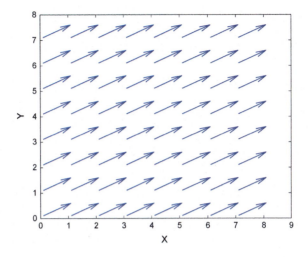

1.8 A wall is held on the plan OXZ and a fluid moves parallel to it. Under some conditions of the fluid stream the velocity may be expressed by the following function

$$\mathbf{v} = by\mathbf{u}_x, \qquad (1.164)$$

b being a constant. Obtain: (a) A graphic representation of the field velocities. (b) The rotational at any point. (c) The divergence.

Solution

(a) Physically (1.164) shows the characteristics of fluid that moves parallel to a solid wall. As it may be seen the fluid velocity on the wall is zero, which means that the fluid close to the solid does not move. The thin layer in contact with the wall is governed by viscous forces and appears as glued to it due to the molecular forces between both the fluid and the wall. A graphic representation of the field velocities is depicted in Fig. 1.22.

In fluid mechanics this kind of flow that moves in parallel layers is said to be laminar flow and, as we can observe, for moving in parallel laminas it is not necessarily the same velocity.

(b) The rotational in cartesian coordinates of the field of velocities may be calculated by (1.41), i.e.,

$$\nabla \times \mathbf{v} = -b\mathbf{u}_z. \qquad (1.165)$$

This result is shown in Fig. 1.23 (blue vectors). We can also calculate (Fig. 1.24a)

$$\oint_\Gamma \mathbf{v}d\mathbf{l} = \int\int_S \nabla \times \mathbf{v}d\mathbf{S} = -b\int\int_S \mathbf{u}_z d S \mathbf{u}_z = -bS. \qquad (1.166)$$

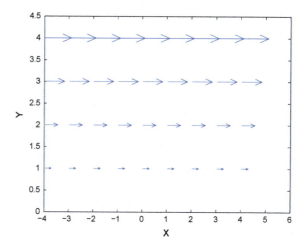

Fig. 1.22 Field of velocities of a fluid moving parallel to a solid wall

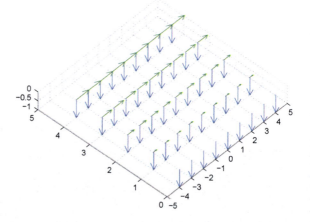

Fig. 1.23 *Green* vectors represent the field velocities and the *blue* one the value of the curl at any point

Taking into account the definition of rotational (1.40) the results obtained (1.165) and (1.166) mean that when a fluid moves with a velocity proportional to the distance to a wall and parallel to it, a net circulation per unit surface appears, i.e., *the density of circulation* at any point of the fluid is $-b$ (see Fig. 1.23). Geometrically, the non-nullity of the rotational shows that there is a net projection of the vector **v** along the closed curve Γ distinct to zero (Fig. 1.24b). Roughly speaking it would mean that, if we would place very small paddle wheels (we can neglect their weights) at different parts of the fluid we would see a rotation at points where the rotational is not zero, being the angular velocity of each wheel a maximum if its rotation axis is parallel to $\nabla \times \mathbf{v}$.[16]

---

[16] A complementary conceptual viewpoint is given in continuum mechanics. In fact, the movement of a point Q in the neighbor of a point P may be expressed as the addition of three terms, namely $\mathbf{v}(Q) = \mathbf{v}(P) + \frac{1}{2}\nabla \times \mathbf{v}(P) \times (Q - P) + \tilde{\zeta}(Q - P)$, where $\tilde{\zeta} = \frac{d\tilde{\varepsilon}}{dt}$ represents the velocity of

Solved Problems

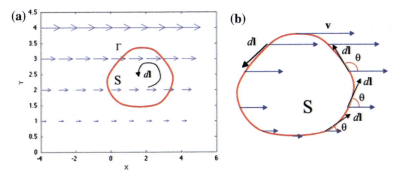

Fig. 1.24 a Closed curve Γ and $d\mathbf{l}$ to perform the curvilinear integral $\oint_\Gamma \mathbf{v} d\mathbf{l}$. b Due to $\nabla \times \mathbf{v} \neq 0$, there exist a circulation of $\mathbf{v}$ per unit surface

(c) Using (1.44) we easily deduce $\nabla \cdot \mathbf{v} = 0$. Remembering the concept of divergence this means that, in the fluid studied there is no net flux of velocity per unit volume.[17]

1.9 Let us consider a rigid solid whose center of gravity coincides with the origin of the coordinate system. This body rotates around one axis that passes through its center with an angular velocity $\omega = \omega \mathbf{u}_z$. Obtain the rotational of the velocity $\mathbf{v}$.

Solution

As it is known, the linear velocity is related with the angular velocity through the equation $\mathbf{v} = \omega \times \mathbf{r}$. If we introduce ω and $\mathbf{r} = x\mathbf{u}_x + y\mathbf{u}_y + z\mathbf{u}_z$ in that equation we have

$$\mathbf{v} = -y\omega\mathbf{u}_x + x\omega\mathbf{u}_y = \omega(-y\mathbf{u}_x + x\mathbf{u}_y). \tag{1.167}$$

This result may be expressed in polar coordinates by changing the variables $x = \rho \cos\phi$ and $y = \rho \sin\phi$, obtaining

$$\mathbf{v} = \omega\rho(-\sin\phi\mathbf{u}_x + \cos\phi\mathbf{u}_y) = \omega\rho\mathbf{u}_\phi, \tag{1.168}$$

where $\mathbf{u}_\phi = -\sin\phi\mathbf{u}_x + \cos\phi\mathbf{u}_y$. Physically it means the longer distance of a point to the rotation axis, the bigger linear velocity. The rotational may be directly calculated by using its definition (Figs. 1.25 and 1.26),

---

(Footnote 16 continued)
the deformation, and $\tilde{\varepsilon} = \frac{1}{2}\left(\frac{\partial u_{ij}}{\partial x_j} + \frac{\partial u_{ji}}{\partial x_i}\right)$ is the strain tensor. Physically this expression says that each small part Q of a continuum system nearby of P translates with the velocity of P, rotates with angular velocity $\omega = \frac{1}{2}\nabla \times \mathbf{v}$ and deforms at velocity of dilatation $\tilde{\zeta}$. As we can intuitively see, the rotational at a point P is in some extent related with the angular velocity of the fluid around P.

[17]In the same way we have explained in the previous footnote for $\nabla \times \mathbf{v}$, we can associate an intuitive understanding to the divergence too. In fact, when $\nabla \cdot \mathbf{v} = \beta \neq 0$, β represents the velocity of volume deformation. Thus, in this problem $\nabla \cdot \mathbf{v} = 0$ means that if we choose a small fluid volume element we cannot observe a deformation of such a volume in time.

Fig. 1.25 Rigid solid turning with constant angular velocity $\omega = \omega \mathbf{u}_z$

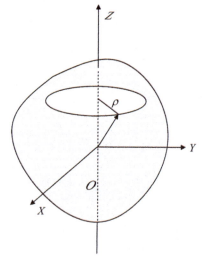

Fig. 1.26 Velocities of the solid on a plane perpendicular to the axis OZ

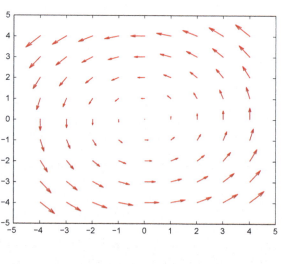

$$\nabla \times \mathbf{v} = 2\omega \mathbf{u}_z = 2\boldsymbol{\omega} \tag{1.169}$$

The rotational is distinct to zero and constant, then we have the same circulation per unit length at any point of the solid. If we remember (see footnote 16) that $\boldsymbol{\omega} = \frac{1}{2}\nabla \times \mathbf{v}$, we obtain $\boldsymbol{\omega} = \boldsymbol{\omega}$. Taking into account that the system only has rotational motion, the result is logical. In fact it means that the neighbor region of any point on a plane perpendicular to the OZ axis rotates at angular velocity ω (Fig. 1.27).

1.10 Obtain the expression of the laplacian operator in cylindrical coordinates.

Solution

The laplacian operator in cartesian coordinates is defined as

Fig. 1.27 Field velocities and rotational on a plane perpendicular to the rotation axis. Note that the rotational (*blue* vectors) is the same at any point

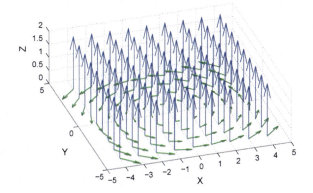

$$\nabla^2 \equiv \frac{\partial^2}{\partial x^2} + \frac{\partial^2}{\partial y^2} + \frac{\partial^2}{\partial z^2}. \quad (1.170)$$

Let us suppose a function $f(x, y, z)$ in this coordinate frame, and we would like to know how such an operator is transformed when working in cylindrical coordinates. This case is sometimes useful in problems with such a symmetry. To calculate the new expression we start with the relation between cartesian and cylindrical coordinates, that is

$$x = \rho \cos \phi, \quad y = \rho \sin \phi, \quad z = z. \quad (1.171)$$

The problem is now to carry out the partial derivatives taking into consideration (1.171). Thus, application of the chain rule leads to

$$\frac{\partial f}{\partial x} = \frac{\partial f}{\partial \rho} \frac{\partial \rho}{\partial x} + \frac{\partial f}{\partial \phi} \frac{\partial \phi}{\partial x}, \quad (1.172)$$

$$\frac{\partial f}{\partial y} = \frac{\partial f}{\partial \rho} \frac{\partial \rho}{\partial y} + \frac{\partial f}{\partial \phi} \frac{\partial \phi}{\partial y}, \quad (1.173)$$

and

$$\frac{\partial f}{\partial z} = \frac{\partial f}{\partial z}. \quad (1.174)$$

We can concentrate our attention on (1.172) and (1.173), because (1.174) does not change therefore may be written

$$\frac{\partial \rho}{\partial x} = \frac{\partial}{\partial x} \sqrt{x^2 + y^2} = \frac{x}{\sqrt{x^2 + y^2}} = \cos \phi, \quad (1.175)$$

$$\frac{\partial \rho}{\partial y} = \frac{\partial}{\partial y} \sqrt{x^2 + y^2} = \frac{y}{\sqrt{x^2 + y^2}} = \sin \phi, \quad (1.176)$$

$$\frac{\partial \phi}{\partial x} = \frac{\partial}{\partial x} \arctan\left(\frac{y}{x}\right) = \frac{-\sin \phi}{\rho}, \qquad (1.177)$$

and

$$\frac{\partial \phi}{\partial y} = \frac{\partial}{\partial y} \arctan\left(\frac{y}{x}\right) = \frac{\cos \phi}{\rho}. \qquad (1.178)$$

Introduction of (1.175), (1.176), (1.177) and (1.178) into (1.172) and (1.173) gives

$$\frac{\partial f}{\partial x} = \frac{\partial f}{\partial \rho} \cos \phi + \frac{\partial f}{\partial \phi} \left(\frac{-\sin \phi}{\rho}\right), \qquad (1.179)$$

$$\frac{\partial f}{\partial y} = \frac{\partial f}{\partial \rho} \sin \phi + \frac{\partial f}{\partial \phi} \frac{\cos \phi}{\rho}. \qquad (1.180)$$

The second partial derivatives are

$$\frac{\partial^2 f}{\partial x^2} = \frac{\partial}{\partial \rho}\left(\frac{\partial f}{\partial \rho} \cos \phi + \frac{\partial f}{\partial \phi}\left(\frac{-\sin \phi}{\rho}\right)\right)\frac{\partial \rho}{\partial x} + \frac{\partial}{\partial \phi}\left(\frac{\partial f}{\partial \rho}\cos\phi + \frac{\partial f}{\partial \phi}\left(\frac{-\sin\phi}{\rho}\right)\right)\frac{\partial \phi}{\partial x} =$$

$$= \left(\cos\phi \frac{\partial^2 f}{\partial \rho^2} + \frac{\sin\phi}{\rho^2}\frac{\partial f}{\partial \phi} - \frac{\sin\phi}{\rho}\frac{\partial^2 f}{\partial \rho \partial \phi}\right)\cos\phi +$$

$$+ \left(-\sin\phi \frac{\partial f}{\partial \rho} + \cos\phi \frac{\partial^2 f}{\partial \rho \partial \phi} - \frac{\cos\phi}{\rho}\frac{\partial f}{\partial \phi} - \frac{\sin\phi}{\rho}\frac{\partial^2 f}{\partial \phi^2}\right)\left(\frac{-\sin\phi}{\rho}\right) =$$

$$= \left(\cos^2\phi \frac{\partial^2 f}{\partial \rho^2} + \frac{2\cos\phi\sin\phi}{\rho^2}\frac{\partial f}{\partial \phi} - \frac{2\cos\phi\sin\phi}{\rho}\frac{\partial^2 f}{\partial \rho \partial \phi} + \frac{\sin^2\phi}{\rho}\frac{\partial f}{\partial \rho} + \frac{\sin^2\phi}{\rho^2}\frac{\partial^2 f}{\partial \phi^2}\right). \qquad (1.181)$$

In a similar way we can calculate the second derivative for the variable y, that is,

$$\frac{\partial^2 f}{\partial y^2} = \frac{\partial}{\partial \rho}\left(\frac{\partial f}{\partial \rho}\sin\phi + \frac{\partial f}{\partial \phi}\frac{\cos\phi}{\rho}\right)\frac{\partial \rho}{\partial y} + \frac{\partial}{\partial \phi}\left(\frac{\partial f}{\partial \rho}\sin\phi + \frac{\partial f}{\partial \phi}\frac{\cos\phi}{\rho}\right)\frac{\partial \phi}{\partial y} =$$

$$= \left(\sin\phi \frac{\partial^2 f}{\partial \rho^2} - \frac{\cos\phi}{\rho^2}\frac{\partial f}{\partial \phi} + \frac{\cos\phi}{\rho}\frac{\partial^2 f}{\partial \rho \partial \phi}\right)\sin\phi +$$

$$+ \left(\cos\phi \frac{\partial f}{\partial \rho} + \sin\phi \frac{\partial^2 f}{\partial \rho \partial \phi} - \frac{\sin\phi}{\rho}\frac{\partial f}{\partial \phi} + \frac{\cos\phi}{\rho}\frac{\partial^2 f}{\partial \phi^2}\right)\left(\frac{\cos\phi}{\rho}\right) =$$

$$= \left(\sin^2\phi \frac{\partial^2 f}{\partial \rho^2} - \frac{2\cos\phi\sin\phi}{\rho^2}\frac{\partial f}{\partial \phi} + \frac{2\cos\phi\sin\phi}{\rho}\frac{\partial^2 f}{\partial \rho \partial \phi} + \frac{\cos^2\phi}{\rho}\frac{\partial f}{\partial \rho} + \frac{\cos^2\phi}{\rho^2}\frac{\partial^2 f}{\partial \phi^2}\right). \qquad (1.182)$$

By adding both expressions for $\dfrac{\partial^2 f(\rho,\phi,z)}{\partial x^2}$ and $\dfrac{\partial^2 f(\rho,\phi,z)}{\partial y^2}$ we find the expression for the laplacian,

$$\nabla^2 \equiv \frac{\partial^2 f}{\partial \rho^2} + \frac{1}{\rho}\frac{\partial f}{\partial \rho} + \frac{1}{\rho^2}\frac{\partial^2 f}{\partial \phi^2} + \frac{\partial^2 f}{\partial z^2}. \qquad (1.183)$$

Solved Problems

These coordinate systems we have seen are some of the most common frames used in basic and intermediate electromagnetics. However, we can have many other problems which, because of their physical characteristics, the aforementioned solutions do not work, and then we must employ other forms for the differential equations, if possible. For example, for calculating the potential and electric field of a metallic or dielectric ellipsoid of revolution in presence of a homogeneous electric field we can use spheroidal coordinates. Some of the solutions for the Laplace equation by means of these coordinates may be also employed for obtaining boundary value problems for geometries bounded by a hyperboloid of revolution. Another example is the computation of the potential created by a charged toroidal conductor. In this case the most appropriate frame seems to be expressed by toroidal coordinates.

1.11 A special situation in physics corresponds to some scalar fields that only depend on the distance r between the source and the point where we are going to study such a field. In this case the field may be represented by a function $\varphi(r)$. Calculate the expression of its gradient.

Solution

$$\nabla\varphi(r) = \left(\frac{\partial\varphi(r)}{\partial r}\right)\left(\frac{\partial r}{\partial x}\right)\mathbf{u}_x + \left(\frac{\partial\varphi(r)}{\partial r}\right)\left(\frac{\partial r}{\partial y}\right)\mathbf{u}_y + \left(\frac{\partial\varphi(r)}{\partial r}\right)\left(\frac{\partial r}{\partial z}\right)\mathbf{u}_z \quad (1.184)$$

$$\frac{\partial\varphi(r)}{\partial r} = \varphi'(r) \quad (1.185)$$

$$\frac{\partial r}{\partial x} = \frac{\partial}{\partial x}\left(\sqrt{x^2+y^2+z^2}\right) = \frac{x}{\sqrt{x^2+y^2+z^2}} \quad (1.186)$$

$$\frac{\partial r}{\partial y} = \frac{\partial}{\partial y}\left(\sqrt{x^2+y^2+z^2}\right) = \frac{y}{\sqrt{x^2+y^2+z^2}} \quad (1.187)$$

$$\frac{\partial r}{\partial z} = \frac{\partial}{\partial z}\left(\sqrt{x^2+y^2+z^2}\right) = \frac{z}{\sqrt{x^2+y^2+z^2}} \quad (1.188)$$

$$\nabla\varphi(r) = \varphi'(r)\frac{x}{\sqrt{x^2+y^2+z^2}}\mathbf{u}_x + \varphi'(r)\frac{y}{\sqrt{x^2+y^2+z^2}}\mathbf{u}_y + \varphi'(r)\frac{z}{\sqrt{x^2+y^2+z^2}}\mathbf{u}_z =$$

$$= \varphi'(r)\left(\frac{x\mathbf{u}_x+y\mathbf{u}_y+z\mathbf{u}_z}{\sqrt{x^2+y^2+z^2}}\right) = \varphi'(r)\frac{\mathbf{r}}{|\mathbf{r}|} \quad (1.189)$$

Functions of these characteristics are, for example, the potentials of electric charges or the potentials generated by point masses. Both examples correspond to the well known newtonian potentials which have the form of $\varphi(r) \sim \frac{1}{r}$.

1.12 Calculate the divergence of a vectorial field of the form

$$\mathbf{F}(\mathbf{r}) = \varphi(r)\frac{\mathbf{r}}{r}.$$

Solution

By using the properties of the nabla operator we have

$$\nabla \cdot (\phi(r)\mathbf{G}) = \phi(r)\nabla \cdot \mathbf{G} + \nabla\phi(r) \cdot \mathbf{G}. \qquad (1.190)$$

If we apply this result to $\mathbf{F}(\mathbf{r})$ it gives

$$\nabla \cdot \mathbf{F}(\mathbf{r}) = \nabla \cdot \left(\varphi(r)\frac{\mathbf{r}}{r}\right) = \nabla \cdot \left(\frac{\varphi(r)}{r}\mathbf{r}\right) = \nabla\left(\frac{\varphi(r)}{r}\right) \cdot \mathbf{r} + \left(\frac{\varphi(r)}{r}\nabla \cdot \mathbf{r}\right), \qquad (1.191)$$

then

$$\nabla\left(\frac{\varphi(r)}{r}\right) = \frac{\partial}{\partial x}\left(\frac{\varphi(r)}{r}\right)\mathbf{u}_x + \frac{\partial}{\partial y}\left(\frac{\varphi(r)}{r}\right)\mathbf{u}_y + \frac{\partial}{\partial z}\left(\frac{\varphi(r)}{r}\right)\mathbf{u}_z. \qquad (1.192)$$

Now employing the chain rule, we get

$$\frac{\partial}{\partial x}\left(\frac{\varphi(r)}{r}\right) = \left(\frac{\partial \varphi(r)}{\partial r}\right)\left(\frac{\partial r}{\partial x}\right)\frac{1}{r} + \varphi(r)\frac{\partial}{\partial r}\left(\frac{1}{r}\right)\frac{\partial r}{\partial x} \qquad (1.193)$$

Taking into account that

$$\frac{\partial r}{\partial x} = \frac{\partial}{\partial x}\left(\sqrt{x^2+y^2+z^2}\right) = \frac{x}{\sqrt{x^2+y^2+z^2}} \qquad (1.194)$$

the aforementioned equality (1.193) leads to

$$\frac{\partial}{\partial x}\left(\frac{\varphi(r)}{r}\right) = \varphi'(r)\frac{x}{\sqrt{x^2+y^2+z^2}}\left(\frac{1}{r}\right) + \varphi(r)\left(\frac{-1}{r^2}\right)\frac{x}{\sqrt{x^2+y^2+z^2}}$$
$$= \frac{x}{r^3}\left(r\varphi'(r) - \varphi(r)\right). \qquad (1.195)$$

Proceeding in the same manner with the variables y and z it is easy to obtain

$$\frac{\partial}{\partial y}\left(\frac{\varphi(r)}{r}\right) = \frac{y}{r^3}\left(r\varphi'(r) - \varphi(r)\right), \qquad (1.196)$$

and

$$\frac{\partial}{\partial z}\left(\frac{\varphi(r)}{r}\right) = \frac{z}{r^3}\left(r\varphi'(r) - \varphi(r)\right). \qquad (1.197)$$

Thus, introducing these results in (1.192) leads to

$$\nabla\left(\frac{\varphi(r)}{r}\right) = \left(\frac{r\varphi'(r) - \varphi(r)}{r^3}\right)(x\mathbf{u}_x + y\mathbf{u}_y + z\mathbf{u}_z) = \left(\frac{r\varphi'(r) - \varphi(r)}{r^3}\right)\mathbf{r}. \tag{1.198}$$

On the other hand, in the second part of (1.191) appears $\nabla \cdot \mathbf{r}$, whose value is

$$\nabla \cdot \mathbf{r} = \frac{\partial x}{\partial x} + \frac{\partial y}{\partial y} + \frac{\partial z}{\partial z} = 3. \tag{1.199}$$

The final expression of (1.191) is

$$\nabla \cdot \mathbf{F}(\mathbf{r}) = \left(\frac{r\varphi'(r) - \varphi(r)}{r^3}\right)\mathbf{r} \cdot \mathbf{r} + \left(\frac{\varphi(r)}{r} \cdot 3\right) = \left(\frac{r\varphi'(r) - \phi(r)}{r}\right) + 3\frac{\varphi(r)}{r}. \tag{1.200}$$

Simplifying that equation we obtain

$$\nabla \cdot \mathbf{F}(\mathbf{r}) = \varphi'(r) + 2\frac{\varphi(r)}{r}. \tag{1.201}$$

1.13 Using the results of the Problems 1.11 and 1.12 obtain the laplacian of a scalar function $\phi(r)$ that only depends on r, that $\nabla^2 \phi(r) = \Delta \phi(r)$.

Solution

The laplacian of function may be written in the following form

$$\nabla^2 \phi(r) = \nabla \cdot (\nabla \phi(r)), \tag{1.202}$$

where $\nabla \phi(r)$ represents the gradient.[18] As we have demonstrated (1.189), the gradient of a function only depending on distance has the form

$$\nabla \phi(r) = \phi'(r)\frac{\mathbf{r}}{|\mathbf{r}|}, \tag{1.203}$$

which is a vector field. At this point we must not calculate the divergence of equality (1.203) again. We have in the preceding exercise demonstrated the value of the divergence of a vector field depending only on distance. In fact, let us put the gradient (1.203) $\mathbf{D}(\mathbf{r})$, and let us name $\phi'(r) = \varphi(r)$, that is

$$\mathbf{D}(\mathbf{r}) = \phi'(r)\frac{\mathbf{r}}{|\mathbf{r}|} = \varphi(r)\frac{\mathbf{r}}{|\mathbf{r}|}. \tag{1.204}$$

As we can see this function is the same that appears in the Problem 1.12. Therefore, by introducing (1.204) into (1.201) we obtain the solution

---

[18] In this exercise we use $\phi(r)$ instead $\varphi(r)$ in order to be clear and do not confuse with the notation the reader.

$$\nabla \cdot \mathbf{D}(\mathbf{r}) = \varphi'(r) + 2\frac{\varphi(r)}{r}, \tag{1.205}$$

but we named $\phi'(r) = \varphi(r)$, then

$$\nabla^2 \phi(r) = \nabla \cdot \mathbf{D}(\mathbf{r}) = \phi''(r) + 2\frac{\phi'(r)}{r}. \tag{1.206}$$

Observe that, in the case of a newtonian potential this identity is zero. In effect, if we introduce in this (1.206) a function of the form $\phi(r) = \frac{a}{r}$, a being a constant, we have

$$\nabla^2 \phi(r) = \phi''(r) + 2\frac{\phi'(r)}{r} = \frac{2a}{r^3} + 2\left(\frac{-a}{r^2}\right)\frac{1}{r} = 0. \tag{1.207}$$

This results tells us that the potentials of spherical symmetry that are inversely proportional to the distance verify the laplace equation where there are no sources. For instance, this is the case of the potential created by point charges excluding the place where the charge is located (in this point Poisson's equation is verified-see Chap. 2).

1.14 Calculate the flux of the vector field

$$\mathbf{F}(x, y, z) = \frac{1}{4\pi} \frac{b\mathbf{r}}{(x^2 + y^2 + z^2)^{3/2}},$$

where b is a constant, across the sphere of radius a centered at the coordinate origin $(0, 0, 0)$.

Solution

$$\phi = \oint_S \mathbf{F} \cdot d\mathbf{S} = \oint_S \frac{1}{4\pi} \frac{b\mathbf{r}}{(x^2 + y^2 + z^2)^{3/2}} \cdot d\mathbf{S} = \frac{b}{4\pi} \oint_S \frac{\mathbf{r}}{r^3} \cdot \mathbf{n} dS. \tag{1.208}$$

The normal $\mathbf{n}$ to the sphere of radius a coincides with the unitary vector $\mathbf{u}_r = \frac{\mathbf{r}}{r}$ in spherical coordinates, then we can write

$$\phi = \frac{b}{4\pi} \oint_S \frac{\mathbf{r}}{r^3} \cdot \mathbf{u}_r dS = \frac{b}{4\pi} \oint_S \frac{\mathbf{r}}{r^3} \cdot \frac{\mathbf{r}}{r} dS = \frac{b}{4\pi} \oint_S \frac{r^2}{r^4} dS = \frac{b}{4\pi} \oint_S \frac{dS}{r^2}. \tag{1.209}$$

As the radius a remains the same for all points of the sphere, we can substitute $r = a$ in the integral above, obtaining

$$\phi = \frac{b}{4\pi} \oint_S \frac{dS}{a^2} = \frac{b}{4\pi a^2} \oint_S dS = \frac{bS}{4\pi a^2}. \tag{1.210}$$

If we calculate this integral (for instance, by using spherical coordinates), we have $S = 4\pi a^2$, and then the flux

$$\phi = b. \tag{1.211}$$

Note: The area of sphere can be calculated as follows:

$$\oint_S dS = \int_0^{2\pi} \int_0^{\pi} J(r, \phi, \theta) d\phi d\theta, \tag{1.212}$$

where $J(r, \phi, \theta) = r^2 \sin \theta$ is the jacobian of the transformation. Thus

$$\oint_S dS = \int_0^{2\pi} \int_0^{\pi} r^2 \sin \theta d\phi d\theta = \int_0^{\pi} r^2 \sin \theta d\theta \int_0^{2\pi} d\phi = 2\pi \int_0^{\pi} r^2 \sin \theta d\theta. \tag{1.213}$$

As $r = R$ for all point on the sphere we have

$$\oint_S dS = 2\pi R^2 \int_0^{\pi} \sin \theta d\theta = 2\pi R^2 (-\cos \theta)_0^{\pi} = 4\pi R^2. \tag{1.214}$$

When we study the electric field we will see that a field like $\mathbf{F}(x, y, z)$ in this problem corresponds to that created by a point charge, and the flux of the electric field throughout any closed surface equals $\frac{q}{\varepsilon_0}$, where q is the net charge[19] enclosed by the surface and ε_0 is a constant. In this example we obtain the same result by substituting the constant b by $\frac{q}{\varepsilon_0}$.

1.15 Obtain the flux of the vector $\mathbf{r}$ throughout the sphere $x^2 + y^2 + z^2 = R^2$.

Solution

Let $\mathbf{F} = \mathbf{r}$. Applying the definition of flux of a vector field we have

$$\phi = \oint_S \mathbf{F}.d\mathbf{S} = \oint_S \mathbf{r}.\mathbf{n}dS. \tag{1.215}$$

The unitary vector to a sphere we have seen is $\mathbf{u}_r = \frac{\mathbf{r}}{r}$, then we can write

$$\phi = \oint_S \mathbf{r}.\frac{\mathbf{r}}{r} dS = \oint_S \frac{r^2}{r} dS = \oint_S r dS. \tag{1.216}$$

As we calculate the flux across the surface of radius R, we substitute $r = R$ in that integral, which leads to

$$\phi = R \oint_S dS = R 4\pi R^2 = 4\pi R^3. \tag{1.217}$$

---

[19] In this problem the expression of $\mathbf{F}$ corresponds to a point charge located at the origin of coordinates, but the final result (known as Gauss's theorem) holds for any interior distribution of charge, that is, the flux across a closed surface S depends only of the total charge enclosed by S.

Another way to calculate the flux is to use the divergence theorem (1.49). As we know the following identity holds

$$\oint_{\partial V} \mathbf{F} \cdot d\mathbf{S} = \int_V \nabla \cdot \mathbf{F} \, dV, \tag{1.218}$$

then employing the divergence of the field, we should obtain the same result. In fact, computing the divergence of $\mathbf{F}$ we have

$$\nabla \cdot \mathbf{F} = 3,$$

and introducing this value in the right side of (1.49), it gives

$$\iiint_V \nabla \cdot \mathbf{F} \, dV = 3 \iiint_V dV = 3 \frac{4}{3} \pi R^3 = 4\pi R^3, \tag{1.219}$$

result that coincides with (1.217).

1.16 A vector field $\mathbf{B} = B \, \mathbf{u}_z$ has constant modulus at any point of space and has the direction of the positive OZ axis. Obtain the flux across the upper hemisphere $x^2 + y^2 + z^2 = R^2$.

Solution

By definition of flux, we have

$$\phi = \oint_S \mathbf{F}.d\mathbf{S} = \oint_S B \, \mathbf{u}_z . \mathbf{n} \, dS. \tag{1.220}$$

From (1.60),

$$\mathbf{n} = \frac{\nabla S(x, y, z)}{|\nabla S(x, y, z)|} = \frac{x \mathbf{u}_x + y \mathbf{u}_y + z \mathbf{u}_z}{\sqrt{x^2 + y^2 + z^2}} = \frac{x \mathbf{u}_x + y \mathbf{u}_y + z \mathbf{u}_z}{R}. \tag{1.221}$$

Introducing this result into (1.220), we get

$$\oint_S \mathbf{F}.d\mathbf{S} = \oint_S B \, \mathbf{u}_z . \left(\frac{x \mathbf{u}_x + y \mathbf{u}_y + z \mathbf{u}_z}{R} \right) dS = \frac{B}{R} \oint_S z \, dS. \tag{1.222}$$

The equation of the hemisphere can be written in explicit form:

$$z = f(x, y) = \sqrt{R^2 - x^2 - y^2}. \tag{1.223}$$

Applying Sect. 1.10.2, we obtain for dS

$$dS = \sqrt{1 + \left(\frac{\partial f(x, y)}{\partial x} \right)^2 + \left(\frac{\partial f(x, y)}{\partial y} \right)^2} \, dx \, dy, \tag{1.224}$$

thus
$$\frac{\partial f(x,y)}{\partial x} = \frac{-x}{\sqrt{R^2 - x^2 - y^2}} \quad (1.225)$$

and
$$\frac{\partial f(x,y)}{\partial y} = \frac{-y}{\sqrt{R^2 - x^2 - y^2}}. \quad (1.226)$$

Employing these results, we have

$$dS = \sqrt{1 + \frac{x^2}{R^2 - x^2 - y^2} + \frac{y^2}{R^2 - x^2 - y^2}} dxdy = \sqrt{\frac{R^2 - x^2 - y^2 + x^2 + y^2}{R^2 - x^2 - y^2}} dxdy$$
$$= \frac{R}{\sqrt{R^2 - x^2 - y^2}} dxdy. \quad (1.227)$$

Introducing these values in (1.222),

$$\oint_S \mathbf{F} \cdot d\mathbf{S} = \frac{B}{R} \int_S \sqrt{R^2 - x^2 - y^2} \frac{R}{\sqrt{R^2 - x^2 - y^2}} dx\,dy = B \int_S dx\,dy = B\pi R^2. \quad (1.228)$$

Problems B

1.17 A vector field in $\Re^3$ has the following form

$$\mathbf{F}(x,y,z) = -\frac{ay}{x^2 + y^2}\mathbf{u}_x + \frac{ax}{x^2 + y^2}\mathbf{u}_y,$$

a being a constant. (a) Discuss the possibility to find a scalar function $V(\mathbf{r})$ so that $\mathbf{F}(r) = -\nabla V(\mathbf{r})$ and obtain the circulation of this field along the circumference $x^2 + y^2 = R^2$. (b) Compute the line integral along the segments shown in Fig. 1.29.

Solution

(a) We have seen in the theory that a necessary condition for finding a potential function is that the rotational of the vector field be zero. Applying (1.81) to $\mathbf{F}(x, y, z)$, holds
$$\nabla \times \mathbf{F}(\mathbf{r}) = 0. \quad (1.229)$$

At first sight we could think that there exists a potential $V(\mathbf{r})$. If this hypothesis is true, the curvilinear integral along any closed curve in the region where we study the problem must be zero. Let us calculate the $\oint_\Gamma \mathbf{F} d\mathbf{l}$, Γ being the circumference of radius R centered at the origin of coordinates on the plane $z = 0$, Fig. 1.28a,

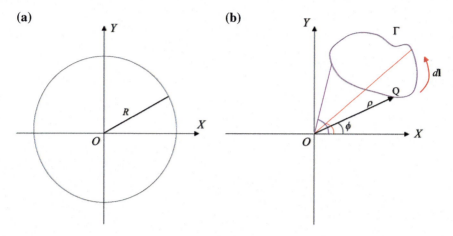

Fig. 1.28 **a** Circumference centered at $P(0, 0, 0)$ contained in the plane $z = 0$. **b** Closed curve Γ that does not intersect the cut and does not pass through the origin

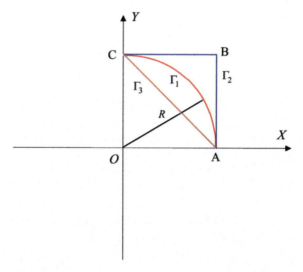

Fig. 1.29 Three different paths with the same ends. The coordinates of A, B, and C are $(R, 0)$, (R, R) and $A(0, R)$, respectively

$$\oint_\Gamma \mathbf{F} \cdot d\mathbf{l} = \oint_\Gamma \left(-\frac{ay}{x^2 + y^2}, \frac{ax}{x^2 + y^2}, 0 \right) \cdot (dx, dy, 0) = a \oint_\Gamma \left(-\frac{y\,dx}{x^2 + y^2} + \frac{x\,dy}{x^2 + y^2} \right). \tag{1.230}$$

Due to the symmetry of the curve Γ we change to polar coordinates by substituting $x = \rho \cos \phi$ and $y = \rho \sin \phi$. Setting these values and those of the differentials $dx = -\rho \sin \phi\, d\phi$, $dy = \rho \cos \phi\, d\phi$ we have

$$a \oint_0^{2\pi} \left(\frac{\rho^2 \sin^2 \phi}{\rho^2} + \frac{\rho^2 \cos^2 \phi}{\rho^2} \right) d\phi = a \oint_0^{2\pi} \left(\frac{R^2 \sin^2 \phi}{R^2} + \frac{R^2 \cos^2 \phi}{R^2} \right) d\phi$$
$$= a \oint_0^{2\pi} d\phi = 2\pi a. \quad (1.231)$$

This non-nullity of the calculation shows that condition (1.81) does not guaranty the existence of a potential function, because the integral calculated does not equal zero. In this case we say that the differential form $\left(-\frac{ydx}{x^2+y^2} + \frac{xdy}{x^2+y^2} \right)$ is closed but not exact (see Sect. 1.12), then the closed line integral depends on the curve chosen. Being $\nabla \times \mathbf{F} = 0$, how can we understand this result? Actually, it is not completely true that the rotational of the field $\mathbf{F}$ is zero, contrary to what seems to be by looking at (1.229). As a matter of fact, in order to be correct we should say that the rotational of $\mathbf{F}$ is zero in all points where this vector field is *well defined*. However, as it may be easily proved, $\mathbf{F}(x, y, z)$ is a discontinuous function at the origin $P(0, 0, 0)$ and therefore any domain containing this point is a multiply connected region, and we have seen that in these cases (1.84) holds. In other words, the calculation of the rotational at $P(0, 0, 0)$ has no mathematical sense because we cannot define the function $\mathbf{F}$ at this point (it tends to infinity), and therefore is not true that $\nabla \times \mathbf{F} = 0$ there.[20] To conclude this discussion we see that the condition curl$(\mathbf{F}(\mathbf{r})) = 0$ is not sufficient for $\mathbf{F}(\mathbf{r})$ to be a gradient in the region $G = (\Re^2 - \{(0,0)\})^{21}$; it would be both necessary and sufficient if G was convex, but this is not the case.

In order to be clear for understanding the idea, we are going to do the same calculation, but now along another closed curve that does not contain the origin of coordinates. Let us suppose we choose the curve Γ represented in the Fig. 1.28b. As the curve is displaced from the origin, not only the angle ϕ varies, but also the distance from O to a point over the curve. We use polar coordinates again but when calculating dx and dy we take into consideration that ρ is not constant (as, for instance, a circumference centered at the origin). By using $x = \rho \cos \phi$ and $y = \rho \sin \phi$ we can write

$$dx = d(\rho \cos \phi) = -\rho \sin \phi \, d\phi + \cos \phi \, d\rho, \quad (1.232)$$

and

$$dy = d(\rho \sin \phi) = \rho \cos \phi \, d\phi + \sin \phi \, d\rho. \quad (1.233)$$

Introducing both expressions into (1.230) we have

$$\oint_\Gamma \mathbf{F} \cdot d\mathbf{l} = a \oint_\Gamma \left(-\frac{\rho \sin \phi(-\rho \sin \phi \, d\phi + \cos \phi \, d\rho)}{\rho^2} + \frac{\rho \cos \phi(\rho \cos \phi \, d\phi + \sin \phi \, d\rho)}{\rho^2} \right). \quad (1.234)$$

---

[20] In the language of the *distributions* we could express the rotational of the field at $(0, 0)$ as $\nabla \times \mathbf{F} = 2\pi a \delta(\rho) \mathbf{u}_z$, $\delta(\rho)$ being the Dirac's delta.

[21] Though the problem is posed in $\Re^3$, the vector field is symmetric with respect the z coordinate, therefore we can refer our study to a two dimensional plane perpendicular to the OZ axis.

By reducing the terms we obtain

$$a \oint_\Gamma \frac{\rho^2 \, d\phi}{\rho^2} = a \oint_\Gamma d\phi = a \int_{\phi_i}^{\phi_f} d\phi, \quad (1.235)$$

where ϕ_i and ϕ_f represent the initial and final angles formed by ρ with the OX axis. But in this case $\phi_i = \phi_f$, because Γ is closed, thus

$$a \int_{\phi_i}^{\phi_f} d\phi = \phi_f - \phi_i = 0. \quad (1.236)$$

This result agrees with the explanations given in Sect. 1.12, and shows that if we do not surround the origin $(0, 0, 0)$ the closed integral is zero. Even though we cannot find a gradient in the domain studied, it is easy to note that **F** is the differential of

$$\phi(x, y) = \tan^{-1}\left(\frac{y}{x}\right). \quad (1.237)$$

This result may be immediately obtained by means of the procedure shown in Sect. 1.13.1. By examining (1.237) we see that this function jumps (Figs. 1.30 and 1.31), and from another point of view this is the reason because the circulation around a closed curve containing the origin is nonzero. However, we may render $\phi(x, y)$ single-valued by choosing a radial cut along one angle (appropiately), for instance, $\phi(x, y) = \phi_0$ (it depends on the function). In this way we restrict valuable values of $\phi(x, y)$ to $\phi_0 \leq \phi(x, y) < \phi_0 + 2\pi$ (Fig. 1.32). Due to the introduction of this cut (observe that the point $(0, 0)$ is included in this restriction), we can consider $\phi(x, y)$ as a gradient in this domain, and therefore, it is logical that the circulation of **F** along any closed simple curve that not contain the origin is not zero. As a corollary, the curvilinear integral over an open curve not passing the origin of coordinates should be only dependent of the initial and end points of that curve (limits of integration). With the aim to see more in deep this possibility, we can examine the next question.

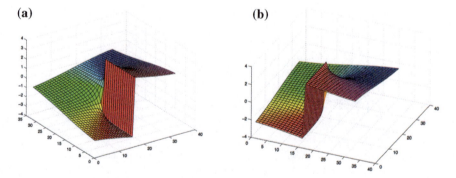

Fig. 1.30 **a** Function $\tan^{-1}\left(\frac{y}{x}\right)$. **b** Idem from another view

Fig. 1.31 Lines of constant values of the function $\tan^{-1}\left(\frac{y}{x}\right)$

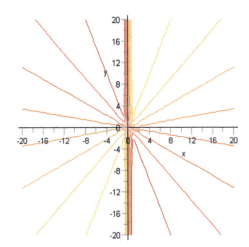

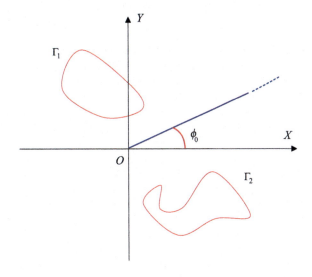

Fig. 1.32 Restriction of the values by means of a cut

(b) Let us calculate the curvilinear integrals along some open curves. Figure 1.29 depicts three open paths with the same ends. The first one Γ_1 corresponds to a quarter of circle; the second one Γ_2 is half of square, and the third Γ_3 is a segment. All of them are different but they have a common characteristic, namely they do not pass across the origin $(0, 0)$. Let us see whether the results depend on the curve chosen.

For the integral along Γ_1 we do not need to repeat the computation in detail. The procedure is the same as shown in (1.231), with the only difference that now the integral extends to $\frac{\pi}{2}$ instead 2π, i.e.,

$$a \int_0^{\frac{\pi}{2}} \left(\frac{R^2 \sin^2 \phi}{R^2} + \frac{R^2 \cos^2 \phi}{R^2} \right) d\phi = a \int_0^{\frac{\pi}{2}} d\phi = a\frac{\pi}{2}. \qquad (1.238)$$

Let us choose Γ_2. The integration along this curve may be divided in two paths: from A to B, and from B to C. In the first one (AB) x remains constant, then $dx = 0$ throughout and only y changes. The opposite occurs for the segment (BC), that is, y is a constant and x varies. Taking this facts into account the line integral gives

$$\int_{\Gamma_2} \mathbf{F} \cdot d\mathbf{l} = \int_A^B \mathbf{F} \cdot d\mathbf{l} + \int_B^C \mathbf{F} \cdot d\mathbf{l} = a \left(0 + \int_0^R \frac{xdy}{x^2+y^2} \right) + a \left(-\int_R^0 \frac{ydx}{x^2+y^2} + 0 \right). \qquad (1.239)$$

The semi-straight lines for the segments (AB) and (BC) are, respectively, $x = R$ and $y = R$, therefore

$$a \int_0^R \frac{Rdy}{x^2+y^2} - a \int_R^0 \frac{Rdx}{x^2+y^2} = a \tan^{-1}\left(\frac{y}{R}\right)\Big|_0^R - a \tan^{-1}\left(\frac{x}{R}\right)\Big|_R^0 = a\frac{\pi}{2}, \qquad (1.240)$$

which is the same as we have obtained before.

In order to be sure, we will calculate the path integral along Γ_3. The segment that lies points A and C is $y = R - x$, thus

$$\int_{\Gamma_3} \mathbf{F} \cdot d\mathbf{l} = a \int_\Gamma \left(-\frac{ydx}{x^2+y^2} + \frac{xdy}{x^2+y^2} \right) = a \left(\int_R^0 -\frac{(R-x)dx}{x^2+(R-x)^2} + \int_R^0 \frac{x(-dx)}{x^2+(R-x)^2} \right) =$$
$$= -aR \int_R^0 \frac{dx}{x^2+(R-x)^2}. \qquad (1.241)$$

With a few calculation this integral converts to

$$-a\left(\frac{2}{R}\right) \int_R^0 \frac{dx}{1 + \left(\frac{2x}{R} - 1\right)^2} = -a \tan^{-1}\left(\frac{2x}{R} - 1\right)\Big|_R^0 = a\frac{\pi}{2}. \qquad (1.242)$$

As we can notice, we obtain the same values for the integral along three different curves with the same ends. It means that, even though we have demonstrated that there is not a gradient in $\Re^2$ excluding only the origin of coordinates, the integral between two point not passing through $(0, 0)$ does not depend on the curve (of course, on the basis that previously $\nabla \times \mathbf{F} = 0$ for these points belonging to $G = (\Re^2 - \{(0,0)\})$, Fig. 1.13). To conclude this problem we can say that, when investigating the possibility of finding a potential function of any vector field we must not only calculate the curl of such a field to prove whether it is zero or not, but we must also study the domain where this field is defined.

To close this problem it is interesting to comment that in Chap. 5, when we study the magnetic field produced by a very large conducting wire currying a constant current I, we will obtain this vector field $\mathbf{F}$ again. There we will have a similar expression like $\mathbf{F}$ where $a = \dfrac{\mu_0 I}{2\pi}$, i.e.,

$$\mathbf{B}(x, y, z) = \dfrac{\mu_0 I}{2\pi \rho} \mathbf{u}_\phi, \qquad (1.243)$$

and we will study Ampère's theorem which asserts that the curvilinear integral of the magnetic field along a closed curve depends only of the net current hooked up by the curve. The point $(0, 0)$ at which we had the discontinuity corresponds to the place where the linear current is located (along the OZ axis), and this is the physical reason because the domain of definition of $\mathbf{F}$ is a multiply connected region.

1.18 Evaluate the flux of the vector field

$$\mathbf{F}(x, y, z) = x\mathbf{u}_x + y\mathbf{u}_y + z\mathbf{u}_y$$

across the cylinder $x^2 + y^2 = R^2$ and limited by the planes $z = \pm a$.

Solution

$$\phi = \oint_S \mathbf{F} \cdot d\mathbf{S} = \oint_S \mathbf{r} \cdot \mathbf{n} dS. \qquad (1.244)$$

The cylinder is formed by three surfaces, the lateral, the basis and the top. For this reason, we can split the surface integral in three parts corresponding to the three surfaces of the cylinder. The unitary vectors of the outward normal to every surface are the following (1.60)

(a) Lateral surface $S_L(x, y, z) = x^2 + y^2 - R^2 = 0$.

$$\mathbf{n}_L = \dfrac{\nabla S(x, y, z)}{|\nabla S(x, y, z)|} = \dfrac{2x\mathbf{u}_x + 2y\mathbf{u}_y}{2\sqrt{x^2 + y^2}} = \dfrac{x\mathbf{u}_x + y\mathbf{u}_y}{R}. \qquad (1.245)$$

(b) Upper basis surface (top) $S_1(x, y, z) = z - a = 0$.

$$\mathbf{n}_1 = \dfrac{\mathbf{u}_z}{1} = \mathbf{u}_z = (0, 0, 1). \qquad (1.246)$$

(c) Lower basis surface $S_2(x, y, z) = z + a = 0$.

$$\mathbf{n}_2 = -\dfrac{\mathbf{u}_z}{1} = -\mathbf{u}_z = (0, 0, -1). \qquad (1.247)$$

In this last equation observe that we have introduced a minus to account that the outward normal at this surface is opposite to calculated in (b).

$$\phi = \oint_S \mathbf{F} \cdot \mathbf{n} dS = \iint_{S_L} \mathbf{F} \cdot \mathbf{n}_L dS + \iint_{S_1} \mathbf{F} \cdot \mathbf{n}_1 dS + \iint_{S_2} \mathbf{F} \cdot \mathbf{n}_2 dS = \qquad (1.248)$$

$$\iint_{S_L} (x, y, z) \cdot \frac{(x, y, 0)}{R} dS + \iint_{S_1} (x, y, z) \cdot (0, 0, 1) dS + \iint_{S_2} (x, y, -z) \cdot (0, 0, -1) dS.$$

In the last integral appears $-z$ as the third component of the field because its projection over OZ is negative (lower space under the plane $z = 0$). Making the scalar products may be obtained

$$\phi = \iint_{S_L} \frac{x^2 + y^2}{R} dS + \iint_{S_1} z\,dS + \iint_{S_2} z\,dS = \iint_{S_L} \frac{R^2}{R} dS + 2\iint_{S_1} z\,dS =$$
$$= R \iint_{S_L} dS + 2z \iint_{S_1} dS. \tag{1.249}$$

The lateral surface of a cylinder is $2\pi Rh$, h being its height, and of a circle πR^2. Considering that $h = 2a$ and $z = \dfrac{h}{2} = \dfrac{2a}{2} = a$ we can write

$$\phi = RS_L + 2zS_1 = R2\pi R\, 2a + 2\pi R^2 a = 6\pi R^2 a. \tag{1.250}$$

The same result should be obtained by means of the divergence theorem

$$\oint_{\partial V} \mathbf{F} \cdot d\mathbf{S} = \int_V \nabla \cdot \mathbf{F}\, dV. \tag{1.251}$$

In order to apply it, let us first calculate the divergence of the field

$$\nabla \cdot \mathbf{F} = 3,$$

and bringing this value in the right side of (1.49) holds

$$\iiint_V \nabla \cdot \mathbf{F}\, dV = 3 \iiint_V dV = 3V. \tag{1.252}$$

Taking into consideration that the volume of a cylinder is $\pi R^2 h$, and that in our case $h = 2a$, it yields

$$\phi = 3\pi R^2 2a = 6\pi R^2 a. \tag{1.253}$$

1.19 Calculate the potential function of the vector field

$$\mathbf{F}(x, y, z) = 2xy^2 z^2 \mathbf{u}_x + 2yx^2 z^2 \mathbf{u}_y + 2zx^2 y^2 \mathbf{u}_z.$$

Solution

Before beginning we must prove that the necessary conditions for the existence of a potential function hold. We can find a function so that its gradient gives $\mathbf{F}(x, y, z)$ if, at least, $\nabla \times \mathbf{F} = 0$. In fact, introducing $\mathbf{F}$ into (1.41) we obtain curl $\mathbf{F} = 0$. Therefore, as $\mathbf{F}(x, y, z)$ is defined in a connected region of $\Re^3$, it is in principle possible to find a scalar function $V(\mathbf{r})$ so that $\mathbf{F}(x, y, z) = \nabla V(\mathbf{r})$. Using the procedure explained

Solved Problems

in Sect. 1.13, let us compute the first integral with respect to the variable x

$$F_x(x, y, z) = -\frac{\partial V}{\partial x} = 2xy^2z^2 \Rightarrow$$

$$V(x, y, z) = \int F_x(x, y, z)dx + C(y, z) = -\int 2xy^2z^2 dx = -x^2y^2z^2 + C(y, z). \tag{1.254}$$

Once we have a general expression of V, where $C(y, z)$ is yet a unknown function, we apply (1.91) as follows

$$F_y(x, y, z) = -\frac{\partial V}{\partial y} = 2yx^2z^2 - \frac{\partial C(y, z)}{\partial y}, \tag{1.255}$$

but we know the second component of **F**, thus introducing F_y into (1.255)

$$2yx^2z^2 - \frac{\partial C(y, z)}{\partial y} = F_y(x, y, z) = 2yx^2z^2 \Rightarrow -\frac{\partial C(y, z)}{\partial y} = 0. \tag{1.256}$$

Because of partial derivative of $C(y, z) = 0$, this function must be a constant (respect to y), which we will label as D, i.e.,

$$C(y, z) = D(z), \tag{1.257}$$

then

$$V(x, y, z) = -x^2y^2z^2 + D(z). \tag{1.258}$$

Operating in the same way as before, we can calculate the value of $D(z)$ by employing (1.92)

$$F_z(x, y, z) = -\frac{\partial V}{\partial z} = -2zx^2y^2 + \frac{\partial D(z)}{\partial z} = -2zx^2y^2 \Rightarrow \frac{\partial D(z)}{\partial z} = 0 \Rightarrow D(z)$$
$$= \text{constant} = R. \tag{1.259}$$

Finally, the potential may be written

$$V(x, y, z) = -x^2y^2z^2 + R. \tag{1.260}$$

By this second procedure we apply (1.98) and we choose at initial point of the finite line element $A = (0, 0, 0)$, then

$$V(x, y, z) = -\int_{x_0}^{x} F_x(x, 0, 0)dx - \int_{y_0}^{y} F_y(x, y, 0)dy - \int_{z_0}^{z} F_z(x, y, z)dz =$$
$$= -\int_{x_0}^{x} 0 dx - \int_{y_0}^{y} 0 dy - \int_{z_0}^{z} 2x^2y^2z dz = -x^2y^2z^2 + R. \tag{1.261}$$

Let us introduce $P_1 = (\lambda x, \lambda y, \lambda z)$ into the function $\mathbf{F}$ of (1.100), then we obtain

$$\mathbf{F}(\lambda \mathbf{r}) = 2xy^2z^2\lambda^5\mathbf{u}_x + 2yx^2z^2\lambda^5\mathbf{u}_y + 2zx^2y^2\lambda^5\mathbf{u}_y = \lambda^5\left(2xy^2z^2\mathbf{u}_x + 2yx^2z^2\mathbf{u}_y + 2zx^2y^2\mathbf{u}_y\right). \tag{1.262}$$

Using $\mathbf{r} = x\mathbf{u}_x + y\mathbf{u}_y + z\mathbf{u}_z$ and calculating the scalar product $\mathbf{F}(\lambda \mathbf{r}) \cdot \mathbf{r}$ in (1.100) we obtain the following integral

$$V(x, y, z) = -\int_0^1 \mathbf{F}(\mathbf{P}_1) \cdot \mathbf{r} \, d\lambda + C = -\int_0^1 \mathbf{F}(\lambda\mathbf{r}) \cdot \mathbf{r} \, d\lambda + C =$$
$$= -\int_0^1 2\lambda^5 \left(xy^2z^2 x + yx^2z^2 y + zx^2y^2 z\right) d\lambda + C. \tag{1.263}$$

The last integral depends only on λ, and therefore we can put the expression depending on x, y, and z outside. Under these conditions (1.263) gives

$$V(x, y, z) = -\int_0^1 2\lambda^5 (x^2y^2z^2 + y^2x^2z^2 + z^2x^2y^2) d\lambda + C$$
$$= -\frac{2}{6}(3x^2y^2z^2) + R = -x^2y^2z^2 + R, \tag{1.264}$$

which agrees with the results obtained before.

1.20 Obtain the vector potential of the vector field $\mathbf{F}(x, y, z) = B\mathbf{u}_z$.

Solution

For calculating the vector potential we employ the fact that

$$\mathbf{F}(\mathbf{r}) = \nabla \times \mathbf{A}(\mathbf{r}) = B\mathbf{u}_z. \tag{1.265}$$

As we have only one component for $\mathbf{F}$ we can write

$$\left(\frac{\partial A_y}{\partial x} - \frac{\partial A_x}{\partial y}\right) = B. \tag{1.266}$$

The easiest possibility to fulfill this condition is to choose $A_x(x, y, z) = 0$, thus

$$\frac{\partial A_y}{\partial x} = B \Rightarrow A_y = \int B dx = Bx + C(y, z), \tag{1.267}$$

where $C(y, z)$ is a function that may depends on y and/or z. We must find the value of the component A_z. To this end we can use the fact that curl $\mathbf{F}(\mathbf{r})$ does not have projections over x and y, then the cross partial derivatives verify

Solved Problems

$$\left(\frac{\partial A_z}{\partial y} - \frac{\partial A_y}{\partial z}\right) = 0, \tag{1.268}$$

and

$$\left(\frac{\partial A_x}{\partial z} - \frac{\partial A_z}{\partial x}\right) = 0. \tag{1.269}$$

Equation (1.268) still contains information about A_y, but also appears A_z. If we choose $A_z(x, y, z) = 0$, a condition for $C(y, z)$ must be found in order to fulfill (1.268). Let us introduce this value that equality

$$\left(0 - \frac{\partial A_y}{\partial z}\right) = -\frac{\partial}{\partial z}(Bx + C(y, z)) = -\frac{\partial C(y, z)}{\partial z} = 0, \tag{1.270}$$

thus

$$C(y, z) = D(y). \tag{1.271}$$

Of course we can set $D(y) = 0$ but, in general (1.271) holds for any continuous function depending on y. As we can observe, as $A_x = A_z = 0$ (1.269) is immediately accomplished. Owing to $C(y, z) = D(y)$, a possibility for the vector potential is the following

$$\mathbf{A}(\mathbf{r}) = (Bx + D(y))\mathbf{u}_y. \tag{1.272}$$

Note that $\mathrm{curl}((Bx + D(y))\mathbf{u}_y) = B\mathbf{u}_z$.

But the vector potential is not single-valued, which means that such a potential is not unique (indeterminate). In effect, we could find other relationship for $\mathbf{A}(\mathbf{r})$, namely $\mathbf{A}'(\mathbf{r})$, which differ each other in the gradient of a scalar function. In other words, if we write

$$\mathbf{A}'(\mathbf{r}) = \mathbf{A}(\mathbf{r}) + \nabla \varphi(\mathbf{r}), \tag{1.273}$$

$\varphi(\mathbf{r})$ being scalar, we get the same $\mathbf{F}(\mathbf{r})$ due to $\nabla \times (\nabla \varphi(\mathbf{r})) = 0$, always. As a consequence an infinite number of vector potentials apply to one vector field.

In order to understand this problem, we will obtain another $\mathbf{A}$ that gives the same $\mathbf{F}$. However, in this case not by adding a gradient, but by directly calculating the potential through (1.116) (or by the procedure demonstrated above setting other values for A_x and A_z). Let us write the solution in the form of the integral

$$\mathbf{A}'(x, y, z) = \int_0^1 [\mathbf{F}(\mathbf{P}_1) \times \mathbf{r}] \lambda d\lambda, \tag{1.274}$$

where $\mathbf{r}$ is the vector (x, y, z) and $\mathbf{F}(\mathbf{P}_1) = \mathbf{F}(\lambda \mathbf{r}) = B\mathbf{u}_z$. The vectorial product gives

$$\mathbf{F}(\mathbf{P}_1) \times \mathbf{r} = xB\mathbf{u}_y - yB\mathbf{u}_x = B(-y\mathbf{u}_x + x\mathbf{u}_y). \tag{1.275}$$

By introducing it into (1.274) yields

$$\mathbf{A}'(x, y, z) = \int_0^1 B(-y\mathbf{u}_x + x\mathbf{u}_y)\lambda\, d\lambda = B(-y\mathbf{u}_x + x\mathbf{u}_y)\int_0^1 \lambda\, d\lambda = \frac{1}{2}B(-y\mathbf{u}_x + x\mathbf{u}_y). \quad (1.276)$$

This result is also valid and shows that **A** is multivalued.

Sometimes we have the opposite question, that is, starting from two potentials, what is the function $\varphi(\mathbf{r})$ that satisfies (1.273)? To solve this problem let us arrange (1.273) as follows

$$\mathbf{A}(\mathbf{r}) - \mathbf{A}'(\mathbf{r}) = -\nabla\varphi(\mathbf{r}) = \mathbf{W}(\mathbf{r}). \quad (1.277)$$

This function $\mathbf{W}(\mathbf{r})$ is the difference of the vector potentials that we have previously calculated, i.e.,

$$\mathbf{W}(\mathbf{r}) = Bx\mathbf{u}_y - \frac{1}{2}B(-y\mathbf{u}_x + x\mathbf{u}_y) = \frac{1}{2}B(y\mathbf{u}_x + x\mathbf{u}_y), \quad (1.278)$$

where, for simplicity, we have chosen $D(y) = 0$. As $\mathbf{W}(\mathbf{r})$ comes from a gradient, $\varphi(\mathbf{r})$ may be computed by means of any of the methods shown in Sect. 1.12. For instance, employing (1.99) yields

$$\varphi(x, y, z) = -\int_{x_0}^{x} W_x(x, y_0, z_0)dx - \int_{y_0}^{y} W_y(x, y, z_0)dy - \int_{z_0}^{z} W_z(x, y, z)dz + R, \quad (1.279)$$

R being an additive constant. Using the components of (1.278) we have

$$\varphi(x, y, z) = -\int_{x_0}^{x} 0\, dx - \int_{y_0}^{y} \frac{1}{2}Bx\, dy - \int_{z_0}^{z} 0\, dz + R = -\frac{1}{2}Bxy + R. \quad (1.280)$$

To test that this potential reproduces the function **W**, simply calculate the gradient of $\varphi(x, y, z)$, giving

$$\mathbf{W} = -\nabla\varphi(\mathbf{r}) = \frac{1}{2}B(y\mathbf{u}_x + x\mathbf{u}_y). \quad (1.281)$$

To conclude a comment about this problem. We have studied two possible vector potentials that generate the same homogeneous field $\mathbf{F}(\mathbf{r}) = B\mathbf{u}_z$. Such a vector field corresponds, for instance, to the magnetic field produced inside a very large solenoid carrying a current I whose revolution axis coincides with OZ (see Chap. 5), or to the magnetic field generated by Helmholtz's coils of radius R at a distance $\frac{R}{2}$ over its symmetry axis with respect to a plane containing any of the coils.

1.21 As we will demonstrate in Chap. 5, the magnetic field generated by a very large conducting wire carrying a current I, and placed along the OZ axis may be expressed by the function

$$\mathbf{B} = \frac{\mu_0 I}{2\pi} \frac{(-y\mathbf{u}_x + x\mathbf{u}_y)}{(x^2 + y^2)}. \quad (1.282)$$

Calculate a vector potential for this field.

Solved Problems

Solution

(a) First procedure

To calculate the vector potential we will use the second method of Sect. 1.14. To do so we must obtain the vectorial product of $\mathbf{B}(\mathbf{P_1})$ and the vector $\mathbf{r}$. In order to express the calculation as easy as possible, let us call the components of the field $(B_x, B_y, 0)$. Using this nomenclature we have

$$\mathbf{B}(\mathbf{P_1}) = \mathbf{B}(\lambda \mathbf{r}) = \frac{\mu_0 I}{2\pi} \frac{(-\lambda y \mathbf{u}_x + \lambda x \mathbf{u}_y)}{\lambda^2(x^2+y^2)} = \frac{\mu_0 I}{2\pi \lambda} \frac{(-y \mathbf{u}_x + x \mathbf{u}_y)}{(x^2+y^2)} = \frac{1}{\lambda}(B_x, B_y, 0), \tag{1.283}$$

and for the product

$$\mathbf{B}(\lambda \mathbf{r}) \times \mathbf{r} = \frac{1}{\lambda} z B_y \mathbf{u}_x - \frac{1}{\lambda} z B_x \mathbf{u}_y + \frac{1}{\lambda}(y B_x - x B_y) \mathbf{u}_z = \frac{1}{\lambda}(z B_y \mathbf{u}_x - z B_x \mathbf{u}_y - \mathbf{u}_z). \tag{1.284}$$

Observe that the component over OZ of the last formula is only $-\mathbf{u}_z$. With this result and (1.116) we conclude

$$\mathbf{A}(x,y,z) = \int_0^1 \frac{1}{\lambda}(z B_y \mathbf{u}_x - z B_x \mathbf{u}_y - \mathbf{u}_z) \lambda d\lambda$$
$$= \int_0^1 (z B_y \mathbf{u}_x - z B_x \mathbf{u}_y - \mathbf{u}_z) d\lambda = z B_y \mathbf{u}_x - z B_x \mathbf{u}_y - \mathbf{u}_z, \tag{1.285}$$

and introducing the projections B_x and B_y the components of $\mathbf{A}$ are

$$A_x(x,y,z) = \frac{\mu_0 I}{2\pi} z \frac{x}{(x^2+y^2)}, \tag{1.286}$$

$$A_y(x,y,z) = \frac{\mu_0 I}{2\pi} z \frac{y}{(x^2+y^2)}, \tag{1.287}$$

and

$$A_z(x,y,z) = -\frac{\mu_0 I}{2\pi}. \tag{1.288}$$

(b) Second procedure

By using the first method of Sect. 1.14 we have many possibilities. For example, choosing $A_z = 0$ or constant we have

$$-\frac{\partial A_y}{\partial z} = F_x(x,y,z), \tag{1.289}$$

$$\frac{\partial A_x}{\partial z} = F_y(x,y,z), \tag{1.290}$$

and
$$\left(\frac{\partial A_y}{\partial x} - \frac{\partial A_x}{\partial y}\right) = F_z(x, y, z). \tag{1.291}$$

Integrating the first equation yields
$$A_y = -\int F_x(x, y, z)dz + C(x, y) = -\int \frac{\mu_0 I}{2\pi} \frac{-y}{(x^2 + y^2)}dz + C(x, y)$$
$$= \frac{\mu_0 I}{2\pi} z \frac{y}{(x^2 + y^2)} + C(x, y), \tag{1.292}$$

and for the second
$$A_x = \int F_y(x, y, z)dz + D(x, y) = \int \frac{\mu_0 I}{2\pi} \frac{x}{(x^2 + y^2)}dz + D(x, y)$$
$$= \frac{\mu_0 I}{2\pi} z \frac{x}{(x^2 + y^2)} + D(x, y). \tag{1.293}$$

Choosing $C(x, y) = 0$, and bringing A_y and A_z into (1.291)
$$\left(\frac{\partial A_y}{\partial x} - \frac{\partial A_x}{\partial y}\right) = F_z(x, y, z) = 0. \tag{1.294}$$

$$\frac{\mu_0 I}{2\pi} zy \frac{-2x}{(x^2 + y^2)^2} - \frac{\mu_0 I}{2\pi} zx \frac{(-2y)}{(x^2 + y^2)^2} - \frac{\partial D(x, y)}{\partial y} = 0, \tag{1.295}$$

thus
$$-\frac{\partial D(x, y)}{\partial y} = 0 \Rightarrow D(x, y) = H(x). \tag{1.296}$$

Summing up the results we can write
$$A_x = \frac{\mu_0 I}{2\pi} z \frac{x}{(x^2 + y^2)} + H(x), \quad A_y = \frac{\mu_0 I}{2\pi} z \frac{y}{(x^2 + y^2)}, \quad A_z = 0. \tag{1.297}$$

Note that if we put $H(x) = 0$ and we would have chosen $A_z = \text{constant} = -\frac{\mu_0 I}{2\pi}$, these results would coincide with the method used previously.

Following this method, we have more possibilities. For instance, if we at first choose $A_x = 0$, we can obtain another form for **A**. In fact, this case corresponds exactly with the development shown in Sect. 1.14. Applying directly (1.108) and (1.107), and setting $D(y, z) = 0$ yields
$$A_y = \int B_z(x, y, z)dx + D(y, z) = 0, \tag{1.298}$$

and

$$A_z = -\int B_y(x, y, z)dx + C(y, z) = -\frac{\mu_0 I}{2\pi}\int \frac{x}{(x^2 + y^2)}dx + C(y, z)$$
$$= -\frac{\mu_0 I}{4\pi}\ln(x^2 + y^2) + C(y, z). \quad (1.299)$$

$$\left(\frac{\partial A_z}{\partial y} - \frac{\partial A_y}{\partial z}\right) = B_x(x, y, z) \Rightarrow -\frac{\mu_0 I}{2\pi}\frac{y}{(x^2+y^2)} + \frac{\partial C(y,z)}{\partial y} = B_x = -\frac{\mu_0 I}{2\pi}\frac{y}{(x^2+y^2)}, \quad (1.300)$$

then

$$\frac{\partial C(y, z)}{\partial y} = 0 \Rightarrow C(y, z) = H(z). \quad (1.301)$$

As a result we can write

$$A_x = 0, \quad A_y = 0, \quad A_z = -\frac{\mu_0 I}{4\pi}\ln(x^2 + y^2) + H(z). \quad (1.302)$$

Usually textbooks of electromagnetism assign to the constant $H(z)$ the value

$$H(z) = \frac{\mu_0 I}{4\pi}\ln(x_0^2 + y_0^2),$$

then the expression for **A** leads to

$$\mathbf{A} = -\frac{\mu_0 I}{4\pi}\ln\left(\frac{x^2 + y^2}{x_0^2 + y_0^2}\right). \quad (1.303)$$

1.22 Let us suppose a scalar field $\varphi(\mathbf{r})$ and a vector field $\mathbf{F}(\mathbf{r})$. Probe the following relations: (a) $\nabla \times (\nabla\varphi(\mathbf{r})) = 0$. (b) $\nabla \cdot (\nabla \times \mathbf{F}(\mathbf{r})) = 0$.

Solution

(a) The gradient of scalar function is given by

$$\nabla\varphi = \frac{\partial\varphi}{\partial x}\mathbf{u}_x + \frac{\partial\varphi}{\partial y}\mathbf{u}_y + \frac{\partial\varphi}{\partial z}\mathbf{u}_z, \quad (1.304)$$

and applying the definition of rotational, we have

$$\nabla \times (\nabla\varphi(\mathbf{r})) = \begin{vmatrix} \mathbf{u}_x & \mathbf{u}_y & \mathbf{u}_z \\ \frac{\partial}{\partial x} & \frac{\partial}{\partial y} & \frac{\partial}{\partial z} \\ \frac{\partial\varphi}{\partial x} & \frac{\partial\varphi}{\partial y} & \frac{\partial\varphi}{\partial z} \end{vmatrix} =$$
$$= \left(\frac{\partial^2\varphi}{\partial y \partial z} - \frac{\partial^2\varphi}{\partial z \partial y}\right)\mathbf{u}_x + \left(\frac{\partial^2\varphi}{\partial z \partial x} - \frac{\partial^2\varphi}{\partial x \partial z}\right)\mathbf{u}_y + \left(\frac{\partial^2\varphi}{\partial x \partial y} - \frac{\partial^2\varphi}{\partial y \partial x}\right)\mathbf{u}_z = 0. \quad (1.305)$$

(b) Let $F_x(x, y, z)$ be the first component of the rotational of the vector field $\mathbf{F(r)}$, whose value is

$$F_x(x, y, z) = \left(\frac{\partial F_z}{\partial y} - \frac{\partial F_y}{\partial z}\right).$$

By introducing this component and the other corresponding to the projections over y and z (1.41) into the definition of divergence, we have

$$\nabla \cdot (\nabla \times \mathbf{F(r)}) = \left(\frac{\partial^2 F_z}{\partial x \partial y} - \frac{\partial^2 F_y}{\partial z \partial x}\right) + \left(\frac{\partial^2 F_x}{\partial z \partial y} - \frac{\partial^2 F_z}{\partial x \partial y}\right) + \left(\frac{\partial^2 F_y}{\partial x \partial y} - \frac{\partial^2 F_x}{\partial y \partial z}\right) = 0. \quad (1.306)$$

Chapter 2
Static Electric Field in Vacuum

Abstract This chapter introduces forces between charges at rest, which are not supposed to be inside any media (Coulomb's law). Concepts such as electric field or electric potential are introduced, as well as its calculus when produced by different charge distributions, including conductive materials. Gauss' law and its use to calculate electric field caused by certain charge distributions is also seen.

2.1 Electric Charge

Charge is a basic and characteristic property of the elementary particles which make up matter. There are two kinds of charges: positive and negative. Every portion of matter contains approximately equal amounts of each type. When speaking about charge, we are referring to the net sum of positive and negative. Therefore, when something is positively charged it is because the amount of positive charges (usually protons) is higher than the negative ones (usually electrons). The electric charge is found in multiples of the elementary charge e (electron or proton charge). It is an experimental fact that charge can be neither created nor destroyed. This is known as **the principle of conservation of charge**: for any process performed in an *isolated system*, net or total charge does not change or in a *non-isolated system* the charge introduced into a system is equal to its increase of charge. Charge is represented by q and its unit in International System (SI) is *coulomb* (C).[1] For continuous distributions of charge, and given the smallness of the elementary charge e, any small element of volume that we consider will be constituted by a large number of electrons and protons. Hence we can consider a charge density function as the limit of the charge per unit volume as the volume becomes infinitesimal, and the corresponding integration will allow us to obtain the overall charge of the object. It is defined as **volume charge density** ρ by

$$\rho = \lim_{\Delta V \to 0} \frac{\Delta q}{\Delta V} \equiv \frac{dq}{dV}, \qquad (2.1)$$

---

[1] Coulomb can be defined as a function of the elementary charge e as $1\,\text{C} = 6.25 \times 10^{18}\,e$. In later chapters, when magnetic experiments are discussed, it will be possible to define it as an ampere's derivative.

which represents the charge per unit volume at each point of a surface. The SI unit is Cm^{-3}. The overall charge q_V in the volume V is obtained as

$$q_V = \int_V \rho dV. \tag{2.2}$$

If the charge is distributed on one surface S, **surface charge density** σ can be defined as

$$\sigma = \lim_{\Delta S \to 0} \frac{\Delta q}{\Delta S} \equiv \frac{dq}{dS}, \tag{2.3}$$

which represents charge per unit area at each point of a surface. Its unit in SI is Cm^{-2}. Total charge q_S on S is obtained as

$$q_S = \int_S \sigma dS. \tag{2.4}$$

When charge is distributed on a material line L, **line charge density** λ can be defined as

$$\lambda = \lim_{\Delta l \to 0} \frac{\Delta q}{\Delta l} \equiv \frac{dq}{dl}, \tag{2.5}$$

that represents charge per unit of length at each point of the line. Its unit in SI is Cm^{-1}. Total charge q_L on L is obtained as:

$$q_L = \int_L \lambda dl. \tag{2.6}$$

2.2 Coulomb's Law

From several observations that took place in 18th century by Coulomb and others, it can be established that force between two electric charges at rest can be mathematically expressed by **Coulomb's law**:

$$\vec{F}_q = k \frac{qq'}{|\vec{r}-\vec{r}'|^2} \frac{\vec{r}-\vec{r}'}{|\vec{r}-\vec{r}'|} = k \frac{qq'}{d^2} \mathbf{u}, \tag{2.7}$$

which states that forces between two point charges q and q' act along the line joining them, and are directly proportional to the product of these charges and inversely proportional to the square of the distance $d = |\mathbf{r}-\mathbf{r}'|$ between them (Fig. 2.1).

Equation (2.7) express the force F_q which acts on q due to q''s action. F_q's direction is determined by the unitary vector $\mathbf{u} = (\mathbf{r}-\mathbf{r}')/|\mathbf{r}-\mathbf{r}'|$, oriented from q' to q, as well as charges' sign. Forces are repulsive if charges have the same sign, and attractive if they have opposite sign. $\mathbf{F}_{q'}$ over q' due to q is the vector $-\mathbf{F}_q$. If $(\mathbf{r}-\mathbf{r}')$ is replaced in (2.7) by $(\mathbf{r}'-\mathbf{r})$, Newton's third Law is obtained.

2.2 Coulomb's Law

Fig. 2.1 Forces between point charges

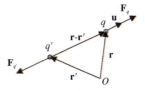

The value of the proportional constant k depends of the units system. In SI, it is

$$k = 10^{-7} c^2, \tag{2.8}$$

where c is the velocity of light in vacuum. It can also be written as

$$k = \frac{1}{4\pi\varepsilon_0}, \tag{2.9}$$

where ε_0, is the *permittivity of free space*. Therefore

$$k \approx 8.9875 \times 10^9 \, \text{Nm}^2/\text{C}^2 \approx 9 \times 10^9 \, \text{Nm}^2/\text{C}^2, \tag{2.10}$$

and

$$\varepsilon_0 \approx 8.8542 \times 10^{-12} \, \text{C}^2/(\text{Nm}^2). \tag{2.11}$$

When several n point charges q_j act on q, it's been experimentally established that the total force acting on a charge is the vector sum of the individual forces which act on it. This is known as **the superposition principle for electrostatic forces**. Therefore, the force is determined by the repeated application of (2.7):

$$\mathbf{F}_q = q \sum_{j=1}^{n} \frac{q_j}{4\pi\varepsilon_0 |\mathbf{r} - \mathbf{r}_j|^2} \frac{\mathbf{r} - \mathbf{r}_j}{|\mathbf{r} - \mathbf{r}_j|} = q \sum_{j=1}^{n} \frac{q_j}{4\pi\varepsilon_0 d_j^2} \mathbf{u}_j, \tag{2.12}$$

where $d_j = |\mathbf{r} - \mathbf{r}_j|$ is the distance between the j-ith charge and q, and $\mathbf{u}_j = (\mathbf{r} - \mathbf{r}_j)/|\mathbf{r} - \mathbf{r}_j|$ is the unitary vector in the direction from q_j to q.

The same principle can be applied in a continuous charge distribution case. If a very small volume dV' is considered at a point in the charge distribution, where the density is ρ, charge inside dV' is, according (2.1), $dq' = \rho dV'$ (Fig. 2.2). If these values are substituted in (2.12) and the sum is substituted by an extended integral to the whole charge, it results:

$$\mathbf{F}_q = \frac{q}{4\pi\varepsilon_0} \int_{V'} \frac{\rho}{|\mathbf{r} - \mathbf{r}'|^2} \frac{\mathbf{r} - \mathbf{r}'}{|\mathbf{r} - \mathbf{r}'|} dV', \tag{2.13}$$

where $\mathbf{r}$ is the point charge position and $\mathbf{r}'$ is the position of each of the volume differentials. Figure 2.2 represents the force $d\mathbf{F}$ of point element $dq' = \rho dV'$ over

Fig. 2.2 Force due to a continuous charge distribution

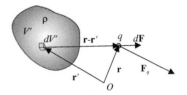

point charge q; $\mathbf{F}_q$'s value comes from adding every element $d\mathbf{F}$. The same expression can be applied when a charge is distributed on a surface or on a line substituting $dq' = \rho dV'$ by $dq' = \sigma dS'$ or $dq' = \lambda dl'$, respectively. The integral in (2.13) is well behaved even in case q falls inside the charge distribution.

2.3 Electric Field

In (2.7), (2.12) and (2.13) it is observed that the force that acts on q is proportional to q. Therefore a vectorial field which is independent from q is introduced. Its dimensions are force per unit of charge. Hence **electric** or **electrostatic field**[2] can be defined as

$$\mathbf{E} = \lim_{q \to 0} \frac{\mathbf{F}}{q}, \qquad (2.14)$$

where the test charge placed at the point goes to zero, so it can be assured that it does not affect the charge distribution which produces $\mathbf{E}$. The electric field unit in SI is NC^{-1}.

For a point charge, the expression of electric field is directly obtained from dividing by q in (2.7):

$$\mathbf{E}(\mathbf{r}) = \frac{q'}{4\pi\varepsilon_0 |\mathbf{r} - \mathbf{r}'|^2} \frac{\mathbf{r} - \mathbf{r}'}{|\mathbf{r} - \mathbf{r}'|} = \frac{q'}{4\pi\varepsilon_0 d^2} \mathbf{u}. \qquad (2.15)$$

When electric field's definition is applied to (2.12) and (2.13) a general equation can be obtained for the electric field due to a given distribution of charge at rest,

$$\mathbf{E}(\mathbf{r}) = \frac{1}{4\pi\varepsilon_0} \sum_{j=1}^{n} \frac{q_j}{|\mathbf{r} - \mathbf{r}_j|^2} \frac{\mathbf{r} - \mathbf{r}_j}{|\mathbf{r} - \mathbf{r}_j|} + \frac{1}{4\pi\varepsilon_0} \int_{V'} \frac{\rho(\mathbf{r}')}{|\mathbf{r} - \mathbf{r}'|^2} \frac{\mathbf{r} - \mathbf{r}'}{|\mathbf{r} - \mathbf{r}'|} dV' +$$
$$+ \frac{1}{4\pi\varepsilon_0} \int_{S'} \frac{\sigma(\mathbf{r}')}{|\mathbf{r} - \mathbf{r}'|^2} \frac{\mathbf{r} - \mathbf{r}'}{|\mathbf{r} - \mathbf{r}'|} dS' + \frac{1}{4\pi\varepsilon_0} \int_{L'} \frac{\lambda(\mathbf{r}')}{|\mathbf{r} - \mathbf{r}'|^2} \frac{\mathbf{r} - \mathbf{r}'}{|\mathbf{r} - \mathbf{r}'|} dl'. \qquad (2.16)$$

---

[2] 'Electrostatic field' is usually used when phenomena are time independent.

2.3 Electric Field

This is the mathematical expression of **superposition principle for electric field**: the electric field created by a number of charges is equal to the sum of the fields produced independently by each of them,[3] where the symbol **r** represents the position vector of the point where the field is calculated (field point), and $\mathbf{r}_j$ or $\mathbf{r}'$ is the vector position of any of the charges or the charge differentials (source point). $\mathbf{r} - \mathbf{r}'$ or $\mathbf{r} - \mathbf{r}_j$ is the vector that goes from each of the source points to the field point, and its magnitude represents distance between them. The sum or integration is calculated over total charge: therefore the variable is not **r**, but $\mathbf{r}_j$ or $\mathbf{r}'$; quantities ρ, σ and λ can also be dependent on variables of position $\mathbf{r}'$.

It is not necessary to apply (2.2)'s formulas to calculate the force that acts on a charged particle q when introduced in a region where exists an electric field **E**, but once **E** is determined it is simple to see from (2.14) that:

$$\mathbf{F} = q\mathbf{E}. \tag{2.17}$$

2.4 Electrostatic Potential

It has been seen in Chap. 1 that if the field's curl is zero, then the vector can be expressed as the gradient of a scalar field. It's easy to demonstrate that $\nabla \times \dfrac{\mathbf{r} - \mathbf{r}'}{|\mathbf{r} - \mathbf{r}'|^3} = 0$, and as every term of (2.16) corresponds to this form, we establish that the electrostatic field is irrotational, and therefore it is derived from a potential:

$$\nabla \times \mathbf{E}(\mathbf{r}) = 0 \Rightarrow \mathbf{E}(\mathbf{r}) = -\nabla V(\mathbf{r}), \tag{2.18}$$

where V is called the **electrostatic potential**. It is important to note that if the phenomena were time-dependent, the electric field's curl would not be zero.

If we bear in mind the gradient's property $dV = \nabla V \cdot d\mathbf{r}$ given by (1.39), and we apply it to (2.18) it results that:

$$V(\mathbf{r}) = -\int \mathbf{E}(\mathbf{r}) \cdot d\mathbf{r}. \tag{2.19}$$

It can be observed that potential represents potential energy per unit charge. So, if we remember potential energy of a conservative force is:

$$E_\mathrm{p}(\mathbf{r}) = -\int \mathbf{F}(\mathbf{r}) \cdot d\mathbf{r}, \tag{2.20}$$

---

[3] The superposition principle has been experimentally checked also for very high field intensities: in engineering practices with fields which reach several millions of volts per meter (accelerators, high voltage discharges), when calculating fields in electron orbits ($E \approx 10^{11} \ldots 10^{17}$ V/m) or when calculating the field of highly weight nucleus ($E \approx 10^{22}$ V/m). For fields over 10^{20} V/m vacuum polarization is introduced and makes the problem non-linear.

and, taking into account from (2.17) that $\mathbf{F} = q\mathbf{E}$, it results that

$$E_p(\mathbf{r})/q = -\int \mathbf{E}(\mathbf{r}) \cdot d\mathbf{r} = V(\mathbf{r}). \qquad (2.21)$$

If two points A and B are taken, it is observed from (2.19) that the *potential difference* between two points is the circulation of the electrostatic field $\mathbf{E}$ between these two points, along any path between them[4]:

$$V_A - V_B = \int_{r_A}^{r_B} \mathbf{E} \cdot d\mathbf{r}. \qquad (2.22)$$

Any convenient point $\mathbf{r}_B = \mathbf{r}_{ref}$ can be chosen as the potential reference, at which $V_B = V_{ref} = 0$ in (2.22). It is common to choose infinity as potential reference. The expression we reach is

$$V_A = \int_{r_A}^{r_{ref}} \mathbf{E} \cdot d\mathbf{r} = \int_{r_A}^{\infty} \mathbf{E} \cdot d\mathbf{r}, \qquad (2.23)$$

which represents the work done by an external agent in transferring the unit of positive charge from infinity to a considered point.

If the $\mathbf{E}$ field is due to a point charge q' (2.15) and (2.19) is integrated, the electrostatic potential is obtained,

$$V(\mathbf{r}) = \frac{q'}{4\pi\varepsilon_0 |\mathbf{r} - \mathbf{r}'|} = \frac{q'}{4\pi\varepsilon_0 d}. \qquad (2.24)$$

The integration constant does not appear because infinity has been chosen as potential reference ($V_\infty = 0 \Rightarrow C = 0$). If the same calculus is applied to (2.16) a general expression for the electrostatic potential is obtained:

$$V(\mathbf{r}) = \frac{1}{4\pi\varepsilon_0} \sum_{j=1}^{n} \frac{q_j}{|\mathbf{r} - \mathbf{r}_j|} + \frac{1}{4\pi\varepsilon_0} \int_{V'} \frac{\rho}{|\mathbf{r} - \mathbf{r}'|} dV' + \frac{1}{4\pi\varepsilon_0} \int_{S'} \frac{\sigma}{|\mathbf{r} - \mathbf{r}'|} dS'$$

$$+ \frac{1}{4\pi\varepsilon_0} \int_{L'} \frac{\lambda}{|\mathbf{r} - \mathbf{r}'|} dl'. \qquad (2.25)$$

This is the mathematical expression of **the superposition principle for electrostatic potential**: the electrostatic potential created by a number of charges is equal to the sum of potentials caused by each of them independently.

---

[4] It must be observed in expression (2.22) that the order of integral limits have been changed. This is due to the negative sign removal.

2.5 Flux of Electric Field. Gauss' Law

Let V be a region in space, bordered by ∂V and let $\mathbf{n}$ be the outward unit normal to ∂V on every point of a surface. The flux Φ_E through ∂V (1.28) of the electric field $\mathbf{E}$ produced by a point charge q located at the origin is

$$\Phi_E = \oint_{\partial V} \frac{q}{4\pi\varepsilon_0 r^3} \mathbf{r} \cdot \mathbf{n}\, dS. \tag{2.26}$$

where $\mathbf{r}$ is a vector that goes from q to a point on ∂V.

Gauss' law states that if V is smooth enough and if $q \notin \partial V$, it is verified

$$\Phi_E = \oint_{\partial V} \frac{q}{4\pi\varepsilon_0} \frac{\mathbf{r} \cdot d\mathbf{S}}{r^3} = \begin{cases} 0 & \text{if } q \notin V, \\ \dfrac{q}{\varepsilon_0} & \text{if } q \in V. \end{cases} \tag{2.27}$$

If we have any charge distribution, and we apply the principle of superposition for electrostatic fields, the previous theorem can be generalized as:

$$\Phi_E = \oint_{\partial V} \mathbf{E} \cdot d\mathbf{S} = \frac{q_{\text{in}}}{\varepsilon_0}, \tag{2.28}$$

where q_{in} represents charge inside surface ∂V, which is usually called a *Gaussian surface*. This is known as **Gauss' law** or Gauss' theorem.

This theorem shows that the flux of the electric field through a closed surface only depends on the charge q_{in} inside the surface. It must be noted that the flux can be zero even though the field is not, as in Fig. 2.3, where the flux is positive in dS_1 and negative in dS_2, and where the direction of field $\mathbf{E}_2$ is towards V's inner part. It must be also observed that the flux (not the field) through one surface is the same as another surface if the charge is in the volume bounded by the surface is the same.

Gauss' law allows calculating an electrostatic field created by charge distributions with different geometric and electric symmetries. This calculus is usually possible if a Gaussian surface with the same electric field (same magnitude and same angle with a normal vector to the surface) can be taken at each of its points.

Fig. 2.3 Flux of an electric field through a closed surface due to a point charge

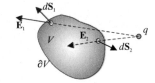

2.6 Electrostatic Equations

The integral expression of Gauss' law makes results dependent on the region we integrate. To avoid this problem, proper application of the divergence theorem results in a differential expression for the law:

$$\nabla \cdot \mathbf{E}(\mathbf{r}) = \frac{\rho(\mathbf{r})}{\varepsilon_0}, \tag{2.29}$$

where $\rho = \rho(\mathbf{r})$ is the charge density on the considered point. This expression represents the *differential form of Gauss' law*,[5] which together with (2.18), compose electrostatics fundamental equations. If we remember the concept of divergence from Chap. 1, it can be observed that electric field sources are positive charge points, and sink ones are points with negative charge.

If we combine the two electrostatic equations, the result is

$$\Delta V(\mathbf{r}) = -\frac{\rho(\mathbf{r})}{\varepsilon_0}, \tag{2.30}$$

which is known as the *Poisson equation*. In regions where charge density is null, we have

$$\Delta V(\mathbf{r}) = 0, \tag{2.31}$$

which is known as the *Laplace equation*.

Equations (2.30) and (2.31) are second order partial differential equations for a scalar field (electrostatic potential). If these equations are integrated and boundary conditions are given by a known charge distribution, the potential can be obtained for particular problems. The problem is simplified if an appropriate coordinate system is chosen. A Poisson (or Laplace) Equation solution is a unique one (*unity theorem*). This property allows us to establish methods to obtain the differential equation solution without specifically solving it (as occurs with the *method of images*, Sect. 7.4). This is because once a solution is obtained, it is unique, independent of the way it is obtained.

2.7 Electric Dipole

A special case of electric charge distribution can be studied: two equal and opposite charges separated by a small distance. This is known as an **electric dipole**. This can occur not only with two charges, but due to more complex charge distributions where

---

[5]This equation is valid even when conditions are not static.

2.7 Electric Dipole

the effective centers of negative and positive charges satisfy dipole characteristics, as it will be seen in Chap. 3. The electric dipole is characterized by its **dipole moment**, expressed by

$$\mathbf{p} \equiv q\mathbf{d}, \tag{2.32}$$

whose SI unit is Cm. The magnitude **d** is equal to the distance between charges, and **d** has the direction from the negative charge to the positive charge. This is especially interesting for the case when the distance d goes to zero (it's very small compared to the other dimensions of the problem): a *point dipole* is formed. It has neither net charge nor space extension, but it is completely characterized by its dipole moment. Polar molecules are an example of a point dipole.[6] The electric field and the potential distribution produced by a point dipole can be calculated with the aid of the formulas of Sects. 2.3 and 2.4. The electric field is

$$\mathbf{E}(\mathbf{r}) = \frac{1}{4\pi\varepsilon_0} \left\{ \frac{3(\mathbf{r} - \mathbf{r}') \cdot \mathbf{p}}{|\mathbf{r} - \mathbf{r}'|^5}(\mathbf{r} - \mathbf{r}') - \frac{\mathbf{p}}{|\mathbf{r} - \mathbf{r}'|^3} \right\}, \tag{2.33}$$

where the point dipole is located at point $\mathbf{r}'$. The potential distribution produced by a point dipole is given by

$$V(\mathbf{r}) = \frac{1}{4\pi\varepsilon_0} \frac{\mathbf{p} \cdot (\mathbf{r} - \mathbf{r}')}{|\mathbf{r} - \mathbf{r}'|^3}. \tag{2.34}$$

2.8 Conductors and Insulators

Materials have charged particles inside them which can move through-out the material under the influence of an outside electric field. These charged particles are called *charge carriers*. Charge carriers are electrons and ions in gases and liquids, electrons in crystalline solids (semiconductors and metals) and pairs of electrons in superconductors. The physical property used to measure the ease of charge movement is the conductivity.[7] According to electric behavior, materials can be divided into conductors, semiconductors and insulators (or dielectric).

Conductors are substances in which charges are free to move throughout the material under the influence of an outside electric field. Metallic conductors are the most characteristic example. The conductivity of metals generally increases with a decrease in temperature. At temperatures near absolute zero ($T \approx 0\,°K$), some conductors exhibit infinite conductivity and are called *superconductors*.

Dielectrics are substances in which charged particles are not free to move (low conductivity). These charges (nucleus and electrons) are strongly linked forming a material's atoms or molecules. In fact, they change position very little.

---

[6][104] can be seen for more detailed information about the electric dipole.
[7]This concept will be studied on Chap. 4.

Semiconductors have electrical properties intermediate between conductors and dielectrics, though in electrostatic fields they behave as conductors.

A material's ability to conduct electricity can be understood with band theory: electrons in solid materials are distributed in bands, each one with a grade of energy. Electrons can change from one to another by absorbing or giving energy. Between bands there can be gaps or forbidden regions where an electron's presence is not possible. In the case of conducting materials, superior bands (conducting ones) are partially full, so the electron can move along them. For the insulating materials, the gap is large; they need a large energy to allow electrons to jump from the highest full band (valence's band) to the next one. In semiconductors, the necessary energy to go from the valence band to the conducting one is small.

Behavior of dielectric materials undergone to electric fields will be studied in next chapter. Charge carriers in conducting materials move until they reach positions where no net force is exerted, so they will have electrostatic balance. Hence, in electrostatic conditions:

- Electric field is null ($\mathbf{E} = 0$) inside conducting materials, because equilibrium implies a null force, and $\mathbf{E}$ is perpendicular on the surface.
- From (2.29), charge density ρ in the interior of the conductor is zero.
- From (2.18), each conductor forms an equipotential region of space.
- If (2.28) is applied, it can be deduced that the field in a very close point from conducting's surface is $E = \sigma/\varepsilon_0$, where σ is the surface charge density of the conductor.

It must be observed that the Laplace equation (2.31) can be applied for conductors problems, because in almost every point the charge density is zero. You can solve the Laplace equation for every point outside the conductor if you know the boundary conditions, which involve the electrostatic potential V. Solution to the problem is already completed, because for the rest of the points (inside the conductor), the solution is the one that follows: as conducting materials are equipotential volumes, the potential inside them is the same as the one on its surface.

2.9 Biot–Savart-like Law in Electrostatics

The Biot–Savart law in one of the most basic relations in electromagnetism. We will study it in later chapters this law which allows us to calculate the total magnetic field $\mathbf{B}$ at a point in space as superposition of $d\mathbf{B}$ produced by the flow of current I through an infinitesimal path segment $d\mathbf{l}$. In [79], we can see an application of Biot–Savart law to obtain the electric field $\mathbf{E}$ produced by a plane charge distribution, bounded by a curve C and kept at a fixed potential V while the rest of the plane is held at zero potential. If $\mathbf{r}'$ locates the source point and $\mathbf{r}$ refers to the field point, it follows that:

$$\mathbf{E}(\mathbf{r}) = \frac{V}{2\pi} \oint_C \frac{(\mathbf{r} - \mathbf{r}') \times d\mathbf{l}'}{|\mathbf{r} - \mathbf{r}'|^3}, \tag{2.35}$$

where $d\mathbf{l}'$ is a length element of the integration path C. The direction of the integration around C is determined by the direction of the outward unit normal via the right-hand rule. Notice that to calculate the electric field we just need to take into account the contributions coming from the boundary contour C.

Solved Problems

Problems A

2.1 In a cartesian coordinate system, with the axis in meters, two point charges are considered, one of them positive of 1 nC, located at the origin of coordinates, and the other one, negative of -20 nC, located at $A(0, 1)$. Determine the resulting field at $B(2, 0)$ and the necessary work to take a positive charge of 3 µC from $B(2, 0)$ to $C(4, 2)$.

Solution

Applying the superposition principle for electrostatic field, field at B will be the vectorial addition of the fields due to each charge (Fig. 2.4), this is:

$$\mathbf{E}_B = \mathbf{E}_{OB} + \mathbf{E}_{AB}.$$

To calculate the field due to each charge, (2.15) is applied. Point B's coordinates are $\mathbf{r} = 2\mathbf{u}_x$. In the case for the charge at O, position vector is $\mathbf{r}' = 0$, and we obtain $\mathbf{r} - \mathbf{r}' = \mathbf{OB} = 2\mathbf{u}_x$ and $|\mathbf{r} - \mathbf{r}'| = 2$. Position vector of charge at A is $\mathbf{r}' = \mathbf{u}_y$, so $\mathbf{r} - \mathbf{r}' = \mathbf{AB} = 2\mathbf{u}_x - \mathbf{u}_y$ and $|\mathbf{r} - \mathbf{r}'| = \sqrt{5}$. If (2.15) is applied to the charge located at O, the result is,

$$\mathbf{E}_{OB} = \frac{1}{4\pi\varepsilon_0} \frac{q'}{|\mathbf{r} - \mathbf{r}'|^2} \frac{\mathbf{r} - \mathbf{r}'}{|\mathbf{r} - \mathbf{r}'|} = 9 \cdot 10^9 \frac{10^{-9}}{4} \mathbf{u}_x = \frac{9}{4}\mathbf{u}_x.$$

And for the charge located at A,

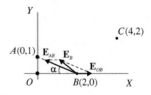

Fig. 2.4 Field produced by point charges

$$\mathbf{E}_{AB} = \frac{1}{4\pi\varepsilon_0} \frac{q'}{|\mathbf{r}-\mathbf{r}'|^2} \frac{\mathbf{r}-\mathbf{r}'}{|\mathbf{r}-\mathbf{r}'|} = 9 \cdot 10^9 \frac{(-20 \cdot 10^{-9})}{5} \left(\frac{2}{\sqrt{5}}\mathbf{u}_x - \frac{1}{\sqrt{5}}\mathbf{u}_y\right)$$
$$= \frac{36}{\sqrt{5}}(-2\mathbf{u}_x + \mathbf{u}_y).$$

The field could be obtained without using position vectors, and bearing in mind Fig. 2.4. Since there is a negative charge at A, field E_{AB} is pointed to A. If we use (2.15) to calculate the electric field magnitude and if we project it, we obtain the field expressed as,

$$\mathbf{E}_{AB} = 9 \cdot 10^9 \frac{20 \cdot 10^{-9}}{5}(-\cos\alpha \mathbf{u}_x + \sin\alpha \mathbf{u}_y) = 36\left(\frac{-2}{\sqrt{5}}\mathbf{u}_x + \frac{1}{\sqrt{5}}\mathbf{u}_y\right).$$

The resulting field will be

$$\mathbf{E}_B = \mathbf{E}_{OB} + \mathbf{E}_{AB} = \left(\frac{9}{4} - \frac{72}{\sqrt{5}}\right)\mathbf{u}_x + \frac{36}{\sqrt{5}}\mathbf{u}_y = -29.95\mathbf{u}_x + 16.10\mathbf{u}_y.$$

Since the electrostatic field is conservative, the work necessary to take a charge from B to C is equal to the variation of potential energy between B and C (2.20), with negative sign. From (2.21) which relates electrostatic potential to the potential energy,

$$W_{BC} = -(E_{PB} - E_{PC}) = -q(V_B - V_C),$$

and potentials at points B and C due to the charge system of the problem must be calculated. To obtain the potential on each point, the superposition principle is applied, this is, the potential to be the added due to each charge, $V_B = V_{OB} + V_{AB}$ and $V_C = V_{OC} + V_{AC}$. If (2.24) (potential produced by a point charge) is applied

$$V = \frac{1}{4\pi\varepsilon_0} \frac{q'}{|\mathbf{r}-\mathbf{r}'|},$$

potentials due to each charge on each point V_{OB}, V_{AB}, V_{OC} and V_{AC} can be determined. It is necessary to define the terms ($|\mathbf{r}-\mathbf{r}'|$), which are the distances from each charge to points B and C. On the case of point B the distances have already been determined for the calculus of the electric field, so potential in B will be

$$V_B = V_{OB} + V_{AB} = 9 \cdot 10^9 \left(\frac{10^{-9}}{2} + \frac{(-20 \cdot 10^{-9})}{\sqrt{5}}\right) = -76 \text{ V}.$$

For point C, its position vector is $\mathbf{r} = 4\mathbf{u}_x + 2\mathbf{u}_y$, and it becomes that $\mathbf{r}-\mathbf{r}'$ values are $\mathbf{OC} = 4\mathbf{u}_x + 2\mathbf{u}_y$ for the charge at O and $\mathbf{AC} = (4\mathbf{u}_x + 2\mathbf{u}_y) - \mathbf{u}_y = 4\mathbf{u}_x + \mathbf{u}_y$ for the charge at A. Potential is

$$V_C = V_{OC} + V_{AC} = 9 \cdot 10^9 \left(\frac{10^{-9}}{\sqrt{20}} + \frac{(-20 \cdot 10^{-9})}{\sqrt{17}} \right) = -41.6 \, \text{V}.$$

And circulation from point B to C is

$$W_{BC} = -q(V_B - V_C) = -3 \cdot 10^{-6}(-76 + 41.6) = 103.2 \cdot 10^{-6} \text{J} = 103.2 \, \mu\text{J}.$$

2.2 In the space region defined by $y > 0$, a charge density $\rho = cy$ exists, with $c = 2 \, \mu\text{C/m}^4$ and y the distance (in meters) from any point to plane XOZ. Calculate:
(a) The flux of the electrostatic field through the prism's surface in Fig. 2.5.
(b) Divergence of electrostatic field in the prism's faces which are parallel to plane XOZ.

Solution

(a) Gauss' law (2.28) states the flux of the electric field due to a charge distribution,

$$\Phi_E = \oint_{\partial V} \mathbf{E} \cdot d\mathbf{S} = \frac{q_{\text{in}}}{\varepsilon_0} = \frac{1}{\varepsilon_0} \int_V \rho \, dV.$$

Therefore we must calculate the charge q_{in} inside the prism in Fig. 2.5. Since charge density ρ only depends on coordinate y, every point located at the same distance y have the same density. Let's consider the infinitesimal volume drawn in Fig. 2.6, $dV = 3 \cdot 3 \, dy = 9 \, dy$. At every point in it the density is cy, so the volume integral becomes a simple integral:

$$q_{\text{in}} = \int_V \rho \, dV = \int_1^5 cy \, 9 \, dy = \left. \frac{9cy^2}{2} \right|_1^5 = 216 \, \mu\text{Cm}^{-3}.$$

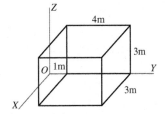

Fig. 2.5 Prism of Problem 2.2

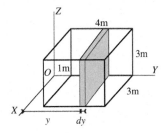

Fig. 2.6 Differential element to calculate internal charge

And therefore the flux is

$$\Phi_E = \frac{q_{in}}{\varepsilon_0} = \frac{216 \cdot 10^{-6}}{8.85 \cdot 10^{-12}} = 24.4 \cdot 10^6 \, \text{NC}^{-1}\text{m}^2.$$

(b) We obtain the divergence of the electric field from (2.29). If we particularize for each of the prism faces, we obtain:

$$\text{div } \mathbf{E} = \frac{\rho}{\varepsilon_0} = \frac{cy}{\varepsilon_0} \begin{cases} y = 1, \text{ div } \mathbf{E} = \dfrac{2 \cdot 10^{-6} \cdot 1}{8.85 \cdot 10^{-12}} = 2.26 \cdot 10^5 \, \text{NC}^{-1}\text{m}^{-1} \\ y = 5, \text{ div } \mathbf{E} = \dfrac{2 \cdot 10^{-6} \cdot 5}{8.85 \cdot 10^{-12}} = 1.13 \cdot 10^6 \, \text{NC}^{-1}\text{m}^{-1} \end{cases}$$

2.3 Determine the electric field produced on any point in space by a very long line (infinite) charged with a uniform density λ.

Solution[8]

This problem has cylindrical symmetry. Hence all the points at the same distance to the line have the same electric field magnitude, and $\mathbf{E}$ is perpendicular to the line.[9] The problem can be solved by applying Gauss' law (2.28). We take as a Gaussian surface ∂V a closed cylindrical surface, with any length L, with the axis on the line and with radius r making the surface to pass through point P, the field desired to be calculated (discontinuous line in Fig. 2.7). The electric field at any point on the Gaussian surface is radial (perpendicular to the cylinder lateral surface), outward pointed, if we suppose the line as positively charged,[10] and has the same magnitude at every point on the lateral surface. It should be observed that the Gaussian surface is a closed surface and therefore to calculate the flux the cylinder lateral surface where the point P is and the cylinder bases, where the magnitude of the field is not constant and is different from the one at the lateral surface, should be considered. However, this is not a problem, since the flux through the bases is null due to the fact that $d\mathbf{S}$ and $\mathbf{E}$ are perpendicular at any point. Calculating the flux through the lateral surface, where $d\mathbf{S}$ and $\mathbf{E}$ are parallel:

$$\Phi_E = \oint_{\partial V} \mathbf{E} \cdot d\mathbf{S} = \int_{S_{lat}} E \, dS = E \int_{S_{lat}} dS = E 2\pi r L,$$

---

[8]Infinite is usually used to express that the element is much longer than the distance r to point P, so the symmetry reasonings can be used for the calculus.

[9]To check this, the field can be considered to be produced at a point by an element dq and its symmetric regarding to the normal to the line by the considered point. Tangential components from one and another have the same magnitude, since they are at the same distance, and opposite directions. Then the result is a radial field.

[10]If the charge were negative, the charge sign on the solution indicates that the field vector has the opposite direction.

Solved Problems

Fig. 2.7 Gaussian surface and field vectors to apply Gauss' law to an infinite line

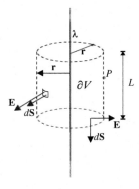

where S_{lat} is the cylinder lateral surface. On another side, if Gauss' law is applied, it is obtained

$$\Phi_E = \frac{q_{in}}{\varepsilon_0} = \frac{\lambda L}{\varepsilon_0}.$$

Equating both expressions,

$$E = \frac{\lambda}{2\pi\varepsilon_0 r},$$

shows that the field varies inversely with the distance to the line. This expression is the same as the one calculated by integration in Problem 2.4. Using vectors:

$$\mathbf{E} = \frac{\lambda}{2\pi\varepsilon_0 r}\frac{\mathbf{r}}{r} = \frac{\lambda}{2\pi\varepsilon_0 r}\mathbf{u}_\rho, \qquad (2.36)$$

where $\mathbf{u}_\rho$ is the radial unit vector for cylindrical coordinates.

Problems B

2.4 Determine the electric field, at any point in space, produced by a line with length L, which has been uniformly charged by a total charge Q: (a) directly; (b) from electrostatic potential. Apply the result for the particular cases of the field produced by an infinitely long line with density λ and to the field produced by a semi-infinite line with density λ at a point located on the perpendicular to its extreme.

Solution

(a) The problem above is drawn in Fig. 2.8, where a cartesian coordinate system has been defined, just to simplify the calculus. To achieve this, we choose a point P where we want to calculate the field (*field point*), and we define a plane XY as drawn. The X axis is on the charged line and the origin is at one extreme of the line.

Fig. 2.8 Field produced by an finite line at any point

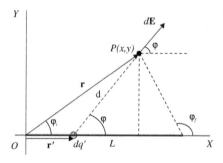

The coordinates of point P will be (x, y) and the electric field $\mathbf{E}$ will be on plane XY.

To solve the problem we calculate the field due to an element dq' from the line, and the superposition principle is applied. The field produced by an element dq' is given by (2.15)

$$d\mathbf{E}(\mathbf{r}) = \frac{dq'}{4\pi\varepsilon_0 |\mathbf{r} - \mathbf{r}'|^2} \frac{\mathbf{r} - \mathbf{r}'}{|\mathbf{r} - \mathbf{r}'|},$$

where $\mathbf{r} = x\mathbf{u}_x + y\mathbf{u}_y$ is the position vector of point P and $\mathbf{r}' = x'\mathbf{u}_x$ is the position of each charge element (*source point*) dq'. Charge dq' is obtained from lineal charge density definition λ (2.5) as

$$dq' = \lambda dl = \lambda dx'.$$

λ is obtained from (2.6), bearing in mind that, since it is a uniform charge, λ is constant:

$$\lambda = \frac{Q}{L}.$$

With $\mathbf{r}$ and $\mathbf{r}'$ values we obtain vector $\mathbf{r} - \mathbf{r}'$ and the distance from dq' to point P:

$$\mathbf{r} - \mathbf{r}' = (x - x')\mathbf{u}_x + y\mathbf{u}_y, \quad d = |\mathbf{r} - \mathbf{r}'| = \sqrt{(x - x')^2 + y^2}.$$

If we substitute the electric field produced by dq' at point P, we have

$$d\mathbf{E}(\mathbf{r}) = \frac{dq'}{4\pi\varepsilon_0 |\mathbf{r} - \mathbf{r}'|^2} \frac{\mathbf{r} - \mathbf{r}'}{|\mathbf{r} - \mathbf{r}'|} = \frac{1}{4\pi\varepsilon_0} \frac{\lambda dx'}{\left[(x - x')^2 + y^2\right]} \frac{(x - x')\mathbf{u}_x + y\mathbf{u}_y}{\sqrt{(x - x')^2 + y^2}}.$$

The total field is obtained by applying the superposition principle (2.16):

$$\mathbf{E} = \frac{\lambda}{4\pi\varepsilon_0} \int_L \frac{dx'}{\left[(x - x')^2 + y^2\right]} \frac{(x - x')\mathbf{u}_x + y\mathbf{u}_y}{\sqrt{(x - x')^2 + y^2}}.$$

Solved Problems

Solving using integrals, where the only variable is x', allows us to obtain the total field's value. A way of solving this integral is by introducing a variable change, depending on the angle φ in Fig. 2.8,

$$\cos\varphi = \frac{x-x'}{d}, \quad \sin\varphi = \frac{y}{d}, \quad \cot\varphi = \frac{\cos\varphi}{\sin\varphi} = \frac{x-x'}{y}.$$

If we take the derivative of the last expression, the result is

$$\frac{d\varphi}{\sin^2\varphi} = \frac{dx'}{y}.$$

If we obtain dx' and d values,

$$dx' = \frac{y\,d\varphi}{\sin^2\varphi}, \quad d = \frac{y}{\sin\varphi},$$

and we substitute into the integral, and substitute $\sqrt{(x-x')^2 + y^2}$ for d, it follows

$$\mathbf{E} = \frac{\lambda}{4\pi\varepsilon_0} \int_L \frac{dx'}{d^2} \frac{(x-x')\mathbf{u}_x + y\mathbf{u}_y}{d}$$

$$= \frac{\lambda}{4\pi\varepsilon_0} \left(\int_\varphi \frac{\sin^2\varphi}{y^2} \frac{y\,d\varphi}{\sin^2\varphi} \cos\varphi \mathbf{u}_x + \int_\varphi \frac{\sin^2\varphi}{y^2} \frac{y\,d\varphi}{\sin^2\varphi} \sin\varphi \mathbf{u}_y \right)$$

$$= \frac{\lambda}{4\pi\varepsilon_0 y} \left(\int_{\varphi_i}^{\varphi_f} \cos\varphi\,d\varphi \mathbf{u}_x + \int_{\varphi_i}^{\varphi_f} \sin\varphi\,d\varphi \mathbf{u}_y \right).$$

where φ_i and φ_f are the angles measured from the initial and final positions of the line, as shown in Fig. 2.8. It should be observed that y, the generic ordinate of a point field, is not a variable in the integral. If we solve it, it results

$$\mathbf{E} = \frac{\lambda}{4\pi\varepsilon_0 y} \left((\sin\varphi_f - \sin\varphi_i)\mathbf{u}_x + (\cos\varphi_i - \cos\varphi_f)\mathbf{u}_y \right). \quad (2.37)$$

The result can be written in Cartesian coordinates:

$$\mathbf{E} = \frac{Q}{4\pi\varepsilon_0 y L} \left[\left(\frac{y}{\sqrt{(L-x)^2 + y^2}} - \frac{y}{\sqrt{x^2 + y^2}} \right) \mathbf{u}_x \right.$$
$$\left. + \left(\frac{x}{\sqrt{x^2 + y^2}} + \frac{L-x}{\sqrt{(L-x)^2 + y^2}} \right) \mathbf{u}_y \right], \quad (2.38)$$

where λ's value has been replaced by Q/L.
(b) To calculate the electric field at P from the potential V, we must obtain a potential generic expression at any point (for example $P(x, y)$) and then obtain the field from

(2.18). It is important to notice that this procedure can be done because the potential at all the points is known, and therefore its gradient can be calculated. To obtain the potential at point P, the potential produced by a charge element dq' is expressed, taking the potential reference at infinity (2.24),

$$dV = \frac{1}{4\pi\varepsilon_0} \frac{dq'}{|\mathbf{r} - \mathbf{r}'|} = \frac{\lambda}{4\pi\varepsilon_0} \frac{dx'}{\sqrt{(x - x')^2 + y^2}},$$

and the superposition principle for electrostatic potential is applied (2.25):

$$V = \frac{\lambda}{4\pi\varepsilon_0} \int_0^L \frac{dx'}{\sqrt{(x - x')^2 + y^2}}.$$

Solving this integral,

$$V = -\frac{\lambda}{4\pi\varepsilon_0} \ln\left(x - x' + \sqrt{(x - x')^2 + y^2}\right)\Big|_0^L = \frac{\lambda}{4\pi\varepsilon_0} \ln \frac{x + \sqrt{x^2 + y^2}}{x - L + \sqrt{(x - L)^2 + y^2}}.$$

If we apply now (2.18),

$$\mathbf{E} = -\nabla V = -(\partial V/\partial x)\mathbf{u}_x - (\partial V/\partial y)\mathbf{u}_y,$$

and if we solve the indicated partial derivative, the total field $\mathbf{E}$ at P is obtained (2.38).

Let's consider the case of an *infinite line* (very long line). Since the line has an infinite length, (2.38) is not obvious. It is easier to use (2.37). It should be observed in Fig. 2.8 that if the line is infinite, initial and final angles are $\varphi_i = 0$ and $\varphi_f = \pi$. If we substitute in (2.37), it results

$$\mathbf{E} = \frac{\lambda}{2\pi\varepsilon_0 y}\mathbf{u}_y.$$

Let's consider that the line begins at O and is very long *(semi-infinite line)* and P is over the perpendicular to the line at O. Point P coordinates are $P(0, y)$. If we consider (2.37), we observe in Fig. 2.8 that, since $P(0, y)$, $\varphi_i = \pi/2$ and $\varphi_f = \pi$. Then,

$$\mathbf{E} = \frac{\lambda}{4\pi\varepsilon_0 y}(-\mathbf{u}_x + \mathbf{u}_y),$$

We should be cautious when applying the used procedure in section (b) to obtain the field when the line is infinitely long. We observe the potential for this charge distribution as

$$V = \lim_{L \to \infty} \frac{\lambda}{4\pi\varepsilon_0} \ln \frac{x + \sqrt{x^2 + y^2}}{x - L + \sqrt{(x - L)^2 + y^2}} \longrightarrow \infty.$$

So it is not possible to obtain the field from this potential. Infinite potential has been obtained because for charge distributions that spread in an infinite region, it can never be certain that this potential converges. However, the field from the potential can be obtained as follows: firstly we calculate the finite line potential, then its gradient, and then we make the line's length to infinity. This difficulty will also appear in the magnetostatic chapter.

2.5 Determine the electric field and the potential at any point in space produced by a spherical crown where the internal radius is R_1 and the external one R_2, with a total charge Q, for the following cases: (a) non conducting and an uniform charge distributed throughout the volume; and (b) metallic and on electrostatic equilibrium. Particularize the results for a solid sphere with radius R.

Solution

(a) In this case the spherical crown has a uniform volume charge density ρ at every point between R_1 and R_2, that can be calculated by applying (2.2):

$$Q = \int_V \rho dV = \rho \int_V dV = \rho V = \rho \frac{4}{3}\pi(R_2^3 - R_1^3) \Rightarrow$$

$$\rho = \frac{Q}{V} = \frac{3Q}{4\pi(R_2^3 - R_1^3)}.$$

Because of its symmetry, Gauss' law (2.28) is applied, considering a spherical Gaussian surface ∂V (Fig. 2.9), concentric with the charge distribution, and passing through the point where we want to calculate the field. At every point inside a sphere whose radius is $r \leq R_1$, field is null since internal charge is zero:

$$E_{(r \leq R_1)} = 0.$$

Fig. 2.9 Gaussian surfaces and field vectors to apply Gauss' law to a spherical crown

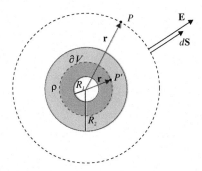

Due to the charge distribution symmetry, the electric field at any other point on the Gaussian surface will be radial,[11] outward[12] and with the same magnitude at every point on the surface. Two different expressions are obtained for the electric field, depending on the point to be studied, if it is outside or inside the spherical crown (P and P' in Fig. 2.9), since internal charge to the Gaussian surface has a different expression.

Let's firstly consider point P, outside the crown. The flux through the Gaussian surface passing through P is

$$\Phi_E = \oint_{\partial V} \mathbf{E} \cdot d\mathbf{S} = \oint_{\partial V} E\, dS = E \oint_{\partial V} dS = E 4\pi r^2,$$

where $4\pi r^2$ is the spherical surface's area, with radius r. We observe that the total charge within the Gaussian surface is all the charge of the spherical crown (grey in figure). If Gauss' law is applied, the flux is

$$\Phi_E = \frac{q_{\text{in}}}{\varepsilon_0} = \frac{Q}{\varepsilon_0} = \frac{\rho 4\pi (R_2^3 - R_1^3)}{3\varepsilon_0}.$$

If we equate the two previous expressions for the flux and we solve it, we obtain the field at an external point P,

$$E_{(r \geq R_2)} = \frac{\rho (R_2^3 - R_1^3)}{3\varepsilon_0 r^2} = \frac{Q}{4\pi \varepsilon_0 r^2},$$

expression that coincides with the field produced by a point charge Q located at the spherical crown centre. In fact, the field produced by a point charge, known from Coulomb's law, can be calculated by applying Gauss' law to a random spherical surface whose centre is on the charge.

To know the field at a point P' (Fig. 2.9), inside the spherical crown, the same procedure is followed. We set up a spherical concentric surface, with radius r, passing through P'. The expression of the flux through a surface of radius r is the same, but since $R_1 \leq r \leq R_2$, internal charge to the Gaussian surface is now not the total charge of the crown charge, but only the dark grey region in the figure. If we calculate

$$\Phi_E = \frac{q_{\text{in}}}{\varepsilon_0} = \frac{\rho 4\pi (r^3 - R_1^3)}{3\varepsilon_0},$$

and then, from $\Phi_E = E 4\pi r^2$, the field at a point P' inside the spherical crown is obtained,

---

[11] It can be checked by taking the field produced by a random element dq and its symmetrical with regard to the diameter that passes through the considered point. Tangential components from one to the other have the same magnitude and opposite directions, and the result is a radial field.

[12] It will be supposed, unless it is stated otherwise, these bodies are positively charged. If the charge is negative, the vector field has the opposite direction.

$$E_{(R_1 \leq r \leq R_2)} = \frac{\rho(r^3 - R_1^3)}{3\varepsilon_0 r^2} = \frac{Q}{4\pi\varepsilon_0 r^2} \frac{(r^3 - R_1^3)}{(R_2^3 - R_1^3)}.$$

Joining both results and expressing the field as a vector, it results

$$\mathbf{E} = \begin{cases} 0 & r \leq R_1, \\ \dfrac{Q}{4\pi\varepsilon_0 r^2} \dfrac{(r^3 - R_1^3)}{(R_2^3 - R_1^3)} \mathbf{u}_r = \dfrac{\rho(r^3 - R_1^3)}{3\varepsilon_0 r^2} \mathbf{u}_r & R_1 \leq r \leq R_2, \\ \dfrac{Q}{4\pi\varepsilon_0 r^2} \mathbf{u}_r = \dfrac{\rho(R_2^3 - R_1^3)}{3\varepsilon_0 r^2} \mathbf{u}_r & r \geq R_2, \end{cases} \quad (2.39)$$

where $\mathbf{u}_r$ is the radial unit vector for spherical coordinates. The field at a point on the outside spherical surface can be calculated by using either one of the expressions, making $r = R_2$:

$$\mathbf{E}_{(r=R_2)} = \frac{Q}{4\pi\varepsilon_0 R_2^2} \mathbf{u}_r = \frac{\rho(R_2^3 - R_1^3)}{3\varepsilon_0 R_2^2} \mathbf{u}_r.$$

If the sphere were solid, with radius R, results can be obtained by replacing $R_1 = 0$ and $R_2 = R$:

$$\mathbf{E} = \begin{cases} \dfrac{Q}{4\pi\varepsilon_0} \dfrac{r}{R^3} \mathbf{u}_r = \dfrac{\rho r}{3\varepsilon_0} \mathbf{u}_r & r \leq R, \\ \dfrac{Q}{4\pi\varepsilon_0 r^2} \mathbf{u}_r = \dfrac{\rho R^3}{3\varepsilon_0 r^2} \mathbf{u}_r & r \geq R, \end{cases} \quad (2.40)$$

And the field at a point on the surface would be:

$$\mathbf{E}_{(r=R)} = \frac{Q}{4\pi\varepsilon_0 R^2} \mathbf{u}_r = \frac{\rho R}{3\varepsilon_0} \mathbf{u}_r.$$

To determine the potential at any point, we calculate the potential difference between that point and infinity, with null potential. Since the circulation of the electrostatic field is path-independent, we take the radial direction from the point, where $\mathbf{E}$ and $d\mathbf{l}$ are parallel,[13] as shown in Fig. 2.10.

For an external point P we have, if we circulate the field $\mathbf{E}$ between P and infinity (Fig. 2.10), that

---

[13] It is not necessary to take a circulation line where $\mathbf{E}$ and $d\mathbf{l}$ are parallel, if we bear in mind the property of any vector $\mathbf{r}$, for which $\mathbf{r} \cdot d\mathbf{r} = |\mathbf{r}|d|\mathbf{r}|$. If we express $\mathbf{E}$ depending on $\mathbf{u}_r$ we would achieve the same calculus expression, but it is explained like this to make the circulation concept comprehension easier.

Fig. 2.10 Potential calculation of a charged spherical crown

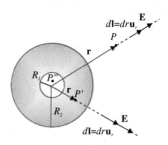

$$V_P - V_\infty = V_P = \int_P^\infty \mathbf{E} \cdot d\mathbf{l} = \int_r^\infty E_{(r \geq R_2)} dr$$
$$= \int_r^\infty \frac{\rho(R_2^3 - R_1^3)}{3\varepsilon_0 r^2} dr =$$
$$= -\left. \frac{\rho(R_2^3 - R_1^3)}{3\varepsilon_0 r} \right|_r^\infty.$$

Then,
$$V_P = \frac{\rho(R_2^3 - R_1^3)}{3\varepsilon_0 r} = \frac{Q}{4\pi\varepsilon_0 r}.$$

It can be observed how the potential given by the charged spherical crown at an external point is the same as a point charge, located at the centre, with the same charge, would create.

If the point P' is inside the spherical crown, from which we need to circulate from P' to infinity, field expressions are different depending on where we circulate, inside or outside the spherical crown. Potential difference is obtained from

$$V_{P'} - V_\infty = V_{P'} = \int_{P'}^\infty \mathbf{E} \cdot d\mathbf{l} = \int_r^{R_2} E_{(r \leq R_2)} dr + \int_{R_2}^\infty E_{(r \geq R_2)} dr$$
$$= \int_r^{R_2} \frac{\rho(r^3 - R_1^3)}{3\varepsilon_0 r^2} dr + \int_{R_2}^\infty \frac{\rho(R_2^3 - R_1^3)}{3\varepsilon_0 r^2} dr =$$
$$= \left. \frac{\rho r^2}{6\varepsilon_0} \right|_r^{R_2} + \left. \frac{\rho R_1^3}{3\varepsilon_0 r} \right|_r^{R_2} - \left. \frac{\rho(R_2^3 - R_1^3)}{3\varepsilon_0 r} \right|_{R_2}^\infty.$$

The potential will be

$$V_{P'} = \frac{\rho(R_2^2 - r^2)}{6\varepsilon_0} + \frac{\rho R_1^3}{3\varepsilon_0}\left(\frac{1}{R_2} - \frac{1}{r}\right) + \frac{\rho(R_2^3 - R_1^3)}{3\varepsilon_0 R_2}.$$

If the point (P'') is in the hole (Fig. 2.10), so $r \leq R_1$, given the field at the hole is null, circulation from point P'' to the inner radius of the crown R_1 is also null. Point P'' potential is the same as the one at point P' located on the inner spherical surface, with $r = R_1$. If the previous result is specified for $r = R_1$ the result is:

$$V_{P''} = \frac{\rho(R_2^2 - R_1^2)}{6\varepsilon_0} + \frac{\rho R_1^3}{3\varepsilon_0}\left(\frac{1}{R_2} - \frac{1}{R_1}\right) + \frac{\rho(R_2^3 - R_1^3)}{3\varepsilon_0 R_2}.$$

If all the results are joined,

$$V = \begin{cases} \dfrac{\rho(R_2^2 - R_1^2)}{6\varepsilon_0} + \dfrac{\rho R_1^3}{3\varepsilon_0}\left(\dfrac{1}{R_2} - \dfrac{1}{R_1}\right) + \dfrac{\rho(R_2^3 - R_1^3)}{3\varepsilon_0 R_2} & r \leq R_1, \\[2ex] \dfrac{\rho(R_2^2 - r^2)}{6\varepsilon_0} + \dfrac{\rho R_1^3}{3\varepsilon_0}\left(\dfrac{1}{R_2} - \dfrac{1}{r}\right) + \dfrac{\rho(R_2^3 - R_1^3)}{3\varepsilon_0 R_2} & R_1 < r \leq R_2, \quad (2.41) \\[2ex] \dfrac{\rho(R_2^3 - R_1^3)}{3\varepsilon_0 r} = \dfrac{Q}{4\pi\varepsilon_0 r} & r \geq R_2. \end{cases}$$

(b) If the spherical crown is metallic and has electrostatic balance, charge is only distributed on its external surface, as it can be deduced from conductor properties seen in Sect. 2.8. To obtain null electric field inside the conductor, the crown must be charged on its external surface with a uniform superficial density. This density can be obtained from (2.4):

$$Q = \int_S \sigma \, dS = \sigma \int_S dS = \sigma 4\pi R_2^2 \Rightarrow \sigma = \frac{Q}{4\pi R_2^2}.$$

The field at internal points ($r < R_2$) is null, due to the charge distribution symmetry. To obtain the electric field produced by the crown at external points to it, and due to the fact that the problem's symmetry is analogous to the one in section (a), we proceed as it was done in that section. We use the same Gaussian surfaces (Fig. 2.9) but just changing that the charge Q is only on the surface. The flux through the spherical surface of radius r is, as it happened with the previous case,

$$\Phi_E = E 4\pi r^2.$$

If Gauss' law is applied to an external point P to the spherical crown,[14] ($r > R_2$), results in

$$\Phi_E = \frac{q_{in}}{\varepsilon_0} = \frac{Q}{\varepsilon_0} = \frac{4\pi R_2^2 \sigma}{\varepsilon_0}.$$

In solving it, the result for an external point P is

$$E_{(r>R_2)} = \frac{Q}{4\pi\varepsilon_0 r^2} = \frac{\sigma R_2^2}{\varepsilon_0 r^2},$$

---

[14] It should be observed that Gauss' theorem cannot be applied to the points exactly located on the sphere's surface, due to the fact that in this case charges would be on the Gaussian surface, breaking the theorem's condition $q \notin \partial V$.

the same expression depending on the total charge Q that in section (a). If it is expressed with vector notation, the results are:

$$\mathbf{E} = \begin{cases} 0 & r < R_2, \\ \dfrac{Q}{4\pi\varepsilon_0 r^2}\mathbf{u}_r = \dfrac{\sigma R_2^2}{\varepsilon_0 r^2}\mathbf{u}_r & r > R_2, \end{cases} \quad (2.42)$$

where $\mathbf{u}_r$ is the radial unit vector for spherical coordinates.

Since the field at external points to the metallic crown is the same as the one in section (a), if the same reference is taken (infinity), the potential is also the same for points outside the crown. For points inside it, given the field is null at these points, potential does not change and has the same value that on the crown's surface. Then,

$$V = \begin{cases} \dfrac{Q}{4\pi\varepsilon_0 R_2} = \dfrac{\sigma R_2}{\varepsilon_0} & r \le R_2, \\ \dfrac{Q}{4\pi\varepsilon_0 r} = \dfrac{\sigma R_2^2}{\varepsilon_0 r} & r \ge R_2. \end{cases} \quad (2.43)$$

2.6 Determine the electric field produced by a very long charged cylinder at any point, with inner radius R_1 and external radius R_2, with a charge per unit of length q, for the following cases: (a) non conductor and uniformly charged; and (b) metallic and in electrostatic balance. Specify the results for a solid cylinder with radius R.

Solution

(a) In this case the cylinder has a uniform volumetric charge density ρ, at every point between R_1 and R_2, that can be calculated by applying (2.2), and considering the finite cylinder length L,

$$Q = \int_V \rho\, dV = \rho \int_V dV = \rho V = \rho \pi (R_2^2 - R_1^2) L \Rightarrow$$

$$\Rightarrow \rho = \frac{Q}{V} = \frac{Q/L}{\pi(R_2^2 - R_1^2)} = \frac{q}{\pi(R_2^2 - R_1^2)}.$$

The problem has cylindrical symmetry: at points inside the cylinder ($r < R_1$), the field is null, due to that symmetry and, for the other zones, all the points at the same distance to the cylinder axis have the same electric field magnitude, with direction perpendicular to that axis.[15] The problem can be solved by applying Gauss' law (2.28), the same as it was done for Problem 2.3. For this we take as a Gaussian surface

---

[15]To check this, consider the field produced at a point by an element dq and its symmetric pair with respect to a perpendicular to the cylinder axis at the considered point. The components parallel to the cylinder axis have the same magnitude, since they are at the same distance, and opposite direction, and we obtain a radial field.

Fig. 2.11 Gaussian surface and field vectors to apply Gauss' law to an infinite cylinder

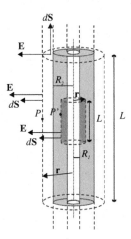

∂V a cylindrical surface, with any length L, with the same axis as the cylinder and radius r, making the surface pass through point P (or P') where we want to calculate the field (discontinuous line in Fig. 2.11).

The flux through the bases of the Gaussian surface, is null, since $d\mathbf{S}$ and $\mathbf{E}$ are perpendicular at any given point. Calculating the flux through the lateral surface, where $d\mathbf{S}$ and $\mathbf{E}$ are parallel:

$$\Phi_E = \oint_{\partial V} \mathbf{E} \cdot d\mathbf{S} = \int_{S_{\text{lat}}} E\, dS = E \int_{S_{\text{lat}}} dS = E 2\pi r L, \qquad (2.44)$$

where S_{lat} is the cylinder lateral surface. Applying Gauss' law we have:

$$\Phi_E = \frac{q_{\text{in}}}{\varepsilon_0},$$

where q_{in}'s value depends on the point where the field is calculated, whether it's inside (P') or outside (P) the charged cylinder. If we calculate the field at an external point P, the entire charged cylinder (with height L, light grey coloured in Fig. 2.11) remains inside the Gaussian surface, and it results

$$\Phi_{E(r>R_2)} = \frac{q_{\text{in}}}{\varepsilon_0} = \frac{1}{\varepsilon_0}\int_{R_1}^{R_2} \rho\, dV = \frac{\rho}{\varepsilon_0}\pi(R_2^2 - R_1^2)L = \frac{qL}{\varepsilon_0},$$

If the point P' is inside the charged cylinder, there is a part of the charge outside the Gaussian surface, which produces no flux. Then the only inner charge that remains is the dark grey coloured one in Fig. 2.11, and it results

$$\Phi_{E(R_1<r<R_2)} = \frac{q_{\text{in}}}{\varepsilon_0} = \frac{1}{\varepsilon_0}\int_{R_1}^{r} \rho\, dV = \frac{\rho}{\varepsilon_0}\pi(r^2 - R_1^2)L.$$

Equating these expressions with (2.44) it results, at external points as P,

$$E(r > R_2) = \frac{\rho\pi(R_2^2 - R_1^2)L}{\varepsilon_0 2\pi r L} = \frac{\rho(R_2^2 - R_1^2)}{2\varepsilon_0 r},$$

And at internal points as P',

$$E(R_1 < r < R_2) = \frac{\rho\pi(r^2 - R_1^2)L}{\varepsilon_0 2\pi r L} = \frac{\rho(r^2 - R_1^2)}{2\varepsilon_0 r}.$$

If we combine both results and we express them using vectors, it results

$$\mathbf{E} = \begin{cases} 0 & r \leq R_1, \\ \dfrac{\rho(r^2 - R_1^2)}{2\varepsilon_0 r}\mathbf{u}_\rho = \dfrac{q}{2\pi\varepsilon_0 r}\dfrac{(r^2 - R_1^2)}{(R_2^2 - R_1^2)}\mathbf{u}_\rho & R_1 \leq r \leq R_2, \\ \dfrac{\rho(R_2^2 - R_1^2)}{2\varepsilon_0 r}\mathbf{u}_\rho = \dfrac{q}{2\pi\varepsilon_0 r}\mathbf{u}_\rho & r \geq R_2, \end{cases}$$

where $\mathbf{u}_\rho$ is the radial unit vector for cylindrical coordinates. The field at any point on the cylinder outside surface can be calculated by using either expressions, and making $r = R_2$.

$$\mathbf{E}_{(r=R_2)} = \frac{\rho(R_2^2 - R_1^2)}{2\varepsilon_0 R_2}\mathbf{u}_\rho = \frac{q}{2\pi\varepsilon_0 R_2}\mathbf{u}_\rho.$$

If the cylinder were solid, with radius R, the field can be obtained by replacing $R_1 = 0$ and $R_2 = R$:

$$\mathbf{E} = \begin{cases} \dfrac{\rho r}{2\varepsilon_0}\mathbf{u}_\rho = \dfrac{q}{2\pi\varepsilon_0 r}\dfrac{r^2}{R^2}\mathbf{u}_\rho & r \leq R, \\ \dfrac{\rho R^2}{2\varepsilon_0 r}\mathbf{u}_\rho = \dfrac{q}{2\pi\varepsilon_0 r}\mathbf{u}_\rho & r \geq R. \end{cases}$$

And at any point on the surface

$$\mathbf{E}_{(r=R)} = \frac{\rho R}{2\varepsilon_0}\mathbf{u}_\rho = \frac{q}{2\pi\varepsilon_0 R}\mathbf{u}_\rho.$$

(b) If the cylinder is metallic (conductor) and in electrostatic balance, charge is only distributed on the external surface and its distribution is uniform.[16] Charge superficial density σ can be calculated from (2.4), considering a cylinder with length L:

---

[16]If the superficial charge density were not uniform, and due to the cylindrical symmetry, the inner field in the conductor would not be null, against the electrostatic balance hypothesis.

$$Q = \int_S \sigma dS = \sigma \int_S dS = \sigma 2\pi R_2 L \Rightarrow \sigma = \frac{Q/L}{2\pi R_2} = \frac{q}{2\pi R_2}.$$

The field at points inside this surface ($r < R_2$) is null, due to the symmetry of the charge distribution. To calculate the field at external points ($r \geq R_2$) we do the same as in section (a). Calculating the flux through the Gaussian surface, the same (2.44) is obtained. Applying Gauss' law, it results:

$$\Phi_{E(r>R_2)} = \frac{q_{in}}{\varepsilon_0} = \frac{1}{\varepsilon_0} \int_S \sigma dS = \frac{\sigma}{\varepsilon_0} S_{lat} = \frac{\sigma}{\varepsilon_0} 2\pi R_2 L = \frac{qL}{\varepsilon_0},$$

where S_{lat} is the lateral surface area of the external cylindrical surface, where the charge is distributed. If we calculate with (2.44) it results

$$E(r > R_2) = \frac{\sigma 2\pi R_2 L}{\varepsilon_0 2\pi r L} = \frac{\sigma R_2}{\varepsilon_0 r} = \frac{q}{\varepsilon_0 2\pi r}.$$

Combining all results,

$$\mathbf{E} = \begin{cases} 0 & r < R_2, \\ \dfrac{\sigma R_2}{\varepsilon_0 r} \mathbf{u}_\rho = \dfrac{q}{\varepsilon_0 2\pi r} \mathbf{u}_\rho & r > R_2, \end{cases}$$

where $\mathbf{u}_\rho$ is the radial unit vector for cylindrical coordinates. The field at a closed point to the external surface of the cylinder can be calculated with $r = R_2$:

$$\mathbf{E}_{(r=R_2)} = \frac{\sigma}{\varepsilon_0} \mathbf{u}_\rho,$$

which is the known value for the field at points near the conductor surface.

2.7 Determine the electric field and the potential produced at any point by a very large plate, with thickness d, on the following cases: (a) non conductor and uniformly charged with density ρ; and (b) metallic and with electrostatic balance, with charge density σ. Particularize these results to the case in which the plate thickness is null (infinite plane). Note: take as the potential reference its central plane.

Solution

(a) Let's firstly consider the case in which charge is uniformly distributed on the plate. Due to symmetry of the charge distribution, every point at the same distance from the central plane of the plate and far away from the ends has the same electric field value,

Fig. 2.12 Gaussian surfaces and field vectors to apply Gauss' law to a uniformly charged plate

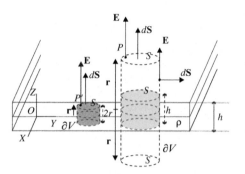

which is also perpendicular to the plate.[17] The problem can be solved by applying Gauss' law (2.28). As a Gaussian surface ∂V, and due to the symmetry, it can be chosen a straight cylinder (any parallelepiped surface would also be valid), with its axis perpendicular to the plate, with one of its bases passing through point where the field is calculated, and the other base symmetric to the previous one, referring to the central plane of the plate (Fig. 2.12). Two different expressions are obtained for the electric field, depending on the point to be studied, if it is outside or inside the plate (P and P' in Fig. 2.12), since charge inside the Gaussian surface has a different expression.

Electric field at any point on the Gaussian surface is perpendicular to the plate, outward if we suppose the plate positively charged, and with the same magnitude at every point of the two cylinder bases. The flux through the lateral surface of either gaussian cylinder ∂V is null, since **E** and $d\mathbf{S}$ are perpendicular at any point of this surface. There is only flux through the bases, it is

$$\Phi_E = \oint_{\partial V} \mathbf{E} \cdot d\mathbf{S} = \int_{B_{\text{upp}}} E dS + \int_{B_{\text{low}}} E dS = E2S. \tag{2.45}$$

It should be noticed that S is the area of the upper base B_{upp} and that of the lower base B_{low}.

If we apply Gauss' law, it is obtained for an external point P,

$$\Phi_E = \frac{q_{\text{in}}}{\varepsilon_0} = \frac{\rho S h}{\varepsilon_0}.$$

Charge inside the Gaussian surface is only inside the cylinder of height h (marked in light grey in Fig. 2.12). That's the reason that the charged volume is just Sh. Equating this expression with (2.45), and solving, it results

---

[17] To check this, consider the field produced at a point by an element and its symmetric pair with respect to a perpendicular to the plate at the considered point. The components parallel to the plate have the same magnitude, since they are at the same distance, and opposite direction, so we need only to add the two normal components of the electric field.

$$E_{\text{ext}} = \frac{\rho h}{2\varepsilon_0}.$$

If we apply Gauss' law for an internal point P', inside the plate, the flux is

$$\Phi_E = \frac{q_{\text{in}}}{\varepsilon_0} = \frac{\rho 2 r S}{\varepsilon_0}$$

The cylinder intersection with the plate is the entire cylinder with height $2r$ (marked in dark grey in Fig. 2.12). Equating this expression with (2.45) and solving, it results

$$E_{\text{in}} = \frac{\rho r}{\varepsilon_0}.$$

If it is expressed with vector notation, using the distance r to the central plane of the plate or Cartesian coordinates, it results

$$\mathbf{E} = \begin{cases} \dfrac{\rho r}{\varepsilon_0} \dfrac{\mathbf{r}}{r} = \dfrac{\rho r}{\varepsilon_0} \operatorname{sgn}(z)\, \mathbf{u}_z & r \leq h/2 \;\; (|z| \leq h/2), \\[1em] \dfrac{\rho h}{2\varepsilon_0} \dfrac{\mathbf{r}}{r} = \dfrac{\rho h}{2\varepsilon_0} \operatorname{sgn}(z)\, \mathbf{u}_z & r \geq h/2 \;\; (|z| \geq h/2), \end{cases} \qquad (2.46)$$

where (x, y, z) are the coordinates of field point P and $\operatorname{sgn}(z)$ the signum function of z, which indicates that the direction of electric field is downward at points under the central plane of the plate.

To determine the potential at any point P, and due to the fact that the potential reference is on the central plane of the plate, the circulation of the electric field must be calculated from the point P to any point of the central plane of the plate. Since circulation is independent of the chosen path, and any point of the central plane has null potential, we take as the circulation line the perpendicular one from the point to the plane, for which $\mathbf{E}$ and $d\mathbf{l} = dr\dfrac{\mathbf{r}}{r}$ are parallel,[18] as shown in Fig. 2.13. If we consider an internal point P' in Fig. 2.13 at a distance r from the central plane of the plate and if O is the point of the central plane perpendicular to P', which has null potential, it results,

$$V_{P'} = V_{P'} - V_O = \int_{P'}^{O} \mathbf{E}_{\text{in}} \cdot d\mathbf{l} = \int_{r}^{0} \frac{\rho}{\varepsilon_0} r\, dr = \frac{\rho}{2\varepsilon_0} r^2 \bigg|_{r}^{0} = -\frac{\rho r^2}{2\varepsilon_0}.$$

To determine the potential at an external point P, since the electric field expression is different depending on whether the point is inside or outside the plate, it is necessary to circulate $\mathbf{E}$ from P to a point P_s on the surface using the expression for an external

---

[18] The same result is achieved without taking a circulation line in which $\mathbf{E}$ and $d\mathbf{l}$ are parallel, as it was already said in Problem 2.5.

Fig. 2.13 Calculation of the potential of a charged plate

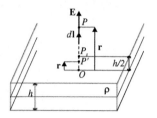

field, and then from this point P_s to the centre of the plate, using the expression for the field at internal points. The result is

$$V_P = V_P - V_O = \int_P^O \mathbf{E} \cdot d\mathbf{l} = \int_P^{P_s} \mathbf{E}_{ext} \cdot d\mathbf{l} + \int_{P_s}^O \mathbf{E}_{in} \cdot d\mathbf{l}$$

$$= \int_r^{h/2} \frac{\rho h}{2\varepsilon_0} dr + \int_{h/2}^0 \frac{\rho r}{\varepsilon_0} dr = \frac{\rho h}{2\varepsilon_0} r \Big|_r^{h/2} + \frac{\rho}{2\varepsilon_0} r^2 \Big|_{h/2}^0$$

$$= \frac{\rho}{2\varepsilon_0} \left(\frac{h^2}{2} - hr - \frac{h^2}{4} \right) = \frac{\rho h}{2\varepsilon_0} \left(\frac{h}{4} - r \right).$$

It can be observed that calculated potentials for point P and for point P' are negative, which coincides with the fact that field $\mathbf{E}$ has the direction of decreasing potentials.
(b) If the plate is a conductor, charge is distributed over the lower and upper surfaces, since interior charge in conductors is null. Charge density σ on the surfaces must be homogeneous, because if not, field inside the plate wouldn't be null, as it has to be a balanced conductor. The field at internal points of the plate is therefore null. To calculate the field at external points, symmetry reasonings are the same as the ones in section (a). The Gaussian surface is the same as before, but now the charge is only in the intersection of the Gaussian surface with the plate surfaces (Fig. 2.14). The flux through the surface is obtained as in (2.45),

$$\Phi_E = E2S.$$

If Gauss' law is applied the result is

$$\Phi_E = \frac{q_{in}}{\varepsilon_0} = \frac{\sigma 2S}{\varepsilon_0}.$$

If both expressions are equated, the result is

$$E_{ext} = \frac{\sigma}{\varepsilon_0},$$

Fig. 2.14 Gaussian surfaces and field vectors to apply Gauss' law to a conducting plate

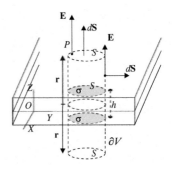

So the field at any external point is constant. If we express it with vectors

$$\mathbf{E} = \begin{cases} 0 & r \leq h/2 \ (|z| \leq h/2), \\ \dfrac{\sigma}{\varepsilon_0}\dfrac{\mathbf{r}}{r} = \dfrac{\sigma}{\varepsilon_0}\,\mathrm{sgn}(z)\mathbf{u}_z & r \geq h/2 \ (|z| \geq h/2), \end{cases}$$

To calculate the potential, since the field is null inside the plate, internal points have the same potential (zero) as the ones on the centre of the plate. To calculate the potential at an external point P, it is enough to circulate $\mathbf{E}$ from P to the plate's surface, since circulation from the surface to the centre of the plate is null. If the circulation path indicated in Fig. 2.13 is followed, the result is

$$V_P = V_P - V_O = \int_P^O \mathbf{E} \cdot d\mathbf{l} = \int_P^{P_s} \mathbf{E}_{\mathrm{ext}} \cdot d\mathbf{l} = \int_r^{h/2} \frac{\sigma}{\varepsilon_0}\,dr = \frac{\sigma}{\varepsilon_0} r \Big|_r^{h/2} = \frac{\sigma}{\varepsilon_0}\left(\frac{h}{2} - r\right).$$

If the plate has no thickness, the problem is the same as the previous one with the only difference that there aren't two charged surfaces with density σ but only one; so if the same calculus is repeated,

$$\Phi_E = \frac{q_{\mathrm{in}}}{\varepsilon_0} = \frac{\sigma S}{\varepsilon_0}.$$

And if we equate with (2.45) the result is

$$E = \frac{\sigma}{2\varepsilon_0}.$$

The resulting potential is

$$V_P = V_P - V_O = \int_P^O \mathbf{E} \cdot d\mathbf{l} = \int_r^0 \frac{\sigma}{2\varepsilon_0}\,dr = -\frac{\sigma}{2\varepsilon_0} r.$$

2.8 We have a wire AB with length l and its line charge density is $\lambda_1 = \lambda(1 + kx)$, where x is the distance of a point of the wire to the central point M of segment OA (Fig. 2.15), and λ and k are two known constants. Perpendicularly to this wire at a distance a of its extreme A, an infinite wire with line charge density $\lambda_2 = \lambda$ is placed. Determine the electric field at point M.

Solution

To solve the problem, the superposition principle for electric fields is applied: fields at point M is the addition of the fields produced by both wires at that point. To calculate the field of wire AB we consider Fig. 2.16, where distance from any element of charge $dq' = \lambda_1 dx$ to point M is expressed by variable x. Field $d\mathbf{E}_1$ produced at point M by the differential element of charge dq' is given by (2.15):

$$d\mathbf{E}_1 = \frac{dq'}{4\pi\varepsilon_0 d^2}\mathbf{u} = \frac{1}{4\pi\varepsilon_0}\frac{\lambda_1 dx}{x^2}(-\mathbf{u}_x) =$$
$$= \frac{1}{4\pi\varepsilon_0}\frac{\lambda(1+kx)dx}{x^2}(-\mathbf{u}_x),$$

To calculate total field due to wire AB, the superposition principle (2.16) is applied:

$$\mathbf{E}_1 = \int_{a/2}^{l+a/2}\frac{\lambda}{4\pi\varepsilon_0}\frac{(1+kx)dx}{x^2}(-\mathbf{u}_x) = \frac{\lambda}{4\pi\varepsilon_0}\left(\int_{a/2}^{l+a/2}\frac{1}{x^2}dx + \int_{a/2}^{l+a/2}\frac{k}{x}dx\right)(-\mathbf{u}_x) =$$
$$= \frac{\lambda}{4\pi\varepsilon_0}\left(\frac{-1}{x} + k\ln x\right)\bigg|_{a/2}^{l+a/2}(-\mathbf{u}_x) = \frac{\lambda}{4\pi\varepsilon_0}\left[\frac{2}{a} - \frac{2}{a+2l} + k\ln\frac{l+a/2}{a/2}\right](-\mathbf{u}_x) =$$
$$= \frac{\lambda}{2\pi\varepsilon_0}\left[-\frac{1}{a} + \frac{1}{a+2l} - \frac{k}{2}\ln\frac{a+2l}{a}\right]\mathbf{u}_x.$$

It should be observed that the limits of the integral are the ends of the charged wire.

To calculate the field produced by the infinite wire with density λ_2 the problem's 2.3 result is applied,

Fig. 2.15 Figure of Problem 2.8

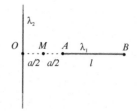

Fig. 2.16 Field produced by the finite wire

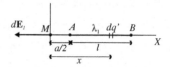

$$\mathbf{E}_2 = \frac{\lambda_2}{2\pi\varepsilon_0 r}\mathbf{u}_\rho = \frac{\lambda}{2\pi\varepsilon_0 a/2}\mathbf{u}_x = \frac{\lambda}{\pi\varepsilon_0 a}\mathbf{u}_x,$$

where distance from the wire to the point field is $a/2$ and the radial unit vector is, in this case, $\mathbf{u}_x$.

The electric field at point M, applying superposition principle, is

$$\mathbf{E} = \mathbf{E}_1 + \mathbf{E}_2 = \frac{\lambda}{2\pi\varepsilon_0}\left(\frac{1}{a} + \frac{1}{a+2l} - \frac{k}{2}\ln\frac{a+2l}{a}\right)\mathbf{u}_x.$$

2.9 Two straight conductors, parallel and infinite, with respective density charge $\lambda_1 = \lambda$ and $\lambda_2 = -2\lambda$ are separated by a distance d. Calculate the potential difference between points A and B in Fig. 2.17.

Solution

To calculate the potential difference between A and B, $V_A - V_B$, it is necessary to know the electrostatic field at every point in a line between these points. For this, the superposition principle is applied, and the resulting field at each point is the addition of the fields produced by each wire independently. From Problem 2.3 it is known that electric field produced by an infinite line is perpendicular to this line, and with the same magnitude at every point of a cylindrical surface whose axis is the line. The electric field at P (Fig. 2.18) is given by (2.36), which for the line of density λ_1 is

$$\mathbf{E}_1 = \frac{\lambda_1}{2\pi\varepsilon_0 r_1}\frac{\mathbf{r}_1}{r_1} = \frac{\lambda}{2\pi\varepsilon_0 r_1}\frac{\mathbf{r}_1}{r_1}.$$

The field produced by the line of density λ_2 is

$$\mathbf{E}_2 = \frac{\lambda_2}{2\pi\varepsilon_0 r_2}\frac{\mathbf{r}_2}{r_2} = \frac{-2\lambda}{2\pi\varepsilon_0 r_2}\frac{\mathbf{r}_2}{r_2} = \frac{\lambda}{\pi\varepsilon_0(d-r_1)}\frac{\mathbf{r}_1}{r_1}.$$

Adding both fields, the total electric field at any point P is obtained

$$\mathbf{E} = \mathbf{E}_1 + \mathbf{E}_2 = \frac{\lambda}{2\pi\varepsilon_0}\left(\frac{1}{r_1} + \frac{2}{d-r_1}\right)\frac{\mathbf{r}_1}{r_1}.$$

To calculate the potential difference between A and B, a circulation of $\mathbf{E}$ from A to B must be done. A line from A to B' has been taken (Fig. 2.19), where $\mathbf{E}$ and

Fig. 2.17 Figure of Problem 2.9

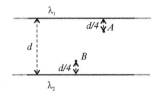

Fig. 2.18 Fields produced by the two infinite wires

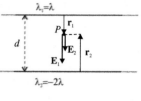

Fig. 2.19 Scheme to calculate the potential difference between A and B

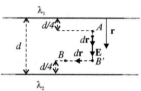

$d\mathbf{r}$ are parallel, and then from B' to B where they are perpendicular, and circulation null.[19] Applying (2.22),

$$V_A - V_B = \int_{r_A}^{r_B} \mathbf{E} \cdot d\mathbf{r} =$$

$$= \int_{d/4}^{3d/4} \frac{\lambda}{2\pi\varepsilon_0}\left(\frac{1}{r} + \frac{2}{d-r}\right) dr =$$

$$= \frac{\lambda}{2\pi\varepsilon_0}\left(\ln r|_{d/4}^{3d/4} - 2\ln(d-r)|_{d/4}^{3d/4}\right) = \frac{\lambda}{2\pi\varepsilon_0} 3\ln 3.$$

2.10 We have an isolated spherical conductor whose radius is $R_1 = 4$ cm, and whose potential is 9000 V referring to ground. After, it is surrounded with a concentric spherical conducting layer, with inner radius $R_2 = 8$ cm and exterior one $R_3 = 10$ cm, isolated and with null total charge. Determine charges and potentials on the inner conductor, as well as the conducting layer, for the following cases: (a) Inner conductor and conducting layer isolated. (b) If the conducting layer is connected to ground. (c) If the layer is once again isolated and the conductor is connected to ground by a conducting wire that goes through a small hole in the layer.

Solution

(a) Note the charge distribution is not known. The isolated spherical conductor will have certain charge q_1, since its potential is not zero. If the expression of potential for a spherical conductor is applied ((2.43) of Problem 2.5), it results

$$V = \frac{q_1}{4\pi\varepsilon_0 R_1}.$$

---

[19] As it was already indicated in Problem 2.5, it is not necessary to consider a specific circulation line, since $\mathbf{r} \cdot d\mathbf{r} = |\mathbf{r}| d|\mathbf{r}|$.

Fig. 2.20 Isolated conducting sphere and isolated conducting layer

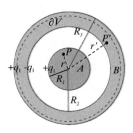

From where conductor's charge q_1 is obtained,

$$q_1 = 4\pi\varepsilon_0 R_1 V = 40\,\text{nC}.$$

If the conductor is surrounded by the conducting layer, Fig. 2.20, charges for both conductors are reorganized, until the balance is reached, all properties in Sect. 2.8 are verified. Since conductor A is isolated, charge q_1 remains, and this charge can only be on its outsider surface. Due to the spherical symmetry of the figure, it will be uniformly distributed over the surface, and thus the electric field inside the conductor is null. On the conducting layer B, charges are distributed so that the field inside it is null. If Gauss' law (2.28) is applied to a Gaussian surface ∂V totally inside the conductor (discontinuous line in the figure),

$$\Phi_E = \oint_{\partial V} \mathbf{E} \cdot d\mathbf{S} = 0 = \frac{q_{\text{in}}}{\varepsilon_0} \Rightarrow q_{\text{in}} = 0 = q_1 + q_{B,\text{in}},$$

where $q_{B,\text{in}}$ is the charge of the inner surface of the conducting layer B. It should be observed that the flux is null because the field at every point of ∂V is zero. Therefore

$$q_{B,\text{in}} = -q_1.$$

If the principle of conservation of charge is applied to conductor B, the external surface charge of B is

$$q_B = 0 = -q_1 + q_{B,\text{ex}} \Rightarrow q_{B,\text{ex}} = +q_1.$$

The charge distribution is the one shown in Fig. 2.20.

Potential at any point is derived from the superposition of potentials created by each of the charge distributions. For every distribution, (2.43) obtained in Problem 2.5 is applied. For conductor A, the distance r from any interior point P to the centre is lower (or equal) than the distance from the centre to the charge distributions ($r \leq R_1, r < R_2, r < R_3$), so it results

$$V_A = \frac{1}{4\pi\varepsilon_0}\left(\frac{q_1}{R_1} - \frac{q_1}{R_2} + \frac{q_1}{R_3}\right) = 9\cdot 10^9 \cdot 40 \cdot 10^{-9}\left(\frac{1}{0.04} - \frac{1}{0.08} + \frac{1}{0.1}\right) = 8100\,\text{V}.$$

Fig. 2.21 Isolated conducting sphere and grounded conducting layer

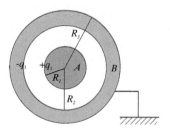

For conductor B, the distance r' from any interior point P' to the centre results in $r' > R_1, r' \geq R_2, r' \leq R_3$, so if (2.43) is applied,

$$V_B = \frac{1}{4\pi\varepsilon_0}\left(\frac{q_1}{r'} - \frac{q_1}{r'} + \frac{q_1}{R_3}\right) = 9 \cdot 10^9 \cdot \frac{40 \cdot 10^{-9}}{0.1} = 3600\,\text{V}.$$

(b) Connecting the conducting layer to ground, Fig. 2.21, we equalize the potentials of conductor B and the ground. Ground potential is taken as a reference (0 V) and, therefore,

$$V_B = 0.$$

Since there is no potential difference between B and ground, there cannot exist any electric field between them (due to (2.18), $\mathbf{E} = -\nabla V$). The field is null in B, as well as outside the conductor B. From (2.29) ($\nabla \cdot \mathbf{E} = \rho/\varepsilon_0$), it is obtained that on the outside surface of conductor B there cannot be charges. The other charge distributions remain the same, as seen by reapplying the reasoning from section (a). It can be observed how grounded conductor B does not have a null charge (it is not an isolated system), but it remains negatively charged. If (2.43) is applied, the result for conductor A is,

$$V_A = \frac{1}{4\pi\varepsilon_0}\left(\frac{q_1}{R_1} - \frac{q_1}{R_2}\right) = 9 \cdot 10^9 \cdot 40 \cdot 10^{-9}\left(\frac{1}{0.04} - \frac{1}{0.08}\right) = 4500\,\text{V}.$$

A device like this, a grounded conductor which surrounds another one, is the base of *electrostatic shields* and it is called a **Faraday cage**. Even though charge or potential inside it is changed, the electric field and potential outside it is always zero. Also, any external electric field would affect neither the electric field nor the potential of the conductors.

(c) If conducting layer B is disconnected from ground, its charge, $-q_1 = -40$ nC, remains and distributes between the inner and outsider surface of B,

$$-q_1 = q_2' + q_3',$$

as seen in Fig. 2.22.

Fig. 2.22 Grounded conducting sphere and isolated conducting layer

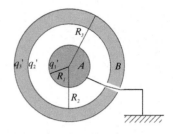

Conductor A does not keep its charge anymore, since it is grounded. New charge is called q'_1. Applying the reasoning from section (a), the charge on the inner surface of conductor B will be

$$q'_2 = -q'_1.$$

It is also known that conductor A potential is zero, since it is grounded. If (2.43) is applied to conductor A the result is

$$V_A = 0 = \frac{1}{4\pi\varepsilon_0}\left(\frac{q'_1}{R_1} + \frac{q'_2}{R_2} + \frac{q'_3}{R_3}\right) \Rightarrow \frac{q'_1}{0.04} + \frac{q'_2}{0.08} + \frac{q'_3}{0.1} = 0.$$

With these three equations the new values of the charges are obtained,

$$q'_1 = \frac{4}{9}q_1 = 17.8\,\text{nC}, \quad q'_2 = -\frac{4}{9}q_1 = -17.8\,\text{nC}, \quad q'_3 = -\frac{5}{9}q_1 = -22.2\,\text{nC}.$$

If (2.43) is applied we obtain conductor B potential,

$$V_B = \frac{1}{4\pi\varepsilon_0}\left(\frac{q'_1}{r'} + \frac{q'_2}{r'} + \frac{q'_3}{R_3}\right) = 9\cdot 10^9 \cdot \frac{-22.2\cdot 10^{-9}}{0.1} = -2000\,\text{V}.$$

2.11 Consider two coaxial conductor cylindrical surfaces, A and B, with infinite length, whose radii are a and b. Outer conductor B is grounded and potential of inner conductor A is V_a. The space between both conductors and outside of conductor B is a vacuum. (a) Calculate surface charge densities on both conductors. (b) If P is a point between A and B, and P' is outside of conductor B, calculate potential difference $V_P - V'_P$.

Solution

(a) Figure 2.23 shows both cylinders. Firstly, since charges can freely move inside conductors, we must study how charges distribute inside both conductors. As it was seen in Sect. 2.8, charges distribute on conductor surfaces, so the field inside them is zero. Due to the symmetry, charge has to be distributed uniformly on the surface. The

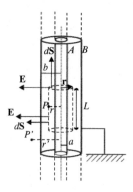

Fig. 2.23 Set of two coaxial hollow cylindrical conductors, the exterior one connected to ground

entire charge of conductor A is distributed on its external surface.[20] Let's suppose that q_a is the charge on conductor A for a finite length L. If Gauss' law for a coaxial cylindrical surface ∂V totally inside conductor B, with length L is applied, the result is, following the same reasoning as Problem 2.10,

$$\Phi_E = \oint_{\partial V} \mathbf{E} \cdot d\mathbf{S} = 0 = \frac{q_{\text{in}}}{\varepsilon_0} \Rightarrow q_{\text{in}} = 0 = q_a + q_{B,\text{in}},$$

where $q_{B,\text{in}}$ is the charge on inner surface of conductor B. It should be observed that the flux is null since the field on every point of the considered Gaussian surface is zero. Therefore

$$q_{B,\text{in}} = -q_a.$$

As conductor B is grounded, its potential is zero. The electric field outside B is zero and charge on the outside surface of the cylinder is also zero, as it was previously explained on Problem 2.10.

To obtain charge densities, we find the electric field's expression at any point P between both conductors. Due to the cylindrical symmetry, the problem can be solved by applying Gauss' law (2.28) in a similar way as it was done in Problem 2.6. To do it, we take as a cylindrical Gaussian surface ∂V, with any length L, with the same axis as the cylinder, and with radius r so that the surfaces passes through point P where we want to calculate the field (discontinuous line in Fig. 2.23). The flux through the bases of the Gaussian surface is zero, since $d\mathbf{S}$ and $\mathbf{E}$ are perpendicular at any point of the bases. So it only remains to calculate the flux through the lateral surface, where $d\mathbf{S}$ and $\mathbf{E}$ are parallel:

$$\Phi_E = \oint_{\partial V} \mathbf{E} \cdot d\mathbf{S} = \int_{S_{\text{lat}}} E \, dS = E \int_{S_{\text{lat}}} dS = E 2\pi r L,$$

---

[20]If there were a charge on the internal surface of conductor A, an electric field inside the conductor should exist, but this would contradict electrostatic equations.

where S_{lat} is the cylinder lateral surface. On the other side, if we apply Gauss' law, we obtain

$$\Phi_E = \frac{q_{\text{in}}}{\varepsilon_0} = \frac{\sigma_a 2\pi a L}{\varepsilon_0},$$

where σ_a is charge density of cylinder A. It should be observed that q_{in} is the charge inside the Gaussian surface, and that there's only charge on the lateral surface of cylinder A, whose radius is a. If both expressions are equalized, electric field is obtained,

$$E = \frac{\sigma_a a}{\varepsilon_0 r} \Rightarrow \mathbf{E} = \frac{\sigma_a a}{\varepsilon_0 r} \mathbf{u}_\rho,$$

where $\mathbf{u}_\rho$ is the radial unit vector for cylindrical coordinates. As additional information, the potential of both conductors is known. The potential difference $V_A - V_B$ between them is $V_a - 0 = V_a$, since B is grounded. If we circulate the electric field by following a line perpendicular to the cylinder axis, so that $\mathbf{E}$ and $d\mathbf{l}$ are parallel, the result is

$$V_A - V_B = V_a = \int_a^b \mathbf{E} \cdot d\mathbf{l} = \int_a^b \frac{\sigma_a a}{\varepsilon_0 r} dr = \frac{\sigma_a a}{\varepsilon_0} \ln\frac{b}{a},$$

from which charge density σ_a of conductor A can be calculated,

$$\sigma_a = \frac{V_a \varepsilon_0}{a \ln\frac{b}{a}}.$$

On a piece of cylinder with length L, A's charge will be $q_a = \sigma_a 2\pi a L$. As charge at inner surface of conductor B is $-q_a$, it results that its density σ_b can be obtained from

$$\sigma_a 2\pi a L = -\sigma_b 2\pi b L \Rightarrow \sigma_b = -\frac{a}{b}\sigma_a = -\frac{V_a \varepsilon_0}{b \ln\frac{b}{a}}.$$

The outer surface of conductor B mentioned previously is not charged due to the fact that it is grounded.

(b) To calculate the potential between points P and P' in Fig. 2.23, at a distance r and r' from the axis, we circulate the electric field between both points,[21]

$$V_P - V_{P'} = V_P = \int_r^{r'} \mathbf{E} \cdot d\mathbf{l} = \int_r^b \frac{\sigma_a a}{\varepsilon_0 r} dr = \frac{\sigma_a a}{\varepsilon_0} \ln\frac{b}{r} = \frac{\ln\frac{b}{r}}{\ln\frac{b}{a}} V_a.$$

---

[21] We will bear on mind the property of every vector $\mathbf{r}$: $\mathbf{r} \cdot d\mathbf{r} = |\mathbf{r}| d|\mathbf{r}|$. It can be reasoned in a similar way by taking the following circulating line: first the perpendicular to the axis from P, where $\mathbf{E}$ and $d\mathbf{l}$ are parallel, until we reach the distance of r' from the center of the cylinders, and then by circulating parallel to the cylinder's axis, whose circulation's value is null, since $\mathbf{E}$ and $d\mathbf{l}$ are perpendicular.

It should be observed that P'''s potential is null: it has ground potential since there is no electric field outside of conductor B. This is also the reason to use b as the superior limit of the integral, and not r': between b and r', the electric field is zero.

Problems C

2.12 Determine the field produced at the coordinate's origin by a circular-shaped arc wire with radius R in Fig. 2.24, symmetrically placed with respect to the X axis, and charged with positive charge density $\lambda = k|\sin\theta|$, where θ is the angle with the horizontal of the position vector of any differential element in the wire, and k is a constant.

Solution

The wire in the problem has a symmetric charge with respect to the X axis. If any charge element dq is taken (Fig. 2.25), field $d\mathbf{E}$ produced by this element of charge has the same magnitude, and forms the same angle θ with X axis as the field produced by its symmetric dq' in reference to this axis. Components parallel to the Y axis of these fields cancel; components along the X axis will be added. The total field produced by a wire has, therefore, a null vertical component, while a horizontal component is twice the one produced by the piece of wire placed in the first quadrant.

Considering the field produced by element $dq = \lambda dl$, which according to (2.15) is

$$d\mathbf{E} = \frac{\lambda dl}{4\pi\varepsilon_0|\mathbf{r}-\mathbf{r}'|^2}\frac{\mathbf{r}-\mathbf{r}'}{|\mathbf{r}-\mathbf{r}'|} = \frac{k|\sin\theta|dl}{4\pi\varepsilon_0 R^2}(-\cos\theta\mathbf{u}_x - \sin\theta\mathbf{u}_y),$$

where $\mathbf{r} = 0$ is field point O position and $\mathbf{r}' = R\cos\theta\mathbf{u}_x + R\sin\theta\mathbf{u}_y$ is the source point dq position.

The total field produced by the wire is

$$\mathbf{E} = \int d\mathbf{E} = \int_L \frac{k|\sin\theta|dl}{4\pi\varepsilon_0 R^2}(-\cos\theta\mathbf{u}_x - \sin\theta\mathbf{u}_y) =$$
$$= \int_{L+} \frac{k\sin\theta dl}{4\pi\varepsilon_0 R^2}(-\cos\theta\mathbf{u}_x - \sin\theta\mathbf{u}_y) + \int_{L-} \frac{k(-\sin\theta)dl}{4\pi\varepsilon_0 R^2}(-\cos\theta\mathbf{u}_x - \sin\theta\mathbf{u}_y),$$

Fig. 2.24 Figure of Problem 2.12

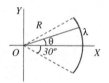

Solved Problems

Fig. 2.25 Field produced at coordinate's origin by the circular-shaped wire

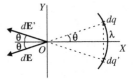

where $L+$ indicates the piece of wire in the first quadrant and $L-$ indicates the piece of wire in the fourth quadrant. Length element dl, arch of circle, can be expressed as a function of angle θ, $dl = Rd\theta$, and the integral results in[22]

$$\mathbf{E} = \int_0^{\frac{\pi}{6}} \frac{k\sin\theta Rd\theta}{4\pi\varepsilon_0 R^2}(-\cos\theta\mathbf{u}_x - \sin\theta\mathbf{u}_y) + \int_{-\frac{\pi}{6}}^0 \frac{k(-\sin\theta)Rd\theta}{4\pi\varepsilon_0 R^2}(-\cos\theta\mathbf{u}_x - \sin\theta\mathbf{u}_y) =$$

$$= -\frac{k}{4\pi\varepsilon_0 R}\left[\left(\int_0^{\frac{\pi}{6}}\sin\theta\cos\theta d\theta - \int_{-\frac{\pi}{6}}^0 \sin\theta\cos\theta d\theta\right)\mathbf{u}_x \right.$$

$$\left. + \left(\int_0^{\frac{\pi}{6}}\sin^2\theta d\theta - \int_{-\frac{\pi}{6}}^0 \sin^2\theta d\theta\right)\mathbf{u}_y\right] =$$

$$= -\frac{k}{4\pi\varepsilon_0 R}\left[\left(\frac{\sin^2\theta}{2}\bigg|_0^{\frac{\pi}{6}} - \frac{\sin^2\theta}{2}\bigg|_{-\frac{\pi}{6}}^0\right)\mathbf{u}_x\right.$$

$$\left. + \left(\left(\frac{2\theta - \sin 2\theta}{4}\right)\bigg|_0^{\frac{\pi}{6}} - \left(\frac{2\theta - \sin 2\theta}{4}\right)\bigg|_{-\frac{\pi}{6}}^0\right)\mathbf{u}_y\right]$$

$$= -\frac{k}{4\pi\varepsilon_0 R}\left[\left(\frac{1}{8} - 0 - 0 + \frac{1}{8}\right)\mathbf{u}_x + \frac{1}{4}(\pi/3 - \sin(\pi/3) - \pi/3 - \sin(-\pi/3))\mathbf{u}_y\right]$$

$$= -\frac{k}{16\pi\varepsilon_0 R}\mathbf{u}_x.$$

The same results can be reached bearing in mind the previous symmetry considerations,

$$\mathbf{E} = \int_L \frac{k|\sin\theta|dl}{4\pi\varepsilon_0 R^2}(-\cos\theta\mathbf{u}_x - \sin\theta\mathbf{u}_y) = 2\int_{L+}\frac{k\sin\theta dl}{4\pi\varepsilon_0 R^2}(-\cos\theta\mathbf{u}_x)$$

$$= -2\int_0^{\frac{\pi}{6}}\frac{k\sin\theta\cos\theta Rd\theta}{4\pi\varepsilon_0 R^2}\mathbf{u}_x =$$

$$= -\frac{k}{2\pi\varepsilon_0 R}\int_0^{\frac{\pi}{6}}\sin\theta\cos\theta d\theta\mathbf{u}_x = -\frac{k}{2\pi\varepsilon_0 R}\frac{\sin^2\theta}{2}\bigg|_0^{\frac{\pi}{6}}\mathbf{u}_x$$

$$= -\frac{k}{2\pi\varepsilon_0 R}\left(\frac{1}{8} - 0\right)\mathbf{u}_x = -\frac{k}{16\pi\varepsilon_0 R}\mathbf{u}_x.$$

---

[22] To solve the integral, remember that $\sin^2\theta = (1 - \cos 2\theta)/2$.

2.13 The spherical crown in Fig. 2.26 (sectioned by plane XY) is shown, whose centre is at point $C(1\,\text{m}, 0, 0)$, with inner radius $R_i = 20\,\text{cm}$ and exterior one $R_e = 50\,\text{cm}$, has a non homogeneous charge density $\rho = k/r$ with $k = 2\,\mu\text{C/m}^2$ and r the distance to the crown's centre measured in meters. The wire in the figure, in XY plane, is infinitely long, makes $45°$ with X axis, and has a charge per unit of length $\lambda = 30\,\text{nC/m}$. Determine the electric field produced at point $A(1\,\text{m}, -1\,\text{m}, 0)$.

Solution

To calculate the field at point A, the superposition principle is applied, adding the fields produced by the spherical crown and the wire at that point. The field produced by each of the distributions can be obtained by applying Gauss' law (2.28).

For the spherical crown, the same procedure as Problem 2.5 is applied: we take a spherical Gaussian surface, concentric with the charged crown, that passes over point A on the field to be calculated (Fig. 2.27). Field $\mathbf{E}_\rho$ is radial and to determine its magnitude, the flux through this Gaussian surface is calculated

$$\Phi_{E_\rho} = \oint_{\partial V} \mathbf{E}_\rho \cdot d\mathbf{S} = E_\rho 4\pi r^2,$$

If Gauss' theorem is applied,

$$\Phi_{E_\rho} = \frac{q_{\text{in}}}{\varepsilon_0} = \frac{1}{\varepsilon_0} \int_{V_{\text{in}}} \rho\, dV = \frac{1}{\varepsilon_0} \int_{V_{\text{in}}} \frac{k}{r} 4\pi r^2\, dr = \frac{4\pi k}{\varepsilon_0} \int_{R_i}^{R_e} r\, dr = \frac{4\pi k}{\varepsilon_0} \frac{R_e^2 - R_i^2}{2},$$

since the volume element in a sphere can be written as $dV = 4\pi r^2 dr$. Equating both flux calculations and taking the data, results are

$$E_\rho = \frac{0.105 k}{\varepsilon_0} = 23729\,\text{N/C}.$$

Fig. 2.26 Figure of Problem 2.13

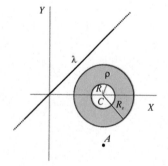

Fig. 2.27 Fields produced by the charged distributions and Gaussian surfaces

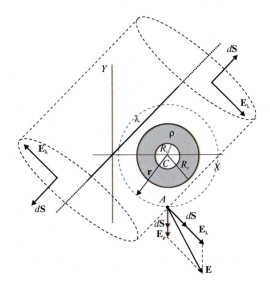

And using vector notation, if Fig. 2.27 is observed:

$$\mathbf{E}_\rho = -23729\,\mathbf{u}_y \text{ N/C}.$$

For the wire, the procedure of Problem 2.4 is applied: a cylindrical Gaussian surface is drawn in Fig. 2.27, whose axis is the wire, with any length L and passing over point A. Field $\mathbf{E}_\lambda$ is perpendicular to the wire, and to obtain its magnitude, the flux through the surface is calculated,

$$\Phi_{E_\lambda} = \oint_{\partial V} \mathbf{E}_\lambda \cdot d\mathbf{S} = \int_{S_{\text{lat}}} E_\lambda\, dS = E_\lambda 2\pi r L\big|_{r=\sqrt{2}\,\text{m}},$$

since the distance from the wire to A, the radius of the cylindrical surface, is $\sqrt{2}$ meters long. And applying Gauss' theorem,

$$\Phi_{E_\lambda} = \frac{q_{\text{in}}}{\varepsilon_0} = \frac{\lambda L}{\varepsilon_0}.$$

Equating the two calculations of the flux, results are

$$E_\lambda = \frac{\lambda}{2\pi\varepsilon_0 r} = \frac{30 \cdot 10^{-9}}{2\pi\varepsilon_0 \sqrt{2}} = 381.5 \text{ N/C}.$$

And using vector notation

$$\mathbf{E}_\lambda = 381.5 \frac{\mathbf{u}_x - \mathbf{u}_y}{\sqrt{2}} = 270\,(\mathbf{u}_x - \mathbf{u}_y)\text{ N/C}.$$

The total field is obtained by adding both fields (vectorially):

$$\mathbf{E} = \mathbf{E}_\rho + \mathbf{E}_\lambda = (270\,\mathbf{u}_x - 23999\,\mathbf{u}_y)\text{ N/C}.$$

2.14 The space region defined by equation $0 < z < 2$ (with Cartesian coordinates in meters) has a charge density $\rho = k|z - 1|$ with $k = 8\,\mu\text{C/m}^4$. (a) Determine the electric field value for the points in the region defined by the sphere with its centre at the coordinates' origin and radius 2 m. (b) Determine the divergence value of the electric field at the previous points. (c) Determine the electric field flux through the previous sphere surface.

Solution

(a) Figure 2.28 represents the region of the problem. Charge distribution is symmetric in reference to the center plane $z = 1$, so Gauss' law (2.28) can be easily applied. A similar procedure as the one used for the infinite plate on Problem 2.7 is followed. Now the points at the field to be calculated are a few specific points defined by the sphere indicated in the statement and light grey coloured in the figure. Points in the higher hemisphere are all the inside points of the charged plate, while points in the lower hemisphere are outside the plate. If Gauss' law is applied as in Problem 2.7, with cylindrical Gaussian surfaces indicated in Fig. 2.28 (which are the ones used before)

$$\Phi_E = \oint_{\partial V} \mathbf{E} \cdot d\mathbf{S} = \int_{B_{upp}} E\,dS + \int_{B_{low}} E\,dS = E2S.$$

To calculate the charge inside the Gaussian surface, it should be taken into account that the charge density is variable. As the charge is symmetric in reference to the

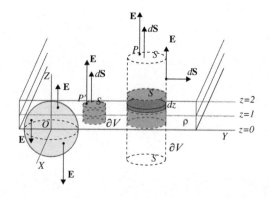

Fig. 2.28 Figure of Problem 2.14

medium plane, we will determine the higher half charge, where $|z - 1| = z - 1$, and we will double the resulting charge. For points outside of the charged region, as point P, it results

$$\Phi_E = \frac{q_{in}}{\varepsilon_0} = \frac{1}{\varepsilon_0}\int_{V_{in}} \rho dV = \frac{1}{\varepsilon_0} 2\int_1^2 k(z-1)Sdz = \frac{2kS}{\varepsilon_0} \frac{(z-1)^2}{2}\bigg|_1^2 = \frac{kS}{\varepsilon_0}.$$

To calculate the integral we have taken a volume element $dV = Sdz$ (in dark grey in the figure), that represents a differential cylinder whose area is S and its height is dz and it is located at any distance from the medium plane. If we compare both flux expressions, the results are

$$E_{ext} = \frac{k}{2\varepsilon_0} = \frac{8 \cdot 10^{-6}}{2\varepsilon_0} = 4.52 \cdot 10^5 \, \text{NC}^{-1},$$

If Gauss' law is applied to an interior point P', it is obtained

$$\Phi_E = \frac{q_{in}}{\varepsilon_0} = \frac{1}{\varepsilon_0} 2\int_1^z k(z-1)Sdz = \frac{2kS}{\varepsilon_0} \frac{(z-1)^2}{2}\bigg|_1^z = \frac{kS}{\varepsilon_0}(z-1)^2.$$

where z is the coordinate in meters of point P'. Comparing both flux expressions, results are

$$E_{in} = \frac{k}{2\varepsilon_0}(z-1)^2 = 4.52 \cdot 10^5 (z-1)^2 \, \text{NC}^{-1}.$$

Vectorially, at points over the center plane ($z = 1$) the sense of field vector is $\mathbf{u}_z$ and for the ones below it is $-\mathbf{u}_z$.

Every point in the lower hemisphere ($z < 0$) is exterior to the charged region, and since they are below the center plane, the sense of the vector field $\mathbf{E}$ is the negative Z-axis, as can be observed in Fig. 2.28. Points in the higher hemisphere are all inside the plate. The vector field has the orientation parallel to the Z-axis in the positive direction at points above the center plane, and it has the opposite orientation at points below the medium plane. If we express the result using vector notation and Cartesian coordinates, it results, for the points in the sphere $x^2 + y^2 + z^2 \leq 4$:

$$\mathbf{E} = \begin{cases} -4.52 \cdot 10^5 \, \mathbf{u}_z \, \text{NC}^{-1} & z < 0, \\ -4.52 \cdot 10^5 (z-1)^2 \, \mathbf{u}_z \, \text{NC}^{-1} & 0 \leq z \leq 1, \\ 4.52 \cdot 10^5 (z-1)^2 \, \mathbf{u}_z \, \text{NC}^{-1} & 1 \leq z \leq 2, \end{cases}$$

with z expressed in meters.

(b) From (2.29) the result is

$$\nabla \cdot \mathbf{E}(\mathbf{r}) = \frac{\rho(\mathbf{r})}{\varepsilon_0} = \begin{cases} 0 & x^2 + y^2 + z^2 \leq 4 \text{ and } z < 0, \\ \dfrac{k|z-1|}{\varepsilon_0} = 9.04 \cdot 10^5 |z-1| \, \text{NC}^{-1}\text{m}^{-1} & x^2 + y^2 + z^2 \leq 4 \text{ and } z \geq 0. \end{cases}$$

It can be observed that the lower hemisphere points are outside of the charged zone ($\rho = 0$) and the electric field's divergence at these points is null.

(c) If Gauss' law (2.28) is applied to the surface defined by the sphere it results

$$\Phi_E = \oint_{\partial V} \mathbf{E} \cdot d\mathbf{S} = \frac{q_{in}}{\varepsilon_0} = \frac{1}{\varepsilon_0} \int_{V_{sph}} \rho \, dV = \frac{1}{\varepsilon_0} \int_{V_{upsph}} k|z-1| dV,$$

where V_{upsph} is the upper hemisphere's volume. To calculate the integral we use spherical coordinates. It can be observed that r varies from 0 to 2 (sphere's radius), coordinate ϕ goes from 0 to 2π because the whole circumference is described, and coordinate θ goes from 0 to $\pi/2$, since only the upper hemisphere is charged:

$$\Phi_E = \frac{1}{\varepsilon_0} \int_{V_{uppsph}} k|z-1| dV = \frac{1}{\varepsilon_0} \int_0^R \int_0^{\pi/2} \int_0^{2\pi} k|r\cos\theta - 1| r^2 \sin\theta \, dr\, d\theta\, d\phi,$$

where we have taken into account that $z = r\cos\theta$. It should be observed that function $|z - 1| = |r\cos\theta - 1|$ has a different value depending on z, whether $z > 1$ or $z < 1$. Where $z = 1$ and the sphere's radius $R = 2$, results are $\cos\theta = 1/2 \to \theta = \pi/3$. Using two integrals to substitute the absolute value expression and solving them, results are

$$\Phi_E = \frac{k}{\varepsilon_0} \left(\int_0^2 \int_0^{\pi/3} \int_0^{2\pi} (r\cos\theta - 1) r^2 \sin\theta \, dr\, d\theta\, d\phi \right.$$
$$\left. - \int_0^2 \int_{\pi/3}^{\pi/2} \int_0^{2\pi} (r\cos\theta - 1) r^2 \sin\theta \, dr\, d\theta\, d\phi \right) =$$
$$= \frac{2\pi k}{\varepsilon_0} \left(\int_0^2 \int_0^{\pi/3} (r\cos\theta - 1) r^2 \sin\theta \, dr\, d\theta - \int_0^2 \int_{\pi/3}^{\pi/2} (r\cos\theta - 1) r^2 \sin\theta \, dr\, d\theta \right) =$$
$$= \frac{2\pi k}{\varepsilon_0} \left(\int_0^{\pi/3} \left(4\cos\theta\sin\theta - \frac{8}{3}\sin\theta \right) d\theta - \int_{\pi/3}^{\pi/2} \left(4\cos\theta\sin\theta - \frac{8}{3}\sin\theta \right) d\theta \right) =$$
$$= \frac{2\pi k}{\varepsilon_0} \left(4\frac{\sin^2\theta}{2} \Big|_0^{\pi/3} + \frac{8}{3}\cos\theta \Big|_0^{\pi/3} - 4\frac{\sin^2\theta}{2} \Big|_{\pi/3}^{\pi/2} - \frac{8}{3}\cos\theta \Big|_{\pi/3}^{\pi/2} \right)$$
$$= \frac{2\pi k}{\varepsilon_0} \left(\frac{3}{2} - \frac{4}{3} - \frac{1}{2} + \frac{4}{3} \right).$$

The flux obtained is

$$\Phi_E = \frac{2\pi k}{\varepsilon_0} = 5.68 \cdot 10^6 \, \text{Nm}^2\text{C}^{-1}.$$

Solved Problems

2.15 Calculate the electric field produced at point P in Fig. 2.29 by the cylinder with volumetric uniform density ρ, whose height is H, inner radius R_i and exterior one R_e.

Solution

The problem can be solved by direct integration using cylindrical coordinates. Any charge element dq (Fig. 2.30), at $\mathbf{r}' = r\cos\phi\mathbf{u}_x + r\sin\phi\mathbf{u}_y + z\mathbf{u}_z$ is expressed by using cylindrical coordinates as follows

$$dq = \rho dV = \rho r\, dr\, d\phi\, dz.$$

The field produced by this element at point P, located at $\mathbf{r} = h\mathbf{u}_z$, is

$$d\mathbf{E} = \frac{dq}{4\pi\varepsilon_0}\frac{\mathbf{r}-\mathbf{r}'}{|\mathbf{r}-\mathbf{r}'|^3} =$$
$$= \frac{\rho}{4\pi\varepsilon_0}\frac{-r\cos\phi\mathbf{u}_x - r\sin\phi\mathbf{u}_y + (h-z)\mathbf{u}_z}{[r^2+(h-z)^2]^{3/2}} r\, dr\, d\phi\, dz.$$

If the superposition principle is applied

Fig. 2.29 Figure of Problem 2.15

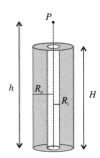

Fig. 2.30 Charge differentials for a finite cylinder

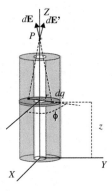

$$\mathbf{E} = \int_{R_i}^{R_e} \int_0^{2\pi} \int_0^H \frac{\rho}{4\pi\varepsilon_0} \frac{-r^2\cos\phi\mathbf{u}_x - r^2\sin\phi\mathbf{u}_y + (h-z)r\mathbf{u}_z}{(r^2 + (h-z)^2)^{3/2}} dr d\phi dz.$$

The sine integral and the cosine integral between 0 and 2π are null, and therefore, components in $\mathbf{u}_x$ and $\mathbf{u}_y$ are null. This can be deduced from the figure's symmetry: the field can only have a component in $\mathbf{u}_z$ as each dq has its symmetric dq' in a horizontal plane whose $d\mathbf{E}'$ makes the same angle with the vertical, and it gives opposite components in $\mathbf{u}_x$ and $\mathbf{u}_y$, and equal components in $\mathbf{u}_z$. The integral to solve is

$$\mathbf{E} = \frac{\rho}{4\pi\varepsilon_0} \int_{R_i}^{R_e} \int_0^{2\pi} \int_0^H \frac{(h-z)r}{(r^2 + (h-z)^2)^{3/2}} dr d\phi dz \mathbf{u}_z$$

$$= \frac{\rho}{2\varepsilon_0} \int_{R_i}^{R_e} \int_0^H \frac{(h-z)r}{(r^2 + (h-z)^2)^{3/2}} dr dz \mathbf{u}_z,$$

because $\int_0^{2\pi} d\phi = 2\pi$. Solving in r, bearing in mind that the derivative of the function $[r^2 + (h-z)^2]$ is $2r$, the result is

$$\mathbf{E} = \frac{\rho}{2\varepsilon_0} \int_0^H \left. \frac{-(h-z)}{(r^2 + (h-z)^2)^{1/2}} \right|_{R_i}^{R_e} dz \mathbf{u}_z =$$

$$= -\frac{\rho}{2\varepsilon_0} \int_0^H \left(\frac{(h-z)}{\sqrt{R_e^2 + (h-z)^2}} - \frac{(h-z)}{\sqrt{R_i^2 + (h-z)^2}} \right) dz \mathbf{u}_z.$$

Hence,

$$\mathbf{E} = \frac{\rho}{2\varepsilon_0} \left(\left. \sqrt{R_e^2 + (h-z)^2} \right|_0^H - \left. \sqrt{R_i^2 + (h-z)^2} \right|_0^H \right) \mathbf{u}_z$$

$$= \frac{\rho}{2\varepsilon_0} \left(\sqrt{R_e^2 + (h-H)^2} - \sqrt{R_i^2 + (h-H)^2} - \sqrt{R_e^2 + h^2} + \sqrt{R_i^2 + h^2} \right) \mathbf{u}_z.$$

2.16 Considere an infinite plate with thickness $h = 1$ m in Fig. 2.31, with uniform charge density $\rho = 20\,\mu\text{C/m}^3$, except inside a spherical cavity with charge density three times that of the plate, with the centre the medium plane, and diameter h. Determine the electric field produced at point $P(1, 2, 2)$, with the coordinates measured in meters referring to the coordinate axis located at the centre of the sphere, as shown in Fig. 2.31.

Solution

To calculate the electric field produced by the charge distribution, it can be observed that the superposition principle can be applied, and therefore it can be calculated as the addition of the fields produced by the two distributions with high symmetry:

Fig. 2.31 Infinite plate with a spherical cavity and higher density

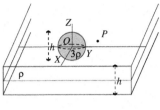

Fig. 2.32 Superposition principle for fields produced by the plate and the sphere

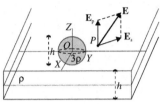

an infinite plate with uniform charge density ρ and a uniformly charged sphere with density 2ρ. In this way the spherical area would have a charge distribution, by adding of the previous ones, of 3ρ indicated on the problem's statement. Fields produced by these distributions (Fig. 2.32) can be obtained from exercises previously worked in this book.

To calculate the field $\mathbf{E}_p$ produced by the whole plate, without the cavity, we apply the obtained results in Problem 2.7, bearing in mind that point P is outside the plate. According to (2.46), the electric field results

$$\mathbf{E}_p = \frac{\rho h}{2\varepsilon_0}\,\mathrm{sgn}(z)\mathbf{u}_z = \frac{20\cdot 10^{-6}\cdot 1}{2\cdot 8.85\cdot 10^{-12}}\mathbf{u}_z = 1.130\cdot 10^6 \mathbf{u}_z \text{ N/C}.$$

To calculate the field produced by the sphere $\mathbf{E}_s$, we apply the result of Problem 2.5, and bear in mind that the sphere is solid (2.40), with density 2ρ and radius $h/2$ and that point P is outside the sphere. The distance r from the sphere's centre to point P is calculated from the vector's position at point P:

$$\mathbf{r} = \mathbf{u}_x + 2\mathbf{u}_y + 2\mathbf{u}_z, \quad r = 3, \quad \mathbf{u}_r = \frac{\mathbf{u}_x + 2\mathbf{u}_y + 2\mathbf{u}_z}{3}.$$

The field results

$$\mathbf{E}_s = \frac{2\rho\left(\frac{h}{2}\right)^3}{3\varepsilon_0 r^2}\mathbf{u}_r = \frac{\rho h^3}{12\varepsilon_0 r^2}\mathbf{u}_r = \frac{20\cdot 10^{-6}\cdot 1^3}{12\cdot 8.85\cdot 10^{-12}\cdot 3^2}\,\frac{\mathbf{u}_x + 2\mathbf{u}_y + 2\mathbf{u}_z}{3}$$
$$= 6975(\mathbf{u}_x + 2\mathbf{u}_y + 2\mathbf{u}_z) \text{ N/C}.$$

The field $\mathbf{E}$ produced by the plate with the cavity is the addition of the fields $\mathbf{E}_p$ and $\mathbf{E}_s$,

$$\mathbf{E} = \mathbf{E}_p + \mathbf{E}_s = (6975\mathbf{u}_x + 13950\mathbf{u}_y + 1.144\cdot 10^6 \mathbf{u}_z) \text{ N/C}.$$

2.17 Consider the bent wire in Fig. 2.33, where coordinates are expressed in meters, with uniform charge density $\lambda = 8\,\mu\text{C/m}$. The horizontal piece is very long (semi-infinite) and the other one, $L = 2$ m long, making an angle of $60°$ with respect to the horizontal axis. Determine the electric field and the electric potential at point $P(2, 1)$.

Solution

To solve the problem, the superposition principle can be applied if the problem is considered as the addition of the fields and potentials produced by a semi-infinite horizontal wire and a finite one ($60°$) in respect to the horizontal axis. For the case of the semi-infinite wire, we consider Fig. 2.34 and (2.37) of Problem 2.4,

$$\mathbf{E}_1 = \frac{\lambda}{4\pi\varepsilon_0 y} \left((\sin\varphi_f - \sin\varphi_i)\mathbf{u}_x + (\cos\varphi_i - \cos\varphi_f)\mathbf{u}_y\right).$$

Angle φ_i can be obtained from its tangent: $\tan\varphi_i = y/x = 1/2 \Rightarrow \varphi_i = 30°$. Angle φ_f is $180°$, since the wire is infinite along the $+X$ direction. Therefore,

$$\mathbf{E}_1 = \frac{8 \cdot 10^{-6}}{4\pi\varepsilon_0 \cdot 1} \left((\sin 180° - \sin 30°)\mathbf{u}_x + (\cos 30° - \cos 180°)\mathbf{u}_y\right)$$
$$= 71.9 \cdot 10^3 \left(-0.5\mathbf{u}_x + 1.87\mathbf{u}_y\right),$$

$$\mathbf{E}_1 = (-36\mathbf{u}_x + 134.2\mathbf{u}_y) \cdot 10^3 \text{N/C}.$$

Considering the finite wire of length $L = 2$ m (Fig. 2.35) and applying (2.37) of Problem 2.4 again, the result is

$$\mathbf{E}_2 = \frac{\lambda}{4\pi\varepsilon_0 h} \left((\sin\varphi_f - \sin\varphi_i)\mathbf{u}_x + (\cos\varphi_i - \cos\varphi_f)\mathbf{u}_y\right).$$

Fig. 2.33 Charged wire of Problem 2.17

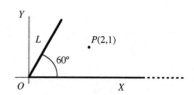

Fig. 2.34 Field produced by the semi-infinite wire

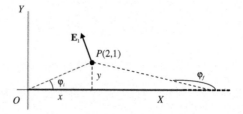

Fig. 2.35 Field produced by the finite wire

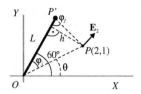

Let's determine the coordinates of the wire's extreme $P'(x', y')$:

$$x' = L\cos 60° = 1, \quad y' = L\sin 60° = \sqrt{3},$$

so

$$\mathbf{PP'} = (1-2)\mathbf{u}_x + (\sqrt{3}-1)\mathbf{u}_y = -\mathbf{u}_x + 0.732\mathbf{u}_y,$$

with $|PP'| = 1.239$.

The vector magnitude $\mathbf{r} = \mathbf{OP}$ is $|OP| = \sqrt{5}$, and angle θ is obtained from

$$\tan\theta = 1/2 \rightarrow \theta = 26.6°.$$

Angle φ_i is, therefore,

$$\varphi_i = 60° - \theta = 33.4°.$$

Distance h from point P to the wire is obtained from

$$h = |OP|\sin\varphi_i = 1.232.$$

The value of φ_f is obtained from

$$\sin\varphi_f = \frac{h}{|PP'|} = 0.994 \rightarrow \varphi_f = 83.9°.$$

If we substitute in $\mathbf{E}_2$ expression

$$\mathbf{E}_2 = \frac{8\cdot 10^{-6}}{4\pi\varepsilon_0 \cdot 1.232}\left((\sin 83.9° - \sin 33.4°)\mathbf{u}_x + (\cos 33.4° - \cos 83.9°)\mathbf{u}_y\right) =$$
$$= 58.44 \cdot 10^3\left(0.44\mathbf{u}_x + 0.73\mathbf{u}_y\right) = (25.9\mathbf{u}_x + 42.6\mathbf{u}_y)\cdot 10^3 \text{N/C}.$$

The total field at point P, if the superposition principle is applied, is

$$\mathbf{E} = \mathbf{E}_1 + \mathbf{E}_2 = \left(-10.1\mathbf{u}_x + 176.8\mathbf{u}_y\right)\cdot 10^3 \text{N/C}.$$

2.18 A thin, flat plate, which has the shape of a regular n-sided polygon, inscribed in a circle of radius a in plane XY, is considered. This plate is maintained at a fixed potential V, while the rest of the plane XY is held at zero potential. Apply Biot–Savart-like law in electrostatics to calculate the electric field produced by this plate at a point P located over the perpendicular to the plate by its centre.

Solution

Figure 2.36 shows the schematic view of the regular n-sided polygon plate, located in the XY-plane, with its centre at the origin, and kept at a potential V ($n = 6$ in Fig. 2.36). In this plane ($z = 0$), the potential is set to zero in the region outside the plate. To solve the problem we apply the (2.35),

$$\mathbf{E}(\mathbf{r}) = \frac{V}{2\pi} \oint_C \frac{(\mathbf{r} - \mathbf{r}') \times d\mathbf{l}'}{|\mathbf{r} - \mathbf{r}'|^3}.$$

$\mathbf{r}$ refers to the field point and $\mathbf{r}'$ locates the source point. $d\mathbf{l}'$ is an element of length of the integration path C. As discussed in Sect. 2.9, we just need to calculate the contributions coming from the boundary contour C.

Looking at Fig. 2.36, it can be seen that electric field at point P can be calculated as the superposition of the field due to n triangles obtained by joining the centre of circle O to the vertices of the polygon: the sense of integration along their common sides is opposite for two adjacent triangles and adding all the contributions from these triangles, we obtain the integration around the contour C. Therefore, the only non-zero contributions come from the n sides that define the contour C. By symmetry, when the n fields are added vectorially, only the components located along Z-axis remains; the XY-plane components add to zero.

Then, considering the point P at Z-axis in the Fig. 2.36 and a point P' and $d\mathbf{l}'$ at the upper straight side,

$$\mathbf{r} = z\mathbf{u}_z, \quad \mathbf{r}' = x'\mathbf{u}_x + a\cos\frac{\pi}{n}\mathbf{u}_y,$$

$$\mathbf{r} - \mathbf{r}' = -x'\mathbf{u}_x - a\cos\frac{\pi}{n}\mathbf{u}_y + z\mathbf{u}_z \quad |\mathbf{r} - \mathbf{r}'| = (x'^2 + a^2\cos^2(\pi/n) + z^2)^{1/2},$$

Fig. 2.36 Schematic view of the regular n-sided polygon plate

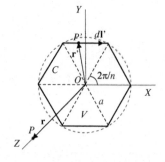

$$d\mathbf{l}' = dx'\mathbf{u}_x \quad (\mathbf{r} - \mathbf{r}') \times d\mathbf{l}' = zdx'\mathbf{u}_y + a\cos\frac{\pi}{n}dx'\mathbf{u}_z.$$

Calling $b^2 = a^2\cos^2(\pi/n) + z^2$ and applying the (2.35) to calculate the contribution E_{iz} along $\mathbf{u}_z$,

$$E_{iz} = \frac{V}{2\pi}\int_{-a\sin(\pi/n)}^{a\sin(\pi/n)} \frac{a\cos(\pi/n)dx'}{(x'^2+b^2)^{3/2}} = \frac{Va\cos(\pi/n)}{2\pi b^2}\left.\frac{x'}{\sqrt{x'^2+b^2}}\right|_{-a\sin(\pi/n)}^{a\sin(\pi/n)} =$$

$$= \frac{Va\cos(\pi/n)}{2\pi b^2}\frac{2a\sin(\pi/n)}{\sqrt{a^2\sin^2(\pi/n)+b^2}} = \frac{V}{2\pi}\frac{a^2\sin(2\pi/n)}{[a^2\cos^2(\pi/n)+z^2]\sqrt{a^2+z^2}}.$$

Adding the n contributions from the n sides that define the contour C, the total electric field at P is obtained,

$$\mathbf{E} = nE_{iz} = \frac{V}{2\pi}\frac{na^2\sin(2\pi/n)}{[a^2\cos^2(\pi/n)+z^2]\sqrt{a^2+z^2}}\mathbf{u}_z.$$

Chapter 3
Static Electric Field in Dielectrics

Abstract In the last chapter, we considered electrostatic fields in free space, produced exclusively by free charges, either by a specified charge distribution or by a free charge on the surface of conductors, but not inside a material media. In this chapter it will be considered the most common case, where materials do not have free charges (*ideal dielectric material*), as well as the case of free charges considered on conductor materials. Actually, a dielectric is composed of charged particles (the atomic nucleus and electrons), which are strongly joined and which form atoms or molecules. They just change their positions lightly, with movements on the order of the radius of an atom, or one angstrom, ($\sim 1\text{Å} = 10^{-10}$ m) as a response to external electric fields. This kind of charge is called *bound charge*, in contrast to free charge found in conducting materials, to express the fact that these charges are not free to move very far or to be extracted from the dielectric material. Strictly speaking, dielectrics do not satisfy this definition, because they have some conductivity, but very little compared to those of metal conductors (more or less 10^{20} times lower). It can be said that dielectrics are non-conductor materials, or *insulators*.

3.1 Polarization

Dielectrics are usually classified as molecularly polar or non-polar. Their behavior when they are placed in an external electric field is different, even though the final result is similar.

Substances with *non-polar* molecules[1] have an electronic negative charge for each molecule symmetrically distributed in reference to its positive nucleus, so its dipole moment **p** (2.32) is null. An external electric field creates a force over the charged particles of the dielectric material, and the positive particles tend to move along the direction of **E** and the negative ones in the opposite direction. When the restoring forces bring the molecule into equilibrium, the centre of positive charge is displaced

---

[1]Monoatomic molecules such as He, Ne or Ar, with spherical symmetry, diatomic with two equal atoms, such as H_2, O_2, and polyatomic with certain symmetries, such as CO_2, C_2H_6 (ethane), C_2H_2 (acetylene) or C_6H_6 (benzene), are not polar, by the fact that the negative and positive charge centres are the same.

© Springer-Verlag Berlin Heidelberg 2017
F. Salazar Bloise et al., *Solved Problems in Electromagnetics*,
Undergraduate Lecture Notes in Physics, DOI 10.1007/978-3-662-48368-8_3

from the centre of negative charge by a very small fraction of an angstrom. So the molecule acquires a dipole moment induced by the external electric field (*electric induced dipoles*). The dielectric is *polarized*.

Substances with *polar* molecules have a permanent polarization caused by the fact that their molecules are asymmetric. These polar molecules can be therefore modeled by an electric dipole formed by the positive and negative charge centres, and with an electric dipole moment **p** which is not null,[2] even without an electric field. It is said that they have a *permanent dipole moment*. However, when there is no electric field these dipole moments will be randomly oriented, and therefore the total dipole moment of the dielectric will be null, as it happened with non-polar dielectrics. When there is an electric field, the opposite movement of the negative and positive charges, whose charge centres are separated by a certain distance, produce a torque which spins every molecule, orienting the positive charges in the direction of the field, and the negative ones in the opposite direction. This produces a dipole moment along the electric field's direction, and the dielectric is polarized.

The polarized dielectric material can be considered as a configuration of electric dipoles, on average neutral, which creates a field at points inside and outside of the dielectric, adding to the external field. As an alternative, it can be supposed that equal charge distributions on the surface and inside the dielectric, called *polarization charges*, create this electric field. An electric field created by a polarized dielectric has opposite direction in relation to the field applied on the dielectric. Hence a complex situation appears: the dielectric's polarization depends on the total electric field in the media, but a part of this field is created by the dielectric itself. Also, the field created by a dielectric can modify the free charge distribution over conductors, which at the same time will modify the field inside the dielectric. It is therefore necessary to create tools to solve this situation.

Let's consider a differential volume dV inside a dielectric material (Fig. 3.1). If the number of polarized molecules in it is N, with dipole moments $\mathbf{p}_i$, dV will also behave as a dipole with net dipole moment equal to the sum of the dipole moments of every molecule, $d\mathbf{p} = \sum_{i=1}^{N} \mathbf{p}_i$. The electric **polarization vector P** is then defined as the electric dipole moment per unit of volume,

$$\mathbf{P}(\mathbf{r}') = \frac{d\mathbf{p}}{dV}, \tag{3.1}$$

whose SI unit is Cm^{-2}. We bear in mind that **P** is a vectorial field defined over the dielectric volume. This concept can be applied to a vacuum by writing $\mathbf{P} = 0$, since there are no molecules that can be polarized. In fact, a vacuum can be considered as a dielectric (it is not a conductor) whose polarization vector is always zero.

---

[2]Polar molecules are almost always diatomic molecules with two different atoms. The order of magnitude of **p** is 10^{-30} Cm, hence a unit for the dipole moment known as the debye exists: 1D = 3.34×10^{-30} Cm. Dipole moments of common bonds for simple diatomic molecules go from 0 to 11D (CO, 0.112D; HI, 0.44D; ClNa 9D; KBr 10.41D). Polyatomic molecules (water, 1.85D; ammonia, 1.47D; formic acid, 1.41D; methanol, 1.70D; formamide, 3.73D; phenol, 1.45D) can also represent electric dipole moments ([77]).

3.2 Polarization Charges

Fig. 3.1 Dipole moments of a polarized dielectric

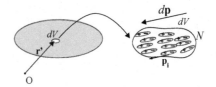

If we calculate the potential produced by the dielectric at any point, as a result of the sum of potentials produced by each of the dipole moments $d\mathbf{p}$, it can be observed that the potential is the same as the one that would create a **polarization volume charge density** ρ_p and a **polarization surface charge density** σ_p given by

$$\rho_p = -\nabla \cdot \mathbf{P}, \tag{3.2}$$

$$\sigma_p = \mathbf{P} \cdot \mathbf{n}, \tag{3.3}$$

where $\mathbf{P}$ represents the electric polarization vector and $\mathbf{n}$ is the outward unit normal for the considered surface at a point on the surface. It can be observed that volumetric density only appears if polarization is not homogeneous, so its divergence is not null. Since polarization charges are a consequence of a molecular charge reorientation, it is always verified that in any finite dielectric, the total polarization charge is null, $q_p = 0$. This can be deduced from the polarization charge densities by applying divergence Theorem 1.49,

$$\begin{aligned} q_p &= \int_V \rho_p \, dV + \int_{\partial V} \sigma_p \, dS = -\int_V \nabla \cdot \mathbf{P} \, dV + \int_{\partial V} \mathbf{P} \cdot \mathbf{n} \, dS \\ &= -\int_{\partial V} \mathbf{P} \cdot \mathbf{n} \, dS + \int_{\partial V} \mathbf{P} \cdot \mathbf{n} \, dS = 0 \,, \end{aligned} \tag{3.4}$$

where V is the dielectric's volume and ∂V its surface.

3.3 The D Field

If we remember the definition of the electric field produced by a charge distribution according to (2.16), it is observed that $\mathbf{E}$ is created by any charge, not depending on its type or origin. A distinction should be made between charges explained in the previous section: the ones that are derived from the dielectric's polarization, also known as *bounded charges* whose density is ρ_p and all the other charges which do not arise from polarization, whose density is ρ_{np}, also known as *free charges*.[3] So the total charge density is

$$\rho = \rho_{np} + \rho_p = \rho_{np} - \nabla \cdot \mathbf{P}. \tag{3.5}$$

And if we apply Gauss' law (2.29)

$$\nabla \cdot \mathbf{E} = \frac{\rho}{\varepsilon_0} = \frac{1}{\varepsilon_0}(\rho_{np} + \rho_p), \tag{3.6}$$

we obtain

$$\nabla \cdot (\varepsilon_0 \mathbf{E} + \mathbf{P}) = \rho_{np}. \tag{3.7}$$

Observing the previous equation, divergence of the vector $\varepsilon_0 \mathbf{E} + \mathbf{P}$ is equal to the free (non-polarization) charge density; it is useful to define a new field vector $\mathbf{D}$ as,

$$\mathbf{D} = \varepsilon_0 \mathbf{E} + \mathbf{P}. \tag{3.8}$$

The vector $\mathbf{D}$ has generally been called **the electric displacement**, but recently it is simply called the $\mathbf{D}$ field. Its units in SI are Cm^{-2}. If we substitute in (3.7), another Gauss' law expression is obtained, and therefore, an alternative form of one of Maxwell's equations,

$$\nabla \cdot \mathbf{D} = \rho_{np}, \tag{3.9}$$

known as Gauss' law for $\mathbf{D}$ field. Equations (3.6) and (3.9) display an essential distinction between $\mathbf{E}$ and $\mathbf{D}$. Any kind of charge density, ρ_p or ρ_{np}, acts as a source for $\mathbf{E}$; but only non-polarization charge density ρ_{np} is a source of $\mathbf{D}$.

If the divergence Theorem 1.49 is used, Gauss' law expression for $\mathbf{D}$ in an integral way can be obtained:

$$\Phi_D = \oint_{\partial V} \mathbf{D} \cdot d\mathbf{S} = q_{np,\text{in}}, \tag{3.10}$$

where ∂V is a closed surface which bounds a volume V and $q_{np,\text{in}}$ is the non-polarization charge inside this volume. An important use of this law is in the determination of $\mathbf{D}$ from non-polarization charge in cases with some degree of symmetry, which mirror the examples and equations done in Chap. 2 to determine $\mathbf{E}$.

---

[3] Even though it is very common to use "free and bounded charges", this can lead to misunderstandings when talking about dielectrics, where non-polarized charges are not free to move. So "free charges" is used to indicate that it is possible to control their distribution, by moving them physically or by introducing them on or inside materials. However, to avoid misunderstandings, we will refer to both charges namely as the ones that come from polarization or the ones that do not.

3.4 The Constitutive Equation

In this chapter it has been studied that the polarization of a dielectric medium occurs in response to the electric field in the medium. The degree of polarization depends not only on the electric field but also on the properties of the dielectric medium. Macroscopically it can be said that a material's behavior is completely specified if the relation between **E** and **P** is known, **P** = **P**(**E**) as determined experimentally. This relation is known as *the constitutive equation* of the material.

Dielectrics can be classified according to this relation. For most dielectrics, **P** vanishes when **E** vanishes; however in some materials the polarization **P** is not null even when the electric field **E** is reduced to zero. The dielectric has permanent polarization and it is called *electret*. When the **P** vector components can be expressed as a function of the first potence of the **E** vector components, the dielectric is known as *linear*, and as *not linear* in the opposite case. The constitutive equation for linear dielectrics is

$$\mathbf{P} = \varepsilon_0 \overline{\overline{\chi_e}} \mathbf{E}, \qquad (3.11)$$

where $\overline{\overline{\chi_e}}$ is a second order tensor (3×3 matrix) called the **electric susceptibility** tensor. The constant ε_0 appears, if χ_e is to be dimensionless, in order to make the equation correct dimensionally. In many cases, at a given point, electric properties of the dielectric do not depend on **E**'s direction. A material is called electrically *isotropic* in this case, and *anisotropic* otherwise. If a material is isotropic, the tensor $\overline{\overline{\chi_e}}$ becomes a scalar, and the constitutive equation results as follows

$$\mathbf{P} = \varepsilon_0 \chi_e \mathbf{E}, \qquad (3.12)$$

where it can be observed that **P** and **E** have the same direction. If electric properties do not depend on the position, the dielectric is called electrically *homogeneous* and susceptibility χ_e will be the same at every point on the material. Many materials are electrically homogeneous and isotropic.

When the electric field in a dielectric is sufficiently large, instead of producing small displacements on dipoles, it begins to pull electrons completely out of the molecules, and the dielectric becomes conducting. This phenomenon is known as **dielectric breakdown**. The maximum field magnitude that can be sustained before a field-stressed material loses its insulating properties is called **dielectric strength**, $E_{\max}$. In SI, the unit of dielectric strength is V/m. Dielectric breakdown can also be described by the **breakdown voltage** of a dielectric, as the maximum voltage difference that can be applied across the material before insulator collapses and conducts.

Substituting (3.12) in (3.8), the result is

$$\mathbf{D} = \varepsilon_0 \mathbf{E} + \mathbf{P} = \varepsilon_0 \mathbf{E} + \varepsilon_0 \chi_e \mathbf{E} = \varepsilon_0 (1 + \chi_e) \mathbf{E}. \qquad (3.13)$$

By setting
$$\varepsilon_r = 1 + \chi_e, \tag{3.14}$$

the **dielectric constant** or *relative permittivity* of the substance,[4] the result is

$$\mathbf{D} = \varepsilon_0 \varepsilon_r \mathbf{E} = \varepsilon \mathbf{E}. \tag{3.15}$$

The multiplying factor $\varepsilon = \varepsilon_0 \varepsilon_r$ that connects $\mathbf{E}$ and $\mathbf{D}$ is called the *absolute permittivity* (or simply permittivity) of the substance and its SI unit is F/m.[5] The same considerations made in regards to isotropy and homogeneity that were made for χ_e can be applied to ε_r.

In problems that only involve linear, homogeneous and isotropic dielectrics, and with some degree of symmetry, Gauss' law (3.10) can be used to determine $\mathbf{D}$, so then the electric field can be determined as

$$\mathbf{E} = \frac{\mathbf{D}}{\varepsilon_0 \varepsilon_r}. \tag{3.16}$$

If we substitute in (3.12), the polarization vector can be obtained from

$$\mathbf{P} = \frac{\varepsilon_r - 1}{\varepsilon_r} \mathbf{D}, \tag{3.17}$$

and polarization charge densities, if we substitute in (3.3) and (3.2), from

$$\sigma_p = \frac{\varepsilon_r - 1}{\varepsilon_r} \mathbf{D} \cdot \mathbf{n}, \tag{3.18}$$

and

$$\rho_p = -\frac{\varepsilon_r - 1}{\varepsilon_r} \nabla \cdot \mathbf{D} = -\frac{\varepsilon_r - 1}{\varepsilon_r} \rho_{np}. \tag{3.19}$$

It can be observed that the three vectors $\mathbf{E}$, $\mathbf{P}$ and $\mathbf{D}$ have the same direction for this kind of dielectric. This relation establishes that if the dielectric is linear, homogeneous and isotropic, there can only be a polarization volume charge if non-polarization charge is distributed in it.

---

[4]In the appendix there is a table with dielectric constants and dielectric strength values of various common materials [66].

[5]Farad (F) is defined in Sect. 3.6. ε unit is also $C^2/(Nm^2)$.

3.5 Boundary Conditions

When we study an electrostatic problem, if many media appear, we must know the field vector's behavior in passing through an interface between two media. The two media may be two different dielectrics, or a dielectric and a conductor. A vacuum can be considered as a dielectric with $\varepsilon_r = 1$.

Let's consider two arbitrary media 1 and 2 in contact (Fig. 3.2). On the interface a certain surface density of charge σ can exist, the sum of the contribution of polarization σ_p and non polarization σ_{np} charges, which may vary from point to point on the interface. Since the electrostatic field is conservative, if the potential difference is calculated along a closed curve *abcda* which goes through both media, as the one in Fig. 3.2a, it can be deduced that the tangential component of field **E** is continuous when going through an interface between two media,

$$E_{1t} = E_{2t}. \tag{3.20}$$

If we apply Gauss' law for **D**, (3.10), to a volume as the one that appears in Fig. 3.2b, it is deduced that the value of the discontinuity on the normal component of electric displacement **D** when going through an interface between two media is the same as the non-polarization charge density on the surface:

$$D_{2n} - D_{1n} = \sigma_{np}. \tag{3.21}$$

Therefore, if non-polarization charge on the surface between two media does not exist, then **D**'s normal component is continuous.

The above results have been obtained for two arbitrary media. If one of the media is a conductor (for example medium 1), $E_1 = 0$ and, by (3.8), also $D_1 = 0$. Thus (3.20) and (3.21) become

$$E_{2t} = 0, \quad D_{2n} = \sigma_{np}. \tag{3.22}$$

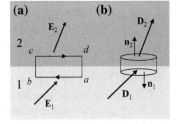

Fig. 3.2 Boundary conditions. Study of: **a** tangential component of **E**; **b** tangential component of **D**

3.6 Coefficients of Potential and Capacitance

In Sect. 2.8 it was mentioned that if we have N conductors with known potential, the problem of solving the potential at every point in space is reduced to solving the Laplace equation for points outside the conductors, while taking into account that boundary conditions are the respective conductor's potentials. Let's now study the case where a linear relationship exists between the potentials and charges on a set of conductors.

Let's consider that all the conductors are discharged, but the j-th. The Laplace equation determines potential V_i of each of the conductors. Since the Laplace equation is linear, if we multiply the charge of j-th conductor by a factor k, potential at every point will be also multiplied by the same factor k. In other words, potential V_{ij} of each conductor i when just conductor j is charged is proportional to its charge q_j:

$$V_{ij} = p_{ij} q_j, \qquad (3.23)$$

where p_{ij} is a constant that only depends on conductor geometries. Using the same reasoning for each of the N conductors and taking into account the Laplace equation linearity, it results that the potential of the i-th conductor, when the N conductors, including itself, are charged, is

$$V_i = \sum_{j=1}^{N} p_{ij} q_j, \qquad (3.24)$$

which represents a linear relationship between potentials V_i and charges q_j. p_{ij} are known as **coefficients of potential** and they only depend on the geometry (shape, size and mutual layout) of conductors. If conductors, instead of being surrounded by a vacuum, were surrounded by a linear dielectric, the analogous result is valid, since the superposition principle of the solutions can also be applied in the Laplace equation, but coefficients p_{ij} will also depend on properties of the dielectric media. In (3.24) it can be observed that potential coefficient p_{ij} is the potential of conductor i when just conductor j is charged and its charge is the unity, which allows its experimental determination. The S.I. unit is $VC^{-1} = F^{-1}$, and **farad** (F) is defined as $1F \equiv 1CV^{-1}$. The farad is a unit too large for most practical applications and it is customary to use either the microfarad ($1\,\mu F = 10^{-6}$ F), the nanofarad (1 nF $= 10^{-9}$ F) or the picofarad (1 pF $= 10^{-12}$ F).

The coefficients of potential verify these relations: $p_{ij} = p_{ji}$ and $p_{ii} \geq p_{ij} > 0$, that is to say, the matrix of the coefficients of potential is symmetric, and all the coefficients are positive.

If the matrix of the coefficients of potential is inverted, it results

$$q_i = \sum_{j=1}^{N} c_{ij} V_j, \qquad (3.25)$$

where c_{ij}, $i \neq j$, are called **coefficients of influence** and c_{ii}, **coefficients of capacitance**. Their units in SI are for both of them the farad (F) and their values depend just on conductor geometries and on dielectric media. It is verified that $c_{ii} > 0$ (coefficients of capacitance are positive) and that $c_{ij} = c_{ji} < 0$ (coefficients of influence are negative and the matrix is symmetric).

3.7 Capacitors

Two conductors which can store equal but opposite charges, independently of whether other conductors in the system are charged, are called a **capacitor**. Let's consider two conductors 1 and 2: number 1 with charge q positive and with potential V_1; number 2 with charge $-q$ negative and potential V_2. The independence of other charges implies that both conductors are placed in a way where all the lines of **E** that originate on the positive body terminate on the negative body, which is known as *total influence*.[6] It will be supposed that between them there is a linear dielectric media or vacuum (in order to be able to use the results obtained in previous sections). The conductors are sometimes referred to as the **plates** of the capacitor. Capacitors are used to store electrical charges and electrostatic energy. When they are introduced in a circuit, they are usually represented by $||$.

Equation (3.24) allows us to express the potentials of both capacitors as a function of their charges, as follows

$$V_1 = p_{11}q + p_{12}(-q) + V_x$$

$$V_2 = p_{21}q + p_{22}(-q) + V_x,$$

where V_x is the common potential contributed by other external charges. If both expressions are subtracted, and bearing in mind that $p_{12} = p_{21}$, we find

$$V_1 - V_2 = (p_{11} - 2p_{12} + p_{22})q, \tag{3.26}$$

this is, the potential difference between the plates of a capacitor is proportional to the charge stored q. Previous expression can be rewritten as

$$q = C(V_1 - V_2), \tag{3.27}$$

where the positive constant of proportionality $C = (p_{11} - 2p_{12} - p_{22})^{-1}$ is called the **capacitance** of the capacitor and represents the electric charge stored in each plate (in absolute value) per unit of the potential difference applied between them. This

---

[6] Since the field is bounded between the conductors it can be deduced, by applying Gauss' theorem to a surface that surrounds both conductors, that charges of both conductors must be equal and opposite.

constant, as well as the coefficients of potential, does not depend on the load or on the potential difference between the conductors, and only depends on the geometry of the conductors and on the media between them. Its SI unit is the farad (F).

The simplest example (but not the only one) of a capacitor consists of a conductor totally surrounded by another one, and between them a vacuum or a linear dielectric. Then it is easy to understand, in this case, that when the inside conductor is charged with a charge $+q$, the outside one remains charged with a charge $-q$ on its internal surface.[7] Any external field would not affect this charge distribution, but it would change both conductor potentials. However, the potential would change by the same amount for both conductors, so the potential difference between the plates would remain constant.

A typical case of a capacitor is the one known as the *parallel-plate capacitor*. It is formed by two parallel plates, with equal and opposite charges and, assuming an ideal case, the plate separation is very small compared with the dimensions of the plate. It can be assumed that the entire electric field remains inside the region between the plates (total influence). The dielectric media between the plates are usually linear, homogeneous and isotropic, distributed by forming parallel layers to the conducting plates. Thus the field between plates is uniform, except for the fringing field at the edge of the parallel plates. This effect may be neglected in this ideal case and it will also be assumed that the plates have a uniform charge distribution.[8]

Capacitors can also be joined by connecting one of the conductors of the first capacitor to a conductor of the second, etc., or in different ways, depending on whether we want more accumulated charge or a reduction of the potential difference. When we join them, the set of capacitors has a certain potential difference ($V_1 - V_2$) and accumulates a charge q. It is called the **equivalent capacitance** of the combination, as

$$C_e = q/(V_1 - V_2), \qquad (3.28)$$

which is equivalent to the capacitance that would have a unique capacitor with the same potential difference and that would store the same amount of charge. When they are connected in parallel (Fig. 3.3a), we match positive plates with negative plates, so all the capacitors have the same potential difference ($V_1 - V_2$) between the plates. Each of the capacitors of the combinations acquires a charge

$$q_i = C_i(V_1 - V_2), \qquad (3.29)$$

---

[7]It is enough to apply Gauss' theorem to a Gaussian surface inside the outsider conductor: since the field is null at every point, the charge inside the Gaussian surface must be zero. So if the inside conductor had charge $+q$, then the outside one should have $-q$.

[8]If charge is not uniformly distributed, electric field would be different on equidistant points to the plates, and when calculating the potential difference between the plates, the results will be different depending on the chosen circulation path.

3.7 Capacitors

Fig. 3.3 Capacitors association **a** in parallel, **b** in series

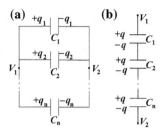

so total stored charge is

$$q = \sum_{i=1}^{n} q_i . \qquad (3.30)$$

This association, therefore, allows storing a higher amount of charge than we would store if we used each capacitor individually. The equivalent capacitance is

$$C_e = \frac{q}{V_1 - V_2} = \frac{1}{V_1 - V_2} \sum_{i=1}^{n} q_i = \sum_{i=1}^{n} \frac{q_i}{V_1 - V_2} = \sum_{i=1}^{n} C_i . \qquad (3.31)$$

The equivalent capacitance of a combination of capacitors in parallel is, therefore, the sum of the capacitances of each capacitor,

$$C_e = \sum_{i=1}^{n} C_i, \qquad (3.32)$$

which is higher than the capacitance of individual capacitors. On the particular case of n capacitors with the same capacitance C, the equivalence capacitance is

$$C_e = nC . \qquad (3.33)$$

If capacitors are connected in series (Fig. 3.3b), a plate from one capacitor is connected with a plate of the next one and so on, and subsequently they are charged by subjecting the extreme plates to a potential difference $(V_1 - V_2)$. All the capacitors take the same charge, but the potential difference is divided in smaller fractions. This combination is usually used when the stored charge produces an electric field inside a capacitor that exceeds the breakdown voltage. In other words, it is used when $V_1 - V_2$ is higher than the breakdown voltage, and therefore it is necessary to divide the potential difference into lower fractions that do not exceed the breakdown voltage of each capacitor. The potential difference ΔV_i of each capacitor will be

$$\Delta V_i = \frac{q}{C_i}, \qquad (3.34)$$

which is lower than $V_1 - V_2$. The equivalent capacitance of the combination is

$$\frac{1}{C_e} = \frac{V_1 - V_2}{q} = \frac{\sum_{i=1}^{n} \Delta V_i}{q} = \sum_{i=1}^{n} \frac{\Delta V_i}{q} = \sum_{i=1}^{n} \frac{1}{C_i}. \quad (3.35)$$

The equivalent capacitance of a capacitor association in series is, therefore,

$$\frac{1}{C_e} = \sum_{i=1}^{n} \frac{1}{C_i}, \quad (3.36)$$

which is smaller than any of the capacitor separately. On the particular case of n capacitors with the same capacitance C, the equivalent capacitance is

$$C_e = \frac{C}{n}. \quad (3.37)$$

Solved Problems

Problems A

3.1 The square plate made of a dielectric material shown in Fig. 3.4 has thickness e and is polarized over its entire volume according to equation $\mathbf{P} = (ay^3 + b)\mathbf{j}$, where a and b are constants. (a) Determine the polarization surface charge density and the polarization volume charge density. (b) Verify explicitly why the total polarization charge is null.

Solution

(a) Polarization surface density σ_p is calculated from (3.3). Since vector $\mathbf{P}$ has a component just in the $\mathbf{u}_y$-direction, it is only necessary to calculate polarization density on the upper and lower surfaces of the plate. On the other ones, the scalar product of (3.3) is zero, because $\mathbf{P}$ and $\mathbf{n}$ are perpendicular at all points. In Fig. 3.5 the orientation of $\mathbf{P}$ at different points can be observed, and also the outward unit normal on surfaces where polarization is not zero. It can be observed on the upper

Fig. 3.4 Square plate of Problem 3.1

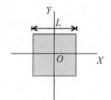

Fig. 3.5 Vectors **P** and **n** in the plate of Problem 3.1

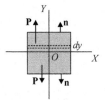

surface of the plate $\mathbf{n} = +\mathbf{u}_y$ and on the lower one, $\mathbf{n} = -\mathbf{u}_y$. If (3.3) is applied it results, for the upper surface, where $y = L/2$,

$$\sigma_{p,\text{upp}} = \mathbf{P} \cdot \mathbf{n} = \left[a \left(\frac{L}{2} \right)^3 + b \right] \mathbf{u}_y \cdot \mathbf{u}_y = a \frac{L^3}{8} + b.$$

And for the lower surface

$$\sigma_{p,\text{low}} = \mathbf{P} \cdot \mathbf{n} = \left[a \left(-\frac{L}{2} \right)^3 + b \right] \mathbf{u}_y \cdot (-\mathbf{u}_y) = a \frac{L^3}{8} - b.$$

To calculate the polarization volume charge density ρ_p, (3.2) is applied:

$$\rho_p = -\nabla \cdot \mathbf{P} = -\left(\frac{\partial P_x}{\partial x} + \frac{\partial P_y}{\partial y} + \frac{\partial P_z}{\partial z} \right) = -3ay^2.$$

(b) To verify that total polarization is null, we calculate this charge:

$$q_p = \int_{\partial V} \sigma_p dS + \int_V \rho_p dV = \int_{S_{\text{upp}}} \left(a \frac{L^3}{8} + b \right) dS$$
$$+ \int_{S_{\text{low}}} \left(a \frac{L^3}{8} - b \right) dS + \int_V -3ay^2 dV.$$

Functions inside the surface integrals are constant. To solve the volume integral, since the function to be integrated only depends on y, the differential volume can be taken as shown in Fig. 3.5, $dV = Ledy$. It results,

$$q_p = \left(a \frac{L^3}{8} + b \right) S_{\text{upp}} + \left(a \frac{L^3}{8} - b \right) S_{\text{low}} + \int_{-L/2}^{L/2} -3ay^2 Ledy =$$
$$= \left(a \frac{L^3}{8} + b \right) Le + \left(a \frac{L^3}{8} - b \right) Le - 3aLe \left. \frac{y^3}{3} \right|_{-L/2}^{L/2} = a \frac{L^4}{4} e - a \frac{L^4}{8} e - a \frac{L^4}{8} e = 0.$$

3.2 The plates of a plane capacitor have $A = 2\,\text{m}^2$ area and they are separated by a distance of $d = 2\,\text{cm}$. Two different cases are established: on one side, space between them is a dielectric medium whose permittivity is $\varepsilon_r = 3$ and, on the other side, space between them is constituted by two different dielectric media: one is $d/4$ thickness with a permittivity of $\varepsilon_{r1} = 3$ and the other is $3d/4$ thickness and its permittivity is $\varepsilon_{r2} = 2$. The capacitor is uniformly charged with $q = 20\,\mu\text{C}$. (a) Determine the electric displacement field at any point of the capacitor. (b) Calculate the capacitance of the capacitor for each configuration.

Solution

(a) Figure 3.6 represents both configurations of the capacitor. The value of **D** does not depend on polarization charges, so its value is the same for both configurations. Charge q of the capacitor, as it was explained in Sect. 3.5, is uniformly distributed over each of the plates with appearing a density on the conducting plates that can be easily calculated from (2.4)

$$q = \int_A \sigma dS = \sigma \int_A dS = \sigma A \Rightarrow \sigma = \frac{q}{A} = 10\,\mu\text{Cm}^{-2}.$$

It can be observed, the same way as it was done in Problem 2.7, that the magnitude of **D** is the same for all the points at the same distance from any of the plates, and the direction is perpendicular (if the fringing field is negligible) from the positive to the negative plate, as shown in Fig. 3.6a. To determine its magnitude, due to the symmetry of the charge distribution, Gauss' law for **D** can be applied. We take as a Gaussian surface ∂V a straight cylinder with its axis perpendicular to the sheet, going through one of the conducting plates (in Fig. 3.6a it goes through the upper plate, whose charge is q) and one of its bases, with section S, inside the dielectric. Flux through the other base (the upper one in the figure) is null, since all the electric field is confined to the plates. If flux through ∂V is calculated, taking into account that on the cylinder's lateral surface **D** and $d\mathbf{S}$ are perpendicular,

$$\Phi_D = \oint_{\partial V} \mathbf{D} \cdot d\mathbf{S} = \int_{S_{\text{low}}} D\,dS = D\int_{S_{\text{low}}} dS = DS,$$

where S is the area of the lower base S_{low} (and of any straight section of the cylinder). If Gauss' law for **D** is applied

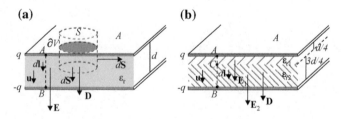

Fig. 3.6 Capacitors of Problem 3.2 and Gaussian surface to determine **D**: **a** with one dielectric medium; **b** with two dielectric media

$$\Phi_D = q_{np,\text{in}} = \sigma S,$$

where $q_{np,\text{in}}$ is the intersected charge by the cylinder when going through the conducting plate (dark grey colored in Fig. 3.6a). If both equations are compared,

$$D = \sigma = 10\ \mu\text{Cm}^{-2}.$$

(b) The capacitance of the capacitor is calculated according to its definition, using (3.27),

$$C = \frac{q}{V_A - V_B},$$

where $V_A - V_B$ is the potential difference between the positive and the negative plates. This potential difference has to be calculated for both of the given configurations. To do it, we calculate the circulation of electric field **E** from the positive plate to the negative one, choosing a circulation path parallel to **E**, which is parallel to **D**, because the dielectrics are linear, homogeneous and isotropic. In the case with only a single dielectric medium, electric field **E** is the same across all the medium, and is obtained by applying (3.16),

$$\mathbf{E} = \frac{\mathbf{D}}{\varepsilon_0 \varepsilon_r} = \frac{\sigma}{\varepsilon_0 \varepsilon_r}\mathbf{u} = \frac{\sigma}{3\varepsilon_0}\mathbf{u},$$

where **u** is the unit normal to the plates, pointing from the positive plate towards the negative one. If the potential difference between the positive plate and the negative one is obtained using the circulation along line AB in Fig. 3.6a, where **E** and $d\mathbf{l}$ are parallel,

$$V_A - V_B = \int_A^B \mathbf{E}\cdot d\mathbf{l} = \int_A^B \frac{\sigma}{\varepsilon_0\varepsilon_r}\mathbf{u}\cdot dl\mathbf{u} = \int_0^d \frac{\sigma}{\varepsilon_0\varepsilon_r}dl = \frac{\sigma d}{\varepsilon_0\varepsilon_r}.$$

Remembering that $q = \sigma A$, and if we substitute in capacitance expression, it results

$$C = \frac{q}{V_A - V_B} = \frac{\sigma A}{\sigma d/(\varepsilon_0\varepsilon_r)},$$

and we obtain what it is known as the *capacitance of a parallel-plate capacitor*,

$$C = \frac{\varepsilon_r\varepsilon_0 A}{d} = \frac{\varepsilon A}{d}. \tag{3.38}$$

Using our numerical values,

$$C = \frac{3\varepsilon_0 A}{d} = \frac{3\varepsilon_0 \cdot 2}{0.02} = 2.655\ \text{nF}.$$

In the second case, with two dielectrics, Fig. 3.6b, it should be observed that even though **D** is the same in both of them, **E** is different in each medium. In medium 1,

$$\mathbf{E}_1 = \frac{\mathbf{D}}{\varepsilon_0 \varepsilon_{r1}} = \frac{\sigma}{\varepsilon_0 \varepsilon_{r1}} \mathbf{u} = \frac{\sigma}{3\varepsilon_0} \mathbf{u}.$$

And in medium 2,

$$\mathbf{E}_2 = \frac{\mathbf{D}}{\varepsilon_0 \varepsilon_{r2}} = \frac{\sigma}{\varepsilon_0 \varepsilon_{r2}} \mathbf{u} = \frac{\sigma}{2\varepsilon_0} \mathbf{u}.$$

If we calculate the potential difference between A and B, line ACB in Fig. 3.6b, we proceed as in the previous case and the result is

$$V_A - V_B = \int_A^C \mathbf{E}_1 \cdot d\mathbf{l} + \int_C^B \mathbf{E}_2 \cdot d\mathbf{l} = \int_0^{d/4} \frac{\sigma}{3\varepsilon_0} dl + \int_{d/4}^d \frac{\sigma}{2\varepsilon_0} dl$$
$$= \frac{\sigma}{3\varepsilon_0} \frac{d}{4} + \frac{\sigma}{2\varepsilon_0} \frac{3d}{4} = \frac{11}{24} \frac{\sigma d}{\varepsilon_0}.$$

And if we substitute this value into the equation of the capacitance,

$$C = \frac{q}{V_A - V_B} = \frac{\sigma A}{11 \sigma d/(24\varepsilon_0)} = \frac{24}{11} \frac{\varepsilon_0 A}{d} = 1.931 \text{ nF}.$$

3.3 A spherical capacitor consists of a conducting sphere 1, whose radius is R_1, surrounded by a conducting spherical crown 2, whose inner radius is R_2. Between both of them there is a dielectric medium, with relative permittivity ε_r. Determine its capacitance.

Solution

Let's suppose that the capacitor plates have electric charges $+q$ and $-q$, as shown in Fig. 3.7. The capacitance of the capacitor is calculated from its definition, by using (3.27),

$$C = \frac{q}{V_A - V_B},$$

Fig. 3.7 Spherical capacitor

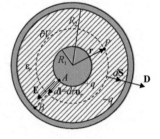

where $V_A - V_B$ is potential difference between the positive plate and the negative one. To determine it, we calculate the circulation of electric field **E** from the positive to the negative one, choosing a line parallel to **E**. So it is necessary to calculate, firstly, **D** and then **E**.

Because of its symmetry, charges $+q$ and $-q$ are uniformly distributed over the outer surface of conductor 1, and the inner surface of conductor 2, respectively, and polarization charges in the dielectric will be also distributed with the same spherical symmetry. Consequently, **E** and **D** have to be radial. Gauss' law (3.10) can be applied to a spherical Gaussian surface ∂V whose radius is r, passing through any point P of the dielectric, as it can be seen in Fig. 3.7 with the discontinuous line. It follows that

$$\Phi_D = \oint_{\partial V} \mathbf{D} \cdot d\mathbf{S} = D 4\pi r^2 = q_{np,in} = q,$$

since charge $+q$ of the inner plate is the only charge inside the Gaussian surface. Solving,

$$D = \frac{q}{4\pi r^2} \Rightarrow \mathbf{D} = \frac{q}{4\pi r^2} \mathbf{u}_r,$$

where $\mathbf{u}_r$ is the radial unit vector for spherical coordinates. The electric field between the plates is obtained by applying (3.16),

$$\mathbf{E} = \frac{\mathbf{D}}{\varepsilon_0 \varepsilon_r} = \frac{q}{4\pi \varepsilon_0 \varepsilon_r r^2} \mathbf{u}_r.$$

If the circulation of **E** between points A and B in Fig. 3.7 is calculated,

$$V_A - V_B = \int_1^2 \mathbf{E} \cdot d\mathbf{l} = \int_{R_1}^{R_2} \frac{q}{4\pi \varepsilon_0 \varepsilon_r r^2} \mathbf{u}_r \cdot dr \mathbf{u}_r = \frac{q}{4\pi \varepsilon_0 \varepsilon_r} \int_{R_1}^{R_2} \frac{dr}{r^2} =$$

$$= -\frac{q}{4\pi \varepsilon_0 \varepsilon_r} \frac{1}{r} \Big|_{R_1}^{R_2} = \frac{q}{4\pi \varepsilon_0 \varepsilon_r} \left(\frac{1}{R_1} - \frac{1}{R_2} \right) = \frac{q}{4\pi \varepsilon_0 \varepsilon_r} \left(\frac{R_2 - R_1}{R_1 R_2} \right).$$

If we introduce this value in (3.27), the capacitance of spherical capacitor is obtained:

$$C = \frac{q}{V_A - V_B} = \frac{4\pi \varepsilon_0 \varepsilon_r R_1 R_2}{R_2 - R_1}.$$

3.4 A cylindrical capacitor consists of a cylindrical conductor with radius R_1 and length L, surrounded by another coaxial cylindrical conductor, with inner radius R_2 and the same length L. Between them a dielectric with permittivity ε_r has been introduced. The capacitor's length is large enough in regards to separation between the conductors, so a fringing field can be neglected. Determine the capacitance of the capacitor.

Fig. 3.8 Cylindrical capacitor

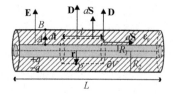

Solution

We follow a procedure similar to that of the preceding problem. We suppose that the conducting cylinders have electric charges $+q$ and $-q$, as shown in Fig. 3.8 and the capacitance of the capacitor is calculated from its definition, using (3.27),

$$C = \frac{q}{V_A - V_B},$$

where $V_A - V_B$ is the potential difference between the positive cylinder and the negative one. To determine it, we calculate the circulation of **E** from the positive to the negative one, by choosing a path parallel to **E**.

If a fringing field is neglected, charges $+q$ and $-q$ are uniformly distributed on the surface of the inner cylindrical conductor and on the inner surface of the outer cylinder, respectively. Their line charge densities are $\lambda = q/L$ and $-\lambda = -q/L$. The electric field **E** and field **D** are radial in regards to the cylinder axis. Gauss' law for **D** (3.10) can be applied to a cylindrical Gaussian surface ∂V with radius r and length l, passing through any point P of the dielectric (discontinuous line in the Fig. 3.8). The flux of **D** through the bases of the Gaussian surface is null, since $d\mathbf{S}$ and **D** are perpendicular at all points (or **D** is null inside the inner conductor). It is only necessary to calculate the flux through the lateral surface, where $d\mathbf{S}$ and **D** are parallel:

$$\Phi_D = \oint_{\partial V} \mathbf{D} \cdot d\mathbf{S} = \int_{S_{\text{lat}}} D\, dS = D \int_{S_{\text{lat}}} dS = D 2\pi r l,$$

where S_{lat} is the lateral surface of the Gaussian cylinder. If (3.10) (Gauss' law for **D**) is applied, it results

$$\Phi_D = q_{np,\text{in}} = \lambda l,$$

since only the charge of the inner conductor is enclosed by the Gaussian surface. If both expressions obtained for the flux are set equal to each other, the result is

$$D = \frac{\lambda}{2\pi r} \Rightarrow \mathbf{D} = \frac{q}{2\pi r L}\mathbf{u}_\rho,$$

where $\mathbf{u}_\rho$ is the radial unit vector for cylindrical coordinates. **E** is obtained from **D** by applying (3.16)

$$\mathbf{E} = \frac{\mathbf{D}}{\varepsilon_0 \varepsilon_r} = \frac{q}{2\pi r L \varepsilon_0 \varepsilon_r}\mathbf{u}_\rho.$$

If we circulate the electric field between points A and B in Fig. 3.8 using the marked path, parallel to **E**,

$$V_A - V_B = \int_A^B \mathbf{E} \cdot d\mathbf{l} = \int_{R_1}^{R_2} \frac{q}{2\pi\varepsilon_0\varepsilon_r rL} \mathbf{u}_\rho \cdot dr\mathbf{u}_\rho = \frac{q}{2\pi\varepsilon_0\varepsilon_r L} \int_{R_1}^{R_2} \frac{dr}{r}$$
$$= \frac{q}{2\pi\varepsilon_0\varepsilon_r L} \ln\left(\frac{R_2}{R_1}\right).$$

which when substituted in (3.27), gives the capacitance of cylindrical capacitor

$$C = \frac{q}{V_A - V_B} = \frac{2\pi\varepsilon_0\varepsilon_r L}{\ln(R_2/R_1)}.$$

Problems B

3.5 A very large metallic sheet with negligible thickness is surrounded by a 2 cm thickness polystyrene layer, whose electric susceptibility is 1.6 and its dielectric strength is 20 kV/mm. (a) Determine the maximum charge density that the sheet can have without reaching the polystyrene dielectric breakdown. (b) Calculate the polarization charge densities in polystyrene for this value.

Solution

(a) In Fig. 3.9 the metallic sheet, whose surface charge density σ has yet to be determined, is grey colored. Polystyrene is over and below the sheet, with $e = 0.02$ m thick, permittivity $\varepsilon_r = \chi_e + 1 = 1.6 + 1 = 2.6$ and dielectric strength $E_{max} = 20$ kV/mm $= 20 \times 10^6$ V/m. Dielectric breakdown has not be reached; therefore the electric field **E** created by this sheet in polystyrene must remain below its dielectric strength E_{max}.

Since there are dielectric materials, it is necessary to calculate the field **D** produced by the sheet to calculate the field **E**. The problem has charge and geometric symmetries (uniformly charged infinite sheet), so Gauss' law for **D** (3.10) can be applied. The problem can be solved according to Sect. 3.4 for linear, homogeneous and isotropic dielectrics. For a uniformly charged sheet, as previously seen in Chap. 2, we take as a Gaussian surface ∂V a cylindrical surface (Fig. 3.9), symmetric according to the sheet, whose bases, with area S, are inside the dielectric, and with height

Fig. 3.9 Gauss' law for **D** in Problem 3.5

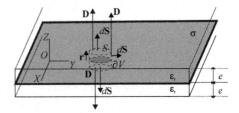

$2r$, where r is the distance from the sheet to any point on the dielectric. The field **D** at any point is normal to the sheet, outwards (if $\sigma > 0$ is supposed) and with the same magnitude at every point of both of the bases of the Gaussian cylinder. The flux through the lateral surface is null, since **D** and $d\mathbf{S}$ are perpendicular at any point of this surface. So there only exists flux through the bases,

$$\Phi_D = \oint_{\partial V} \mathbf{D} \cdot d\mathbf{S} = \int_{S_{\text{upp}}} DdS + \int_{S_{\text{low}}} DdS = D2S.$$

It should be observed that S is the area of both the highest S_{upp} and lowest S_{low} bases of the Gaussian cylinder. If Gauss' law for **D** is applied (3.10) the result is

$$\Phi_D = q_{np,\text{in}} = \sigma S,$$

where the charge $q_{np,\text{in}}$ within ∂V equals the intersection between the sheet and the cylinder (darker grey coloured in Fig. 3.9). If we equate both equations, the result is

$$D = \sigma/2.$$

If (3.16) is applied, we have for field **E**, with the same direction of **D**,

$$E = \frac{D}{\varepsilon_0 \varepsilon_r} = \frac{\sigma}{2\varepsilon_0 \varepsilon_r}.$$

And if we make this field to be lower than the dielectric strength, σ can be obtained,

$$\frac{\sigma}{2\varepsilon_0 \varepsilon_r} < E_{\max} \Rightarrow \sigma < 2\varepsilon_0 \varepsilon_r E_{\max} = 920.4 \, \mu\text{Cm}^{-2}.$$

(b) To obtain the polarization charge densities in polystyrene, we follow what was previously said in Sect. 3.4 for linear, homogeneous and isotropic dielectrics. The polarization volume charge density ρ_p is obtained from (3.19),

$$\rho_p = -\frac{\varepsilon_r - 1}{\varepsilon_r} \nabla \cdot \mathbf{D} = -\frac{\varepsilon_r - 1}{\varepsilon_r} \rho_{np} = 0,$$

since inside the dielectric there are no non-polarization charges. The polarization surface charge density σ_p is obtained from (3.18),

$$\sigma_p = \frac{\varepsilon_r - 1}{\varepsilon_r} \mathbf{D} \cdot \mathbf{n}.$$

for every surface of the dielectric. It should be firstly considered that, due to the symmetry of the figure, density of the upper layer of polystyrene is the same that the one of lower layer. If we consider the upper layer (Fig. 3.10), there are two surfaces: one in contact with the sheet, where the angle between unitary $\mathbf{n}_1$ and **D** is 180°; and

Fig. 3.10 Vectors to calculate polarization surface charge densities on the *upper* layer

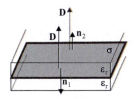

the other in contact with the vacuum, where the angle between unitary $\mathbf{n}_2$ and $\mathbf{D}$ is $0°$. If (3.18) is applied for each of them, taking into account that $D = \sigma/2$, where $\sigma = 920.4\,\mu\text{Cm}^{-2}$ was calculated in the previous section, it results

$$\sigma_{p1} = \frac{\varepsilon_r - 1}{\varepsilon_r}\mathbf{D}\cdot\mathbf{n}_1 = -\frac{\varepsilon_r - 1}{\varepsilon_r}D = -\frac{1.6}{2.6}\sigma/2 = -283.2\,\mu\text{Cm}^{-2},$$

$$\sigma_{p2} = \frac{\varepsilon_r - 1}{\varepsilon_r}\mathbf{D}\cdot\mathbf{n}_2 = \frac{\varepsilon_r - 1}{\varepsilon_r}D = \frac{1.6}{2.6}\sigma/2 = 283.2\,\mu\text{Cm}^{-2}.$$

3.6 A point charge $q = -4\,\text{nC}$ is located at the centre of a spherical crown dielectric material, whose permittivity $\varepsilon_r = 2$, and its inner and outsider radii are $R_1 = 20\,\text{cm}$ and $R_2 = 60\,\text{cm}$ respectively. (a) Determine the polarization charge densities in the spherical crown. (b) Calculate the potential difference between the spherical surfaces.

Solution

(a) The polarization volume charge density ρ_p is obtained from (3.19),

$$\rho_p = -\frac{\varepsilon_r - 1}{\varepsilon_r}\nabla\cdot\mathbf{D} = -\frac{\varepsilon_r - 1}{\varepsilon_r}\rho_{np} = 0,$$

since there are no non-polarization charges inside the dielectric. The polarization surface charge density σ_p is obtained from (3.18),

$$\sigma_p = \mathbf{P}\cdot\mathbf{n} = \frac{\varepsilon_r - 1}{\varepsilon_r}\mathbf{D}\cdot\mathbf{n}.$$

for both surfaces of the dielectric crown. To obtain the field $\mathbf{D}$ created by the point charge at any point, distance r from the centre, Gauss' law (3.10) can be applied considering a spherical surface ∂V, whose centre is charge q (discontinuous line in Fig. 3.11). Since $\mathbf{D}$ only depends on non-polarization charges, and the only one is q, its value is the same for every point in space:

$$\Phi_D = \oint_{\partial V}\mathbf{D}\cdot d\mathbf{S} = q_{np,in} \Rightarrow \oint_{\partial V} -DdS = q \Rightarrow -D4\pi r^2 = q.$$

Fig. 3.11 Vectors for calculus in Problem 3.6

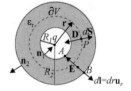

It should be observed that since q is negative, field $\mathbf{D}$ points towards charge q and makes 180° with $d\mathbf{S}$. If we solve it

$$D = \frac{-q}{4\pi r^2} = \frac{4 \cdot 10^{-9}}{4\pi r^2} \Rightarrow \mathbf{D} = \frac{10^{-9}}{\pi r^2}(-\mathbf{u}_r).$$

where $\mathbf{u}_r$ is the radial unit vector for the spherical coordinate system. If we substitute $\mathbf{D}$ in (3.18), and considering that $r = R_1 = 0.2$ m and $\mathbf{n}_1 = -\mathbf{u}_r$ for the inner surface, and $r = R_2 = 0.6$ m and $\mathbf{n}_2 = \mathbf{u}_r$ for the outer surface, it results

$$\sigma_p|_{r=R_1} = \frac{\varepsilon_r - 1}{\varepsilon_r} \mathbf{D} \cdot \mathbf{n}_1|_{r=R_1} = \frac{1}{2} \cdot \frac{10^{-9}}{\pi \cdot (0.2)^2}(-\mathbf{u}_r) \cdot (-\mathbf{u}_r) = 3.98 \cdot 10^{-9}\,\text{Cm}^{-2}.$$

$$\sigma_p|_{r=R_2} = \frac{\varepsilon_r - 1}{\varepsilon_r} \mathbf{D} \cdot \mathbf{n}_2|_{r=R_2} = \frac{1}{2} \cdot \frac{10^{-9}}{\pi \cdot (0.6)^2}(-\mathbf{u}_r) \cdot \mathbf{u}_r = -4.42 \cdot 10^{-10}\,\text{Cm}^{-2}.$$

(b) To determine the potential difference between both surfaces of the spherical crown, and since circulation does not depend on the path, we circulate the electric field along the radial direction, where $\mathbf{E}$ and $d\mathbf{l}$ are parallel, from A on the inner surface to B on the outer one, as shown in Fig. 3.11. $\mathbf{E}$ is obtained[9] from $\mathbf{D}$ by applying (3.16),

$$\mathbf{E} = \frac{\mathbf{D}}{\varepsilon_0 \varepsilon_r} = \frac{10^{-9}}{\pi r^2 \varepsilon_0 \varepsilon_r}(-\mathbf{u}_r).$$

All points along the circulation path are inside the dielectric crown, where $\varepsilon_r = 2$. So the potential difference between A and B is

$$V_A - V_B = \int_A^B \mathbf{E} \cdot d\mathbf{l} = \int_A^B \frac{10^{-9}}{\pi r^2 \varepsilon_0 \varepsilon_r}(-\mathbf{u}_r)\mathbf{u}_r dr = -\int_{0.2}^{0.6} \frac{10^{-9}}{2\pi \varepsilon_0 r^2} dr$$

$$= \left.\frac{10^{-9}}{2\pi \varepsilon_0 r}\right|_{0.2}^{0.6} = -59.95\,\text{V}.$$

3.7 A coaxial cable consists of a cylindrical copper conductor whose radius is $a = 2$ mm surrounded by a coaxial cylindrical dielectric (polyethylene) of exterior

---

[9] It can be observed that $\mathbf{E}$ could be directly obtained by applying (2.15) for the electric field produced by a point charge by merely substituting ε_0 by $\varepsilon_0 \varepsilon_r$. Field $\mathbf{P}$, necessary for Section (a), could be directly obtained from the constitutive equation $\mathbf{P} = \varepsilon_0 \chi_e \mathbf{E}$, being $\chi_e = \varepsilon_r - 1$.

Solved Problems

radius $b = 8\,\text{mm}$, whose relative dielectric constant $\varepsilon_r = 2.3$, protected by a coaxial cylindrical outer conductor whose radius is also b. The inner conductor is maintained at a potential of $300\,\text{kV}$ and the outer conductor at null potential. Determine the line charge density of the conductor.

Solution

The charged conductor produces an electric field at points in the dielectric (which depends on the charge of the conductor). Line charge density of the conductor will therefore appear in the mathematical expression of this electric field.

The problem has cylindrical symmetry: every point of the dielectric located at the same distance of the cylinder's axis has the same magnitude of electric field, which is perpendicular to this axis. The problem can be solved by applying Gauss' law for field $\mathbf{D}$ (3.10) and then by obtaining $\mathbf{E}$, bearing in mind that the dielectric is linear, homogeneous and isotropic. So we take as the Gaussian surface ∂V (discontinuous line in Fig. 3.12) a cylindrical surface, whose length is L, coaxial with the conductor and with radius r, passing through a point P of the dielectric.

The flux of $\mathbf{D}$ through the bases of the Gaussian surface is null, since $d\mathbf{S}$ and $\mathbf{D}$ are perpendicular at all points (or $\mathbf{D}$ is null, at any point inside the conductor). It is only necessary to calculate the flux through the lateral surface, where $d\mathbf{S}$ and $\mathbf{D}$ are parallel:

$$\Phi_D = \oint_{\partial V} \mathbf{D} \cdot d\mathbf{S} = \int_{S_{\text{lat}}} D\, dS = D \int_{S_{\text{lat}}} dS = D 2\pi r L,$$

where S_{lat} is the lateral surface of the cylinder. Furthermore, if Gauss' law for field $\mathbf{D}$ (3.10) is applied, it results

$$\Phi_D = q_{np,\text{in}} = \lambda L,$$

where λ is the line charge density of the conductor whose radius is a, which we want to know. If both expressions obtained for the flux are equated, then the result is

$$D = \frac{\lambda L}{2\pi r L} \Rightarrow \mathbf{D} = \frac{\lambda}{2\pi r}\mathbf{u}_\rho,$$

where $\mathbf{u}_\rho$ is the radial unit vector for cylindrical coordinates. $\mathbf{E}$ is obtained from $\mathbf{D}$ by applying (3.16)

$$\mathbf{E} = \frac{\mathbf{D}}{\varepsilon_0 \varepsilon_r} = \frac{\lambda}{2\pi r \varepsilon_0 \varepsilon_r}\mathbf{u}_\rho.$$

Fig. 3.12 Electric cable of Problem 3.7 and Gaussian surface to calculate $\mathbf{D}$

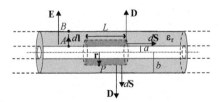

Additional data in the problem are the potentials of both conductors, so we can calculate the potential difference between them. Since the circulation of **E** does not depend on the path, we use a radial path between two points A and B, one for each conductor (Fig. 3.12). Therefore $d\mathbf{l} = dr\mathbf{u}_\rho$ is parallel to the field **E** between the conductors:

$$W_{AB} = \int_A^B \mathbf{E} \cdot d\mathbf{l} = \int_A^B \frac{\lambda}{2\pi r \varepsilon_0 \varepsilon_r} \mathbf{u}_\rho \mathbf{u}_\rho dr = \frac{\lambda}{2\pi \varepsilon_0 \varepsilon_r} \int_a^b \frac{dr}{r}$$

$$= \frac{\lambda}{2\pi \varepsilon_0 \varepsilon_r} \ln r \Big|_a^b = \frac{\lambda}{2\pi \varepsilon_0 \varepsilon_r} \ln \frac{b}{a}.$$

The circulation calculated between points A and B is the potential difference $V_A - V_B$ between both conductors, which according to the problem's statement is 300 kV. If we equalize it and solve it, the charge density λ of the inner conductor is obtained:

$$3 \cdot 10^5 = \frac{\lambda}{2\pi \varepsilon_0 \varepsilon_r} \ln \frac{b}{a} \Rightarrow \lambda = \frac{6 \cdot 10^5 \pi \varepsilon_0 \varepsilon_r}{\ln \frac{b}{a}}$$

$$= \frac{6 \cdot 10^5 \pi \varepsilon_0 \cdot 2.3}{\ln 4} = 2.77 \cdot 10^{-5}\,\mathrm{C/m} = 27.7\,\mu\mathrm{C/m}.$$

3.8 Two charged spherical shells, with radius R, are placed in a linear, homogeneous and isotropic dielectric liquid, as shown in Fig. 3.13. Their charges stay uniformly distributed. The sphere S_1 on the left side has a charge q, while the sphere S_2 on the right side has a charge $-q$. The potential difference between points A and B is the same as the potential difference between A and C, if the spheres were in a vacuum. Determine the relative permittivity of the liquid.

Solution

Since the problem's statement refers to a potential difference between points, it is necessary to calculate the electric field produced by the charge distribution of the figure at any point in the dielectric medium. Since there are two spheres, we firstly calculate the field created by any spherical shell with charge q and then we apply the superposition principle for electric fields.

To calculate the electric field **E** we follow the procedure of Problem 3.6: **D** is calculated by applying Gauss' law for **D** (3.10) and then **E** is obtained by applying (3.16). The Gaussian spherical surface ∂V is discontinuously plotted in Fig. 3.14, concentric with the charged spherical shell, passing through any point P of the dielectric.

Fig. 3.13 Figure of Problem 3.8

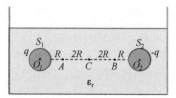

Fig. 3.14 Field produced by a charged sphere inside a dielectric medium

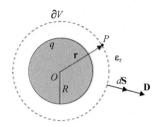

Applying (3.10),

$$\Phi_D = \oint_{\partial V} \mathbf{D} \cdot d\mathbf{S} = q_{np,in} \Rightarrow \oint_{\partial V} DdS = q \Rightarrow D 4\pi r^2 = q,$$

bearing in mind that the charge enclosed by the Gaussian surface is the charge q of the sphere. If we solve it

$$D = \frac{q}{4\pi r^2} \Rightarrow \mathbf{D} = \frac{q}{4\pi r^2}\frac{\mathbf{r}}{r} = \frac{q}{4\pi r^2}\mathbf{u}_r.$$

where $\mathbf{u}_r$ is radial unit vector for spherical coordinates. $\mathbf{E}$ is obtained from $\mathbf{D}$ by applying (3.16),

$$\mathbf{E} = \frac{\mathbf{D}}{\varepsilon_0 \varepsilon_r} = \frac{q}{4\pi \varepsilon_0 \varepsilon_r r^2}\mathbf{u}_r.$$

In Fig. 3.13 it is observed that all points referred to the statement are placed over line $\overline{O_1 O_2}$ which joins both spheres' centre. The circulation of E may be calculated along this line to obtain the potential differences, so we should know the expression of the electric field at every point on this line. Let's consider O_1 as the origin of the coordinate system. Fields produced by spheres at any point P on this line, at a distance r of O_1 (Fig. 3.15) are

$$\mathbf{E}_1 = \frac{q}{4\pi \varepsilon_0 \varepsilon_r r^2}\frac{\mathbf{r}}{r} = \frac{q}{4\pi \varepsilon_0 \varepsilon_r r^2}\mathbf{u}_r,$$

for sphere S_1 and

$$\mathbf{E}_2 = \frac{-q}{4\pi \varepsilon_0 \varepsilon_r r_2^2}\frac{\mathbf{r}_2}{r_2} = \frac{q}{4\pi \varepsilon_0 \varepsilon_r (8R-r)^2}\mathbf{u}_r,$$

for sphere S_2. It is considered that $r_2 = O_2 O_1 - r = 8R - r$ and that $\frac{\mathbf{r}_2}{r_2} = -\mathbf{u}_r$. The total field $\mathbf{E}$ at a point on the line $\overline{O_1 O_2}$ is

$$\mathbf{E} = \mathbf{E}_1 + \mathbf{E}_2 = \frac{q}{4\pi \varepsilon_0 \varepsilon_r}\left(\frac{1}{r^2} + \frac{1}{(8R-r)^2}\right)\mathbf{u}_r.$$

Fig. 3.15 Electric field due to both charged spheres

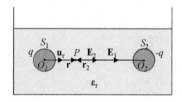

To obtain the potential differences indicated in the statement, we circulate the previous electric field[10] between points A and B as indicated in Fig. 3.13:

$$V_A - V_B = \int_A^B \mathbf{E} \cdot d\mathbf{l} = \int_A^B \frac{q}{4\pi\varepsilon_0\varepsilon_r} \left(\frac{1}{r^2} + \frac{1}{(8R-r)^2} \right) \mathbf{u}_r dr \mathbf{u}_r$$

$$= \frac{q}{4\pi\varepsilon_0\varepsilon_r} \int_{2R}^{6R} \left(\frac{1}{r^2} + \frac{1}{(8R-r)^2} \right) dr.$$

If the integral is solved,

$$V_A - V_B = \frac{q}{4\pi\varepsilon_0\varepsilon_r} \left(-\frac{1}{r} + \frac{1}{8R-r} \right) \bigg|_{2R}^{6R} = \frac{q}{4\pi\varepsilon_0\varepsilon_r} \left(-\frac{1}{6R} + \frac{1}{2R} + \frac{1}{2R} - \frac{1}{6R} \right)$$

$$= \frac{q}{4\pi\varepsilon_0\varepsilon_r} \frac{2}{3R}.$$

The statement says that the potential difference between points A and C is the same as the previous one, if the spheres were in a vacuum. We follow the same procedure, but making $\varepsilon_r = 1$, and considering the new integration limits,

$$V_A - V_C = \int_A^C \mathbf{E} \cdot d\mathbf{l} = \frac{q}{4\pi\varepsilon_0} \int_{2R}^{4R} \left(\frac{1}{r^2} + \frac{1}{(8R-r)^2} \right) dr$$

$$= \frac{q}{4\pi\varepsilon_0} \left(-\frac{1}{r} + \frac{1}{8R-r} \right) \bigg|_{2R}^{4R} = \frac{q}{4\pi\varepsilon_0} \frac{1}{3R}.$$

Since both potential differences have to be the same,

$$\frac{q}{4\pi\varepsilon_0\varepsilon_r} \frac{2}{3R} = \frac{q}{4\pi\varepsilon_0} \frac{1}{3R}.$$

Hence

$$\varepsilon_r = 2.$$

3.9 The conducting plate, which is gray coloured in Fig. 3.16 and whose thickness is $d = 20$ cm, is large enough in regards to this dimension. It is charged with a uniform surface charge density $\sigma = 200\,\mu\text{C/m}^2$. It is completely covered

---

[10] Potential differences could also be calculated for each field separately and then the superposition principle applied.

Fig. 3.16 Figure of Problem 3.9.

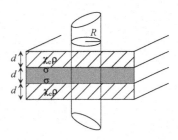

by layers of dielectric material whose electric susceptibility is $\chi_e = 3$ and its thickness is d. The dielectric has a uniform charge density, that does not come from polarization, and whose value is $\rho = 500\,\mu\text{C/m}^3$. (a) Determine the flux of the electric displacement **D** through the cylinder with oblique bases in the figure, whose straight section has a radius of $R = 30\,\text{cm}$. (b) Determine the polarization surface charge densities and the polarization volume charge density in the dielectric layers. (c) Determine the minimum value of dielectric strength to avoid the breakdown of the dielectric layers.

Solution

(a) To calculate the flux of D through the cylindrical surface of Fig. 3.16, Gauss' law for **D** (3.10) is applied. Non-polarization interior charge $q_{np,\text{in}}$ is determined by intersecting the cylinder with the charged areas. The conductor is charged with density σ on both surfaces. The dielectric is entirely charged with density ρ. Then

$$\Phi_D = \oint_{\partial V} \mathbf{D} \cdot d\mathbf{S} = q_{np,\text{in}} = \int_{\text{cond}} \sigma dS + \int_{\text{diel}} \rho dV = 2\sigma\pi R^2 + \rho\pi R^2 2d = 170\,\mu\text{C}.$$

(b) The dielectric is linear, homogeneous and isotropic, so the polarization volume charge density ρ_p is obtained from (3.19),

$$\rho_p = -\frac{\varepsilon_r - 1}{\varepsilon_r}\rho_{np} = -\frac{3}{4} \cdot 500\,\mu\text{C/m}^3 = -375\,\mu\text{C/m}^3,$$

where $\varepsilon_r = \chi_e + 1 = 4$. The polarization surface charge density σ_p is obtained from (3.18),

$$\sigma_p = \mathbf{P} \cdot \mathbf{n} = \frac{\varepsilon_r - 1}{\varepsilon_r}\mathbf{D} \cdot \mathbf{n}.$$

Calculus should be applied to each of the four surfaces of the dielectric. From the symmetry of the problem, Gauss' law for D can be applied to obtain the field **D** at any point of the dielectric. We take as a Gaussian surface ∂V a cylindrical surface (Fig. 3.17), whose bases are inside the dielectrics at the same distance in regards to the centre of the conducting plate, with area S and height $2r$, where r is the distance from the centre of the plate to any point of the dielectric. Field **D** at any point is normal to the plate, outward, with the same magnitude at every point on both bases

Fig. 3.17 Field **D** and unit normals to calculate polarization surface densities

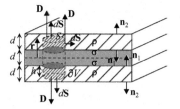

of the Gaussian cylinder. Flux of **D** through the cylinder lateral surface is null, since **D** and $d\mathbf{S}$ are perpendicular at any point on the surface. So there only exists flux through the bases,

$$\Phi_D = \oint_{\partial V} \mathbf{D} \cdot d\mathbf{S} = \int_{B_{\text{upp}}} D dS + \int_{B_{\text{low}}} D dS = D2S.$$

Applying Gauss' law for **D**, the result is

$$\Phi_D = q_{np,\text{in}} = 2\sigma S + \rho S 2h = 2S(\sigma + \rho h),$$

where $h = r - d/2$, is the dielectric's height inside either of the Gaussian cylinder halves. The charge inside the Gaussian surface is the result of intersecting the cylinder with the conducting plate (dark grey coloured in Fig. 3.17) and with the dielectric (lined in Fig. 3.17). If both expressions are equated, the result is

$$D = \sigma + \rho h.$$

If we substitute this value in (3.18), and we consider that **D** and **n** make 180° on dielectric surfaces in contact with the conductor (surface 1) and 0° on the other ones (as it can be seen in Fig. 3.17), it results

$$\sigma_{p1} = \frac{\varepsilon_r - 1}{\varepsilon_r} \mathbf{D}_1 \cdot \mathbf{n}_1 = -\frac{3}{4} D_1 = -\frac{3}{4}\sigma = -150\,\mu\text{Cm}^{-2}.$$

$$\sigma_{p2} = \frac{\varepsilon_r - 1}{\varepsilon_r} \mathbf{D}_2 \cdot \mathbf{n}_2 = \frac{3}{4} D_2 = \frac{3}{4}(\sigma + \rho d) = 225\,\mu\text{Cm}^{-2}.$$

It should be observed that $h = 0$ for the surfaces in contact with the conductor, and $h = d$ for the furthest surfaces.

(c) To obtain the dielectric strength, it is necessary to calculate the maximum electric field which the dielectric can withstand without breaking down. Since field **D** is known at every point of the dielectric, field **E** can be obtained by applying (3.16),

$$\mathbf{E} = \frac{\mathbf{D}}{\varepsilon_0 \varepsilon_r}.$$

The maximum value of the electric field **E** occurs when **D** has the highest value. If previous results for **D** are observed, this occurs on the most outlying surface of the dielectric ($h = d$). If we substitute, the minimum dielectric strength is obtained,

$$E_{max} = \frac{D_{max}}{\varepsilon_0 \varepsilon_r} = \frac{\sigma + \rho d}{\varepsilon_0 \varepsilon_r} = 8.47 \, \text{kV/mm}.$$

3.10 A point charge $q = 24 \, \mu\text{C}$ is placed at the centre of an aluminium spherical crown, that is also charged with $q = 24 \, \mu\text{C}$, whose interior radius is $R_1 = 2$ cm and the exterior one $R_2 = 10$ cm. The sphere is surrounded by a concentric plastic sphere, whose relative permittivity is $\varepsilon_r = 4$ and its thickness $e = 5$ cm. (a) Determine the induced charge densities in the plastic. (b) Determine the potential difference between the conducting sphere and a point P placed 20 cm away from the centre.

Solution

(a) The polarization volume charge density ρ_p is obtained from (3.19),

$$\rho_p = -\frac{\varepsilon_r - 1}{\varepsilon_r} \nabla \cdot \mathbf{D} = -\frac{\varepsilon_r - 1}{\varepsilon_r} \rho_{np} = 0,$$

since there are non-polarization charges inside the dielectric. The polarization surface charge density σ_p is obtained from (3.18),

$$\sigma_p = \mathbf{P} \cdot \mathbf{n} = \frac{\varepsilon_r - 1}{\varepsilon_r} \mathbf{D} \cdot \mathbf{n},$$

for both surfaces of the plastic crown. It is necessary, therefore, to calculate **D** at points of the dielectric.

The charges of the conductor (aluminium) move until they find positions in which the field inside the conductor is null (Sect. 2.8 and Problem 2.10). If Gauss' law 2.28 is applied to a Gaussian surface ∂V_1 totally inside the conductor (discontinuous line in Fig. 3.18),

$$\Phi_E = \oint_{\partial V_1} \mathbf{E} \cdot d\mathbf{S} = 0 = \frac{q_{in}}{\varepsilon_0} \Rightarrow q_{in} = 0 = q + q_{cond, in},$$

Fig. 3.18 Fields **D** and **E**, normal unit vectors and path to calculate circulation in Problem 3.10

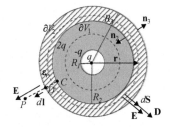

where $q_{\text{cond,in}}$ is the charge on the inner surface of the conductor. It should be observed that the flux is null because the field at every point of ∂V_1 is null. Therefore

$$q_{\text{cond,in}} = -q.$$

If the principle of conservation of charge is applied to the conductor, with charge q, charge of its outer surface is obtained,

$$q_{\text{cond}} = q = -q + q_{\text{cond,ex}} \Rightarrow q_{\text{cond,ex}} = 2q$$

and the charge distribution of Fig. 3.18 is obtained.

Due to spherical symmetry, charge will be uniformly distributed over the conductor surface, making the field inside it null. As a consequence, the field outside the conductor will be radial. Due to the existence of a dielectric (plastic), we need to calculate **D** before calculating the electric field **E**. Due to spherical symmetry, Gauss' law for **D** (3.10) is applied to a Gaussian spherical surface ∂V_2, discontinuous in Fig. 3.18, concentric with the conductor, and passing through any point of the dielectric, at a distance r from the centre,

$$\Phi_D = \oint_{\partial V_2} \mathbf{D} \cdot d\mathbf{S} = q_{np,\text{in}} \Rightarrow \oint_{\partial V_2} DdS = 2q \Rightarrow D4\pi r^2 = 2q,$$

where the charge $q_{np,\text{in}}$ inside the Gaussian surface is the sum of all the charges: $q - q + 2q$. Solving,

$$D = \frac{q}{2\pi r^2} \Rightarrow \mathbf{D} = \frac{q}{2\pi r^2} \frac{\mathbf{r}}{r} = \frac{q}{2\pi r^2} \mathbf{u}_r.$$

where $\mathbf{u}_r$ is the radial unit vector for spherical coordinates. Since **D** only depends on non-polarization charges, and there are no charges of this kind beyond the conductor, the calculated value is also valid outside the dielectric.

If we substitute it into (3.18), bearing in mind that for the inner surface, $r = R_2 = 0.1$ m and $\mathbf{n}_2 = -\mathbf{u}_r$, and for the outer one, $r = R_3 = R_2 + e = 0.15$ m and $\mathbf{n}_3 = \mathbf{u}_r$, it results

$$\sigma_p|_{r=R_2} = \frac{\varepsilon_r - 1}{\varepsilon_r} \mathbf{D} \cdot \mathbf{n}_2|_{r=R_2} = \frac{3}{4} \cdot \frac{24 \cdot 10^{-6}}{2\pi \cdot (0.1)^2} \mathbf{u}_r \cdot (-\mathbf{u}_r) = -286 \,\mu\text{Cm}^{-2}.$$

$$\sigma_p|_{r=R_3} = \frac{\varepsilon_r - 1}{\varepsilon_r} \mathbf{D} \cdot \mathbf{n}_3|_{r=R_3} = \frac{3}{4} \cdot \frac{24 \cdot 10^{-6}}{2\pi \cdot (0.15)^2} \mathbf{u}_r \cdot \mathbf{u}_r = 127 \,\mu\text{Cm}^{-2}.$$

(b) To determine the potential difference we circulate the electric field along the radial direction, where **E** and $d\mathbf{l}$ are parallel, from point C at the conductor surface (since the conductor is equipotential) to point P, as shown in Fig. 3.18,

$$V_C - V_P = \int_C^P \mathbf{E} \cdot d\mathbf{l} = \int_C^D \mathbf{E}_{\text{plas}} \cdot d\mathbf{l} + \int_D^P \mathbf{E}_{\text{vac}} \cdot d\mathbf{l}.$$

The path CP goes through two different media, plastic and vacuum, so the electric field $\mathbf{E}$ will be different, even though field $\mathbf{D}$ is the same. $\mathbf{E}$ is obtained from $\mathbf{D}$ by applying (3.16),

$$\mathbf{E}_{\text{plas}} = \frac{\mathbf{D}}{\varepsilon_0 \varepsilon_r} = \frac{q}{8\pi\varepsilon_0 r^2}\mathbf{u}_r,$$

$$\mathbf{E}_{\text{vac}} = \frac{\mathbf{D}}{\varepsilon_0 \varepsilon_r} = \frac{q}{2\pi\varepsilon_0 r^2}\mathbf{u}_r.$$

If we calculate the potential difference, taking into account that $\mathbf{E}$ is parallel to $d\mathbf{l}$,

$$V_C - V_P = \int_{0.1}^{0.15} \frac{q}{8\pi\varepsilon_0 r^2} dr + \int_{0.15}^{0.2} \frac{q}{2\pi\varepsilon_0 r^2} dr$$
$$= -\frac{q}{2\pi\varepsilon_0}\left(\frac{1}{4r}\Big|_{0.1}^{0.15} + \frac{1}{r}\Big|_{0.15}^{0.2}\right) = 1.08 \cdot 10^6 \text{ V}.$$

3.11 An empty spherical cavity with radius R is made in a linear, homogeneous and isotropic dielectric of infinite extent, with relative permittivity $\varepsilon_r = 2$. The dielectric has a non-polarization charge density $\rho = a/r^4$, with a as a constant and r the distance from the centre of the spherical cavity to any point of the dielectric. Determine: (a) the polarization charge densities in the dielectric; (b) the potential at the centre of the spherical cavity, taking infinity as the reference potential.

Solution

(a) The polarization volume charge density ρ_p at inner points of the dielectric is obtained from (3.19),

$$\rho_p = -\frac{\varepsilon_r - 1}{\varepsilon_r}\nabla \cdot \mathbf{D} = -\frac{\varepsilon_r - 1}{\varepsilon_r}\rho_{np} = -\frac{a}{2r^4}.$$

For points of the dielectric material in contact with the cavity, the polarization surface charge density σ_p is obtained from (3.18),

$$\sigma_p = \mathbf{P} \cdot \mathbf{n} = \frac{\varepsilon_r - 1}{\varepsilon_r}\mathbf{D} \cdot \mathbf{n}.$$

Since there are no charges in the hollow, and due to spherical symmetry of outer charges, the electric field in the cavity is null, and therefore $\mathbf{D} = 0$. Hence

$$\sigma_p = 0.$$

Fig. 3.19 Fields **D** and **E** and path to calculate circulation in Problem 3.11

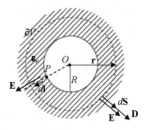

(b) To calculate the potential at the centre of the cavity, it is necessary to calculate the potential difference between the centre of the sphere and infinity (reference potential). We need to know the electric field **E** at every point in space. Due to spherical symmetry of charges, **E** is null at every point in cavity ($r < R$) and radial at every point in dielectric. Therefore,

$$V_O - V_\infty = V_O = \int_O^\infty \mathbf{E} \cdot d\mathbf{l} = \int_P^\infty \mathbf{E} \cdot d\mathbf{l},$$

where P is any point on dielectric surface (Fig. 3.19).

To calculate **E** in the dielectric, we firstly determine **D**. Due to the spherical symmetry of the charge distribution, Gauss' law for **D** (3.10) can be applied considering a spherical Gaussian surface ∂V concentric with the cavity (discontinuous line in Fig. 3.19), and passing through any point of the dielectric, at a distance r from the centre,

$$\Phi_D = \oint_{\partial V} \mathbf{D} \cdot d\mathbf{S} = q_{np,\text{in}} \Rightarrow \oint_{\partial V} D dS = D 4\pi r^2 = q_{np,\text{in}}.$$

Since charge density depends on distance r from the centre, to calculate non-polarization charge inside the volume, $q_{np,\text{in}}$, we take into account that the volume element in a sphere can be written as $dV = 4\pi r^2 dr$,

$$q_{np,\text{in}} = \int_{V_{\text{in}}} \rho dV = \int_R^r \frac{a}{r^4} 4\pi r^2 dr = 4\pi a \int_R^r \frac{1}{r^2} dr = 4\pi a \left. \frac{-1}{r} \right|_R^r = 4\pi a \left(\frac{1}{R} - \frac{1}{r} \right).$$

Therefore, substituting the expression for $q_{np,\text{in}}$ into the flux expression, we obtain **D**,

$$D = \frac{q_{np,\text{in}}}{4\pi r^2} = \frac{a}{r^2} \left(\frac{1}{R} - \frac{1}{r} \right) \Rightarrow \mathbf{D} = \frac{a}{r^2} \left(\frac{1}{R} - \frac{1}{r} \right) \mathbf{u}_r.$$

where $\mathbf{u}_r$ is the radial unit vector for spherical coordinates. **E** is obtained from **D** by applying (3.16),

$$\mathbf{E} = \frac{\mathbf{D}}{\varepsilon_0 \varepsilon_r} = \frac{a}{2\varepsilon_0 r^2} \left(\frac{1}{R} - \frac{1}{r} \right) \mathbf{u}_r.$$

To calculate the potential difference, we circulate the electric field along the radial direction, in which **E** and $d\mathbf{l}$ are parallel,

$$V_O = \int_P^\infty \mathbf{E} \cdot d\mathbf{l} = \int_R^\infty \frac{a}{2\varepsilon_0 r^2}\left(\frac{1}{R} - \frac{1}{r}\right)dr = \frac{a}{2\varepsilon_0}\left(\frac{-1}{Rr} + \frac{1}{2r^2}\right)\bigg|_R^\infty = \frac{a}{4\varepsilon_0 R^2}.$$

3.12 A parallel-plate capacitor consists of two squared shells whose side measures L, separated by a distance d, much smaller than L. There are two dielectric media, whose relative permittivities are $\varepsilon_{r1} = 2$ and $\varepsilon_{r2} = 3$. They can be placed between plates in two ways, (a) and (b), shown in Fig. 3.20. Each dielectric occupies half the space between the shells in both cases. Capacitors are charged with a charge q and they remain isolated. Make a comparison between the capacitance of both capacitors.

Solution

The capacitance is calculated from its definition, using (3.27),

$$C = \frac{q}{V_A - V_B},$$

where $V_A - V_B$ is the potential difference between the positive and the negative plate. We calculate it by circulation of the electric field **E** from the positive shell to the negative one. Since there is a dielectric medium, **D** must be calculated first.

Let's first consider the dielectrics of Fig. 3.20a. We follow a procedure similar to that of Problem 3.2. Because of the symmetry, charge q is uniformly distributed, with a density

$$\sigma = q/L^2.$$

D is perpendicular to the plates (if the fringing field at the edge is neglected) and with direction from the positive shell to the negative one. To obtain its magnitude, Gauss' law for **D** (3.10) can be applied, by taking as Gaussian surface ∂V a cylindrical surface, drawn with a discontinuous line in Fig. 3.21a. The flux of **D** through the upper base is null, since the electric field is confined between the plates. Flux through the lateral surface is also zero, since **D** and $d\mathbf{S}$ are perpendicular. We only need to calculate the flux through the lower base, and if Gauss' law for **D** is applied, the result is

$$\Phi_D = \oint_{\partial V} \mathbf{D} \cdot d\mathbf{S} = \int_{S_{\text{low}}} D\, dS = DS = q_{np,\text{in}} = \sigma S \Rightarrow D = \sigma,$$

Fig. 3.20 Parallel-plate capacitors with two different layouts for the dielectrics

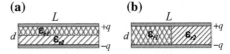

Fig. 3.21 Gaussian surface to calculate **D** and **E**

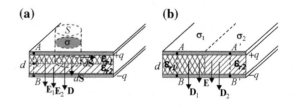

where S is the area of the lower base S_{low} of the Gaussian cylinder (and of any straight section of it). Electric field **E** for each dielectric medium is

$$E_1 = \frac{D}{\varepsilon_0 \varepsilon_{r1}} = \frac{\sigma}{2\varepsilon_0}, \quad E_2 = \frac{D}{\varepsilon_0 \varepsilon_{r2}} = \frac{\sigma}{3\varepsilon_0}.$$

Calculating the potential difference between A and B, line ACB of Fig. 3.21a, perpendicular to the plates, the result is

$$V_A - V_B = \int_A^B \mathbf{E} \cdot d\mathbf{l} = \int_A^C \mathbf{E}_1 \cdot d\mathbf{l} + \int_C^B \mathbf{E}_2 \cdot d\mathbf{l} = \int_0^{d/2} \frac{\sigma}{2\varepsilon_0} dl + \int_{d/2}^d \frac{\sigma}{3\varepsilon_0} dl$$

$$= \frac{\sigma}{2\varepsilon_0} \frac{d}{2} + \frac{\sigma}{3\varepsilon_0} \frac{d}{2} = \frac{5}{12} \frac{\sigma d}{\varepsilon_0}.$$

And if it is substituted in the definition of capacitance,

$$C_a = \frac{q}{V_A - V_B} = \frac{\sigma L^2}{5\sigma d/(12\varepsilon_0)} = \frac{12}{5} \frac{\varepsilon_0 L^2}{d}.$$

If we consider the layout of Fig. 3.20b, charge q does not uniformly distribute on the plates; it now depends on permittivity of the dielectric between them. Let σ_1 be the charge density on the plate over dielectric 1 and σ_2 the charge density on plate over dielectric 2, Fig. 3.21b. The result is,

$$q = \sigma_1 L \frac{L}{2} + \sigma_2 L \frac{L}{2}.$$

Since the circulation of the electrostatic field is path-independent (AB through dielectric 1 or $A'B'$ through dielectric 2), the potential difference between the plates of capacitor is the same. If we calculate these potential differences, and taking into account previous calculations made for layout (a), the result is, through dielectric 1

$$V_A - V_B = \int_A^B \mathbf{E}_1 \cdot d\mathbf{l} = \int_0^d \frac{\sigma_1}{2\varepsilon_0} dl = \frac{\sigma_1 d}{2\varepsilon_0},$$

and through dielectric 2

$$V'_A - V'_B = \int_{A'}^{B'} \mathbf{E}_2 \cdot d\mathbf{l} = \int_0^d \frac{\sigma_2}{3\varepsilon_0} dl = \frac{\sigma_2 d}{3\varepsilon_0}.$$

Since both potential differences are the same, it can be observed that both fields $\mathbf{E}_1$ and $\mathbf{E}_2$ must be the same ($\mathbf{E}$ in Fig. 3.21b). It follows that

$$V_A - V_B = V'_A - V'_B \Rightarrow \frac{\sigma_1 d}{2\varepsilon_0} = \frac{\sigma_2 d}{3\varepsilon_0} \Rightarrow \frac{\sigma_1}{\sigma_2} = \frac{2}{3}.$$

It should be observed that the relation between charge densities is the same as the relation of permittivities. If the capacitance is calculated, by using either expression for potential difference,

$$C_b = \frac{q}{V_A - V_B} = \frac{\sigma_1 \frac{L^2}{2} + \sigma_2 \frac{L^2}{2}}{\frac{\sigma_1 d}{2\varepsilon_0}} = \frac{\sigma_1 \frac{L^2}{2} + \frac{3}{2}\sigma_1 \frac{L^2}{2}}{\frac{\sigma_1 d}{2\varepsilon_0}} = \frac{5}{2}\frac{\varepsilon_0 L^2}{d}.$$

The same results could be reached if we considered each layout a combination of separate capacitors. In case (a), the capacitor is equivalent to the one shown in Fig. 3.22a. It is necessary to introduce a metallic plate between both dielectrics, but the capacitance does not change. This capacitor is therefore equivalent to two parallel-plate capacitors connected in series. The equation of capacitance for each capacitor has been calculated in Problem 3.2 (3.38). The equivalent capacitance is obtained by (3.36),

$$\frac{1}{C_a} = \frac{1}{C_1} + \frac{1}{C_2} = \frac{d/2}{\varepsilon_0 \varepsilon_{r1} L^2} + \frac{d/2}{\varepsilon_0 \varepsilon_{r2} L^2},$$

from where

$$C_a = \frac{12}{5}\frac{\varepsilon_0 L^2}{d}.$$

In case (b), the capacitor is equivalent to the one shown in Fig. 3.22b: two capacitors connected in parallel. Its capacitance can be calculated, according to (3.32), as the sum of the capacitances of both parallel-plate capacitors:

Fig. 3.22 Equivalence of plane capacitors of Fig. 3.20

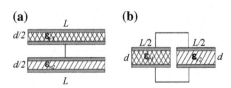

$$C_b = C_1 + C_2 = \frac{\varepsilon_0 \varepsilon_{r1} L^2/2}{d} + \frac{\varepsilon_0 \varepsilon_{r2} L^2/2}{d}.$$

Therefore
$$C_b = \frac{5}{2} \frac{\varepsilon_0 L^2}{d}.$$

Problems C

3.13 A very long dielectric cylinder, with length L, radius $R_1 = 2$ m and permittivity $\varepsilon_r = 4$, is charged with a non-polarization charge density $\rho = cr^2$, where $c = 6\,\mu\text{Cm}^{-5}$ and r is the distance from the cylinder axis in meters. It is surrounded by a grounded conducting coaxial cylinder, also very long, with radii $R_2 = 4$ m and $R_3 = 6$ m. The space between them is a vacuum. (a) Determine the charge distribution in the conductor. (b) Determine the flux of the electric field through the cylindrical surface whose radius is $R = 3$ m and length is $l = 5$ m, coaxial with the very long cylinders. (c) Determine the charge densities in the conductor and in the dielectric. (d) Determine the potential at a point placed 1 m from the axis of the cylinders.

Solution

(a) Figure 3.23 shows the layout of the problem. Charges in the conductor (grey coloured) are reorganized in order to make the field **E** null inside the conductor (Sect. 2.8). Gauss' law (2.28) is applied to a Gaussian cylindrical surface ∂V_1, length L, totally inside conductor (discontinuous line in the figure)

$$\Phi_E = \oint_{\partial V_1} \mathbf{E} \cdot d\mathbf{S} = 0 = \frac{q_{\text{in}}}{\varepsilon_0} \Rightarrow q_{\text{in}} = 0 = q + q_{\text{cond,in}},$$

where $q_{\text{cond,in}}$ is the charge on the inner surface of the conductor, and q is the charge of the dielectric, which is inside the cylinder ∂V_1. It should be observed that the flux is null, since field **E** is null at every point of ∂V_1. Therefore

$$q_{\text{cond,in}} = -q.$$

We have to determine charge q of the dielectric. Since the charge distribution has cylindrical symmetry, taking as a volume element a cylinder with any radius r, length L and thickness dr ($dV = 2\pi r L dr$):

Fig. 3.23 Charged dielectric cylinder surrounded by a conductor

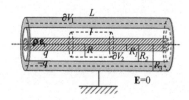

$$q = \int_{V_{in}} \rho dV = \int_0^{R_1} cr^2 2\pi rL dr = 2\pi cL \frac{r^4}{4}\bigg|_0^{R_1} = \frac{\pi c R_1^4 L}{2} = 1.5 \cdot 10^{-4} L \, \text{C}.$$

To determine charge on the outer surface of the conductor, it is taken into account that there is no potential difference between ground and the conductor, since the conductor is grounded. There cannot exist any field between them (because of (2.18), $\mathbf{E} = -\nabla V$). Field $\mathbf{E}$ is null in the conductor and outside it and, from (2.29) ($\nabla \cdot \mathbf{E} = \rho/\varepsilon_0$), there cannot be any charges on the outer surface of the conductor. Therefore,

$$q_{\text{cond,in}} = -1.5 \cdot 10^{-4} L \, \text{C}, \quad q_{\text{cond,ex}} = 0.$$

(b) To calculate the flux through the cylindrical surface whose radius is $R = 3$ m and length is $l = 5$ m (discontinuous line in Fig. 3.23), Gauss' law (2.28) can be applied

$$\Phi_E = \oint_{\partial V_2} \mathbf{E} \cdot d\mathbf{S} = \frac{q_{\text{in}}}{\varepsilon_0}.$$

q_{in} is the charge of the dielectric, whose value can be calculated the same way as it was done in section (a), but inside a cylinder of length l:

$$\Phi_E = \frac{q_{\text{in}}}{\varepsilon_0} = \frac{1.5 \cdot 10^{-4} l}{\varepsilon_0} = 84.7 \cdot 10^6 \, \text{Vm}.$$

(c) Charge in the conductor is located on the inner surface, with value $-q$. Charge density is obtained from (2.3). Since the cylinder is very long, a fringing field at the edges can be neglected and therefore the charge density is uniform,

$$\sigma_{\text{cond}} = \frac{dq}{dS} = \frac{-q}{S_{\text{cond,in}}} = \frac{-q}{2\pi R_2 L} = -6 \, \mu\text{Cm}^{-2},$$

where $S_{\text{cond,in}}$ is the area of the inner surface of the conductor. The polarization volume charge density ρ_p at inner points of the dielectric is obtained from (3.19),

$$\rho_p = -\frac{\varepsilon_r - 1}{\varepsilon_r} \nabla \cdot \mathbf{D} = -\frac{\varepsilon_r - 1}{\varepsilon_r} \rho_{np} = -\frac{3}{4} \cdot 6 \cdot 10^{-6} r^2 = -4.5 \cdot 10^{-6} r^2 \, \text{Cm}^{-3},$$

with r in meters.

The polarization surface charge density σ_p at points on the dielectric's surface is obtained from (3.18),

$$\sigma_p = \mathbf{P} \cdot \mathbf{n} = \frac{\varepsilon_r - 1}{\varepsilon_r} \mathbf{D} \cdot \mathbf{n}\big|_{r=R_1}.$$

with $\mathbf{n} = \mathbf{u}_\rho$, the outward unit normal to the dielectric, as can be seen in Fig. 3.24. It is necessary to calculate $\mathbf{D}$ on the surface of the dielectric. Since in section (d) we

Fig. 3.24 Polarization densities and potential differences calculus of Problem 3.13

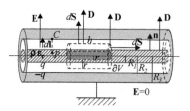

will need to calculate field **E** (and therefore **D**) at every point in space, we will make this calculus and then we will take $r = R_1$.

Field is null inside the conductor, and also for distances $r > R_3$, since the conductor is grounded. For $r < R_2$, because of the cylindrical symmetry of charge distribution, every point at the same distance from the cylinder's axis has the same field magnitude **D**, which is perpendicular to this axis (neglecting the fringing field at the edges). To obtain field **D** Gauss' law for **D** (3.10) can be applied. As Gaussian surface ∂V we construct a cylindrical surface, which is coaxial with the actual cylinders, with length h and radius r, passing through the point where the field is calculated (discontinuous lines in Fig. 3.24). Flux through the bases of the Gaussian surface is null, since $d\mathbf{S}$ and **D** are perpendicular at every point of them. We only need to calculate the flux through the lateral surface, where $d\mathbf{S}$ and **D** are parallel:

$$\Phi_D = \oint_{\partial V} \mathbf{D} \cdot d\mathbf{S} = \int_{S_{lat}} D\, dS = D \int_{S_{lat}} dS = D 2\pi r h, \qquad (3.39)$$

where S_{lat} is the lateral surface of the cylinder.

On the other hand, if Gauss' law (3.10) is applied for **D**, the result is

$$\Phi_D = q_{np,in}.$$

The value of $q_{np,in}$ depends on the position of the point where the field is calculated (inside or outside the dielectric cylinder). If the point is outside, the dielectric cylinder of radius R_1 and length h is totally inside the Gaussian surface, as it can be seen in Fig. 3.24. Charge is determined, as in section (a), taking as a volume element a cylinder with any radius r, length h and thickness dr ($dV = 2\pi r h dr$):

$$\Phi_{D(R_1 \leq r < R_2)} = q_{np,in} = \int_{V_{in}} \rho dV = \int_0^{R_1} cr^2 2\pi r h dr = \frac{\pi c R_1^4 h}{2}.$$

If the point is inside the dielectric, part of the dielectric cylinder stays outside the Gaussian surface, not producing any flux. Interior charge is only in the dark grey colored region in Fig. 3.24. Therefore, the result is

$$\Phi_{D(r \leq R_1)} = q_{np,in} = \int_{V_{in}} \rho dV = \int_0^r cr^2 2\pi r h dr = \frac{\pi c r^4 h}{2}.$$

If these expressions are equated to the one of the calculated flux (3.39), and replacing the values of $R_1 = 2$ m and $R_2 = 4$ m, **D** is obtained:

$$\mathbf{D} = \begin{cases} \dfrac{\pi c r^4 h}{4\pi r h}\mathbf{u}_\rho = \dfrac{cr^3}{4}\mathbf{u}_\rho & r \leq 2, \\[2mm] \dfrac{\pi c R_1^4 h}{4\pi r h}\mathbf{u}_\rho = \dfrac{4c}{r}\mathbf{u}_\rho & 2 \leq r < 4 \\[2mm] 0 & r \geq 4, \end{cases}$$

where $\mathbf{u}_\rho$ is the radial unit vector for cylindrical coordinates. **E** is obtained from **D** by applying (3.16) ($\mathbf{E} = \dfrac{\mathbf{D}}{\varepsilon_0 \varepsilon_r}$) where $\varepsilon_r = 4$ if the point is inside the dielectric and $\varepsilon_r = 1$ if the point is in the vacuum,

$$\mathbf{E} = \begin{cases} \dfrac{cr^3}{16\varepsilon_0}\mathbf{u}_\rho & r \leq 2, \\[2mm] \dfrac{4c}{r\varepsilon_0}\mathbf{u}_\rho & 2 \leq r < 4 \\[2mm] 0 & r \geq 2. \end{cases}$$

Substituting the values of **D** into (3.18), with $r = R_1 = 2$ m and $\mathbf{n} = \mathbf{u}_\rho$, the result is

$$\sigma_p = \mathbf{P} \cdot \mathbf{n} = \dfrac{\varepsilon_r - 1}{\varepsilon_r}\mathbf{D} \cdot \mathbf{n}|_{r=R_1} = \dfrac{3}{4} \cdot \dfrac{cR_1^3}{4}\mathbf{u}_\rho \cdot \mathbf{u}_\rho = 9\,\mu\text{Cm}^{-2}.$$

(d) To determine the potential difference we circulate the electric field along the normal direction to the cylinder axis, where **E** and $d\mathbf{l}$ are parallel (Fig. 3.24), from point P (1 m from the axis, inside the dielectric) to a point C on the conductor surface, with a known potential (0 V, since the conductor is grounded),

$$V_P - V_C = V_P = \int_P^C \mathbf{E} \cdot d\mathbf{l} = \int_P^D \mathbf{E}_{\text{diel}} \cdot d\mathbf{l} + \int_D^C \mathbf{E}_{\text{vac}} \cdot d\mathbf{l}.$$

D is a point on the dielectric surface, at a distance R_1 of the axis. The path PDC goes through two different media, dielectric and vacuum. Therefore the electric field **E** will be different. Calculating, taking into account that that the angle between the vectors **E** and $d\mathbf{l}$ is $0°$,

$$V_P = \int_1^2 \dfrac{cr^3}{16\varepsilon_0}dr + \int_2^4 \dfrac{4c}{r\varepsilon_0}dr = \dfrac{c}{\varepsilon_0}\left(\dfrac{r^4}{64}\bigg|_1^2 + 4\ln r|_2^4\right) = 2.04 \cdot 10^6 \text{ V}.$$

3.14 A very large plate of a dielectric material, whose dielectric constant is ε_r and its thickness e, is charged with a volume charge density $\rho = ax$, being a a positive constant and x the distance from a point of the dielectric to one of the surfaces. Determine fields **D**, **E** and **P** at any point placed at distance x, as well as the polarization charge densities.

Solution

Figure 3.25 shows the dielectric plate. The vertical direction (perpendicular to the plate) is taken as axis X, with the origin on the upper surface. **D** can be directly obtained by applying (3.9),

$$\nabla \cdot \mathbf{D} = \rho_{np} \Rightarrow \frac{\partial D_x}{\partial x} = ax,$$

because **D** only changes along direction x. If the equation is integrated,

$$D_x = \frac{1}{2}ax^2 + C,$$

where C is a constant to be determined. Since there is the same charge beneath the upper face as over the lower one, **D** must be the same at $x = 0$ and at $x = e$, but with opposite direction,

$$\frac{1}{2}a0^2 + C = -(\frac{1}{2}ae^2 + C) \Rightarrow C = -\frac{ae^2}{4}.$$

And therefore,

$$\mathbf{D} = \frac{a}{2}\left(x^2 - \frac{e^2}{2}\right)\mathbf{u}_x.$$

E is obtained from **D** by applying (3.16),

$$\mathbf{E} = \frac{\mathbf{D}}{\varepsilon_0 \varepsilon_r} = \frac{a}{2\varepsilon_0 \varepsilon_r}\left(x^2 - \frac{e^2}{2}\right)\mathbf{u}_x.$$

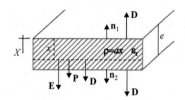

Fig. 3.25 Dielectric plate with variable charge

P is obtained from (3.8) (definition of field **D**) or by using (3.17),

$$\mathbf{D} = \varepsilon_0 \mathbf{E} + \mathbf{P} \Rightarrow \mathbf{P} = \mathbf{D} - \varepsilon_0 \mathbf{E} = \frac{\varepsilon_r - 1}{\varepsilon_r} \mathbf{D} = \frac{a}{2} \frac{\varepsilon_r - 1}{\varepsilon_r} \left(x^2 - \frac{e^2}{2} \right) \mathbf{u}_x.$$

The polarization volume charge density ρ_p is

$$\rho_p = -\nabla \cdot \mathbf{P} = -\frac{\partial P_x}{\partial x} = -\frac{\varepsilon_r - 1}{\varepsilon_r} ax.$$

The same result can be obtained if (3.19) is applied.
Polarization surface charge densities are

$$\sigma_p|_{x=0} = \mathbf{P} \cdot \mathbf{n}_1 = \frac{a}{2} \frac{\varepsilon_r - 1}{\varepsilon_r} \left(0^2 - \frac{e^2}{2} \right) \mathbf{u}_x \cdot (-\mathbf{u}_x) = \frac{\varepsilon_r - 1}{\varepsilon_r} \frac{ae^2}{4},$$

and

$$\sigma_p|_{x=e} = \mathbf{P} \cdot \mathbf{n}_2 = \frac{a}{2} \frac{\varepsilon_r - 1}{\varepsilon_r} \left(e^2 - \frac{e^2}{2} \right) \mathbf{u}_x \cdot \mathbf{u}_x = \frac{\varepsilon_r - 1}{\varepsilon_r} \frac{ae^2}{4},$$

taking into account the unit normals drawn in Fig. 3.25.

3.15 Determine the capacitance of a parallel-plate capacitor whose plates are circular, separated by a distance e, and whose radius is R, very large compared to e. There is an isotropic dielectric between the plates, whose relative permittivity is $\varepsilon_r = a + br$, with a and b constants and r the distance from the revolution axis of the plates. Neglect the fringing field at the edge.

Solution

The capacitance of the capacitor is calculated from its definition, using (3.27),

$$C = \frac{q}{V_A - V_B},$$

where $V_A - V_B$ is the potential difference between the positive and negative plates. To solve it, the circulation of electric field **E** is calculated from the positive plate to the negative one. To obtain **E**, it is necessary to know **D**, since there is a dielectric medium. As the dielectric permittivity depends on distance r to the axis, the charge q of the capacitor does not distribute uniformly; the charge density on conducting plates also depends on the distance ($\sigma = \sigma(r)$). Due to rotational symmetry, field **D** will be perpendicular to the plates (if fringing field at the edge is neglected) and with direction from the positive to the negative plate. As **D** magnitude depends on the charges, D will also depend on r.

Let's consider the set of points located at any distance r from the axis, and the ones at a $r + dr$ (hollow cylinder, darker in Fig. 3.26). These sets of points form a differential capacitor whose area is $dS = 2\pi r dr$ and the distance between plates is

Fig. 3.26 Parallel-plate capacitor of circular plates, with a dielectric of variable permittivity

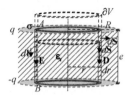

e, with a dielectric medium whose permittivity is $\varepsilon_r = a + br$, the same at every point. Surface charge density σ on the plates is, therefore, the same at every point in the differential volume. To determine the magnitude of **D**, Gauss' law for **D** (3.10) can be applied, taking as a Gaussian surface ∂V (discontinuous line in Fig. 3.26) the surfaces of a cylinder with the same axis and the same radii r and $r + dr$ as the differential capacitor. ∂V only goes through one of the conducting plates (the upper one in Fig. 3.26) so one of the bases of ∂V (the lower one) is inside the dielectric. The flux of **D** through the upper base is null, since all the electric field is confined between the capacitor plates. Flux through the lateral surfaces is also zero, since **D** and $d\mathbf{S}$ are perpendicular on the lateral surfaces of the cylinder. If the flux through surface ∂V is calculated, the result is

$$\Phi_D = \oint_{\partial V} \mathbf{D} \cdot d\mathbf{S} = \int_{S_{\text{low}}} D\, dS = D \int_{S_{\text{low}}} dS = D 2\pi r\, dr,$$

where $S_{\text{low}} = 2\pi r\, dr$ is the area of the lower base. If Gauss' law for **D** is applied

$$\Phi_D = q_{np,\text{in}} = \sigma 2\pi r\, dr,$$

which is the charge intersected by the cylinder when going through the conducting plate, (black colored in Fig. 3.26). If both equations are compared,

$$\mathbf{D} = \sigma \mathbf{u},$$

where **u** is the unit normal to the plates, with direction from the positive to the negative one.

In the differential capacitor, since $\varepsilon_r = a + br$ is a constant value in it, the electric field **E**, parallel to **D**, is obtained from (3.16),

$$\mathbf{E} = \frac{\mathbf{D}}{\varepsilon_0 \varepsilon_r} = \frac{\sigma}{\varepsilon_0 \varepsilon_r}\mathbf{u} = \frac{\sigma}{\varepsilon_0(a + br)}\mathbf{u}.$$

To calculate the circulation of the electric field **E** from the positive plate to the negative one, AB is the chosen path (Fig. 3.26), parallel to **E**, where the angle between **E** and $d\mathbf{l} = dl\mathbf{u}$ is zero,

$$V_A - V_B = \int_A^B \mathbf{E} \cdot d\mathbf{l} = \int_0^e \frac{\sigma}{\varepsilon_0(a+br)} dl = \frac{\sigma e}{\varepsilon_0(a+br)},$$

where $\sigma = \sigma(r)$ depending on the distance r to the axis. It can be observed that we use dl instead of dr because the circulation is calculated along the vertical direction. Since the potential difference does not depend on the chosen path, even though σ and ε_r are not constant, this quotient is the same for any defined differential capacitor. From this expression σ can be obtained, dependent on r,

$$\sigma = \frac{(V_A - V_B)\varepsilon_0(a+br)}{e}.$$

From this value, the charge q of the capacitor can be obtained,

$$q = \int_S \sigma dS = \int_0^R \sigma 2\pi r dr = \int_0^R \frac{(V_A - V_B)\varepsilon_0(a+br)}{e} 2\pi r dr$$
$$= \frac{2\pi(V_A - V_B)\varepsilon_0}{e}\left(\frac{aR^2}{2} + \frac{bR^3}{3}\right).$$

The resulting capacitance of the capacitor is

$$C = \frac{q}{V_A - V_B} = \frac{2\pi\varepsilon_0}{e}\left(\frac{aR^2}{2} + \frac{bR^3}{3}\right).$$

Similarly to Problem 3.12, this problem could have been solved as a capacitor association. The differential capacitor of area $dS = 2\pi r dr$ (dark in Fig. 3.26) is a parallel-plate capacitor whose capacitance is (Problem 3.2, (3.38))

$$dC = \frac{\varepsilon_r \varepsilon_0 dS}{e} = \frac{(a+br)\varepsilon_0}{e} 2\pi r dr.$$

The capacitor of the problem can be considered as a set of capacitors connected in parallel, with r from 0 to R. Therefore, by applying (3.32) for a capacitor association in parallel, if the sum becomes an integral, the result is

$$C = \int_0^R \frac{(a+br)\varepsilon_0}{e} 2\pi r dr = \frac{2\pi\varepsilon_0}{e}\left(\frac{aR^2}{2} + \frac{bR^3}{3}\right).$$

Chapter 4
Electric Current

Abstract In the previous chapters we saw problems with charge at rest. In this chapter we will study charges in motion, which is known as an electric current. Conducting materials will be studied, in which charge carriers are free to move under electric field action. This movement leads to two types of currents: conduction currents and convection currents. In conduction currents, free electric charges, which are known as current carriers, move on a neutral media. The most characteristic example is the current in metals, where valence (or conductive) electrons (which can move from one atom to another randomly) move in an organized way when an external electric field is applied, and they generate electric current. This type of current can also be seen in electrolytes or in ionized gases, where conduction is by positive and negative ions, which travel in opposite directions. However, they produce a current in the same direction. By convention, the direction in which the positive carrier moves is taken as the direction, or *sense*, of the current. In convection currents, negatively or positively charged media (liquid or solid) suffer a hydrodynamic movement which implies mass transport. This is how an electric current is generated. Some examples of this are atmospheric electricity or electron beams in a cathode ray tube.

4.1 Current Density. The Current

In Fig. 4.1 a metal conductor is shown with electric current in it. Electric current is characterized by the **current density**, defined for every point of the conductor as

$$\mathbf{j}(\mathbf{r}, t) \equiv q n(\mathbf{r}, t) \mathbf{v}(\mathbf{r}, t), \tag{4.1}$$

where $\mathbf{v}(\mathbf{r}, t)$ is a velocity field (current carriers velocity on each point $\mathbf{r}$ of the material for each point in time t), q is every carrier charge (for a metal, $q = -e = -1.6 \times 10^{-19}$ C) and n is the number of carriers per unit volume. The current density is a vector point function. Observing Fig. 4.1, we see current density and carriers velocity have opposite direction, due to the electron negative sign. From Expression (4.1) it is deduced that the current density vector unit in SI is $Cs^{-1}m^{-2}$. Cs^{-1} is called

Fig. 4.1 Conduction current in a metal

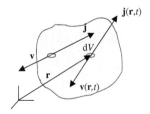

ampere, with symbol A. So the unit of **j** is Am^{-2}. The ampere is considered as one of the fundamental units in SI, instead of the coulomb.[1]

When several types of carriers exist (in electrolytes or in ionized gases), the expression (4.1) converts into

$$\mathbf{j}(\mathbf{r}, t) \equiv \sum_i q_i n_i(\mathbf{r}, t) \mathbf{v}_i(\mathbf{r}, t), \qquad (4.2)$$

where index i refers to different carriers types.

Expression (4.1) can also be applied to convection currents, where q is a free moving charge. If it is observed that Nq represents free moving charges per volume unit ρ, the same equation can be written as

$$\mathbf{j}(\mathbf{r}, t) = \rho(\mathbf{r}, t) \mathbf{v}(\mathbf{r}, t). \qquad (4.3)$$

Current density is a point vector function. If total current through a macroscopic arbitrary surface with area S is studied, it will be necessary to determine current density flux **j** through S. This calculus is known as **current** or **current intensity** I, expressed by:

$$I = \int_S \mathbf{j} \cdot d\mathbf{S}. \qquad (4.4)$$

The unit of I in SI is ampere (A). Current density may have positive or negative sign depending on the scalar product between **j** and $d\mathbf{S}$ or, equivalently, on surface orientation relative to the carrier velocity.

It can be demonstrated that the current I is the charge that goes through surface S per time unit. This is the reason why I is also defined as:

$$I \equiv \lim_{\Delta t \to 0} \frac{\Delta q}{\Delta t} \equiv \frac{dq}{dt}, \qquad (4.5)$$

which shows that, generally, current is time dependent.

---

[1] It is necessary to use the magnetic force concept (Chap. 5) to define ampere as a fundamental unit.

4.2 The Equation of Continuity

Current density **j** and charge density ρ are not independent quantities. They are related at each point by a differential equation called **the equation of continuity**. This equation can be easily deduced from the principle of charge conservation, by applying (4.4) to an arbitrary *closed* surface. If a current I flows towards the *inside* region, charge in the volume must *increase*. If flux through this surface is calculated and the divergence theorem is applied to volume V enclosed by S, the result is:

$$I(t) = -\oint_S \mathbf{j} \cdot d\mathbf{S} = -\int_V \operatorname{div} \mathbf{j}\, dV. \tag{4.6}$$

The negative sign appears because $d\mathbf{S}$ is outward oriented, and we want to consider positive current when **j** moves into the volume. If we take into account that I represents the rate at which the charge is transported into V, the result also is,

$$I = \frac{dq}{dt} = \frac{d}{dt}\int_V \rho\, dV = \int_V \frac{\partial \rho}{\partial t} dV. \tag{4.7}$$

Since the volume is fixed, the derivative operates just over ρ and it can be introduced in the integral as a partial derivative. If (4.6) and (4.7) are equated,

$$\int_V \left(\frac{\partial \rho}{\partial t} + \operatorname{div} \mathbf{j} \right) dV = 0. \tag{4.8}$$

But V is arbitrary, and the only way that previous expression can hold for an arbitrary volume is

$$\operatorname{div} \mathbf{j} + \frac{\partial \rho}{\partial t} = 0, \tag{4.9}$$

to express the equation of continuity.

4.3 Direct Current

Electric current is said to be **direct**, **constant** or **stationary**, if none of the magnitudes previously seen are time-dependent: $\mathbf{v} = \mathbf{v}(\mathbf{r})$, $\mathbf{j} = \mathbf{j}(\mathbf{r})$. Therefore, current I, given by the flux of **j**, is also time independent. For direct current, $\partial \rho / \partial t = 0$, and the equation of continuity (4.9) states that

$$\operatorname{div} \mathbf{j} = 0. \tag{4.10}$$

This means that the current density is a solenoidal vector field. Its current lines have neither sources nor sinks. Since these lines must be inside a finite volume in space,

Fig. 4.2 Scheme for the equation of continuity deduction in a metal straight wire

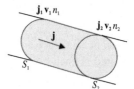

they have to be closed curves. The volume integral of this equation, if the divergence theorem is applied, is

$$\oint_S \mathbf{j} \cdot d\mathbf{S} = 0. \quad (4.11)$$

The equation of continuity for direct current is usually particularized to current in a conducting straight wire. Let's consider a wire segment between two straight sections, with areas S_1 and S_2, as shown in Fig. 4.2. $\mathbf{j}$ can be assumed to be parallel to the wire and uniform on every section. Outward current from the volume defined by these two sections and lateral surface is, from (4.11)

$$0 = \oint_S \mathbf{j} \cdot d\mathbf{S} = \int_{S_1} \mathbf{j} \cdot d\mathbf{S} + \int_{S_2} \mathbf{j} \cdot d\mathbf{S} + \int_{S_{lat}} \mathbf{j} \cdot d\mathbf{S}$$
$$= -j_1 S_1 + j_2 S_2 = -e(n_2 v_2 S_2 - n_1 v_1 S_1),$$

The third integral is null, due to the fact that $d\mathbf{S}$ is perpendicular to $\mathbf{j}$ (wire tangent) on its lateral surface. If it is solved, we obtain

$$j_1 S_1 = j_2 S_2, \quad (4.12)$$

the equation of continuity for a wire with direct current, which expresses current equality on every section from the wire. If the electron density is the same on every section, $n_1 = n_2$, the result is

$$v_1 S_1 = v_2 S_2, \quad (4.13)$$

which states that electron velocity on every section is inversely proportional to its cross-section.

4.4 Ohm's Law

It can be analytically justified, and it has been experimentally verified that, on almost every conducting material, including metals, electrolytes and ionized gases, when temperature is constant, current is proportional to the electric field, with same direction and sense, and null when it does not exist:

$$\mathbf{j} = \sigma \mathbf{E}. \tag{4.14}$$

This relation is known as **Ohm's law**, and conductors that follow this law are called *ohmic conductors* or *isotropic lineal conductors*.[2]

The constant of proportionality $\sigma > 0$ that appears in Ohm's law is called the **conductivity**, and its inverse

$$\rho = \frac{1}{\sigma}, \tag{4.15}$$

the **resistivity**. From (4.14), it is obtained that resistivity is measured in $VA^{-1}m$. An *ohm* (Ω) is defined as a VA^{-1}. Therefore, resistivity is measured in Ωm. From (4.15) it is obtained that conductivity is measured in $\Omega^{-1}m^{-1}$. Ω^{-1} is known as *siemens* (S).

Sometimes, atoms or molecule disposition may lead to dependence between resistivity and the direction which an electric field is applied on. These materials are known as *anisotropic*. For these, Ohm's law (4.14) is expressed as $\mathbf{j} = \overline{\overline{g}}\mathbf{E}$, where $\overline{\overline{g}}$ is a second order tensor, represented as a 3×3 matrix. For other materials, resistivity and conductivity depend not only on material or temperature, but also on the applied field. These materials are *non linear*, and Ohm's law (4.14) should be replaced by $\mathbf{j} = g(E)\mathbf{E}$.

Conductivity and resistivity values are specific for each material, but change significantly with temperature. This variation is usually empirically expressed by

$$\rho(T) = \rho_0[1 + \alpha(T - T_0) + \beta(T - T_0)^2 + \cdots], \tag{4.16}$$

where T is temperature and ρ_0 resistivity to $T = T_0$. α is known as the *temperature coefficient*[3] and is measured with K^{-1}.

4.5 Ohm's Law in a Conducting Straight Wire. Resistance

Let's now consider an ohmic conductor, straight wire-shaped, with section S and resistivity ρ not necessarily uniform along it. Two sections of the wire are maintained at a constant potential difference $V_1 - V_2$ (Fig. 4.3). There will be a static electric field, with decreasing potential direction, and due to Ohm's law, a direct current in the same direction. According to (4.14), the electric field and the current density $\mathbf{j}$ can be supposed approximately perpendicular and uniform on each straight section of the wire.

---

[2] Field E which appears on the expression is force per unit time, independent from its origin, electrostatic or not, and is represented sometimes as $\mathbf{E}_{ef}$. If the unique origin is electrostatic, as occurs with conductors, it will be necessary to maintain the conductor out of the balance by using right devices (generators), so the field is not null.

[3] See Appendix for a table with resistivity values and temperature coefficients for common materials [34, 66].

Fig. 4.3 Vectors for Ohm's law application to a conducting straight wire

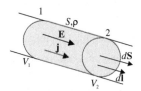

Potential difference between sections 1 and 2 (Fig. 4.3) can be written, using Ohm's law (4.14) and (4.15), as

$$V_1 - V_2 = \int_1^2 \mathbf{E} \cdot d\mathbf{l} = \int_1^2 \rho \mathbf{j} \cdot d\mathbf{l} = \int_1^2 \rho \frac{dl}{S} jS, \qquad (4.17)$$

where $d\mathbf{l}$ and $\mathbf{j}$ have the same direction. The product jS is the current I on every wire section (from (4.4), if $d\mathbf{S}$ is taken with $d\mathbf{l}$ direction), with constant value due to direct current. Therefore it can be removed from the integral, and the result is

$$V_1 - V_2 = I \int_1^2 \rho \frac{dl}{S}. \qquad (4.18)$$

The **resistance** of the wire is defined as

$$R = \int_1^2 \rho \frac{dl}{S}. \qquad (4.19)$$

The equation may be rewritten

$$V_1 - V_2 = RI, \qquad (4.20)$$

which is the familiar form of **Ohm's law**. This law states that *the potential difference (voltage drop) between two points in a conducting wire is proportional to the current that goes through it*. In Sect. 4.11 it will be shown that (4.14) implies (4.20), independently of the shape of the conductor.

Resistance $R \geq 0$ is the proportionality constant, and is a characteristic property of each conductor that depends on its material resistivity (and from temperature), on its shape, and dimensions. If, as is common, resistance and section do not vary along the wire, (4.19) becomes

$$R = \frac{\rho L}{S}, \qquad (4.21)$$

where L is the wire length. Resistance from a uniform wire section increases with its resistivity and length, and decreases when its cross-sectional area is increased.

4.6 Power Supplied by Electric Field. Joule's Law

When conduction charges are accelerated by an electric field, they continually gain energy from it. This energy is transferred to the bulk of the material by collisions, increasing the thermal motions of the atoms. If the metal is not thermally isolated, energy per unit of time P is transferred to the surrounding media as heat. This phenomenon is called the **Joule's effect**.

Work done by an electric field $\mathbf{E}$ to move charge q_i a distance Δl_i is $q_i \mathbf{E} \cdot (\Delta \mathbf{l}_i)$. Power is

$$p_i = \lim_{\Delta t \to 0} \frac{\Delta w_i}{\Delta t} = q_i \mathbf{E} \cdot \mathbf{v}_i, \qquad (4.22)$$

where $\mathbf{v}_i = \dfrac{\Delta \mathbf{l}_i}{\Delta t}$ is the velocity of the carrier i. Power given to all the carriers in a volume dV is

$$dP = \sum_i p_i = \mathbf{E} \cdot \left(\sum_i n_i q_i \mathbf{v}_i \right) dV, \qquad (4.23)$$

n_i being the number of carriers per volumen unit. If (4.2) is applied, the result is

$$dP = \mathbf{E} \cdot \mathbf{j}\, dV, \qquad (4.24)$$

or

$$\frac{dP}{dV} = \mathbf{E} \cdot \mathbf{j}, \qquad (4.25)$$

which represents the **power density** in a direct current. It is a point function measured in Wm^{-3}. For a macroscopic volume V, total electric power given by an electric field and transformed into heat, is

$$P = \int_V \mathbf{E} \cdot \mathbf{j}\, dV, \qquad (4.26)$$

measured in W. Equation (4.26) is known as **Joule's law**.

If a wire conductor is considered, (4.26) can be rewritten, if $\mathbf{j}$ is taken with the same direction as $d\mathbf{l}$

$$P = \int_L E\, dl \int_S j\, dS = (V_1 - V_2) I, \qquad (4.27)$$

where I is the current in the conductor and $V_1 - V_2$ is the potential difference established between the conductor ends. If Ohm's law (4.20) is applied the result is

$$P = R I^2, \qquad (4.28)$$

a well known equation from elementary circuit theory. The equation represents heat dissipated in resistance R per time unit.

4.7 Direct Current Generators

It's been seen that if we want an electric field to appear and to maintain direct current in a conducting wire (on grey in Fig. 4.4) it is necessary to maintain a potential difference $V_1 > V_2$ between two of its points. Besides, due to (4.10), **j**'s vector lines must be closed, setting up a *circuit*. But with electric fields, $\oint \mathbf{E} \cdot d\mathbf{l} = 0$, which implies $\oint \mathbf{j} \cdot d\mathbf{l} = 0$, applying Ohm's law (4.14). In other words, a pure electrostatic force cannot make the current circulate in the same direction around a closed path. It is necessary to apply another type of force (mechanic, chemical, etc.) so, in a part of the circuit, charge moves in an opposite direction to electrostatic field **E**. This is done by an **electric generator** (also known as **voltage source**), that allows current to pass and maintains the electric potential difference, preventing it from reaching equilibrium. A generator produces non electrostatic fields due to changing magnetic fields, different concentrations, etc.

Potential difference that applies to an ideal electric generator between its terminals is called the **electromotive force** $\mathcal{E}$ (or simply the emf) of the generator, and is measured, therefore, in volts (V). An ideal generator has an extreme (pole) with higher potential V_1 or *positive terminal*, and another extreme with lower potential V_2 or *negative terminal*. Current *exits* the generator via the *positive* terminal.

To understand the physic meaning of emf the scheme in Fig. 4.4 should be observed. In the generator two electric fields appear: electrostatic field **E**, due to electric potential difference between the terminals, and field $\mathbf{E}_{ns}$ (non electrostatic field). The force acting over a charge q is $\mathbf{F} = q(\mathbf{E} + \mathbf{E}_{ns})$, where $\mathbf{E}_{ef} = \mathbf{E} + \mathbf{E}_{ns}$ is known as the *effective field*. The **electromotive force (emf)** $\mathcal{E}$ **around a closed path** is defined as the circulation of the effective field along this line.

$$\mathcal{E} = \oint_\Gamma \mathbf{E}_{ef} \cdot d\mathbf{l} = \oint_\Gamma \mathbf{E} \cdot d\mathbf{l} + \oint_\Gamma \mathbf{E}_{ns} \cdot d\mathbf{l} = \oint_\Gamma \mathbf{E}_{ns} \cdot d\mathbf{l}, \qquad (4.29)$$

which depends solely on the non electrostatic field circulation, since circulation of the electrostatic field along a closed path is null. If the only source of non electrostatic field included in the closed path is the generator, hence that between its terminals there will only exist $\mathbf{E}_{ns}$, so

$$\mathcal{E} = \int_{gen} \mathbf{E}_{ns} \cdot d\mathbf{l}, \qquad (4.30)$$

Fig. 4.4 Scheme of a generator's behaviour

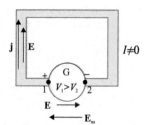

4.7 Direct Current Generators

which is the **electromotive force (emf)** of the source. The integral is taken around the circuit from one terminal of the source to the other, with a *positive value if we circulate with the same direction as the non electrostatic field* (and therefore, with the same direction as the current). It represents the energy per unit of circulating charge dq that the generator should supply to move the charge against the electrostatic field, carrying this charge from the negative terminal to the positive one (from 2 to 1 in Fig. 4.4), that is,

$$\mathcal{E} \equiv \frac{dW_G}{dq}. \tag{4.31}$$

An ideal generator is the one where all the work $dW_G = dq\mathcal{E}$ is expended to increase the energy of charge dq. As dq goes from point 2, with lower potential V_2, to 1, with higher potential V_1, this increase is $dq(V_1 - V_2)$. So, for an ideal generator, we have

$$dq\,\mathcal{E} = dq(V_1 - V_2) \Rightarrow \mathcal{E} = (V_1 - V_2). \tag{4.32}$$

In real generators, from all the developed work $dq\mathcal{E}$, a part is lost by the generator itself and the rest is used to increase the energy of charge dq when it passes through the generator. The simplest supposition is that losses per unit of charge depend linearly on current. So, for a real generator, we have

$$dq\,\mathcal{E} = dq(V_1 - V_2) + \text{losses} \Rightarrow V_1 - V_2 = \mathcal{E} - rI, \tag{4.33}$$

where coefficient $r > 0$ is the **internal resistance of the generator**, measured in ohms (Ω). It should be observed that the voltage drop in terminals of a real generator where current circulates is always lower than its emf. Emf is the same as the electric potential difference between the generator terminals when no current circulates through it (due, for example, to a circuit where the switch is open). It is also called the *open-circuit voltage*. **Power supplied** to the carriers (also known as the power at the generator terminals) is obtained applying (4.27) and (4.33),

$$P_{\text{sup}} = (V_1 - V_2)I = \mathcal{E}I - rI^2. \tag{4.34}$$

$\mathcal{E}I$ represents actual power produced by the generator, considering zero losses (or **power generated** by the generator):

$$P_{\text{gen}} = \frac{dq}{dt}\mathcal{E} = \mathcal{E}I = (V_1 - V_2)I + rI^2. \tag{4.35}$$

Power supplied by the generator is always lower than power generated by it in the quantity rI^2, which represents the part of the power generated that is dissipated by the generator itself, as heat (*losses caused by Joule's effect*).

4.8 Direct Current Motors

Besides generators and electric resistances, elements generically known as **motors** are usually included in circuits. They take electric energy from the circuit to convert into other types of energy.

A motor has an extreme or *positive terminal* where potential is higher (V_3 in Fig. 4.5), and another extreme or *negative terminal* with lower potential (V_4 on the figure). Current *enters* the motor via the *positive* terminal. It is characterized by its **counter-electromotive force** $\mathcal{E}$ (cemf), which is the energy that the motor converts (on mechanical energy, chemical energy, etc.) per unit of circulating charge. With the same reasoning used in the previous section, it is observed that cemf coincides with the circulation of non electrostatic field inside the motor. It is a positive magnitude that is measured in volts (V).

When a charge dq circulates through the motor, from 3 to 4, it loses an energy $dq(V_3 - V_4)$. An ideal motor converts this energy integrally: $dW_{\text{transf}} = dq\mathcal{E}$. So, for an ideal motor, we have

$$dq(V_3 - V_4) = dq\mathcal{E} \Rightarrow V_3 - V_4 = \mathcal{E}, \tag{4.36}$$

this means that the electric potential difference between terminals 3 and 4 from an ideal motor is the same as the counter-electromotive force. In real motors, from all the energy $dq(V_3 - V_4)$ that reaches the motor, a part of it is converted by it, and the rest is lost in it; this is

$$dq(V_3 - V_4) = dq\mathcal{E} + \text{losses} \Rightarrow V_3 - V_4 = \mathcal{E} + rI, \tag{4.37}$$

where it has been supposed that losses depend linearly on the current, as it occurs in most motors. Proportionality constant $r > 0$ is known as the **internal resistance of the motor** and is measured in ohms (Ω). The voltage drop in terminals of a real motor is, therefore, higher than its electromotive force.

From (4.36) it is deduced that energy converted by a motor per unit of time (power) is

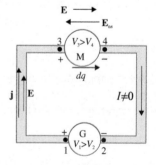

Fig. 4.5 Motor's behavior scheme

4.8 Direct Current Motors

$$P_{\text{conv}} = \frac{dq}{dt}\mathcal{E} = \mathcal{E}I, \quad (4.38)$$

and applying (4.37), the result is

$$P_{\text{conv}} = \mathcal{E}I = (V_3 - V_4)I - rI^2. \quad (4.39)$$

The term $(V_3 - V_4)I$ is power consumed (absorbed) or power extracted from the circuit (also known as power at the motor terminals),

$$P_{\text{cons}} = (V_3 - V_4)I = \mathcal{E}I + rI^2 \quad (4.40)$$

The power converted by a motor is always rI^2 less than power extracted from the circuit. rI^2 represents power dissipated by the motor itself (losses because of Joule's effect).

4.9 Ohm's Law in Circuits

A simple direct current circuit can consist of one or several generators, with electromotive forces $\mathcal{E}_i$ and internal resistances r_i, of one or several motors, with counter-electromotive forces $\mathcal{E}'_i$ and internal resistance r'_i, joined by conductors with resistance R_i.

Let's consider an example of a simple circuit (Fig. 4.6), that consists of a generator G_1 and two motors M_1 and M_2 joined by conductor wires with resistance R_i, represented by ⋎⋎⋎. The symbol used to represent polarized devices (generators and motors) is ⊢, where the longest stroke represents the positive pole and the shortest the negative one. It should be observed that the current exits the generators and enters

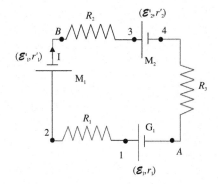

Fig. 4.6 Example of simple circuit

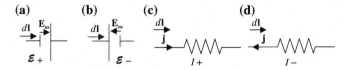

Fig. 4.7 Sign criteria to apply Ohm's law

the motors via the positive pole.[4] Let's calculate the circulation of electrostatic field along all the closed circuit (we know its value is null). This circulation will be the sum of voltage drops between the different marked point on the figure, if we start and finish in the same point (A, for example). From (4.20), (4.33) and (4.37) the result is

$$\sum_i \mathcal{E}_i = I \sum_i R_i , \qquad (4.41)$$

which is known as *Ohm's law for a circuit*. $\sum_i R_i$ includes every resistance in the circuit: conductor ones and internal ones in generators and motors. $\sum_i \mathcal{E}_i$ includes all emf and cemf, with positive sign in the generator's case and a negative one in motors. If the obtained value for current that circulates through the circuit were negative, it would mean that motors couldn't work with that cemf in the given circuit.[5]

It should be observed, as it was previously said when Ohm's law was deduced, that assigned direction to $d\mathbf{S}$ (perpendicular to the conductor's cross section used to calculate I), should be the same as the one that the circulation of electric field has (marked by $d\mathbf{l}$), so (4.20), (4.33) and (4.37) are right. Current, electromotive force and counter-electromotive force signs are a consequence of scalar product sign that appears in its definitions. Bearing in mind that the direction coincidence between $d\mathbf{l}$ and $d\mathbf{S}$, it is convenient to emphasize the sign criteria for $\mathcal{E}$ and for I only depend on $d\mathbf{l}$, as it can be observed in Fig. 4.7.

If when circulating we first go through the negative pole (Fig. 4.7a), $\mathcal{E}$ is positive, because according to expression $\mathcal{E} = \oint_\Gamma \mathbf{E}_{ns} \cdot d\mathbf{l}$, directions of $\mathbf{E}_{ns}$ and $d\mathbf{l}$ are the same. Current I is positive when we circulate with the same current direction because $\mathbf{j}$ and $d\mathbf{l}$ (or $\mathbf{j}$ and $d\mathbf{S}$ in the expression $I = \int \mathbf{j} \cdot d\mathbf{S}$) have the same direction (Fig. 4.7c).

To balance the energy it is sufficient to multiply the terms of the (4.41) by current I, separating on different sides generators' and motors' power, and the result is

$$\sum_i \mathcal{E}_i I = \sum_i \mathcal{E}'_i I + \sum_i R_i I^2 . \qquad (4.42)$$

---

[4] In practice some devices, known as reversible ones, can work both as generators or motors (releasing or absorbing power). The way to know their working mode is analyzing the direction of the current when going through them.

[5] In fact, motor or motors could work with a lower regime, with a cemf value lower than the nominal one.

4.9 Ohm's Law in Circuits

This equation expresses the energy conservation principle in a direct current circuit. Therefore, generated energy per unit time is used in converting energy in the motors and in dissipating energy as heat.

If all this is applied to a circuit portion (branch) in Fig. 4.6, the result is in what is usually known as *Ohm's law for a branch*:

$$V_A - V_B + \sum_i \mathcal{E}_i = \sum_i R_i I, \qquad (4.43)$$

with sign criteria previously mentioned for I and $\mathcal{E}_i$.[6]

If both sides in (4.43) are multiplied by current I and we rearrange them, the energy balance for a branch is:

$$(V_B - V_A)I = \sum_i \mathcal{E}_i I - \sum_i \mathcal{E}'_i I - \sum_i R_i I^2, \qquad (4.44)$$

which shows that energy given by a branch to the rest of the circuit (if $(V_B - V_A)I > 0$) or absorbed from it (if $(V_B - V_A)I < 0$), is the difference between the one generated by generators in the branch and the one consumed by its resistance and motors.

4.10 Direct Current Networks

It has been previously stated that current that circulates through each element in a circuit is the same, because elements are connected one after the other. However, current is not necessarily the same because several conductors can be joined at any point of the circuit. They form a conductor **network** where all the other studied devices are included (generators and motors). The point where two or more conductors are linked is known as a **node**. Each circuit portion between two consecutive nodes is known as a **branch**, and every closed path defined in the network, without going through the same branch twice, is known as a **mesh** (or loop[7]).

4.10.1 Kirchhoff's Circuit Laws

In simple circuits, Ohm's law (4.41) allows us to determine a circulating current if the device characteristics are known (resistance, electromotive forces and counter-electromotive ones). If we have a conductor network, the basic problem is: given the

---

[6]For this case, the equation is as follows: $V_A - V_B + \mathcal{E}_1 - \mathcal{E}'_1 = (r_1 + r'_1 + R_1)I$.
[7]Strictly speaking a mesh is a loop in a planar circuit. Planar circuits are circuits that can be drawn on a plane surface with no wires crossing each other.

resistance and emf of each circuit element, find the current in each of these elements. To solve this problem, nodes and branches are firstly numbered and then a sense is assigned to the current on each branch, due to the fact that at this time it is not that easy to determine the sense or to know which device works as a generator or as a motor (if we have reversible ones). Let's call I_i the current by branch i and R_i the total resistance in the branch, including internal generators and motors. To solve the problem, two rules known as **Kirchhoff's circuit laws** are applied:

1. *(Kirchhoff's current law (KCL))*. The algebraic sum of the currents flowing towards a node is zero,

$$\sum_i I_i = 0. \qquad (4.45)$$

2. *(Kirchhoff's voltage law (KVL))*. The algebraic sum of the electric potential differences around any closed path is zero,

$$\sum_i \Delta V_i = 0.$$

If we bear expression (4.43) in mind for each branch in the mesh, the result is $\Delta V_i = R_i I_i - \sum_i \mathcal{E}_i$, where $\sum_i \mathcal{E}_i$ is the summed emfs, with its sign, from branch i. Therefore it can be written that in a mesh, the *algebraic* sum of electromotive forces is equal to the algebraic sum of the voltage drop in resistances. That is,

$$\sum_i \mathcal{E}_i = \sum_i R_i I_i. \qquad (4.46)$$

KCL is an immediate consequence of the equation of continuity for direct current (4.10), $\text{div}\,\mathbf{j} = 0$, which implies that there are neither sources nor sinks of direct current. If a closed surface, which includes the node, is taken and considering that the equation is integrated for the closed volume, the result is that the total flux through the surface is null. This means that the algebraic sum of all currents leaving and entering the given node is also null. The number of independent equations that can be obtained from the first law is the same as the number of nodes minus one.

KVL is an immediate consequence of the fact that the electric field is conservative, determined by the gradient of the potential. Therefore, its circulation along a closed path is null. If we bear in mind the equations previously obtained with KCL, the application of KVL allows us to obtain the necessary amount of independent equations to reach the number of branches (and therefore, of unknown values) in the network. In other words, the number of equations is the number of branches minus the number of nodes plus one. Bear in mind that when selecting meshes to complete the number of equations, every branch must be picked up at least once. Sign criteria is the same as the one indicated in Fig. 4.7. If the numerical solution of these equations yields a negative value for a particular current, the correct sense of this current is opposite to that assumed.

4.10.2 Mesh Analysis

If the number of conductors in a network is large, the number of equations resulting from Kirchhoff's laws will be so large that solving the problem becomes a cumbersome task. The mesh analysis (also known as mesh current method) is used to solve planar circuits and to find the currents (and indirectly the voltages) at any place in the circuit. By applying this method fewer equations are needed to solve the problem.

This method is based on supposing that one continuous current, with current i_j, is flowing in each mesh j of a circuit. These currents are called *mesh currents*. The current in each element of the circuit is the algebraic sum of all mesh currents flowing through that element. The number of meshes to consider is the same as the one obtained if KVL is applied.

To solve an electric circuit problem by means of mesh analysis, we select a set of meshes such that at least one mesh passes through each branch. One mesh current is assigned to each mesh. The direction chosen for each such current is arbitrary. KVL is applied around each mesh, applying the usual sign criteria (Fig. 4.7): positive sign is assigned to mesh current i_j when we circulate with the same current direction, and negative sign if we circulate in the opposite direction. Then, N equations (as many as meshes) are obtained,

$$\mathcal{E}_i = \sum_{j=1}^{N} R_{ij} i_j . \qquad (4.47)$$

The NxN resistances matrix R_{ij}, which includes generator and motor internal resistances, is a symmetric matrix; an element R_{ii} in the main diagonal is the sum of all the resistances on mesh i; an element R_{ij}, with $i \neq j$, is the sum of the resistances that have in common meshes i and j, with a negative sign if circulating directions are opposite.[8] $\mathcal{E}_i$ and i_j are column vectors. $\mathcal{E}_i$ represents the sum of electromotive and counter-electromotive forces in mesh i, applying the usual sign criteria (Fig. 4.7). Mesh currents i_j, always with a positive sign, are the unknown values to obtain. Currents I_i through each branch are obtained as the algebraic sum of all mesh currents through this branch.

4.11 Passive Network Equivalence

Two-terminal networks are called *terminally equivalent* if the same current flows into both networks when the potential difference between their terminals (or terminal voltages) is equal, and/or if the same electric potential difference appears between their terminals when identical currents are forced into both networks.

Resistance is a property of a material object and it depends on both the nature of the material from which the object is composed and on its geometry. A conducting object

---

[8] If the direction chosen for each mesh current is the same in all meshes, this sign is always negative.

Fig. 4.8 Passive network equivalence

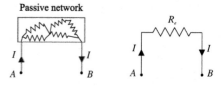

of convenient shape, primarily characterized by its resistance, is called a *resistor*. It is known as a **passive network** (or resistance network) in direct current to a network just formed by resistors.

Let's consider a random passive network (Fig. 4.8). Furthermore, let's suppose that when an electric potential difference $V_A - V_B$ between two of its points is established, a certain current I enters by point A, and by point B it leaves the same one. It is called the **equivalent resistance** of the passive network between two points A and B to a resistance R_e that when the same potential difference $V_A - V_B$ is established between these two points, the same current I flows.

4.11.1 Resistances in Series, Parallel, Triangle and Star Associations

N resistors connected as shown in Fig. 4.9a are said to be connected in **series**. Current is the same in all of them, but the potential difference across each resistor is generally different. If Ohm's law (4.20) is applied to the potential difference between the terminals of the equivalent resistance R_e (Fig. 4.9b), $V_A - V_B = R_e I$. If this value is compared to the one obtained with all the resistances $V_A - V_B = V_A - V_C + \cdots + V_D - V_B = R_1 I + \cdots + R_n I$, the result is that the equivalent resistance R_e for N resistances R_i in series verifies

$$R_e = \sum_{i=1}^{N} R_i, \qquad (4.48)$$

which is larger than any of individual resistance.

N resistors connected as shown in Fig. 4.10a are said to be connected in **parallel**. The potential difference across each resistor is the same, but generally each one with a different current. If Ohm's law (4.20) is applied to determine the current I that goes through an equivalent resistance (Fig. 4.10b), $I = (V_A - V_B)/R_e$. If we do the

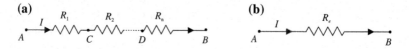

Fig. 4.9 a Resistors connected in series and b their equivalent resistance

4.11 Passive Network Equivalence

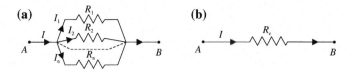

Fig. 4.10 a Resistors connected in parallel and b their equivalent resistance

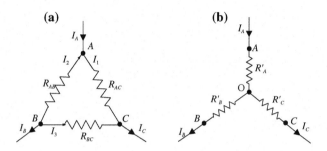

Fig. 4.11 a Triangle and b star associations

same for every current I_i on every branch, $I_i = (V_A - V_B)/R_i$; if these currents I_i are summed, this addition is equal to current I. Therefore equivalent resistance R_e between points A and B for N parallel resistances R_i verifies

$$\frac{1}{R_e} = \sum_{i=1}^{N} \frac{1}{R_i}, \qquad (4.49)$$

which is smaller than any of the individual resistances.

Figure 4.11a shows another way of connecting resistances, and it is known as a **triangle** association. In this case there's not an equivalent resistance to the three of them, but another one called a **star** association does (Fig. 4.11b). Both associations will be equivalent, if the potential differences between its points A, B and C are the same the currents that go through the other circuit branches are the same. Every resistance value in the star association can be determined from the resistances of the associated triangle association by the **Kennelly's theorem**

$$R'_A = \frac{R_{AB} R_{AC}}{R_{AB} + R_{AC} + R_{BC}}, \qquad (4.50)$$

$$R'_B = \frac{R_{AB} R_{BC}}{R_{AB} + R_{AC} + R_{BC}}, \qquad (4.51)$$

$$R'_C = \frac{R_{AC} R_{BC}}{R_{AB} + R_{AC} + R_{BC}}. \qquad (4.52)$$

Practically every resistance in the star is the product of resistances that converge on each node, divided by the sum of resistances in the triangle association.

By solving the system of equations (4.50)–(4.52), each resistance in the triangle association can be written in terms of the resistances of the star association:

$$R_{BC} = \frac{R'_A R'_B + R'_A R'_C + R'_B R'_C}{R'_A}, \tag{4.53}$$

$$R_{AC} = \frac{R'_A R'_B + R'_A R'_C + R'_B R'_C}{R'_B}, \tag{4.54}$$

$$R_{AB} = \frac{R'_A R'_B + R'_A R'_C + R'_B R'_C}{R'_C}. \tag{4.55}$$

Every passive network can be easily reduced to an equivalent resistance by applying previous equivalences.

4.11.2 Resistance in a Conductor with Any Shape

Previous results can be applied to obtain a non linear conductor's equivalent resistance. So let's consider the conductor block with constant thickness (parallelepiped) in Fig. 4.12. As it was supposed in (4.21) where current entered perpendicular to the wire's section, here it can be also considered to enter perpendicularly, through the shaded left face and to leave it perpendicularly through the right one, parallel to the rest of the block faces. This way we could consider a problem similar to the wire. The boundary conditions are therefore that the block has two equipotential surfaces (the shaded faces), perpendicular to current, while the rest of the faces are impermeable, and through them no current passes. So (4.21) can be equally applied, if S is the shaded section area and L the block's length (Fig. 4.12).

If we want to apply all this to any conductor shape, the block in Fig. 4.12 can be divided in both ways as shown in Fig. 4.13. Figure 4.13a shows the block shown as a set of conductors in parallel (they are usually known as tubes of current), since extremes in contact with shaded faces have the same potential and are separated by ideal insulating sheets (light grey in figure). Figure 4.13b shows the block as a set of conductors in series (equipotential slices), all of them traversed by the same current and with separations of infinite conductivity (dark grey in figure).

Fig. 4.12 Conducting block

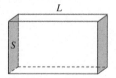

4.11 Passive Network Equivalence

Fig. 4.13 a Tubes of current. **b** Equipotential slices

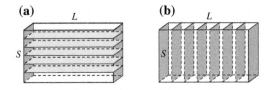

Equivalent resistance from a block formed by n tubes of current is obtained by applying (4.49)

$$\frac{1}{R_e} = \sum_{i=1}^{n} \frac{1}{R_{ti}}, \tag{4.56}$$

where R_{ti} is every tube resistance. Since there are n tubes, each straight section is $S_i = S/n$, the same for all of them, and the result for R_{ti} (4.21):

$$R_{ti} = \frac{\rho L}{S_i} = \frac{n\rho L}{S} = R_t, \tag{4.57}$$

And then (4.56) results

$$\frac{1}{R_e} = \frac{n}{R_t}. \tag{4.58}$$

Equivalent resistance from a block formed by m equipotential slices is obtained by applying (4.48)

$$R_e = \sum_{i=1}^{m} R_{si}, \tag{4.59}$$

where R_{si} is the resistance of each of the slices. Since there are m slices, the length of each one is $L_i = L/m$, the same for all of them, and the result is for R_{si}:

$$R_{si} = \frac{\rho L_i}{S} = \frac{\rho L}{mS} = R_s, \tag{4.60}$$

with (4.59) as

$$R_e = mR_s. \tag{4.61}$$

It is obvious that nothing is earned by the subdivision of a parallelepiped into tubes and slides. The power of this method becomes apparent only if the shape of the conductor is more complicated. Previous expressions allow us to separate a block in any form in different lengths and sections of tubes and slices, so its equivalent resistance can be obtained. Lengths and sections from slices and tubes can be as small as wanted, so they can be adjusted appropriately to the form of the considered block.

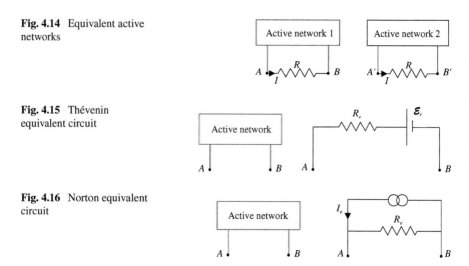

Fig. 4.14 Equivalent active networks

Fig. 4.15 Thévenin equivalent circuit

Fig. 4.16 Norton equivalent circuit

4.12 Thévenin's and Norton's Theorems

We define an **active network** as a network with a direct current that which is formed by resistances, generators and motors. A passive network is obtained by removing generators and motors and substituting them solely with their internal resistances. Two active networks (Fig. 4.14) are equivalent when having equal resistances R between their terminals, the current I that circulates through them is the same for both networks.

Thévenin's theorem: A circuit between two points A and B which has an electric potential difference $V_A - V_B$ and an equivalent passive resistance R_e between these points is equal to an ideal generator with emf $\mathcal{E}_e = V_A - V_B$ and a series resistance R_e (Fig. 4.15).

Norton's theorem: A circuit between two points A and B which has an electric potential difference $V_A - V_B$ and an equivalent passive resistance R_e between these points is equal to an ideal current source[9] with current $I_e = (V_A - V_B)/R_e$ and a parallel resistance R_e (Fig. 4.16).

To explain the previous theorems, let's consider with subscripts T all values in the Thévenin equivalent circuit and with subscripts N all values in the Norton equivalent circuit. If a resistor of value $R = \infty$ (an open-circuit) is connected between terminals A and B (Fig. 4.14) and we measure the potential difference (the voltage) between A and B, the open-circuit voltage V_{oc} is determined. In the Thévenin equivalent circuit, Fig. 4.15, $V_A - V_B = V_{oc} = \mathcal{E}_T$. In the Norton equivalent circuit, Fig. 4.16, $V_A - V_B = V_{oc} = I_N R_N$. Thus,

---

[9] A current source (ideal) is a device that supplies constant current, regardless of what is connected to its terminals. It is represented by –∞–.

4.12 Thévenin's and Norton's Theorems

$$V_{oc} = \mathcal{E}_T = I_N R_N.$$

If the value of external resistance in Fig. 4.14 is changed to $R = 0$, (a short-circuit), the current I that flows out of the network is the short-circuit current I_{sc}. In Fig. 4.15, $I_{sc} = \mathcal{E}_T/R_T$. In Fig. 4.16, $I_{sc} = I_N$. Thus,

$$I_N = \frac{\mathcal{E}_T}{R_T},$$

and therefore

$$R_T = R_N.$$

It should be noted that the value of R_T (or R_N) can be observed from the terminals if and only if the internal voltage source in the Thévenin circuit (or the current source in the Norton circuit) is set to zero. In other words, its value is the equivalent resistance R_e of the passive network.

Summarizing,

- The voltage source in the Thévenin equivalent circuit is the open-circuit voltage.
- The current source in the Norton equivalent circuit is the short-circuit current.
- The series resistor in the Thévenin equivalent is identical to the parallel resistor in the Norton equivalent ($R_T = R_N$), and its value is the equivalent resistance R_e of the passive network.
- The previous values are interrelated by Ohm's law: $\mathcal{E}_T = I_N R_T$.

Solved Problems

Problems A

4.1 Let's consider two cylindrical threads with length 4 m and radius 4 mm. The first thread is made of tungsten (the first half) and aluminum (the other half). The other thread is made by surrounding a 2 mm radius tungsten cylinder with a cylindrical aluminum layer, also with 2 mm thickness. Each thread is subjected to a potential difference between their extremes of 1.5 V. Find the current that goes through each line. **Conductivities**: tungsten $1.81 \times 10^7 \, \Omega^{-1}\mathrm{m}^{-1}$; aluminum $3.77 \times 10^7 \, \Omega^{-1}\mathrm{m}^{-1}$.

Solution

The first of the wires can be considered as a resistance R_1 resulting from joining two resistors in series, one with resistance R_{11} constituted by a 2 m length and 4 mm radius tungsten wire, and the other one with resistance R_{12} constituted by a 2 m length and 4 mm radius aluminum wire. Thread and its equivalence are shown in Fig. 4.17a.

The second wire can be considered as a resistance R_2 resulting from joining two parallel resistors, one with resistance R_{21} constituted by a 4 m length and 2 mm radius

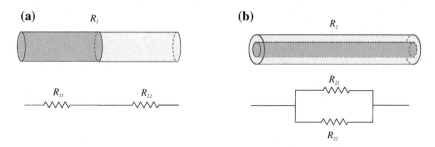

Fig. 4.17 Cylindrical wires: **a** series tungsten (*dark grey*) and aluminum (*light grey*) resistor; **b** parallel tungsten and aluminum resistor

tungsten wire, and the other one with resistance R_{22} constituted by a 4 m length, inner radius 2 mm and outer one 4 mm, aluminum wire. Thread and its equivalence are shown in Fig. 4.17b.

If we bear in mind that conductivity and section do not vary for every resistor, each one's value can be obtained by applying (4.21),

$$R_{ij} = \frac{\rho_j L_{ij}}{S_{ij}}.$$

Resistivities ρ_j can be obtained from (4.15)

$$\rho_j = \frac{1}{\sigma_j},$$

where σ_j are conductivity data given for the problem. So, for tungsten,

$$\rho_1 = \frac{1}{\sigma_1} = \frac{1}{1.81 \times 10^7} = 5.52 \times 10^{-8}\,\Omega\mathrm{m}.$$

For aluminum,

$$\rho_2 = \frac{1}{\sigma_2} = \frac{1}{3.77 \times 10^7} = 2.65 \times 10^{-8}\,\Omega\mathrm{m}.$$

with these values all resistances can be obtained:

$$R_{11} = \frac{\rho_1 L_{11}}{S_{11}} = \frac{\rho_1 L_{11}}{\pi r^2} = \frac{5.52 \times 10^{-8} \cdot 2}{\pi (4 \times 10^{-3})^2} = 2.20\,\mathrm{m}\Omega,$$

$$R_{12} = \frac{\rho_2 L_{12}}{S_{12}} = \frac{\rho_2 L_{12}}{\pi r^2} = \frac{2.65 \times 10^{-8} \cdot 2}{\pi (4 \times 10^{-3})^2} = 1.05\,\mathrm{m}\Omega,$$

$$R_{21} = \frac{\rho_1 L_{21}}{S_{21}} = \frac{\rho_1 L_{21}}{\pi r_{\mathrm{int}}^2} = \frac{5.52 \times 10^{-8} \cdot 4}{\pi (2 \times 10^{-3})^2} = 17.57\,\mathrm{m}\Omega,$$

$$R_{22} = \frac{\rho_2 L_{22}}{S_{22}} = \frac{\rho_2 L_{22}}{\pi[r_{ext}^2 - r_{int}^2]} = \frac{2.65 \times 10^{-8} \cdot 4}{\pi[(4 \times 10^{-3})^2 - (2 \times 10^{-3})^2]} = 2.81\,\text{m}\Omega.$$

For the first of the wires, since resistors are in series, the result is,

$$R_1 = R_{11} + R_{12} = 2.20 + 1.05 = 3.25\,\text{m}\Omega.$$

For the second one, since resistors are in parallel, the result is,

$$\frac{1}{R_2} = \frac{1}{R_{21}} + \frac{1}{R_{22}} = \frac{1}{17.57} + \frac{1}{2.81} = 0.41 \Rightarrow R_2 = 2.42\,\text{m}\Omega.$$

Current is obtained by applying Ohm's law (4.20), $V_1 - V_2 = RI \Rightarrow I = (V_1 - V_2)/R$, for every resistance, where 1 and 2 are each of the studied wire extremes. For the first wire,

$$I_1 = (V_1 - V_2)/R_1 = \frac{1.5}{3.25 \times 10^{-3}} = 461.5\,\text{A}.$$

For the second one,

$$I_2 = (V_1 - V_2)/R_2 = \frac{1.5}{2.42 \times 10^{-3}} = 619.8\,\text{A}.$$

4.2 If any of three equal bulbs are connected to a 220 V voltage source, it absorbs 100 W. How should these three bulbs be connected, so that the power dissipated by the three bulbs is the highest? (It is assumed that temperature variation over bulb's resistance value is negligible.)

Solution

Power dissipated by each bulb individually allows us to determine its resistance. This is the only parameter that does not depend on how we connect them (with the condition of discounting temperature's effect). Due to this fact, (4.28) is used, and if in it we replace (4.20) (Ohm's law), the result is

$$P = RI^2 = V^2/R \Rightarrow R_1 = R_2 = R_3 = R = V^2/P = \frac{220^2}{100} = 484\,\Omega.$$

The four different interconnections of bulbs linked to a 220 V source are shown in Fig. 4.18. To determine the power dissipated by them, the current through each bulb should be calculated for every case.

In series interconnection (Fig. 4.18a), the current, since it is direct, is the same on the whole circuit,

$$V = R_e I \Rightarrow I = \frac{V}{R_e} = \frac{V}{R_1 + R_2 + R_3} = \frac{220}{3 \cdot 484} = 0.15\,\text{A}.$$

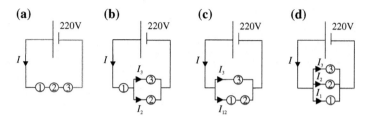

Fig. 4.18 Three equal bulbs interconnection. **a** Three of them in series. **b** 1 in series with 2 and 3 in parallel. **c** 1 and 2 in series with 3 in parallel. **d** Three of them in parallel

Note that since resistors are in series, equivalent resistance is the sum of the individual resistances.

Power dissipated by each bulb is obtained from (4.28) (Joule's effect losses),

$$P_1 = P_2 = P_3 = RI^2 = 484 \cdot 0.15^2 = 11.11\,\text{W}.$$

Total power is the sum of the powers dissipated by each bulb,

$$P = P_1 + P_2 + P_3 = 33.33\,\text{W}.$$

The same result can be obtained by using the equivalence resistance:

$$P = \frac{V^2}{R_e} = \frac{220^2}{3 \cdot 484} = 33.33\,\text{W}.$$

In Fig. 4.18b interconnection, the current I through the circuit (and, therefore, through the bulb 1) is firstly determined. To do this, equivalent resistance from 2 and 3 is calculated,

$$\frac{1}{R_{23}} = \frac{1}{R_2} + \frac{1}{R_3} = \frac{2}{R} \Rightarrow R_{23} = \frac{R}{2},$$

in series with bulb 1 resistance. The current through the circuthe result is:

$$I = \frac{V}{R_e} = \frac{V}{R_1 + R_{23}} = \frac{V}{R + R/2} = \frac{220}{1.5 \cdot 484} = 0.30\,\text{A}.$$

To obtain the current through bulbs 2 and 3, KCL is applied. Since bulbs are equal, the current through both is the same:

$$I = I_2 + I_3 = 2I_2 \Rightarrow I_2 = I_3 = I/2 = 0.15\,\text{A}.$$

So power dissipated by each bulb is:

$$P_1 = RI^2 = 484 \cdot 0.30^2 = 44.44\,\text{W} \quad P_2 = P_3 = RI_2^2 = RI_3^2 = 484 \cdot 0.15^2 = 11.11\,\text{W}.$$

Total power, as a sum of individual powers, is

$$P = P_1 + P_2 + P_3 = 66.66 \, \text{W}.$$

The same result can be obtained by using the equivalence resistance:

$$P = \frac{V^2}{R_e} = \frac{220^2}{1.5 \cdot 484} = 66.67 \, \text{W}.$$

For interconnection in Fig. 4.18c, we can obtain the current through the bulbs by applying Ohm's law (4.20) to every branch, due to the fact that each one has the same potential difference of 220 V:

$$V = (R_1 + R_2)I_{12} = R_3 I_3 \Rightarrow I_{12} = \frac{V}{R_1 + R_2} = \frac{220}{2R} = 0.23 \, \text{A},$$

$$I_3 = \frac{V}{R_3} = \frac{220}{R} = 0.45 \, \text{A}.$$

Power dissipated by each bulb is:

$$P_1 = P_2 = RI_{12}^2 = 484 \cdot 0.23^2 = 25 \, \text{W} \quad P_3 = RI_3^2 = 484 \cdot 0.45^2 = 100 \, \text{W}.$$

Total power, which is the sum of the three individual powers, is

$$P = P_1 + P_2 + P_3 = 150 \, \text{W}.$$

The same result can be obtained by using the equivalence resistance

$$\frac{1}{R_e} = \frac{1}{R_1 + R_2} + \frac{1}{R_3} = \frac{3}{2R} \Rightarrow R_e = \frac{2R}{3} = 322.67 \, \Omega.$$

Power dissipated is

$$P = \frac{V^2}{R_e} = \frac{220^2}{322.67} = 150 \, \text{W}.$$

In parallel interconnection (Fig. 4.18d), we can obtain the current through the bulbs by applying Ohm's law to every branch, since each one has the same potential difference of 220 V:

$$V = R_1 I_1 = R_2 I_2 = R_3 I_3 \Rightarrow I_1 = I_2 = I_3 = \frac{V}{R} = \frac{220}{484} = 0.45 \, \text{A}.$$

Power dissipated by each bulb is:

$$P_1 = P_2 = P_3 = RI_1^2 = 484 \cdot 0.45^2 = 100 \, \text{W},$$

and total power
$$P = P_1 + P_2 + P_3 = 300\,\text{W}.$$

The same result can be obtained by using the equivalence resistance
$$\frac{1}{R_e} = \frac{1}{R_1} + \frac{1}{R_2} + \frac{1}{R_3} = \frac{3}{R} \Rightarrow R_e = \frac{R}{3} = 161.33\,\Omega.$$

Power dissipated is
$$P = \frac{V^2}{R_e} = \frac{220^2}{161.33} = 300\,\text{W}.$$

From these results it can be observed how the three bulbs in parallel interconnection is the one that dissipates the highest power value: the voltage source is the same in all the connections but the equivalent resistance has the lowest value in parallel interconnection.

4.3 A generator with emf $\mathcal{E}$ supplies current to a motor with cemf $\mathcal{E}' = 100$ V. Both of them have the same internal resistance r. Determine the generator's emf if the power converted by the motor into mechanical power is half the power supplied to the network by the generator.

Solution

Figure 4.19 represents the circuit, with a generator of emf $\mathcal{E}$ that supplies current to a motor with cemf $\mathcal{E}'$. Direct current circulates following the indicated sense: it exits the generator and enters the motor via the positive pole. Let's apply Ohm's law (4.20) to determine the current that goes through the circuit:

$$\mathcal{E} - \mathcal{E}' = 2rI \Rightarrow I = \frac{\mathcal{E} - \mathcal{E}'}{2r}.$$

If we bear in mind (4.34) and (4.39), which give us power supplied by the generator, and the one converted by the motor, it is obtained

$$\mathcal{E}'I = 0.5\left(\mathcal{E}I - rI^2\right) \Rightarrow \mathcal{E} - rI = 2\mathcal{E}'.$$

Fig. 4.19 Direct current generator supplying current to a motor

Substituting the current's I value, emf is obtained:

$$\mathcal{E} - r\frac{\mathcal{E} - \mathcal{E}'}{2r} = 2\mathcal{E}' \Rightarrow \mathcal{E} = 3\mathcal{E}' \Rightarrow \mathcal{E} = 300\,\text{V}.$$

4.4 In Fig. 4.20, determine the voltage of point A, the power supplied by generators to the circuit and total losses due to Joule's effect.

Solution

At least one generator is needed in a circuit, therefore both devices behave as generators. Since current exits the generator via its positive pole, the current direction must be the one indicated in Fig. 4.21.

To calculate voltage in A, potential difference between A and B is calculated, bearing in mind that, as B is grounded, it is considered to be at zero potential. If Expression (4.43) is applied, and if we circulate from A to B following the current direction, the result is:

$$V_A - V_B + \sum_i \mathcal{E}_i = \sum_i R_i I \Rightarrow$$

$$\Rightarrow V_A - V_B + 110 = (40 + 5)I \Rightarrow V_A = 45I - 110.$$

Note that 110 V emf is introduced with positive sign, because if we circulate following the mentioned direction we firstly pass through its negative pole. The current I is positive because the direction of circulation is the same as the one of the current. Every resistance is taken into account, including internal resistances of the generators between A and B.

Expression (4.41) (Ohm's law for a circuit) is applied to determine current I through the circuit. If we circulate around a closed path following the same direction of current, the result is:

$$\sum_i \mathcal{E}_i = I \sum_i R_i \Rightarrow 220 + 110 = (40 + 5 + 25 + 50 + 5 + 40)I \Rightarrow I = 2\,\text{A}.$$

And substituting in voltage's expression:

Fig. 4.20 Circuit of Problem 4.4

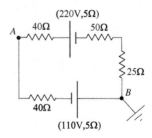

Fig. 4.21 Circuit of Problem 4.4, with current direction

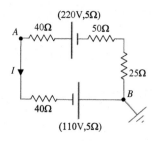

$$V_A = 45I - 110 = -20\,\text{V}.$$

We could check the result by calculating the same potential difference, but this time from A to B by the opposite direction as the current. The result is

$$V_A - V_B - 220 = -(40 + 5 + 50 + 25)I \Rightarrow V_A = 220 - 120I = -20\,\text{V}.$$

Emf has been introduced with a negative sign because we firstly pass through the positive pole. I is introduced with a negative sign, because it is opposite to the direction of circulation. The result is the same as the one obtained before.

To determine the power supplied by generators, Expression (4.34) is applied to both generators in the network. For the 220 V generator:

$$P_{\text{sup}} = \mathcal{E}I - rI^2 = 220 \cdot 2 - 5 \cdot 2^2 = 420\,\text{W}.$$

For the 110 V generator:

$$P_{\text{sup}} = \mathcal{E}I - rI^2 = 110 \cdot 2 - 5 \cdot 2^2 = 200\,\text{W}.$$

The losses due to Joule's effect in the circuit are obtained if (4.28) is applied to every resistance in the network, including the generators' internal ones:

$$P = \sum_i R_i I^2 = (40 + 25 + 50 + 40) \cdot 2^2 + (5 + 5) \cdot 2^2 = 620 + 40 = 660\,\text{W}.$$

Internal resistance losses have been separated from the rest of the ones in the circuit, so it can be noted that the power supplied by generators is the same as the losses of the network's resistances.

4.5 A generator with emf $\mathcal{E}$ and internal resistance r, connected to a resistance line R, is shown in Fig. 4.22. We want to connect to poles a and b a device, with emf $\mathcal{E}'$ and internal resistance r', that is able to work as a generator or as a motor. Determine in which case (generator or motor) losses in the circuit, due to Joule's effect, are the highest.

Fig. 4.22 Generator and supplying line of Problem 4.5

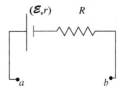

Solution

Any device connected between a and b results in a current leaving the generator by its positive pole. Losses due to Joule's effect in the circuit (power dissipated by resistances) are given by (4.28):

$$P = (R+r+r')I^2.$$

where r' is the resistance of the connected device. Since resistances are the same no matter how the element is connected, losses will be higher as the current I increases.

In Fig. 4.23 current enters by the negative pole of the device with emf $\mathcal{E}'$, that works as a generator. To determine current I through the circuit, (4.41) (Ohm's law for a circuit) is applied, circulating with current direction:

$$\sum_i \mathcal{E}_i = I \sum_i R_i \Rightarrow \mathcal{E} + \mathcal{E}' = (R+r+r')I,$$

$$I = \frac{\mathcal{E}+\mathcal{E}'}{R+r+r'}.$$

Note that emfs are both positive, because when circulation is done, we pass first through the negative poles of the devices. Substituting in the previous expression of power, it is obtained, in this case:

$$P = (R+r+r')\left(\frac{\mathcal{E}+\mathcal{E}'}{R+r+r'}\right)^2 = \frac{(\mathcal{E}+\mathcal{E}')^2}{R+r+r'}.$$

In the other case, Fig. 4.24, current enters through the positive pole of the device with emf $\mathcal{E}'$, therefore it works as a motor. The new current I is obtained applying (4.41). If we circulate in the direction of the current, the result is

Fig. 4.23 Circuit of Problem 4.5, with the device working as a generator

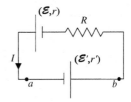

Fig. 4.24 Circuit of Problem 4.5, with the device working as a motor

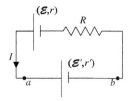

$$\sum_i \mathcal{E}_i = I \sum_i R_i \Rightarrow \mathcal{E} - \mathcal{E}' = (R + r + r')I,$$

$$I = \frac{\mathcal{E} - \mathcal{E}'}{R + r + r'}.$$

Note the negative sign of emf $\mathcal{E}'$, due to the fact that we firstly pass through its positive pole. And if we substitute it in previous expression of power, losses in this case are obtained:

$$P = (R + r + r') \left(\frac{\mathcal{E} - \mathcal{E}'}{R + r + r'} \right)^2 = \frac{(\mathcal{E} - \mathcal{E}')^2}{R + r + r'},$$

lower than the one previously calculated.

As it can be observed, when the device is connected as a generator, current is higher, therefore losses due to Joule's effect are higher.

Problems B

4.6 A generator with emf $\mathcal{E}$ and internal resistance r supplies current to a set of two resistors in parallel, each one with value R. Determine value R, in order to obtain the maximum power dissipated.

Solution

Figure 4.25a represents the circuit's scheme. The circuit is firstly simplified by obtaining the equivalent resistance of the two parallel resistances (Fig. 4.25b):

$$\frac{1}{R_e} = \frac{1}{R} + \frac{1}{R} = \frac{2}{R} \Rightarrow R_e = R/2$$

Power dissipated by resistance R_e is obtained from (4.28),

$$P = I^2 R_e = I^2 R/2.$$

It is necessary to determine the current I through the circuit, which is the same that flows through resistance R_e. If (4.41) is applied (Ohm's law for a circuit), the result is:

Fig. 4.25 a Generator supplying current to two resistors in parallel. b Equivalent circuit

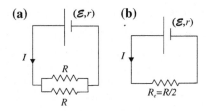

$$\sum_i \mathcal{E}_i = I \sum_i R_i \Rightarrow \mathcal{E} = I(r + R_e) \Rightarrow I = \frac{\mathcal{E}}{r + R_e}.$$

Note that in $\sum_i R_i$, the generator's internal resistance r and $R_e = R/2$ are included. If this current value is substituted in the power's expression, the power dissipated by resistance R is obtained,

$$P = \frac{\mathcal{E}^2 R_e}{(r + R_e)^2}.$$

This power should be maximum. Therefore the derivative of P in regards to the variable (in this case R_e) is computed and equaled zero:

$$\frac{dP}{dR_e} = \frac{\mathcal{E}^2 (R_e + r)^2 - 2(R_e + r)\mathcal{E}^2 R_e}{(R_e + r)^4} = 0.$$

Solving this equation,

$$(R_e + r)^2 = 2(R_e + r)R_e \Rightarrow R_e + r = 2R_e,$$

and then

$$R_e = r \Rightarrow R = 2r.$$

is the resistance value.

From the result it might be deduced that we should not use higher resistance to increase the losses by Joule's effect (for example, to generate more heat). Ohm's law explains it: if the circuit's resistance increases, current through it decreases. This affects losses due to Joule's effect. It must be observed that equivalent resistance of the network should have the same value as the generator's internal resistance.

4.7 A direct current line is 1 km long and has a resistance of 0.2 Ω. This line, due to a bad isolation, has a leakage current to earth ground, so current at the entrance is 53 A and at the end 45 A. Voltage at the entrance is 230 V, and at the end 220 V. Determine where the breakdown has occurred and the value of the leakage resistance to ground.

Solution

Figure 4.26 represents the problem's line. Resistance R_f represents the leakage resistance to ground at the breakdown point A. If the line is $L = 1000$ m long, point A splits the line in two stretches $A1$ and $A2$ with respective lengths of x and $L - x$, where x is the value to be determined. Point O (ground) is at 0 V. Entering voltage, electric potential difference between point 1 and ground is $V_1 = 230$ V. Ending voltage, potential difference between point 2 and ground is $V_2 = 220$ V. Through the first stretch, current is $I_{1A} = 53$ A, and through the second one $I_{A2} = 45$ A. Leakage current through resistance R_f is easily determined, applying KCL (4.45) at point A:

$$I_{1A} = I_{A2} + I_f \Rightarrow I_f = I_{1A} - I_{A2} = 8 \text{ A}.$$

To determine resistance on every stretch, we apply Ohm's law between point 1 and ground (point O), and the result is

$$V_1 - 0 = R_{1A} I_{1A} + R_f I_f,$$

and between point 2 and ground,

$$V_2 - 0 = -R_{A2} I_{A2} + R_f I_f,$$

because in this case current I_{A2} direction is opposite to the chosen direction for the circulation (from point 2 to ground). Total resistance of the line $R = 0.2 \, \Omega$ is the sum of both resistance R_{1A} and R_{A2} of both stretches, so $R_{A2} = R - R_{1A}$. Substituting into the second equation,

$$V_2 - 0 = (R_{1A} - R) I_{A2} + R_f I_f,$$

and substracting both equations, we obtain

$$V_1 - V_2 = R_{1A}(I_{1A} - I_{A2}) + R I_{A2} \Rightarrow R_{1A} = \frac{V_1 - V_2 - R I_{A2}}{I_{1A} - I_{A2}}$$

$$= \frac{230 - 220 - 0.2 \cdot 45}{53 - 45} = 0.125 \, \Omega.$$

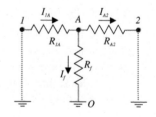

Fig. 4.26 Leakage current to ground

To calculate length x of stretch 1A, (4.21) (resistance of a wire with constant resistivity and cross-section) is applied to the resistance of the complete line and to the one of stretch 1A, whose length x is the unknown value in the problem:

$$R = \rho \frac{L}{S}; \quad R_{1A} = \rho \frac{x}{S},$$

where ρ is conductor's resistivity and S the section's line.

If both expressions are divided, the value of x is obtained

$$\frac{R_{1A}}{R} = \frac{x}{L} \Rightarrow x = L\frac{R_{1A}}{R} = 1000 \frac{0.125}{0.2} = 625 \text{ m}.$$

So the breakdown is 625 m away from the beginning point.

The leakage resistance value can be determined by finding the value in any of previous expressions where Ohm's law is applied, and the result is

$$V_1 - 0 = R_{1A}I_{1A} + R_f I_f \Rightarrow R_f = \frac{V_1 - R_{1A}I_{1A}}{I_f} = \frac{230 - 0.125 \times 53}{8} = 27.9\,\Omega.$$

4.8 A generator with emf $\mathcal{E}$ and internal resistance $r = 0.2\,\Omega$ supplies current through a line with resistance $R = 0.4\,\Omega$ to a lighting installation. It is constituted by 10 lamps, each one consuming 550 W at a voltage of 220 V. Determine: (a) Equivalent resistance of the lighting installation. (b) Power supplied by the generator. (c) Line's efficiency. (d) Generator's emf.

Solution

Figure 4.27 represents the network's scheme of the problem.

(a) If (4.27) is applied to each lamp, current I_i through each of them can be determined

$$P_i = (V_A - V_B)I_i \Rightarrow 550 = 220 I_i,$$

$$I_i = 2.5\,A,$$

which will be the same for all the lamps, since they are equal. The total current through the circuit is obtained by applying KCL, (4.45),

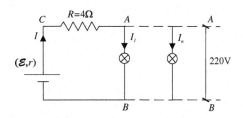

Fig. 4.27 Direct current generator supplying a lighting installation

$$I = \sum_i I_i = 10 I_i = 25 \text{ A}.$$

The circuit equivalent resistance is obtained by applying Ohm's law (4.20):

$$V_A - V_B = R_e I \Rightarrow 220 = 25 R_e \Rightarrow R_e = 8.8 \, \Omega.$$

(b) The electric potential difference (voltage drop) at the generator terminals, $(V_C - V_B)$, is the sum of voltage drops at the lamps and the voltage drop in the line. If Ohm's law is applied:

$$V_C - V_B = (V_C - V_A) + (V_A - V_B) = 0.4 I + 220 = 230 \text{ V}.$$

The power supplied by the generator is obtained by applying (4.34):

$$P_{\text{sup}} = (V_C - V_B) I = 230 \cdot 25 = 5750 \text{ W}.$$

(c) Since the power consumed by each lamp is 550 W, the total power consumed by the installation is

$$P = 10 P_i = 5500 \text{ W}.$$

Efficiency η of the line is obtained by dividing power consumed by the installation between power supplied by it:

$$\eta = \frac{5500}{5750} = 0.96.$$

(d) If (4.33) is applied, the generator's emf can be obtained:

$$V_C - V_B = \mathcal{E} - rI \Rightarrow \mathcal{E} = 230 + 0.2 \cdot 25 = 235 \text{ V}.$$

4.9 The potential difference measured by voltmeter V between terminals C and D (Fig. 4.28), is supplied to motor M and resistance R. Its value is 380 V. The current measured by the ammeter A is 6 A. The resistance value is 2 kΩ and the motor has a cemf of 300 V, and an internal resistance with unknown value. Determine this resistance and study the power in the branches. (Note: ammeter

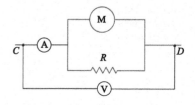

Fig. 4.28 Circuit of Problem 4.9

Solution

and voltmeter are supposed to be ideal: zero resistance for the ammeter and infinite resistance for the voltmeter.)

Solution

Let's suppose that current I measured by the ammeter splits in two currents, as shown in Fig. 4.29. Through the voltmeter no current circulates, because it has infinite resistance. If Ohm's law (4.20) is applied to the lower branch, the result is

$$V_C - V_D = RI_1 \Rightarrow I_1 = \frac{V_C - V_D}{R} = \frac{380}{2000} = 0.19 \text{ A}.$$

If KCL (4.45) is applied, current I_2 is obtained:

$$I = I_1 + I_2 \Rightarrow I_2 = I - I_1 = 6 - 0.19 = 5.81 \text{ A}.$$

If Ohm's law for a branch (4.43) is applied to the upper branch, the internal resistance r is obtained:

$$V_C - V_D - 300 = rI_2 \Rightarrow r = \frac{(V_C - V_D) - 300}{I_2} = \frac{380 - 300}{5.81} = 13.77 \, \Omega.$$

Note that the motor's cemf of 300 V is introduced in the expression with negative sign because if we circulate from C to D we firstly pass through the motor's positive pole.

Power converted by the motor is obtained by applying (4.39):

$$P_{\text{conv}} = \mathcal{E}I_2 = 300 \cdot 5.81 = 1743 \text{ W}.$$

Power extracted (or consumed) from the circuit by the motor is obtained by applying (4.40):

$$P_{\text{cons}} = (V_C - V_D)I_2 = \mathcal{E}I_2 + rI_2^2 = 380 \cdot 5.81 = 2207.8 \text{ W}.$$

Power dissipated by Joule's effect, in resistance R and in the motor's internal resistance r, is obtained by applying (4.28):

$$P_{\text{Joule}} = RI_1^2 + rI_2^2 = 2000 \cdot 0.19^2 + 13.77 \cdot 5.81^2 = 537 \text{ W}.$$

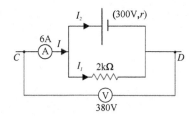

Fig. 4.29 Currents of Problem 4.9

4.10 In the circuit shown in Fig. 4.30, devices have electromotive forces $\mathcal{E}_1 = 24\,V$, $\mathcal{E}_2 = 6\,V$, $\mathcal{E}_3 = 12\,V$, and internal resistances $r = 1\,\Omega$. Calculate: (a) If switch C is open, the potential difference between A and B. If switch C is closed, (b.1) the potential difference between A and B; (b.2) total power in each element.

Solution

(a) When switch C is open, closed paths cannot be found. Therefore, the current is null through every branch. To determine the potential difference between A and B, (4.43) (Ohm's law for a branch) is applied. Circulating from A to B through the lower branch, we obtain

$$V_A - V_B + \sum_i \mathcal{E}_i = \sum_i R_i I \Rightarrow V_A - V_B - 12 - 6 = 0 \Rightarrow V_A - V_B = 18\,V.$$

Emfs are introduced with negative sign, because circulating from A to B, we firstly go through the positive poles.

(b.1) When switch C is closed (Fig. 4.31), we find a unique closed path in the network, formed by its external lines. To calculate the potential difference between A and B using (4.43), we need to determine the currents through the circuit branches. Through the branch with generator of emf $\mathcal{E}_2$ there's no current, because the branch is open. To determine the current through the external circuit, we suppose a current direction (showed in Fig. 4.31), and we apply Ohm's law for a circuit, (4.41):

$$\sum_i \mathcal{E}_i = I \sum_i R_i \Rightarrow 24 - 12 = (8 + 1 + 1 + 20)I \Rightarrow I = 0.4\,A,$$

including in $\sum_i R_i$ every resistance of the resistors in the external circuit, also including generator internal resistances. Circulation has been done counterclockwise, therefore current is positive (the same direction as the circulation), and the emf of device 1 has a positive sign (we firstly go through the negative pole), and a negative sign for $\mathcal{E}_3$ (we firstly go through the positive one).

Once the current is obtained through every branch in the circuit, (4.43) can be applied to obtain the potential difference between A and B. Circulation is done through the lower branch, bearing in mind the sign criteria explained previously:

Fig. 4.30 Circuit of Problem 4.10

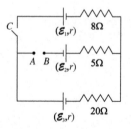

Fig. 4.31 Circuit of Problem 4.10. Switch C is closed

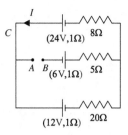

$$V_A - V_B + \sum_i \mathcal{E}_i = \sum_i R_i I \Rightarrow V_A - V_B - 12 - 6 = (1 + 20) \cdot 0.4.$$

Hence
$$V_A - V_B = 26.4 \, \text{V}.$$

If we do the calculation through the upper branch, we reach the same result:

$$V_A - V_B - 24 - 6 = (1 + 8) \cdot (-0.4) \Rightarrow V_A - V_B = 26.4 \, \text{V}.$$

Note the current's negative sign, because the direction of the circulation is opposite to the current.

(b.2) Since no current passes through the device whose emf is $\mathcal{E}_2 = 6\,\text{V}$, power supplied or consumed by it is zero. The current exits the device with emf $\mathcal{E}_1$ via the positive pole, so it behaves as a generator. The current enters the device with emf $\mathcal{E}_3$ via the positive pole, therefore it behaves as a motor.

If (4.34) is applied to the generator we obtain the power supplied by it:

$$P_{1\text{sup}} = \mathcal{E}_1 I - rI^2 = 24 \cdot 0.4 - 1 \cdot 0.4^2 = 9.44 \, \text{W},$$

where $\mathcal{E}_1 I = 24 \cdot 0.4 = 9.6\,\text{W}$ is the power generated by it, and $rI^2 = 1 \cdot (0.4)^2 = 0.16\,\text{W}$ are Joule's effect losses.

If (4.40) is applied to the motor, the power consumed by it is obtained:

$$P_{3\text{cons}} = \mathcal{E}_3 I + rI^2 = 12 \cdot 0.4 + 1 \cdot 0.4^2 = 4.96 \, \text{W},$$

where $\mathcal{E}_3 I = 12 \cdot 0.4 = 4.8\,\text{W}$ is the power converted by the motor and $rI^2 = 1 \cdot (0.4)^2 = 0.16\,\text{W}$ the loss due to Joule's effect.

If we calculate losses due to the resistors in the circuit by applying (4.28) (without bearing in mind internal resistances), and taking into account that no current passes through resistance $5\,\Omega$, the result is:

$$P = \sum_i R_i I^2 = (8+20) \cdot 0.4^2 = 4.48\,\text{W},$$

equal to the difference between the power supplied by generator 1 and the power consumed by generator 3.

4.11 For the network in Fig. 4.32, find the value that variable resistance R_x has if no current goes through resistance R. In this case, determine the potential difference between points a and b in the figure, as well as power supplied by the generator or generators to the network.

Solution

In Fig. 4.33 the three nodes in the network have been labeled. But since the branch connected between nodes 2 and 3 is open and its current is null, it can be considered only a branch through node 3, whose current is I_1. Therefore, the network has only two nodes, *1* and *2*. In Fig. 4.33, the direction chosen for each current is arbitrary. Note that no direction has been assigned to the open branch, because the current is null.

If KCL (4.45) is applied to any of them (for example *1*), the result is:

$$\sum_i I_i = 0 \Rightarrow I_1 = I_2 + I_3.$$

Since the problem's statement shows that current through resistance R is null, the previous equation becomes as follows:

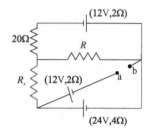

Fig. 4.32 Network of Problem 4.11

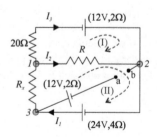

Fig. 4.33 Network of Problem 4.11 with the directions of currents and circulations

$$I_2 = 0 \quad I_1 = I_3.$$

Considering these results, and applying KVL (4.46) to specified meshes in the figure, the result is:

Mesh (I): $12 = 22I_3 - RI_2 \Rightarrow I_3 = I_1 = \frac{6}{11}$ A.
Mesh (II): $24 = (R_x + 4)I_1 + RI_2 \Rightarrow 24 = (R_x + 4)\frac{6}{11} \Rightarrow R_x = 40\,\Omega$.

To calculate the electric potential difference between a and b, we apply Ohm's law for a branch (4.43). If we circulate from a to b by the lower branch, and taking into account the sign criteria in Fig. 4.7, the result is:

$$V_a - V_b + 12 - 24 = -4I_1 \Rightarrow V_a - V_b = 9.82\,\text{V}.$$

Devices of 12 V (in upper branch) and 24 V behave as generators, since the current exits both of them via the positive pole. Power supplied by them is obtained by applying (4.34):

$$P_{\text{sup}} = \mathcal{E}I - rI^2 = (24I_1 - 4I_1^2) + (12I_3 - 2I_3^2) = 17.85\,\text{W}.$$

4.12 Two generators with the same electromotive force $\mathcal{E}$ and internal resistance r, are connected in parallel, as shown in Fig. 4.34. A passive network is connected to terminals A and B, whose equivalent resistance is R_e. Discuss how the generators should be connected in order to obtain the maximum power dissipated by the passive network.

Solution

Generator terminals can be connected by joining the same sign poles (Fig. 4.35) or by joining the positive sign of one with the negative one of the other (Fig. 4.36).

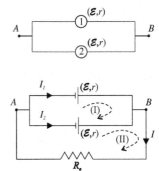

Fig. 4.34 Generators of Problem 4.12

Fig. 4.35 Generators connected by joining the poles with the same sign

Fig. 4.36 Generators connected by joining opposite sign poles

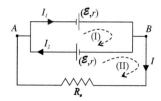

If Ohm's law (4.20) is applied to the ends of the equivalent resistance R_e, and the I direction is taken into account, the result for both cases is:

$$V_B - V_A = R_e I \rightarrow I = (V_B - V_A)/R_e. \tag{4.62}$$

The power consumed by resistance R_e, according to (4.28), is

$$P = R_e I^2 = R_e[(V_B - V_A)/R_e]^2 = (V_B - V_A)^2/R_e.$$

From the previous expression, since R_e is a constant value, it follows that the highest value of $(V_B - V_A)$ makes the maximum power. Thus $(V_B - V_A)$ should be determined for both settings.

Considering the interconnection in Fig. 4.35 and applying KCL (4.45), the result is

$$I = I_1 + I_2.$$

Applying KVL (4.46), with the directions for the circulations shown in Fig. 4.35, and taking into account the previous node's equation, the result is:

Mesh (I): $\quad \mathcal{E} - \mathcal{E} = rI_1 - rI_2 \quad \Rightarrow I_1 = I_2 = I/2$.

Mesh (II): $\mathcal{E} = R_e I + rI_2 = (R_e + r/2)I \Rightarrow I = \dfrac{\mathcal{E}}{R_e + r/2}$.

If this result for I is compared to the one obtained in (4.62), the result is:

$$I = \frac{\mathcal{E}}{R_e + r/2} = \frac{V_B - V_A}{R_e} \Rightarrow V_B - V_A = \frac{\mathcal{E} R_e}{R_e + r/2}.$$

The same procedure is done for the interconnection in Fig. 4.36. Applying KCL (4.45),

$$I_1 = I + I_2.$$

Applying KVL (4.46) with the directions shown in Fig. 4.36,

Mesh (I): $\quad \mathcal{E} + \mathcal{E} = rI_1 + rI_2$
$\quad\quad 2\mathcal{E} = r(I_2 + I) + rI_2 = rI + 2rI_2.$
Mesh (II): $\quad -\mathcal{E} = R_e I - rI_2.$

Solving
$$0 = (r + 2R_e)I \Rightarrow I = 0.$$

The potential difference is
$$V_B - V_A = R_e I = 0.$$

So for the circuit in Fig. 4.36 the power consumed by the passive network is null, since no current passes through it.

Maximum power, therefore, corresponds to the interconnection in Fig. 4.35 and its value is
$$P = (V_B - V_A)^2 / R_e = \frac{\mathcal{E}^2 R}{(R + r/2)^2}.$$

4.13 For the circuit in Fig. 4.37, determine the value of emf $\mathcal{E}$ that will cause the power converted by the motor to be 90 % of the value of the power absorbed by it. Motor values: 220 V cemf and 20 Ω internal resistance. Resistance value: $R = 200 \, \Omega$.

Solution

To make the calculus easier, it is better to simplify the circuit by obtaining the equivalent resistance of the ones that allow it. Firstly, the two resistances R on the central branch AC are in series and are added, so we obtain a resistance $2R$ (Fig. 4.38a). In this figure a triangle association between nodes A, B and C can be seen. It is simplified to star abc, according to Kennelly's theorem (4.50)–(4.52):

$$a = \frac{R \cdot 2R}{R + R + 2R} = \frac{R}{2} = 100 \, \Omega \quad b = \frac{R \cdot R}{R + R + 2R} = \frac{R}{4} = 50 \, \Omega$$
$$c = \frac{R \cdot 2R}{R + R + 2R} = \frac{R}{2} = 100 \, \Omega.$$

The simplified circuit is shown in Fig. 4.38b. There are three branches and two nodes. Applying KCL, one node equation is obtained, but two more mesh equations to solve the three unknown currents are needed. The direction chosen for each current has been indicated, as well as the direction for the circulation assigned to the two meshes. If KCL (4.45) is applied to any of the nodes C or D in Fig. 4.38b,

$$I_1 = I_2 + I_3.$$

Fig. 4.37 Network of Problem 4.13

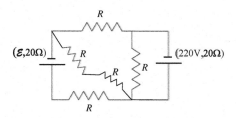

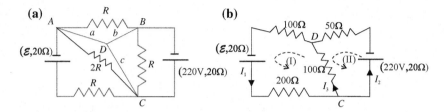

Fig. 4.38 a Simplified network of Problem 4.13. **b** Scheme to apply Kirchhoff's laws

If KVL (4.46) is applied to the two indicated meshes, bearing in mind the sign criteria shown at Fig. 4.7, the result is:

$$\text{Mesh (I): } \mathcal{E} = (100 + 20 + 200)I_1 + 100I_3$$
$$\text{Mesh (II): } -220 = (20 + 50)I_2 - 100I_3,$$

a system of three-equations with four unknown values.

The additional equation is obtained from the condition, $P_{\text{conv}} = 0.9 P_{\text{cons}}$, and the result is, from (4.39) and (4.40):

$$220I_2 = 0.9(220I_2 + 20I_2^2) \Rightarrow I_2 = 1.22 \, \text{A}.$$

If this value is substituted into the previous Kirchhoff's laws equations, and we solve it, the result is:

$$I_3 = 3.06 \, \text{A}, \quad I_1 = 4.28 \, \text{A}, \quad \mathcal{E} = 1675.6 \, \text{V}.$$

Problems C

4.14 Extreme 1 of the copper wire in Fig. 4.39 is in contact with water at a temperature of 80 °C, while extreme 2 is in contact with boiling water, so a linear distribution of temperatures along the wire is generated. The wire, of 20 cm length, has tronco-conical shape, with its straight section radius of 2 mm on extreme 1, and 6 mm on extreme 2. Its resistivity varies with temperature according to expression $\rho = \rho_0(1 + \alpha(t - t_0))$, where copper resistivity is $\rho_0 = 1.71 \times 10^{-8} \, \Omega\text{m}$ at $t_0 = 20\,°\text{C}$ and its temperature coefficient $\alpha = 4 \times 10^{-3} \, \text{K}^{-1}$. Determine wire resistance.

Fig. 4.39 Copper wire with non-constant section submitted to a temperature gradient

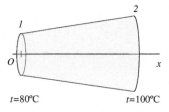

Solution

If (4.19), resistance of a wire, is applied,

$$R = \int_1^2 \rho \frac{dl}{S} = \int_1^2 \rho_0 \left[1 + \alpha(t - t_0)\right] \frac{dl}{S}.$$

It is indicated that on the wire, a linear temperature relation has been established,

$$(x - x_0) \Rightarrow t - 80 = a(x - 0).$$

As values on the wire's extremes are known, its slope value a can be determined:

$$(100 - 80) = a(0.2 - 0) \Rightarrow a = 100 \, \text{km}^{-1}.$$

If we bear in mind the temperature value depends on the position, $t = 80 + 100x$ and $t_0 = 20$, the expression for the resistance can be integrated, with $dl = dx$,

$$R = \int_1^2 \rho_0 \left[1 + \alpha(t - 20)\right] \frac{dl}{S} = \int_0^L \rho_0 \left(1 + \alpha(60 + 100x)\right) \frac{dx}{S}.$$

Since the wire's section S is not constant, its value should be obtained as x-dependent. If the slope of the cone's generatrix is obtained, the x-dependent radius r can be obtained (both in meters) (Fig. 4.40):

$$m = \tan \varphi = \frac{r_2 - r_1}{L} = \frac{6 - 2}{200} = 0.02$$

$$r = r_1 + mx = 2 \times 10^{-3} + 0.02x.$$

If this value is introduced to obtain the wire's circular section S, the result is:

Fig. 4.40 Scheme to obtain radius' equation depending on its position

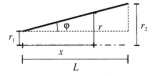

$$R = \int_0^L \frac{\rho_0 \left(1 + \alpha(60 + 100x)\right)}{\pi (r_1 + mx)^2} dx.$$

To solve the integral, the polynomial is decomposed:

$$R = \frac{\rho_0}{\pi} \int_0^L \frac{(1 + \alpha(60 + 100x))}{(r_1 + mx)^2} dx = \frac{\rho_0}{\pi} \int_0^L \left(\frac{A}{(r_1 + mx)^2} + \frac{B}{r_1 + mx} \right) dx =$$

$$= \frac{\rho_0}{\pi} \left[\frac{-A}{m(r_1 + mx)} + \frac{B}{m} \ln(r_1 + mx) \right]_0^L = \frac{\rho_0}{\pi} \left(\frac{-A}{m(r_1 + mL)} + \frac{A}{mr_1} + \frac{B}{m} \ln \frac{r_1 + mL}{r_1} \right).$$

When decomposing the polynomial the result is $A = 1 + 60\alpha - 100\alpha r_1/m = 1.2$ and $B = 100\alpha/m = 20$. If values are substituted, the resistance is obtained:

$$R = \frac{1.71 \times 10^{-8}}{\pi}(20000 + 1000 \ln 3) = 1.15 \times 10^{-4}\, \Omega.$$

4.15 Figure 4.41 represents the straight section of a trapezoidal block with resistivity ρ, whose dimensions are measured in meters. Both lateral striped surfaces are equipotential. The other surfaces are isolated. Determine, approximately, resistance per unit depth of the block.

Solution

In Fig. 4.41, the current flux would enter perpendicular to the left surface, and would leave it perpendicular to the right side. Therefore we can suppose that equipotential surfaces are vertical, and we can decompose the block into slices connected in series by highly conductive thin sheets, as we did in Fig. 4.13b (Sect. 4.11.2). The figure could be also decomposed in the way shown in Fig. 4.42. The narrower the trapezoids of the decomposition are, the more similar to the rectangle in Fig. 4.13b they will be, and therefore the more accurate the calculation will be.

It is true that the white trapezoid i whose thickness is l_i and height is h_i is not a rectangular block, but since $h_i \gg l_i$ the approximation of using the mean height h_i is justified. Resistance of the block i (4.21) is:

$$R_i = \frac{l_i}{h_i} \rho.$$

Fig. 4.41 Conductor block of Problem 4.14

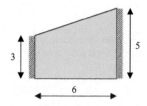

Fig. 4.42 Block as slices

Fig. 4.43 Block as a simple tube

The total block's resistance, if we apply (4.59), is

$$R_e = \sum_{i=1}^{m} R_i.$$

If, for example, we decompose it in only two trapezoids, their thickness will be of 3 m and their medium heights will respectively be 3.5 and 4.5 m, and resistance

$$R_- = \frac{3}{3.5}\rho + \frac{3}{4.5}\rho = 1.524\rho,$$

which corresponds to the lowest resistance value that can be obtained. If we decompose it, for example, in 10 trapezoids (slices), the result would be $R = 1.5321\rho$, more accurate than the previous result.[10]

Figure 4.41 is decomposed now into current tubes. The simplest way (Fig. 4.43) would be to suppose that current I enters perpendicular through the left surface, and flows horizontally until the right one. The effect would be to ignore the upper triangle's material conductivity, and therefore a higher level of the resistance is obtained. If we apply (4.21) the result is:

$$R_+ = \frac{\rho L}{S} = \frac{6}{3}\rho = 2\rho.$$

If we calculate the media of the lowest (R_-) and highest (R_+) values obtained for the resistance,

$$\bar{R} = \frac{R_- + R_+}{2} = \frac{1.524 + 2}{2}\rho = 1.762\rho.$$

and we determine the error

$$(R_+ - \bar{R})/\bar{R} = 0.135,$$

---

[10] See Appendix, Matlab programm to calculate the resistance using slices.

Fig. 4.44 Block as tubes

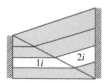

we have the range of values for the resistance:

$$R = 1.762\rho \pm 13.5\,\%.$$

Better schemes of tubes can be built, as for example the one in Fig. 4.44, where one of the diagonals has been drawn and tubes parallel to every face of the trapezoid have been obtained. The two white tubes ($1i$ and $2i$) are connected in series. If its lengths and heights are respectively l_{1i}, h_{1i} and l_{2i}, h_{2i}, the white tube's resistance is, applying (4.59) and (4.21):

$$R_i = \frac{l_{1i}}{h_{1i}}\rho + \frac{l_{2i}}{h_{2i}}\rho.$$

The whole set of tubes, as the white one, are connected at the same time in parallel, obtaining the equivalent resistance of the block if (4.49) is applied:

$$\frac{1}{R} = \sum_{i=1}^{n} \frac{1}{R_i}.$$

To apply this method it is necessary to determine the lengths and heights of each tube.[11] For example, in case of dividing the block in 10 tubes, it results in $R = 1.645\rho$.

4.16 In the network in Fig. 4.45, every device has 10 V emf and 1 Ω internal resistance. (a) Calculate the Thévenin and Norton equivalents between terminals A and B. (b) If a small motor, with 6 V cemf and 0.5 Ω internal resistance, is connected to these terminals, calculate the power consumed by this motor.

Solution

(a) To determine Thévenin and Norton equivalents it is necessary to calculate the potential difference and equivalent resistance between the asked points. To calculate the potential difference it is always necessary to know the currents that go through the network between these points. Firstly, it can be observed that there is a triangle association abc (Fig. 4.46a) that can be simplified by Kennelly's theorem (4.50)–(4.52), obtaining a star-association. If equations of this theorem are applied, the values of the resistances in the star are obtained. In practice these equations represent the product of the resistances that converge in the node divided by the sum of the resistances of the association:

---

[11] See Appendix, Matlab programm to calculate the resistance using tubes.

Fig. 4.45 Network of Problem 4.15

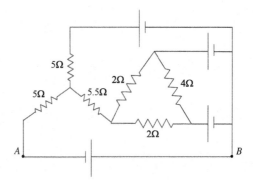

$$R_a = \frac{2 \cdot 2}{2+2+4} = 0.5\,\Omega \quad R_b = \frac{2 \cdot 4}{2+2+4} = 1\,\Omega \quad R_c = \frac{2 \cdot 4}{2+2+4} = 1\,\Omega.$$

The resulting network is shown in Fig. 4.46b, where the resistance of 5.5 Ω from the original star and the one of $R_a = 0.5\,\Omega$, which are in series, have also been summed. This network has a lower number of nodes and branches than the original one, so its solution is easier.

Since there are so many nodes and branches, it is better to solve it by applying mesh analysis (4.47), with the mesh currents shown in Fig. 4.46b. The only real current which is necessary to be known to determine the potential difference between A and B is current I that goes through the branch between A and B. This current I is the same as mesh current i_3. Following the indicated criteria for the mesh current method (Sect. 4.10.2), the matrix of emfs, mesh resistances and currents results:

$$\begin{pmatrix} -10+10 \\ -10+10 \\ 10+10 \end{pmatrix} = \begin{pmatrix} 1+1+1+5+6 & -1-1 & -6 \\ -1-1 & 1+1+1+1+4 & -1-1 \\ -6 & -1-1 & 1+1+6+5+1 \end{pmatrix} \begin{pmatrix} i_1 \\ i_2 \\ i_3 \end{pmatrix}$$

$$= \begin{pmatrix} 14 & -2 & -6 \\ -2 & 8 & -2 \\ -6 & -2 & 14 \end{pmatrix} \begin{pmatrix} i_1 \\ i_2 \\ i_3 \end{pmatrix}$$

Emfs have been introduced following the sign criteria indicated in Fig. 4.7. Values on the principal diagonal of the resistance matrix are the sum of the resistances on every mesh. Common resistances to the different meshes are introduced with negative sign because circulations through them are opposite depending on the mesh. The internal resistance of the generators and motors should not be forgotten. Solving the system of equations,

$$i_3 = I = 1.93\,\text{A}.$$

To obtain the potential difference between A and B, Ohm's law for a branch (4.43) is applied to the lower branch in the figure,

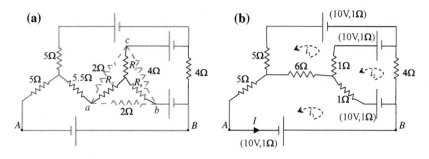

Fig. 4.46 a Simplified network of Problem 4.15. b Scheme to apply Mesh analysis

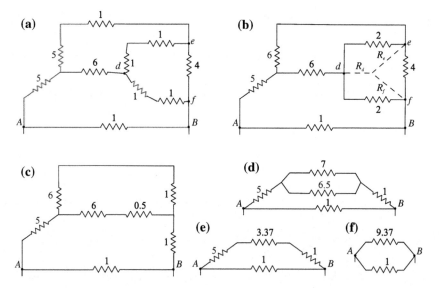

Fig. 4.47 Scheme for the calculation of the passive network equivalent resistance

$$V_A - V_B + \sum_i \mathcal{E}_i = \sum_i R_i I \Rightarrow V_A - V_B + 10 = 1 \cdot 1.93 \Rightarrow V_A - V_B = -8.07\,\text{V},$$

$$\mathcal{E}_e = V_B - V_A = 8.07\,\text{V}.$$

To calculate equivalent resistance we follow the scheme in Fig. 4.47, with every value in ohms. It should be remembered that points A and B cannot be changed, because we are calculating equivalent resistance between them. Figure 4.47a is the passive network of the circuit in Fig. 4.46b. 1 Ω resistances of branches de and df on the right side in Fig. 4.47a are in series and have to be summed, as well as resistances of 5 and 1 Ω in the higher branch. 2 Ω resistances in branches de and df, and the one of 4 Ω in branch ef in Fig. 4.47b are in triangle configuration. If they are simplified to a star association, each resistance has a value of

$$R_d = \frac{2 \times 2}{2+2+4} = 0.5\,\Omega \quad R_e = \frac{4 \times 2}{2+2+4} = 1\,\Omega \quad R_f = \frac{4 \times 2}{2+2+4} = 1\,\Omega.$$

In the central branch of Fig. 4.47c, 6 and 0.5 Ω resistances are in series, as well as the 6 and 1 Ω in the upper branch. They are summed and the result is Fig. 4.47d. 7 and 6.5 Ω resistances are in parallel. Its equivalent is obtained by

$$\frac{1}{R_p} = \frac{1}{7} + \frac{1}{6.5} \Rightarrow R_p = 3.37\,\Omega.$$

In Fig. 4.47e resistances in the upper branch are in series. They are summed resulting Fig. 4.47f, with two resistances in parallel. Equivalent resistance is

$$\frac{1}{R_e} = \frac{1}{9.37} + \frac{1}{1} \Rightarrow R_e = 0.9\,\Omega,$$

which is the resistance we are asked for.

If we consider the scheme in Fig. 4.15, the Thévenin equivalent is the one in Fig. 4.48a. It should be observed that the negative terminal is on the side of point A, since this is the one with the lowest potential ($V_A - V_B$ is negative).

To obtain the Norton equivalent it is necessary to calculate the current source,

$$I_e = \mathcal{E}_e/R_e = 8.07/0.9 = 8.97\,\text{A}.$$

According to the scheme in Fig. 4.16, the Norton equivalent is the one in Fig. 4.48b. The current source supplies current in the sense of lower to higher potential, from A to B, which explains the direction of current I_e.

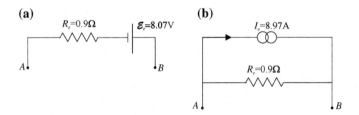

Fig. 4.48 a Thévenin equivalent. b Norton equivalent

Fig. 4.49 Circuit to calculate power consumed by the motor

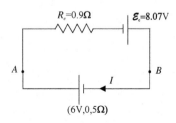

(b) To calculate power consumed by the motor, we connect it to terminals A and B from Thévenin equivalent (Fig. 4.49), since its behavior is the same as the one of the complete network. If Ohm's law (4.41) is applied to the circuit in the current direction,

$$\sum_i \mathcal{E}_i = I \sum_i R_i \Rightarrow 8.07 - 6 = (0.9 + 0.5)I ,$$

whence

$$I = 1.48 \,\text{A} .$$

Power consumed by the motor (taken from the circuit) is, according to Expression (4.40),

$$P = (\mathcal{E} + rI)I = (6 + 0.5 \cdot 1.48) \cdot 1.48 = 9.975 \,\text{W} .$$

4.17 For the circuit in Fig. 4.50, the current through generator G is four times the one that passes through motor M. It is also known that motor converts 1000 W into mechanical energy. Determine Thévenin and Norton equivalents between points A and B, and power supplied to the network by the generator. Lines' resistance values: $R = 100 \,\Omega$. Generator's and motor's internal resistance: $r = 10 \,\Omega$.

Solution

To determine Thévenin and Norton equivalents it is necessary to calculate the potential difference and equivalent resistance between the asked points. To calculate the potential difference it is necessary to know the currents that circulate through the branches of the network between these points.

The circuit can be simplified, since there are associations of resistances that can be substituted by their equivalent resistance. Both vertical resistances with value $2R$ are in parallel. If (4.49) is applied, the result is:

$$\frac{1}{R_p} = \frac{1}{2R} + \frac{1}{2R} = \frac{1}{R} \Rightarrow R_p = R ,$$

as shown in Fig. 4.51a. Resistance R and $2R$ between C and D are in series and have been summed. In this figure, once these simplifications are done, it can be observed that between points C, D and E a triangle association exists that can be simplified to

Fig. 4.50 Network of Problem 4.17

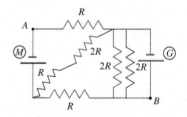

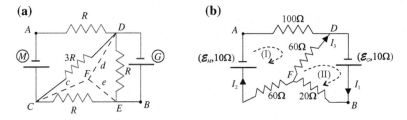

Fig. 4.51 **a** Simplified network of Problem 4.17. **b** Scheme to apply Kirchhoff's laws

the star association *cde* shown in the figure. Applying Kennelly's theorem (4.50)–(4.52):

$$c = \frac{R \cdot 3R}{R+R+3R} = \frac{3R}{5} = 60\,\Omega \quad d = \frac{R \cdot 3R}{R+R+3R} = \frac{3R}{5} = 60\,\Omega$$

$$e = \frac{R \cdot R}{R+R+3R} = \frac{R}{5} = 20\,\Omega.$$

Figure 4.51b shows the circuit already simplified. Note that there are three branches and two nodes (D and F). If KCL is applied, only one node equation is obtained. We need two more mesh equations to solve the three unknown values, which are currents through the branches. Supposed currents have been drawn through the different branches in the network, as well as the direction of circulation for the two meshes. If KCL (4.45) is applied to any of the two nodes D or F in Fig. 4.51b, the result is:

$$I_1 = I_2 + I_3.$$

If KVL (4.46) is applied to both of the indicated meshes, taking into account the sign criteria indicated in Fig. 4.7, the result is:

$$\text{Mesh (I):} \quad \mathcal{E}_G = 30I_1 + 60I_3$$
$$\text{Mesh (II):} \quad -\mathcal{E}_M = 170I_2 - 60I_3.$$

Since current through the generator is four times the one through the motor, it is obtained:

$$I_1 = 4I_2.$$

Since power converted by the motor to mechanical energy is 1000 W, if (4.39) is applied, we obtain:

$$P_M = \mathcal{E}_M I_2 = 1000 \Rightarrow \mathcal{E}_M = 1000/I_2.$$

If these two equations are introduced in the three equations that we have obtained from Kirchhoff's laws, the following values are obtained:

$$I_1 = 40\,\text{A}, \quad I_2 = 10\,\text{A}, \quad I_3 = 30\,\text{A}, \quad \mathcal{E}_G = 3000\,\text{V}, \quad \mathcal{E}_M = 100\,\text{V}.$$

With these results, power supplied by the generator to the network can be determined, if we apply (4.34):

$$P_{\text{sup}} = \mathcal{E}_G I_1 - rI_1^2 = 3000 \cdot 40 - 10 \cdot 40^2 = 104\,\text{kW}.$$

The potential difference between points A and B is obtained by applying Ohm's law for a branch (4.43). If circulation is done using the path ADB, the result is:

$$V_A - V_B + 3000 = 100 I_2 + 10 I_1 \Rightarrow V_A - V_B = -1600\,\text{V}.$$

$\mathcal{E}_G$ is introduced with positive sign, due to the fact that it goes firstly through the negative pole of the generator. We can check the result by circulating through the the path AFB:

$$V_A - V_B + 100 = -70 I_2 - 20 I_1 \Rightarrow V_A - V_B = -1600\,\text{V}.$$

To obtain the equivalent resistance between A and B, the passive network of the circuit (Fig. 4.51b) is obtained by following the scheme in Fig. 4.52, with all the values in ohms. In the passive network in Fig. 4.52a, resistances of 10 and 60 Ω are in series and are summed.

There is also a triangle association BDF (also the triangle association ADF could be taken) which simplifies the star association bdf according to Kennelly's theorem (4.50)–(4.52):

$$b = \frac{20 \cdot 10}{20 + 10 + 60} = \frac{20}{9}\,\Omega \quad d = \frac{60 \cdot 10}{20 + 10 + 60} = \frac{20}{3}\,\Omega \quad b = \frac{20 \cdot 60}{20 + 10 + 60} = \frac{40}{3}\,\Omega.$$

Note that A and B are nodes, because they are the endings of the equivalent. The result obtained is Fig. 4.52b, where resistances AD and DG are in series, being summed, as

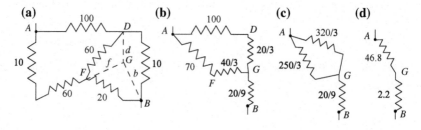

Fig. 4.52 Scheme for equivalent resistance in the passive network calculus

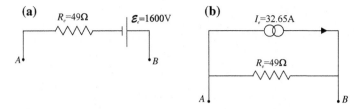

Fig. 4.53 a Thévenin equivalent. b Norton equivalent

well as resistances AF and FG, resulting in Fig. 4.52c. Resistances in both branches AG are in parallel, whose equivalent is (4.49):

$$\frac{1}{R_p} = \frac{3}{320} + \frac{3}{250} \Rightarrow R_p = 46.8\,\Omega.$$

Figure 4.52d shows the final result with two resistances in series, whose sum is the equivalent resistance that is asked,

$$R_e = 49\,\Omega.$$

If we consider the scheme in Fig. 4.15, the Thévenin equivalent is the one in Fig. 4.53a. Note that the negative pole is on point A side, since this is the one with the lowest potential ($V_A - V_B$ is negative).

To obtain the Norton equivalent it is necessary to calculate the current source,

$$I_e = \mathcal{E}_e/R_e = 1600/49 = 32.65\,\text{A}.$$

According to the scheme in Fig. 4.16, the Norton equivalent is the one in Fig. 4.53b. The current source supplies current from lower to higher potential (from A to B), which explains the current I_e direction.

4.18 The connection of Fig. 4.54 between four resistance R_x, R_v, R_1 and R_2, the generator G and the galvanometer W is used to measure resistance R_x and it is called the Wheatstone bridge. Resistances are known: $R_1 = 10\,\Omega$, $R_2 = 5\,\Omega$. R_v is a variable resistance which can be adjusted until the galvanometer W shows that current through branch BC is zero. The generator G has 24 V emf and $2\,\Omega$ internal resistance, and the galvanometer detects no current when $R_v = 8\,\Omega$. (a) Determine the value of R_x. (b) By using the Thévenin equivalent, find the cemf value of a motor, whose internal resistance is $2\,\Omega$, so that if the motor is connected between points B and D, maximum power is converted.

Solution

(a) Since resistance R_v can be adjusted so that no current passes through the galvanometer W, current through the two upper branches of the Wheatstone bridge is

Fig. 4.54 Network of Problem 4.18

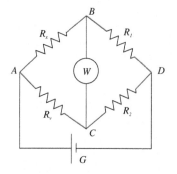

Fig. 4.55 Currents scheme of Problem 4.18 to calculate R_x

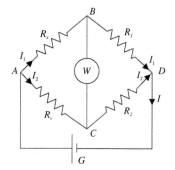

equal (Fig. 4.55). The same occurs with the two lower branches. Since no current passes through branch BC, the voltage drop is zero and points B and C have the same potential:

$$V_B - V_C = R_W I_W = 0 \Rightarrow V_B = V_C,$$

where R_W is the galvanometer resistance and I_W is the current through it. If the potential difference between points A, B, C, D of the circuit is calculated through the different branches of the circuit, the result is:

$$\left. \begin{array}{l} V_A - V_B = R_x I_1 \\ V_A - V_C = R_v I_2 \end{array} \right\} \quad \begin{array}{l} V_C = V_B \\ R_x I_1 = R_v I_2 \end{array} \Rightarrow \frac{R_x}{R_v} = \frac{I_2}{I_1} \qquad (4.63)$$

In a similar way,

$$\left. \begin{array}{l} V_B - V_D = R_1 I_1 \\ V_C - V_D = R_2 I_2 \end{array} \right\} \quad \begin{array}{l} V_C = V_B \\ R_1 I_1 = R_2 I_2 \end{array} \Rightarrow \frac{R_1}{R_2} = \frac{I_2}{I_1} \qquad (4.64)$$

From (4.63) and (4.64) it follows that:

$$R_x = \frac{R_1}{R_2} R_v, \qquad (4.65)$$

Substituting the problem values,

$$R_x = 16\,\Omega.$$

(b) To obtain the Thévenin equivalent it is necessary to know the potential difference and equivalent resistance between points B and D. With previous equations and with the value of R_x, the potential difference can be determined. First, from (4.65), the result is

$$R_v = \frac{R_2}{R_1} R_x = 8\,\Omega.$$

From (4.63) or from (4.64) it is also obtained that

$$I_2 = 2I_1.$$

The same result can be obtained by applying KVL (4.46) to mesh (I) in Fig. 4.56, where resistance values have already been substituted, and the current is considered null through the galvanometer. Taking into account the indicated senses in the figure, the result is

Mesh (I): $0 = (16 + 10)I_1 - (8 + 5)I_2 \Rightarrow I_2 = 2I_1$.

One of the other two equations is obtained from KCL (4.45), by applying it, for example, to node D:

$$I = I_1 + I_2 = 3I_1,$$

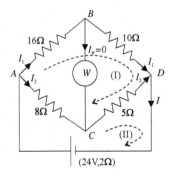

Fig. 4.56 Current scheme of Problem 4.18 to calculate potential difference between B and D

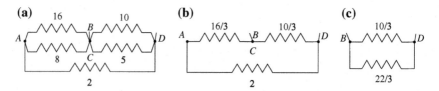

Fig. 4.57 Scheme to calculate the passive network equivalent resistance

Fig. 4.58 Thévenin equivalent

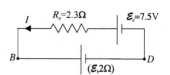

and the other one by applying KVL to mesh (II), with the indicated senses,

Mesh (II): $24 = (8+5)I_2 + 2I = 26I_1 + 6I_1 = 32I_1 \Rightarrow I_1 = 0.75$ A.

The potential difference between B and D (and therefore the Thévenin equivalent's electromotive force) is obtained by applying Ohm's law (4.20) between these points

$$\mathcal{E}_e = V_B - V_D = 10I_1 = 7.5 \text{ V}.$$

To determine the equivalent resistance of the circuit between points B and D, its passive network should be solved (Fig. 4.57a, with all the values in ohms). B and C have the same potential. Points B and D should stay the same, since the equivalent has to be calculated between these two points. In the figure, resistance between A and B and the ones between B and D are in parallel. If (4.49) is applied, the result is

$$\frac{1}{R_{p1}} = \frac{1}{16} + \frac{1}{8} \Rightarrow R_{p1} = \frac{16}{3} \Omega,$$

and

$$\frac{1}{R_{p2}} = \frac{1}{10} + \frac{1}{5} \Rightarrow R_{p2} = \frac{10}{3} \Omega,$$

as it can be seen in Fig. 4.57b. It seems that resistance on the upper branches are in series, but this is not true: point B behaves as a node and therefore cannot disappear, as well as D. However, resistance of $16/3\,\Omega$ and $2\,\Omega$ are in series, because point A is not a node, and they can be summed (Fig. 4.57c). These two resistances are in parallel (B and D are nodes), and the equivalent resistance results

$$\frac{1}{R_e} = \frac{3}{10} + \frac{3}{22} \Rightarrow R_e = 2.3\,\Omega.$$

If we consider the scheme in Fig. 4.15, the Thévenin equivalent between B and D is the one in Fig. 4.58. Note that the negative pole is on the side of point D, since this is the one with the lowest potential.

A motor is connected to the equivalent, with unknown $\mathcal{E}$ cemf and $2\,\Omega$ internal resistance. Current I has the indicated direction, leaving the generator through the positive pole and entering to the motor by its positive one. Applying Ohm's law (4.41),

$$7.5 - \mathcal{E} = (2.3 + 2)I \Rightarrow I = \frac{7.5 - \mathcal{E}}{4.3}.$$

The motor must convert the maximum power. The expression of power converted is (4.39)

$$P_{\text{conv}} = \mathcal{E}I = \mathcal{E}\frac{7.5 - \mathcal{E}}{4.3},$$

which has to be its maximum. Applying the maximum condition:

$$\frac{dP}{d\mathcal{E}} = \frac{7.5 - 2\mathcal{E}}{4.3} = 0 \Rightarrow \mathcal{E} = 3.75\,\text{V},$$

which is the value of the counter electromotive force we are asked for.

Chapter 5
Magnetostatics

Abstract This chapter deals with the basic characteristics of the magnetostatic field. Even though such a field is a special case of the general electromagnetic field described by the Maxwell equations (Chap. 10), it is important for many actual problems that occur in applied science and technology. For this reason, some procedures for calculating it are presented, and in detail explained.

5.1 Differential Equation of the Magnetostatic Field

When we studied the electrostatic field we saw that through some differential equations it is possible to determine the electric field **E** at any point in space for stationary phenomena. In addition, these equations allowed us to understand what were the sources of the electric field. In the same way we are now interested in finding out what are the general laws governing the magnetic field independent of time. The study of this particular case is the purpose of magnetostatics.

The basic equations of the magnetostatic field are the following:

$$\nabla \times \mathbf{B} = \mu_0 \mathbf{j}, \tag{5.1}$$

$$\nabla \cdot \mathbf{B} = 0. \tag{5.2}$$

In addition, by using the law of charge conservation (see Chap. 2) for stationary currents:

$$\nabla \cdot \mathbf{j} = 0. \tag{5.3}$$

Equation (5.1) indicates that a possible source of the magnetic field **B** is the electric current, i.e. moving electric charges.[1] The second equality, being equal to zero, means that there are no sources or sinks for **B**, which implies that magnetic charges do not exist (monopoles). In short, the magnetic field **B** has vector sources (current density)

---

[1] Commonly this field is referred in many textbooks as *magnetic induction*, but we will refer to it as the magnetic field **B**.

but no scalar sources (magnetic charge density). If magnetic charges would exist similarly to the electric charges, the equation $\nabla \cdot \mathbf{B}$ would be different from zero, in a way similar to Gauss law for electric field.[2] The non-existence of magnetic charge restricts the topological characteristics of the magnetic field lines. As a consequence of it being divergence free, we can distinguish three possibilities for the geometry of the magnetic field lines (MFL): (a) They may be closed; (b) They can be born and die at infinity, i.e. they extend infinitely. (c) They can ergodicaly recover a surface in a bounded region, unable to establish origin or end.

In this chapter we shown the differential system of equations we can employ for determining the of the vector lines of a vector field (1.19). In the same way we can write for the magnetic field

$$\frac{dx}{B_x} = \frac{dy}{B_y} = \frac{dz}{B_z}, \quad (5.4)$$

which define a set of trajectories in $\Re^3$. Geometrically it has a simple interpretation. In fact, the integrals of (5.4) represent two surfaces in the space, whose intersection corresponds to the field lines of the magnetic field $\mathbf{B}$. Magnetic fields, in general, do not form magnetic surfaces. Such surfaces arise in magnetohydrostatic equilibria and for some highly symmetric field configurations.[3]

The unit of measurement in the SI of the magnetic field $\mathbf{B}$ is the Tesla, and is symbolized by the letter T. However, due to the fact that the the Tesla quite large, on many occasions, a smaller unit than the Tesla is used, namely the gauss. The equivalence between the two is: 1 Gauss = 10^{-4} T. With the aim of providing some idea about these quantities we show an example.

As it is known Earth's magnetic field is not uniform over all parts of the Earth. However the order of magnitude can be estimated at about $0.2 - 0.6 \cdot 10^{-4}$ T, and the field generated by the coils often used in laboratory practices is around 10^{-3} T, for an intensity that ranges from 0.3 to 0.5 A. This means that the Tesla is a large unit. There are other examples of practical importance where the magnitude of the magnetic field may differ from the aforementioned values. This is the case of the nuclear magnetic resonance (NRM)[4] for medical diagnosis in which the field employed is of the order of one Tesla. But in other cases it exceeds many times that magnetic field. For

---

[2] There is a theory about magnetic monopoles, although, to date, has not been experimentally proven.

[3] When studying the characteristics of the magnetic field lines and surfaces, an important concept is the magnetic helicity. The helicity $\mathcal{H}$ of a magnetic field $\mathbf{B}$ is defined as $\mathcal{H} = \int \int \int_V \mathbf{A} \cdot \mathbf{B} dV$ (see Sect. 5.3). This concept plays an important role in Magnetohydrodynamics. The physical meaning is complex, but sometimes is related with the linkage of the field lines.

[4] This non-invasive technique, also called magnetic resonance imaging (MRI), employs several coils that produce different kinds of magnetic fields. One of them creates a quasi-homogeneous static $\mathbf{B}$ and the other ones produce variable fields in different directions. By using radiowaves the atoms of the body can absorb energy at different frequencies, and depending on the way the absorbtion takes place, and by means of computer technology, makes it possible to reconstruct three-dimensional images of the body. The data obtained through these images provides structural and biochemical information about tissue, and also the possibility of detecting spine abnormalities, cerebral edema, and early-stage cancer.

5.1 Differential Equation of the Magnetostatic Field

example, the magnetic field existing in nuclear fusion experimental reactors ranges from 3.5 (JET) to 5.3 Tesla (ITER). In stellerators, such as the Wendelstein 7-X the magnetic field can reach 3 Tesla and at CERN, the highest field recently obtained has been 13.5 Tesla.

5.2 Integral Form of the Equations

The equations of the previous section give information about the magnetostatic field at each point in the space, i.e. they are pointlike equations. However, these expressions can be represented in an integral form, which under certain circumstances are more useful to address problems.

The integral representation of (5.1) is called Ampère's law, which states that circulation of the magnetic field $\mathbf{B}$ along a closed Γ trajectory, depends only on the net current passing through the open surface S whose boundary is Γ (∂S) (Fig. 5.1), i.e.

$$\oint_\Gamma \mathbf{B}(\mathbf{r}) d\mathbf{l} = \mu_0 \int_S \mathbf{j} \cdot d\mathbf{S} = \mu_0 I. \tag{5.5}$$

Ampère's theorem is important because under certain conditions it easily allows us to calculate the magnetic field components known the current density. Theoretically this law is always true if the conditions of Stokes's theorem are satisfied (see Chap. 1, (1.48)). However, from a practical point of view Ampère's theorem is only useful when the system under study possesses symmetries. In this case the basic idea for applying (5.5) is to find a curve containing the symmetry of the component of the magnetic field to be analyzed and performing the integral. In this case, the scalar product of $\mathbf{B}(\mathbf{r}) \cdot d\mathbf{l}$ is a constant along the path, then the corresponding component i[5] of $\mathbf{B}$ may be move outside of the integral like a scalar, i.e.

$$B_i \oint_\Gamma dl = \mu_0 \int_S \mathbf{j} \cdot d\mathbf{S}. \tag{5.6}$$

To apply Ampere's law some caution must be taken into account. Rewriting (5.5), we have:

$$\oint_{\partial S} \mathbf{B} \cdot d\mathbf{l} = \mu_0 \int_S \mathbf{j} \cdot d\mathbf{S} = \mu_0 \int_S \mathbf{j} \cdot \mathbf{n} dS = \mu_0 I, \tag{5.7}$$

where $\mathbf{n}$ is the unit normal vector at each point of the open surface. This equation shows that the choice of the direction over Γ to perform the contour integral determines the orientation of the surface $\mathbf{S}$ and vice versa. In short, Γ is directly related to S, and determines the sign of the scalar product of $\mathbf{j} \cdot \mathbf{n} dS$. Regarding the double integral of the second member in (5.7), in principle it extends to the surface S chosen,

---

[5] This component i may be the projection of $\mathbf{B}(\mathbf{r})$ over x, y or z in a cartesian coordinate frame, or if we work with cylindrical coordinates ρ, ϕ or z.

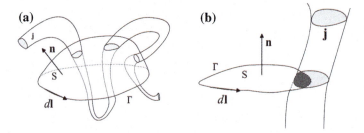

Fig. 5.1 Open surface crossed by a density current **j**. The integral along the closed curve Γ depend only of the net intensity I. **a** The total intensity crossing the surface S is calculated by $\int_S \mathbf{j} \cdot d\mathbf{S}$. **b** In this case only a part of the current density crosses the surface

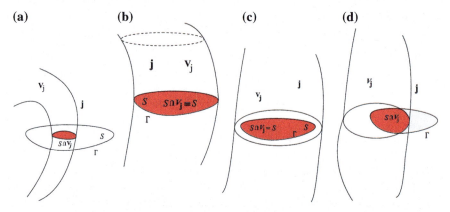

Fig. 5.2 a The actual surface of integration is smaller than S. **b** Here the limits of integration coincide with $S = S \cap V_j$. **c** In this case the intersection also corresponds with S, although j extends beyond S. **d** The intersection corresponds only to a part of the surface S chosen to perform the integral

however the actual surface of integration is only the intersection between S and the volume V_j corresponding to the current density **j**, since the scalar product $\mathbf{j} \cdot \mathbf{S}$ is zero where S or j are zero (see Fig. 5.2). Accordingly, we can write:

$$\int_S \mathbf{j} \cdot d\mathbf{S} = \int_{S \cap V_j} \mathbf{j} \cdot \mathbf{n} dS = I. \tag{5.8}$$

In relation to (5.2), its integral form is what is known the flux law for the magnetic field **B**. It shows that the magnetic field flux **B** through a closed surface is always zero, which is a direct consequence of the non-existence of magnetic charge

$$\int_V \nabla \cdot \mathbf{B} dV = \oint_{\partial V} \mathbf{B} \cdot d\mathbf{S} = 0, \tag{5.9}$$

where ∂V denotes the boundary of V, i.e., the surface S.

5.3 Vector Potential

As discussed in Chap. 2 (see also Chap. 1), electrostatic conditions meant that $\nabla \times \mathbf{E} = 0$. This means that, in principle, we can find a potential function $V(x, y, z)$ such that $\mathbf{E} = -\nabla V$. The importance of this is that, to calculate the electric field corresponding to an arbitrary charge distribution it is not necessary to use the integral expression (2.16), rather finding it from the gradient of this potential function. This is simpler because $V(x, y, z)$ is a scalar, then we must perform only one integration. In the same way the question that arises is whether there is a similar function from which it is possible to obtain field $\mathbf{B}$.

To answer this question one need only examine the (5.2). Taking into account some of the properties of the differential operators seen in Chap. 1, the divergence of the curl is always zero. Consequently, because the divergence of $\mathbf{B}$ is always zero, we can express the magnetic field as the curl of a vectorial function, that is,

$$\nabla \cdot \mathbf{B} = \nabla \cdot (\nabla \times \mathbf{A}) = 0, \tag{5.10}$$

thus we can write

$$\mathbf{B}(\mathbf{r}) = \nabla \times \mathbf{A}(\mathbf{r}) \tag{5.11}$$

This new vector field is called vector potential $\mathbf{A}(\mathbf{r})$. This potential has some differences and similarities to that established for the electric field. $V(x, y, z)$ is a scalar function while $\mathbf{A}(\mathbf{r})$ is vectorial and generally more complicated. However, like the electrostatic potential the vector potential is multivaluated. As we have seen in Chap. 2, if we add a constant c to the potential V the new function $V' = V + c$ is a potential too. It means both potentials provide the same electric field. In a similar way, the vector potential suffers some indetermination. To show this let us consider a scalar function $\varphi(x, y, z)$. If we calculate the rotational of this results

$$\nabla \times (\nabla \varphi) = 0. \tag{5.12}$$

This means that if we define a new field say $\mathbf{A}'$ such that

$$\mathbf{A}' = \mathbf{A} + \nabla \varphi, \tag{5.13}$$

this new vector potential reproduces the same field $\mathbf{B}$. In fact, introducing $\mathbf{A}'$ in (5.11) we have

$$\nabla \cdot \mathbf{B} = \nabla \times \mathbf{A}' = \nabla \times \mathbf{A} + \nabla \times (\nabla \varphi) = \nabla \times \mathbf{A}, \tag{5.14}$$

and then for a magnetic field $\mathbf{B}$ given we have infinite fields $\mathbf{A}$. This might suggest that $\mathbf{A}$ is only a mathematical procedure for calculating the field without physical meaning. This, however, is not what experience shows, since particles are capable of detecting the vector potential $\mathbf{A}$ in a region of space where $\mathbf{B} = 0$ (see Aharonov–Bohm effect in [28, 48, 97, 112]). With the aim of obtaining an expression for calculating the vector potential, we have

$$\nabla \times \nabla \times \mathbf{A}(\mathbf{r}) = \nabla(\nabla \cdot \mathbf{A}(\mathbf{r})) - \nabla^2 \mathbf{A}(\mathbf{r}) = \mu_0 \mathbf{j}. \tag{5.15}$$

The solution to this differential equation provides $\mathbf{A}$, which allows us to determine the magnetic field $\mathbf{B}$ by means of the rotational of $\mathbf{A}$. At first sight it seems to be very difficult and contrary to the first idea for calculating $\mathbf{B}$, however as we will show (5.15) may be simplified.

If we examine in more detail this equation we see that, if we did not have $\nabla(\nabla \cdot \mathbf{A})$, its structure would be similar to Poisson's equation seen in Chaps. 2 and 7. In this case, we could directly obtain the solution to our problem. With the aim to resolve (5.15) as easily as possible, we will impose the condition that the divergence of the potential vector be zero, i.e.

$$\nabla \cdot \mathbf{A} = 0, \tag{5.16}$$

hence obtaining the result we are seeking. However, this election must be justified.[6] In fact, in order to demonstrate that this condition may be always chosen, we apply the nabla operator on both sides of (5.13)

$$\nabla \cdot \mathbf{A}' = \nabla \cdot \mathbf{A} + \nabla^2 \varphi. \tag{5.17}$$

Let us suppose that (5.16) does not hold and we choose, for instance, $\nabla \cdot \mathbf{A} = f(r)$, $f(r)$ being a scalar smooth function. Introducing it in (5.17) we have

$$\nabla \cdot \mathbf{A}' = f(r) + \nabla^2 \varphi. \tag{5.18}$$

Remembering that all vector potentials differing in $\nabla \varphi$ give the same magnetic field $\mathbf{B}$, we can use the potential $\mathbf{A}'$ instead of $\mathbf{A}$. If we impose the aforementioned condition on $\mathbf{A}'$ we can write

$$0 = \nabla \cdot \mathbf{A}' = f(r) + \nabla^2 \varphi \Rightarrow \nabla^2 \varphi = -f(r). \tag{5.19}$$

This result tell us that we can always choose $\nabla \cdot \mathbf{A} = 0$, and if this condition is not chosen, we can resolve $\nabla^2 \varphi = -f(r)$ and then work with $\mathbf{A}'$, whose divergence is zero (5.19). In this case we have

$$\nabla^2 \mathbf{A}'(\mathbf{r}) = -\mu_0 \mathbf{j}. \tag{5.20}$$

---

[6]This election only on the basis of simplifying (5.15) is confusing and a priori non comprehensible. Otherwise, why don't we choose $\nabla^2 \mathbf{A} = 0$?

5.3 Vector Potential

This differential equation is similar, for each of its components, to the Poisson equation for the electrostatic field (Chaps. 2 and 7). It follows that the solution must be formally the same, so we can write[7]

$$\mathbf{A}(\mathbf{r}) = \frac{\mu_0}{4\pi} \int_{V'} \frac{\mathbf{j}_v(\mathbf{r}')dV'}{|\mathbf{r}-\mathbf{r}'|}, \quad (\text{Tm}) \tag{5.21}$$

where $\mathbf{j}_v$ represents the volume current density. In the case that the currents extend over a surface the equalities are:

$$\mathbf{A}(\mathbf{r}) = \frac{\mu_0}{4\pi} \int_{S'} \frac{\mathbf{j}_s(\mathbf{r}')dS'}{|\mathbf{r}-\mathbf{r}'|}, \quad (\text{Tm}) \tag{5.22}$$

where $\mathbf{j}_s$ represents the surface current density. Finally, when the system may be considered mathematically one-dimensional, taking into consideration that $\mathbf{j}_v dV' = \mathbf{j}_v S' \cdot d\mathbf{l}' = I d\mathbf{l}'$, we obtain

$$\mathbf{A}(\mathbf{r}) = \frac{\mu_0}{4\pi} \oint_{\Gamma'} \frac{I d\mathbf{l}'}{|\mathbf{r}-\mathbf{r}'|}, \quad (\text{Tm}) \tag{5.23}$$

where Γ' is the line where it extends the current. Note that the domain of integration of all expressions for the vector potential extends over the respective current elements, i.e. integration variables dV', dS' and $d\mathbf{l}'$, respectively (variables have $'$).

There is also the called magnetic scalar potential, but we will talk a little about it in Sect. 5.7 and in Chap. 7 (Problem 7.20).

5.4 The Biot–Savart Law

In this section we are interested in finding an expression for directly obtaining the magnetic field **B** created by an arbitrary current distribution. To this aim, calculating the rotational (5.21) we have

$$\mathbf{B}(\mathbf{r}) = \frac{\mu_0}{4\pi} \int_{V'} \frac{\mathbf{j}_v(\mathbf{r}') \times (\mathbf{r}-\mathbf{r}')}{|\mathbf{r}-\mathbf{r}'|^3} dV', \quad (\text{T}). \tag{5.24}$$

This expression is known as the Biot–Savart law.

As in the preceding section, it may occur that the current distribution extends in a two-dimensional or one-dimensional domain. For these cases we use,

$$\mathbf{B}(\mathbf{r}) = \frac{\mu_0}{4\pi} \int_{S'} \frac{\mathbf{j}_s(\mathbf{r}') \times (\mathbf{r}-\mathbf{r}')}{|\mathbf{r}-\mathbf{r}'|^3} dS', \tag{5.25}$$

---

[7] In the following we use **A** instead **A**$'$. The reasoning employed to justify the election for the divergence of the vector potential does not depend on the nomenclature chosen.

and
$$\mathbf{B}(\mathbf{r}) = \frac{\mu_0}{4\pi} \oint_{\Gamma'} \frac{Id\mathbf{l}' \times (\mathbf{r} - \mathbf{r}')}{|\mathbf{r} - \mathbf{r}'|^3}. \tag{5.26}$$

5.5 Forces on Currents

Another important problem is the effect of the magnetic field over currents. Let us imagine a region Ω in the space $\mathbb{R}^3$ in which we have a magnetic field $\mathbf{B}$. If in such a region a density current $\mathbf{j}_v$ extends over V', such that $V' \subseteq \Omega$, by using the Lorentz force (Chap. 9), we can find that

$$\mathbf{F}(\mathbf{r}) = \int_{V'} \mathbf{j}_v(\mathbf{r}') \times \mathbf{B}(\mathbf{r}') dV'. \tag{5.27}$$

If the current flows into a filament, i.e. is a one-dimensional current, the expression for the force exerted is

$$\mathbf{F}(\mathbf{r}) = I \int_{\Gamma'} d\mathbf{l}' \times \mathbf{B}. \tag{5.28}$$

Of particulary interest is if the magnetic field present in the region is homogeneous. In this case we have,

$$\mathbf{F}(\mathbf{r}) = I \left[\int_{\Gamma'} d\mathbf{l}' \right] \times \mathbf{B}(\mathbf{r}). \tag{5.29}$$

From (5.29) we can deduce that the force exert by a homogeneous magnetic field over a closed filament crossed by a current I' is zero since $\oint_{\Gamma'} d\mathbf{l}' = 0$.

On the other hand, the moment of the force $\mathbf{N}$ is defined as

$$\mathbf{N}(\mathbf{r}) = \int_{\Gamma'} \mathbf{r}' \times d\mathbf{F}(\mathbf{r}'). \tag{5.30}$$

In some circumstances we have two or more circuits for which currents flow. Taking into account that each circuit produces a magnetic field, they exert a force on each other. For simplicity, let us suppose that we have two circuits represented by Γ_1 and Γ_2 whose currents are respectively, I_1 and I_2. If we look at one of the circuits, for instance Γ_2, it is affected by the presence of the magnetic field $\mathbf{B}_1$. By using (5.29) it may be demonstrated that the force over the current I_2 is

$$\mathbf{F}_{1-2} = \frac{\mu_0}{4\pi} \oint_{\Gamma_2} \oint_{\Gamma_1} \frac{I_2 d\mathbf{l}_2 \times (I_1 d\mathbf{l}'_1 \times [\mathbf{r} - \mathbf{r}'])}{|\mathbf{r} - \mathbf{r}'|^3}. \tag{5.31}$$

where $\mathbf{r}'$ represents the vector joining the origin of reference with a generic point on the first circuit. In case we had calculated the force on the first circuit due to the magnetic field generated for I_2, we would have obtained the same results provide the changes of the subindexes in (5.31) were done adequately.

5.6 Magnetic Dipole

The magnetic field created on a point in the space by a circuit of an arbitrary shape is, in general, of huge complexity. However, when calculating **B** produced by any circuit of magnetic moment **m** far away from its location, it can be demonstrated that this field does not depend on the specific geometrical characteristics of such a circuit but only of **m**. This means that all circuits with the same magnetic moment **m** produce the same **B** in a point away from the source. Of course the question of the actual meaning of *far away* arises. In our context the words far or near are meaningless without a reference for comparison. For this reason, when we say that one point is very far from a system it should be understood that the distance from the origin of the circuit to the point where we calculate the field is much greater than the linear dimensions of such a circuit. The expressions for the magnetic vector potential and the magnetic field **B** created by a dipole are, respectively,

$$\mathbf{A} = \frac{\mu_0}{4\pi} \frac{\mathbf{m} \times \mathbf{r}}{r^3}, \qquad (5.32)$$

and,

$$\mathbf{B} = \frac{\mu_0}{4\pi} \left(-\frac{\mathbf{m}}{r^3} + 3 \cdot \frac{\mathbf{m} \cdot \mathbf{r}}{r^5} \cdot \mathbf{r} \right). \qquad (5.33)$$

Sometimes it is necessary to study the mechanical behavior of a dipole in the presence of an external uniform magnetic field $\mathbf{B}_e$. In this case we can use the notion of the potential energy of the dipole, which is defined as follows:

$$E_p = -\mathbf{m} \cdot \mathbf{B}_e = -mB_e \cos\theta, \qquad (5.34)$$

where θ is the angle formed by both vectors. From this equation it is inferred that magnetic moments tend to orient in the same direction as the external magnetic field $\mathbf{B}_e$, because for this disposition the potential energy is a minimum. This result may be also obtained by applying (5.30) i.e. $\mathbf{N} = \mathbf{m} \times \mathbf{B}_e$.

5.7 Off-Axis Magnetic Field for Axisymmetric Systems

Let us suppose a system with rotational symmetry with respect to the OZ axis. The solution for the axial and radial components of the magnetic field **B** for this system with axisymmetric geometry is given by

$$B_z = \sum_{n=0}^{\infty} (-1)^n \frac{B_z^{(2n)}(0, z)}{2^{2n}(n!)^2} \rho^{2n}, \qquad (5.35)$$

and for the radial component

$$B_\rho = \sum_{n=0}^{\infty} (-1)^n \frac{B_z^{(2n)}(0, z)}{2^{2n+1} n!(n+1)!} \rho^{2n+1}, \tag{5.36}$$

where $B_z(0, z)$ is the z-component of the magnetic field along the revolution axis of the system. Due to the rotational symmetry the magnetic vector potential has only the angular component distinct to zero

$$A_\phi = \sum_{n=0}^{\infty} (-1)^{n+1} \frac{B_z^{(2n+1)}(0, z)}{2^{2n+1} n!(n+1)!} \rho^{2n+1}. \tag{5.37}$$

Another possibility is to use the magnetic scalar potential. In a current-free region $\nabla \times \mathbf{B} = 0$ we can derive the magnetic field through a potential function, i.e. $\mathbf{B} = -\mu_0 \nabla U_m(\mathbf{r})$.[8] On the other hand from the Maxwell equations $\nabla \cdot \mathbf{B} = 0$ resulting in $\nabla^2 U_m = 0$, which is Laplace's equation for the magnetic potential in any region free of currents (see Chap. 7). Due to the axial symmetry of the system we can use spherical coordinates to express the general solution of the Laplace differential equation. Our solution for the axisymmetric scalar potential is

$$U(r, \theta) = \sum_{n=0}^{\infty} \left(a_l r^l + \frac{b_l}{r^{l+1}} \right) P_l(\cos \theta), \tag{5.38}$$

where a_l and b_l are unknown coefficients to be determined and $P_l(cos\theta)$ represent a Legendre polynomial of order l. In order to avoid singularities inside the system to be analyzed, we set $b_l = 0$ in of (5.38) obtaining the following development

$$U(r, \theta) = a_0 + a_1 r P_1(\cos \theta) + a_2 r^2 P_2(\cos \theta) + \cdots + a_n r^n P_n(\cos \theta) + \cdots \tag{5.39}$$

Particularizing for points over the revolution axis of the system (OZ-axis) we introduce $\cos \theta = 1$ in (5.39) obtaining

$$U(r, \theta) = a_0 + a_1 z + a_2 z^2 + \cdots + a_n z^n + \cdots \tag{5.40}$$

The problem we have is determining the values of the coefficients a_l. With this aim we employ the relation between the magnetic scalar potential and the magnetic field $\mathbf{B}$. For the component z of the field this relation is

$$B_z(z) = -\mu_0 \frac{\partial U_m}{\partial z}, \tag{5.41}$$

---

[8]In the next chapter we will define the magnetic field $\mathbf{H}$ with which we can express this equation in a equivalent form as $\mathbf{B} = -\mu_0 \nabla U_m(\mathbf{r}) = \mu_0 \mathbf{H}$, thus $\mathbf{H} = -\nabla U_m(\mathbf{r})$. See Chap. 7, Problem 20.

5.7 Off-Axis Magnetic Field for Axisymmetric Systems

and integrating (5.41) we have

$$U_m(z) = -\frac{1}{\mu_0} \int B(z)dz + C, \qquad (5.42)$$

C being a constant. Once $U_m(z)$ is known, we have a way to determine the coefficients a_l of (5.40). In fact, expanding the magnetic potential (5.42) in Taylor series around $z = 0$ [9] for our specific system (solenoid, paraboloid, etc.), (5.42) leads to

$$U_m(z) = U_m(0) + \frac{\partial U_m}{\partial z}z + \frac{1}{2!}\frac{\partial^2 U_m}{\partial z^2}z^2 + \cdots + \frac{1}{n!}\frac{\partial^n U_m}{\partial z^n}z^n + \vartheta(z^{(n+1)}), \qquad (5.43)$$

whose coefficients are the same that appear in (5.40), thus it holds

$$a_l = \frac{1}{l!}\frac{\partial^l U_m}{\partial z^l}. \qquad (5.44)$$

From these data we have the general expression (5.40) for the potential $U_m(\mathbf{r})$. Thus, taking its gradient, the components of the magnetic field **B** take the form

$$B_r(r, \theta) = -\mu_0 \frac{\partial U_m}{\partial r}, \qquad (5.45)$$

and

$$B_\theta(r, \theta) = -\mu_0 \frac{1}{r}\frac{\partial U_m}{\partial \theta}. \qquad (5.46)$$

This magnetostatic potential technique gives good analytical results everywhere inside of the system to be studied, if the number of terms of the Taylor expansion is enough. However, it does not accurately describe the magnetic field outside the system.

The most important conclusion of this technique explained, is that for calculating the magnetic field inside a system with rotational symmetry we only need to know the field **B** along its symmetry axis (in our development z). This procedure is very useful for analyzing electromagnetic systems, in which the computation of the off-axis magnetic field through the current distributions is very difficult.

---

[9] We have expanded U_m around $z = 0$ for simplicity. In general, we can take the Taylor expansion around any point of interest.

Solved Problems

Problems A

5.1 Calculate the magnetic field created by a very long straight conducting wire carrying a homogeneous current density $\mathbf{j} = j\mathbf{u}_z$, in any point of the space exterior to the wire.

Solution

The magnetic field created for a very long conducting wire may be solved in different ways. One of them is by directly applying the Biot–Savart law, and the other one is by following Ampère's law. Both procedures are equivalent but we will first resolve it through the second method leaving the (5.26) for another exercise. We will explain the solution in detail because the reasoning and results we are going to see are applicable to many other problems.

As we have mentioned in the theory, the Ampère law procedure is specially useful when the system that creates the magnetic field has symmetries. For this reason we will begin with the study of geometrical characteristics of our system. Let us suppose the current extends along the OZ axis. This election facilitates the calculation without loss of generality. In principle, as we have seen in (5.1), the existence of a density current may generate a magnetic field. Let us fix a generic point P of coordinates (x, y, z) at which we want to study such a field. As we do not know a priori the structure of $B(\mathbf{r})$, we work with the usual components in three dimensions (Fig. 5.3). With the aim to study the symmetries, let us consider another point, say P', with the same projection on the OXY plane but different coordinate over OZ, i.e. $(x, y, z + z_0)$, z_0 being a constant (Fig. 5.4). From a geometrical point of view both points are equivalent. In fact, an observer located at P translates to P' cannot realize the difference between both positions because, due to the infinite length of the wire, nothing is changed upwards nor downwards. This means that the system is invariant under a translation along OZ and, as a consequence the magnetic field at P and P' must be equal.

Fig. 5.3 Metallic wire

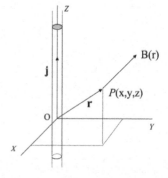

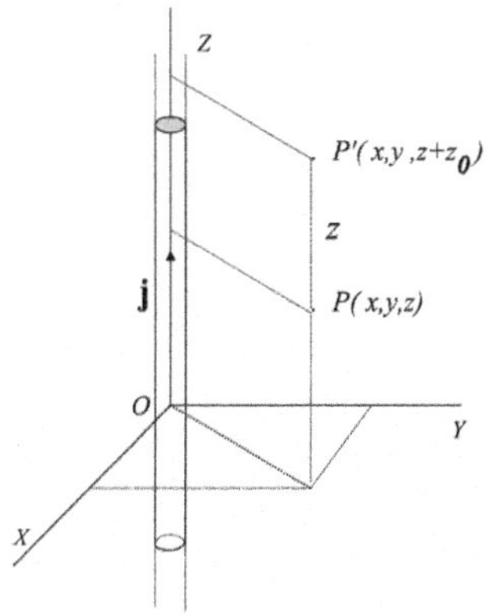

Fig. 5.4 The points P and P' have the same projections on the plane OXY but different coordinate z

Fig. 5.5 Point $P*$ is reached by a rotation of angle ξ around OZ

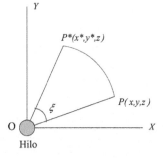

Now let us imagine a parallel plane to OXY and on it a $P(x, y, z)$ (Fig. 5.5). If we rotate this point an angle ξ around the OZ axis we obtain another point $P*$ of coordinates $(x*, y*, z)$. This new point has different projections over OX and OY, however both points are again equivalent because we are not able to distinguish the magnetic field detected at both positions. Thus the system is invariant under rotations around one axis passing through the wire. Therefore, rotational symmetry is inferred and the magnetic field must be equivalent as well. Strictly speaking, when rotating the reference system $S(OXYZ)$ to another $S'(OX'Y'Z')$, the cylindrical components of **B** in the new system S' are the same as the projections of the magnetic field in S.

Due to the symmetries of the system we will employ a cylindrical system of coordinates ($\{\mathbf{u}_\rho, \mathbf{u}_\phi, \mathbf{u}_z\}$). In order to apply Ampère's law for obtaining the components of the magnetic field, we must choose different curves for performing the integral (5.5). The curve chosen in every case must be such that when performing the scalar product $\mathbf{B} \cdot d\mathbf{l}$ there appears the component of the field to be calculated. But, on the other hand, this projection over $d\mathbf{l}$ needs to be constant, otherwise we cannot reach an expression like (5.6).

Let us begin by calculating the tangential component of **B**. With this aim, because of the rotational symmetry of the system, it seems logical using a curve with the same characteristics of such a symmetry. For this reason we choose a circle of radius ρ whose center is on the wire and contained in a plane perpendicular to the current (Fig. 5.6).

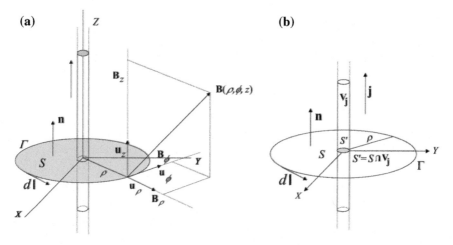

Fig. 5.6 **a** Three components of the magnetic field in cylindrical coordinates. **b** Actual area of integration. Observe that the intersection between the volume density current and the surface S is surface S' on which we perform the integration (case (**a**) in Fig. 5.2)

Together with the curve we need a differential $d\mathbf{l}$ along the curve and a surface (whose boundary is the curve) to perform the right side of (5.7). In principle, we can chose any well defined open surface whose contour coincides with the curve Γ. In order to resolve the problem as easy as possible, we employ the plane where the circle is contained. We also select the sense (or direction) of the differential length to be anticlockwise, then the surface S is oriented in a direction parallel to the positive OZ axis. The vector $d\mathbf{l}$ tell us the way we go along the curve and its direction (clockwise or anticlockwise) determines the surface orientation, and the sign of the integral, but do not change the final result of the magnetic field produced. As aforesaid, taking into consideration the symmetries of the system, we will use cylindrical coordinates. Since the circle lies entirely in the OXY plane, and the density current over OZ, we have $\mathbf{j} = j\mathbf{u}_z$ and $d\mathbf{l} = dl\mathbf{u}_\phi = (0, dl, 0)$. On the other hand, as we chose a plane surface parallel to OXY, the element differential of surface is $d\mathbf{S} = dS\,\mathbf{u}_z = (0, 0, 1)\,dS$. Introducing these values in (5.7) we obtain:

$$\oint_{\partial S} (B_\rho, B_\phi, B_z) \cdot (0, dl, 0) = \mu_0 \int_S (0, 0, j) \cdot (0, 0, 1) dS.$$

and operating the scalar products

$$\oint_{\partial S} B_\phi dl = \mu_0 \int_{S'=S \cap V_j} j dS.$$

Solved Problems

By analyzing the integral on the left we observe that now B_ϕ is a constant magnitude (due to the scalar product). It means that its value is the same for any point on the circumference, then we can place it outside the integral. With respect to the term on the right side, we must calculate the intensity by evaluating the scalar product of $j\mathbf{u}_z$ through the surface S. However, as we commented in the theory, such a surface does not always coincide with the surface S bounded by Γ. In fact, this integral is taken over the surface $S' = S \cap V_j$ ($S' \subset S$). When calculating the scalar product we see that it is positive and the current density is constant, thus

$$\int_0^L B_\phi dl = \mu_0 \int_S j dS \Rightarrow B_\phi \int_0^L dl = \mu_0 j \int_S dS, \qquad (5.47)$$

where $L = 2\pi\rho$ and $jS = I$, I being the current through the wire. Then

$$B_\phi L = \mu_0 j S' \Rightarrow B_\phi 2\pi\rho = \mu_0 I,$$

obtaining the result

$$B_\phi = \frac{\mu_0 I}{2\pi\rho}. \qquad (5.48)$$

Since I is positive, $\mathbf{B}_\phi$ is the tangential component of the field $\mathbf{B}$, which is positive over the coordinate system chosen,

$$\mathbf{B}_\phi = \frac{\mu_0 I}{2\pi\rho}\mathbf{u}_\phi. \qquad (5.49)$$

If we had taken the differential $d\mathbf{l}$ in the opposite direction, we would have obtained the scalar products with the signs changed, but on both sides of (5.47), thus not varying the result.

By following this procedure, we only obtain the tangential component of the field. This does not mean the other components are zero because they do not appear. It just means that with the curve chosen we were able to compute the tangential component of $\mathbf{B}$. Ultimately, the geometry of the curve determines the information about the magnetic field we can get from Ampère's law. To calculate the components over z and ρ, it is necessary to use other integration paths.

In effect, if we wish to calculate the field over OZ, we need a curve that can account for the symmetry related to this coordinate axis. Taking into account the analysis employed at the beginning of this exercise, let us consider a rectangle-shaped curve γ (Fig. 5.7). This rectangle is coplanar with the wire, and its opposite sides are parallel to each other. Subdividing the curve into smaller segments γ_i ($i = 1, 2, 3, 4$) we see that γ_1 and γ_3 are parallel to OZ, where we have translation symmetry. Observe that the distances from the segments γ_1 and γ_3 to OZ are not the same (we shall explain the reason later). On the other hand, following in the same way shown for the tangential component of the field, we choose a plane surface for integrating the right side of (5.5), and $d\mathbf{l}$ counterclockwise. Applying again Ampère's theorem over

Fig. 5.7 Rectangular curve. Throughout the segments γ_1 and γ_3 only appears the component B_z, once we have computed the scalar product $\mathbf{B} \cdot d\mathbf{l}$. In the same way, over the curves γ_2 and γ_4 we have information about B_ρ

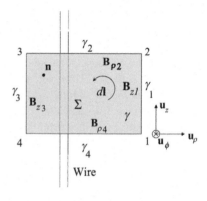

γ ($\gamma_1 + \gamma_2 + \gamma_3 + \gamma_4$), and on the surface Σ, we find by subdividing γ into its four segments

$$\oint_\gamma (B_\rho, B_\phi, B_z) \cdot d\mathbf{l} = \int_{\gamma_1} \mathbf{B}_1 \cdot d\mathbf{l}_1 + \int_{\gamma_2} \mathbf{B}_2 \cdot d\mathbf{l}_2 + \int_{\gamma_3} \mathbf{B}_3 \cdot d\mathbf{l}_3 + \int_{\gamma_4} \mathbf{B}_4 \cdot d\mathbf{l}_4 = \mu_0 \int_\Sigma (0,0,j) \cdot d\mathbf{S}. \tag{5.50}$$

To develop (5.50) we introduce different $\mathbf{B}_i$ for each segment, because we only know $\mathbf{B}_\phi$. By setting the differential element of length for every path corresponding to the curve γ, i.e. $d\mathbf{l}_1 = dz\mathbf{u}_z = (0,0,dz)$, $d\mathbf{l}_2 = d\rho\mathbf{u}_\rho = (d\rho,0,0)$, $d\mathbf{l}_3 = dz\mathbf{u}_z = (0,0,dz)$ y $d\mathbf{l}_4 = d\rho\mathbf{u}_\rho = (d\rho,0,0)$, into (5.50) we obtain

$$\int_1^2 (B_{\rho 1}, B_{\phi 1}, B_{z1}) \cdot (0,0,dz) + \int_2^3 (B_{\rho 2}, B_{\phi 2}, B_{z2}) \cdot (d\rho,0,0) + \int_3^4 (B_{\rho 3}, B_{\phi 3}, B_{z3}) \cdot (0,0,dz) + \tag{5.51}$$

$$+ \int_4^1 (B_{\rho 4}, B_{\phi 4}, B_{z4}) \cdot (d\rho,0,0) = \int_1^2 B_{z1} dz + \int_2^3 B_{\rho 2} d\rho + \int_3^4 B_{z3} dz + \int_4^1 B_{\rho 4} d\rho.$$

By analyzing this last expression we notice that because of the symmetry of translation along the OZ direction, the value of the magnetic field over all points of the curve γ_2 (path 2–3) must be the same as the field over γ_4 (path 2–3), then $B_{\rho 2}(\rho,\phi,z)_{\gamma_2} = B_{\rho 4}(\rho,\phi,z)_{\gamma_4}$, and then

$$\int_2^3 B_{\rho 2} d\rho = \int_1^4 B_{\rho 4} d\rho. \tag{5.52}$$

However, the integral over γ_4 in (5.51) goes from point 4 to 1 and not vice versa, then their signs must be different. As a consequence we can write

$$\int_2^3 B_{\rho 2} d\rho = -\int_4^1 B_{\rho 4} d\rho. \tag{5.53}$$

By introducing it into (5.50), and knowing that $d\mathbf{S} = -dS\mathbf{u}_\phi = (0, -1, 0)dS$, we have

$$\int_1^2 B_{z1}dz + \int_3^4 B_{z3}dz = \mu_0 \int_\Sigma (0, 0, j) \cdot (0, -1, 0)dS = 0. \quad (5.54)$$

Due to the aforementioned symmetry we cannot distinguish between two points located over the same curve γ_1, or γ_3. This means that independently of the point belonging to γ_1 the value of B_{z1} is the same for all points of this segment, and must not depend on z, otherwise we would have different values of B_{z1} for the same distance to the wire, breaking the translation symmetry in the OZ direction. The same reasoning applies to B_{z3} over the segment γ_3. However, B_{z1} must not be necessarily the same as B_{z3} because curves γ_1 and γ_3 are placed at distinct distances to OZ. This is the reason why we have chosen a non-symmetrical square for applying the Ampère law. Otherwise the integrals corresponding to this paths would have cancelled each other out due to symmetry. Therefore we can write,

$$B_{z1}\int_1^2 dz + B_{z3}\int_3^4 dz = 0 \Rightarrow B_{z1}L - B_{z3}L = 0 \Rightarrow B_{z1} = B_{z3}, \quad (5.55)$$

where the minus sign appears due to the fact that the integrals over γ_1 and γ_3 are calculated in opposite direction, and L is the length of the segment. This result shows that the z component of the magnetic field is identical for all points, independently of the distance to the wire, then it cannot depend on the ρ. Neither shall it depend on z, nor ϕ, because it contradicts the argument of symmetries previously done. From this it follows that $B_{z1} = B_{z3} = C$, where C is a constant undetermined. This result is ambiguous, as it shows that B_z is constant for every point but does not say what is its value. In order to determine the constant C let us imagine that we repeat the last calculations but not over the original curve γ, but for another rectangle γ^*, similar to γ in shape, but with Sects. 2.3 and 1.4 larger (Figs. 5.8 and 5.9). If this is done we observe that the result obtained for B_z is identical to the first one. Then, apparently we are not able to determine the value of the constant C.

As a last attempt let us consider the same calculation again, but choosing a rectangular curve in which the lengths γ_2 and γ_4 are as large as possible (Fig. 5.9). The

Fig. 5.8 Rectangular curve with longer sides

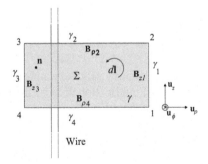

Fig. 5.9 Rectangular curve composed of very long segments

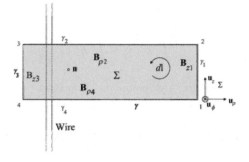

Fig. 5.10 Plane curve for applying the theorem

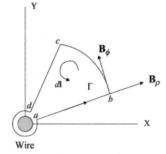

result is again the same, but there is a difference. In effect, it is known that a general property of the magnetic field is that it must be a well defined function. It means that the field must converge from far away from the currents, i.e. the wire in our case. Therefore, theoretically, the field must tend to zero at infinity. Using this argument we arrive at the conclusion that C has to be zero. As the physical reasoning followed for all the curves chosen is the same, this result obtained must be valid for all of them, thus we infer that the constant C is zero in any case, and then we can write $B_z = 0$. To calculate the radial part of the field through (5.7), we must first find a path over which B_ρ appears when doing $\mathbf{B} \cdot d\mathbf{l}$ and, moreover, that is constant for this radial component. To do so we look at the symmetries studied. In principle, a possible curve may be a sector as showing in Fig. 5.10. This curve

$$\int_a^b (B_{\rho 1}, B_{\phi 1}, B_{z1}) \cdot (d\rho, 0, 0) + \int_b^c (B_{\rho 2}, B_{\phi 2}, B_{z2}) \cdot (0, \rho d\phi, 0) + \int_c^d (B_{\rho 3}, B_{\phi 3}, B_{z3}) \cdot (d\rho, 0, 0) + \tag{5.56}$$

$$\int_d^a (B_{\rho 4}, B_{\phi 4}, B_{z4}) \cdot (0, \rho d\phi, 0) = \int_a^b B_{\rho 1} d\rho + \int_b^c B_{\phi 2} \rho d\phi + \int_c^d B_{\rho 3} d\rho + \int_d^a B_{\phi 4} \rho d\phi.$$

However, when closely examining this closed path it is observed that it can not give us the desired information because the integrals along (a–b) and (c–d) are the same but in opposite directions, then they cancel each other, losing the information about B_ρ. For this reason, it is necessary to consider a different point of view for determining the radial component.

Fig. 5.11 Closed surface used for applying flux theorem

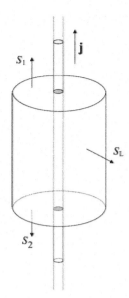

To calculate the radial component we may try (5.2) in its integral form, which accounts for the flux of the magnetic field throughout a closed surface (see Chap. 10). The application of this equality differs from (5.7) because the integral must be taken over a closed surface instead a closed curve. However, we can continue to use the symmetries.

The only surface that retains some of the symmetries of the wire is a cylinder. Thus, let us consider a cylinder-shaped surface of height h (Fig. 5.11) and radius ρ, whose axis of revolution coincides with the wire. The closed surface may be divided in three parts: the upper and lower bases, S_1, S_2, and the lateral surface S_L. Applying (5.2) we obtain

$$\oint_S \mathbf{B} \cdot d\mathbf{S} = \int_{S_1} \mathbf{B}_1 \cdot d\mathbf{S}_1 + \int_{S_2} \mathbf{B}_2 \cdot d\mathbf{S}_2 + \int_{S_L} \mathbf{B}_L \cdot d\mathbf{S}_L = 0 \,, \tag{5.57}$$

Setting, $d\mathbf{S}_1 = dS_1\mathbf{u}_z = (0, 0, 1)dS_1$, $d\mathbf{S}_2 = dS_2\mathbf{u}_z = (0, 0, -1)dS_2$ y $d\mathbf{S}_L = dS_L\mathbf{u}_\rho = (1, 0, 0)dS_L$, we have:

$$\int_{S_1}(B_{\rho_1}, B_{\phi_1}, B_{z_1}) \cdot (0, 0, 1)dS_1 + \int_{S_2}(B_{\rho_2}, B_{\phi_2}, B_{z_2}) \cdot (0, 0, -1)dS_2$$

$$+ \int_{S_L}(B_{\rho_L}, B_{\phi_L}, B_{z_L}) \cdot (1, 0, 0)dS_L = 0.$$

Owing to the translational symmetry, the field that exists at a point $P_1(\rho, \phi, z)$ on the surface S_1, has to be equal to another point P_2 on S_2 translated an amount h parallel to the OZ axis, i.e. $P_2(\rho, \phi, z - h)$. The same does not apply a point on the lateral surface S_L, because the distance to the rotation axis is different, so the magnetic field does not have to be identical to the other ones. However, due to the rotational symmetry, all points on S_L are equivalent since the distance of such a point to OZ is the same, and therefore the modulus of the field is the same too. Putting $dS_1 = dS_2$ in the above equation we have

$$\int_{S_1} B_{z_1} S_1 - \int_{S_1} B_{z_1} dS_1 + \int_{S_L} B_\rho dS_L = 0 \Rightarrow B_\rho \int_{S_L} dS_L = 0 \Rightarrow B_\rho S_L = 0 \Rightarrow B_\rho = 0, \tag{5.58}$$

and then,

$$\mathbf{B}_\rho = 0. \tag{5.59}$$

In conclusion we observe after all the calculus for each of the components of **B**, the magnetic field created by a very long wire which carries a current I, only has tangential component, thus the final expression is

$$\mathbf{B} = \frac{\mu_0 I}{2\pi\rho} \mathbf{u}_\phi. \tag{5.60}$$

5.2 A current of $I = 2$ A flows through a large cylindrical conductor of radius $R = 10$ cm. Obtain: (a) The magnetic field B for $\rho < R$. (b) Idem for $\rho > R$.

Solution

(a) Let us suppose we have chosen a reference system in which the cylinder is placed coinciding its revolution axis with the OZ axis. Due to the rotational symmetry of the conductor, it seems appropriate to use (5.7) instead the Biot–Savart Law. To calculate the magnetic field we have to define a curve for performing the integral. Because of the aforementioned symmetry a circle is chosen (Fig. 5.12). Using polar coordinates we can write for $d\mathbf{l} = (0, dl, 0)$, and for the current density $(0, 0, j)$. By using Ampère's theorem we have

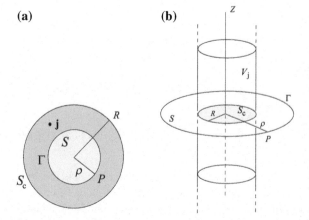

Fig. 5.12 a Plane view of the current in direction of OZ axis. The surface S is delimited by the curve Γ of radius $\rho < R$ ($S < S_c$). **b** For a point P exterior to the cylinder the effective surface is the same as the surface of the conductor S_c, and it corresponds to the intersection between the volume where the current **j** exists and the surface S whose border is the curve Γ ($S \cap V_j = S_c$)

Solved Problems

$$\oint_\Gamma \mathbf{B} \cdot d\mathbf{l} = \oint_\Gamma (B_\rho, B_\phi, B_z) \cdot (0, dl, 0) = \mu_0 \int_S \mathbf{j} \cdot d\mathbf{S} = \mu_0 \int_S (0, 0, j) \cdot (0, 0, dS), \tag{5.61}$$

and making the scalar products

$$\oint_\Gamma B_\phi dl = \mu_0 \int_S j dS \Rightarrow \int_0^{2\pi} B_\phi \rho d\phi = \mu_0 \int_0^{2\pi} \int_0^\rho j\rho d\rho d\phi. \tag{5.62}$$

Due to its symmetry, the system is invariant under a rotation around the OZ axis, then B_ϕ and j do not change over the curve Γ of radius ρ. For this reason we may write

$$B_\phi \rho \int_0^{2\pi} d\phi = \mu_0 j \int_0^{2\pi} d\phi \int_0^\rho \rho d\rho \Rightarrow 2\pi \rho B_\phi = 2\pi \mu_0 j \frac{1}{2}\rho^2, \tag{5.63}$$

that is

$$B_\phi = \frac{1}{2}\mu_0 j \rho. \tag{5.64}$$

This result depends on the current density j which is not given in the problem. We must transform this expression to another where I appears. With this aim let us employ the definition of intensity, i.e.

$$I = \int_S \mathbf{j} \cdot d\mathbf{S}. \tag{5.65}$$

In this problem we know the total intensity I through the conductor. Thus, applying (5.65) to a surface of the conductor perpendicular to the current density $\mathbf{j}$, it results

$$I = \int_0^{2\pi} d\phi \int_0^R \rho d\rho = j\pi R^2 \Rightarrow j = \frac{I}{\pi R^2}, \tag{5.66}$$

in the direction of the OZ axis. Combining this result with (5.64) we obtain

$$B_\phi = \frac{\mu_0 I}{2\pi R^2} \rho. \tag{5.67}$$

(b) The calculation for an exterior point may be made in a similar form as shown in the preceding paragraph, but choosing a curve that passes for the point to be analyzed. In fact, applying the Ampère theorem to the curve Γ in Fig. 5.12b we have

$$\oint_\Gamma B_\phi dl = \mu_0 \int_{S \cap V_j} j dS \Rightarrow \int_0^{2\pi} B_\phi \rho d\phi = \mu_0 \int_0^{2\pi} \int_0^R j\rho d\rho d\phi. \tag{5.68}$$

Note that the integral of the right extends over the surface of radius R of the cylinder and not ρ as seen in section (a). The reason for this is that, although the surface of integration extends to ρ, the scalar product $\mathbf{j} \cdot d\mathbf{S}$ is zero outside the conductor, where

j is zero (Fig. 5.12). Therefore the integral extends to R, which corresponds to the intersection between the surface S and the volume V_j where the current density flows ($S \cap V_j$). Integrating this latter expression we have

$$B_\phi = \frac{\mu_0 I}{2\pi \rho}, \tag{5.69}$$

result similar to the obtained for the wire. For the other components of the magnetic field can proceed similarly as we saw in Problem 5 obtaining the same result, i.e. $B_\rho = B_z = 0$.

5.3 Find the magnetic field created by a toroidal solenoid of N turns if the density current that circulates is j.

Solution

A torus is the geometry that is the result of rotating a circle of radius R around an axis coplanar with it (Fig. 5.13). The final surface may be mathematically described by giving its minor and major radius, which are labelled in the figure by R_1 and R_2, respectively. Roughly speaking it appears to be like a donut or a smooth wheel. In our problem what we physically have is a wire wound around an imaginary torus, obtaining a geometry as depicted in Fig. 5.14. As it may be observed, the curve formed by the wire is like a spring screw (Fig. 5.15a). To calculate the tangential component of the magnetic field we suppose that the wire is closely wound, then

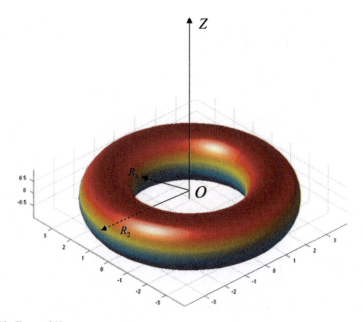

Fig. 5.13 Torus of N turns

Solved Problems

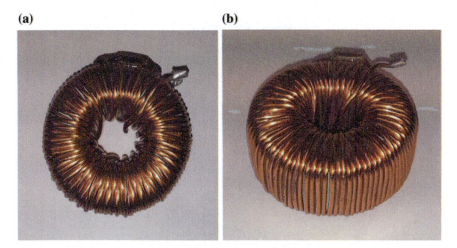

Fig. 5.14 a Plane view of a toroidal solenoid. **b** The same figure in perspective. Observe that the wire is closely wound

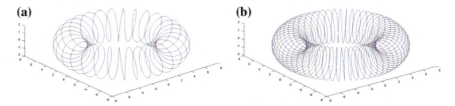

Fig. 5.15 a Winding of wire in form of a screw helix. **b** In this case the wire is closely wound, then the solenoid may be illustrated as a set of N independent coils

the winding does not have space (Fig. 5.15b). Examining the symmetries of the torus we can see that it is invariant under rotations around an axis that crosses its geometrical center (OZ axis in Fig. 5.13). If we remember the explanations done for the infinite wire, the symmetries show a way to solve problems simply. In the present case we are dealing with the computation of the tangential component of the field, where the system has symmetry. For this reason we will apply Ampère's law in lieu of the Biot–Savart integral. Due to the aforementioned rotational symmetry, we choose cylindrical coordinates for the calculation. On the other hand for studying the different parts of the torus, we divide the problem in three parts. The first one corresponding to the inner solenoid ($R_1 < \rho < R_2$), and the other when ($0 < \rho < R_1$) and ($R_2 < \rho < \infty$), respectively. (a) ($R_1 < \rho < R_2$). For working this problem we choose a curve Γ circular in shape, centered in the geometrical center of the solenoid (Fig. 5.16). For calculating the flux of $\mathbf{j}$ corresponding to the right side of the Ampère integral we choose an open plane surface and the element $d\mathbf{l}$ in clockwise direction, then the normal to the surface is oriented in the negative direction of the OZ axis, i.e. $d\mathbf{S}_1 = -dS\,\mathbf{n}_1 = (0, 0, -1)\,dS$. If we suppose, for instance, that the current density

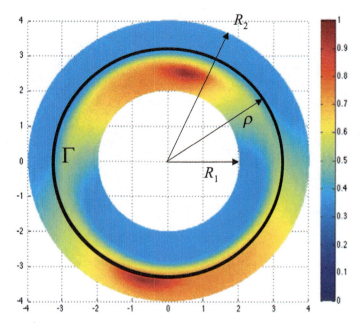

Fig. 5.16 Plan view of the torus. A circular curve Γ of radius ρ is chosen ($R_1 < \rho < R_2$)

follows the direction entering the surface, its expression at the intersection with the surface S is $(0, 0, -j)$.[10] Introducing all these data in (5.7) we have

$$\oint_\Gamma (B_\rho, B_\phi, B_z) \cdot (0, dl, 0) = \mu_0 \int_{S_1} (0, 0, -j) \cdot (0, 0, -1) dS \Rightarrow \int_0^{2\pi} B_\phi dl = \int_{S_1} j dS.$$
(5.70)

As we have mentioned previously, due to curve used for integration, the only component that appears after the scalar product is B_ϕ which, on the other hand, must be the same for all points on the curve Γ because of the symmetry of revolution. For the surface integral of the second member, the domain of integration for the intensity is, in principle, the surface S. However the integrand is zero except at intersections of S with each one of the turns,[11] then the intensity of current through the surface S, $\int_S \mathbf{j} \cdot d\mathbf{S}$, is NI. Considering this, and putting the differential element of length dl in polar coordinates, we have:

---

[10] The density current is tangent a each point of the toroidal solenoid, hence this expression is only valid for points of intersection between $\mathbf{j}$ and $d\mathbf{S}$. If the surface chosen for calculating the flux is not planar, the components of $\mathbf{j}$ over $d\mathbf{S}$ would be different to $(0, 0, -j)$ but the final result of the Ampère law does not change.

[11] In this case $S \cap V_j$ corresponds to the section of the wire used in the coil (see Sect. 5.2).

$$\int_0^{2\pi} B_\phi \rho d\phi = \int_{S \cap V_j} j dS = NI \Rightarrow B_\phi \int_0^{2\pi} \rho d\phi = NI \Rightarrow 2\pi \rho B_\phi = NI \Rightarrow B_\phi = \frac{NI}{2\pi \rho}, \quad (5.71)$$

that is,

$$\mathbf{B}_\phi = \frac{NI}{2\pi \rho} \mathbf{u}_\phi \quad (5.72)$$

(b) $\rho < R_1$. To analyze this case we follow the same procedure as before. Let us consider a circular curve Γ of radius ρ but now $\rho < R_1$ (Fig. 5.17). As we have seen when explaining Ampère's law, the curvilinear integral of B over the curve Γ depends on the net intensity crossing the open surface S. Bearing in mind this idea, it is obvious that there is no net current traversing S, then

$$\int_0^{2\pi} B_\phi \rho d\phi = 0 \Rightarrow B_\phi \int_0^{2\pi} \rho d\phi = 0 \Rightarrow 2\pi \rho B_\phi = 0 \Rightarrow B_\phi = 0, \quad (5.73)$$

that expressed in vectorial form gives

$$\mathbf{B}_\phi = 0. \quad (5.74)$$

(c) $\rho > R_2$. Let us now choose a curve Γ of radius ρ larger than R_2 (see Fig. 5.18). Again applying Ampère's law, no differences are found in the integral of the field. Following the same procedure as the preceding sections (a) and (b), a similar result is obtained. However this is not the case of the intensity (right hand of (5.70)). In

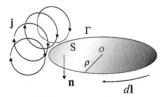

Fig. 5.17 In this case **j** does not cross the surface S

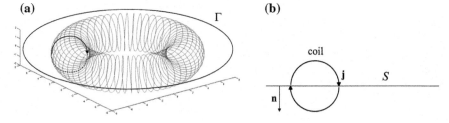

Fig. 5.18 **a** Curve Γ for $\rho > R_2$. **b** Side view. The density **j** cross S once down and another up, then the net intensity is zero

effect, the density current **j** crosses the surface S once down and again up, the net intensity crossing S is zero. For this reason we can write

$$\int_0^{2\pi} B_\phi \rho d\phi = NI - NI \Rightarrow B_\phi = 0 \Rightarrow \mathbf{B}_\phi = 0. \tag{5.75}$$

In short we can say that a toroidal solenoid of N coils carrying a current I produces a tangential magnetic field $\mathbf{B}_\phi$ only inside, being zero at any point outside it.

5.4 Find the tangential component of the magnetic field corresponding to a closely wound finite solenoid of length L, radius a, and n turns per unit length, carrying a current I for $\rho < a$ and $\rho > a$ (Fig. 5.29).

Solution

To address the problem different ways may be used, and the difficulty of the resolution depends, to some extent, on the initial suppositions about the geometry of the system. If we observe an actual solenoid it consists of a wire of diameter d wound in a helix. If the pitch of the helix coincides with d the coil is said to be closely wound (Fig. 5.19b). On the contrary, if we augment the pitch it does not (Fig. 5.19a). However, even though it occurs the calculation of the field may be not easy. In effect, in both cases, due to the finiteness of the solenoid and the form of the turn, we do not have either translational symmetry nor rotational invariance, hence the use of Ampère's law does not apply (it is not of practical use). For this reason the first possibility is to employ the Biot–Savart integral and directly see the tangential field component at the end of the calculation. Another similar method, however tedious too, would be to calculate the vector potential (5.26) and its curl. However, these procedures are usually of high complexity, for it requires the integration along the parametric equations of the helical curve.

But the problem actually may be simplified if we suppose that, besides that the coil is closely wound, the diameter of the wire is very thin. In this case we might neglect the pith of the helix, thus the solenoid would be equivalent to a set of independent rings carrying a current I (Fig. 5.20). The advantage of this new configuration is that it has rotational symmetry, hence allowing us to use Ampere's law as we have seen previously. But in order to do this simplification, what does thin mean? 0.5 cm? If not, 1 mm? 0.1 mm? To intuitively answer this question it is enough to look at Fig. 5.19. In effect, if the angle formed for the coils with respect to the plane perpendicular to the OZ axis is very small, then it means that the wire is thin. However, we need to make this intuitive idea more specific for applying it to actual problems. Even though each problem may require different conditions, a possible criterium could be the following (Fig. 5.21)

$$\tan(\alpha) = \frac{d}{2a} << 0.0025, \tag{5.76}$$

which corresponds to a solenoid of $D = 2a = 4$ cm and diameter of the wire $d = 1$ mm.

Supposing condition (5.76) holds, let us first begin with the calculation of the field for a point $\rho < a$. To this aim for performing the line integral of $\mathbf{B}(\mathbf{r})$ we choose a curve

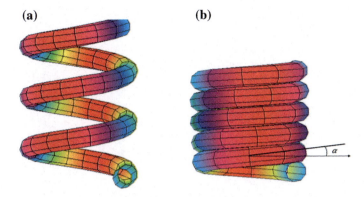

Fig. 5.19 **a** Finite solenoid of radius a for a pitch greater than the diameter of the wire. **b** Closed wound solenoid. Due to the dimension of the diameter, although the helix pitch coincides with d, the turn wire is inclined with an angle α with respect to a plane perpendicular to the OZ axis

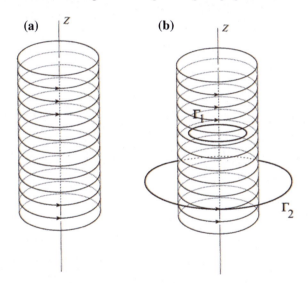

Fig. 5.20 **a** Finite solenoid of radius a. **b** Curves Γ_1 and Γ_2 for $\rho < a$ and $\rho > a$

Γ_1 circular in shape, concentric with the coil and contained on a plane perpendicular to the OZ axis (Fig. 5.22a and b). With respect to the open surface for calculating the flux of $\mathbf{B}(\mathbf{j})$, as we have seen in the problem (5.1) we can use a plane, due to its simplicity. In the present case we have $\mathbf{j}(\mathbf{r}') = (0, j_\phi, 0)$, $\mathbf{n} = (0, 0, 1)$ and $d\mathbf{l}' = (0, dl', 0)$. Introducing them into (5.7) we get

$$\oint_{\Gamma_1} (B_\rho, B_\phi, B_z) \cdot (0, dl, 0) = \mu_0 \int_S (0, j_\phi, 0) \cdot (0, 0, 1) dS = 0, \quad (5.77)$$

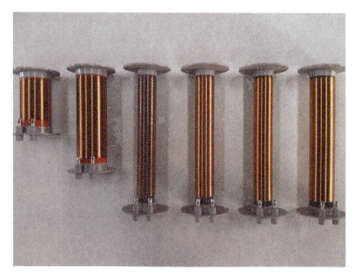

Fig. 5.21 Solenoids of different lengths and radii

Fig. 5.22 **a** Plane view of the curve Γ_1 and $\mathbf{j}(\mathbf{r}') = (0, j_\phi, 0)$. **b** Side view of the finite solenoid of radius a. Observe that there is no flux of $\mathbf{j}$ through the surface S

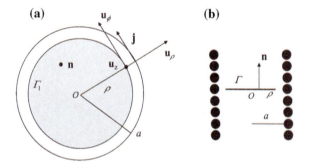

then

$$\int_0^L B_\phi dl = 0 \Rightarrow B_\phi = 0. \tag{5.78}$$

Therefore, the tangential magnetic potential of a finite coil is zero. For a point $\rho > a$ the procedure is the same, but by choosing a curve of a larger radius. In that case we again obtain identical result.

Observe that the results obtained does not depend on the high z at which the curve Γ_i ($i = 1, 2$) is located with respect to the solenoid.

5.5 A circular loop of radius a and constant current I is located at the coordinate origin as shown in Fig. 5.23. Obtain the magnetic field $\mathbf{B}$ at a point $P(0, 0, z)$.

Fig. 5.23 Circular loop carrying a current I

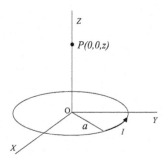

Solution

As the current flows through a filament the Biot–Savart law (5.26) may be used directly. To evaluate the field at P, we must first find the coordinates of the sources (primed) and the field coordinates (unprimed). The radius vector from the coordinate origin to the source (current) is,

$$\mathbf{r}' = (x', y', 0),$$

and for the field point

$$\mathbf{r} = (0, 0, z).$$

Besides the differential element of curve $d\mathbf{l}'$ is

$$d\mathbf{l}' = (dx', dy', 0),$$

and the radius vector from the source to the field point

$$(\mathbf{r} - \mathbf{r}') = (-x', -y', z).$$

With these data we can perform the vectorial product appearing in (5.26) and the integral. In fact, introducing these vectors in (5.26) we obtain

$$d\mathbf{l}' \times (\mathbf{r} - \mathbf{r}') = z\,dy'\mathbf{u}_x - z\,dx'\mathbf{u}_y + (x\,dy' - y'\,dx')\mathbf{u}_z, \qquad (5.79)$$

thus,

$$\mathbf{B} = \frac{\mu_0 I}{4\pi} \left\{ \oint_\Gamma \frac{z\,dy'\mathbf{u}_x - z\,dx'\mathbf{u}_y + (x\,dy' - y'\,dx')\mathbf{u}_z}{(x'^2 + y'^2 + z^2)^{3/2}} \right\}, \qquad (5.80)$$

and separating the integral into components

$$\mathbf{B} = \frac{\mu_0 I}{4\pi} \left\{ \oint_\Gamma \frac{z\,dy'}{(x'^2 + y'^2 + z^2)^{3/2}} \mathbf{u}_x - \oint_\Gamma \frac{z\,dx'}{(x'^2 + y'^2 + z^2)^{3/2}} \mathbf{u}_y + \oint_\Gamma \frac{(x\,dy' - y'\,dx')}{(x'^2 + y'^2 + z^2)^{3/2}} \mathbf{u}_z \right\}. \qquad (5.81)$$

Due to the symmetry it is convenient to write the coordinates of the source in term of polar coordinates, i.e.

$$\mathbf{B} = \frac{\mu_0 I}{4\pi} \left\{ \int_0^{2\pi} \frac{za\cos\phi' d\phi'}{(a^2+z^2)^{3/2}} \mathbf{u}_x + \int_0^{2\pi} \frac{za\sin\phi' d\phi'}{(a^2+z^2)^{3/2}} \mathbf{u}_y + \int_0^{2\pi} \frac{a^2(\cos^2\phi'+\sin^2\phi')d\phi'}{(a^2+z^2)^{3/2}} \mathbf{u}_z \right\}. \tag{5.82}$$

For an arbitrary point over the OZ axis, the denominator of each summand of the above expression is a constant magnitude, so it can be taken outside the integral,

$$\mathbf{B} = \frac{\mu_0 I}{4\pi} \frac{za}{(a^2+z^2)^{3/2}} \int_0^{2\pi} \cos\phi' d\phi' \mathbf{u}_x + \frac{za}{(a^2+z^2)^{3/2}} \int_0^{2\pi} \sin\phi' d\phi' \mathbf{u}_y + \frac{a^2}{(a^2+z^2)^{3/2}} \int_0^{2\pi} d\phi' \mathbf{u}_z. \tag{5.83}$$

The first and second integrals are zero, then we have

$$\mathbf{B} = \frac{\mu_0 I}{4\pi} \frac{a^2}{(a^2+z^2)^{3/2}} \int_0^{2\pi} d\phi' \mathbf{u}_z = \frac{\mu_0 I a^2}{2(a^2+z^2)^{3/2}} \mathbf{u}_z, \tag{5.84}$$

and the final expression for the loop field

$$\mathbf{B}(z) = \frac{\mu_0 I a^2}{2(a^2+z^2)^{3/2}} \mathbf{u}_z. \tag{5.85}$$

5.6 A metallic ring of radius a is centered at the origin and lying in the z plane. If the circulating current I is constant, obtain the magnetic vector potential at point $(0, 0, z)$. Do these results hold for calculating the magnetic field over the rotation axis OZ?

Solution

This problem may be solved by directly applying (5.23). With this aim we start by setting the field and source coordinates, and the differential vector line element, i.e. $\mathbf{r} = (0, 0, z)$, $\mathbf{r}' = (x', y', 0)$ and $d\mathbf{l}' = (dx', dy', 0)$. Introducing these data in (5.23) it results,

$$\mathbf{A}(\mathbf{r}) = \frac{\mu_0 I}{4\pi} \oint_\Gamma \frac{dx' \mathbf{u}_x + dy' \mathbf{u}_y}{(x'^2+y'^2+z^2)^{1/2}} = \frac{\mu_0 I}{4\pi} \left\{ \oint_\Gamma \frac{dx'}{(x'^2+y'^2+z^2)^{1/2}} \mathbf{u}_x + \oint_\Gamma \frac{dy'}{(x'^2+y'^2+z^2)^{1/2}} \mathbf{u}_y \right\}.$$

Due to symmetry we may use polar coordinates to compute the integrals obtaining

$$\mathbf{A}(\mathbf{r}) = \frac{\mu_0 I}{4\pi} \left\{ \int_0^{2\pi} \frac{-a\sin\phi' d\phi'}{(a^2+z^2)^{1/2}} \mathbf{u}_x + \int_0^{2\pi} \frac{a\cos\phi' d\phi'}{(a^2+z^2)^{1/2}} \mathbf{u}_y \right\}$$

$$= \frac{\mu_0 I}{4\pi} \left\{ \frac{-a}{(a^2+z^2)^{1/2}} \int_0^{2\pi} \sin\phi' d\phi' \mathbf{u}_x + \frac{a}{(a^2+z^2)^{1/2}} \int_0^{2\pi} \cos\phi' d\phi' \mathbf{u}_y \right\} = 0.$$

(b) At first sight this results seems to be useful to calculate the magnetic field at any point P over the OZ axis. However, as we shall show this assumption leads to a contradiction. In fact, introducing the value of the magnetic vector potential obtained into (5.11) we have

$$\mathbf{A} = 0 \Rightarrow \mathbf{B} = \nabla \times \mathbf{A} = \nabla \times 0 = 0, \tag{5.86}$$

then the magnetic field created by a circular loop of radius a, at any point of coordinates $(0, 0, z)$ is zero, which contradicts (5.85). This result leads to two possibilities either (5.85) is wrong and the new value for $\mathbf{B}$ is valid, or we have not proceeded properly in some of our explanation when calculating the magnetic field through $\mathbf{A}$ (5.86).

The reason for this contradiction is that we have applied the (5.11) incorrectly. If we examine (5.11), we see that to determine the magnetic field at a generic point P we have to do certain partial derivatives of the vector potential. A derivative gives us information about how a function varies in the neighbor of the point. This, from a mathematical point of view requires not only the knowledge at P of the function to be derived but also the values that it takes around this point. In our case we only know the vector potential at point $(0, 0, z)$, then it is not possible to calculate the curl. Thus, the result (5.86) does not mean that the magnetic field is zero, simply shows that the procedure employed is wrong. To calculate $\mathbf{B}$ from $\mathbf{A}$ we have to calculate first the magnetic vector potential at a generic point $P(x, y, z)$ of space. As a result we have the value of the function at P and its surroundings (in a ball $B_\epsilon(P)$ of radius ϵ centered at P), and we can compute the rotational. To finish, particularize the value of the rotational (field B) for the point under study (in our case $(0, 0, z)$) (Figs. 5.24 and 5.25).

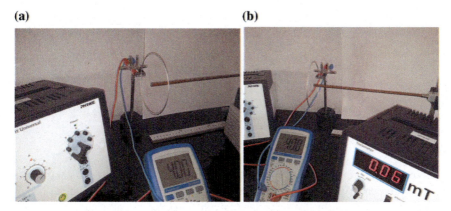

Fig. 5.24 Coil of radius $a = 4$ cm. **a** Lateral view with a magnetometer located at its center. **b** Magnetic field registered for an intensity $I = 4$ A. The field of $6 \cdot 10^{-5}$ T coincides with the field calculated theoretically

Fig. 5.25 Finite wire of length $L = L_1 + L_2$

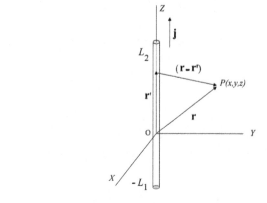

Fig. 5.26 This complex system may be divided into two parts, one of them corresponding to the segment from $-L_1$ to L_2, and the other one to the rest of the circuit

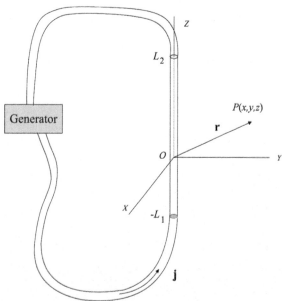

5.7 Find the magnetic field produced by a finite straight wire of length L carrying a steady current I, in a point $P(x, y, z)$.

Solution

Before solving the problem it is worthy to make a clarification. If we analyze the exercise rigourously we find that, in principle, what we want to calculate makes no sense. The reason for this lies in the fact that an open circuit cannot circulate a stationary current, and therefore the computation of the field is irrelevant. Observe that the Biot–Savart integral must be taken over a closed curve (5.26). In order to make sense, the problem needs to be reformulated. In fact, what we should actually say when calculating fields of open conducting filaments is: assuming a system as in Fig. 5.26 (or similar), obtain $\mathbf{B}(\mathbf{r})$ at $P(x, y, z)$, belonging only to a straight segment of length L. Physically it means that we are interested in only in the field contribution owing to this segment, even though the rest of the circuit produces a magnetic field, too. Bearing in mind this idea, let us apply first the Biot–Savart law to all the conductor of Fig. 5.26. We divide the interval of integration in two parts, one of them corresponding to segment of length L, and the other one to the rest,

$$\mathbf{B}(\mathbf{r}) = \frac{\mu_0 I}{4\pi} \oint_\Gamma \frac{d\mathbf{l}' \times (\mathbf{r} - \mathbf{r}')}{|\mathbf{r} - \mathbf{r}'|^3} = \frac{\mu_0 I}{4\pi} \int_{-L_1}^{L_2} \frac{d\mathbf{l}' \times (\mathbf{r} - \mathbf{r}')}{|\mathbf{r} - \mathbf{r}'|^3} + \frac{\mu_0 I}{4\pi} \int_{\text{rest of the circuit}} \frac{d\mathbf{l}' \times (\mathbf{r} - \mathbf{r}')}{|\mathbf{r} - \mathbf{r}'|^3}. \tag{5.87}$$

Observe that the separation of the integral in two summands (5.87) is but the principle of superposition of fields. From these two parts we only calculate the integral corresponding to the straight wire, which is the only part that interests us,

$$\frac{\mu_0 I}{4\pi} \int_{-L_1}^{L_2} \frac{d\mathbf{l}' \times (\mathbf{r} - \mathbf{r}')}{|\mathbf{r} - \mathbf{r}'|^3} \tag{5.88}$$

Proceeding in this way, the initial contradiction disappears, allowing us to solve the problem (the procedure explained has general validity, and can be applicable to any problem).

In order to compute (5.26), we must write the value of $\mathbf{r}$ and $\mathbf{r}'$. The variables with prime represent source points, which are our variables of integration. Points denoted without prime are called field points, and they show places at which we calculate $\mathbf{B}$. In our case we have: $\mathbf{r} = (x, y, z)$, $\mathbf{r}' = (0, 0, z')$ and $d\mathbf{l}' = (0, 0, dz')$, thus $d\mathbf{l}' \times (\mathbf{r} - \mathbf{r}') = -y dz' \mathbf{u}_x + x dz' \mathbf{u}_y$, and introducing them into (5.26),

$$\mathbf{B}(\mathbf{r}) = \frac{\mu_0 I}{4\pi} \int_{-L_1}^{L_2} \frac{-y dz' \mathbf{u}_x + x dz' \mathbf{u}_y}{[x^2 + y^2 + (z - z')^2]^{3/2}}. \tag{5.89}$$

This integral may be decomposed into two parts,

$$\mathbf{B}(\mathbf{r}) = \frac{\mu_0 I}{4\pi} \left\{ -y \int_{-L_1}^{L_2} \frac{dz'}{[x^2 + y^2 + (z - z')^2]^{3/2}} \mathbf{u}_x + x \int_{-L_1}^{L_2} \frac{dz'}{[x^2 + y^2 + (z - z')^2]^{3/2}} \mathbf{u}_y \right\}$$

$$= (-y \mathbf{u}_x + x \mathbf{u}_y) \frac{\mu_0 I}{4\pi} \int_{-L_1}^{L_2} \frac{dz'}{[x^2 + y^2 + (z - z')^2]^{3/2}}$$

$$= (-y \mathbf{u}_x + x \mathbf{u}_y) \frac{\mu_0 I}{4\pi} \left[\frac{-(z - z')}{(x^2 + y^2)[x^2 + y^2 + (z - z')^2]^{1/2}} \right]_{-L_1}^{L_2}.$$

Remembering the study of the symmetries done in the problem of the metallic wire, we can also see that the finite wire has symmetry of revolution as well, thus it seems useful to introduce cylindrical coordinates for the field points (it does not affect the variable of integration). Therefore we can write $x = \rho \cos \phi$ and $y = \rho \sin \phi$, where ρ represents the distance from point $P(x, y, z)$ to the OZ.

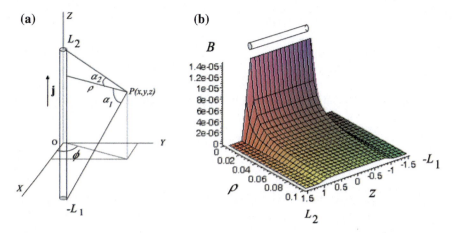

Fig. 5.27 **a** α_1 y α_2 are measured with respect ρ, ρ being the segment perpendicular to the OZ axis (it does not depend on the length of the finite wire). **b** Plot of $B = B(\rho, z)$ for symmetric wire of $L = 2\,\mathrm{m}$

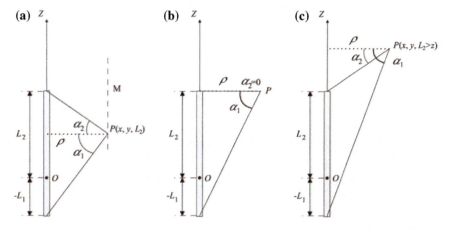

Fig. 5.28 **a** $P(x, y, z < L_2)$. **b** $P(x, y, z = L_2)$. **c** $P(x, y, z > L_2)$. Observe that ρ is always perpendicular to an imaginary line across the wire

$$\mathbf{B}(\mathbf{r}) = \frac{\mu_0 I}{4\pi} \left[\frac{-(z-z')}{\rho^2 \sqrt{\rho^2 + (z-z')^2}} \right]_{-L_1}^{L_2} (-\rho \sin\phi \mathbf{u}_x + \rho \cos\phi \mathbf{u}_y) = \quad (5.90)$$

$$\frac{\mu_0 I}{4\pi} \left[\frac{-(z-L_2)}{\rho \sqrt{\rho^2 + (z-L_2)^2}} - \frac{-(z+L_1)}{\rho \sqrt{\rho^2 + (z+L_1)^2}} \right] \mathbf{u}_\phi .$$

This result may be expressed in an easier form if we look at the geometrical significance of its two terms. From Fig. 5.27, we have:

$$\mathbf{B}(\mathbf{r}) = \frac{\mu_0 I}{4\pi \rho}(\sin\alpha_1 + \sin\alpha_2)\mathbf{u}_\phi. \quad (5.91)$$

where α_1 and α_2 are the angles between ρ and the straight lines that join such a point $P(x, y, z)$ with the top and the bottom of the wire, respectively. Taking into account the discussion at the beginning of the problem, the approach of the problem may appear to not make sense, because a finite wire cannot have a stationary current flowing through it. The result is important because many electromagnetic systems are composed of electrical elements, some of them are straight segments, requiring knowledge of the magnetic field created for the entire system. In this case, by applying the superposition principle of linear fields, we can find an easy way for calculating the magnetic field. Before finishing the exercise, it is worthy to comment about the application of (5.91), when the coordinate z of point P at which the magnetic field must be calculated is greater than L_2 or smaller than $-L_1$. To understand the procedure let us begin with the geometry shown in Fig. 5.28a. In that case, the coordinate z belongs to the interval $(-L_1 < z < L_2)$, and the angles α_1 and α_2 are both positives over the distance ρ. If we displace point P parallel to the OZ axis over the line M, the angle α_1 increases and α_2 decreases to zero, when the coordinate $z = L_2$ (Fig. 5.28b). If we go higher still, we reach the picture of Fig. 5.28c. Here α_1 continues increasing and α_2 becomes negative respect to ρ. The same result may be obtained from (5.90) by putting any value of $z > L_2$. On the other hand it should be noted that ρ must be drawn perpendicular to the axis of symmetry of the wire (in our case the OZ axis), even though there is no filament (see Fig. 5.28c).

5.8 Find the magnetic field produced by a very large straight wire carrying a steady current I, in a point $P(x, y, z)$, by using (5.91).

Solution

As we will demonstrate, we can solve the problem in an easy way, by applying (5.91) directly. In effect, when extending the segment of length L at its ends, a very large wire is achieved. Geometrically, the larger the wire, the larger the angles and its limit should hold $\alpha_1 = \alpha_2 = \alpha = \pi/2$. By introducing these values into (5.91), we have

$$\mathbf{B}(\mathbf{r}) = \frac{\mu_0 I}{4\pi\rho}(2\sin\alpha)\mathbf{u}_\phi = \frac{\mu_0 I}{2\pi\rho}\mathbf{u}_\phi, \quad (5.92)$$

result which agrees with (5.60).

The use of the Biot–Savart law has the advantage that its application gives the solution to the three components of the magnetic field. We do not need symmetries or something else to apply it. However, the disadvantage is that its calculation may be highly difficult needing the use of numerical techniques. On the contrary, Ampère's law is usually easier, but is application is restricted to systems with a certain degree of symmetry, and we need different integration curves for each component (Fig. 5.29).

5.9 A solenoid of length L, radius a and n turns per unit length, is carrying a current I. Find the magnetic $\mathbf{B}$ at a generic point $P(0, 0, z)$ over the OZ axis.

Fig. 5.29 Finite solenoid of length L and radius a

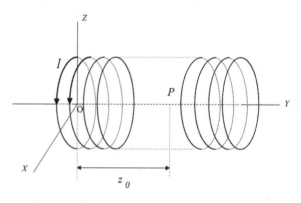

Fig. 5.30 Section of the solenoid. Note how the loops dN are displaced a distance z', with respect to the origin O. Mathematically it corresponds to a shift over the OZ axis

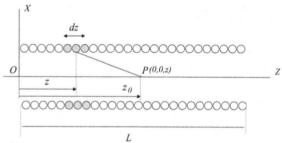

Solution

To resolve this exercise, we can adopt two different points of view. The first one is to assume that the solenoid is closely wound, so that it is equivalent to a set of independent coils with radius a. The second, completely equivalent, is to see the solenoid not as a set of independent rings but rather as a conducting sheet on which a tangential density current circular in shape flows. Starting from the first model for the solenoid, we are dealing with the calculation of the magnetic field at a point z over the axis of revolution of the system. To do so, we first compute the field produced by a differential loop of current located at z', and then we add the contributions of all rings by integrating over the total length of the solenoid. Using the reference frame of Fig. 5.30, the magnetic field $\mathbf{B}$ at point $P(0.0, z)$ due to dN coils situated at z' with respect to the origin of the coordinates is

$$d\mathbf{B}(P) \approx \frac{\mu_0 dN I a^2}{2\left[a^2 + (z-z')^2\right]^{3/2}} \mathbf{u}_z. \tag{5.93}$$

As we can see in the denominator of this equality appears $(z - z')$, and not z' which is a consequence of the location of the element dN. At first glance it seems easier to choose the current loops dN at the origin of coordinates and then integrate. However this is not possible, because if we take a differential element at the origin $(0, 0, 0)$, the variable of integration z' in the denominator disappears automatically, making an

Solved Problems

error in the calculus. We must always take the current elements (or charges) located at generic points.

Now we must integrate (5.93), however we have a problem because in this expression two variables appear, z' and N. For this integration we should have one independent variable, then it is necessary to find the function that relates to both variables. To do this we can use the definition of number of turns per unit length, $n = \frac{N}{l}$, which is a known quantity. However, taking into account the viewpoint followed at the beginning of our reasoning, we need a relationship between the two differential quantities. For this reason we may employ the definition of n in differential form,

$$n = \frac{dN}{dz'}, \tag{5.94}$$

which represents the number of turns per unit length referred to the symmetry axis of the solenoid. This result is important because it allows us to obtain dN as a function of dz'. In fact, by using the concept of differential of a function we have[12]

$$dN = \left(\frac{dN}{dz'}\right) dz' = n dz'. \tag{5.95}$$

Introducing this result into (5.93) we obtain:

$$dB(P) = \frac{\mu_0 I a^2 n dz'}{2\left[a^2 + (z - z')^2\right]^{3/2}}, \tag{5.96}$$

which depends on the variable z' only. Once the field produced by a differential of loops is known, the total magnetic field B produced by the system is computed by integrating (5.96),

$$B(P) = \int_0^L \frac{\mu_0 I a^2 n dz}{2\left[a^2 + (z-z')^2\right]^{3/2}} = \frac{\mu_0 I a^2 n}{2} \int_0^L \frac{dz}{\left[a^2 + (z-z')^2\right]^{3/2}} = \tag{5.97}$$

$$\frac{\mu_0 I a^2 n}{2}\left[\frac{-(z-z')}{a^2\left[a^2+(z-z')^2\right]^{1/2}}\right]_0^L = \frac{\mu_0 I n}{2}\left[\frac{-(z-L)}{\left[a^2+(z-L)^2\right]^{1/2}} - \frac{-z}{\left[a^2+z^2\right]^{1/2}}\right].$$

This is the expression for the magnetic field **B** produced by a solenoid at a point over its revolution axis (in our case the OZ axis). The sense of the field depends on the sign of the intensity current. Equation (5.97) may be simplified if we take into consideration the geometry of the solenoid (Fig. 5.31).

---

[12] Notice how dN is not obtained directly from (5.94) by multiplying n by dz' from the member on the right. This is not possible since *mathematically it is not correct to multiply and divide by differential quantities*; for this reason we have used the definition of differential on dN directly.

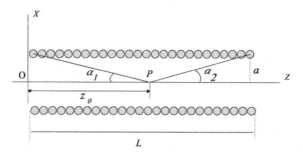

Fig. 5.31 Final scheme for calculating the magnetic field along the symmetry axis. α_i ($i = 1, 2$) corresponds to the angle between the segment joining the point $P(0, 0, z)$ and the last coil on the *right* (*left*) of the solenoid and the OZ axis

In fact, by analyzing the two parts that appear in (5.97) we see that the quotients in the bracket correspond to the cosines of the angles formed by the semi-straight line joining the upper extreme of the solenoid and the Oz axis, thus it holds

$$\cos \alpha_2 = \frac{-(z-L)}{[a^2 + (z-L)^2]^{1/2}}, \quad \cos \alpha_1 = \frac{z}{[a^2 + z^2]^{1/2}}, \tag{5.98}$$

and for the magnetic field

$$\mathbf{B}(z) = \frac{\mu_0 I n}{2} (\cos \alpha_1 + \cos \alpha_2) \mathbf{u}_z, \tag{5.99}$$

where α_1 and α_2 are measured positively with respect to OZ.

This equation, therefore, allows the magnetic field calculation of a finite solenoid at any point over its axis of revolution, but care should be taken outside the length of the coil. In fact, as we explained before (see problem for the finite wire), there are locations in the space where the angles may change their sign (see Fig. 5.28), and this case can also occur in a solenoid. To demonstrate this, let us consider a point $P(0, 0, z)$ for which $z > L$. For this situation α_1 decreases and α_2 increases exceeding $\frac{\pi}{2}$ (Fig. 5.32). It means we can write $\alpha_2 = (\pi - \beta)$ in (5.99), thus

$$\mathbf{B}(z) = \frac{\mu_0 I n}{2} (\cos \alpha_1 + \cos(\pi - \beta)) \mathbf{u}_z = \frac{\mu_0 I n}{2} (\cos \alpha_1 - \cos \beta) \mathbf{u}_z, \tag{5.100}$$

where $\beta = \arctan \frac{a}{(z-L)}$ (Fig. 5.33).

As commented at the beginning, we can model the solenoid in another way. In fact, as the coils are closely wound, we can face the solenoid as a conducting sheet over which a tangential current density j_s flows. Physically this density current represents the charge per unit time and per unit length (located on the surface) perpendicular to movement of charges, i.e. (Fig. 5.34)

$$j_s = \frac{dI}{dl} = \frac{IN}{L}. \tag{5.101}$$

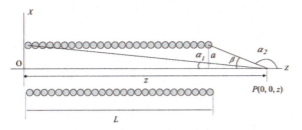

Fig. 5.32 Point $P(0, 0, z)$ outside of the solenoid. Observe that the angle α_2 exceeds $\frac{\pi}{2}$, then $cos(\alpha_2)$ changes its sign

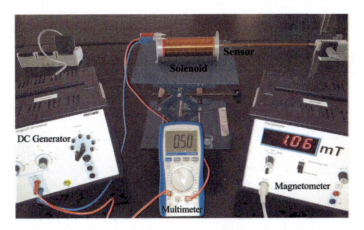

Fig. 5.33 Set-up for measuring the magnetic field created by a finite solenoid. The device is composed by a DC generator, a multimeter, and a magnetometer. In the attached figure the head sensor is located at a point over the revolution axis of the coil

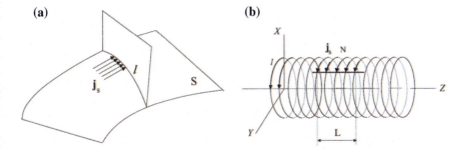

Fig. 5.34 **a** Density j_s is tangent to the surface at every point. It represents the charge per unit time and per unit length perpendicular to the movement of charges. Observe that j_s has dimensions of A/m. **b** Model of solenoid. The current now circulates over the surface of a conducting sheet

Taking into consideration that $(N/L) = n$ and that in this case the length perpendicular to the current is the OZ axis, we can write

$$dI = \left(\frac{dI}{dz}\right) dz = j_s \, dz = \frac{IN}{L} dz = nI \, dz, \qquad (5.102)$$

then

$$d\mathbf{B}(P) \approx \frac{\mu_0 a^2 dI}{2 \left[a^2 + (z-z')^2\right]^{3/2}} \mathbf{u}_z = \frac{\mu_0 a^2 nI dz}{2 \left[a^2 + (z-z')^2\right]^{3/2}} \mathbf{u}_z, \qquad (5.103)$$

which coincides with (5.93).

5.10 Through a solenoid of radius a very long length, flows a current I, find: (a) The magnetic field B at a point P over its revolution axis. (b) The magnetic field in any interior point ($\rho < a$). (c) Idem for ($\rho > a$).

Solution

(a) In a previous exercise we have demonstrated the magnetic field created by a finite solenoid along the OZ axis. Now the problem deals with an analogous situation, but when the length of the coil is very large. For finding the field it is not necessary to repeat the calculation, on the contrary it is enough to apply (5.99) for our special case. In fact, for a very large solenoid we can suppose the extremes of the solenoid are far away from point P, then $\alpha_1 = \alpha_2 \approx 0$ (see Fig. 5.31). Introducing these values in (5.99) we have (Fig. 5.35)

$$\mathbf{B}(z) = \frac{\mu_0 In}{2}(1+1)\mathbf{u}_z = \mu_0 In \, \mathbf{u}_z. \qquad (5.104)$$

(b) The result obtained (5.104) represents only the magnetic field at any point over OZ, but it is not valid for points out of the symmetry axis (OZ). Thus, (5.104), in principle, does not solve the question, because we need to calculate the magnetic field at any point inside of the system, i.e. at points that are not only located along the OZ axis. For calculating the field at an inner point of the solenoid, we may use Ampère's Law. In fact, supposing that the the system may be treated as a set of independent rings, it has symmetry of translation, and symmetry of rotation with respect to OZ.

Let us suppose we choose a rectangular curve Γ as shown in Fig. 5.36a, and a planar surface for calculating the flux of $\mathbf{j}$. By dividing the integral of Ampère in four paths, corresponding to each segment of the curve, and choosing $d\mathbf{l}$ in the clockwise direction, we can write

$$\oint_\Gamma \mathbf{B} \cdot d\mathbf{l} = \int_1^2 B_{z1} dz + \int_2^3 B_{\rho 2} d\rho + \int_3^4 B_{z3} dz + \int_4^1 B_{\rho 4} d\rho = \mu_0 NI. \qquad (5.105)$$

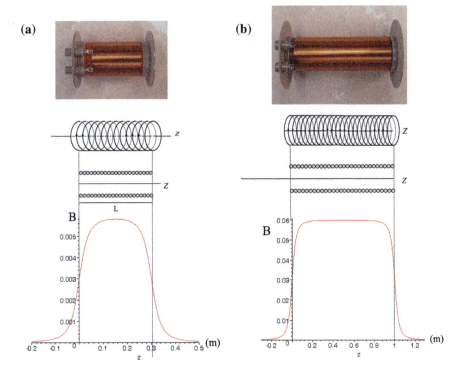

Fig. 5.35 **a** Magnetic field over the OZ axis for a solenoid of $L = 30\,\text{cm}$, $n = 10{,}000$, $a = 4\,\text{cm}$ and $I = 0.5\,\text{A}$, as function of distance z. Observe that the field is quasi-homogeneous in the geometrical center and for a small interval around of $L/2$. **b** The same calculation but for $L = 1\,\text{m}$. In this case the region where the field is practically uniform grows. From it we can imagine that if the lengths of the solenoids are very large we reach a homogeneous field along the OZ axis (see (5.104))

A similar procedure was explained in the problem of the infinite wire. There we saw that, due to the translation symmetry of the system, the integrals over the segments (2–3) and (4–1) cancel each other out, because the path integrals are taken in opposite directions. Then, the calculation refers only to the other segments,

$$\int_1^2 B_{z1} dz + \int_3^4 B_{z3} dz = B_{z1} \int_1^2 dz + B_{z3} \int_3^4 dz = \mu_0 NI \Rightarrow B_{z1} l - B_{z3} l = \mu_0 NI, \tag{5.106}$$

where B_{z1} is the field along the path (1–2) and B_{z3} is the field along the path (3–4). Therefore,

$$B_{z1} - B_{z3} = \frac{\mu_0 NI}{l} \Rightarrow B_{z1} = \mu_0 nI + B_{z3}. \tag{5.107}$$

The above equation depends on a constant to be determined. To obtain B_{z3}, let us suppose we repeat the same calculation but over a similar rectangular curve Γ,

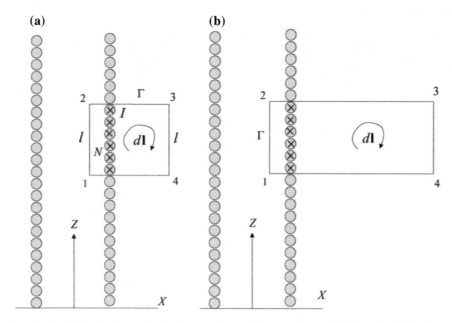

Fig. 5.36 a Curve Γ. b Curve Γ with two of its parallel segments elongated ((2–3) and (4–1))

whose segments perpendicular to the solenoid ((2–3) and (4–1)) are much longer (Fig. 5.36b). The result we will obtain is identical. Hence, it is not possible to distinguish between them. In other words, the result does not depend on the curves we have chosen. However, if the path (3–4) is far from the solenoid, as we have seen, the magnetic field must decrease rapidly because of its convergence. Thus, as B_{z3} represents the magnetic field over the segment (3–4), we conclude that $B_{z3} = 0$, and then

$$B_{z1} = \mu_0 n I. \tag{5.108}$$

This result means that the component z of the magnetic field is the same for any point interior to a large solenoid. The procedure for finding B_ϕ is identical as explained in the problem of the infinite metallic wire. The final result is that the tangential component for the infinite solenoid is zero. With respect to B_ρ for the infinite solenoid, we can integrate the Maxwell equation $\nabla \cdot \mathbf{B} = 0$ over a closed cylindrical surface with radius $\rho < R$ (Fig. 5.37a), as we made for the infinite wire. The result we obtain is the same as there (5.59). The radial component of B is zero as well.

(c) For points exterior to the spool let us begin with B_z. We proceed as we have seen, but now the curve Γ is located outside of the solenoid (Fig. 5.38). Employing Ampère's law we have

$$\int_1^2 B_{z1} dz + \int_3^4 B_{z3} dz = B_{z1} \int_1^2 dz + B_{z3} \int_3^4 dz = 0 \Rightarrow B_{z1} l - B_{z3} l = 0. \tag{5.109}$$

Fig. 5.37 a Cylindrical closed surface of radius $\rho < a$. Idem for $\rho > a$

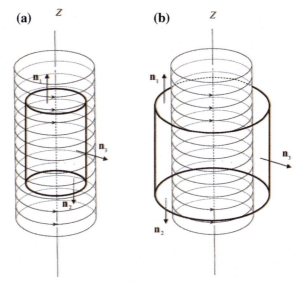

Fig. 5.38 Curve Γ for applying Ampère's law. If we enlarge the curve so that points 3 and 4 extend to $3'$ and $4'$, far away from the solenoid, we obtain the same result

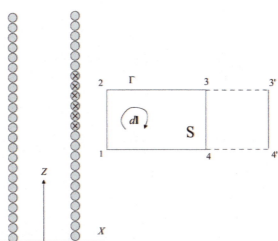

Observe that in this case the integral must vanishes because no current crosses the open surface S, whose delimitation is Γ. From (5.109) follows that

$$B_{z1}l - B_{z3}l = 0 \Rightarrow B_{z1} = B_{z3}, \tag{5.110}$$

that is, the B_z is a constant. However, if we enlarge Γ to infinity, the result must be equivalent, as we have shown in the section above. Then, $B_{z1} = B_{z\infty}$ which shows that B_z is zero for any point exterior to the solenoid. For the components ϕ and ρ, there are no differences in the way to calculate such projections. In effect, let us look at Fig. 5.37b. Applying the equation of the flux of B for a cylinder of radius greater than the radius of the solenoid, we find that $B_\rho = 0$. The same occurs with the tangential component, $B_\phi = 0$. The only difference is the radius of the curve Γ to perform the integration.

The final result viewed in this exercise is that, a very large circular solenoid whose coils may be considered as independent rings carrying a current I creates only a magnetic field in its interior, and the exterior field is zero. Note that the field for $\rho < a$ is uniform (the same for all interior points), which is not true for a finite solenoid.

Problems B

5.11 Using the the characteristics of the vector potential demonstrate the formula (5.37).

Solution

As we have seen, the magnetic field **B** can be derived from a vector potential through the relation

$$\mathbf{B} = \nabla \times \mathbf{A}. \tag{5.111}$$

Due to the circular symmetry of the system, we choose a plane surface delimited by a circular curve Γ, whose revolution axis coincides with the symmetry axis of the physical system to be studied (solenoid, hyperboloid, hemispherical cap, etc.). By integrating over the open surface S (5.111) we obtain

$$\int\int_S \mathbf{B} \cdot d\mathbf{S} = \int\int_S \nabla \times \mathbf{A} \cdot d\mathbf{S}, \tag{5.112}$$

an expression that may be expressed by a curvilinear integral along the curve Γ by applying the Stokes theorem,

$$\int\int_S \mathbf{B} \cdot d\mathbf{S} = \oint_\Gamma \mathbf{A} \cdot d\mathbf{l}, \tag{5.113}$$

$d\mathbf{l}$ being the differential curve length. Equation (5.113) states that the magnetic flux of the field B through an open surface S is equal to the circulation of the magnetic potential vector along the delimiting curve of the surface. Employing (5.113) and (5.35) and (5.36) together we have

$$\int\int_S (B_\rho, B_\phi, B_z) \cdot (0, 0, 1) dS = \oint_\Gamma (A_\rho, A_\phi, A_z) \cdot (0, 1, 0) dl, \tag{5.114}$$

thus,

$$\int\int_S B_z dS = \oint_\Gamma A_\phi dl. \tag{5.115}$$

Substituting (5.35) into (5.115) it yields the relation between the potential vector and the magnetic field

$$\int\int_S \sum_{n=0}^{\infty}(-1)^n \frac{B_z^{(2n)}(0,z)}{2^{2n}(n!)^2}\rho^{2n}dS = \oint_\Gamma A_\phi dl. \qquad (5.116)$$

Because of the symmetry the modulus of the component A_ϕ of the vector potential takes the same values for all points belonging to the curve chosen Γ, then we can set it outside of the integral

$$\sum_{n=0}^{\infty}(-1)^n \frac{B_z^{(2n)}(0,z)}{2^{2n}(n!)^2} \int\int_S \rho^{2n}dS = A_\phi \oint_\Gamma dl, \qquad (5.117)$$

and integrating

$$A_\phi 2\pi\rho = \sum_{n=0}^{\infty}(-1)^n \frac{B_z^{(2n)}(0,z)}{2^{2n}(n!)^2} \int_0^{2\pi}d\phi \int_0^\rho \rho^{2n}\rho d\rho = 2\pi\sum_{n=0}^{\infty}(-1)^n \frac{B_z^{(2n)}(0,z)}{2^{2n}(n!)^2} \frac{\rho^{2n+2}}{(2n+2)}. \qquad (5.118)$$

Taking into account that $(n!)^2(2n+2) = n! \cdot n! \cdot ((n+1)+(n+1)) = n! \cdot (n! \cdot (n+1) + n! \cdot (n+1)) = 2n! \cdot (n+1)!$ expression (5.118) may be written

$$A_\phi = \sum_{n=0}^{\infty}(-1)^n \frac{B_z^{(2n)}(0,z)}{2^{2n+1}n!(n+1)!}\rho^{2n+1}, \qquad (5.119)$$

equation which coincides with (5.37).

5.12 In $\Re^3$ there is a vector potential whose value may be represented by

$$\mathbf{A} = \begin{cases} \frac{a}{2}(x^2+y^2)\mathbf{u}_z & \text{if } \rho < R_1 \\ 0 & \text{if } \rho > R_1, \end{cases}$$

where a is a constant. Find: (a) The current density for any point of $\Re^3$. (b) The magnetic field B_ϕ for $\rho < R_1$ using Ampère's law. (c) The differential equation of the magnetic field lines and its geometry for $\rho < R_1$.

Solution

(a) For calculating the density current we can use (5.20), which gives the relation between the laplacian of the vector potential and $\mathbf{j}$

$$\Delta\mathbf{A} = -\mu_0 \mathbf{j} = 2a\,\mathbf{u}_z \Rightarrow \mathbf{j} = -\frac{2a}{\mu_0}\mathbf{u}_z, \qquad (5.120)$$

for $\rho < R_1$. Notice that the density obtained has the same direction and symmetry as the vector potential.

For $\rho > R_1$, the application of the same formula as before (5.120) leads to

$$\Delta \mathbf{A} = -\mu_0 \mathbf{j} = 0 \Rightarrow \mathbf{j} = 0. \tag{5.121}$$

(b) The current density calculated has cylindrical symmetry with respect to the OZ axis. On the other hand we already know the mathematical expression of the density current, so we try to calculate the tangential component of the magnetic field $\mathbf{B}$ by means of Ampère's theorem. By choosing a circular curve of radius $\rho < R_1$ we have

$$\oint_\Gamma \mathbf{B} \cdot d\mathbf{l} = \mu_0 \int_S \mathbf{j} \cdot d\mathbf{S} = \mu_0 I \Rightarrow \oint_\Gamma (B_\rho, B_\phi, B_z) \cdot (0, dl, 0) = -\mu_0 \int_S \frac{2a}{\mu_0} \mathbf{u}_z \cdot d\mathbf{S}\, \mathbf{u}_z. \tag{5.122}$$

Making the scalar products of both sides it leads to

$$2\pi\rho B_\phi = -\mu_0 \frac{2a}{\mu_0} \pi \rho^2 \Rightarrow B_\phi = -a\rho. \tag{5.123}$$

(c) The differential equation of the field lines can be found by applying the following equality (1.19)

$$\frac{dx}{B_x} = \frac{dy}{B_y} = \frac{dz}{B_z}. \tag{5.124}$$

Regrouping terms

$$\frac{dx}{B_x} = \frac{dy}{B_y} \Rightarrow B_x dy = B_y dx. \tag{5.125}$$

To resolve this equation we need to introduce the field projections over OX and OY axis. To do this, we transform (5.123) from cylindrical into cartesian coordinates

$$\mathbf{B}_\phi = -a\rho \mathbf{u}_\phi = -a\rho(-\sin\phi\, \mathbf{u}_x + \cos\phi\, \mathbf{u}_y)$$
$$= -a(-\rho\sin\phi\, \mathbf{u}_x + \rho\cos\phi\, \mathbf{u}_y) = -a(-y\mathbf{u}_x + x\mathbf{u}_y). \tag{5.126}$$

Introducing these components into (5.125) and integrating we obtain

$$B_x dy = B_y dx \Rightarrow aydy = -axdx \Rightarrow \int ydy = -\int xdx \Rightarrow \frac{1}{2}y^2 = -\frac{1}{2}x^2 + C, \tag{5.127}$$

C being a constant. Joining the variables x and y on a side

$$\frac{1}{2}x^2 + \frac{1}{2}y^2 = C \Rightarrow x^2 + y^2 = 2C, \tag{5.128}$$

which is the equation of a family of circles centered at the reference frame OXY. This result agrees with previous problems where there was rotational symmetry.

5.13 Calculate the magnetic field **B** of a very large wire starting from the magnetic vector potential.

Solution

The idea of this problem is to obtain the vector potential and thereafter the magnetic field **B** by means of the relation $\mathbf{B} = \nabla \times \mathbf{A}$.

In previous problems we have calculated the magnetic field of a wire by using Ampère's law and the Biot–Savart integral. Now we will obtain **B** by using (5.11). To do so, we first need the vector potential **A** and then we take its rotational. To apply (5.23) we have to identify all quantities appearing in the integral. From Fig. 5.25 we see that $\mathbf{r} = (x, y, z)$, $\mathbf{r}' = (0, 0, z')$, and $d\mathbf{l}' = (0, 0, dz')$. Introducing them in (5.23), and choosing the sense of the curve from below to above, we have

$$\mathbf{A} = \frac{\mu_0 I}{4\pi} \int_{-L_1}^{L_2} \frac{dz'}{[x^2 + y^2 + (z-z')^2]^{1/2}} \mathbf{u}_z = \frac{\mu_0 I}{4\pi} \int_{-L_1}^{L_2} \frac{dz'}{[\rho^2 + (z-z')^2]^{1/2}} \mathbf{u}_z$$

$$= \frac{\mu_0 I}{4\pi} \ln\left[(z'-z) + \sqrt{\rho^2 + (z'-z)^2}\right]\Big|_{-L_1}^{L_2}$$

$$= \frac{\mu_0 I}{4\pi} \left\{ \ln\left[(L_2 - z) + \sqrt{\rho^2 + (L_2 - z)^2}\right] - \ln\left[(-L_1 - z) + \sqrt{\rho^2 + (L_1 + z)^2}\right] \right\}$$

$$= \frac{\mu_0 I}{4\pi} \ln \frac{\left[(L_2 - z) + \sqrt{\rho^2 + (L_2 - z)^2}\right]}{\left[(-L_1 - z) + \sqrt{\rho^2 + (L_1 + z)^2}\right]}.$$

The expression found corresponds to the vector potential of a finite wire. To calculate the magnetic field of the complete conducting filament we extend the limit of integration L_1 and L_2 to infinity and over the final mathematical expression, we take the curl. However, this statement even though correct, sets up an important question, as we will see. In fact, when computing the limit of (5.129) we find

$$\lim_{L_1, L_2 \to \infty} \frac{\mu_0 I}{4\pi} \left[\ln \frac{(z + L_1) + \sqrt{\rho^2 + (z + L_1)^2}}{(z - L_2) + \sqrt{\rho^2 + (z - L_2)^2}} \right] = \infty. \qquad (5.129)$$

This result does not make sense, because we already know that the magnetic field of a wire is perfectly defined. It might suggest that the procedure followed is unsuitable for calculating the field, however this is not true. Actually sometimes it occurs when resolving some electromagnetic problems. The reason may be understood by looking at the basic hypothesis on which the electromagnetic theory is found. If we do not impose any restriction about the geometrical location of charges and currents in the space (infinite), it is not possible to ensure a priori the convergence of the potentials like **A** or $V(\mathbf{r})$. On the contrary, a more realistic point of view for this situation is to consider that the currents are confined to *finite regions of the space*, then conceptually difficulties as shown in (5.129) disappear. However, even though we could adopt a non-restricting start point like this, and the singularity of the vector potential remains, it is possible to avoid the problem for the calculation of the magnetic field.

In effect, let us again consider (5.129) for the finite thread. The trick is to evaluate first $\nabla \times \mathbf{A}$ on this wire of total length L, obtaining the magnetic field of a finite filament, and then to take the limit when L_1 and L_2 tend to infinity over this expression. Using cylindrical coordinates, we have:

$$\mathbf{B} = \nabla \times \mathbf{A} = -\frac{\partial A_z}{\partial \rho} \mathbf{u}_z$$

$$= \frac{\mu_0 I}{4\pi\rho} \left[\frac{\frac{\rho}{\sqrt{\rho^2 + (L_2 - z)^2}}}{(L_2 - z) + \sqrt{\rho^2 + (L_2 - z)^2}} - \frac{\frac{\rho}{\sqrt{\rho^2 + (z + L_1)^2}}}{(-z - L_1) + \sqrt{\rho^2 + (z + L_1)^2}} \right] \tag{5.130}$$

$$= \frac{\mu_0 I}{4\pi\rho} \left[\frac{L_1 + z}{\sqrt{\rho^2 + (L_1 + z)^2}} + \frac{L_2 - z}{\sqrt{\rho^2 + (L_2 - z)^2}} \right] \mathbf{u}_\phi \tag{5.131}$$

Now by making L_1 and L_2 very large, yields

$$\frac{\mu_0 I}{2\pi\rho} \mathbf{u}_\phi, \tag{5.132}$$

a result that agrees with (5.92).

The conclusion we can highlight from this problem is that when having singularities for the potentials like (5.129), due to the large regions over which currents extend (infinite), calculate first the potential for a finite part of the system and then take the limit when the variable becomes infinity. By means of this procedure the difficulty disappears allowing us the calculation of the magnetic field.

5.14 In Fig. 5.39, a toroidal solenoid with radius $R = 10$ cm, main radius $R_m = 50$ cm and $n = 100$ is represented. The current circulating is $I_s = 0.1$ A. Together with the torus, a finite conducting wire of length $L = 1$ m is placed symmetrically along its revolution axis. If throughout the wire circulates a current $I_h = 2$ A, find: (a) The magnetic field $\mathbf{B}$ at $P_1(R_m, 0, 0)$. (b) The magnetic field $\mathbf{B}$ at point $P_2(1/2, 0, 1/2)$ m.

Solution

(a) Supposing that the currents across the wire and solenoid have no influence on each other, we can figure out this problem as the superposition of two independent problems consisting of two systems: the wire and the torus. Let us first focus our attention on the finite wire. The filament is located symmetrically with respect to the solenoid (see Fig. 5.39), then the angles α_1 and α_2 are the same for this configuration. By applying (5.91) for this case we have

$$\mathbf{B}_h = \frac{\mu_0 I_h}{4\pi\rho} (\sin \alpha_1 + \sin \alpha_2) \mathbf{u}_\phi = \frac{\mu_0 I_h}{4\pi\rho} 2 \sin \alpha_1 \mathbf{u}_\phi = \frac{\mu_0 I_h}{2\pi R_m} \sin \alpha_1 \mathbf{u}_\phi. \tag{5.133}$$

Fig. 5.39 System composed by a torus and a finite wire

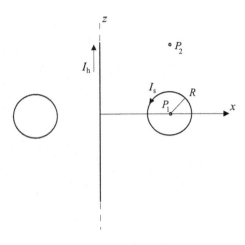

On the other hand, the field produced by the torus at point P may be calculated directly through the (5.72). In fact, by setting $\rho = R_m$ in this formula we can write

$$\mathbf{B}_s = \frac{\mu_0 N I_s}{2\pi \rho} \mathbf{u}_\phi = -\frac{\mu_0 N I_s}{2\pi R_m} \mathbf{u}_\phi = -\mu_0 n I_s \mathbf{u}_y. \tag{5.134}$$

Applying now the superposition principle of fields we find $\mathbf{B}$, i.e.

$$\mathbf{B} = \mathbf{B}_h + \mathbf{B}_s = \frac{\mu_0 I_h}{2\pi R_m} \sin \alpha_1 \mathbf{u}_\phi - \mu_0 n I_s \mathbf{u}_y = \frac{\mu_0 I_h}{2\pi R_m} \sin \alpha_1 \mathbf{u}_y - \mu_0 n I_s \mathbf{u}_y = 1.2 \cdot 10^{-5} \mathbf{u}_y. \tag{5.135}$$

Observe that at point $P_1(R_m, 0, 0)$ the unitary vector $\mathbf{u}_\phi$ in (5.133) coincides with $\mathbf{u}_y$.

(b) For this second part of the problem, the idea is the same but not the point P where the magnetic field must be calculated. If we look at the data, point $P_2(1/2, 0, 1/2)$ corresponds to the case shown in Fig. 5.28b. By setting $L = 1$ m and $\rho = R_m = 0.5$ we have $\alpha_1 = 0$ and $\alpha_2 = \arctan(L/R_m) = 63.4°$, then

$$\mathbf{B}_h = \frac{\mu_0 I_h}{4\pi R_m} (0 + \sin \alpha_2) \mathbf{u}_y = 3.6 \cdot 10^{-7} \mathbf{u}_y. \tag{5.136}$$

5.15 Figure 5.40 represents a long region with rotational symmetry about the OZ axis that consists of two zones. The first one is a cylinder of radius $R_1 = 5$ cm and altitude h, very large, and the second region is a cylinder with inner and external radius $R_1 = 5$ cm and $R_2 = 10$ cm, respectively. For the first region flow $n_p = 10^{19}$ m^{-3} protons with velocity $\mathbf{v}_p = 1000\,\vec{u}_z$ m/s, and for the second ($R_1 < \rho < R_2$) $n_e = 10^{17}$ m^{-3} electrons with velocity $\mathbf{v}_e = 500\,\mathbf{u}_z$ m/s. Find: (a) The magnetic field $\vec{B}_\phi$ for any point such that $(0 < \rho < R_1)$. (b) The magnetic field $\mathbf{B}_\phi$ for a generic interior point to the hollow cylinder $(R_1 < \rho < R_2)$. (c) Supposing the conditions of the inner region $(0 < \rho < R_1)$,

Fig. 5.40 Cylindrical region

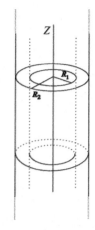

determine the increment of the electron density Δn_e in $(R_1 < \rho < R_2)$ in order for the magnetic field $\mathbf{B}_\phi$ to be zero in any point $P(\rho > R_2)$. Consider the electrons are homogeneously distributed in the volume.

Solution

(a) The system represented in the figure attached is symmetric under translations and rotations. This means that the magnetic field produced must be the same when translating the reference system parallel to the OZ axis or when rotating the reference frame about OZ. For this reason we can try to solve the exercise by means of Ampère's theorem (see problem of the infinite wire). With this goal we have to use (5.7), however, as we can notice in the statement of the problem, no density current is given, then we cannot perform the right hand of (5.7). Nonetheless we are able to calculate $\mathbf{j}$ for each region of the space with the numerical values corresponding to the particles velocities. In fact, remembering the definition of density current, we can write for the protons

$$\mathbf{j}_p = n_p q_p \mathbf{v}_p = 1062 \, \mathbf{u}_z, \tag{5.137}$$

and for the electrons

$$\mathbf{j}_e = n_e q_e \mathbf{v}_e = -8 \, \mathbf{u}_z. \tag{5.138}$$

For the calculation of the field for a distance smaller than R_1, only the protons contribute to generate $\mathbf{B}$. In effect, as we explained Ampère's law in the introduction, the circulation of the magnetic field over a closed curve Γ only depends on the net intensity crossing the open surface whose delimiting curve is Γ, therefore, protons and electrons outside the curve do not play any role. Applying the Ampère theorem for a curve Γ and a surface $S_p = S \cap V_j$ as shown in Fig. 5.41 we obtain

$$\oint_\Gamma \mathbf{B} \cdot d\mathbf{l} = \oint_\Gamma (B_\rho, B_\phi, B_z) \cdot (0, dl, 0) = \mu_0 \int_S \mathbf{j}_p \cdot d\mathbf{S} = \mu_0 \int_S (0, 0, j_p) \cdot (0, 0, dS), \tag{5.139}$$

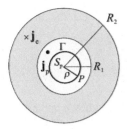

Fig. 5.41 Plane view of the current in direction of OZ axis. The surface S is delimited by the curve Γ of radius $\rho < R_1$. The effective surface S_p for performing the flux of $\mathbf{j}_p$ corresponds to the intersection between the volume, where the current $\mathbf{j}_p$ exists and the surface S whose border is the curve Γ ($S \cap V_j = S_p$)

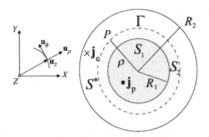

Fig. 5.42 a Plane view of the current of electrons and protons. The surface S^* is delimited by the curve Γ. For a point P in the region $R_1 < \rho < R_2$ the effective surface of integration S^* is divided in two parts. The first one S_1 and the second S_2 that corresponds to the intersection between the volume where the electrons move and the total surface S^* ($S_2 = S^* \cap V_{j_e} = (\pi \rho^2 - \pi R_1^2)$)

$$\oint_\Gamma B_\phi dl = \mu_0 \int_{S_p = S \cap V_j} j_p dS \Rightarrow \int_0^{2\pi} B_\phi \rho d\phi = \mu_0 \int_0^{2\pi} \int_0^{\rho < R_1} j_p \rho d\rho d\phi. \quad (5.140)$$

Due to the symmetries mentioned, the system is invariant under a rotation about the OZ axis, then B_ϕ and j_p do not change over the curve Γ of radius ρ. For this reason we may write

$$B_\phi \rho \int_0^{2\pi} d\phi = \mu_0 j_p \int_0^{2\pi} d\phi \int_0^{\rho < R_1} \rho d\rho \Rightarrow 2\pi \rho B_\phi = 2\pi \mu_0 j_p \frac{1}{2} \rho^2, \quad (5.141)$$

hence

$$B_\phi = \frac{1}{2} \mu_0 j_p \rho. \quad (5.142)$$

This result tells us that the tangential component of the magnetic field varies linearly with distance to the OZ axis (Fig. 5.42).

(b) The magnetic field for ($R_1 < \rho < R_2$) may be computed in the same way, because the symmetries apply to this case as well. When applying Ampère's law we must divide the surface of integration S^* in two parts, one corresponding to $S_1 = \pi R_1^2$ where protons move, and the other S_2 where electrons displace (surface between S^*

and S_1), i.e. $S_2 = \pi\rho^2 - \pi R_1^2$ (see Fig. 5.2b). In fact, setting $S^* = S_1 + S_2$ in (5.7), we have

$$\oint_{\partial S} \mathbf{B} \cdot d\mathbf{l} = \mu_0 I = \mu_0 \int_{S^*} \mathbf{j} \cdot d\mathbf{S} = \mu_0 \int_{S_1} \mathbf{j}_p \cdot d\mathbf{S} + \mu_0 \int_{S_2} \mathbf{j}_e \cdot d\mathbf{S}. \quad (5.143)$$

Observe that with the nomenclature shown in Sect. 5.2 (see Fig. 5.2) $S_1 = S^* \cap V_{j_p} = \pi R_1^2$ and $S_2 = S^* \cap V_{j_e} = (\pi\rho^2 - \pi R_1^2)$. Integrating (5.143) yields

$$2\pi\rho B_\phi = \mu_0 j_p \int_0^{2\pi} d\phi \int_0^{R_1} \rho d\rho + \mu_0 j_e \int_0^{2\pi} d\phi \int_{R_1}^{\rho < R_2} \rho d\rho = \mu_0 [j_p \pi R_1^2 + j_e(\pi\rho^2 - \pi R_1^2)], \quad (5.144)$$

and then the tangential component of the field

$$B_\phi = \frac{\mu_0}{2\rho}[j_p R_1^2 + j_e(\rho^2 - R_1^2)] = \frac{6.2 \cdot 10^{-7}}{\rho}[4 - 8(\rho^2 - 0.01)]. \quad (5.145)$$

(c) Before beginning to make calculations, we try to understand the question. If we would compute B_ϕ for a point $P(\rho > R_2)$ using density currents (5.137) and (5.138), we would obtain, in principle, a magnetic field distinct from zero. The question that arises is, how many electrons must be added in the hollow cylinder in order to have no magnetic field in the exterior region of the system. To investigate it, we first have to calculate the field outside R_2 supposing an unknown density current j'_e for the electrons, and later imposing the condition of zero magnetic field. With this aim, let us determine B_ϕ for a generic exterior point by means of Ampère's theorem. In this case we do not have density current outside R_2, then the effective surface of integration $S*$ is $S_1 + S_2$, distinct to S (see Fig. 5.43). As a result, proceeding in the same way as shown in (5.144) we obtain

$$B_\phi = \frac{\mu_0}{2\rho}[j_p R_1^2 + j'_e(R_2^2 - R_1^2)] = 0 \Rightarrow j_p R_1^2 = -j'_e(R_2^2 - R_1^2). \quad (5.146)$$

Notice that in this expression we have R_2^2 instead of ρ^2 (5.145). On the other hand, employing the definition of current density we can write

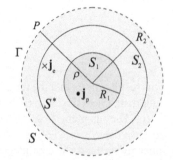

Fig. 5.43 For a point P belonging to $\rho > R_2$, the effective surface of integration S^* coincides with $S_1 + S_2$ because outside of this region there is not current density (see Fig. 5.2c)

$$\mathbf{j}'_e = n'_e q \mathbf{v}_e = (n_e + \Delta n_e) q \mathbf{v}_e, \quad (5.147)$$

where Δn_e represents the increment of electrons in the hollow cylinder we must to augment for obtaining zero exterior magnetic field. Resolving this equality we have

$$j_p R_1^2 = -(n_e + \Delta n_e) q_e v_e (\rho^2 - R_1^2) \Rightarrow \Delta n_e = 6.6 \cdot 10^{18}. \quad (5.148)$$

5.16 Figure 5.44 represents a finite solenoid of length $L = 80$ cm, radius $r = 3$ cm and $n = 1000$ turns per unit length. The coil is crossed by a very large conducting wire. The symmetry axis of the solenoid coincides with the axis OY, and the wire is perpendicular to it and located on the OYZ plane. If the currents circulating through the solenoid and wire are $I_s = 0.5$ A $I_w = 0.2$ A, respectively, find: (a) The magnetic field **B** at the origin of the coordinate frame. (b) The force on the wire.

Solution

(a) To calculate the magnetic field at point $P(0, 0, 0)$ we apply the principle of superposition of linear fields.

Let us first focus our attention to the field generated by the solenoid. For obtaining its field we will directly use (5.99) because the point P is located on the line corresponding to its symmetry axis (remember that (5.99) is only valid for points belonging to the revolution axis of the coil). Thus, taking into consideration that the revolution axis of the solenoid lies along the OY coordinate axis, we can write

$$\mathbf{B}_y = \frac{\mu_0 n I}{2}(\cos \alpha_1 + \cos \alpha_2)\mathbf{u}_y. \quad (5.149)$$

On the other hand, as the point P is located outside of the coils, one of the angles of the cosine function is negative. In fact, from Fig. 5.45 we can see that $\alpha_1 = (\pi - \beta)$, and therefore

$$\mathbf{B}_s = \frac{\mu_0 n I}{2}(-\cos \beta + \cos \alpha_2)\mathbf{u}_y = 9.2 \cdot 10^{-5} \mathbf{u}_y. \quad (5.150)$$

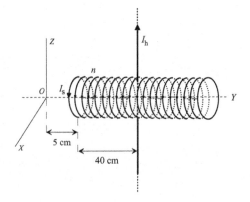

Fig. 5.44 Finite solenoid and very large conducting wire

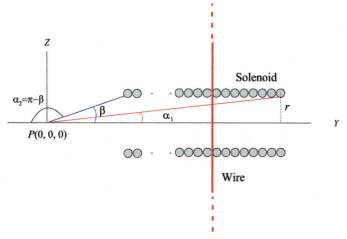

Fig. 5.45 Due to the point $P(0, 0, 0)$ is outside of the solenoid one of the angles is greater than $\frac{\pi}{2}$

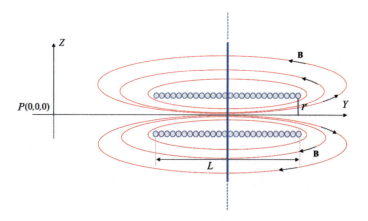

Fig. 5.46 Magnetic field lines of a solenoid

Once the magnetic field of the solenoid is obtained, the field **B** of the conducting wire may be easily computed by means of (5.49)

$$\mathbf{B}_h = \frac{\mu_0 I}{2\pi \rho} \mathbf{u}_x = 8.9 \cdot 10^{-8} \mathbf{u}_x. \tag{5.151}$$

Now, the resulting field at P is

$$\mathbf{B}_s = \mathbf{B}_h + \mathbf{B}_s = 8.9 \cdot 10^{-8} \mathbf{u}_x + 9.2 \cdot 10^{-5} \mathbf{u}_y. \tag{5.152}$$

(b) The force exerted on the wire may be calculated by means of (5.28). However, some comment about it is needed before. In fact, when trying to apply (5.28) we

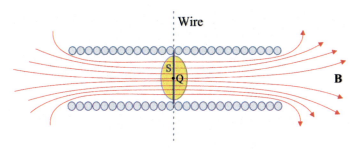

Fig. 5.47 Observe that the in the central region the magnetic field is almost uniform

immediately see that we are not able to calculate it easily. In order to compute the integral in (5.28), we must first know the magnetic field created by the solenoid along the wire.

In Problem 5 we obtained the field generated by a finite solenoid at any point on its symmetry axis, but not in space regions off of this axis. Therefore, if we would like to determine the force exerted by the magnetic field of the coils along the conducting wire, we have to calculate the field generated by the solenoid at a generic point $P(x, y, z)$ (Fig. 5.46), and then introducing it into (5.28) for computing the force. This procedure, although correct, is very hard to come by. The calculation requires the application of the Biot–Savart law by direct integration, or other more complicated techniques. However, by carefully examining the pose of the problem, we can avoid the complicated calculus. In fact, if we look at the geometrical characteristics of the system we can observe that we have a slender solenoid. By this noun we mean that the coil is longer than wide, i.e. its length $L \gg r$, r being the radius. As a consequence, the field inside the solenoid where the wire is placed is almost homogeneous, and outside of the coils the field strength is low. Thus, at first sight, we can neglect the influence of the external field on the wire outside of the coils. Ultimately, even though we do not accurately describe the magnetic field along all parts of the metallic wire, nevertheless our simplification provides a reasonable way for solving the problem. Taking into account the reasoning commented, we can consider that there exist a uniform magnetic field along the segment of the wire inside of the solenoid (central region in Fig. 5.47), and outside it no field. For the calculation we will suppose that the internal field over the segment S (see Fig. 5.47) is the same, approximately, as the magnetic field at the centre of the coil (point Q). Introducing $\alpha_1 = \alpha_2 = \alpha = \arctan(\frac{2r}{L}) = 4.3°$ into (5.149) we obtain

$$\mathbf{B}_y(Q) = \frac{\mu_0 n I}{2}(\cos \alpha_1 + \cos \alpha_2)\mathbf{u}_y = \mu_0 n I \cos \alpha \, \mathbf{u}_y, \quad (5.153)$$

and then

$$\mathbf{F}(\mathbf{r}) = I_h \int_{\Gamma'} d\mathbf{l}' \times \mathbf{B}(\mathbf{r}) \approx I_h \mathbf{L} \times \mathbf{B} = I_h(L\mathbf{u}_z \times (\mu_0 n I \cos \alpha \, \mathbf{u}_y)) = -I_h L B \mathbf{u}_x = 7.5 \cdot 10^{-6} \mathbf{u}_x \, N. \quad (5.154)$$

5.17 In the region between two very large concentric cylinders of radius R_1 and R_2 ($R_1 < R_2$), whose revolution axis coincides with OZ, flow $n_e = 10^{21}$ electrons per cm^{-3} ($\mu_r \approx 1$) (Fig. 5.48). Their velocities may be represented by $\mathbf{v} = b\rho \mathbf{u}_\phi$, where b is a constant and ρ is the distance from a generic point within the current to the OZ axis. Find: (a) The magnetic field B_ϕ in $\rho < R_1$. (b) The component B_z in the interval $R_1 < \rho < R_2$. (c) Idem if $\rho > R_2$. Data: $R_1 = 10$ mm; $R_2 = 11$ mm; $b = 10$ s^{-1}

Solution

(a) For ($\rho < R_1$). Taking into account that the system is invariant under rotations around the OZ axis and translations parallel to the generatrix of the hollow cylinder, the use of Ampère law seems to be adequate. To calculate the tangential component of the field in the interior of the system, we choose a circular curve of radius $\rho < R_1$ and a plane surface S. Proceeding in the same way as we have seen in other exercises, we may write

$$\oint_\Gamma \mathbf{B} \cdot d\mathbf{l} = \mu_0 \int_S \mathbf{j} \cdot d\mathbf{S} = \mu_0 I \Rightarrow \oint_\Gamma (B_\rho, B_\phi, B_z) \cdot (0, dl, 0) = \mu_0 I. \quad (5.155)$$

For determining B_ϕ we need the intensity through the surface S, but we do not know its value. However, it is possible to calculate it by means of the particle velocities. By using the definition of current density, we have

$$\mathbf{j} = nq\mathbf{v} = nqb\rho \mathbf{u}_\phi. \quad (5.156)$$

Taking into account that $d\mathbf{S} = (0, 0, dS)$

$$\int_S \mathbf{j} \cdot d\mathbf{S} = \int_S (0, j_\phi, 0) \cdot (0, 0, dS) = 0, \quad (5.157)$$

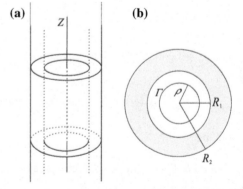

Fig. 5.48 a Current density in a cylindrical region between the radii delimited by $R_1 < \rho < R_2$. **b** Surface S and circular curve Γ of radius $\rho < R_1$

then,
$$2\pi\rho B_\phi = 0 \Rightarrow B_\phi = 0. \tag{5.158}$$

(b) Let us again apply Ampère's law over a rectangular curve directed counterclockwise, and a plane surface as shown in Fig. 5.49 (see Problem 5.1). Dividing the curve Γ in four paths, we have

$$\oint_\Gamma (B_\rho, B_\phi, B_z) \cdot d\mathbf{l} = \int_1^2 \mathbf{B}_{12} \cdot d\mathbf{l}_{12} + \int_2^3 \mathbf{B}_{23} \cdot d\mathbf{l}_{23} + \int_3^4 \mathbf{B}_{34} \cdot d\mathbf{l}_{34} + \int_4^1 \mathbf{B}_{41} \cdot d\mathbf{l}_{41}. \tag{5.159}$$

In order to obtain information for the z component of the field we suppose distinct values for $B_{\rho 1}$ and $\mathbf{B}_z$ over each part of Γ, that is, $\mathbf{B}_i$, for $i = 1, 2, 3, 4$. Introducing $d\mathbf{l}_{12} = dz\mathbf{u}_z = (0, 0, dz)$, $d\mathbf{l}_{23} = d\rho\mathbf{u}_\rho = (d\rho, 0, 0)$, $d\mathbf{l}_{34} = dz\mathbf{u}_z = (0, 0, dz)$ and $d\mathbf{l}_{41} = d\rho\mathbf{u}_\rho = (d\rho, 0, 0)$ in (5.159), we have

$$\int_1^2 (B_{\rho 1}, B_{\phi 1}, B_{z1}) \cdot (0, 0, dz) + \int_2^3 (B_{\rho 2}, B_{\phi 2}, B_{z2}) \cdot (d\rho, 0, 0) + \int_3^4 (B_{\rho 3}, B_{\phi 3}, B_{z3}) \cdot (0, 0, dz) +$$

$$\int_4^1 (B_{\rho 4}, B_{\phi 4}, B_{z4}) \cdot (d\rho, 0, 0) = \int_1^2 B_{z1} dz + \int_2^3 B_{\rho 2} d\rho + \int_3^4 B_{z3} dz + \int_4^1 B_{\rho 4} d\rho. \tag{5.160}$$

In the same way we have seen in Problem 1, the integrals along the line segment (2–3) and (4–1) cancel each other out because are run in opposite directions, then

$$\int_2^3 B_{\rho 2} d\rho = -\int_4^1 B_{\rho 4} d\rho, \tag{5.161}$$

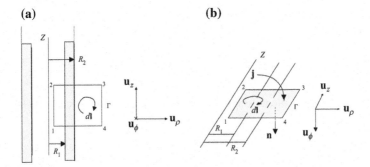

Fig. 5.49 a Plane view of the current of electrons and protons. The surface S is delimited by the curve Γ. For a point P in the region $R_1 < \rho < R_2$ the effective surface of integration S_2 corresponds to the intersection between the volume where the electrons moves and the total surface S ($S_2 = S \cap V_{j_e} = (\pi\rho^2 - \pi R_1^2)$)

which is also true if $B_\rho = B_\rho(\rho)$. By setting this expression into the general Ampère law (5.7), and $d\mathbf{S} = dS\mathbf{u}_\phi = (0, 1, 0)dS$ for the differential surface element, we have:

$$\int_1^2 B_{z1}dz + \int_3^4 B_{z3}dz = \mu_0 \int_\Sigma (0, j, 0) \cdot (0, 1, 0)dS = \mu_0 I. \quad (5.162)$$

Let us calculate the intensity current through the plane surface S. Remembering that $\mathbf{j} = nqb\rho\,\mathbf{u}_\phi$, the flux of the density current through the surface S is

$$I = \int_S \mathbf{j} \cdot d\mathbf{S} = \int_0^h \int_{R_1}^{R_2} nqb\rho\, d\rho\, dz = nqb \int_{R_1}^{R_2} \rho\, d\rho \int_0^h dz = \frac{1}{2}nqbh(R_2^2 - R_1^2) \quad (5.163)$$

and introducing it into (5.162) results

$$\int_1^2 B_{z1}dz + \int_3^4 B_{z3}dz = \mu_0 I = \frac{1}{2}\mu_0 nqbh(R_2^2 - R_1^2). \quad (5.164)$$

As a consequence of the translation symmetry of the system, the magnetic fields B_{z1} and B_{z3} have a constant value over their respective integration paths, hence we can locate them outside of the integrals,

$$B_{z1}h + B_{z3}h = \mu_0 I = \frac{1}{2}\mu_0 nqbh(R_2^2 - R_1^2) \Rightarrow B_{z1} = \mu_0 I = \frac{1}{2}\mu_0 nqb(R_2^2 - R_1^2) - B_{z3}. \quad (5.165)$$

In this expression the constant B_{z3} is not known yet. However its value may be determined by using the same argument showed in the Problem 1 when calculating the longitudinal component of the field for the conducting wire. In effect, if we would have chosen a similar curve with the segments (2–3) and (4–1) very large, we would not have observed any change in the final result. This means that this new rectangular curve Γ has the same validity for obtaining the magnetic field. The advantage of employing it is that we know that the magnetic field must be convergent, then at infinity it must vanish. For this reason we conclude that $B_{z3} = 0$, hence

$$B_{z1} = \mu_0 I = \frac{1}{2}\mu_0 nqb(R_2^2 - R_1^2) = 0,22T. \quad (5.166)$$

(c) If $\rho > R_2$, from the reasoning of the above section (see also Problem 5.1) we see that $B = 0$.

5.18 A very large solenoid with radius a and n turns per unit length, is carrying a current I_1. In its interior two parallel conducting wires A y B are located. If the distance between the wires is d, and their currents are I_A and I_B, respectively, find: (a) The magnetic field generated by the solenoid and wire A at any point over the filament B. (b) The force exerted over the segment L of the wire B between the points 1 and 2.

Solution

(a) For calculating the magnetic field exerted at any point on wire A, we can apply the superposition principle of fields proceeding in two steps. In the first one we determine the field generated by the solenoid and in the second we obtain the magnetic contribution due to the wire B (Figs. 5.50 and 5.51).

As the wire A is located over the OZ axis of the reference frame, coinciding with the symmetry axis of the solenoid, we can use the formula (5.99) for the field along the revolution axis of a spool. Taking into consideration that the coil is very large we can introduce $\alpha_1 = \alpha_1 \approx 0$ in (5.99), thus

$$\mathbf{B}_s = \frac{\mu_0 n I_1}{2}(\cos \alpha_1 + \cos \alpha_2)\mathbf{u}_z = \mu_0 n I_1 \mathbf{u}_z. \tag{5.167}$$

On the other hand, the magnetic strength of a large conducting wire carrying an intensity I_A is (5.49)

$$\mathbf{B}_A = \frac{\mu_0 I_A}{2\pi \rho}\mathbf{u}_\phi = \frac{\mu_0 I_A}{2\pi d}\mathbf{u}_\phi, \tag{5.168}$$

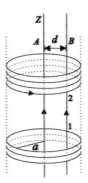

Fig. 5.50 System of wires and solenoid

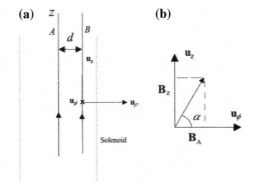

Fig. 5.51 a Lateral section of the system. b Magnetic fields corresponding to the wire and coils

ρ being the distance between the two wires, and considering $\mathbf{u}_\phi$ with respect a coordinate system located on wire A. With these expressions the total field at any point of wire B is

$$\mathbf{B} = \mathbf{B}_s + \mathbf{B}_A = \mu_0 n I_1 \mathbf{u}_z + \frac{\mu_0 I_A}{2\pi d} \mathbf{u}_\phi. \tag{5.169}$$

Notice that the resulting magnetic field has two components, one of them over $\mathbf{u}_\phi$ and the other one over $\mathbf{u}_z$. The direction of the magnetic field forms an angle α with the direction denoted by $\mathbf{u}_\phi$ whose value is

$$\tan\alpha = \frac{\mu_0 n I_1}{\frac{\mu_0 I_A}{2d\pi}} = \frac{2nd\pi I_1}{I_A} \Rightarrow \alpha = \arctan\left(\frac{2nd\pi I_1}{I_A}\right). \tag{5.170}$$

(b) To calculate the force on wire B we must know the expression of the magnetic field generated by filament A, but this is the same as we have seen in the above section for I_B. The only difference is that in this new case we locate the reference system on the wire A. By introducing (5.169) into (5.29)

$$\mathbf{F}(\mathbf{r}) = I_B \int_{\Gamma'} d\mathbf{l}' \times \mathbf{B}(\mathbf{r}) = I_B \int_1^2 dl' \mathbf{u}_z \times \left[\mu_0 n I_1 \mathbf{u}_z + \frac{\mu_0 I_A}{2\pi d} \mathbf{u}_\phi\right] =$$

$$-I_B \int_z^{z+L} \frac{\mu_0 I_A}{2\pi d} dz' \mathbf{u}_\rho = -\frac{\mu_0 I_A I_B}{2\pi d} \int_z^{z+L} dz' \mathbf{u}_\rho = -\frac{\mu_0 I_A I_B L}{2\pi d} \mathbf{u}_\rho. \tag{5.171}$$

From this result we notice that the field produced by the solenoid has no influence on the magnetic forced exerted over the segment L, because the direction of the field $\mathbf{B}$ of the coil and the direction of the element $d\mathbf{l}'$ are parallel.

5.19 The system in Fig. 5.52 is formed by two semi-infinite solenoids, of radii $r = 5$ cm and $R = 10$ cm, and $n_1 = 500$ and $n_2 = 1000$ turns per unit length, respectively. The distance between the right side of the first coil to the origin of coordinates O is $a = 10$ cm, and the left top of the second solenoid is placed at an unknown distance b. If the current through the solenoid is $I_1 = 1$ A, and through coil 2 the absolute value of the current is $I_2 = 2$ A, find: (a) The sign of I_2 and the value of b for obtaining a magnetic field zero at point $P(0, 0, 0)$. (b) The value of the field at a point $P(0, y > 0, 0)$ placed at a very large distance from the origin O.

Solution

(a) Due to the dispositions of the solenoids and the current of coil 1, we can deduce the sign of the current I_2 for obtaining zero magnetic field. In fact, applying the rule of the right hand we see that the solenoid 1 creates a field at P over the positive direction of the OY axis. Hence the magnetic field generated by 2 must be in opposite direction, which happens if the current I_2 flows from the bottom to top (clockwise). Once we know the sign of this current, we can calculate the magnitude of I_2. To do this we will calculate the mathematical expressions of the magnetic fields generated

Fig. 5.52 System of two solenoids

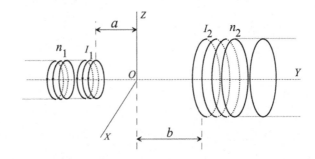

for both solenoids at point P. As P belongs to the symmetry axis of the coils, we can use (5.99), i.e.

$$\mathbf{B}_y = \frac{\mu_0 n I}{2}(\cos \alpha_1 + \cos \alpha_2)\mathbf{u}_y. \tag{5.172}$$

As the solenoids are very large by one of their extremes and the loci where we want to obtain the field is outside the coil (see Fig. 5.32) α_2 excedes $\frac{\pi}{2}$ and $\cos \alpha_2$ has a negative values (5.100), then we have

$$\mathbf{B}_1 = \frac{\mu_0 n I_1}{2}(\cos \alpha_1 - \cos \beta)\mathbf{u}_y \approx \frac{\mu_0 n I_1}{2}\left(\cos 0 - \frac{a}{\sqrt{a^2 + r^2}}\right)\mathbf{u}_y. \tag{5.173}$$

Let us call α_1' and α_2' the angles corresponding to the coil 2. Proceeding in the same way as shown for the solenoid 1 it holds

$$\mathbf{B}_2 = \frac{\mu_0 n I_2}{2}(-\cos \beta_1' + \cos \alpha_2')\mathbf{u}_y \approx \frac{\mu_0 n I_2}{2}\left(-\frac{b}{\sqrt{b^2 + r^2}} + \cos 0\right)\mathbf{u}_y. \tag{5.174}$$

Now applying the superposition principle of the linear fields we add both at P, and we impose the condition that the total field be zero (5.174), resulting in

$$\mathbf{B} = \mathbf{B}_1 + \mathbf{B}_2 = \frac{\mu_0 n I_1}{2}\left(\cos 0 - \frac{a}{\sqrt{a^2+r^2}}\right)\mathbf{u}_y + \frac{\mu_0 n(-I_2)}{2}\left(-\frac{b}{\sqrt{b^2+r^2}} + \cos 0\right)\mathbf{u}_y = 0 \Rightarrow b = 42.5 \text{ cm}. \tag{5.175}$$

(b) As point $P(0, y, 0)$ is located at a very large distance from the origin of the coordinate frame $\alpha_1 = 0$, but α_2 does not equal zero, even though it is very small. Thus we can write (Fig. 5.53)

$$\mathbf{B}_1 \approx \frac{\mu_0 n I_1}{2}\left(\cos 0 - \left(1 - \frac{r^2}{a^2}\right) + \cdots\right)\mathbf{u}_y = \frac{\mu_0 n I_1}{2}\frac{r^2}{a^2}\mathbf{u}_y, \tag{5.176}$$

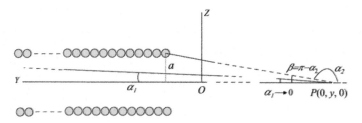

Fig. 5.53 Point $P(0, y, 0)$ located a very large distance on the *right* of the first solenoid. Notice that $\alpha_1 \approx 0$ and $\beta = \pi - \alpha_2$ is very small

Fig. 5.54 Point $P(0, y, 0)$ placed at the end of the coil 2

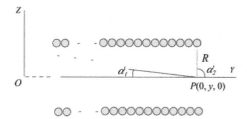

which tends to zero when a is very large. For the coil 2 the point P is viewed as placed at the end of its right side (see Fig. 5.54), hence the magnetic field is

$$\mathbf{B}_2 = \frac{\mu_0 n I_2}{2}(\cos \alpha_1' + \cos \alpha_2')\mathbf{u}_y = \frac{\mu_0 n I_2}{2}\left(1 + \cos\frac{\pi}{2}\right)\mathbf{u}_y. \tag{5.177}$$

Considering these approximations, the total field at P is the following

$$\mathbf{B} = \mathbf{B}_1 + \mathbf{B}_2 = \frac{\mu_0 n I_1}{2}\frac{r^2}{a^2}\mathbf{u}_y + \frac{\mu_0 n I_2}{2}\mathbf{u}_y \approx \frac{\mu_0 n I_2}{2}\mathbf{u}_y = -1.25 \cdot 10^{-3}\mathbf{u}_y,\ T. \tag{5.178}$$

5.20 A metallic ring of radius R holds a free charge Q. The ring begins to spin around its revolution axis with an angular velocity $\Omega = (0, 0, \omega)$. Find: (a) The component B_z of the field at $P(0, 0, z)$. (b) The magnetic field B_ϕ at $P_1 = (0, y, z)$.

Solution

(a) The effect of the ring rotation around the OZ axis (see Fig. 5.55) is to create a current, due to the charge movement. On the other hand, as a consequence of the circular geometry of the ring, the system is equivalent to a coil of radius R carrying a current I generated by the velocity of the charges. Accepting this physical model, the first step for determining the magnetic field at a point over the OZ axis is to calculate the current, and then the field corresponding to such a current. By employing the definition of current, we have

Fig. 5.55 Circular loop of radius R. Observe that the rotation makes the ring equivalent to a coil of the same radius carrying a current $I = \lambda \omega R$

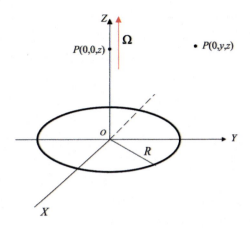

$$I = \frac{Q}{T} = \frac{Qv}{2\pi R} = \frac{Q\omega R}{2\pi R} = \frac{Q\omega}{2\pi}, \qquad (5.179)$$

T being the period of the rotation. This equation can be expressed in terms of the linear density of charge. In effect, remembering its definition we may write

$$\lambda = \frac{Q}{L} = \frac{Q}{2\pi R}. \qquad (5.180)$$

Thus introducing (5.180) into (5.179) it holds

$$I = \frac{2\pi R \lambda \omega}{2\pi} = \lambda \omega R. \qquad (5.181)$$

Once the intensity is obtained, the magnetic field may be computed by (5.85) directly,

$$\mathbf{B} = \frac{\mu_0 I R^2}{2(R^2 + z^2)^{3/2}} \mathbf{u}_z = \frac{\mu_0 \lambda \omega R^3}{2(R^2 + z^2)^{3/2}} \mathbf{u}_z. \qquad (5.182)$$

(b) Due to the circular symmetry of the system with respect to the OZ axis, it is advisable to use Ampère's Law. To this end, we choose a circular curve Γ parallel and concentric to the metallic ring, passing at point P of coordinates $(0, y, z)$ (see Fig. 5.56). As no currents flow through this curve, we can write

$$\oint_\Gamma \mathbf{B} \cdot d\mathbf{l} = \mu_0 \int_S \mathbf{j} \cdot d\mathbf{S} = \mu_0 I \Rightarrow \oint_\Gamma (B_\rho, B_\phi, B_z) \cdot (0, dl, 0) = 0 \Rightarrow 2\pi \rho B_\phi = 0, \qquad (5.183)$$

then

$$B_\phi = 0. \qquad (5.184)$$

Fig. 5.56 Circular curve Γ of radius ρ for applying Ampère's law

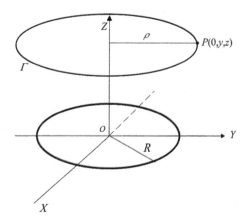

This means that the tangential magnetic field generated by circular current is zero.

5.21 A hollow cylinder of radius R, length L, surface charge density σ_s, and neglected thickness, rotates around its symmetry axis with an angular velocity ω constant. Find: (a) The surface density current $\mathbf{j}_s$. (b) The magnetic field $\mathbf{B}$ at any point over its revolution axis.

Solution

Let us consider the OZ axis coinciding with the axis of revolution of the hollow cylinder. In order to find the magnetic field, we must first calculate the density current generated by the system as a result of the rotation of the free charge, here represented by σ_s. Employing the definition of $\mathbf{j}_s$ and the relation between linear and the angular velocity, it follows

$$\mathbf{j}_s = \sigma_s \mathbf{v} = \sigma_s(\boldsymbol{\omega} \times \mathbf{r}'). \tag{5.185}$$

Taking into consideration that $\boldsymbol{\omega} = (0, 0, \omega)$ and $\mathbf{r}' = (x', y', z')$, the vectorial product is

$$\boldsymbol{\omega} \times \mathbf{r}' = \omega x' \mathbf{u}_y - \omega y' \mathbf{u}_x. \tag{5.186}$$

Using cylindrical coordinates $x' = R\cos\phi'$ $y' = R\sin\phi'$, and introducing them into (5.186) results

$$\boldsymbol{\omega} \times \mathbf{r}' = \omega R \cos\phi' \mathbf{u}_y - \omega R \sin\phi' \mathbf{u}_x = \omega R(-\sin\phi' \mathbf{u}_x + \cos\phi' \mathbf{u}_y) = \omega R \mathbf{u}_\phi, \tag{5.187}$$

hence,

$$\mathbf{j}_s = \sigma_s \omega R \mathbf{u}_\phi, \tag{5.188}$$

which has dimensions of $[j_s] = [\text{Cm}^{-2}\text{s}^{-1}\text{m}] = [\text{Cm}^{-1}\text{s}^{-1}] = [\text{Am}^{-1}]$. Examining the physical significance of this result, we observe that the effect of the constant

Solved Problems

charge movement tangential to the cylinder surface is similar to a solenoid of the same radius carrying a current per unit length done by (5.101). Thus, identifying $nI = j_s$, we have

$$\mathbf{B} = \frac{\mu_0 nI}{2}(\cos\alpha_1 + \cos\alpha_2)\mathbf{u}_z = \frac{\mu_0 j_s}{2}(\cos\alpha_1 + \cos\alpha_2)\mathbf{u}_z = \frac{\mu_0 \sigma_s \omega R}{2}(\cos\alpha_1 + \cos\alpha_2)\mathbf{u}_z. \tag{5.189}$$

5.22 A metallic wire is closely wound on a truncated conical surface of radii R and r, respectively (Fig. 5.57). If the number of turns per unit length is n, considered with respect to its symmetry axis: (a) Determine the magnetic field B at point $P(0, 0, 0)$. (b) The vector potential A at the same point.

Solution

Since the radius of the wire is very small and closely wound, in the same way as shown for the finite solenoid, we can model our problem as a conical coil carrying a current per unit length nI, where n is defined as the following expression (Fig. 5.58)

$$n = \frac{dN}{dz'}. \tag{5.190}$$

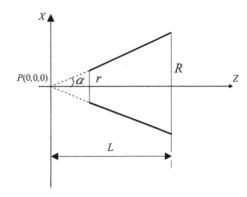

Fig. 5.57 Conical surface. The density current surface is like a sheet current $j_s = nI$

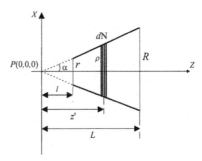

Fig. 5.58 Magnetic field created by a set of dN coils carrying a current I

z' being the coordinate denoting a source point. Let us suppose we focus our attention to a set of dN coils of the solenoid. As dN is very small, the radius associated to these rings may be considered the same and equal to ρ. For this reason the magnetic field produced by these differential number of coils at a point P over its revolution axis is

$$d\mathbf{B}(P) = \frac{\mu_0 dNI\rho^2}{2\left[\rho^2 + (z-z')^2\right]^{3/2}} \mathbf{u}_z. \tag{5.191}$$

From (5.190) and using the definition of differential of a function we have

$$dN = \left(\frac{dN}{dz'}\right)dz' = n(z')dz'. \tag{5.192}$$

By introducing this values into (5.191)

$$d\mathbf{B}(P) = \frac{\mu_0 I \rho^2 n dz'}{2\left[\rho^2 + (z-z')^2\right]^{3/2}}. \tag{5.193}$$

This last expression depends on two variables, z' and ρ, hence we cannot perform the integration directly. With the aim to integrate (5.191) we need only one variable, which may be obtained by means of the geometrical relation between z' and ρ,

$$\frac{\rho}{z'} = \tan\alpha \Rightarrow \rho = z'\tan\alpha.$$

Setting $\rho = f(z')$ in (5.191), we have

$$d\mathbf{B}(P) = \frac{\mu_0 I (z'\tan\alpha)^2 n dz'}{2\left[(z'\tan\alpha)^2 + (z-z')^2\right]^{3/2}} \mathbf{u}_z. \tag{5.194}$$

Actually, the point at which we must compute the field is $P(0, 0, 0)$, then we calculate (5.194) for $z = 0$

$$d\mathbf{B}(P) = \frac{\mu_0 I (z'\tan\alpha)^2 n dz'}{2\left[(z'\tan\alpha)^2 + z'^2\right]^{3/2}} \mathbf{u}_z = \frac{\mu_0 I (z'\tan\alpha)^2 n dz'}{2|z'|^3 \left(1+\tan^2\alpha\right)^{3/2}} \mathbf{u}_z = \frac{\mu_0 n I z'^2 dz' \sin^2\alpha \cos\alpha}{2|z'|^3} \mathbf{u}_z. \tag{5.195}$$

Now, integrating it holds

$$\mathbf{B}(P) = \frac{\mu_0 n I \sin^2\alpha \cos\alpha}{2} \int_l^L \frac{dz'}{|z'|} \mathbf{u}_z. \tag{5.196}$$

We do not known the values of L and l, but of the radii R and r. However, we can express these limits in the integral as function of the radii as follows

$$\frac{R}{L} = \tan \alpha \Rightarrow L = \frac{R}{\tan \alpha},$$

and

$$\frac{r}{l} = \tan \alpha \Rightarrow l = \frac{r}{\tan \alpha}.$$

Setting these values in the limits of the integral it leads to

$$\mathbf{B}(P) = \frac{\mu_0 n I \, \sin^2 \alpha \, \cos \alpha}{2} \int_{\frac{r}{\tan \alpha}}^{\frac{R}{\tan \alpha}} \frac{dz'}{|z'|} \mathbf{u}_z = \frac{\mu_0 n I \, \sin^2 \alpha \, \cos \alpha}{2} \ln \frac{R}{r} \mathbf{u}_z. \quad (5.197)$$

5.23 Find the magnetic field at the center of a polygon of N identical segments carrying a current I and prove that when the number of sides is large, the magnetic field created agrees with the field of a circular coil.

Solution

We can face this problem as the superposition of N finite wires disposed as shown in the Fig. 5.59. Let us first focus our analysis on only a metallic segment of the polygon. In order to apply (5.91) directly, let us assume that the OZ axis is perpendicular to the plane where the polygon lies, then the magnetic field of one of the wires which forms the polygon may be expressed as

$$\mathbf{B}_z = \frac{\mu_0 I}{4\pi \rho}(\sin \alpha_1 + \sin \alpha_2)\,\mathbf{u}_z. \quad (5.198)$$

Remembering that ρ is the distance between the center of the polygon and the wire chosen, we realize that $\alpha_1 = \alpha_2 = \alpha$. Introducing these angles into (5.91), we have

$$\mathbf{B}_z = 2\frac{\mu_0 I}{4\pi \rho} \sin \alpha_1 \,\mathbf{u}_z = \frac{\mu_0 I}{2\pi \rho} \sin \alpha \,\mathbf{u}_z. \quad (5.199)$$

From Fig. 5.59 we see that $\frac{a}{2} = R \sin \alpha$, and $\rho = R \cos \alpha$, then

Fig. 5.59 Polygon formed by N wires of length a

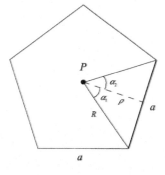

$$\mathbf{B}_z = \frac{\mu_0 I}{2\pi R \cos \alpha} \sin \alpha \, \mathbf{u}_z = \frac{\mu_0 I}{2\pi R} \tan \alpha \, \mathbf{u}_z. \tag{5.200}$$

On the other hand, for a polygon of N sides it holds that $2\alpha \cdot N = 2\pi$, thus

$$\mathbf{B}_z = \frac{\mu_0 I}{2\pi R} \tan\left(\frac{\pi}{N}\right) \mathbf{u}_z. \tag{5.201}$$

Applying the principle of superposition of fields, the magnetic field created by the polygon is N times (5.201)

$$\mathbf{B}_z = \frac{\mu_0 I N}{2\pi R} \tan\left(\frac{\pi}{N}\right) \mathbf{u}_z. \tag{5.202}$$

When the number of wires composing the polygon is large, the field produced should be the same as that produced by a circular loop carrying an intensity I. In order to demonstrate this, we calculate the limit of (5.202) for $N \to \infty$

$$\lim_{N \to \infty} \frac{\mu_0 I N}{2\pi R} \tan\left(\frac{\pi}{N}\right) = \infty \cdot 0, \tag{5.203}$$

which is an indetermination. But this result may be expressed in another form, so that we may employ L'Hôpital rule

$$\lim_{N \to \infty} \frac{\mu_0 I}{2\pi R} \frac{\tan\left(\frac{\pi}{N}\right)}{\frac{1}{N}} = \frac{0}{0}. \tag{5.204}$$

In fact, the quotient $\frac{0}{0}$ allows us to use such a rule as follows

$$\lim_{N \to \infty} \frac{\mu_0 I}{2\pi R} \frac{\tan\left(\frac{\pi}{N}\right)}{\frac{1}{N}} = \lim_{N \to \infty} \frac{\mu_0 I}{2\pi R} \frac{\frac{-\pi \left(\frac{1}{N}\right)^2}{1+\left(\frac{\pi}{N}\right)^2}}{-\left(\frac{1}{N}\right)^2} = \lim_{N \to \infty} \frac{\mu_0 I}{2\pi R} \frac{\pi}{\left(1+\left(\frac{\pi}{N}\right)^2\right)} = \frac{\mu_0 I}{2R}. \tag{5.205}$$

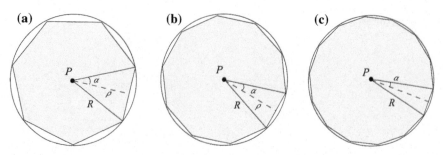

Fig. 5.60 a Polygon of 7 sides. **b** Polygon of 10 sides. **c** Polygon of 14 sides. Observe that if we increase the number of sides, when $N \to \infty$ a circle is obtained

Fig. 5.61 System formed by a semi-infinite conducting wire and a semi-infinite hollow metallic cylinder

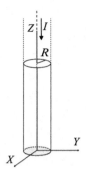

This result is the same reached by directly applying (5.85) with $z = 0$ (Fig. 5.60).

Another possibility for obtaining the same result is to substitute $n = \frac{\pi}{\alpha}$ into (5.202), i.e.

$$\mathbf{B}_z = \frac{\mu_0 I \pi}{2\pi R \alpha} \tan \alpha \, \mathbf{u}_z. \tag{5.206}$$

When the number of sides grows to infinity, the angle α tends to zero, then may be written

$$\lim_{\alpha \to 0} \frac{\mu_0 I}{2R\alpha} \tan \alpha = \frac{0}{0}. \tag{5.207}$$

Again, by applying L'Hôpital's rule we have

$$\lim_{\alpha \to 0} \frac{\mu_0 I}{2R} \frac{\frac{1}{\cos^2 \alpha}}{1} = \frac{\mu_0 I}{2R}, \tag{5.208}$$

which agrees with (5.205).

Problems C

5.24 The system of the figure attached (Fig. 5.61) is built by a semi-infinite conducting wire joined to one of the bases of a metallic hollow cylinder of the same length and with radius R. If a current I flows along the wire, find: (a) The density current circulating by the lateral side of the cylinder. (b) The tangential component of the magnetic field at a generic point for $z > 0$, and $\rho > R$.

Solution

(a) Let us apply the equation of the conservation of the electric charge in integral form. As the current is stationary ($\frac{\partial \rho(\mathbf{r})}{\partial t} = 0$), we can write

$$\int_V \nabla \cdot \mathbf{j}\, dV = \oint_S \mathbf{j} \cdot d\mathbf{S} = 0. \tag{5.209}$$

In order to obtain the intensity circulating by the lateral surface of the conducting cylinder, we choose a *closed* cylindrical surface S with basis Σ' and height $2h$, as shown in Fig. 5.62. Since the current flows on the surface of the basis Σ' of the cylinder, $\mathbf{j}$ may be expressed by means of a Dirac's delta, i.e. $\mathbf{j} = j\delta(z)\mathbf{u}_\rho$ (note that $\mathbf{j}$ is a current density of free charge). As a consequence, the flux of the density current through the closed surface S actually represents the flux across a circular line Γ (intersection between the lateral surface of the cylinder and the closed surface S-see Figs. 5.62b and 5.63). In fact, the intersection of $\mathbf{j}$ and $\mathbf{S}$ is a circular line when

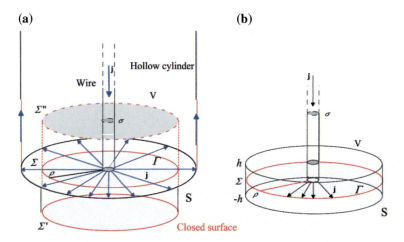

Fig. 5.62 a Closed surface S together with the system. Observe that, as we have chosen the closed surface S it contains a segment of wire and part of the basis of the hollow cylinder. Observe that Σ, Σ', and Σ' are three identical circular parallel surfaces. Σ', and Σ' correspond to the closed surface S and Σ is the intersection of the actual basis of the hollow cylinder. b Closed surface for applying the theorem of the charge conservation

Fig. 5.63 Density current on the basis of the conducting cylinder. Note that the flux is across the curve Γ. This curve corresponds to the intersection of the closed surface S and the base of the hollow cylinder

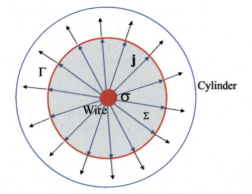

the basis of the cylinder is considered (Fig. 5.63). On the other hand, the intersection of the current flow along the wire $\mathbf{j} = -j\mathbf{u}_z$ and the closed surface S is σ, direct application of (5.209) gives

$$\int\int_\sigma \mathbf{j} \cdot d\mathbf{S} + \int\int_\Sigma \mathbf{j} \cdot d\mathbf{S} = -I + \int\int_\Sigma \mathbf{j} \cdot d\mathbf{S} = -I + \int\int_\Sigma j\,\delta(z)\,\mathbf{u}_\rho \cdot d\mathbf{S}\,\mathbf{u}_\rho =$$
$$-I + \int_0^{2\pi}\int_{-h}^{h} j_s\,\delta(z)\,\rho\,d\phi\,dz = -I + j_s\,\rho\int_0^{2\pi} d\phi\int_{-h}^{h}\delta(z)\,dz = -I + 2\pi\,\rho j_s = 0,$$
(5.210)

then,

$$j_s = \frac{I}{2\pi\rho} \Rightarrow \mathbf{j}_s = \frac{I}{2\pi\rho}\mathbf{u}_\rho.$$ (5.211)

If we examine this result obtained, we observe that the intensity depends on the distance ρ from a point placed on the basis of cylinder to the axis of revolution of the system. This current density diminishes when ρ increases. When the charge goes out of the wire it must expand over the entire base of the cylinder, then its magnitude should decrease with distance in such a manner.

(b) For calculating the magnetic field at a point $P(x, y, z > 0)$ outside of the cylinder we may use Ampère's law. In fact, due to the rotational symmetry of the system, we choose a circle Γ of radius $\rho > R$ for applying this theorem (see Fig. 5.64) and a differential element of curve $d\mathbf{l} = (0, dl, 0)$, all in cylindrical coordinates. Introducing these data in (5.7) we have for the first integral

$$\int_\Gamma \mathbf{B} \cdot d\mathbf{l} = \int_0^L B_\phi dl = \int_0^{\rho > R} B_\phi \rho\,d\phi = 2\pi\rho B_\phi.$$ (5.212)

The second part corresponding to the net current crossing the open surface S can be obtained by employing (5.211). For calculating the flux of $\mathbf{j}$ through S we must take into account that $\mathbf{j}$ extends on the lateral surface of the cylinder. This means that the product $\mathbf{j} \cdot d\mathbf{S}$ actually represents the flux of the density current across a line (see Fig. 5.34). As we have seen $\mathbf{j}$ depends on ρ, but when the current reaches the lateral surface the distance remains constant, then we can write for this case

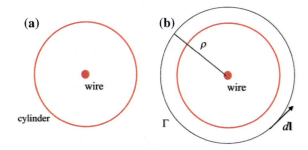

Fig. 5.64 a Plane view of the system. b Closed curve Γ for applying Ampère's theorem

$$\mathbf{j}_s = \frac{I}{2\pi R}\mathbf{u}_\rho. \tag{5.213}$$

By setting (5.213) into the integral of the intensity, it yields

$$\int_S \mathbf{j}\cdot d\mathbf{S} = \int\int_S \mathbf{j}_{wire}\cdot d S \mathbf{u}_z + \int\int_S j_s(R)\,\delta(\rho-R)\,\mathbf{u}_z\cdot d S \mathbf{u}_z. \tag{5.214}$$

As the density current corresponding to the wire goes in the direction of the negative OZ axis, and the element $d\mathbf{S} = (0,0,dS)$, the intensity due to the filament is negative, then it follows

$$\int_S \mathbf{j}\cdot d\mathbf{S} = -I + \int\int_S j_s(R)\delta(\rho-R)\cdot\rho\,d\rho\,d\phi =$$

$$-I + \int_0^{2\pi} d\phi\int_0^{\rho>R}\frac{I}{2\pi R}\delta(\rho-R)\,\rho\,d\rho = -I + 2\pi\frac{I}{2\pi R}\int_0^{\rho>R}\delta(\rho-R)\,\rho\,d\rho =$$

$$-I + 2\pi\frac{I}{2\pi R}R = -I + I = 0. \tag{5.215}$$

Using this value with (5.212) it results

$$B_\phi = 0.$$

5.25 A very large hollow metallic cylinder of negligible thickness, is located coinciding its symmetry axis with OZ. Along its surface flows a density current $\mathbf{j}_s = 100\,\mathbf{u}_z$ (A/m). If the radius of the cylinder is $R = 10$ cm, find: (a) The magnetic field B_ϕ if $\rho < R$ and $\rho > R$. (b) The component A_z of the vector potential for an exterior point ($\rho > R$). (c) Prove that for $\rho > R$ the general equation $\mathbf{B} = \nabla\times\mathbf{A}$ applies.

Solution

(a) In the same way we have seen in other problems, we can use Ampère's law for answering this first question. In fact, due to the rotational symmetry of the system we may use integral (5.7) for calculating the tangential component of the field. With this aim let us apply the law to a circular curve of radius $\rho < R$ and delimiting plane surface S. By making the scalar product $\mathbf{B}\cdot d\mathbf{l}$ and taking into consideration that no density current across the surface S exists, we have

$$\oint_{\partial S} B_\phi dl = \mu_0\int_S j dS = 0 \Rightarrow B_\phi = 0. \tag{5.216}$$

This result shows that no magnetic field B_ϕ is present in the interior of a hollow cylinder.

For an exterior point we follow the same procedure, but now the flux of the density current through the surface S is not zero. When calculating such a flux, we observe a difference from other problems in which we have used Ampère's theorem. We have

Fig. 5.65 **a** Close curve Γ for a point P interior to the cylinder. **b** Plane view of the density current over the lateral surface of the cylinder

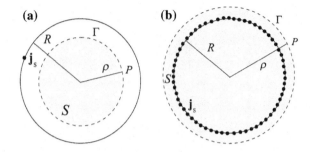

shown that the effective surface of integration corresponded to the intersection of the volume where $\mathbf{j}_v$ flows with the surface S whose delimitation is the curve Γ. However in this case the density current $\mathbf{j}_s$ extends on the surface of the cylinder, but not in a volume. Hence the actual region of integration is a curve.

As the current is located over the lateral surface of the cylinder, as shown in Fig. 5.65, we may represent it by means of Dirac's delta distribution as follows

$$\mathbf{j}_s = j_s\,\delta(\rho - R)\,\mathbf{u}_z = 100\,\delta(\rho - R)\,\mathbf{u}_z. \tag{5.217}$$

By introducing this expression into the right side of Ampère's law, we have

$$\oint_{\partial S} B_\phi dl = \mu_0 \int_S \mathbf{j}_s \cdot d\mathbf{S} = \mu_0 \int_0^{2\pi} \int_0^\rho j_s\,\delta(\rho - R)\,\rho\,d\rho\,d\phi, \tag{5.218}$$

and integrating

$$B_\phi 2\pi\rho = \mu_0 j_s \int_0^{2\pi} d\phi \int_0^\rho \delta(\rho - R)\,\rho\,d\rho = \mu_0 j_s 2\pi \int_0^\rho \delta(\rho - R)\,\rho\,d\rho = 2\pi\mu_0 j_s R. \tag{5.219}$$

From this last result we find for the tangential component of the magnetic field

$$\mathbf{B}_\phi = \frac{\mu_0 j_s R}{\rho}\mathbf{u}_\phi. \tag{5.220}$$

This equation may be written in another form. In effect, the product $2\pi R j_s$ represents the flux of j_s across the circular line of length $2\pi R$, then we can call it the surface intensity I_s. Setting it into (5.220) we have

$$\mathbf{B}_\phi = \frac{\mu_0 I_s}{2\pi\rho}\mathbf{u}_\phi, \tag{5.221}$$

whose form is identical to those of the very large wire carrying intensity I (see Problem 1).

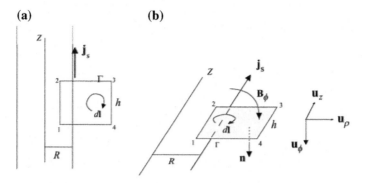

Fig. 5.66 a Closed curve Γ for a point P interior to the cylinder. b Plane view of the current density over the lateral surface of the cylinder

(b) For the calculation of the vector potential we may use the following relation

$$\oint_\Gamma \mathbf{A} \cdot d\mathbf{l} = \int\int_S \mathbf{B} \cdot d\mathbf{S}. \tag{5.222}$$

In this equation we find a relation between the magnetic field **B** and **A**. It tells us that the circulation of the vector potential A along a closed curve Γ is the same as the flux of the magnetic field through the open surface S whose boundary is Γ. As we know the magnetic field, by choosing an adequate curve and surface it should be possible to determine any of the components of **A**. In fact, due to the translational symmetry we employ a curve rectangular in shape as depicted in Fig. 5.66. By dividing such a curve in four paths, we have

$$\oint_\Gamma (A_\rho, A_\phi, A_z) \cdot d\mathbf{l} = \int_1^2 \mathbf{A}_{12} \cdot d\mathbf{l}_{12} + \int_2^3 \mathbf{A}_{23} \cdot d\mathbf{l}_{23} + \int_3^4 \mathbf{A}_{34} \cdot d\mathbf{l}_{34} + \int_4^1 \mathbf{A}_{41} \cdot d\mathbf{l}_{41}, \tag{5.223}$$

and setting the components of **A**

$$\int_1^2 (A_{\rho 1}, A_{\phi 1}, A_{z1}) \cdot (0, 0, dz) + \int_2^3 (A_{\rho 2}, A_{\phi 2}, A_{z2}) \cdot (d\rho, 0, 0) +$$
$$\int_3^4 (A_{\rho 3}, A_{\phi 3}, A_{z3}) \cdot (0, 0, dz) + \int_4^1 (A_{\rho 4}, A_{\phi 4}, A_{z4}) \cdot (d\rho, 0, 0) =$$
$$\int_1^2 A_{z1} dz + \int_2^3 A_{\rho 2} d\rho + \int_3^4 A_{z3} dz + \int_4^1 A_{\rho 4} d\rho. \tag{5.224}$$

Solved Problems

As we explained in the problem of the infinite wire, the integrals from (2–3) and (4–1) take the same values but opposite sign, the they cancel each other out,

$$\int_2^3 A_{\rho 2} d\rho = -\int_4^1 A_{\rho 4} d\rho, \tag{5.225}$$

hence we can write

$$\oint_\Gamma \mathbf{A} \cdot d\mathbf{l} = \int_1^2 A_{z1} dz + \int_3^4 A_{z3} dz = A_{z1} h - A_{z3} h. \tag{5.226}$$

For the plane surface shown in Fig. 5.66, the lines of the tangential magnetic field are perpendicular to S. Due to the election of $d\mathbf{l}$, surface S has its normal in the direction of $\mathbf{u}_\phi$. Thus introducing its value on right side of (5.222), we have

$$\int\int_S \mathbf{B} \cdot d\mathbf{S} = \int\int_S (B_\rho, B_\phi, B_z) \cdot (0, dS, 0) = \int\int_S B_\phi dS = \int_0^h dz \int_R^\rho \frac{\mu_0 j_s R}{\rho} d\rho =$$

$$\mu_0 j_s R h \ln \frac{\rho}{R}, \tag{5.227}$$

and the component z of the vector potential is

$$A_{z1} h - A_{z3} h = \mu_0 j_s R h \ln \frac{\rho}{R} \Rightarrow A_{z3} = A_{z1} - \mu_0 j_s R \ln \frac{\rho}{R}, \tag{5.228}$$

thus

$$\mathbf{A} = \left(A_{z1} - \mu_0 j_s R \ln \frac{\rho}{R}\right) \mathbf{u}_z, \tag{5.229}$$

A_{z1} being a constant.

(c) If we know the vector potential $\mathbf{A}$, we can obtain the magnetic field $\mathbf{B}$ by means of $\mathbf{B} = \nabla \times \mathbf{A}$. Due to $\mathbf{A}$ only having non-zero projections over the OZ axis, (5.11) becomes

$$\nabla \times \mathbf{A} = -\frac{\partial A_z}{\partial \rho} \mathbf{u}_\phi. \tag{5.230}$$

Introducing (5.229) into (5.230) leads to

$$\mathbf{B} = -\frac{\partial}{\partial \rho}\left(A_{z1} - \mu_0 j_s R \ln \frac{\rho}{R}\right) \mathbf{u}_\phi = \frac{\mu_0 j_s R}{\rho} \mathbf{u}_\phi, \tag{5.231}$$

results that agrees with (5.220).

5.26 A cylindrical solenoid of length $L = 20$ cm and constant radius $R = 5$ cm is located with respect to the coordinate axes as shown in Fig. 5.30. The wire is closely wound in such a way that the number of turns per unit length is not a constant, but the function $n(z) = bz$, $b = 1000$ m^{-2}. (a) Find the component

z of the magnetic field B at a generic point $P(0, 0, z)$. (b) What is the value of the field over the OZ axis if $z \gg L$, and $z \gg R$?

Solution

(a) As the solenoid is closely wound, we can consider this solenoid equivalent to a ensemble of independent loops together where the current forms a sheet of current per unit length

$$n(z) = 1000z.$$

For this reason the calculus of the magnetic field at the point P, may be regarded as the addition of the fields produced by many circular loops. In that sense, this problem can solved as explained in the problem of the solenoid.

To start we calculate the magnetic field due to a differential set of coils located at a generic position $z = z'$ only. As the number of loops dN is small enough, we can write

$$d\mathbf{B}(P) \approx \frac{\mu_0 dNIR^2}{2\left[R^2 + (z - z')^2\right]^{3/2}} \mathbf{u}_z. \tag{5.232}$$

As known, the relation between dN and dz' is

$$dN = \left(\frac{dN}{dz'}\right) dz' = n(z')dz'. \tag{5.233}$$

By introducing this expression into (5.232), we have

$$d\mathbf{B}(P) \approx \frac{\mu_0 IR^2 n(z')dz'}{2\left[R^2 + (z - z')^2\right]^{3/2}} \mathbf{u}_z = \frac{\mu_0 IR^2 1000z'dz'}{2\left[R^2 + (z - z')^2\right]^{3/2}} \mathbf{u}_z. \tag{5.234}$$

Once we know the field produced by these differential currents, the total magnetic field is found by integrating over the complete length of the solenoid,

$$\mathbf{B}(P) = \int_0^L \frac{\mu_0 IR^2 1000z'dz'}{2\left[R^2 + (z - z')^2\right]^{3/2}} \mathbf{u}_z = \frac{1000\mu_0 IR^2}{2} \int_0^L \frac{z'dz'}{\left[R^2 + (z - z')^2\right]^{3/2}} \mathbf{u}_z. \tag{5.235}$$

By substituting $(z - z') = t$, we have

$$\mathbf{B}(P) = \frac{1000\mu_0 IR^2}{2} \int_z^{(z-L)} \frac{(z - t)(-dt)}{\left[R^2 + t^2\right]^{3/2}} \mathbf{u}_z = \tag{5.236}$$

$$\frac{1000\mu_0 IR^2}{2} \left(-z \int_z^{(z-L)} \frac{dt}{\left[R^2 + t^2\right]^{3/2}} + \int_z^{(z-L)} \frac{tdt}{\left[R^2 + t^2\right]^{3/2}}\right) \mathbf{u}_z.$$

Integrating we obtain

$$\mathbf{B}(P) = \frac{1000\mu_0 I R^2}{2} \left\{ z\left(\frac{z}{R^2\left[R^2+z^2\right]^{1/2}} - \frac{(z-L)}{R^2\left[R^2+(z-L)^2\right]^{1/2}}\right) \right.$$
$$\left. + \left(\frac{1}{\left[R^2+z^2\right]^{1/2}} - \frac{1}{\left[R^2+(z-L)^2\right]^{1/2}}\right) \right\} \mathbf{u}_z, \quad (5.237)$$

and introducing R^2 in the brackets it follows

$$\mathbf{B}(P) = \frac{1000\mu_0 I}{2} \left\{ z\left(\frac{z}{\left[R^2+z^2\right]^{1/2}} - \frac{(z-L)}{\left[R^2+(z-L)^2\right]^{1/2}}\right) \right.$$
$$\left. + \left\{ R\left(\frac{R}{\left[R^2+z^2\right]^{1/2}} - \frac{R}{\left[R^2+(z-L)^2\right]^{1/2}}\right) \right\} \mathbf{u}_z. \quad (5.238)$$

If we look at the geometry of the solenoid, it is easy to identify

$$\frac{z}{\left[R^2+z^2\right]^{1/2}} = \cos\alpha_1, \quad \frac{-(z-L)}{\left[R^2+(z-L)^2\right]^{1/2}} = \cos\alpha_2, \quad (5.239)$$

and

$$\frac{R}{\left[R^2+z^2\right]^{1/2}} = \sin\alpha_1, \quad \frac{R}{\left[R^2+(z-L)^2\right]^{1/2}} = \sin\alpha_2. \quad (5.240)$$

With all these equalities, the mathematical expression for the magnetic field may be written as

$$\mathbf{B}(0,0,z) = \frac{1000\mu_0 I}{2} \left[z(\cos\alpha_1 + \cos\alpha_2) + R(\sin\alpha_1 - \sin\alpha_2)\right] \mathbf{u}_z. \quad (5.241)$$

(b) To analyze the behavior of the magnetic field **B** at distances z large compared to the length L and the radius of the solenoid, we can expand the trigonometric functions that appear in (5.241). In fact, by using the series for the square root, we have

$$\frac{z}{\left[R^2+z^2\right]^{1/2}} \approx 1 - \frac{1}{2}\left(\frac{R}{z}\right)^2 + \cdots, \quad \frac{(z-L)}{\left[R^2+(z-L)^2\right]^{1/2}} \approx 1 - \frac{1}{2}\left(\frac{R}{z-L}\right)^2 + \cdots, \quad (5.242)$$

and

$$\frac{R}{\left[R^2+z^2\right]^{1/2}} \approx \frac{R}{z}\left(1 - \frac{1}{2}\left(\frac{R}{z}\right)^2 + \cdots\right),$$

$$\frac{R}{\left[R^2+(z-L)^2\right]^{1/2}} \approx \frac{R}{(z-L)}\left(1 - \frac{1}{2}\left(\frac{R}{z-L}\right)^2 + \cdots\right), \quad (5.243)$$

thus

$$[z(\cos\alpha_1 + \cos\alpha_2) + R(\sin\alpha_1 - \sin\alpha_2)] \approx \frac{R^2L^2}{2z(z-L)^2} \approx \frac{R^2L^2}{2z^3}. \quad (5.244)$$

Introducing this expression into (5.241) it results

$$\mathbf{B}(0,0,z) = \frac{1000\mu_0 I R^2 L^2}{4z^3}\mathbf{u}_z. \quad (5.245)$$

5.27 Let us suppose a solenoid of variable cross circular section, and length $L = (z_2 - z_1)$. If the number of turns per unit arc s is η and the current circulating through each coil is I, obtain a general expression for calculating the field at a point on its axis of revolution.

Solution

Before beginning the calculation of the magnetic field, we draw a solenoid of variable radius as shown in Fig. 5.67. As the cross section of the system is circular in shape, we can apply the procedure used in the Problems 5.4 and 5.5 to compute the magnetic field $\mathbf{B}$. With this aim, first we calculate the field B due to a set of dN rings located at a generic distance z' of the origin O, and then we add the contributions of all loops of the solenoid. In fact, as dN is a differential quantity, these circular wires have the same radius, approximately, then the following expression holds

$$d\mathbf{B}(P) \approx \frac{\mu_0 dN I R^2}{2\left[R^2 + (z-z')^2\right]^{3/2}}\mathbf{u}_z, \quad (5.246)$$

where R is constant only for the rings contained in the distance dz'. If we follow the same way as the Problem 5 – 3 it may be seen that, in order to integrate (5.246), we need to have one variable only (say z'). However, two difficulties arise. The first one

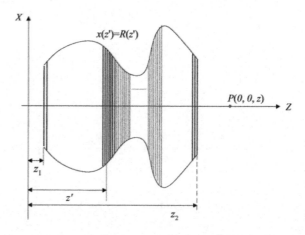

Fig. 5.67 Solenoid of variable cross-circular section

is that R depends on the distance z' (see Fig. 5.67); and the second is that the number of turns per unit *length* is unknown. Thus, to solve the problem it is necessary to know the functions that relate to radius R, and the value of dN, with distance z'. To find dN we use the definition of number of turns per unit length in (5.94),

$$dN = \left(\frac{dN}{dz'}\right) dz' = n(z')dz',$$

where $n(z')$ is in our case unknown. To obtain the expression that depends on the number of turns per unit arc it is enough to apply the chain rule,

$$n(z') = \left(\frac{dN}{dz'}\right) = \left(\frac{dN}{ds}\right)\left(\frac{ds}{dz'}\right), \tag{5.247}$$

then

$$dN = \left(\frac{dN}{dz'}\right) dz' = \left(\frac{dN}{ds}\right)\left(\frac{ds}{dz'}\right) dz'. \tag{5.248}$$

On the other hand, if the function representing the outline of the longitudinal cross-section of the solenoid is $x(z')$ (really, $x(z') = R(z')$), the expression for $\left(\frac{ds}{dz'}\right)$ is

$$\frac{ds}{dz'} = \sqrt{1 + \left(\frac{dx}{dz'}\right)^2}. \tag{5.249}$$

By introducing (5.249) into (5.248), we have

$$dN = n(z')dz' = \left(\frac{dN}{dz'}\right) dz' = \eta\sqrt{1 + \left(\frac{dx}{dz'}\right)^2} dz', \tag{5.250}$$

η being the turns per unit arc s. Hence, by using (5.250) and adding the contribution of all loops we have

$$\mathbf{B}(P) = \frac{\mu_0 I}{2} \int_{z_1}^{z_2} \frac{\eta(z')[R(z')]^2}{[[R(z')]^2 + (z_0 - z')^2]^{3/2}} \sqrt{1 + \left(\frac{dR(z')}{dz'}\right)^2} dz' \mathbf{u}_z. \tag{5.251}$$

Sometimes the equation of the curve may be obtained in parametric form. In this case, by parameterizing $z = z(t)$ and $x = x(t)$, (5.251) takes the form

$$\mathbf{B}(P) = \frac{\mu_0 I}{2} \int_{t_1}^{t_2} \frac{\eta(t)[x(t)]^2}{[[x(t)]^2 + (z - z(t))^2]^{3/2}} \sqrt{\dot{x}(t)^2 + \dot{z}(t)^2} dt \, \mathbf{u}_z, \tag{5.252}$$

where t is the parameter and z is the point at which the magnetic field must be calculated.

Fig. 5.68 Solenoid spherical in shape

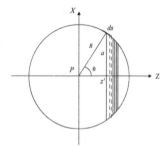

5.28 A wire is closely wound forming a spherical surface with radius R (Fig. 5.68). The solenoid is equivalent to N turns per unit arc s, η constant. If current I circulates through the turns, calculate the magnetic field B in the center of the sphere.

Solution

As we have seen in previous problems, the calculus of the magnetic field at the center of the sphere may be regarded as the addition of the fields produced by many circular loops. However, in this case, the circular wires have different radii.

With the same idea we compute the magnetic field $\mathbf{B}$ at point P due to a differential set of coils located at position $z = z'$, and then we integrate for all the coils. For this purpose we use the formula corresponding to dN circular coils with radius a,

$$d\mathbf{B}(P) \approx \frac{\mu_0 I a^2 dN}{2\left[a^2 + (z-z')^2\right]^{3/2}} \mathbf{u}_z = \frac{\mu_0 I a^2 dN}{2\left[a^2 + (z')^2\right]^{3/2}} \mathbf{u}_z. \qquad (5.253)$$

This expression has three unknowns, a, dN and z', but we need only one to carry out the integration. In the problem of the solenoid, this difficulty was solved by using the expression of dN versus dz', but it is not possible due to dependence of a in this case. To solve this problem we employ first the relation between a and z'. For a generic loop located at z' holds

$$\tan\theta = \frac{a}{z'} \Rightarrow a = z'\tan\theta.$$

Introducing this result in $d\mathbf{B}$, we have

$$d\mathbf{B}(P) = \frac{\mu_0 I (z'\tan\theta)^2 dN}{2\left[(z'\tan\theta)^2 + (z')^2\right]^{3/2}} \mathbf{u}_z = \frac{\mu_0 I z'^2 \tan^2\theta dN}{2|z'|^3 \left[1 + \frac{\sin^2\theta}{\cos^2\theta}\right]^{3/2}} \mathbf{u}_z = \qquad (5.254)$$

$$\frac{\mu_0 I z'^2 \tan^2\theta dN}{2|z'|^3 \frac{1}{\cos^3\theta}} \mathbf{u}_z = \frac{\mu_0 I \sin^2\theta \cos\theta dN}{2|z'|} \mathbf{u}_z.$$

Two variables z' and dN remain even with these changes, thus the functional relation between these unknowns must be found. With attention to the geometry of the

Solved Problems

problem we can write $z' = R\cos\theta$, then

$$d\mathbf{B}(P) = \frac{\mu_0 I \sin^2\theta dN}{2R}\mathbf{u}_z. \tag{5.255}$$

Now, we need only to express $dN = f(\theta)$. If we look at the data of the problem, we know the number of turns per unit arc η, and we can write

$$\eta = \frac{dN}{ds} \Rightarrow dN = \eta ds = \eta R d\theta.$$

Using this expression, (5.255) yields

$$d\mathbf{B}(P) = \frac{\mu_0 I \sin^2\theta \eta d\theta}{2}\mathbf{u}_z, \tag{5.256}$$

and summing all the wires

$$\mathbf{B}(P) = \frac{\mu_0 I \eta}{2}\int_0^\pi \sin^2\theta d\theta = \frac{\pi\mu_0 I \eta}{4}\mathbf{u}_z. \tag{5.257}$$

The same result may be obtained if (5.251) is employed. In fact, from (5.250) we have

$$n(z')' = \eta\sqrt{1 + \left(\frac{df}{dz'}\right)^2}.$$

There is another possibility to resolve this exercise. It consist of computing the magnetic field directly by means of (5.252). Taking into account the symmetry of the ball (in two dimensions) we may choose polar coordinates as follows,

$$z'(\theta) = R\cos\theta$$

$$x'(\theta) = R\sin\theta,$$

where θ is the parameter. Introducing these coordinates and their derivatives in (5.252) we have for $\mathbf{B}$

$$\mathbf{B}(P) = \frac{\mu_0 I}{2}\int_0^\pi \frac{\eta R^2 \sin^2\theta}{[R^2\sin^2\theta + R^2\cos^2\theta]^{3/2}}\sqrt{R^2\sin^2\theta + R^2\cos^2\theta}\,d\theta\,\mathbf{u}_z =$$
$$= \frac{\mu_0 I \eta}{2}\int_0^\pi \sin^2\theta d\theta = \frac{\pi\mu_0 I \eta}{4}\mathbf{u}_z, \tag{5.258}$$

Fig. 5.69 Conducting wire and a finite solenoid of length L

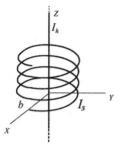

which is the same result obtained above.

5.29 The Fig. 5.69 represents a system formed by a rectilinear conducting wire, very large, located on the OZ axis and a finite solenoid of length $L = 10$ cm and radius $a = 5$ cm, closely wound and concentric with the filament. The coil is constructed by a metallic wire in form of helix of pitch $b = 1$ mm. If the current circulating through the wire is $I_h = 0.1$ A and for the solenoid $I_s = 0.05$ A, find: (a) The force on the solenoid. (b) The momentum.

Solution

(a) For calculation the forced exerted on the solenoid due to the magnetic field created by the infinite wire, we directly use (5.28)

$$\mathbf{F}(\mathbf{r}) = I \int_{\Gamma'} d\mathbf{l}' \times \mathbf{B}(\mathbf{r}), \tag{5.259}$$

where I is the intensity circulating through the system upon which the force is calculated (in this case the solenoid, $I = I_s$), and $\mathbf{B}$ is the magnetic field produced by the wire. As the wire is very large we can use (5.49),

$$\mathbf{B} = \mathbf{B}_\phi = \frac{\mu_0 I_h}{2\pi\rho}\mathbf{u}_\phi. \tag{5.260}$$

Expressing the value of the unitary vector $\mathbf{u}_\phi$ with its cartesian components we may write

$$\mathbf{B} = \frac{\mu_0 I}{2\pi\rho}(-\sin\phi\,\mathbf{u}_x + \cos\phi\,\mathbf{u}_y). \tag{5.261}$$

Taking into account that $d\mathbf{l}'$ is tangential to the curve representing the system upon which we will compute the force, we put it in a general form and then we express it in the appropriate coordinates for this geometry, i.e. $d\mathbf{l}' = (dx', dy', dz')$. By using this last expression we can calculate the vectorial product appearing in (5.28),

$$d\mathbf{l}' \times \mathbf{B}(\mathbf{r}) = -B_y dz'\mathbf{u}_x + B_x dz'\mathbf{u}_y + (B_y dx' - B_x dy')\mathbf{u}_z. \tag{5.262}$$

With the aim to perform the integral as easiest as possible, we may try to express the curve Γ' in its parametric form. Thus a circular helix of radius R and pitch b may be described by the following equations

$$x(\phi) = R\cos\phi$$
$$y(\phi) = R\sin\phi$$
$$z(\phi) = \frac{b}{2\pi}\phi. \tag{5.263}$$

With this parameterization, the differential element of the curve has the form $dl' = (dx', dy', dz') = (-R\sin\phi, R\cos\phi, \frac{b}{2\pi})d\phi$, and introducing it into (5.262) and (5.28), gives

$$\mathbf{F}(\mathbf{r}) = \frac{\mu_0 I_h I'}{2\pi R} \int_{\Gamma'} -\cos\phi\, dz'\mathbf{u}_x + \sin\phi\, dz'\mathbf{u}_y + (\cos\phi\, dx' - \sin\phi\, dy')\mathbf{u}_z$$

$$= \frac{\mu_0 I_h I'}{2\pi R} \int_{\Gamma'} -\cos\phi\left(\frac{b}{2\pi}\right)d\phi\, \mathbf{u}_x + \sin\phi\left(\frac{b}{2\pi}\right)d\phi\, \mathbf{u}_y$$
$$+ (\cos\phi(-R\sin\phi) - \cos\phi(R\sin\phi))d\phi\, \mathbf{u}_z, \tag{5.264}$$

where I' is the intensity circulating by the solenoid. The limits of integration may be obtained by using the diameter of the wire. In fact, as the wire is closely bounded in each turn the filament advances a distance equal to its diameter, therefore $L = Nb$, b and N being the diameter of the wire (pitch) and the number of turns, respectively. From this, it yields

$$z = 0 \Rightarrow \phi = 0$$
$$z = L = \frac{b}{2\pi}\phi \Rightarrow \phi = \frac{2\pi L}{b} = 2\pi N. \tag{5.265}$$

Introducing this result into the limits of integration, we have

$$\mathbf{F}(\mathbf{r}) = \frac{\mu_0 I_h I'}{2\pi R} \left\{ -\int_0^{\frac{2\pi L}{b}} \cos\phi\left(\frac{b}{2\pi}\right)d\phi\, \mathbf{u}_x + \int_0^{\frac{2\pi L}{b}} \sin\phi\left(\frac{b}{2\pi}\right)d\phi\, \mathbf{u}_y \right.$$
$$\left. -\int_0^{\frac{2\pi L}{b}} (\cos\phi\sin\phi + \sin\phi\cos\phi)d\phi\, \mathbf{u}_z \right\}$$

$$= \frac{\mu_0 I_h I'}{2\pi R} \left\{ -\left(\frac{b}{2\pi}\right)\int_0^{\frac{2\pi L}{b}} \cos\phi\, d\phi\, \mathbf{u}_x + \left(\frac{b}{2\pi}\right)\int_0^{\frac{2\pi L}{b}} \sin\phi\, d\phi\, \mathbf{u}_y \right.$$
$$\left. -2\int_0^{\frac{2\pi L}{b}} \cos\phi\sin\phi\, d\phi\, \mathbf{u}_z \right\} = 0. \tag{5.266}$$

(b) Once the force upon the solenoid is obtained, the calculus for the moment of the force may be computed by employing (5.30)

$$\mathbf{N}(\mathbf{r}) = \int_{\Gamma'} \mathbf{r}' \times d\mathbf{F}(\mathbf{r}'). \tag{5.267}$$

Using the results obtained in (a) we can write

$$dF_x = -I'B_y dz' = -\frac{\mu_0 I_h I'}{2\pi R} \cos\phi \, dz' = -\frac{\mu_0 I_h I'}{2\pi R} \cos\phi \frac{b}{2\pi} d\phi$$

$$dF_y = I'B_x dz' = -\frac{\mu_0 I_h I'}{2\pi R} \sin\phi \, dz' = -\frac{\mu_0 I_h I'}{2\pi R} \sin\phi \frac{b}{2\pi} d\phi, \tag{5.268}$$

and

$$dF_z = (B_y dx' - B_x dy') = \frac{\mu_0 I_h I'}{2\pi R} (\cos\phi \, dx' - \sin\phi \, dy')$$

$$= \frac{\mu_0 I_h I'}{2\pi R}(-R\cos\phi\sin\phi + R\sin\phi\cos\phi) = 0, \tag{5.269}$$

that is, the component z of the force disappears. Introducing these expressions into the vectorial product of (5.30) it results

$$\mathbf{r}' \times d\mathbf{F}(\mathbf{r}') = \left(-zdF_y \mathbf{u}_x + zdF_x \mathbf{u}_y + (xdF_y - ydF_x) \mathbf{u}_z\right) =$$
$$\frac{\mu_0 I_h I'}{2\pi R} \left\{ \left(\frac{b}{2\pi}\phi\right) \sin\phi \left(\frac{b}{2\pi}\right) d\phi \mathbf{u}_x + \left(\frac{b}{2\pi}\phi\right) \cos\phi \left(\frac{b}{2\pi}\right) d\phi \mathbf{u}_y + \right.$$
$$\left. \left(-R\cos\phi\sin\phi \left(\frac{b}{2\pi}\right) d\phi + R\sin\phi\cos\phi \left(\frac{b}{2\pi}\right) d\phi\right) \mathbf{u}_z \right\}, \tag{5.270}$$

and integrating by parts over the curve Γ', it holds (Fig. 5.70)

$$\mathbf{N}(\mathbf{r}) = \frac{\mu_0 I_h I'}{2\pi R} \left(\frac{b}{2\pi}\right)^2 \int_0^{\frac{2\pi L}{b}} \phi \sin\phi d\phi \mathbf{u}_x + \left(\frac{b}{2\pi}\right)^2 \int_0^{\frac{2\pi L}{b}} \phi \cos\phi d\phi \mathbf{u}_y = -\frac{\mu_0 I_h I' b L}{4\pi^2 R} \mathbf{u}_x. \tag{5.271}$$

5.30 A very thin flat disk with inner and outer radii R_1 cm and R_2 cm, respectively, has a uniform surface charge density σ_s Cm^{-2}. The disk is coplanar with the OXY plane and its axis of revolution coinciding with the axis OZ (Fig. 5.70). If the disk rotates with a constant angular velocity $\omega = \omega \mathbf{u}_z$, find: (a) The magnetic field $\mathbf{B}$ at point $P(0, 0, z)$. (b) Idem at point $P(0, 0, 0)$ cm. (c) The field B_ϕ at $P(x > R_2, 0, 0)$ cm.

Solution

(a) To calculate the magnetic field at a point on the OZ axis we first need to know the current circulating in the ring. As the charge per unit surface on the disc is σ_s,

Solved Problems

when the ring rotates about its axis of symmetry the charges on it move and generate a surface current density. Thus, the first step is to obtain $\mathbf{j}_s$ and introduce it in (5.25). With this aim let us use the definition of $\mathbf{j}_s$, i.e.

$$\mathbf{j}_s = \sigma_s \mathbf{v} = \sigma_s (\boldsymbol{\omega} \times \mathbf{r}') = \sigma_s (x'\omega \mathbf{u}_y - y'\omega \mathbf{u}_x). \tag{5.272}$$

Changing to cylindrical coordinates $x' = \rho' \cos \phi'$ and $y' = \rho' \sin \phi'$ we can express (5.272) as the following

$$\mathbf{j}_s = \sigma_s (\rho' \cos \phi' \omega \mathbf{u}_y - \rho' \sin \phi' \omega \mathbf{u}_x) = \sigma_s \omega \rho' \mathbf{u}_\phi \tag{5.273}$$

Now we will employ (5.25)

$$\mathbf{B}(\mathbf{r}) = \frac{\mu_0}{4\pi} \int_{S'} \frac{\mathbf{j}_s(\mathbf{r}') \times (\mathbf{r} - \mathbf{r}')}{|\mathbf{r} - \mathbf{r}'|^3} dS'. \tag{5.274}$$

The vectorial product in (5.25), where the field and source points are $\mathbf{r} = (0, 0, z)$ and $\mathbf{r}' = (x', y', 0)$, respectively, yields

$$\mathbf{j}_s(\mathbf{r}') \times (\mathbf{r} - \mathbf{r}') = \sigma_s \omega \rho' (z \cos \phi' \mathbf{u}_x + y' \sin \phi' \mathbf{u}_z + x' \cos \phi' \mathbf{u}_z + z \sin \phi' \mathbf{u}_y). \tag{5.275}$$

Introducing into (5.25) these results together with $x' = \rho' \cos \phi'$ and $y' = \rho' \sin \phi'$ and the jacobian $J(\rho', \phi') = \rho'$, we have

$$\mathbf{B}(\mathbf{r}) = \frac{\mu_0 \sigma_s \omega}{4\pi} \left(z \int_0^{2\pi} \cos \phi' d\phi' \int_{R_1}^{R_2} \frac{\rho'^2 d\rho'}{(\rho'^2 + z^2)^{\frac{3}{2}}} \mathbf{u}_x + z \int_0^{2\pi} \sin \phi' d\phi' \int_{R_1}^{R_2} \frac{\rho'^2 d\rho'}{(\rho'^2 + z^2)^{\frac{3}{2}}} \mathbf{u}_y \right.$$
$$\left. + \int_0^{2\pi} d\phi' \int_{R_1}^{R_2} \frac{\rho'^3 (\sin^2 \phi' + \cos^2 \phi') d\rho'}{(\rho'^2 + z^2)^{\frac{3}{2}}} \mathbf{u}_z \right). \tag{5.276}$$

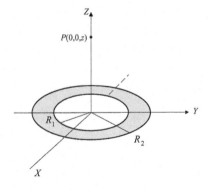

Fig. 5.70 Thin flat disc with a homogeneous surface charge density

When integrating this expression, the components of the field over OX and OY disappear. The reason is that the integrals of the circular functions $\cos\phi'$ and $\sin\phi'$ over the interval $[0, 2\pi]$ are zero. Thus, we can write[13]

$$\mathbf{B}(\mathbf{r}) = \frac{\mu_0 \sigma_s \omega}{4\pi} \int_0^{2\pi} d\phi' \int_{R_1}^{R_2} \frac{\rho'^3}{(\rho'^2 + z^2)^{\frac{3}{2}}} \mathbf{u}_z = \frac{\mu_0 \sigma_s \omega}{2} \int_{R_1}^{R_2} \frac{\rho'^3}{(\rho'^2 + z^2)^{\frac{3}{2}}} \mathbf{u}_z =$$

$$\frac{\mu_0 \sigma_s \omega}{2} \left\{ \left[\frac{z^2}{\sqrt{\rho'^2 + z^2}} \right]_{R_1}^{R_2} + \left[\sqrt{\rho'^2 + z^2} \right]_{R_1}^{R_2} \right\} \mathbf{u}_z =$$

$$\frac{\mu_0 \sigma_s \omega}{2} \left\{ \frac{z^2}{\sqrt{R_2^2 + z^2}} + \sqrt{R_2^2 + z^2} - \frac{z^2}{\sqrt{R_1^2 + z^2}} - \sqrt{R_1^2 + z^2} \right\} \mathbf{u}_z \quad (5.277)$$

(b) For obtaining the magnetic field at the origin of coordinates we may directly use the last result from section (a). In fact, by setting $z=0$ in (5.277), we obtain

$$\mathbf{B}(\mathbf{r}) = \frac{\mu_0 \sigma_s \omega}{2}(R_2 - R_1)\mathbf{u}_z \quad (5.278)$$

(c) Since the disc has symmetry of rotation around the OZ axis, and that the component of the field to be investigated is B_ϕ, we can use Ampère's theorem. As we have seen in the exercise of the solenoid, tangential component of the magnetic field at $P(x > R_2, 0, 0)$ is zero.

5.31 Find the off-axis magnetic field produced by a finite solenoid of length L and radius a, carrying a current I.

Solution

As we have studied in the Problem 5, a finite solenoid of constant radius is a system with rotational symmetry. Thus, for determining the off-axis magnetic field, we can apply the procedure explained in Sect. 1.6. The first step of the calculation is to know the field along the symmetry axis, which in the present case coincides with OZ. Using directly Fig. 5.29, we have

$$B(P) = \frac{\mu_0 I n}{2}\left[\frac{-(z-L)}{\left[a^2 + (z-L)^2\right]^{1/2}} + \frac{z}{\left[a^2 + z^2\right]^{1/2}}\right]. \quad (5.279)$$

Once we have $B(z)$, we may calculate the magnetic potential $U_m(z)$ by means of (5.42), i.e.

---

[13] For calculating the following integral you can consult [36], p. 101 or [91], p. 156 integral 114.

$$U_m(z) = -\frac{In}{2} \int \left[\frac{-(z-L)}{[a^2+(z-L)^2]^{1/2}} + \frac{z}{[a^2+z^2]^{1/2}} \right] dz + C$$
$$= \frac{In}{2} \left[(a^2+(z-L)^2)^{1/2} - (a^2+z^2)^{1/2} \right] + C. \tag{5.280}$$

Now we must obtain the coefficients of the expansion (5.40). To do this, it is necessary to develop our formula (5.280) for obtaining the coefficients a_l. Due to the magnetic field of a solenoid being symmetric with respect to its center of symmetry, we expand the magnetic potential around $z = L/2$, obtaining

$$a_0 = U_m\left(\frac{L}{2}\right) = 0 \tag{5.281}$$

$$a_1 = \left(\frac{\partial U_m}{\partial z}\right)_{z=\frac{L}{2}} = \frac{nI}{2}\left[-\frac{L}{(a^2+(\frac{L}{2})^2)^{\frac{1}{2}}}\right], \tag{5.282}$$

$$a_2 = \frac{1}{2!}\left(\frac{\partial^2 U_m}{\partial z^2}\right)_{z=\frac{L}{2}} = 0, \tag{5.283}$$

$$a_3 = \frac{1}{3!}\left(\frac{\partial^3 U_m}{\partial z^3}\right)_{z=\frac{L}{2}} = \frac{nI}{2}\left[\frac{L}{2(a^2+(\frac{L}{2})^2)^{\frac{3}{2}}} - \frac{L^3}{8(a^2+(\frac{L}{2})^2)^{\frac{5}{2}}}\right], \tag{5.284}$$

$$a_4 = \frac{1}{4!}\left(\frac{\partial^4 U_m}{\partial z^4}\right)_{z=\frac{L}{2}} = 0, \tag{5.285}$$

$$a_5 = \frac{1}{5!}\left(\frac{\partial^5 U_m}{\partial z^5}\right)_{z=\frac{L}{2}} = \frac{nI}{2}\left[-\frac{21}{384}\frac{L^5}{(a^2+(\frac{L}{2})^2)^{\frac{9}{2}}} + \frac{15}{48}\frac{L^3}{(a^2+(\frac{L}{2})^2)^{\frac{7}{2}}} - \frac{3}{8}\frac{L}{(a^2+(\frac{L}{2})^2)^{\frac{5}{2}}}\right], \tag{5.286}$$

and

$$a_6 = \frac{1}{6!}\left(\frac{\partial^6 U_m}{\partial z^6}\right)_{z=\frac{L}{2}} = 0. \tag{5.287}$$

Observe that all the even derivatives are zero.

By introducing (5.281)–(5.287) into the general solution for axisymmetric systems (5.40), the expression of the scalar magnetic potential $U_m(r, \theta)$ is directly obtained. Therefore, the components of the field in spherical coordinates may be calculated by means of the gradient of the potential, (see (5.45) and (5.46)), i.e.

$$B_r(r, \theta) = -\mu_0 \frac{\partial U_m(r,\theta)}{\partial r} = \frac{\mu_0 nI}{2}\left[\frac{L}{(a^2+(\frac{L}{2})^2)^{\frac{1}{2}}}\right]\cos\theta \tag{5.288}$$

$$-\frac{3}{4}r^2(5\cos^2\theta - 3\cos\theta) \cdot \left[\frac{\mu_0 nIL}{2(a^2 + \frac{L^2}{4})^{\frac{3}{2}}} - \frac{\mu_0 nIL^3}{8(a^2 + \frac{L^2}{4})^{\frac{5}{2}}}\right] + \vartheta(r^3),$$

and for the θ component,

$$B_\theta(r,\theta) = -\mu_0 \frac{1}{r}\frac{\partial U_m(r,\theta)}{\partial \theta} = -\frac{\mu_0 nI}{2}\left[\frac{L}{(a^2 + (\frac{L}{2})^2)^{\frac{1}{2}}}\right]\sin\theta - \left(\frac{3}{4}\sin\theta - \frac{15}{4}\cos^2\theta\sin\theta\right) \cdot \tag{5.289}$$

$$\left[\frac{\mu_0 nIL}{2(a^2 + \frac{L^2}{4})^{\frac{3}{2}}} - \frac{\mu_0 nIL^3}{8(a^2 + \frac{L^2}{4})^{\frac{5}{2}}}\right] + \vartheta(r^3).$$

5.32 In some atomic problems special solenoids must be designed. One of them is the well known Zeeman slowing technique. By this method, using a solenoid as shown in Fig. 5.71, a parabolic axial magnetic field is produced to keep the radiation pressure constant throughout the coil. Such a field is modeled by the following expression

$$B(z) = B_1 + B_0\sqrt{1 - \beta z} \tag{5.290}$$

where B_1, B_0 and β are constants. Find the magnetic field inside the solenoid by using the magnetostatic potential method.

Solution

This solenoid is used in different fields of science such as atomic and molecular physics. The aim of this non-uniform coil is to reduce the velocity of a beam of atoms or molecules that cross the solenoid in the direction of its axis of symmetry. This geometry is known as a Zeeman-slower, and by using it in conjunction with the radiation pressure of a laser light, it is able to slow down and cool neutral atoms. The design consist of an inhomogeneous conoidal-like current distribution (see Fig. 5.71), that leads to a non-uniform magnetic field along the symmetry axis of the system. At the same time, a pumped laser beam is directed through the cavity of the coil in the direction opposite to the particle motion. If the transition of the atoms (or molecules) are nearly resonant with the laser beam, they could absorb a photon, then the particle reaches an excited state. However, this new state is unstable and the atom will reach

Fig. 5.71 Solenoid with inhomogeneous current distribution

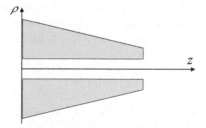

Solved Problems

its ground state by emitting spontaneously a photon in a direction compatible with the momentum conservation law, but random, which leads to an increases of the atoms velocity, whose directions are randomly distributed. When studying this process for a large number of events, statistically, it is observed that the absorption of the photon diminishes the speed of the atoms, thus reducing its temperature (cooling). By this procedure it is possible to diminish the temperature up to mK.

To calculate the magnetic field produced by this system at any point inside of the coil, we will apply the equations of Sect. 5.7.

In this case we know the value of the magnetic field produced by the solenoid along its symmetry axis, which is given by (5.290)

$$B(z) = B_1 + B_0\sqrt{1 - \beta z}, \tag{5.291}$$

then introducing it into (5.42) we have

$$U_m(z) = -\frac{1}{\mu_0}\int B(z)dz + C = -\frac{1}{\mu_0}\int (B_1 + B_0\sqrt{1-\beta z})dz + C = -\frac{1}{\mu_0}\left(B_1 z - \frac{2}{3}\frac{B_0}{\beta}(1-\beta z)^{\frac{3}{2}}\right). \tag{5.292}$$

By using (5.43), expanding in Taylor series around $z = 0$ (5.292) we obtain

$$a_0 = U_m(0) = \left(\frac{2B_0}{3\mu_0\beta}\right), \tag{5.293}$$

$$a_1 = \left(\frac{\partial U_m}{\partial z}\right)_{z=0} = -\frac{(B_0 + B_1)}{\mu_0}, \tag{5.294}$$

$$a_2 = \frac{1}{2!}\left(\frac{\partial^2 U_m}{\partial z^2}\right)_{z=0} = \frac{B_0\beta}{2!2\mu_0}, \tag{5.295}$$

$$a_3 = \frac{1}{3!}\left(\frac{\partial^3 U_m}{\partial z^3}\right)_{z=0} = \frac{B_0\beta^2}{3!4\mu_0}, \tag{5.296}$$

$$a_4 = \frac{1}{4!}\left(\frac{\partial^4 U_m}{\partial z^4}\right)_{z=0} = \frac{3B_0\beta^3}{4!8\mu_0}, \tag{5.297}$$

and so on. Taking into account that the gamma function[14] verifies that $\Gamma(n + 1) = n\Gamma(n)$, we can express the magnetic scalar potential in a closed form as follows

$$U_m(z) = \frac{2}{3}\frac{B_0}{\mu_0\beta} - \frac{(B_0 + B_1)}{\mu_0}r\cos\theta + \frac{B_0}{\mu_0\sqrt{\pi}}\sum_{n=2}^{\infty}\frac{\beta^{(n-1)}\Gamma(n-\frac{1}{2})}{n!(2n-3)}r^n P_n(\cos\theta). \tag{5.298}$$

---

[14]See, for example [29] pp. 204 and 219.

The magnetic field may be calculated by means of (5.45) and (5.46)

$$B_r(r, \theta) = (B_1 + B_0) \cos \theta - \frac{B_0}{\sqrt{\pi}} \sum_{n=2}^{\infty} \frac{\beta^{(n-1)} \Gamma(n - \frac{1}{2})}{(n-1)!(2n-3)} r^{(n-1)} P_n(\cos \theta), \quad (5.299)$$

and for the θ component

$$B_\theta(r, \theta) = -(B_1 + B_0) \sin \theta - \frac{B_0}{\sqrt{\pi} \sin \theta} \sum_{n=2}^{\infty} \frac{\beta^{(n-1)} \Gamma(n - \frac{1}{2})}{(n-1)!(2n-3)} r^{(n-1)}$$
$$\cdot (\cos \theta P_n(\cos \theta) - P_{n-1}(\cos \theta)). \quad (5.300)$$

Due to the rotational symmetry, sometimes it is useful to manipulate the result in cylindrical coordinates, even though we started with spherical coordinates. In this case we can express the result in a cylindrical reference frame by using the following change

$$B_\rho = B_r \sin \theta + B_\theta \cos \theta \quad (5.301)$$

$$B_z = B_r \cos \theta - B_\theta \sin \theta.$$

Now calling $\tilde{z} = \dfrac{z}{\sqrt{\rho^2 + z^2}}$, we obtain

$$B_\rho(\rho, z) = -\frac{B_0}{\sqrt{\pi}} \sum_{n=2}^{\infty} \frac{\beta^{(n-1)} \Gamma(n - \frac{1}{2})}{(n-1)!(2n-3)} \frac{(\sqrt{\rho^2 + z^2})^n}{\rho} (P_n(\tilde{z}) - \tilde{z} P_{n-1}(\tilde{z})), \quad (5.302)$$

and

$$B_z(\rho, z) = (B_1 + B_0) - \frac{B_0}{\sqrt{\pi}} \sum_{n=2}^{\infty} \frac{\beta^{(n-1)} \Gamma(n - \frac{1}{2})}{(n-1)!(2n-3)} (\sqrt{\rho^2 + z^2})^n P_{n-1}(\tilde{z})). \quad (5.303)$$

Chapter 6
Static Magnetic Field in Presence of Matter

Abstract In this chapter we will study the behavior of matter in the presence of magnetic fields. A comprehensive explanation of magnetism of substances requires employing quantum physics, which is beyond the scope of this book. In this chapter we will use classical physics to give a qualitative and conceptually simple explanation of the phenomenon of magnetism. This will be not very rigorous, but it will give us an idea about the phenomenon of magnetization.

6.1 Magnetization

From a microscopic point of view, gases, liquids, solids and plasmas, are constituted by atoms. According to Bohr's model, a possible construction of the atom consist of a nucleus of positive charge (protons) and neutrons without charge surrounded by electrons with negative charge which move in definite closed orbits around it. If we assume that the number of electrons and protons is the same, we obtain a model of an atom that is electrically neutral. In this way, matter can be thought to be formed by circuits (micro-coils) of the order of atomic dimensions by which a determined current circulates, producing, therefore, a magnetic field. Taking into account that such circuits are very small compared with the distance to which the magnetic field is measured, its behavior may be approximated to that of a magnetic dipole (see Chap. 5). In short, from this point of view we will consider magnetic matter to be equivalent to an ensemble of magnetic dipoles. This is the basic idea to understand the classical concept of magnetism.

We will define the vectorial field magnetization, and we will denote it as **M**, like the magnetic dipolar moment per unit volume,

$$\mathbf{M}(\mathbf{r}') = \lim_{\Delta V \to 0} \frac{\Delta \mathbf{m}}{\Delta V'} \equiv \frac{d\mathbf{m}}{dV'}, \tag{6.1}$$

the unit which is in the S.I. as Am^{-1}. Conceptually, the magnetization represents the density of dipole moments at each point in the material.

Thus, magnetic fields may be produced both by electric currents (Chap. 5) and by magnetized materials.

6.2 Magnetic Current Densities

Assuming that, from a magnetic point of view, matter behaves like many magnetic dipoles together, it seems logical that if we add all contributions of the dipoles that compose a body, sometimes we find a zero magnetic field at a given point in space. In this case two microscopic possibilities may occur. One of them is that the magnetic moment of each micro-coil is zero, and then the addition for all atoms $\sum \mathbf{m}_i = 0$. Another possibility is that the magnetic moment m_i of the individual atoms is not zero, but due to their atomic spacial locations, and the their type of atoms, the net contributions of the atomic currents is also zero, $\sum \mathbf{m}_i = 0$. However, this is not the only case that may occur. In fact, a non-complete cancellation of the bounded atomic currents can occur, giving a net magnetization $\mathbf{M}$ in the material. In this case, it is interesting to know what is the magnetic field produced. To this aim, we can calculate the vector potential at a point $P(x, y, z)$ and, from it the magnetic field $\mathbf{B}$ by means of the curl. Let us suppose we have a magnetized substance occupying the volume V' (see Fig. 6.1).

If we choose a differential volume element dv', whose magnetic moment is $d\mathbf{m}$, its contribution to the vector potential $\mathbf{A}$ at P is

$$d\mathbf{A} = \frac{\mu_0}{4\pi} \frac{d\mathbf{m} \times \mathbf{r}}{r^3}. \tag{6.2}$$

Now, to calculate the total vector potential of the body, we sum over all differential volume elements of moment $d\mathbf{m}$, obtaining

$$\mathbf{A}(\mathbf{r}') = \frac{\mu_0}{4\pi} \left\{ \int_{V'} \frac{\nabla' \times \mathbf{M}(\mathbf{r}')}{|\mathbf{r} - \mathbf{r}'|} dV' + \oint_{S'} \frac{\mathbf{M}(\mathbf{r}') \times \mathbf{n}}{|\mathbf{r} - \mathbf{r}'|} dS' \right\}. \tag{6.3}$$

If we compare this expression with those given to the vector potential in Chap. 5, we see that instead of $\nabla' \times \mathbf{M}(\mathbf{r}')$ and $\mathbf{M}(\mathbf{r}') \times \mathbf{n}$ we have the volumetric current density $\mathbf{j}_v$ and the surface current density $\mathbf{j}_s$, respectively. For this reason we can extract a simile, that the vector magnetic potential $\mathbf{A}$ created by the magnetized matter is

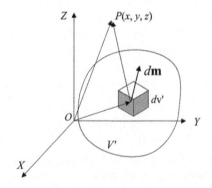

Fig. 6.1 Vector potential at point $P(x, y, z)$ produced by a differential volume element dv' of magnetic moment $d\mathbf{m}$

6.2 Magnetic Current Densities

equivalent to that would be created in vacuum volume and surface density currents of values

$$\mathbf{j}_m = \nabla' \times \mathbf{M}, \tag{6.4}$$

and

$$\mathbf{j}_{ms} = \mathbf{M} \times \mathbf{n}, \tag{6.5}$$

where **n** represents the outward normal vector to that surface. These currents are called volumetric magnetization current density and surface magnetization current density, respectively. Note that these currents are *bounded currents* different from the free currents seen in Chap. 5. In this way we can rewrite the expression of the following potential form:

$$\mathbf{A}(\mathbf{r}) = \frac{\mu_0}{4\pi} \left\{ \int_{V'} \frac{\mathbf{j}_m}{|\mathbf{r}-\mathbf{r}'|} dV' + \oint_{S'} \frac{\mathbf{j}_{ms}}{|\mathbf{r}-\mathbf{r}'|} dS' \right\}. \tag{6.6}$$

In short, what this result shows is that the effect provoked by a magnetized material is the same as that produced by these currents.

To obtain the magnetic field **B** we use (5.11)

$$\mathbf{B}(\mathbf{r}) = \frac{\mu_0}{4\pi} \left\{ \int_{V} \frac{\mathbf{j}_m \times (\mathbf{r}-\mathbf{r}')}{|\mathbf{r}-\mathbf{r}'|^3} dV + \oint_{S} \frac{\mathbf{j}_{ms} \times (\mathbf{r}-\mathbf{r}')}{|\mathbf{r}-\mathbf{r}'|^3} dS \right\}. \tag{6.7}$$

As we have commented $\mathbf{j}_m$ and $\mathbf{j}_{ms}$ are not free current densities. The free current $\mathbf{j}_f$ is actually the only current that appears in the problems of the previous chapter, however we wrote **j** instead of $\mathbf{j}_f$. The reason for this is twofold. Firstly, at that time, we did not have the magnetization vector defined yet. Making this distinction in Chap. 5 would be confusing to the reader. Secondly, and perhaps more importantly, in the Ampère theorem explained in Chap. 5 appears **j**, which includes all density currents, and not only $\mathbf{j}_f$. If we would write there only the free currents it could make us to think that the magnetic field **B** can be produced only by conduction currents $\mathbf{j}_f$, which is obviously *false*.

6.3 The Magnetic Field H

As we have seen, magnetized matter can create a magnetic field **B**, which can be explained through the currents $\mathbf{j}_m$ and $\mathbf{j}_{ms}$. On the other hand, free moving electric charges, also called conduction currents, are also source of **B**. As a consequence, the (5.1) can be written as

$$\nabla \times \mathbf{B}(\mathbf{r}) = \mu_0 \mathbf{j} = \mu_0 \left(\mathbf{j}_f + \mathbf{j}_m \right). \tag{6.8}$$

Note that we have not included $\mathbf{j}_{ms}$ in the expression (6.8), which is also one of the bound current densities defined in the previous section due to the existence of the field $\mathbf{M}$. There are several reasons. Firstly, the differential partial equation (6.8) is defined on a mathematically open region of space,[1] given the information concerning to the surface and their currents with the boundary conditions. Secondly, to calculate the curl of $\mathbf{B}$ in (6.8) we need that the magnetic field be differentiable in all points of the region to be investigated. However this condition is violated on the surface of the body where the field changes abruptly. Thirdly, it is not possible to write $\mathbf{j} = \mathbf{j}_f + \mathbf{j}_m + \mathbf{j}_{ms}$ because the unities of the surface current density are ampère per meter, and of the other currents Am^{-2}.

By introducing (6.4) into (6.8), we have

$$\nabla \times \left[\frac{\mathbf{B}}{\mu_0} - \mathbf{M} \right] = \mathbf{j}_f. \tag{6.9}$$

This new magnitude in the brackets is defined as the magnetic field $\mathbf{H}$, i.e.

$$\mathbf{H} = \frac{\mathbf{B}(\mathbf{r})}{\mu_0} - \mathbf{M}. \quad [Am^{-1}] \tag{6.10}$$

The physical units of $\mathbf{H}$ in the S.I. are the same as $\mathbf{M}$, that is, Am^{-1}.

With this definition, the differential equation (6.9) may be written as follows

$$\nabla \times \mathbf{H} = \mathbf{j}_f. \tag{6.11}$$

Equation (6.11) states that a current of free charge creates a magnetic field $\mathbf{H}$, or equivalently: the current $\mathbf{j}_f$ is a possible source of magnetic field $\mathbf{H}$. It is important to note that we do not exclude other possible sources for this field. In fact, from (6.10) we have

$$\mathbf{B} = \mu_0(\mathbf{H} + \mathbf{M}), \tag{6.12}$$

and by using (5.2), we get

$$\nabla \cdot \mathbf{H} + \nabla \cdot \mathbf{M} = 0 \Rightarrow \nabla \cdot \mathbf{H} = -\nabla \cdot \mathbf{M}. \tag{6.13}$$

This result is very important and it shows another source of magnetic field $\mathbf{H}$. Those points of space at which the divergence of $\mathbf{M}$ is distinct from zero, are points of non-zero divergence of $\mathbf{H}$ and, consequently, we will have a source or sink (depending on the sign) of lines of field H.

---

[1] If the region is infinite (from a mathematical viewpoint) we have no boundary conditions nor surface density currents.

6.3 The Magnetic Field H

In addition, there would be a third source of field **H**, which is the variation of electric displacement **D** with time, although variable fields have not yet been introduced, this possibility does not affect us in this chapter (see the Maxwell equations in Chap. 10).

6.4 The Ampère Law of the Magnetic Field H

Ampère's theorem is the integral expression of (6.11) and provides that the circulation of the magnetic field **H** along a closed path Γ depends only on the net intensity due to the free current that crosses any open surface S whose boundary is Γ,

$$\oint_{\partial S} \mathbf{H} \cdot d\mathbf{l} = \int_S \mathbf{j}_f \cdot d\mathbf{S} = \int_S \mathbf{j}_f \cdot \mathbf{n} dS = I_f. \tag{6.14}$$

where I_f corresponds to the conduction currents.

Note that in (6.14) the conduction currents only appear, whereas the expression (5.7) includes all currents. To explain this fact, let us write Ampère's theorem for the field **B** again considering all density currents

$$\oint_{\partial S} \mathbf{B} \cdot d\mathbf{l} = \mu_0 \int_S \mathbf{j} \cdot d\mathbf{S} = \int_S (\mathbf{j}_f + \mathbf{j}_m) \cdot d\mathbf{S} + \oint_\Gamma \mathbf{j}_{ms} \cdot \mathbf{n} dl = I_f + I_m + I_{ms} \tag{6.15}$$

Observe that in this case we have included $\mathbf{j}_{ms}$, which on the other hand, seems to be a contradiction if we compare it with (6.8) and remember the arguments given in the previous section for not considering the surface density current in the differential equation. The reason of this deals with the boundary conditions in (6.8). In fact, when the partial differential equation is converted into the integral form (6.15), the boundary conditions disappear, thus we would lose this information. Actually, the information retained in the BC is transformed into a surface integral. However, accepting this conclusion, another problem arises. In effect, taking into consideration the units of $\mathbf{j}_{ms}$ (A/m), we have seen that this current cannot be added to volumetric densities $\mathbf{j}_f$ and $\mathbf{j}_{ms}$, which apparently contradicts (6.15), however correct it may be.

To explain the idea, let us consider Fig. 6.2. As we can observe, although the $\mathbf{j}_{ms}$ flows over the surface Σ, the product (intersection) $\mathbf{j}_{ms} \cdot d\mathbf{S}$ (it has no sense; see the physical units) extends over a one-dimensional manifold (the curve Γ), which implies that the integration domain for $\mathbf{j}_{ms}$ is not a surface, but actually the curve Γ (see Problem 6.23). Thus, dS behaves as if it were an element of length dl, making the units of this product $j_{ms} dl$ in the integrand ampère (A), like $(\mathbf{j}_c + \mathbf{j}_m) \cdot d\mathbf{S}$.

$$\int_S \mathbf{j}_{ms} \cdot d\mathbf{S} = \int_\Gamma \mathbf{j}_{ms} \cdot \mathbf{n} dl = I_{ms}. \tag{6.16}$$

To some extent this equation represents the flux of j_{ms} throughout the line Γ (the number of lines j_{ms} crossing this curve).

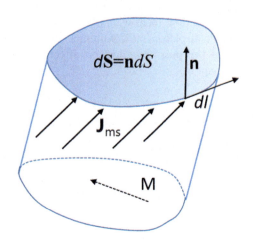

Fig. 6.2 Flux of the surface current density $\mathbf{j}_{ms}$ across the surface S. Observe that the product $\mathbf{j}_{ms} \cdot \mathbf{n} dl$ extends along the curve Γ. This situations differs from that corresponding to $(\mathbf{j}_c + \mathbf{j}_m) \cdot d\mathbf{S}$ in which the intersection is a surface (see also Fig. 5.2 in Chap. 5)

6.5 Basic Kinds of Magnetic Materials

From a magnetic point of view, substances existing in nature can be divided into diamagnetic, paramagnetic, ferromagnetic, antiferromagnetic, ferrimagnetic and metamagnetic. In reality this classification refers to different types of magnetism, since, for example, a substance that at a temperature is ferromagnetic becomes paramagnetic if a certain temperature is exceeded, namely the Curie temperature. Actually matter at high temperatures behaves as paramagnetic or diamagnetic.

In order to classify magnetic materials we must study the relation between the magnetization M and the magnetic field H. To this aim, we define the susceptibility χ_m as

$$\chi_m = \frac{\mathbf{M}}{\mathbf{H}}. \tag{6.17}$$

Thus, it is possible to characterize magnetic materials by giving the functional dependence $\mathbf{M} = \mathbf{M}(\mathbf{H})$. On the other hand, we can divide the magnetism of materials in two types: weak magnetism and intense magnetism. The first one includes diamagnetism and paramagnetism, and the second one is represented basically by the ferro- and ferrimagnetism. The fundamental characteristics of all of them are manifested by (6.17) or, equivalently, by the functional dependence of B versus H. In general, we can write

$$\mathbf{M} = \chi_m(\mathbf{H})\mathbf{H}, \tag{6.18}$$

where χ_m may be, in general, a function of H. Therefore, introducing (6.18) into (6.12) we have

$$\mathbf{B} = \mu_0[1 + \chi_m(\mathbf{H})]\mathbf{H} = \mu_0\mu_r(\mathbf{H})\mathbf{H} = \mu(\mathbf{H})\mathbf{H}, \tag{6.19}$$

where μ is the magnetic permeability of the material and μ_r the relative magnetic permeability. Bearing this in mind, we can make the following classification:

6.5 Basic Kinds of Magnetic Materials

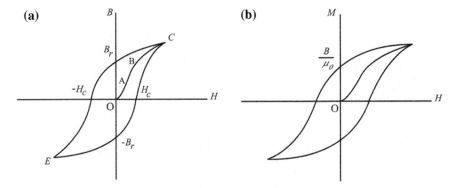

Fig. 6.3 a Hysteresis curve on plane (H–B). b Idem on (M, H)

(A) Diamagnetic materials. χ_m is constant, small and negative, then μ_r is smaller than 1. The relations $\mathbf{M} = \mathbf{M}(\mathbf{H})$ and $\mathbf{B} = \mathbf{B}(\mathbf{H})$ are linear.

(B) Paramagnetic materials. $\chi_m > 0$ and constant and $\mu_r > 1$, also functions $\mathbf{M} = \mathbf{M}(\mathbf{H})$ and $\mathbf{B} = \mathbf{B}(\mathbf{H})$ are linear.

(C) Ferromagnetic materials. Here the relationship $\mathbf{M} = \mathbf{M}(\mathbf{H})$ is not linear (Fig. 6.3), being, therefore, χ_m and μ_r functions of H. In these substances the relation $B = B(H)$ and $M = M(H)$ are called hysteresis curve on the plane BH and on the plane MH, respectively. In order to know at which point in the hysteresis curve we are, we will call the path OC, the first magnetization curve, and the stretch CE and EC, the second and third magnetization curves, respectively. The MH diagram can be obtained from cycle BH using the general expression $\mathbf{B} = \mu_0(\mathbf{M} + \mathbf{H})$. In these materials the magnetization appears spontaneously. The reason of that is the existence of magnetic domains each of which behaves like magnets of magnetic moment $\mathbf{m}_i$. The magnetization inside each microscopic region (domain)[2] is made up of the sum of the atomic magnetic moments which are aligned in a definite direction due to the exchange force. On the other hand, between magnetic domains are boundaries in which the direction of magnetization vector rotates from one domain to the next; this transition region is called the Bloch wall.[3]

6.6 Description of the Magnetization Curve

As we have mentioned in the preceding section, the magnetization process of a ferromagnetic material may be divided in three segments. The first magnetization

---

[2] In general for solid bodies, they are different factors that may contribute to the distribution of the spontaneous magnetization such as magnetostriction, magnetostatic energy and magnetic anisotropy.

[3] The boundary between the magnetic regions is thin but not sharp (discontinuous) when regarding on atomic scale. Actually, the wall spreads over a thickness which depends on the material (from 100 to 1000 atomic lattice constants). In this transition layer between domains the spin directions of the atoms change gradually.

Fig. 6.4 First magnetization curve. The part OA is reversible. The second segment goes from A to B, and is irreversible. The third zone BC corresponds to the magnetization rotation, and in CD the saturation is reached

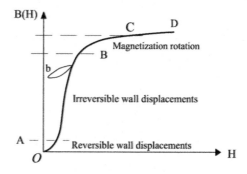

curve starts at point O (Fig. 6.4) where $H = 0$ and $B = 0$, and represents the demagnetized state. Starting from this point the magnetization increases along the curve OAB with a increase of the magnetic field H, and if we continue increasing H, the saturation magnetization at point C is reached. For greater values of H, the magnetic field B scarcely varies (point D). To describe in more detail this first part we distinguish three intervals.

The first one corresponds to the path OA and it is almost reversible. In this stage, the application of a magnetic field H leads to an increase in volume of the domains whose magnetic moments are substantially located in the same direction of the applied field. On the contrary, the domains with magnetization directions very different from H tend to decrease the size. This variation in volume of the domains is possible because of the displacement of the Bloch wall.

At the beginning in the neighborhood of the origin O the curve has a finite slope. We define the initial permeability as

$$\mu_i = \left(\frac{dB}{dH}\right)_{H \to 0}. \tag{6.20}$$

This path goes from O to A and it may be fitted to

$$B = \mu_i H + \nu H^2. \tag{6.21}$$

Taking into account that $B = \mu(H)H$ we can write (6.21) as follows

$$B = (\mu_i + \nu H)H = \mu(H)H, \tag{6.22}$$

and then,

$$\frac{d\mu(H)}{dH} = \nu,$$

is a constant. The second part extends from A to B with a steep slope. In this interval the process is not reversible. If the field H is decreased, the magnetization also changes but not along the path BA, but rather following the segment b (Fig. 6.4). The

6.6 Description of the Magnetization Curve

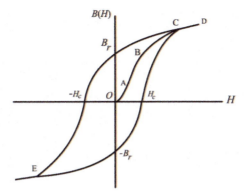

Fig. 6.5 Hysteresis loop. Observe that by increasing the magnetic field H between C and D, the magnetic field B is almost the same (saturation). When we go back the fields H and B follow the paths CE (second magnetization curve) and EC (third magnetization curve), respectively, and not the curve CO. This means that the magnetization process is not reversible. In some extent we can say that these kind of materials have *memory*

reason for this irreversibility is that when the magnetic field diminishes, Bloch's walls cannot move back by the same paths to their initial positions. As a consequence, the domains do not acquire their original size and then the sum of all magnetic moments corresponding to each magnetic domain is not zero. On the other hand, over this segment AB the permeability $\mu(H)$ reaches its maximum. The third part covers the segment from B to C. In this interval most of the domains have reached a stable configuration, then when magnetic field H increases, the only possibility is to rotate the magnetization in each domain. For greater values of H the material follows the straight CD in which the practical totality of the domains have their magnetization vector in the direction of the magnetic field H applied. This last segment is reversible, that is, if H decreases a little, the systems goes back through the straight DC. If the magnetic field is now reduced to zero the curve is not retracted (see Fig. 6.5). On the contrary, the magnetic field B follows the line CE and stops at the B-axis (where $H = 0$). This point of the curve is the remanet magnetic field B_r. If we continue reducing H the point $(-H_c, 0)$ is obtained, in which B is zero (observe that in the second quadrant H is negative, then considering its sign we can say "reducing". However in modulus we increase the magnetic field H). This value of H is called the coercitive magnetic field H_c. By increasing the modulus of H, we reach E, where the material is again saturated. From this point diminishing H to zero $-B_r$ is reached, whose value coincides in modulus with the remanet magnetic field aforementioned. If we continue augmenting it, positively, more and more, we holds C again.

For experimentally obtaining the hysteresis curve, a device as shown in Fig. 6.6 may be used. By changing the value of the variable resistance VR the current circulating in the circuit is controlled. For low values of the current we cannot obtain a saturated hysteresis curve. We obtain a smaller loop (Fig. 6.7) within the major

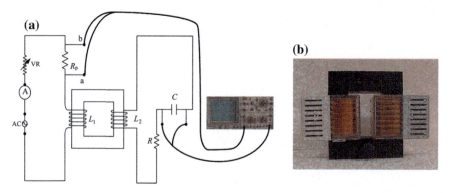

Fig. 6.6 a Experimental set-up for obtaining the hysteresis loop. In this picture appear: AC, alternating current source; R_p, standard resistance; VR, variable resistance; A, amperimeter; R, resistance of the secondary circuit; L_1 and L_2, autoinductances (see Chap. 8); C, capacitor. On the right the oscilloscope. **b** Subsystem formed by the ferromagnetic core and the coils L_1 and L_2 (transformer). See next section

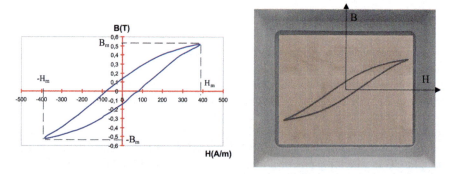

Fig. 6.7 Experimental hysteresis curve obtained in the laboratory for a steel sample. Observe that the first magnetization curve does not appear

loop as shown in Fig. 6.8, where H_m and B_m are the maximum values for H and B, respectively (observe the first loops in the Fig. 6.8).[4]

However, by increasing the intensity enough, it is possible to cycle the material between symmetrical fields H to achieve saturation (see the bigger hysteresis loop in Fig. 6.8). If we vary the magnitude of the field strength H following the above process,

---

[4]The first magnetization curve does not appear in this figure, due to the experimental procedure we have used for measuring this hysteresis loop. When an AC current is employed, the fields vary rapidly (50 or 60 times per second), incapable for the human eye to observe the initial part of the hysteresis loop. For detecting directly this segment of the curve we should take a photograph with an exposure time on the order of $\left(\frac{1}{5\times 50(60)}\right)$ s^{-1}.

6.6 Description of the Magnetization Curve

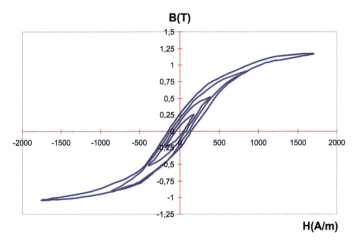

Fig. 6.8 Family of hysteresis loops

and we draw all curves together, thus we get a family of loops whose appearance is displayed in Fig. 6.8. This enables the reconstruction of the first magnetization curve by joining the maxima (H_m, B_m) for each loop.[5]

6.7 Magnetic Circuits and Electromagnets

A magnetic circuit is, in general, a system formed by magnetic materials, often with presence of electric currents, although not necessarily (e.g., the permanent magnet). The most representative circuit is, perhaps, the denominated electromagnet, shown in Fig. 6.9a. This is formed by two parts, one corresponding to the material, we will call core or yoke, and an air gap.[6] Furthermore, it has one or several insulated copper coils around the core. The main objective in the problems is the determination of the fields H, M and B, in all parts of the eletromagnet, i.e. in the core and the air gap. The exact calculation of the fields is a very difficult problem which requires special techniques. However, it is possible to find an approximate solution by making some approximations. With the aim of answering these questions, the resolution of a magnetic circuit may be carried out by following three steps: (1) Application of the Ampère law for H. To this end we choose a closed curve Γ passing through the mean lengths of the electromagnet arms. Depending on the geometry of the problem one or more curves may be used (see Figs. 6.9a, 6.10). Let us suppose, for simplicity, that we have a simple electromagnet with an air gap and constant cross section (Fig. 6.11). For this case it holds that

---

[5] In some cases the form of the hysteresis loop may be very different as we have seen for ferromagnetic materials. See for example [94].

[6] In fact, an electromagnet can be formed by more than one material. We start with this scheme from simplicity. The same procedure holds for other more complex systems.

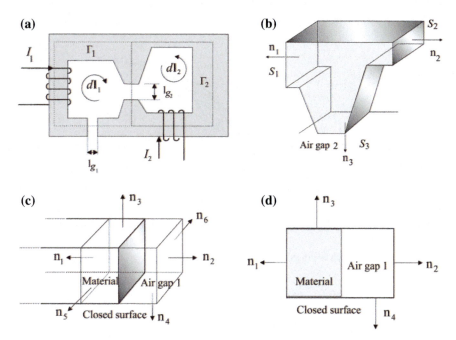

Fig. 6.9 **a** Electromagnet. **b** Closed surface. Observe that this surface encloses three parts of the magnetic material. **c** Closed surface which contains a part of the material and a zone of the air gap. **d** Lateral view of (**c**). By calculating the magnetic flux through these closed surfaces we can find equations which relate the magnetic field B in the regions of interest

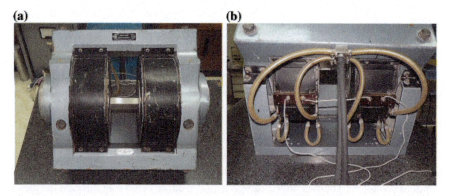

Fig. 6.10 **a** Front view of an electromagnet that can reach 1.1 Tesla when cone poles are inserted. **b** Back view. In order to refrigerate the electromagnet, water flows through a system of tubes that have been located properly

$$\oint_\Gamma \mathbf{H} dl = \int\int_S \mathbf{j}_c \cdot d\mathbf{S} = I_f \Rightarrow \int_{core} H \, dl + \int_{gap} H \, dl = H_c l_c + H_g l_g = NI, \quad (6.23)$$

6.7 Magnetic Circuits and Electromagnets

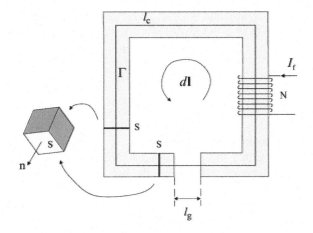

Fig. 6.11 Electromagnet of constant cross-section S and mean lengths l_c and l_g

where NI is known as the magnetomotive force (m.m.f.). This result is valid under the hypothesis that the mean magnetic field H is the same along the core where the cross section does not vary. For a more general case we can write

$$\sum_1^n \int_{\Gamma_i-(core)} H_i\, dl + \int_{gap} H\, dl = NI, \tag{6.24}$$

and if the magnetic field H_i for each part of the electromagnet is constant, (6.24) yields

$$\sum_1^n H_i l_i + H_g l_g = NI. \tag{6.25}$$

(2) Calculation of the flux of B through closed surfaces. In (6.23) a field appears that corresponds to the core and air gap, simultaneously. Hence this equation is not sufficient to calculate the fields in the different parts of the system; we need more equations. In order to have equalities, we use a general property of the magnetic field, namely its divergence is always zero. We need to relate parts of the electromagnet in which either the magnetic properties of the materials change or the sections vary (Fig. 6.9b), since both may cause a variation of the magnetic fields. In the case of Fig. 6.11 the electromagnet has a constant cross section, thus the zone of interest lies in the transition from the magnetic material to the air gap. By choosing a closed surface for this electromagnet as shown in Fig. 6.9c, d, and integrating $\nabla \cdot \mathbf{B}$ throughout, we have

$$\oint_S \mathbf{B} d\mathbf{S} = \int_{S_1} \mathbf{B}_1 d\mathbf{S} + \int_{S_2} \mathbf{B}_2 d\mathbf{S} + \int_{S_3} \mathbf{B}_3 d\mathbf{S} + \int_{S_4} \mathbf{B}_4 d\mathbf{S} + \int_{S_5} \mathbf{B}_5 d\mathbf{S} + \int_{S_6} \mathbf{B}_6 d\mathbf{S} = 0. \tag{6.26}$$

If we suppose that the field lines follow the geometry of the core arms, and that no flux leakage at the air gap exists, the magnetic flux crossing the surfaces 3, 4, 5, and 6 are zero, and then (6.26) leads to

$$-\int_{S_1} B_1 dS + \int_{S_2} B_2 dS = 0 \Rightarrow B_c S_c = B_g S_g, \qquad (6.27)$$

where $B_1 = B_c$ in the core and $B_2 = B_g$ in the air gap. However, since the cross sections are identical, $B_c = B_g$.

(3) Using the material equations including the gap properties. The magnetic characteristics of the materials that form the core are given by the equation $\mathbf{B} = \mu_0 \mu_r(\mathbf{H})\mathbf{H} = \mu(\mathbf{H})\mathbf{H}$. If the yoke is made of a ferromagnetic material, this corresponds to the hysteresis curve. For the gap it holds that $\mathbf{B} = \mu_0 \mathbf{H}$.

6.8 Operating Straight Line and Operating Point

By using the general equation of B in the case of the air gap, we can substitute H_g as a function of B_g,

$$B_g = \mu_0(H_g) \Rightarrow H_g = \frac{B_g}{\mu_0}, \qquad (6.28)$$

and introducing it into (6.23),

$$H_c l_c + H_g l_g = NI \Rightarrow H_c l_c + \left(\frac{B_g}{\mu_0}\right) l_g = NI. \qquad (6.29)$$

This equation has two unknowns corresponding to the core and the air gap, respectively. The first step to solve it is to express this equality as a function of fields only corresponding to the core. To this aim, we employ (6.27) together with (6.29), obtaining

$$H_c l_c + \left(\frac{B_c}{\mu_0}\right) l_g = NI \Rightarrow B_c = \frac{\mu_0 NI}{l_g} - \frac{\mu_0 l_c}{l_g} H_c, \qquad (6.30)$$

which is known as operating straight line (OSL), and depends on the fields in the yoke. To calculate H_c and B_c we still need another equation. This is the material equation, which in case of ferromagnetic bodies is the hysteresis loop. The solution of both equations simultaneously corresponds to the intersection between the operating straight line and the material equation, and is said to be the operating point of the system. Because of the hysteresis curve, for ferromagnetic materials the intersection gives three (or two in the third quadrant) different solutions as depicted in Fig. 6.12a. It means that to calculate the intersection in a problem it is necessary to know the history of the material. So if the specimen starts from a demagnetized state (point O), by increasing the current we run throughout the first magnetization curve, then

6.8 Operating Straight Line and Operating Point

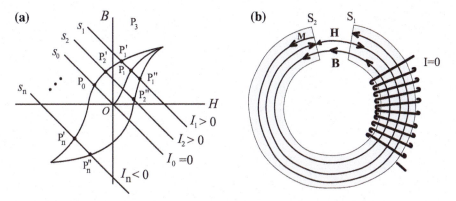

Fig. 6.12 **a** The operating straight line *OSL* intersects at three points on the hysteresis curve (P_1, P'_1, and P''_1). The solution depends on the previous history of the material. If the body started from a demagnetized state and the intensity grows positively, the operating point is P_1. If the saturation is reached and the intensity is diminished, the solution is P'_1. If we continue in this direction we get P_0. At this operating point ($I = 0$) the magnetic internal field B is opposite to H, then it tends to demagnetize the material. This field is called demagnetizing field H_d. In the case we cross $I = 0$ and the direction of the intensity is inverted, we have a generic point P'_n. By increasing the absolute value of I, we obtain the saturation again (third quadrant), and reducing $|I|$ the operating point $P''n$ is obtained. **b** Electromagnet when $I = 0$. Observe that the points where $\nabla \cdot \mathbf{M} \neq 0$ act as sources or sinks of magnetic field H

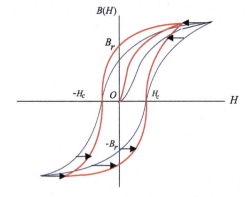

Fig. 6.13 Observe that owing the demagnetizing field H_d, the effective field inside of the specimen is less than the external field H, and therefore the actual hysteresis curve is sheared

the operation point must be P_1. If we would have reached material saturation, the state of the system is described by the second curve of magnetization. As a result, the operation point belongs to this upper or the lower curve depending on the value of the intensity (point P'_1 or P'_n), passing through P_0 which corresponds to $I = 0$. This last point is of some interest and it should be commented in more detail. In fact, at this point no intensity is flowing through the coils, but we have magnetic fields H and B. To understand this, we have to remember that a magnetostatic field H may be generated not only by free currents, but also by the existence of a nonzero divergence of magnetization (6.13). If we examine the electromagnet of Fig. 6.12b we see that

Fig. 6.14 Operating straight line *OSL* for a paramagnetic material. Here the intersection is single valued. Observe that the operating straight lines *OSL* for each problem are parallel (same slope)

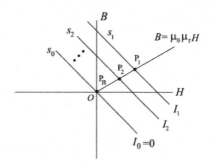

on the surface S_1 magnetization M is born ($\nabla \cdot \mathbf{M} > 0$), then the magnetic field H dies ($\nabla \cdot \mathbf{H} < 0$), and on S_2 the contrary occurs ($\nabla \cdot \mathbf{M} < 0$ and $\nabla \cdot \mathbf{H} > 0$).

This result shows two important characteristics of H for the present case. The first one is that the magnetic field H is discontinuous on the surfaces delimiting the material and air-gap; secondly, that H inside the electromagnet (for $I = 0$) has an opposite direction to the magnetic field B and M. As a result inside the material B is less than M ($B = \mu_0(-H + M)$), hence it tends to demagnetize the sample. For this reason this field is called *demagnetizing field* H_d. This field must be taken into account in device design, however its calculation is a task of high difficulty (see Problem 6.2). Due to the fact this demagnetizing field is opposite to magnetization, when studying the hysteresis loop this fact must be taken into account (Fig. 6.13).

If the material of the core is paramagnetic we simply use $B = \mu_0 \mu_r H$. The geometrical interpretation is shown in Fig. 6.14.

6.9 The Permanent Magnet

If we look at Fig. 6.12a we see that for $I = 0$, the magnetic fields H and B are non-zero. This means that in absence of electric currents magnetic fields may arise from magnetized matter. When that occurs, the material is said to be a permanent magnet. The magnetic circuit for a magnet may be calculated in the same way shown in the preceding section. The only difference is that the right side of (6.29) is zero (Fig. 6.15),

$$H_i l_i + \left(\frac{B_i}{\mu_0}\right) l_g = 0 \Rightarrow B_i = -\frac{\mu_0 l_i}{l_g} H_i. \tag{6.31}$$

Observe that the value of B_i depends directly of the quotient $\frac{l_i}{l_g}$. Although the air gap is an important part for many devices (recording heads, filters, etc.), it may not necessarily be for some applications. In this case, the magnet must be designed without an air gap (see Fig. 6.16). In this case, a toroidal core geometry may be used. The key idea is to bring the ferromagnetic core to saturation by using

6.9 The Permanent Magnet

Fig. 6.15 Permanent magnet with air-gap

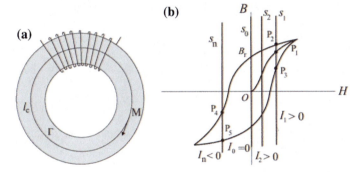

Fig. 6.16 a Permanent magnet of mean length l_c with tangential magnetization $\mathbf{M} = M\mathbf{u}_\phi$. **b** Operating straight line and operating point. For a magnet without air gap the *OSL* has an infinite slope. As a consequence, the intersection between this line and the hysteresis curve coincides with the remanent magnetic field B_r

multiple-layer solenoidal coils (Fig. 6.16a), and then reducing the electric current to zero. In applying this procedure the magnetic field in the material coincides with B_r. The reason is that the operating straight line has an infinite slope, i.e. is parallel to the *B*-axis (Fig. 6.16b), thus the intersection between the operating straight line and the second magnetizing curve of the material, when $I = 0$, corresponds to the remanent magnetic field.[7]

---

[7] When we explained the hysteresis loop of ferromagnetic materials in Sect. 6.5, we simply described it without commenting about the *OSL*, because we had not yet defined it.

6.10 The Demagnetizing Field

As we have commented in Sect. 6.8, when a material of finite size is magnetized, as a consequence of the existence of a divergence of its magnetization, on their ends appears a magnetic field in a direction opposite that of the magnetization **M**. We will call this field that tends to demagnetize a body the *demagnetizing field*. As a result, the effective magnetic field **H** inside of the body is smaller than the field if we do not consider this effect. Therefore the magnetic field **H** in the interior of the material must be corrected for obtaining its correct value. Considering a first order demagnetizing field H_d, it is, in general, proportional to the magnetization, it may be expressed as follows[8]

$$H_{d_p} = N_{d_{pq}} M_q, \qquad (6.32)$$

$N_{d_{pq}} = \bar{\bar{N}}_d$ being the demagnetizing tensor.[9]

The demagnetizing correction is usually very difficult to calculate. Only in the case of ellipsoidal materials with a uniform magnetization may an analytical solution of the demagnetizing factor be obtained, also resulting in a homogeneous demagnetizing field. Even in the case of uniform magnetization of a specimen, if the body is irregular in shape, its demagnetizing field will be non-homogeneous. In general, for non-ellipsoidal bodies the computation is highly difficult to determine, with it being necessary to employ numerical techniques.

For ellipsoidal samples, the internal field may be expressed by the following formula

$$\mathbf{H}_{in} = \mathbf{H}_{ex} + \mathbf{H}_d = \mathbf{H}_{ex} - \bar{\bar{N}}_d \mathbf{M}, \qquad (6.33)$$

where $\mathbf{H}_{ex}$ is the external field applied. For this geometry N is a diagonal tensor whose components verify the relation

$$N_{xx} + N_{yy} + N_{zz} = 1. \qquad (6.34)$$

$$N_{xx} = \frac{\cos\varphi \cos\theta}{\sin^3\theta \sin^2\alpha}(F(k,\theta) - E(k,\theta)), \qquad (6.35)$$

$$N_{yy} = \frac{\cos\varphi \cos\theta}{\sin^3\theta \sin^2\alpha \cos^2\alpha}\left(E(k,\theta) - \cos^2\alpha F(k,\theta) - \frac{\sin^2\alpha \sin\theta \cos\theta}{\cos\varphi}\right), \qquad (6.36)$$

---

[8] We will comment on this definition latter, when we deal with non-ellipsoidal samples.
[9] For very large one-dimensional samples, N is only a number, thus employing the formula $H_d = N_d M$. This is the normal case in solving problems for students. In this case the noun *demagnetizing factor* instead demagnetizing tensor is usually employed.

6.10 The Demagnetizing Field

$$N_{zz} = \frac{\cos\varphi \cos\theta}{\sin^3\theta \sin^2\alpha}\left(\frac{\cos\varphi \sin\theta}{\cos\theta} - E(k,\theta)\right), \quad (6.37)$$

where

$$\cos\theta = \frac{c}{a}, \quad \left(0 \le \theta \le \frac{\pi}{2}\right) \quad (6.38)$$

$$\cos\varphi = \frac{b}{a}, \quad \left(0 \le \varphi \le \frac{\pi}{2}\right) \quad (6.39)$$

and

$$\sin\alpha = \left(\frac{1 - \left(\frac{b}{a}\right)^2}{1 - \left(\frac{c}{a}\right)^2}\right), \quad \left(0 \le \alpha \le \frac{\pi}{2}\right). \quad (6.40)$$

Functions $F(k,\theta)$ and $E(k,\theta)$ are elliptic integrals of the first and second kinds, respectively, k being the modulus and θ the amplitude of these integrals. In (6.38) and (6.39), a, b and c are the lengths of the semiaxis of the ellipsoid.

These equations may be simplified in some cases. For instance, if the ellipsoid is elongated ($c \gg a, b$) and has one axis of symmetry we can work with only one coefficient of the tensor. In this case, supposing the magnetization parallel to the long axis we have

$$N_{zz} = \frac{1}{\kappa^2 - 1}\left(\frac{\kappa}{\sqrt{\kappa^2 - 1}}\ln(\kappa + \sqrt{\kappa^2 - 1} - 1)\right), \quad (6.41)$$

κ being the ratio of its long length to the diameter of the ellipsoid (cross-section perpendicular to c). In the case of an oblate ellipsoid, if the magnetization is parallel to its circular section we can use

$$N = \frac{1}{2}\left(\frac{\kappa^2}{(\kappa^2 - 1)^{\frac{3}{2}}}\sin^{-1}\frac{\sqrt{\kappa^2 - 1}}{\kappa} - \frac{1}{\kappa^2 - 1}\right), \quad (6.42)$$

where here κ is the quotient between diameter and thickness.

If the body is non-ellipsoidal in shape the above definition of the demagnetizing factor must be changed because, in general, the relation between the demagnetizing field and the magnetization is not linear. In fact, when it occurs, supposing that the specimen is located in a homogeneous magnetic field, the demagnetizing factor is defined as the ratio of the average demagnetizing magnetic field **H** to the average magnetization of the whole body, i.e.

$$\int\int_S \mathbf{H}_d(\mathbf{r}) \cdot d\mathbf{S} = -N \int\int_S \mathbf{M}(\mathbf{r}) \cdot d\mathbf{S}. \quad (6.43)$$

By this type of non-ellipsoidal shaped materials the magnetization is usually non-uniform which leads to demagnetizing factors depending on the point examined of the sample. It means that the components of the tensor will be a function of the coordinates, i.e. $N_{pq} = N_{pq}(x, y, z)$, where $p, q = x, y, x$.[10]

Solved Problems

Problems A

6.1 A finite bar of radius R and length L is magnetized homogeneously in the direction of its symmetry axis (Fig. 6.17). Find the volumetric and surface density currents of magnetization, and give the expression of the magnetic field **B** at any point over its symmetry axis.

Solution

Due to the symmetry of the bar let us choose a cylindrical system of coordinates, in which its revolution axis coincides with OZ. In this case we can write the magnetization field as $\mathbf{M} = M\,\mathbf{u}_z$. For determining the volumetric magnetization current density we have to apply its definition (6.4),

$$\mathbf{j}_m = \nabla' \times \mathbf{M}. \qquad (6.44)$$

Substituting $\mathbf{M} = (0, 0, M)$ into this equation we obtain

$$\mathbf{j}_m = \nabla' \times (0, 0, M) = 0, \qquad (6.45)$$

which is logical because of the homogeneity of **M** in the volume.

With respect to the surface magnetization current density, we follow the same process. However in this case it is necessary to distinguish the different parts of the bar, the two bases and the lateral face. Applying definition (6.5) we have

$$\mathbf{j}_{ms} = \mathbf{M} \times \mathbf{n}, \qquad (6.46)$$

For the right side of the cylinder its normal vector to its surface is $\mathbf{n}_1 = (0, 0, 1)$. Using definition (6.5), we have (Fig. 6.17)

$$\mathbf{j}_{ms1} = \mathbf{M}(\mathbf{r}') \times \mathbf{n}_1 = (0, 0, M) \times (0, 0, 1) = 0. \qquad (6.47)$$

For the another base the normal is $\mathbf{n}_2 = (0, 0, -1)$, and $\mathbf{j}_{ms}$:

$$\mathbf{j}_{ms2} = \mathbf{M}(\mathbf{r}') \times \mathbf{n}_2 = (0, 0, M) \times (0, 0, -1) = 0. \qquad (6.48)$$

---

[10] The knowledge of the local distribution of the demagnetizing factors $N_{pq}(x, y, z)$ is very important for some technological applications, for instance in plane ferrite elements. These type of materials are the constituents of many devices such as phase shifters, magnetic circuits, isolators, and microwave systems.

Fig. 6.17 Rod homogeneously magnetized

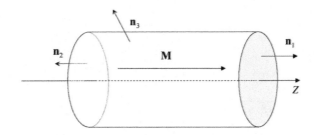

The density corresponding to the lateral surface may by calculated in the same way, but changing the normal vector. As we have chosen cylindrical coordinates,[11] such a vector has only one component which corresponds to the direction $\mathbf{u}_\rho$, then it holds that $\mathbf{n}_3 = \mathbf{u}_\rho = (1, 0, 0)$. Introducing it into (6.5), it yields

$$\mathbf{j}_{ms3} = (0, 0, M) \times (1, 0, 0) = (0, M, 0) = M\mathbf{u}_\phi. \tag{6.49}$$

If we analyze the preceding results we conclude that the bar only has surface magnetization current density on the lateral surface. Geometrically we can represent this result as shown in Fig. 6.18, which remeinds us of a solenoid (Chap. 5). By using (5.101) and (5.102) we can write,

$$dI_{m3} = \left(\frac{dI_{m3}}{dz}\right) dz = j_{ms3}\, dz, \tag{6.50}$$

where I_{m3} represents the flux of $\mathbf{j}_{ms3}$ through a line perpendicular to this current density,

$$\int_S \mathbf{j}_{ms3} \cdot d\mathbf{S} = \int_\Gamma j_{ms3}\, dl = I_{ms3}. \tag{6.51}$$

As we have studied in Chap. 5 when obtaining the magnetic field of a solenoid closely wound, we were able to calculate **B** by computing first the field produced by a set of dN coils, and then integrating for the total length of the solenoid. In our case we try to use the same idea, but with the current I_m due to magnetization because we do not have free charge currents. To confront this problem, let us suppose we wish to calculate first the magnetic field produced by a slice dz' of the bar at a point $P(0, 0, z)$ (Fig. 6.19)

$$d\mathbf{B}(P) \approx \frac{\mu_0 dI_m a^2}{2\left[a^2 + (z - z')^2\right]^{3/2}} \mathbf{u}_z = \frac{\mu_0 a^2 j_{ms3} dz}{2\left[a^2 + (z - z')^2\right]^{3/2}} \mathbf{u}_z. \tag{6.52}$$

---

[11] If we would have chosen cartesian coordinates the normal to the surface would be $\mathbf{n}_3 = cos\phi\, \mathbf{u}_x + sin\phi\, \mathbf{u}_y$, obtaining the same result.

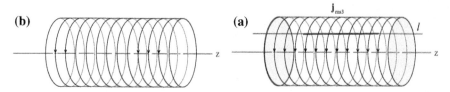

Fig. 6.18 a Surface density current on the lateral side S_3. Observe that j_{ms3} is perpendicular to the line l. I_m represents the flux of j_{ms3} across that line (see Fig. 5.31). **b** Equivalent Solenoid to a homogeneous magnetized bar of the same length and radius

Fig. 6.19 Magnetized bar and surface magnetization currents

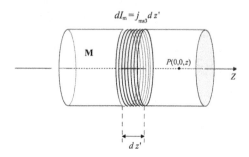

If we compare this expression with (5.96) we observe that both equations are similar, however the only difference is that in (6.52) j_{ms3} appears instead nI. In other words, the computation of the magnetic field for the magnetized bar from (6.52) is *formally* the same as that performed for the coil of n turns per unit length, but now appears j_{ms3}. For this reason when making calculations with a homogeneous magnetized bar, we may work with an *equivalent solenoid* if we identify j_{ms3} with nI. Physically it means that the magnetic field **B** produced by a cylinder of uniform magnetization M along its revolution axis, is the same as the field **B** created by an actual solenoid of the same geometry with a flux current of nI. Notice that we identify the effects of both systems, but not the causes that create such a field. By integrating (6.52), we conclude that the magnetic field B may be expressed as follows (see (5.99) in Chap. 5)

$$\mathbf{B} = \frac{\mu_0 j_{ms}}{2}(\cos\alpha_1 + \cos\alpha_2)\mathbf{u}_z = \frac{\mu_0 M}{2}(\cos\alpha_1 + \cos\alpha_2)\mathbf{u}_z. \qquad (6.53)$$

Ultimately, if we construct a coil of length L and radius R so that the product nI coincides with the magnetization of a bar of the same shape, we cannot distinguish between the magnetic fields **B** produced by both systems, because their effects are the same. In this regard we come to understand the term *equivalent solenoid* (Fig. 6.19).

6.2 A magnetic bar of circular cross-section has a length and radius $\ell = 0.2$ m and $r = 0.03$ m, respectively. The material is homogeneously magnetized in the direction of its axis of revolution reaching magnetization $M = 15{,}000\,\text{Am}^{-1}$. Find: (a) The magnetic fields B and H over its symmetry axis at its center and on the top. (b) A graphical representation of B, M and H. (c) A sketch of the field lines of **B**, **M** and **H**.

Solution

(a) As we have shown in the preceding exercise, for calculating the magnetic field B of a homogeneously magnetized bar, we can use the expression of a finite solenoid if we put $nI = j_{ms}$, nI being the current per unit length circulating through the actual coil. By applying (6.53) we have

$$B = \frac{\mu_0 nI}{2}(\cos\alpha_1 + \cos\alpha_2) = \frac{\mu_0 M}{2}(\cos\alpha_1 + \cos\alpha_2). \quad (6.54)$$

For a point at the center of the bar holds $\alpha_1 = \alpha_2 = \alpha$. On the other hand $\alpha = \left[\text{arctg}\left(\frac{r}{\ell/2}\right)\right]$, then we can write

$$B = \frac{\mu_0 M}{2} 2\cos\left[\text{arctg}\left(\frac{r}{\ell/2}\right)\right] = \frac{4\pi \times 10^{-7} \times 15{,}000}{2} 2\cos\left[\text{arctg}\left(\frac{0.03}{0.2/2}\right)\right] = 0.018 \text{ T}.$$

To obtain B on the center of one base of the bar (point P) we proceed in the same way, but now $\alpha_1 \neq \alpha_2$. For this case we introduce $\alpha_1 = \left[\text{arctg}\left(\frac{r}{\ell}\right)\right]$ and $\alpha_2 = \frac{\pi}{2}$, then we have (Fig. 6.20)

$$B = \frac{\mu_0 M}{2} \cos\left[\text{arctg}\left(\frac{r}{\ell}\right)\right] = \frac{4\pi \times 10^{-7} \times 15{,}000}{2} \cos\left[\text{arctg}\left(\frac{0.03}{0.2}\right)\right] = 0.009 \text{ T}.$$

For determining the magnetic field H we use its definition, i.e.

$$\mathbf{H} = \frac{\mathbf{B}}{\mu_0} - \mathbf{M}, \quad (6.55)$$

where **M** is the magnetization, which is known. By employing the above results for B, we can find the values of H at the center and on the base. In fact, for $L/2$ over its revolution axis, we can write

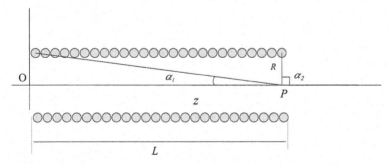

Fig. 6.20 Point P at the extreme of the solenoid. Notice that $\alpha_2 = \frac{\pi}{2}$

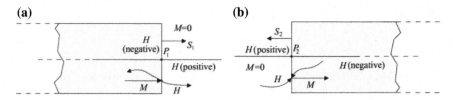

Fig. 6.21 The magnetic field H changes its sigh for points immediately outside of the rod. **a** The right and **b** the left end of the thin bar

$$H = \frac{0.018}{4\pi \times 10^{-7}} - 15{,}000 \text{ A/m} = -6300 \text{ A/m}. \quad (6.56)$$

The negative sign means that **H** and **B** have opposite directions in the material. On the top of the bar we should distinguish between a point on the inner face of the material, and outside of the material (the point very near of the base).

Since the points we are going to study are very close to each other, we can put for both cases the angles corresponding to calculation at the edge on the top. The only difference when applying (6.54) is that there is magnetization if we are inside of the material, but there is no magnetization outside the bar, then for a point close to the end and inside of the system we have

$$H = \frac{0.0093}{4\pi \times 10^{-7}} - 15{,}000 \text{ A/m} = -7600 \text{ A/m}, \quad (6.57)$$

on the left of the surface S (see Fig. 6.21), but very near P. For P outside on the right,

$$H = \frac{0.0093}{4\pi \times 10^{-7}} \text{ A/m} = 7400 \text{ A/m}. \quad (6.58)$$

(b) Taking into account the values of **H**, **M** and **B** obtained, the graphical representation of these fields has the forms shown in Fig. 6.22.

(c) To sketch the field lines inside and outside of the bar it is necessary to understand the physical significance of the results obtained in section (a). These results mean that at the center of the circular base H changes sharply, because its sign changes (Fig. 6.21). The magnetic field H is negative just on the inner side where there is material (left), and on the same face but outside on the point exterior to the material the resulting H field is positive (right). As this result is conceptually important, we will devote some discussion to it. To gain some physical insight into its meaning we are going to use (6.13). From this equation we saw that we can have a source or a sink of magnetic field H at any point where the divergence of the magnetization is nonzero. Let us suppose we examine the surface S_1 corresponding to the right side end of the bar (Fig. 6.21a). On this surface the lines of **M** die on S_1 then $\nabla \cdot \mathbf{M} < 0$. It means that at P_1 $\nabla \cdot \mathbf{H} > 0$, hence lines of magnetic field H must be born (Fig. 6.21a). A similar analysis holds on the surface S_2, but in this case

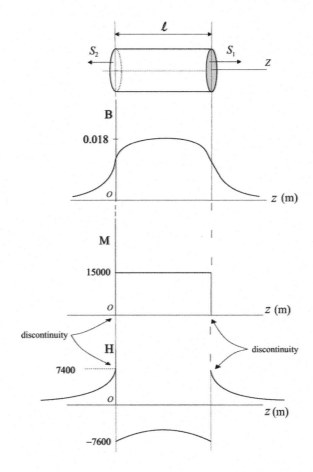

Fig. 6.22 This graphic shows the fields B, M and H versus the distance z, inside and outside of the magnetic bar

the physical result differs. In fact, at P_2 we have a source of magnetization and then $\nabla \cdot \mathbf{M} > 0$ (Fig. 6.21b). At the same time from (6.13) it leads to $\nabla \cdot \mathbf{H} < 0$, thus lines of $\mathbf{H}$ die at this point. Definitely, point P_1 behaves like a sink of magnetic field $\mathbf{H}$, and P_2 like a source of H. In the interior of the rod there are neither sources nor sinks because $\nabla \cdot \mathbf{M} = 0$.

By analyzing the results obtained we see that each field behaves differently. So, the magnetic field $\mathbf{B}$ is formed by closed lines (Fig. 6.22(1)), which is a consequence of the non-existence of monopoles. However, the magnetic field $\mathbf{H}$ is discontinuous on the base of the rod and the directions inside and outside the material are opposite (Fig. 6.23(3)). This field is going from north to south outside of the bar, and inside too (contrary to what occurs with the magnetic field B). This interior field H opposite the magnetization M is the *demagnetizing field* H_d, already commented in the theoretical introduction. In this case, we can mathematically express this field by means of a scalar instead of a tensor as follows

$$H_d = N_d M, \tag{6.59}$$

Fig. 6.23 In this picture a qualitative sketch of the fields **B**, **M** and **H** are presented. Notice that as a consequence of the existence of $\nabla \cdot \mathbf{M} \neq 0$, the magnetic field H is discontinuous on the surfaces S_1 and S_2

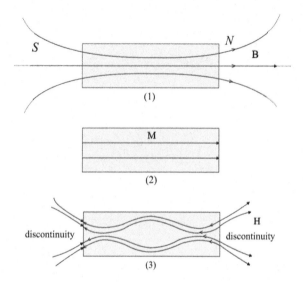

where N_d is the *demagnetizing factor*. Observe that, because H_d goes in the opposite direction to M, applying (6.10) we can write for the magnetized bar $B = \mu_0(-H_d + M)$, which means that the magnetic field inside of the specimen is less than M, but in the same direction. However, the magnitude of H_d does not surpasses M.

As the system is a slender bar (see (6.41)), an easy expression for N_d may be found. In fact, a slender rod may be approximated to an ellipsoid whose revolution axis c is greater than its diameter, then $r(\kappa = \frac{c}{a} = \frac{r}{l} \gg 1)$. In this case, by using (6.41) we obtain

$$N_{zz} \approx \frac{(\ln(2\kappa) - 1)}{\kappa^2}. \tag{6.60}$$

As the rod is homogenously magnetized along OZ and its cross-section is circular then $N_{xx} = N_{yy}$, and the value of N_{xx} may be easily deduced by applying (6.34).

6.3 A magnetic bar of radius $a = 5$ cm extends from the origin of coordinates to a very far point on the OY axis. If the rod has a homogeneous magnetization $\mathbf{M} = 1000\,\mathbf{u}_y$, find: (a) The magnetic field **B** at $P(0, -10, 0)$ cm. (b) The magnetic field **H** at the same point P. (c) The magnetic field **H** at $Q(0, 5, 0)$ cm.

Solution

(a) To answer this question we may directly use the basic results obtained in the Problems 6.1 and 6.2. This means we identify the magnetic field B produced by a homogeneous magnetized bar with the field of a solenoid of the same radius and length in which the identity $nI = M$ holds. From (6.53) it follows

$$\mathbf{B} = \frac{\mu_0 M}{2}(\cos\alpha_1 + \cos\alpha_2)\mathbf{u}_y. \tag{6.61}$$

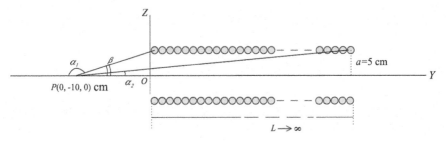

Fig. 6.24 Magnetic field B outside of the bar. Notice that α_1 exceeds $\pi/2$

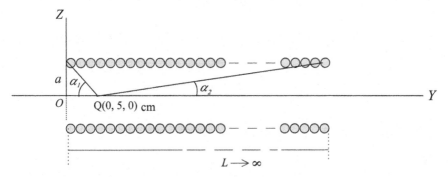

Fig. 6.25 Point Q over the OY axis inside of the rod. In this case the angle $\alpha_1 < \frac{\pi}{2}$, but α_2 does not change because the right side of the bar (equivalent solenoid) is far away from Q

As point P is located over the OY axis outside of the rod, the angle α_1 exceeds $\pi/2$. Actually $\alpha_1 = (\pi - \beta)$, then we can introduce it into (6.61) $\cos \alpha_1 = \cos(\pi - \beta) = -\cos \beta$. On the other hand, as the bar is very large over the OY axis, the angle α_2 is very small, and at the limit where the length of the right side of the solenoid tends to infinity $\alpha_2 \to 0$, i.e. (Fig. 6.24)

$$\mathbf{B} = \frac{\mu_0 n I}{2}(-\cos \beta + \cos \alpha_2)\mathbf{u}_y = \frac{\mu_0 M}{2}(-\cos \beta + 1)\mathbf{u}_y = 6.6 \cdot 10^{-5} \text{ T}, \quad (6.62)$$

where we have substituted $\beta = \arctan(\frac{5}{10}) = 26.6°$ (Fig. 6.25).

(b) Once we know the magnetic field B, the calculation of H may be computed by directly applying (6.12). In fact, introducing B into that equation we have

$$\mathbf{H} = \frac{\mathbf{B}}{\mu_0} - \mathbf{M} = \frac{\mathbf{B}}{\mu_0} - 0 = \frac{\mathbf{B}}{\mu_0} = 52.9 \text{ A/m}. \quad (6.63)$$

Observe that outside the rod $M = 0$, because we do not have matter.

(c) For determining H at point Q inside of the bar, we must first calculate the magnetic field B. The reason for this (see Exercise 6.1) is that the physical effects we are going to equalize refer to B, but not to H or M. In other words, the magnetized bar and a solenoid of the same dimensions in which $nI = M$ are only equivalent when regarding their magnetic fields B produced.

Starting again from (6.53) we can write

$$\mathbf{B} = \frac{\mu_0 M}{2}(\cos\alpha_1 + \cos\alpha_2)\mathbf{u}_y = \frac{\mu_0 M}{2}(\cos\alpha_1 + 1)\mathbf{u}_y. \qquad (6.64)$$

For the point Q holds $\alpha_1 = \arctan(\frac{5}{3}) = 45°$, and then $\mathbf{B} = 1.1 \cdot 10^{-3}$ T. Unlike in section (b) point Q is located inside of the material, then magnetization $M \neq 0$. This means that for calculating H we must use its value, i.e.

$$\mathbf{H} = \frac{\mathbf{B}}{\mu_0} - \mathbf{M} = \frac{\mathbf{B}}{\mu_0} - 1000\,\mathbf{u}_y = -146.5 \ \mathrm{A/m}. \qquad (6.65)$$

6.4 Figure 6.26 represents a system composed by a finite solenoid of length $L = 50$ cm and radius $R = 20$ cm, and a paramagnetic cylindrical slender bar of $l = 2$ cm, radius $r = 1$ mm, and $\mu_r = 10$. The current circulating through the coil is $I = 2$ A and the number of turns per unit length is $n = 10,000$. If the distance between the solenoid and the left top of the rod is $d = 1$ m, find: (a) The magnetization M of the bar, approximately. (b) The magnetic field $\mathbf{B}$ in the magnetic bar. (c) The magnetic current densities of the rod.

Solution

(a) The magnetization in the bar may be obtained by means of its definition, that is,

$$\mathbf{M} = \chi_m \mathbf{H},$$

where $\chi_m = (\mu_r - 1)$. Then, magnetization of the slender rod depends on the intensity of $\mathbf{H}$, which, on the other hand, is produced by the solenoid. For calculating the field generated by the coil we can use the (5.99), demonstrated in the previous chapter, that is,

$$\mathbf{B} = \frac{\mu_0 n I}{2}(\cos\alpha_1 + \cos\alpha_2)\mathbf{u}_z.$$

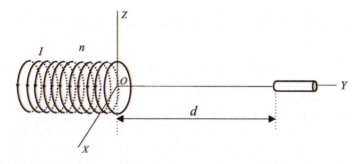

Fig. 6.26 System formed by a finite solenoid and a magnetic slender rod

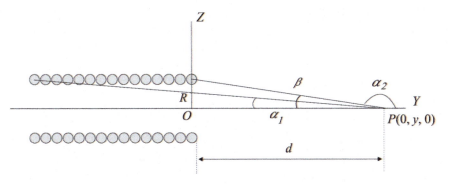

Fig. 6.27 Finite solenoid. Observe as the point where the field must be calculated is located outside of the coil

However, for calculating **M** we need **H** and not **B**. We can easily determine it by using (6.12)

$$\mathbf{B} = \mu_0(\mathbf{H} + \mathbf{M}),$$

but in a vacuum $\mathbf{M} = 0$, hence

$$\mathbf{B} = \mu_0(\mathbf{H} + \mathbf{M}) \Rightarrow \mathbf{H} = \frac{\mathbf{B}}{\mu_0}.$$

Now we must apply (5.99), but some approximations are needed. In fact, this equation gives the magnetic field **B** at any point over the symmetry axis of the coil. In the present case the magnetic rod has a finite extent, thus the angles α_1 and α_2 do not have a constant value along OY (for each point of the rod). However, taking into account that the distance between the solenoid and the slender rod is much greater than the length of said bar ($L + d \gg l$), and that its diameter is larger than the diameter of the rod ($R \gg r$), we can suppose that the field created by the solenoid in the volume where the rod is placed is homogeneous, approximately. If so, we can put (see Fig. 6.27) the same angles for all parts of the bar, i.e. $\alpha_1 \approx \arctan\left(\frac{0.2}{1.5}\right) = 7.59°$ and $\beta \approx \arctan\left(\frac{0.2}{1}\right) = 11.31°$, and then the magnetic field H

$$\mathbf{H} = \frac{nI}{2}(\cos\alpha_1 + \cos\alpha_2)\mathbf{u}_y = \frac{10{,}000 \cdot 2}{2}(\cos\alpha_1 - \cos\beta)\mathbf{u}_y = 106.6\,\mathbf{u}_y \text{ A/m}. \tag{6.66}$$

For the calculation of **M** we employ its definition, however some comments must be said. Firstly, the specimen is small compared to the solenoid and the distance between both systems is very long; secondly, as we have seen in the theory that when a sample is magnetized, a demagnetized field appears. In our case we have a slender rod, which means that $r \ll l$, that is, its diameter is much smaller that its length, then we can neglect the demagnetized field produced. For this reason we can work with the field **H** generated by the solenoid without introducing corrections. Acceptingthis statement we can write

$$\mathbf{M} = \chi_m \mathbf{H} = (\mu_r - 1)\mathbf{H} \approx (\mu_r - 1)\mathbf{H}(P) = 9\mathbf{H} = 959.6\,\mathbf{u}_y. \tag{6.67}$$

(b) For obtaining the magnetic field we apply (6.12)

$$\mathbf{B} = \mu_0(\mathbf{H} + \mathbf{M}) \approx \mu_0(\mathbf{H} + \mathbf{M}) = 1.3 \times 10^{-3}\,\text{T}. \tag{6.68}$$

(c) Considering that the fields in the bar are homogeneous, approximately, we can compute the magnetization current densities directly through (6.4) and (6.5). Using a cylindrical coordinate frame were the symmetry axis coincides with the OY axis, we obtain,

$$\mathbf{j}_m = \nabla' \times \mathbf{M} = 0, \tag{6.69}$$

and

$$\mathbf{j}_{ms} = \mathbf{M} \times \mathbf{n} = M\,\mathbf{u}_y \times \mathbf{n} = M\,\mathbf{u}_\phi = 937.6\,\mathbf{u}_\phi\ (\text{A/m}). \tag{6.70}$$

6.5 In the interior of a torus with mean radius R_m and N turns of wire closely wound, a ferromagnetic material is introduced, whose curve of first magnetization is given by

$$B = aH/(b + cH),$$

where a, b and c are constants. Find: (a) The magnetic field B if a current I flows throughout the entire wire. (b) The value of $\chi_m(H)$. (c) The magnetic field B at saturation.

Solution

(a) A magnetic circuit may be solved by following three steps:
  (1) Application of Ampère's Law for the magnetic field H. (2) The flux of B. (3) The magnetic material equations.

The Ampère theorem for H only depends on the free current of charge, i.e.

$$\oint_\Gamma \mathbf{H} d\mathbf{l} = \int\int_S \mathbf{j}_f \cdot d\mathbf{S} = I_f. \tag{6.71}$$

For applying this law we take a circular curve Γ of radius R_m, concentric with the magnetic torus, and a plane surface S. As we explained in Chap. 5 (Problem 4), a flat surface is the easiest to calculate the flux of $\mathbf{j}_f$ throughout. Introducing the cylindrical coordinates for all components of $\mathbf{H}$ and the current density, we can write

$$\oint_\Gamma H_\phi \mathbf{u}_\phi\, dl\mathbf{u}_\phi = \int\int_S j_f \mathbf{u}_z \cdot d\mathbf{S}\,\mathbf{u}_z = I_f(\text{total}) \Rightarrow \int_0^L H_\phi\, dl = H_\phi \int_0^L dl = NI. \tag{6.72}$$

$$H_\phi L = NI \Rightarrow H_\phi = H = \frac{NI}{L} = nI, \tag{6.73}$$

Fig. 6.28 Hysteresis curve and operating straight line (OSL). Observe that in this case the *OSL* is parallel to the *OB* axis because the electromagnet has no air-gap

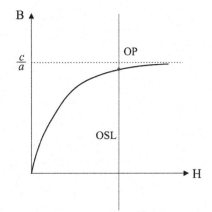

where $L = 2\pi R_m$ and $n = \frac{N}{L}$. To obtain the magnetic field B we must find the operating point of the system, which corresponds to the intersection of the operating straight line (OSL) with the hysteresis curve of the electromagnet. In this case, since that the system has no air-gap, the OSL is simply given by (6.73), and the material equation is

$$B = aH/(b + cH). \tag{6.74}$$

To determine such an intersection (see Fig. 6.28) we introduce the value of H in (6.74)

$$B = \frac{anI}{(b + cnI)}. \tag{6.75}$$

(b) To obtain the magnetic susceptibility, we can use (6.19). In fact,

$$B = \mu_0[1 + \chi_m(H)]H = \mu_0\mu_r(H)H = \frac{a}{(b + cH)}H, \tag{6.76}$$

then

$$\chi_m = \mu_r - 1 = \frac{a}{\mu_0(b + cnI)} - 1. \tag{6.77}$$

(c) The magnetic field B in saturation may be obtained by increasing H in the hysteresis curve (Fig. 6.28). As we have the function of the first magnetization curve, the limit of $B(H)$ for extremely high values of H,

$$\lim_{H \to \infty} B = \lim_{H \to \infty} \frac{aH}{(b + cH)} = \frac{a}{c}. \tag{6.78}$$

6.6 Over an electromagnet of lengths l_i and l_g for the iron and the air-gap, respectively, N coils are closely wound. Starting from the demagnetized state, the current I circulating through the coils is increased progressively to reach saturation. From this state, the intensity is diminished to an unknown value. If the second magnetization curve corresponding to the two first quadrants is given by $B = a\sqrt{H + |H_c|}$ (for $-H_c \leq H \leq H_{max}$), where a is a constant and H_c is the coercitive field, find the condition for the current circulating through each coil in order that the magnetic field H_i in the iron core always takes positive values.

Solution

As we have reached saturation, and considering the current is diminishing, the operating point for the material must correspond to the intersection of the operating straight line and the second magnetization curve (see Fig. 6.29). In order to find the aforementioned condition, we proceed as usual, i.e. we begin with the Ampère law, then the flux of the magnetic field B is evaluated, and finally the material equation is included in the resulting equation for the operating straight line.

Applying Ampère's theorem we have

$$\oint_\Gamma \mathbf{H} d\mathbf{l} = \int\int_S \mathbf{j}_f \cdot d\mathbf{S} = I_f \Rightarrow H_i l_i + H_g l_g = NI. \qquad (6.79)$$

in which we have substituted the intensity of free charge I_f by I, in order to simplify the notation. On the other hand, the section is the same for all parts of the electromagnet, hence $B_i = B_g$. The material equation for the air gap is $B_g = \mu_0 H_g$, and in the material is the hysteresis loop, whose mathematical expression is given in the statement. Introducing in (6.79) $H_g = \frac{B_g}{\mu_0} = \frac{B_i}{\mu_0}$, we obtain the operating straight line (OSL)

$$B_i = \frac{\mu_0 NI}{l_g} - \frac{\mu_0 l_i}{l_g} H_i. \qquad (6.80)$$

As we know, any operating point corresponds to the intersection between the OSL and the hysteresis loop, thus we have to calculate such a point by means of (6.80) and

$$B_i = a\sqrt{H_i + |H_c|}. \qquad (6.81)$$

Introducing (6.81) into the left side of (6.80), we obtain

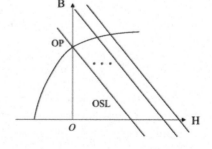

Fig. 6.29 The operating point corresponds to the intersection of the (OSL) for $H = 0$ and the hysteresis curve

$$a\sqrt{H_i + |H_c|} = \frac{\mu_0 NI}{l_g} - \frac{\mu_0 l_i}{l_g} H_i. \quad (6.82)$$

The magnetic field in the iron core must always be positive, then all possible values of H and B allowed are in the first quadrant. We can find the condition by calculating just the limit when $H_i \to 0$, guaranteeing that the magnetic field H is still positive,

$$a\sqrt{|H_c|} = \frac{\mu_0 NI}{l_g} \Rightarrow I = \frac{a\sqrt{|H_c|}\, l_g}{\mu_0 N}. \quad (6.83)$$

This current corresponds to the operation point $(B_r, 0)$, but in our case $H_i > 0$, hence the current must verify the condition

$$I > \frac{a\sqrt{|H_c|}\, l_g}{\mu_0 N}. \quad (6.84)$$

This unequality means that all operating points corresponding to values of the intensity given by (6.84) verify that the magnetic field H is always positive.

6.7 An electromagnet with iron core of 3.0 m in length, and an air-gap of length 1 cm and 1000 turns is constructed. For the material employed $B = K\sin(H/C)$ in a first magnetization curve, where $K = 1.2$ T and $C = 2800$ A/m. A field of 0.6 T in the air-gap is required. Calculate the intensity that must flow along the coil. Disregard the flux leakage.

Solution

As explained previously, for resolving the question we will follow three steps. To begin, we use Ampère's law for the magnetic field H. In order to calculate the integral we employ a curve passing over the midline of the electromagnet (see Fig. 6.11), obtaining

$$\oint_\Gamma \mathbf{H} d\mathbf{l} = \int\int_S \mathbf{j}_c \cdot d\mathbf{S} = I_f \Rightarrow H_c l_c + H_g l_g = NI, \quad (6.85)$$

l_c and l_g being the middle lengths corresponding to the iron core and air-gap, respectively. Since the flux leakage of B is not considered, magnetic field B is limited to the zone of the air gap and core, and hence, due to the law of the flux applied to a pole of the electromagnet, the value of B must be the same in the iron core as in the air-gap, thus $B_c = B_g$. On the other hand, in the air gap it holds that

$$B_g = \mu_0 H_g \Rightarrow H_g = \frac{B_g}{\mu_0} = \frac{B_c}{\mu_0}, \quad (6.86)$$

and then we can write the operating straight line

$$H_c l_c + \frac{B_c}{\mu_0} l_g = NI. \quad (6.87)$$

Fig. 6.30 Hysteresis loop corresponding to the first magnetization curve. The maximum value of B is 1.2 T. Notice that the operating straight line (OSL) cross the curve at $B = 0.6$ T

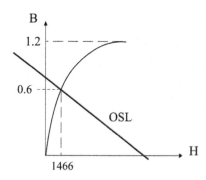

The operating point of the electromagnet corresponds to the intersection between (OSL) and the material equation for the iron core, the magnetization curve (Fig. 6.30). In this problem we know the value of B we will reach, then in (6.87) we can substitute $B_c = 0.6$ T. The equation remains with two unknowns, namely H_c and I. Easily we can calculate H_c. In fact, as the first magnetization loop is done analytically, we can obtain the magnetic field H_c as follows

$$B = 1.2 \sin\left(\frac{H}{2800}\right) = 0.6 \Rightarrow H_c = 2800 \arcsin\left(\frac{0.6}{1.2}\right) = 1466 \, \text{A/m}. \quad (6.88)$$

Now, substitution of the two above results into (6.87) yields

$$I = \frac{1}{N}\left\{H_c l_c + \frac{B_c}{\mu_0} l_g\right\} = \frac{1}{1000}\left\{1466 \cdot 3 + \frac{0.6}{\mu_0} 0.001\right\} = 9.7 \, \text{A}. \quad (6.89)$$

6.8 In an electromagnet of constant cross section $S = 16 \, \text{cm}^2$, the length of the iron core and the air-gap are $l_c = 44$ cm and $l_g = 5$ mm, respectively. A multiturn of 1000 coils are wound around the core. The material employed for the core is ferromagnetic and its first magnetization curve is $B = 0.001 H^2$, for $0 < H < 10{,}000 \, \text{Am}^{-1}$. Starting from the demagnetizing state the carrying current through the coils is increased to a unknown value I, in which the magnetization is $M = 10^5 \, \text{A m}^{-1}$. Find the intensity circulating for each coil. There is no flux leakage.

Solution

Following the procedure described in the theory, and as stated in the previous problem, applying first Ampère's law yields

$$\oint_\Gamma \mathbf{H} d\mathbf{l} = \int\int_S \mathbf{j}_f \cdot d\mathbf{S} = I_f \Rightarrow H_c l_c + H_g l_g = NI. \quad (6.90)$$

As we do not have flux leakage, the divergence theorem gives

because the electromagnet has constant cross section. Introducing it in (6.90), and using the fact that $B_g = \mu_0 H_g$, we have

$$H_c l_c + \frac{B_c}{\mu_0} l_g = NI, \qquad (6.92)$$

which corresponds to the operating straight line. The idea for calculating the intensity is to obtain the magnetic fields H and B and substitute them in (6.92). Contrary to what usually occurs, we know the material magnetization, then when calculating the operation point (H, B) by means of the intersection between the (OSL) and the material equation (hysteresis loop), such a point of coordinates must be what corresponds to a magnetization $M = 10^5 \, \text{A m}^{-1}$. In order to determine H and B we use (6.12)

$$B = \mu_0(H + M) = 0.001 \, H^2 \Rightarrow \mu_0(H + 10^5) = 0.001 \, H^2, \qquad (6.93)$$

that is

$$795.7 \, H^2 - H - 10^5 = 0. \qquad (6.94)$$

The solution of this equation gives two values for H. One of them is negative, which is physically incorrect because we are in the first magnetization curve, therefore the value of H must be positive. For this reason the only magnetic field possible is

$$H = 11.2 \, \text{A/m}.$$

Once we know H, the field B may be calculated directly through the hysteresis curve $B = 0.001 \, H^2$, obtaining

$$B = 0.13 \, T.$$

Now we know the operating point (H, B) where the electromagnet works, hence such values must also satisfy (6.95). For this reason, the current being carried through each coil I is

$$I = \frac{1}{N}\left\{H_c l_c + \frac{B_c}{\mu_0} l_g\right\} = \frac{1}{1000}\left\{11.2 \cdot 0.44 + \frac{0.13}{\mu_0} 0.005\right\} = 0.5 \, \text{A}. \qquad (6.95)$$

6.9 Over an electromagnet of total mean length $L = 0.4$ m, and an unknown air gap l_g, a set of $N = 10{,}000$ turns are closely wound around. The susceptibility $\chi_m = \chi_m(H)$ in the interval corresponding to the first magnetization curve may be approximated by the following function

$$\chi_m = \left(9.5 \times 10^{-4} \, H^2 - 1\right), \quad 100 \leq H \leq 1500 \, \text{(A/m)}.$$

$$\oint_S \mathbf{B} d\mathbf{S} = 0 \Rightarrow B_c S_c = B_g S_g \Rightarrow B_c = B_g, \qquad (6.91)$$

Obtain l_g if we would reach a magnetic field 0.3 T in the air gap, when the intensity circulating through the coils is 1 A.

Solution

(a) We begin with the Ampère theorem for the magnetic field **H**,

$$\oint_\Gamma \mathbf{H} d\mathbf{l} = \int\int_S \mathbf{j}_c \cdot d\mathbf{S} = I_c \Rightarrow H_c l_c + H_g l_g = NI. \tag{6.96}$$

Taking into consideration the absence of flux leakage, we can write

$$\oint_S \mathbf{B} d\mathbf{S} = 0 \Rightarrow B_c S_c = B_g S_g, \tag{6.97}$$

but due to $S_c = S_g$ the fields B_c and B_g are equal. In the air gap $B_g = \mu_0 H_g$ and, on the other hand, the cross sections of the electromagnet are equal. As a consequence, the operating straight line has the form

$$H_c l_c + \frac{B_c}{\mu_0} l_g = NI. \tag{6.98}$$

We know the total length L of the electromagnet, but neither the length of the air gap l_g nor l_c. However, it is always true that $L = l_c + l_g$, thus we can substitute l_c by $(L - l_g)$ in (6.98) obtaining

$$H_c(L - l_g) + \frac{B_c}{\mu_0} l_g = NI. \tag{6.99}$$

This is the operating straight line of the magnetic system, and for obtaining the operating point we need to use the material equation of the iron, i.e.

$$B = \mu_0(H + M) = \mu_0 \mu_r H = \mu_0(1 + \chi_m(H))H, \tag{6.100}$$

and by setting the value of $\chi_m = \chi_m(H)$, we have

$$B = \mu_0(9.5 \times 10^{-4} H^2)H = 9.5 \times 10^{-4} \mu_0 H^3. \tag{6.101}$$

To determine the point of the hysteresis curve at which the electromagnet works, we calculate the intersection between (6.101) and (6.99). In this case we know the magnetic field B to be obtained, hence we clear H away as function of B,

$$\left(\frac{B_i}{9.5 \times 10^{-4} \mu_0}\right)^{\frac{1}{3}}(L - l_g) + \frac{B_i}{\mu_0} l_g = NI. \tag{6.102}$$

Introducing now $B = 0.3$ T, $N = 10{,}000$ and $I = 1$ A, we obtain

Solved Problems

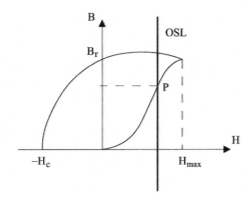

Fig. 6.31 Hysteresis loop and operating point P. Notice that the operating straight line is vertical since the system has no air gap

$$631(0.4 - l_g) + 238732\, l_g = 10{,}000 \Rightarrow l_g \approx 4.1\,\text{cm}. \qquad (6.103)$$

6.10 A torus of mid average radius R_m and N turns, is built with a ferromagnetic material whose first and second magnetization curves are represented by the following function (see Fig. 6.31):

$$B = a\,H^3,$$
$$(H - b)^2 + B^2 = c^2 \qquad (c > b \text{ and } -H_c \leq H \leq H_{\text{máx}}),$$

where a, b, and c are constants. Find: (a) The magnetic field H in the material when starting in the demagnetized state when the current circulating through the coils reaches the value I. (b) Idem for the magnetic field B. (c) If the current increases so that the saturation point is reached and then the current is brought to zero, find the magnetic field B_r.

Solution

(a) Let us apply Ampère law for **H** over a circular curve Γ of length $2\pi R_m$

$$\oint_\Gamma \mathbf{H}d\mathbf{l} = \int\int_S \mathbf{j}_f \cdot d\mathbf{S} = I_f \Rightarrow H_i l_i = NI, \qquad (6.104)$$

and introducing $l = 2\pi R_m$ we have

$$H = \frac{NI_f}{2\pi R_m}, \qquad (6.105)$$

which corresponds to the intersection between the operating straight line parallel to the OB axis and the first magnetization curve. This result is logical because the electromagnet has a constant cross section and does not have an air-gap. As a consequence, the field H over the mean curve Γ must be the same inside the material. Notice that in this case there is no demagnetizing field (we have neglected the flux leakage).

(b) The magnetic field B is found by using the magnetization curve. In effect, we know the coordinate H of the operating point (6.105), hence another coordinate of the point P is

$$B = a \left(\frac{N I_c}{2\pi R_m} \right)^3. \tag{6.106}$$

(c) If the intensity is diminished to zero from saturation we reach the remanent magnetic field. In this case the operating straight line coincides with the OB axis, hence B_r is found by introducing in the second magnetization curve $H = 0$, that is

$$B = \sqrt{c^2 - b^2} \tag{6.107}$$

Problems B

6.11 A hollow bar of length $L = 20$ cm is built of a ferromagnetic material. The rod is magnetized along its revolution axis which coincides with OX (see Fig. 6.32). If the minimum and maximum radii are $R_1 = 2$ cm and $R_2 = 5$ cm, respectively, and the magnetization $\mathbf{M} = 10{,}000\, \mathbf{u}_x$ A/m, find: (a) The magnetic field B at $P_1(10, 0, 0)$ cm. (b) The vector M at the same point. (c) The magnetic field H at $P_2(20, 0, 0)$ cm.

Solution

Before making calculations, we should focus our attention on the geometry of the system. In effect, this problem seems to be very different from the preceding exercises of magnetized bars, however as we will see, the present problem may easily be resolved using the same ideas we explained there. So, by means of the principle of superposition of linear fields,[12] the hollow magnetized bar is equivalent to two rods of opposite magnetization, whose geometries coincide with the two cylindrical surfaces (exterior and interior), and thus delimiting the hollow bar (Fig. 6.33).

This means that, in order to compute the magnetic field B, we can divide the problem in two. The first one would correspond to the field produced by a rod of length L, radius R_1 and magnetization $\mathbf{M} = M\,\mathbf{u}_x$, and the second one to a bar of the same length but with radius R_2 and magnetization $\mathbf{M} = -M\,\mathbf{u}_x$. Denoting by B_2 the field produced by the rod of radius R_2, and by B_1 the corresponding field of R_1, it follows

---

[12] It is important to note that, even though the process of the magnetization of the ferromagnetic bar is non-linear (see Sect. 6.6) we can apply the principle of superposition for calculating the magnetic field. The reason is that we examine the bar when it has already reached its magnetization **M**, but we do not account for the previous process of magnetization. In other words, we work with a *final physical state* of rod regarding only the effect of its magnetization (the creation of the fields **B** and **H**). To some extent (in the language of the system theory) the bar behaves like a *black box* of magnetization **M**.

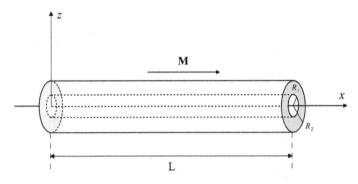

Fig. 6.32 Hollow magnetized ferromagnetic bar

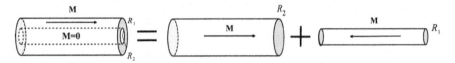

Fig. 6.33 The hollow magnetized cylinder behaves like two massive bars of radius R_1 and R_2, respectively, whose magnetizations are in opposite directions

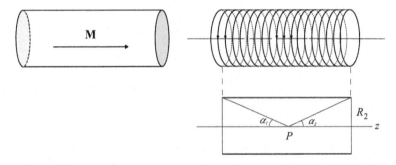

Fig. 6.34 Massive bar of radius R_1 and its equivalent solenoid. Notice that for this case point P is located at the middle point of the rod, then $\alpha_1 = \alpha_2$

$$\mathbf{B}_2 = \frac{\mu_0 M}{2}(\cos\alpha_1 + \cos\alpha_2)\mathbf{u}_x = \frac{\mu_0 M}{2}(2\cos\alpha_1)\mathbf{u}_x, \quad (6.108)$$

where we have written $\alpha_1 = \alpha_2$ because P_1 is at the mid point (Fig. 6.34). For the small bar we proceed in the same way, but we must change the angles,

$$\mathbf{B}_1 = -\frac{\mu_0 M}{2}(\cos\alpha_1' + \cos\alpha_2')\mathbf{u}_x = -\frac{\mu_0 M}{2}(2\cos\alpha_1')\mathbf{u}_x. \quad (6.109)$$

Now, setting $\alpha_1 = \arctan(\frac{5}{10})$ and $\alpha_1' = \arctan(\frac{2}{10})$ into (6.109), we have (Fig. 6.35)

$$\mathbf{B} = \mathbf{B}_2 + \mathbf{B}_1 = \mu_0 M \cos\alpha_1 \mathbf{u}_x - \mu_0 M \cos\alpha_1' \mathbf{u}_x = -1.1 \times 10^{-3}\mathbf{u}_x \text{ (T)}. \quad (6.110)$$

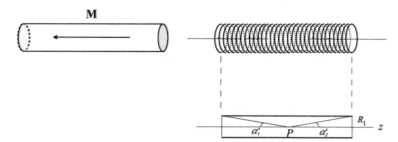

Fig. 6.35 Massive bar of radius R_2 with its corresponding equivalent solenoid. Here the angles α'_1 and α'_2 are different than α_1 and α_2 of the big bar

(b) Point P_1 is placed over the OX axis of the system, in the hollow region of the cylinder, where there is no matter present, then **M** must be zero.

(c) For obtaining the magnetic field H at $P_2(20, 0, 0)$, we previously calculate the magnetic field B. As point P_2 is at the end of the hollow rod the angles are $\alpha_1 = \arctan(\frac{5}{20})$, $\alpha_2 = \frac{\pi}{2}$, and $\alpha'_1 = \arctan(\frac{2}{20})$, $\alpha'_2 = \frac{\pi}{2}$, resulting in

$$\mathbf{B} = \mathbf{B}_1 + \mathbf{B}_2 = \frac{\mu_0 M}{2}(\cos\alpha_1 - \cos\alpha_2)\mathbf{u}_x - \frac{\mu_0 M}{2}(\cos\alpha'_1 - \cos\alpha'_2)\mathbf{u}_x = \quad (6.111)$$

$$= \mu_0 M(\cos\alpha_1 - \cos\alpha'_1)\mathbf{u}_z = -1.56 \times 10^{-4}\mathbf{u}_x \text{ (T)}.$$

Introducing this result into (6.10) leads to

$$\mathbf{H} = \frac{\mathbf{B}}{\mu_0} - \mathbf{M} = \frac{\mathbf{B}}{\mu_0} - 0 = -124.4\,\mathbf{u}_x \text{ (A/M)}. \quad (6.112)$$

6.12 A very long hollow cylinder of radii $R_1 = 5$ cm and $R_2 = 8$ cm as shown in Fig. 6.36, is constructed of a magnetic material whose magnetization is $\mathbf{M} = 10{,}000\,\mathbf{u}_z$ (A/m). Find: (a) The magnetic field **B** for every point belonging $\Re^3$. (b) The magnetic field **H**.

Solution

(a) $(0 < \rho < R_1)$ In the Problem 6.11 of this chapter, we have calculated the magnetic field **B** produced by a finite magnetized hollow cylinder at any point of its symmetry axis. There we saw that, for homogeneous magnetization the field **B** produced by this system was equivalent to that produced by two solenoids of length L and radii R_1 and R_2, respectively. Now we are dealing with the field produced at any region of space when the hollow bar is very long. At first glance it seems to be not possible to solve the problem in the same way, because the formula obtained there was found for the special case of a cylinder of finite length, however such a formula does not apply here (in our case the hollow cylinder is very long). However, even though the latter is true, we will explain how to determine **B** based on the equivalence of the magnetic fields **B** produce from different geometries (Fig. 6.37).

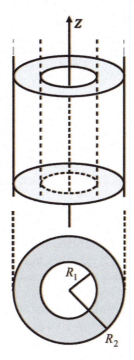

Fig. 6.36 Magnetized hollow cylinder

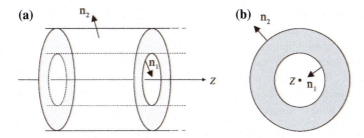

Fig. 6.37 a Normal vectors to the inner and outer surfaces. **b** Plan view cross section of the magnetized hollow cylinder

In effect, let us first calculate the magnetization currents of the system. Beginning with the volumetric current density, using (6.4) it holds that

$$\mathbf{j}_m = \nabla' \times \mathbf{M} = 0. \tag{6.113}$$

We regard to the surface we apply (6.5) to both the inner and outer surfaces of the cylinder, obtaining

$$\mathbf{j}_{ms_1} = \mathbf{M} \times \mathbf{n_1} = -M\mathbf{u}_\phi, \tag{6.114}$$

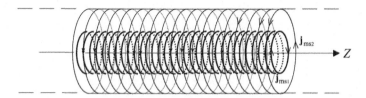

Fig. 6.38 The field **B** produced by a very long homogeneous magnetized hollow cylinder is equivalent to the magnetic field **B** generated by two long solenoids of the same dimensions

and
$$\mathbf{j}_{ms_2} = \mathbf{M} \times \mathbf{n_2} = M\mathbf{u}_\phi. \tag{6.115}$$

Taking into account the explanation made in the Problems 6.1, 6.2 and 6.11, we can assert that, from the point of view of the magnetic field **B**, the hollow slender magnetized bar behaves like a system composed of two very long coils of the same length and radii of the inner and outer cylinders, through which circulates an intensity per unit length of $n_1 I_1 = M$ and $n_2 I_2 = -M$, respectively. Therefore, for calculating **B** we can use the formula studied in Chap. 5 referring to the solenoid. In fact, we can apply the general expression, only valid over the axis of revolution, for the infinite solenoid by setting $\alpha_1 \to 0$ and $\alpha_2 \to 0$, giving

$$\lim_{\alpha_1=\alpha_2\to 0} \mathbf{B}(\alpha_1, \alpha_2) = \frac{\mu_0 nI}{2}(\cos\alpha_1 + \cos\alpha_2)\mathbf{u}_z = \mu_0 nI\, \mathbf{u}_z. \tag{6.116}$$

In principle this result applies over the OZ axis (see Fig. 6.38), but the problem refers to any point of the space, thus we have found only a partial solution, right? Wrong. Actually, the result obtained holds for any point inside the hole, or what is the same, inside of a solenoid with radius R_1. The reason for this is that when the solenoid is very long, the magnetic field inside of the coil is homogeneous (the same value anywhere) and its value coincides with (6.116) (see Problem 5.10). Thus we may write for the field produced by j_{ms} inside of the solenoids with radius R_1

$$\mathbf{B}_1 = \mu_0 n_1 I_1\, \mathbf{u}_z = -\mu_0 M\, \mathbf{u}_z, \tag{6.117}$$

and for the solenoid with radius R_2

$$\mathbf{B}_2 = \mu_0 n_2 I_2\, \mathbf{u}_z = \mu_0 M\, \mathbf{u}_z. \tag{6.118}$$

The total magnetic field **B** for $\rho < R_1$ is

$$\mathbf{B}_1 = \mathbf{B}_1 + \mathbf{B}_2 = -\mu_0 M\, \mathbf{u}_z + \mu_0 M\, \mathbf{u}_z = 0. \tag{6.119}$$

This result differs from (6.110) as a consequence of the non-finite length of the hollow bar.

$(R_1 < \rho < R_2)$

Outside of a very large solenoid it was demonstrated that a magnetic field does not exist, and so in the region between solenoids only the coil of radius R_2 generates a magnetic field **B**, hence

$$\mathbf{B} = 0 + \mathbf{B}_2 = \mu_0 M \mathbf{u}_z. \tag{6.120}$$

$(\rho > R_2)$
For any point outside both long solenoids there is no magnetic field, then

$$\mathbf{B} = 0. \tag{6.121}$$

(b) $(0 < \rho < R_1)$
Referring to the magnetic field **H**, once **B** is known it may be calculated by using (6.12), i.e.

$$\mathbf{B} = \mu_0(\mathbf{H} + \mathbf{M}) \Rightarrow \mathbf{H} = \frac{\mathbf{B}}{\mu_0} - \mathbf{M} = 0,$$

where we have used the fact that in a vacuum the magnetization **M** equals zero.
$(R_1 < \rho < R_2)$
For this region we also employ the value of **B** calculated in the previous section

$$\mathbf{H} = \frac{\mathbf{B}}{\mu_0} - \mathbf{M} = \frac{\mathbf{B}_2}{\mu_0} - \mathbf{M} = \frac{\mu_0 M}{\mu_0} \mathbf{u}_z - M \mathbf{u}_z = 0. \tag{6.122}$$

$(\rho > R_2)$
For this part we have neither magnetization **M** nor magnetic field **B**, and therefore

$$\mathbf{H} = \frac{\mathbf{B}}{\mu_0} - \mathbf{M} = 0.$$

6.13 A very long conducting cylinder of radius $R = 10$ cm, whose revolution axis coincides with the OZ axis, is carrying a density current $\mathbf{j} = (1000/\pi)\mathbf{u}_z$ (A/m^2). The bar has a hole also cylindrical in shape of radius $r = 2$ cm, whose symmetry axis is parallel to the other one. The distance between the axes of both cylinders is $d = 4$ cm. If the metallic bar is surrounded by a homogeneous linear magnetic material of relative permittivity $\mu_r = 100$, find: (a) The magnetic field $\mathbf{B}_\phi$ at $P(8, 0, 0)$ cm. (b) The magnetic field $\mathbf{B}_\phi$ at $P(15, 0, 0)$ cm. (c) The magnetization **M** at $P(15, 0, 0)$ cm.

Solution

(a) At first sight this problem seems to be very difficult to resolve because we do not have enough symmetries to apply Ampère's law. In fact the hollow cylinder has translational symmetry along OZ, but it does not have rotational symmetry, then, in principle, it may be not useful. As a consequence, we should try to employ other methods for calculating the magnetic field. However, as we will demonstrate, another simple method may be applied (Fig. 6.39).

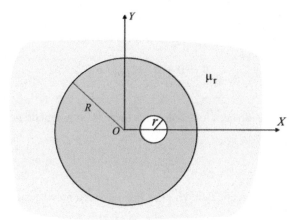

Fig. 6.39 Hollow cylinder carrying a density current **j**

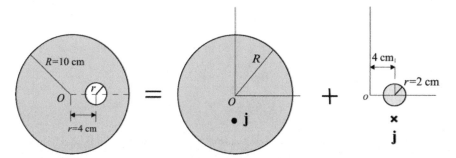

Fig. 6.40 Decomposition of the hollow cylinder into two bars whose density currents flow in opposite directions

As we have commented in other parts of the book, for linear fields there is an important principle, namely the principle of superposition of linear fields. In our case, we should divide the problem into parts so that the addition of each contribution equals the original problem. If this applies, by looking at the geometry of the hollow bar we observe that the system should behave like the sum of two separated cylinders of radii R and r, respectively, carrying the same density current but in opposite directions (see Fig. 6.40). This means we can compute each cylinder separately, and then add the magnetic fields obtained. When examining the first bar we notice that now the system has revolution symmetry and, as a consequence, we may directly use Ampère's theorem. As in this first question we are inside the cylinder where $\mu_r \approx 1$, thus we can directly employ the Ampère law for the magnetic field **B** (see Fig. 6.41). In effect, choosing a plane surface S and circular delimiting curve Γ_1 of radius $\rho_1 = 8$ cm ($\rho_1 < R$) we have

Fig. 6.41 Curve Γ_1 of radius $\rho_1 = 8$ cm, where the magnetic field B must be calculated

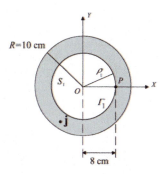

Fig. 6.42 Curve Γ_2 with radius $\rho_4 = 4$ cm, passing through point P where the field B must be calculated

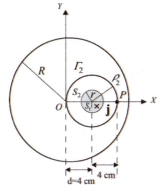

$$\oint_{\Gamma_1} (B_\rho, B_\phi, B_z)(0, dl, 0) = \mu_0 \int\int_{S_1} \mathbf{j} \cdot d\mathbf{S} = \mu_0 \int_0^{2\pi} \int_0^{\rho_1 < R_1} j\mathbf{u}_z \, dS \, \mathbf{u}_z$$

$$= \mu_0 \int_0^{2\pi} \int_0^{\rho_1 < R_1} j\rho \, d\rho \, d\phi, \qquad (6.123)$$

and integrating both terms of the equality

$$B_\phi = \frac{1}{2}\mu_0 j \rho_1, \qquad (6.124)$$

which represents the field contribution at point P due to the cylinder of radius R. With the same idea we calculate the field B at the same point, created by the bar of radius r located at $(4, 0, 0)$ cm, carrying a current $\mathbf{j} = -j\,\mathbf{u}_z$. Taking a circular curve Γ_2 and a flat surface S_2, we have (Fig. 6.42)

$$\oint_{\Gamma_2} (B_\rho, B_\phi, B_z)(0, dl, 0) = \mu_0 \int\int_{S_2} \mathbf{j} \cdot d\mathbf{S} = \mu_0 \int_0^{2\pi} \int_0^{\rho_2 > r} (-j)\mathbf{u}_z \, dS \, \mathbf{u}_z$$

$$= -\mu_0 \int_0^{2\pi} \int_0^{\rho_2 > r} j\rho \, d\rho \, d\phi. \qquad (6.125)$$

Fig. 6.43 Curve Γ_1 with radius $\rho = 15$ cm, and surface S. Notice that in this region $\mu_r = 100$, then it is better to begin with the Ampère law for **H**

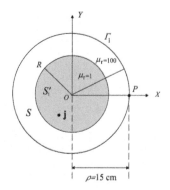

However, though the last integral extends to $\rho_2 > r$, between r and ρ there is no current density, and thus the product of $\mathbf{j} \cdot d\mathbf{S}$ is zero. This means that the effective surface of integration must correspond to $S \cap V_j = S$ (see Chap. 5). In our case $S_2 \cap V_j = S_r$, S_r being the surface of the circle with radius r. By performing the integrations we obtain

$$2\pi \rho_2 B_\phi = -\mu_0 \int_0^{2\pi} \int_0^r j \, \rho \, d\rho \, d\phi = -\pi \mu_0 j \, r^2 \Rightarrow B_\phi = -\frac{\mu_0}{2\rho_2} j \, r^2, \qquad (6.126)$$

where $\rho_2 = (\rho_1 - d) = 4$ cm for the point $(8, 0, 0)$ cm, and $\rho_1 = 8$ cm. Now, applying the superposition principle of fields, the total magnetic field produced by the hollow cylinder is

$$B_\phi = \mu_0 j \left(\frac{1}{2} \rho_1 - \frac{1}{2|\rho_1 - d|} r^2 \right) = 1.2 \cdot 10^{-5} \, \text{T}. \qquad (6.127)$$

(b) As we have explained in other exercises, if magnetic matter is present, then it is preferable to use Ampère's theorem for the field **H** rather than for **B**, because in magnetostatics, when the demagnetizing fields are negligible, H only depends of the free current density. On the contrary, if we start with Ampère's law for **B**, we must take into account the possible contributions of the magnetization currents, if any. Applying such a law for the field **H**, and considering a circular curve of radius $\rho_1 = 15$ cm concentric with the cross section of the bar, it results (Fig. 6.43)

$$\oint_{\Gamma_1} (H_\rho, H_\phi, H_z)(0, dl, 0) = \int\int_{S_1} \mathbf{j}_c \cdot d\mathbf{S} = \int_0^{2\pi} \int_0^{\rho > R} j \mathbf{u}_z \, dS \, \mathbf{u}_z = \int_0^{2\pi} \int_0^R j \, \rho \, d\rho \, d\phi, \qquad (6.128)$$

and then

$$2\pi \rho_1 H_\phi = 2\pi \frac{1}{2} j R^2 \Rightarrow H_\phi = \frac{1}{2\rho_1} j R^2. \qquad (6.129)$$

Observe that, in this case, the effective surface of integration is $S \cap V_j = S'_1 = \pi R^2$.

Fig. 6.44 Curve Γ_2 centered at a point located at distance d from the origin of the coordinate frame

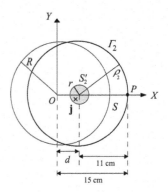

This field corresponds to the whole cylinder without a hole and it fails to determine the magnetic field **H** produced by the thin bar carrying a current $\mathbf{j} = -j\mathbf{u}_z$. To do this, we also choose a circular curve, but with radius $\rho_2 = 11$ cm, centered at point $(4, 0, 0)$ cm, where the cross section of that cylinder has its revolution axis (see Fig. 6.44). Bear in mind that the effective surface of integration is $S_2' = \pi r^2$, we may write

$$2\pi\rho_2 H_\phi = -2\pi\frac{1}{2}j r^2 \Rightarrow H_\phi = -\frac{1}{2\rho_2}j r^2. \tag{6.130}$$

Once we know the fields produced by each part of the equivalent system, using the superposition principle it follows

$$H_\phi = \frac{1}{2\rho_1}j R^2 - \frac{1}{2\rho_2}j r^2. \tag{6.131}$$

Considering the data of the statement $\rho_2 = (\rho_1 - d) = 11$ cm, then

$$H_\phi = \frac{j}{2}\left(\frac{1}{\rho_1}R^2 - \frac{1}{|\rho_1 - d|}r^2\right) = 10 \text{ A/m}. \tag{6.132}$$

Once we have H, we calculate the magnetic field B by means of (6.19),

$$B_\phi = \mu_0(1 + \chi_m)H = \mu_0\mu_r H_\phi = \frac{\mu_0\mu_r j}{2}\left(\frac{1}{\rho_1}R^2 - \frac{1}{|\rho_1 - d|}r^2\right) = 1.2 \cdot 10^{-3} \text{ T}. \tag{6.133}$$

(c) The magnetization may be directly obtained by using (6.12)

$$B = \mu_0(H + M) \Rightarrow M_\phi = \frac{B_\phi}{\mu_0} - H_\phi = \chi_m H_\phi = 990 \text{ Am}^{-1}. \tag{6.134}$$

Fig. 6.45 Magnetic system formed by a big region in which there is a hollow cylinder of radius $R_2 = 15$ cm. Inside of this hole a very long wire of radius $R_1 = 5$ cm is introduced

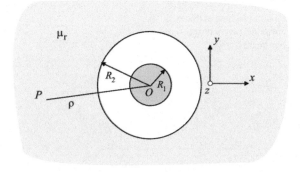

Fig. 6.46 Curve Γ with radius ρ used to calculate the field in region of $\mu_r = 10$

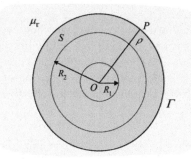

6.14 Throughout a big region of space of relative permittivity $\mu_r = 10$, a current density $\mathbf{j} = \frac{-0.32}{\rho^2}\mathbf{u}_z$ Am^{-2} is circulating, where ρ is the distance from the origin of coordinates O to a generic point P in cylindrical coordinates (see Fig. 6.45). A cavity cylindrical in shape with radius $R_2 = 15$ cm is made, whose rotational axis passes by O. A very long cylindrical metallic wire of radius $R_1 = 5$ cm is introduced into the hole parallel to the cavity, and coinciding with the symmetry axis of the system. If the density current flowing through the wire is $\mathbf{j} = 510\,\mathbf{u}_z$ Am^{-2}, find: (a) The magnetic field $\mathbf{B}_\phi$ for $\rho = 20$ cm. (b) The magnetization $\mathbf{M}$ at the same point. (c) The values of ρ for which the magnetic field H_ϕ is zero.

Solution

(a) The system shown in Fig. 6.45 is symmetric under a translation parallel to OZ and also under rotations about this axis. The point at which we must calculate the field belongs to the region where the material has a relative permittivity 10, then we will employ Ampère's law for the magnetic field $\mathbf{H}$. Choosing a circular curve of radius $\rho > R_2$, centered at O, and a plane surface S whose boundary is such a curve, we have (Fig. 6.46)

$$\oint_\Gamma (H_\rho, H_\phi, H_z)(0, dl, 0) = \int\int_S \mathbf{j} \cdot d\mathbf{S}, \tag{6.135}$$

and taking into consideration that in the region $R_1 < \rho < R_2$ there is no current density, Ampère's theorem leads to

$$\oint_\Gamma H_\phi dl = \int_0^{2\pi}\int_0^{R_1} 510\mathbf{u}_z d\mathbf{S}\mathbf{u}_z + \int_0^{2\pi}\int_{R_1}^{R_2} 0 dS - \int_0^{2\pi}\int_{R_2}^{\rho>R_2} \frac{0.32}{\rho^2}\mathbf{u}_z d\mathbf{S}\mathbf{u}_z =$$
$$510\int_0^{2\pi}\int_0^{R_1} \rho d\rho d\phi - \int_0^{2\pi}\int_{R_2}^{\rho>R_2} \frac{0.32}{\rho^2} \rho d\rho d\phi =$$
$$2\pi \cdot 510 \int_0^{R_1} \rho d\rho - 2\pi \cdot 0.32 \int_{R_2}^{\rho>R_2} \frac{1}{\rho} d\rho = 2\pi 510 \frac{1}{2}R_1^2 - 2\pi \cdot 0.32 \ln \frac{\rho}{R_2}.$$
(6.136)

As we have seen in other problems the first integral of the left side is

$$\int_0^{2\pi} H_\phi \rho d\phi = 2\pi \rho H_\phi,$$
(6.137)

thus,

$$H_\phi = \frac{255}{\rho}R_1^2 - \frac{0.32}{\rho} \ln \frac{\rho}{R_2} = 2.7 \, \text{Am}^{-1}.$$
(6.138)

Once the magnetic field **H** is known, we get the field **B** by (6.12)

$$\mathbf{B} = \mu_0(\mathbf{H} + \mathbf{M}) = \mu_0\mu_r\mathbf{H} = \mu_0\mu_r H_\phi \mathbf{u}_\phi = 3.4 \times 10^{-5} \mathbf{u}_\phi \, \text{T}$$
(6.139)

(b) By using the aforementioned equation, we can write

$$\mathbf{B} = \mu_0(\mathbf{H} + \mathbf{M}) \Rightarrow \mathbf{M} = \frac{\mathbf{B}}{\mu_0} - \mathbf{H} = (\mu_r - 1)\mathbf{H} = 24.5\mathbf{u}_\phi \, \text{Am}^{-1}.$$
(6.140)

(c) By making (6.138) equal to zero, we obtain the points for which the magnetic field H disappear,

$$\frac{255}{\rho}R_1^2 - \frac{0.32}{\rho} \ln \frac{\rho}{R_2} = 0 \Rightarrow \rho \approx 1.1 \, \text{m}.$$
(6.141)

This result tells us that over all points corresponding to a circumference of radius $\rho \approx 1.1$ m ($\rho > R_2$) the field **H** equals zero. But it is possible that other points also have the same characteristics. In fact, applying the Ampère theorem again to the cross section of the wire of radius R_1, we get

$$\int_0^{2\pi} H_\phi \rho d\phi = 2\pi \rho H_\phi = \int_0^{2\pi}\int_0^{\rho<R_1} 510\mathbf{u}_z . d\mathbf{S}\mathbf{u}_z,$$
(6.142)

and calculating the double integral as shown in preceding sections

$$H_\phi = 510\,\pi\rho^2 \Rightarrow H_\phi = 255\,\rho\,(\text{A/m}), \tag{6.143}$$

that is, the magnetic field H_ϕ in the interior of the cylindrical conductor of radius R_1 is a linear function of the distance ρ to the symmetry axis OZ. From this result we can conclude that at point $\rho = 0$ the magnetic field H_ϕ is zero as well.

6.15 A magnetic cylinder of length $L = 10$ cm and circular cross section of radius $R = 2$ cm, is located with its axis of symmetry coinciding with the OZ coordinate axis. The cylinder is magnetized heterogeneously in such a manner that $\mathbf{M} = az'\mathbf{u}_z$, where a is a constant of value $13{,}000\,\text{A}\,\text{m}^{-2}$. Calculate: (a) The magnetic field H at the point $P(0, 0, 8)$ cm. (b) The magnetic field B at $P(0, 0, 1)$ m.

Solution

As we have seen in Exercise 4, it is not necessary to always employ (6.6) and its curl for solving a problem of magnetization. Sometimes it is possible to construct a simple model that allows us to obtain the field. The idea is to calculate the magnetization currents and analyze if there is an equivalent system from the viewpoint of the effects with respect to the magnetic field $\mathbf{B}$ produced.

The currents $\mathbf{j}_m$ and $\mathbf{j}_{ms}$ are

$$\mathbf{j}_m = 0, \tag{6.144}$$

and

$$\mathbf{j}_{ms} = az'\mathbf{u}_\phi. \tag{6.145}$$

This result shows that a magnetization current only exists over the surface of the bar. As these currents are tangent in the circumferential direction to the cylinder surface, the system seems to be equivalent to a solenoid in which the current I is variable with distance z'. In Chap. 5 we have studied a solenoid whose turns per unit length were also proportional to z'. When we analyzed the homogeneously magnetized bar we saw that there was an equivalence between nI of the actual solenoid and the modulus of the magnetization M. Now, following the same process and taking into consideration the physical dimensions of $\mathbf{j}_{ms}$, we can write

$$M(z') = az' = j_{ms} = n(z')I, \tag{6.146}$$

but $n(z') = bz'$ (see Problem 5.26), hence

$$az' = bz'I \Rightarrow a = bI. \tag{6.147}$$

This means that the mathematical expression for the magnetic field B at a point P over the axis of symmetry of the cylinder is the same as (5.241) but changing bI by a (i.e. $1000I$ by $13{,}000$).

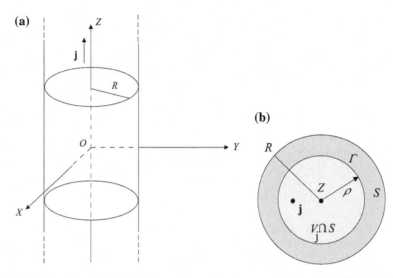

Fig. 6.47 a Magnetic slender bar. Observe the rotational symmetry around the OZ axis. **b** Plane section of the bar. A curve Γ of radius $\rho < R$ is chosen. In this figure $V_j \cap S$ represents the intersection of the volumetric current density **j** and the surface S chosen for applying Ampère's Law

$$\mathbf{B}(0,0,z) = \frac{13{,}000\mu_0}{2}[z(\cos\alpha_1 + \cos\alpha_2) + R(\sin\alpha_1 - \sin\alpha_2)]\,\mathbf{u}_z. \quad (6.148)$$

This equivalence means that the magnetic field B produced by the heterogeneously magnetized bar is the same as that generated by a solenoid carrying a current per unit length $bz'I$ (we identify effects but not causes!). Once we know B, the value of the magnetic field H is obtained by (6.10) (Fig. 6.47)

$$\mathbf{H} = \frac{\mathbf{B}}{\mu_0} - \mathbf{M} = \left\{\frac{a}{2}[z(\cos\alpha_1 + \cos\alpha_2) + R(\sin\alpha_1 - \sin\alpha_2)] - az\right\}\mathbf{u}_z$$
$$= -228\,\mathbf{u}_z\ (\mathrm{A/m}). \quad (6.149)$$

(b) In the same way as in the former section, employing (6.10) and taking into account that for $P(0, 0, 1)$ meter there is no matter, we have

$$\mathbf{H} = \frac{\mathbf{B}}{\mu_0} - \mathbf{M} = \frac{\mathbf{B}}{\mu_0} \Rightarrow \mathbf{B}(0,0,z) = \frac{13{,}000\mu_0}{2}[z(\cos\alpha_1 + \cos\alpha_2) + R(\sin\alpha_1 - \sin\alpha_2)]\,\mathbf{u}_z$$
$$= 3.2 \times 10^{-7}(\mathrm{T}). \quad (6.150)$$

6.16 A slender cylindrical bar of radius $R = 8\,\mathrm{cm}$ is made of a conducting ferromagnetic material whose initial magnetization curve is

$$M(H) = 600H + 101H^2 \quad (6.151)$$

A constant current $I = 2\,\text{A}$ corresponding to a current density $\mathbf{j}$ along its axis of revolution flows through the bar. Calculate: (a) The magnetic field B_ϕ for $\rho = 6\,\text{cm}$. (b) The magnetic flux through a surface perpendicular to the revolution axis of the bar.

Solution

(a) Since we have matter it is convenient to employ the Ampère law for the magnetic field H. In this problem the only sources for H are the moving free charges resulting from the current I. Applying (6.14) we can write

$$\oint_\Gamma \mathbf{H} d\mathbf{l} = \int\int_S \mathbf{j} \cdot d\mathbf{S} = I_c. \tag{6.152}$$

The bar is symmetric with respect to a rotation around the OZ axis, then it seems appropriate to use cylindrical coordinates. Rewriting (6.152) for this coordinate system it yields to

$$\oint_\Gamma (H_\rho, H_\phi, H_z)(0, dl, 0) = \int\int_S \mathbf{j} \cdot d\mathbf{S} \tag{6.153}$$

If we look at the point where the magnetic field must be calculated, we see that it belongs to the interior region of the bar. It means that the effective surface for computing the integral of the right side is not the total cross-section surface S of the cylinder given in the problem. Actually, the surface of integration corresponds to the intersection of the volumetric current density $\mathbf{j}$ and the surface S. As a result this effective surface is smaller than the cross-section S, ρ and Γ being the radius and delimiting curve, respectively. Thus, it follows

$$\oint_\Gamma H_\phi dl = j \int\int_{V_j \cap S} dS \Rightarrow H_\phi 2\pi\rho = jS = j\pi\rho^2, \tag{6.154}$$

and this leads to

$$H_\phi = \frac{1}{2} j\rho. \tag{6.155}$$

We only know the value of I but not j. However, we can find the relation between both magnitudes by means of the definition of intensity. In fact, the intensity through the total circular surface of the bar is

$$\int\int_S \mathbf{j} \cdot d\mathbf{S} = I_c \Rightarrow j\pi R^2 = I, \tag{6.156}$$

then

$$j = \frac{I}{\pi R^2}, \tag{6.157}$$

and introducing this result in (6.155)

$$H_\phi = \frac{I\rho}{2\pi R^2}.\tag{6.158}$$

Now, to calculate the magnetic field **B** the hysteresis curve must be used

$$B_\phi(H) = \mu_0(H + M) = \mu_0(H + 600H + 101H^2) = \mu_0(1 + 600 + 101H)H =$$
$$\tag{6.159}$$
$$\mu_0\left(1 + 600 + 101\frac{I\rho}{2\pi R^2}\right)\frac{I\rho}{2\pi R^2} = 1.7 \cdot 10^{-3}\,\text{T}.$$

(b) Due to the circular symmetry of the bar and the fact that its diameter is much smaller than its length, the magnetic field produced is tangential only, hence the flux of this field through a circular surface perpendicular to its revolution axis is zero (the field is perpendicular to the surface S).

6.17 A very large magnetic hollow cylinder of radii R_1 and R_2, carries a homogeneous current I. If the magnetic permeability is $\mu_r = \rho/R_1$, ρ being the distance from the revolution axis of the system to a generic point, find: (a) The magnetic field B_ϕ for $R_1 < \rho < R_2$. (b) The flux of B_ϕ across a section of the cylinder of high h as shown in Fig. 6.48.

Solution

(a) The system has translational symmetry along the OZ axis, and also rotational symmetry about OZ. Hence the best way for calculating B_ϕ is to apply Ampère's law. As we have magnetic matter, in principle, we cannot control the magnetization density currents, so we employ Ampère's theorem for H

$$\oint_\Gamma \mathbf{H} \cdot d\mathbf{l} = \int_S \mathbf{j}_f \cdot d\mathbf{S}.\tag{6.160}$$

Fig. 6.48 Very large hollow cylinder carrying current I

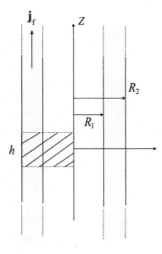

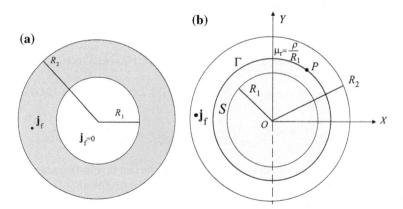

Fig. 6.49 a Cross-section of the hollow cylinder carrying a current density $\mathbf{j}_f$. b Application of the Ampère theorem over an open plane surface S of radius ρ ($R_1 < \rho < R_2$) and boundary Γ

For applying this law, we choose a plane surface whose boundary is a circle of radius ρ such that $R_1 < \rho < R_2$ (Fig. 6.49). Using cylindrical coordinates it yields

$$\oint_\Gamma \mathbf{H} \cdot d\mathbf{l} = \oint_\Gamma (H_\rho, H_\phi, H_z) \cdot (0, dl, 0) = \int_S \mathbf{j}_f \cdot d\mathbf{S} = \int_S (0, 0, j_c) \cdot (0, 0, dS) \quad (6.161)$$

and performing the scalar products

$$\oint_\Gamma H_\phi dl = \int_S j_f dS \Rightarrow \int_0^{2\pi} H_\phi \rho d\phi = \int_0^{2\pi} \int_{R_1}^{\rho < R_2} j_f \rho d\rho d\phi \Rightarrow H_\phi \rho \int_0^{2\pi} d\phi$$
$$= j_f \int_0^{2\pi} d\phi \int_{R_1}^{\rho < R_2} \rho d\rho. \quad (6.162)$$

Notice that the variable ρ in the double integral extends from R_1 to a generic ρ inside the crown $R_1 < \rho < R_2$, because in the region $0 < \rho < R_1$ there is no density current j_f. Operating the integrals it follows

$$2\pi \rho H_\phi = \pi j_f (\rho^2 - R_1^2) \Rightarrow H_\phi = \frac{j_f(\rho^2 - R_1^2)}{2\rho}. \quad (6.163)$$

This last expression does not depend on I but on j_f, which is unknown (the data given in the statement is the total current I). However, we can calculate the value of the current density of free charge as a function of I. In fact, employing the definition of intensity we can write

$$I = \int_S \mathbf{j}_f \cdot d\mathbf{S}. \quad (6.164)$$

Now, applying (6.164) over a surface of the hollow conductor perpendicular to the current density $\mathbf{j}_f$ leads to (Fig. 6.49b)

$$I = \int_0^{2\pi} d\phi \int_{R_1}^{R_2} j_f \rho d\rho = j_f \pi (R_2^2 - R_1^2) \Rightarrow j_f = \frac{I}{\pi(R_2^2 - R_1^2)}, \qquad (6.165)$$

and introducing it into H_ϕ, we have

$$H_\phi = \frac{I}{2\pi\rho} \frac{(\rho^2 - R_1^2)}{(R_2^2 - R_1^2)}. \qquad (6.166)$$

Once we know **H**, the magnetic field **B** may be obtained using (6.19)

$$\mathbf{B} = \mu_0 \mu_r \mathbf{H} \Rightarrow B_\phi = \mu_0 \mu_r H_\phi = \frac{\mu_0 \mu_r I}{2\pi\rho} \frac{(\rho^2 - R_1^2)}{(R_2^2 - R_1^2)}, \qquad (6.167)$$

where μ_r depends on ρ. Now substituting the value of μ_r we have

$$B_\phi = \frac{\mu_0 I}{2\pi R_1} \frac{(\rho^2 - R_1^2)}{(R_2^2 - R_1^2)} \qquad (6.168)$$

(b) In the previous section we have determined the field **B** in the shaded region (Fig. 6.48), but for calculating the flux of **B** through the entire surface we also need to know the magnetic field in the region $[R_1, 0] \times [0, h]$ ($\rho < R_1$ in Fig. 6.48; see also Fig. 6.50). For this reason, we are going to compute such a field in the hole of the bar. Applying Ampère's theorem again over a circular curve of radius $\rho < R_1$

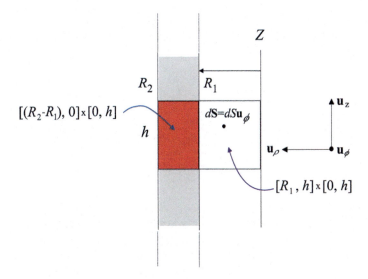

Fig. 6.50 Lateral section of the hollow cylinder for calculating the flux across the surface S. Observe that the normal to this surface has the direction of the unitary vector $\mathbf{u}_\phi$. The right part of S ($\rho < R_1$) does not contribute to the flux integral because $B_\phi = 0$ in this region

Fig. 6.51 Circular curve of radius $\rho < R_1$ and planar surface S to apply Ampère's law. Notice that this surface (*blue*) is not crossed by any current

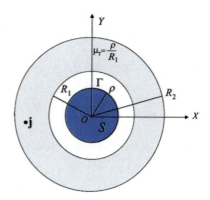

(Fig. 6.51), we have

$$\oint_\Gamma H_\phi dl = \int_S j_f dS = \int_0^{2\pi} \int_0^{\rho<R_1} j_f \rho d\rho d\phi = 0 \Rightarrow H_\phi = 0, \quad (6.169)$$

because there is no density current in the hole of the cylinder, and therefore $B_\phi = 0$. The calculation of the flux may be performed directly by its definition, i.e.

$$\phi = \int_S \mathbf{B} \cdot d\mathbf{S}. \quad (6.170)$$

Taking the normal outward to S, the unit vector $\mathbf{u}_\phi$ of the magnetic field coincides with $\mathbf{n}$ on this surface (Fig. 6.50). On the other hand, due to the fact that there is no magnetic field for $\rho < R_1$, the integral for the variable ρ extends only from R_1 to R_2, thus

$$\phi = \int_S \mathbf{B} \cdot d\mathbf{S} = \int\int_S B\mathbf{u}_\phi \cdot d\mathbf{S}\,\mathbf{u}_\phi = \int_0^h \int_{R_1}^{R_2} \frac{\mu_0 I}{2\pi R_1} \frac{(\rho^2 - R_1^2)}{(R_2^2 - R_1^2)} d\rho\,dz =$$

$$\frac{\mu_0 I}{2\pi R_1 (R_2^2 - R_1^2)} \int_0^h dz \int_{R_1}^{R_2} (\rho^2 - R_1^2)\,d\rho =$$

$$\frac{\mu_0 I h}{2\pi R_1 (R_2^2 - R_1^2)} \left\{ \frac{1}{3}(R_2^3 - R_1^3) - R_1^2(R_2 - R_1) \right\} \quad (6.171)$$

6.18 A very large cylindrical conductor whose revolution axis coincides with OZ, has an unknown radius R. A density current of free charge $\mathbf{j}_f = (-\rho^2 + 0.005)\mathbf{u}_z\,\mathrm{A\,m^{-2}}$ circulates along the OZ axis. If the relative permeability is $\mu_r = 2$, find: (a) The value of R in order that for any external point to the cylinder ($\rho > R$), the resulting magnetic field $\mathbf{B}$ be zero. (b) The magnetic field $\mathbf{B}$ at $\rho = 0.05$ m.

Solution

(a) To know the radius R, we need to first calculate the magnetic field B for a generic external point and then we will impose the condition that such a field be zero.

As the system has translational and rotational symmetry, we proceed as shown in the former exercises, that is, by applying Ampère's law for $\mathbf{H}$ over a circular curve Γ of radius $\rho > R$, which is the boundary of an open plane surface S

$$\oint_\Gamma \mathbf{H} \cdot d\mathbf{l} = \int_S \mathbf{j}_f \cdot d\mathbf{S} \qquad (6.172)$$

$$\oint_\Gamma \mathbf{H} \cdot d\mathbf{l} = \oint_\Gamma (H_\rho, H_\phi, H_z) \cdot (0, dl, 0) = \int_S \mathbf{j}_f \cdot d\mathbf{S} = \int_S (0, 0, j_f) \cdot (0, 0, dS), \qquad (6.173)$$

and making the scalar products

$$\oint_\Gamma H_\phi dl = \int_S j_f dS \Rightarrow \int_0^{2\pi} H_\phi \rho d\phi = \int_0^{2\pi}\int_0^R (-\rho^2 + 0.005)\rho d\rho d\phi. \qquad (6.174)$$

Due to the rotational symmetry around the axis of the cylinder, the modulus of the magnetic field $\mathbf{H}_\phi$ is the same for all points along the curve Γ used for making the integral, hence we can put it outside

$$H_\phi \rho \int_0^{2\pi} d\phi = \int_0^{2\pi} d\phi \int_0^R (-\rho^2 + 0.005)\rho d\rho = 2\pi \left[-\int_0^R \rho^3 d\rho + \int_0^R 0.005 \rho d\rho \right] =$$
$$2\pi \left[-\frac{R^4}{4} + 0.005 \frac{R^2}{2} \right] \Rightarrow 2\pi \rho H_\phi = 2\pi \left[-\frac{R^4}{4} + 0.005 \frac{R^2}{2} \right], \qquad (6.175)$$

then

$$H_\phi = \frac{R^2}{2\rho}\left[-\frac{R^2}{2} + 0.005\right]. \qquad (6.176)$$

Once the magnetic field H is known the field B may be calculated by (6.12),

$$B = \mu_0(H + M) \Rightarrow B_\phi = \mu_0 \mu_r H_\phi = \frac{\mu_0 \mu_r R^2}{2\rho}\left[-\frac{R^2}{2} + 0.005\right]$$
$$= \frac{2\mu_0 R^2}{2\rho}\left[-\frac{R^2}{2} + 0.005\right]. \qquad (6.177)$$

This expression depends on R and gives us the value of B_ϕ for any point outside of the conductor. Now we must impose the condition that the field is zero

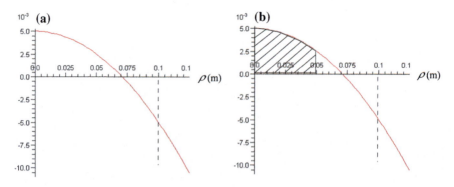

Fig. 6.52 a Density current of free charge versus distance ρ to the symmetry axis of the cylinder. Observe that j_f is not homogeneous inside the conductor. **b** For point $R = 0.05$ m, the density current that contributes to the field flows in the direction of the positive OZ axis. At this point j_f has not reached zero yet ($\rho = 0.7$ m)

$$B = \mu_0(H + M) \Rightarrow B_\phi = \frac{2\mu_0 R^2}{2\rho}\left[-\frac{R^2}{2} + 0.005\right] = 0 \Rightarrow R = 0.1 \text{ m}. \quad (6.178)$$

To understand the result obtained, Fig. 6.52a helps us. Bearing in mind $R = 0.1$ m, we see that the density current has a negative part in the interval $[0.07, 0.1]$ m. This is logical; if we do not have a part of the density current in an opposite direction inside of the cylinder, we can not reach a zero field at any exterior point.

(b) Point $\rho = 0.05$ m belongs to the interior of the conductor. For calculating B_ϕ inside, we proceed in the same way used in Section (a). The only difference is that the integral over ρ extends to a specific ρ, i.e.

$$2\pi \rho H_\phi = \int_0^{2\pi} d\phi \int_0^{\rho < R = 0.05} (-\rho^2 + 0.005)\rho d\rho = 2\pi \left[-\frac{\rho^4}{4} + 0.005\frac{\rho^2}{2}\right]_0^\rho, \quad (6.179)$$

thus

$$H_\phi = \frac{\rho}{2}\left[-\frac{\rho^2}{2} + 0.005\right], \quad (6.180)$$

and B

$$B = \mu_0(H + M) \Rightarrow B_\phi = \frac{2\mu_0 \rho}{2}\left[-\frac{\rho^2}{2} + 0.005\right]. \quad (6.181)$$

For $\rho = 0.05$ m, (6.181) yields $B = 2.3 \times 10^{-10}$ T.

6.19 In the interior of a very large solenoid with radius $R_2 = 6$ cm and $n = 1000$ m^{-1} turns per unit length, a cylindrical paramagnetic bar, also very large, of radius $R_1 = 4$ cm is placed with their symmetry axes coinciding. If the magnetic permeability of the slender rod is $\mu_r = 10$ and the current through the solenoid is 1 A, obtain: (a) The magnetic field **B** in the bar. (b) The component A_ϕ of the vector potential at a generic point inside of the rod ($0 < \rho < R_1$). (c) The value of A_ϕ between the rod and the coil ($R_1 < \rho < R_2$). (d) Idem for any point outside of the system ($\rho > R_2$).

Solution

(a) The magnetic field **H** created by a very long solenoid is

$$\mathbf{H} = nI\,\mathbf{u}_z, \tag{6.182}$$

which is homogeneous and confined inside of the coils (remember that outside of a very long solenoid (mathematically infinite), the field produced is zero-see Problem 5.10). This field acts on the bar and as a result the rod reaches a magnetization **M**, which in turn produces a magnetic field **B** inside of value

$$\mathbf{B} = \mu_0(\mathbf{H} + \mathbf{M}) = \mu_0\mu_r\mathbf{H} = \mu_0\mu_r nI\,\mathbf{u}_z = 0.013\,\mathbf{u}_z\ T. \tag{6.183}$$

The same result may be obtained by considering the magnetization reached by the slender bar due to the magnetic field **H**. In fact, the presence of **H** leads to a homogeneous magnetization **M** in the direction of the symmetry axis of the cylindrical bar. As we have studied, the magnetic field **B** of a bar magnetized along its revolution axis is equivalent to a solenoid of the same dimensions in which $nI = j_{ms} = M$. If we apply this result to our problem, the magnetic field inside the cylindrical bar may be calculated as the superposition of the magnetic field $\mathbf{B}_s$ of the solenoid plus the magnetic field $\mathbf{B}_b$ of the equivalent solenoid corresponding to the bar. The existence of **H** leads to a magnetization $\mathbf{M} = \chi_m \mathbf{H} = (\mu_r - 1)\mathbf{H}$. With this magnetization the equivalent solenoid creates the magnetic field

$$\mathbf{B}_b = \mu_0 j_{ms}\,\mathbf{u}_z = \mu_0 M\,\mathbf{u}_z = \mu_0(\mu_r - 1)H\,\mathbf{u}_z = \mu_0(\mu_r - 1)nI\,\mathbf{u}_z. \tag{6.184}$$

then, the resulting magnetic field **B** is

$$\begin{aligned}\mathbf{B} &= \mathbf{B}_s + \mathbf{B}_b = \mu_0 nI\,\mathbf{u}_z + \mu_0(\mu_r - 1)nI\,\mathbf{u}_z = \mu_0\,(nI + (\mu_r - 1)nI)\\\mathbf{u}_z &= \mu_0\mu_r nI\,\mathbf{u}_z = 0.013\,\mathbf{u}_z\ T,\end{aligned} \tag{6.185}$$

which is the same result shown in (6.183).

(b) For obtaining the potential vector inside of the rod ($\rho < R_1$) we can use the relation between the magnetic field **B** and **A** (remember that $\mathbf{B} = \nabla \times \mathbf{A}$-see Chap. 5), i.e.

$$\int\int_S \mathbf{B}\cdot d\mathbf{S} = \int\int_S \nabla\times\mathbf{A}\cdot d\mathbf{S} = \oint_\Gamma \mathbf{A}\cdot d\mathbf{l}, \tag{6.186}$$

and introducing the corresponding vectors

$$\int\int_S (B_\rho, B_\phi, B_z) \cdot (0, 0, 1) dS = \oint_\Gamma (A_\rho, A_\phi, A_z) \cdot (0, 1, 0) dl, \tag{6.187}$$

thus,

$$\int\int_S B_z dS = \oint_\Gamma A_\phi dl \Rightarrow \int_0^{2\pi}\int_0^{\rho<R_1} B_z \rho d\rho d\phi = \int_0^{2\pi} A_\phi \rho d\phi, \tag{6.188}$$

$$2\pi \frac{\rho^2}{2} B_z = 2\pi A_\phi \rho \Rightarrow A_\phi = \frac{1}{2}\mu_0 \mu_r nI \, \rho = 0.0063 \, \rho \, T \, m. \tag{6.189}$$

(c) For a generic point between the bar and the solenoid ($R_1 < \rho < R_2$) we have

$$\int_0^{2\pi}\int_0^{\rho>R_1} B_z \rho d\rho d\phi = \int_0^{2\pi} A_\phi \rho d\phi \tag{6.190}$$

Since the magnetic field **B** is different in the bar and in the region between the solenoid and the magnetic material, the integral on the left hand side of (6.190) must be separated in two parts. The first one corresponds to the bar radius R_1, and the second region goes from R_1 to ρ belonging to a point P in ($R_1 < \rho < R_2$)

$$\int_0^{2\pi}\int_0^{R_1} \mu_0 \mu_r nI \rho d\rho d\phi + \int_0^{2\pi}\int_{R_1}^{\rho} \mu_0 nI \rho d\rho d\phi = \int_0^{2\pi} A_\phi \rho d\phi. \tag{6.191}$$

Integrating this last expression it yields

$$2\pi \frac{R_1^2}{2}\mu_0 \mu_r nI + 2\pi \frac{(\rho^2 - R_1^2)}{2}\mu_0 nI = 2\pi A_\phi \rho \Rightarrow A_\phi = \frac{\mu_0 \mu_r nI R_1^2}{2\rho} + \frac{\mu_0 nI(\rho^2 - R_1^2)}{2\rho} = \tag{6.192}$$

$$= \frac{6.3 \cdot 10^{-5}}{\rho} + \frac{6.3 \cdot 10^{-3}(\rho^2 - 0.0016)}{\rho}.$$

(d) For a exterior point to the system ($\rho > R_2$), by means of the same procedure employed in (b) and (c), and denoting as $B(i)$ for $i = 1, 2, 3$ the magnetic fields in the three different regions of the system, we have

$$\int_0^{2\pi}\int_0^{R_1} B_z(1) \rho d\rho d\phi + \int_0^{2\pi}\int_{R_1}^{R_2} B_z(2) \rho d\rho d\phi + \int_0^{2\pi}\int_{R_2}^{\rho} B_z(3) \rho d\rho d\phi = \int_0^{2\pi} A_\phi \rho d\phi. \tag{6.193}$$

However, the magnetic field outside of a very large solenoid ($\rho > R_2$) is zero, so the preceding equality has the following form

$$\int_0^{2\pi}\int_0^{R_1} B_z(1) \rho d\rho d\phi + \int_0^{2\pi}\int_{R_1}^{R_2} B_z(2) \rho d\rho d\phi + 0 = \int_0^{2\pi} A_\phi \rho d\phi, \tag{6.194}$$

Solved Problems 373

or

$$\int_0^{2\pi}\int_0^{R_1} \mu_0\mu_r nI \rho d\rho d\phi + \int_0^{2\pi}\int_{R_1}^{R_2} \mu_0 nI \rho d\rho d\phi = \int_0^{2\pi} A_\phi \rho d\phi, \quad (6.195)$$

which leads to

$$2\pi \frac{R_1^2}{2}\mu_0\mu_r nI + 2\pi \frac{(R_2^2 - R_1^2)}{2}\mu_0 nI = 2\pi A_\phi \rho \Rightarrow A_\phi = \frac{\mu_0\mu_r nIR_1^2}{2\rho} + \frac{\mu_0 nI(R_2^2 - R_1^2)}{2\rho} = \quad (6.196)$$

$$= \frac{6.3 \cdot 10^{-5}}{\rho} + \frac{1.3 \cdot 10^{-6}}{\rho}.$$

6.20 The attached figure represents an iron core built from two different materials with two distinct coils closely wound of $N_h = 1000$ and $N_p = 15{,}000$ turns, respectively. The material of the lower part is paramagnetic and its permeability is $\mu_r = 25$. The upper is built with a ferromagnetic material, whose second magnetization curve is given by

$$B_h(H) = \begin{cases} B_h(H) = 1, & 0 < H_h < 40{,}000 \\ B_h(H) = 1 + \dfrac{H_h}{20{,}000}, & -20{,}000 < H_h \leq 0 \end{cases}$$

The middle lengths of both parts are the same, approximately, and of value $l_h \approx l_p = 20$ cm. The cross sections are $S_h = 25$ cm^2 and $S_p = 16$ cm^2. From the demagnetized state the system reaches saturation by increasing the current circulating through the coils. Under these circumstances, find: (a) The magnetic fields H_h and B_h, if starting from the saturation state the currents vary to $I_h = 2$ A and $I_p = 1$ A. (b) The fields H_p and B_p if both currents are reduced to zero (Fig. 6.53).

Solution

Unlike all previous problems, in this case the core system has two different materials. However, the steps for calculating the fields H_h and B_h in the iron core are the same

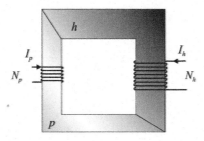

Fig. 6.53 Electromagnet composed by two parts

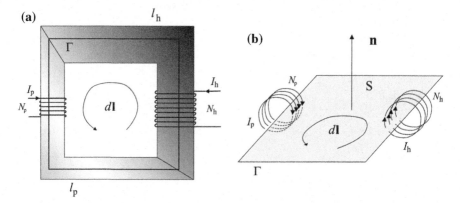

Fig. 6.54 **a** Curve Γ chose for applying Ampère's law. **b** Observe that for the $d\mathbf{l}$ counterclockwise the normal is outward of the surface S, and as a consequence the $N_p I_p$ has a negative sign and $N_h I_h$ a positive one

as those followed for electromagnets. Choosing a mean curve Γ passing through the system it results (Fig. 6.54)

$$\oint_\Gamma \mathbf{H} d\mathbf{l} = I_f \Rightarrow H_h l_h + H_p l_p = N_h I_h - N_p I_p. \tag{6.197}$$

As we do not have flux leakage and the cross sections of the electromagnet are different for each part, we may write

$$\oint_S \mathbf{B} d\mathbf{S} = 0 \Rightarrow B_h S_h = B_p S_p. \tag{6.198}$$

Introducing $B_h S_h = B_p S_p$ into (6.23), and substituting the equation for the paramagnetic material $B_p = \mu_0 H_p$

$$H_h l_h + \frac{B_p}{\mu_0 \mu_r} l_p = N_h I_h - N_p I_p \Rightarrow H_h l_h + \frac{B_h S_h}{\mu_0 \mu_r S_p} l_p = N_h I_h - N_p I_p. \tag{6.199}$$

Taking into consideration that $l_h \approx l_p$, the above equation takes the form

$$H_h + \frac{B_h S_h}{\mu_0 \mu_r S_p} \approx \frac{N_h I_h - N_p I_p}{l_p}, \tag{6.200}$$

and then the operating straight line is

$$B_h = \frac{\mu_0 \mu_r S_p}{S_h} \left(\frac{N_h I_h - N_p I_p}{l_p} \right) - \frac{\mu_0 \mu_r S_p}{S_h} H_h. \tag{6.201}$$

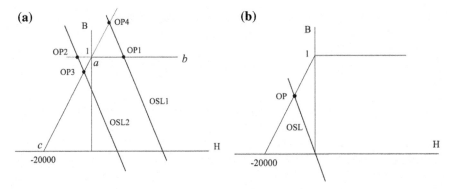

Fig. 6.55 **a** Different possibilities for the position of the (OSL). Observe that if at the beginning we had chosen the curve (c–a), the intersection with the (OSL) corresponding to the values $I_h = 2$ A and $I_p = 1$ A would be the point OP4, which is outside of the hysteresis loop; for this reason the only possible values is OP1. If due to the values of I_h and I_p the operating straight line would be (OSL1), we also have two possibilities corresponding to the intersections of the (OSL) with (a–b) and with (c–a). In this case only one of the operating points, i.e. either OP2 or OP3, would be valid. **b** Operating point when the intensities are zero

We do not know whether the operating point (OP) is in the first quadrant (a–b) or in the second quadrant (c–a) of the hysteresis curve (HC). Hence for calculating the intersection between the straight line (OSL) and (HC) we must try with the two parts. In principle, four possibilities for the operating point arise from the graphic represented in Fig. 6.55a. Firstly, let us consider the function $B(H) = 1$ corresponding to the interval $0 < H < 40{,}000$ (segment (a–b)). If we calculate the intersection with the operating straight line whose equation is (6.201), we have two possibilities. In fact, the intersection can be the point OP1 or OP2 depending if the (OSL) is OSL1 or OSL2. In order to know the point, we calculate the intersection of both curves and then we examine the result. It reduces to solve the following equation systems

$$\begin{cases} B_h(H) = 1, & 0 < H_h < 40{,}000 \\ B_h(H) = \dfrac{\mu_0 \mu_r S_p}{S_h}\left(\dfrac{N_p I_p - N_h I_h}{l_p}\right) - \dfrac{\mu_0 \mu_r S_p}{S_h} H_h. \end{cases}$$

Introducing the values of the lengths, turns, sections and intensities, it yields $H_h = 15264\,\text{A}\,\text{m}^{-1}$, and $B_h = 1$ T. This point corresponds to (OP1) in Fig. 6.55a, and belongs to the curve, so this is the desired intersection. However if we had calculated the intersection with the curve segment (c–a) the intersection would be the point (OP4), which is outside of the hysteresis curve, and therefore is invalid.

(b) If the intensities are reduced to zero, the operating point must be in the second quadrant of the hysteresis loop (Fig. 6.55b). In fact, making $I_p = I_h = 0$ in (6.201) we obtain

$$B_h = -\frac{\mu_0 \mu_r S_p}{S_h} H_h, \tag{6.202}$$

which is the equation of a straight line passing through the origin of coordinates, whose slope is $-\frac{\mu_0\mu_r S_p}{S_h}$. The intersection with the hysteresis curve corresponds to the point represented in Fig. 6.55, for which $(-20,000 < H < 0)$. Hence to find the solution we resolve the system of equations composed by (6.202) and

$$B_h(H) = 1 + \frac{H_h}{20,000}. \tag{6.203}$$

The result of that intersection is $H_h = -14,260\,\text{A}\,\text{m}^{-1}$ and $B_h = 0.28\,\text{T}$. However, these values of the fields refer to the ferromagnetic material, but not the paramagnetic on the lower part of the system. For obtaining the magnetic fields in this part we need to use (6.198), and the equation for the paramagnetic material $B_p = \mu_0 H_p$,

$$B_h S_h = B_p S_p \Rightarrow B_p = \frac{B_h S_h}{S_p} = 0.4\,\text{T}, \tag{6.204}$$

and

$$H_p = \frac{B_p}{\mu_0} = 14,000\,\text{Am}^{-1}. \tag{6.205}$$

Notice that in this electromagnet we do not have an air-gap, however we obtain an operating straight line that is not parallel to the axis of the magnetic field B (Fig. 6.16b). At first sight it may seem contradictory with that explained in section (6.9), where the permanent magnet without an air-gap had an OSL parallel to the vertical axis. Thus, the same result could be expected in this case, but it is not what we see and the reason is the following.

From the point of view of the equation employed for solving the problem, the electromagnet of Fig. 6.53 behaves like a system with an air-gap (compare (6.197) and (6.23); there is no difference). In effect, to some extent it is like an electromagnet in which the air-gap has been filled up with the paramagnetic material,[13] then it is logical that the OSL has the form shown in (6.202). For this reason this equation is a straight line of finite slope, as we obtained for systems with an air-gap (Fig. 6.12a). Observe that the magnet presented in Fig. 6.16 was built only of one material. In conclusion we may say that, even though we do not have an air gap, we can have an operating straight line of finite slope if the electromagnet is formed with several specimens of distinct physical characteristics.

Problems C

6.21 The figure attached shows a truncated magnetized cone of radii $r = 5\,\text{cm}$ and $R = 10\,\text{cm}$, respectively. The angle formed by the generatrix and the OZ axis

---

[13] In this reasoning we do not take into account the length of the air-gap, which usually is very small; this is only one way for explaining the form of the OSL.

Fig. 6.56 Truncated cone solenoid

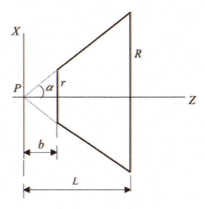

is $\alpha = 30°$, and the magnetization of the material is $\mathbf{M} = -10{,}000\mathbf{u}_z$ A/m. Find: (a) The volumetric magnetization current. (b) The surface magnetization current. (c) The magnetic field **B** created by the magnet at point $P(0, 0, 0)$.

Solution

(a) To determine the volumetric magnetization current we apply its definition (Fig. 6.56)

$$\mathbf{j}_m = \nabla' \times \mathbf{M} = \nabla' \times (0, 0, -M) = 0. \tag{6.206}$$

This result is logical if we consider that the magnetization of the cone is homogeneous.

(b) For calculating the surface magnetization current we divide the sample into three parts, which correspond to the faces of the truncated cone. The perpendicular unitary vector to the left base is $\mathbf{n}_1 = (0, 0, -1)$, and therefore

$$\mathbf{j}_{ms_1} = \mathbf{M} \times \mathbf{n}_1 = 0. \tag{6.207}$$

For the right side

$$\mathbf{n}_2 = (0, 0, 1)$$

which leads to

$$\mathbf{j}_{ms_2} = \mathbf{M} \times \mathbf{n}_2 = 0, \tag{6.208}$$

that is, we do not have magnetization currents on both bases. In the same way, the normal to the lateral surface of the specimen is $\mathbf{n}_3$, and then $\mathbf{j}_{ms_3} = \mathbf{M} \times \mathbf{n}_3$. However, in this case the unitary normal vector depends on the point on the surface, i.e. $\mathbf{n}_3 = \mathbf{n}_3(x, y, z)$. For obtaining the value of $\mathbf{n}_3$ we can use the gradient of the surface (see Chap. 1) as follows

$$\mathbf{n}_3 = \frac{\nabla S(x, y, z)}{|\nabla S(x, y, z)|}, \tag{6.209}$$

$S(x, y, z)$ being the equation of the surface.

A conical surface may be represented in cartesian coordinates by

$$x^2 + y^2 = \tan^2 \alpha \, z^2 \tag{6.210}$$

or in cylindrical coordinates

$$\rho^2 = \tan^2 \alpha \, z^2. \tag{6.211}$$

By applying the nabla operator over $S(x, y, z)$, we obtain

$$S(x, y, z) \equiv x^2 + y^2 - \tan^2 \alpha \, z^2 \Rightarrow \nabla S(x, y, z) = 2x \, \mathbf{u}_x + 2y \, \mathbf{u}_y - 2 \tan^2 \alpha \, z \, \mathbf{u}_z, \tag{6.212}$$

and dividing by the its modulus we have the normal vector $\mathbf{n}_3$

$$\mathbf{n}_3 = \frac{2x \, \mathbf{u}_x + 2y \, \mathbf{u}_y - 2 \tan^2 \alpha \, z \, \mathbf{u}_z}{2\sqrt{x^2 + y^2 + (\tan^2 \alpha)^2 \, z^2}}. \tag{6.213}$$

Introducing the equation of the surface (6.210) into the denominator of (6.209) we get

$$\mathbf{n}_3 = (n_x, n_y, n_z) = \frac{x \, \mathbf{u}_x + y \, \mathbf{u}_y - \tan^2 \alpha \, z \, \mathbf{u}_z}{\sqrt{x^2 + y^2 + (\tan^2 \alpha)(x^2 + y^2)}} = \frac{1}{\sqrt{x^2 + y^2}} \frac{x \, \mathbf{u}_x + y \, \mathbf{u}_y - \tan^2 \alpha \, z \, \mathbf{u}_z}{\sqrt{1 + \tan^2 \alpha}} =$$

$$\frac{x \, \mathbf{u}_x + y \, \mathbf{u}_y - \tan^2 \alpha \, z \, \mathbf{u}_z}{\sqrt{x^2 + y^2}} \cos \alpha. \tag{6.214}$$

Once we know the expression of $\mathbf{n}_3$ we can calculate the surface magnetization current

$$\mathbf{j}_{ms_3} = \mathbf{M} \times \mathbf{n}_3 = M \, n_y \, \mathbf{u}_x - M \, n_x \, \mathbf{u}_y = \frac{y \, \mathbf{u}_x - x \, \mathbf{u}_y}{\sqrt{x^2 + y^2}} M \cos \alpha. \tag{6.215}$$

Due to the rotational symmetry of the truncated cone is seems to be logical to change to polar coordinates. In fact, substituting $x = \rho \cos \phi$ and $y = \rho \sin \phi$ leads to

$$\mathbf{j}_{ms_3} = \frac{M \cos \alpha}{\rho} (\rho \sin \phi \, \mathbf{u}_x - \rho \cos \phi \, \mathbf{u}_y)$$
$$= -M \cos \alpha (-\sin \phi \, \mathbf{u}_x + \cos \phi \, \mathbf{u}_y) = -M \cos \alpha \, \mathbf{u}_\phi. \tag{6.216}$$

From the definition of the modulus of the surface magnetization current density (6.50) we can write

$$j_{ms_3} = \frac{dI_m}{dl} = -M \cos \alpha \tag{6.217}$$

Considering a very thin slice of material, the magnetic field B created is (6.52) (Fig. 6.57)

Fig. 6.57 Slice of the Truncated cone

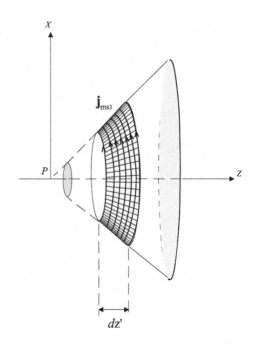

$$d\mathbf{B}(P) = \frac{\mu_0 a^2 dI_m}{2\left[a^2 + (z_0 - z)^2\right]^{3/2}} \mathbf{u}_z, \qquad (6.218)$$

where dI_m is the differential surface magnetization intensity of such a slice. From (6.217) we can write

$$dI_m = \left(\frac{dI_m}{dl}\right) dl = j_{ms_3} dl = -M \cos \alpha \, dl, \qquad (6.219)$$

and substituting into (6.218)

$$d\mathbf{B}(P) = \frac{\mu_0 a^2 j_{ms_3} dl}{2\left[a^2 + (z_0 - z)^2\right]^{3/2}} \mathbf{u}_z. \qquad (6.220)$$

As we can observe, to this formula three variables appear, namely a, z' and l. For performing the calculation we need to express it in terms of only one variable. To reduce the number of unknowns we first see the relation between a and z'

$$\frac{a}{z'} = \tan \alpha \Rightarrow a = z' \tan \alpha, \qquad (6.221)$$

and therefore

$$d\mathbf{B}(P) = \frac{\mu_0 (z' \tan \alpha)^2 j_{ms_3} \, dl}{2\left[(z' \tan \alpha)^2 + (z - z')^2\right]^{3/2}} \mathbf{u}_z. \qquad (6.222)$$

Taking into account that the point where we wish to calculate the magnetic field is the origin of coordinates, we can introduce the value 0 for z. In this case (6.222) reduces to

$$d\mathbf{B}(P) = \frac{\mu_0 (z' \tan \alpha)^2 j_{ms_3} \, dl}{2 \left[(z' \tan \alpha)^2 + z'^2 \right]^{3/2}} \mathbf{u}_z = \frac{\mu_0 (z' \tan \alpha)^2 j_{ms_3} \, dl}{2|z'|^3 (1 + \tan^2 \alpha)^{3/2}} \mathbf{u}_z = \frac{\mu_0 (z' \tan \alpha)^2 j_{ms_3} \, dl}{2|z'|^3 \left(\frac{1}{\cos^3 \alpha} \right)} \mathbf{u}_z.$$
(6.223)

By introducing (6.219) in (6.223)

$$d\mathbf{B}(P) = -\frac{\mu_0 \, z'^2 \sin^2 \alpha \, \cos \alpha \, M \, \cos \alpha \, dl}{2|z'|^3} \mathbf{u}_z = -\frac{\mu_0 \sin^2 \alpha \, M \, \cos^2 \alpha \, dl}{2|z'|} \mathbf{u}_z.$$
(6.224)

Observe that the variable of integration is the length l and not z', then we cannot compute the integral. In order to express (6.224) as a function of one of them, we use the relation to the arc of length,

$$dl = \sqrt{1 + \left(\frac{dx}{dz'} \right)^2} \, dz'.$$
(6.225)

In our case $x = \tan \alpha \, z'$, thus $\left(\frac{dx}{dz'} \right) = \tan \alpha$, and (6.224) converts to

$$\frac{\mu_0 \sin^2 \alpha \, \cos^2 \alpha \, M \, \sqrt{1 + \tan^2 \alpha} \, dz'}{2|z'|} \mathbf{u}_z = -\frac{\mu_0 \sin^2 \alpha \, \cos^2 \alpha \, M \, \left(\frac{1}{\cos \alpha} \right) dz'}{2|z'|} \mathbf{u}_z =$$
$$= -\frac{\mu_0 \sin^2 \alpha \, \cos \alpha \, M \, dz'}{2|z'|} \mathbf{u}_z.$$
(6.226)

Integrating this expression we obtain,

$$\mathbf{B}(P) = -\frac{\mu_0 \, M \, \sin^2 \alpha \, \cos \alpha}{2} \int_b^L \frac{dz'}{|z'|} \mathbf{u}_z.$$
(6.227)

We do not know the values of L and l, however the radii R and r are known. Hence, we can express these limits in the integral as functions of the radii as follows

$$\frac{R}{L} = \tan \alpha \Rightarrow L = \frac{R}{\tan \alpha},$$

and

$$\frac{r}{b} = \tan \alpha \Rightarrow b = \frac{r}{\tan \alpha}.$$

Setting these values in the limits of the integral it leads to

$$\mathbf{B}(P) = -\frac{\mu_0 M \sin^2 \alpha \cos \alpha}{2} \int_{\frac{r}{\tan \alpha}}^{\frac{R}{\tan \alpha}} \frac{dz'}{|z'|} \mathbf{u}_z = -\frac{\mu_0 M \sin^2 \alpha \cos \alpha}{2} \ln \frac{R}{r} \mathbf{u}_z.$$
(6.228)

(b) Second method.

Taking into consideration that $l = \sqrt{a^2 + z'^2}$, and setting into (6.218) $z = 0$, may be written

$$d\mathbf{B}(P) = \frac{\mu_0 a^2 dI_m}{2l^3} \mathbf{u}_z = \frac{\mu_0 a^2 j_{ms_3} \, dl}{2l^3} \mathbf{u}_z.$$
(6.229)

In this equation the denominator does not depend on z' but on l directly. Thus, another possibility to integrate (6.229) is to express it as function of l. However, this equation also depends on a, so l is not the only variable. It therefore remains to find the relation between a and l. Knowing that $a = l \sin \alpha$, we have,

$$d\mathbf{B}(P) = -\frac{\mu_0 (l \sin \alpha)^2 M \cos \alpha \, dl}{2l^3} \mathbf{u}_z = -\frac{\mu_0 M \sin^2 \alpha \cos \alpha \, dl}{2l} \mathbf{u}_z.$$
(6.230)

The integration of this last result between l_1 l_2 leads to

$$\mathbf{B}(P) = -\frac{\mu_0 M \sin^2 \alpha \cos \alpha}{2} \int_{l_1}^{l_2} \frac{dl}{l} \mathbf{u}_z = -\frac{\mu_0 M \sin^2 \alpha \cos \alpha}{2} \ln \frac{l_2}{l_1} \mathbf{u}_z.$$
(6.231)

On the other hand, by introducing $l_1 = \frac{r}{\sin \alpha}$ and $l_2 = \frac{R}{\sin \alpha}$ we have

$$\mathbf{B}(P) = \frac{\mu_0 M \sin^2 \alpha \cos \alpha}{2} \ln \frac{R}{r} \mathbf{u}_z,$$
(6.232)

which coincides with (6.228).

6.22 A ball with radius $R = 10$ cm of a ferromagnetic isotropic material has a homogeneous magnetization $\mathbf{M} = 100{,}000 \, \mathbf{u}_z$ A/m. If the center of gravity of the sphere coincides with origin of the coordinate frame, find: (a) The magnetic field $\mathbf{B}$ at $P(0, 0, 0)$. (b) The magnetic field $\mathbf{H}$.

Solution

(a) A first approach to solve this problem could be to calculate the magnetization density currents and then to apply (6.3). However, this procedure is difficult because of the calculation of the second integral that appears (the first over the volume is zero). Perhaps the easiest way to calculate $\mathbf{B}$ is to try finding an equivalent solenoid of the same geometrical characteristics to the ball, in which we can identify their effects with respect to the magnetic field $\mathbf{B}$. In other words, we follow the same idea shown in the Exercise 6.2 when the field produced by a magnetized finite bar was studied. To do so, we obtain first the $\mathbf{j}_m$ and $\mathbf{j}_{ms}$, i.e.

$$\mathbf{j}_m = \nabla' \times \mathbf{M} = \nabla' \times (0, 0, M) = 0.$$
(6.233)

For the surface magnetization current density we have

$$\mathbf{j}_{ms} = \mathbf{M} \times \mathbf{n}. \tag{6.234}$$

In this case, to calculate it, we need to know the normal $\mathbf{n}$, which will depend on the coordinates (x, y, z). We can determine this vector by means of the gradient of the spherical surface as follows

$$\mathbf{n} = \frac{\nabla S(x,y,z)}{|\nabla S(x,y,z)|}. \tag{6.235}$$

Taking into consideration that a spherical surface of radius R is

$$x^2 + y^2 + z^2 = R^2, \tag{6.236}$$

its gradient has the form

$$S(x,y,z) \equiv x^2 + y^2 + z^2 - R^2 \Rightarrow \nabla S(x,y,z) = 2x\,\mathbf{u}_x + 2y\,\mathbf{u}_y + 2z\,\mathbf{u}_z. \tag{6.237}$$

Dividing by its modulus, we obtain the normal at any point of the sphere

$$\mathbf{n} = \frac{2x\,\mathbf{u}_x + 2y\,\mathbf{u}_y + 2z\,\mathbf{u}_z}{2\sqrt{x^2+y^2+z^2}} = \frac{x\,\mathbf{u}_x + y\,\mathbf{u}_y + z\,\mathbf{u}_z}{\sqrt{x^2+y^2+z^2}} = \frac{x\,\mathbf{u}_x + y\,\mathbf{u}_y + z\,\mathbf{u}_z}{R}. \tag{6.238}$$

Introducing that into (6.234), it leads to

$$\mathbf{j}_{ms} = -M\,n_y\,\mathbf{u}_x + M\,n_x\,\mathbf{u}_y = \frac{M}{R}(-y\,\mathbf{u}_x + x\,\mathbf{u}_y). \tag{6.239}$$

Due to the symmetry of the problem, it seems to be adequate to introduce spherical coordinates, thus we have

$$\mathbf{j}_{ms} = \frac{M}{R}(-R\,\sin\phi\,\sin\theta\,\mathbf{u}_x + R\,\cos\phi\,\sin\theta\,\mathbf{u}_y)$$
$$= M\,\sin\theta(-\sin\phi\,\mathbf{u}_x + \cos\phi\,\mathbf{u}_y) = M\,\sin\theta\,\mathbf{u}_\phi. \tag{6.240}$$

This result tells us that the surface density current is distributed on the spherical surface. By employing (6.50), we can determine dI_m

$$j_{ms} = \frac{dI_m}{dl} = M\,\sin\theta \Rightarrow dI_m = \left(\frac{dI_m}{dl}\right)dl = j_{ms}\,dl = M\,\sin\theta\,dl, \tag{6.241}$$

where dl is the differential arc length on the surface of the sphere. Once we have obtained the value of dI_m, we may try to find the magnetic field at the center of the system.

Fig. 6.58 Homogeneous magnetized sphere

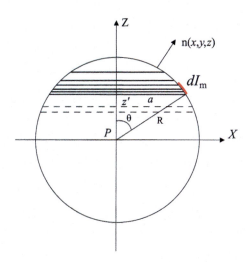

If we look at (6.240) we observe that the geometry of the currents $\mathbf{j}_{ms}$ is similar to that corresponding to coils of different radii located on a sphere of radius R (see Problem 5.28). For this reason, for computing the field $\mathbf{B}$ we can pose the problem as we had a set of coils of variable radii carrying a current that depends on the angle θ (6.241). Formally, the mathematical procedure is as shown in Chap. 5 (however conceptually it is very different). In this sense, we first calculate the field produced by a set of surface magnetization currents dI_m corresponding to a small slice (see Fig. 6.58), and then we integrate for all the system (6.52). By using the expression of the magnetic field created by a coil of radius R (5.85), the differential field produced at the origin of the coordinate frame by the aforementioned slice corresponding to dI_m is

$$d\mathbf{B}(P) = \frac{\mu_0 a^2 dI_m}{2\left[a^2 + z'^2\right]^{3/2}} \mathbf{u}_z = d\mathbf{B}(P) = \frac{\mu_0 a^2 j_{ms} \, dl}{2\left[a^2 + z'^2\right]^{3/2}} \mathbf{u}_z = \frac{\mu_0 a^2 M \sin\theta \, dl}{2\left[a^2 + z'^2\right]^{3/2}} \mathbf{u}_z, \quad (6.242)$$

where we have assumed the same radii for this set of fictitious currents, approximately. Due to the spherical symmetry of the problem we can express z' as a function of θ, then reducing the number of integration variables (Fig. 6.58). Therefore we can write

$$\frac{a}{z'} = \tan\theta \Rightarrow a = z' \tan\theta, \quad (6.243)$$

and introducing it into (6.242) it yields

$$d\mathbf{B}(P) = \frac{\mu_0 (z' \tan\theta)^2 M \sin\theta \, dl}{2\left[(z' \tan\theta)^2 + z'^2\right]^{3/2}} \mathbf{u}_z = \frac{\mu_0 (z' \tan\theta)^2 M \sin\theta \, dl}{2|z'|^3 \left(1 + \tan^2\theta\right)^{3/2}} \mathbf{u}_z =$$

$$\frac{\mu_0 (z' \tan\theta)^2 M \sin\theta \, dl}{2|z'|^3 \left(\frac{1}{\cos^3\theta}\right)} \mathbf{u}_z = \frac{\mu_0 \sin^3\theta M \cos\theta \, dl}{2|z'|} \mathbf{u}_z, \quad (6.244)$$

but $z' = R \cos \theta$, hence

$$d\mathbf{B}(P) = \frac{\mu_0 \sin^3 \theta M \cos \theta \, dl}{2R \cos \theta} \mathbf{u}_z = \frac{\mu_0 \sin^3 \theta M \, dl}{2R} \mathbf{u}_z. \tag{6.245}$$

The arc length may be expressed as a function of the angle θ by using polar coordinates, i.e. $dl = R \, d\theta$, thus

$$d\mathbf{B}(P) = \frac{\mu_0 \sin^3 \theta M \, d\theta}{2} \mathbf{u}_z. \tag{6.246}$$

Introducing it into (6.244) and integrating, we have

$$\mathbf{B}(P) = \int_0^\pi \frac{\mu_0 \sin^3 \theta M \, d\theta}{2} \mathbf{u}_z = \frac{\mu_0 M}{2} \int_0^\pi \sin^3 \theta \, d\theta \mathbf{u}_z = \frac{\mu_0 M}{2} \int_0^\pi \sin \theta (1 - \cos^2 \theta) \, d\theta \mathbf{u}_z =$$
$$\frac{\mu_0 M}{2} \left\{ \int_0^\pi \sin \theta \, d\theta - \int_0^\pi \sin \theta \cos^2 \theta \, d\theta \right\} \mathbf{u}_z = \frac{2}{3} \mu_0 M \mathbf{u}_z. \tag{6.247}$$

Observe that this problem can also be directly solved by using methods shown in Chap. 5, yielding the same result. If we remember, when we studied the homogeneous magnetized bar (Problems 6.1 and 6.2) we saw that the field created by the rod was the same as the field created by an equivalent solenoid of the same dimensions as the rod, in which the free current per unit length verifies that $nI = M$. To some extent the basic idea of the procedure consisted of finding a system of currents (solenoid) that produces identical effects regarding $\mathbf{B}$. In that case, we were able to calculate the field produced by the magnetized body in a simple way, without resolving differential equations or difficult integrals.

This idea is not restricted to a rod. On the contrary, as we are going to see, this method may also be used in this problem. In general, if we construct a model in which it is possible to identify the magnetization currents of the matter with the free currents of its equivalent system, we will be able to solve the problem as we had no matter. Though we could find such a model, as we can suppose, the procedure is only easy if we have some symmetries. In our case we have studied a magnetized sphere, thus, we can likely find an easy model.

In fact, in Problem 5.28 we calculated the magnetic field created by a coil of N turns per unit arc η (constant) wound on a spherical surface. There we saw that, setting dN as a function of η, we obtained the field $\mathbf{B}$ at the center of the spherical solenoid. If we compare (5.256) with (6.246) we see that both expressions are identical if $I\eta = M \sin \theta = j_{ms}$. Therefore, introducing the values of $I\eta$ into (5.256) we have

$$d\mathbf{B}(P) = d\mathbf{B}(P) = \frac{\mu_0 \sin^2 \theta I \eta d\theta}{2} \mathbf{u}_z = \frac{\mu_0 \sin^2 \theta M \sin \theta d\theta}{2} \mathbf{u}_z, \tag{6.248}$$

and integrating it gives the same value as shown in (6.248). This result means that if we construct a spherical solenoid or radius R carrying an intensity *per unit arc*

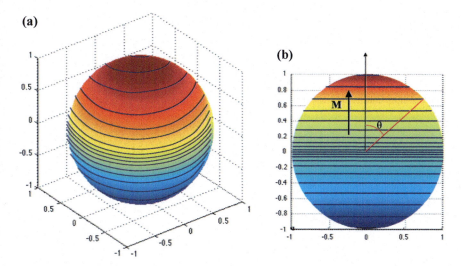

Fig. 6.59 a Homogeneous magnetized sphere. Notice that due to $j_{ms}(\theta) = M \sin\theta$ the fictitious current per unit length accumulates more near the equator $\theta = \frac{\pi}{2}$. **b** Front view

$I\eta = M \sin\theta$, it will produce the same magnetic field **B** that those corresponding to a sphere homogeneously magnetized.

It is important to note that, in this case, contrary to what occurred with the homogeneous magnetized bar, we cannot directly substitute $I\eta = M \sin\theta$ into the final result (5.257). The reason is that in the case of the rod the surface magnetization current density *is constant* ($j_{ms} = M$), and therefore we can replace nI by M in the final equation (5.99). However, in the present problem j_{ms} is a function of the angle θ ($j_{ms}(\theta) = M \sin\theta$), then it must be introduced in (6.242) in order to be integrated.[14] Physically, this dependence of the surface magnetized currents on the direction, that is $j_{ms}(\theta) = \frac{dI_m}{dl} = M \sin\theta$, means that the fictitious currents per unit length are not homogeneously distributed over the surface of the ball. They concentrate more in the proximity of $\theta = \frac{\pi}{2}$ (zero meridian-equator) where the sine function is one, and decreases when $\theta \to 0$ and $\theta \to \pi$ (the poles). See Fig. 6.59.

6.23 A very large cylinder of radius R, is magnetized $\mathbf{M} = K\rho^2\,\mathbf{u}_\phi$ A/m, where K is a constant and ρ represents the distance between a generic point and the symmetry axis of the cylinder. Find: (a) The volumetric and surface magnetization currents. (b) The fields $\mathbf{H}_\phi$ and $\mathbf{B}_\phi$ at any point of space. (c) Sketch a graph of the fields as a function of distance to the revolution axis of the system.

Solution

(a) We can get the magnetization density currents by applying definitions (6.4) and (6.5), i.e.

---

[14]Note that the introduction of $I\eta = M \sin\theta$ into (5.257) gives $\mathbf{B}(P) = \frac{\pi\mu_0 I\eta}{4}\mathbf{u}_z = \frac{\pi\mu_0 M \sin\theta}{4}\mathbf{u}_z$ (*false!*), a result very different from what we obtained in (6.247).

Fig. 6.60 a Non-homogeneous magnetized cylinder. Notice that M increases with the distance to the axis of symmetry (OZ). **b** Magnetization density currents

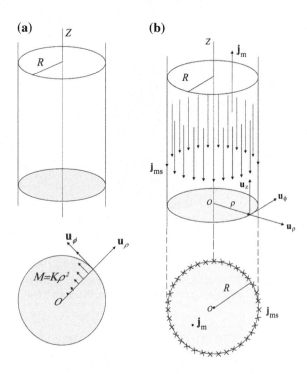

$$\mathbf{j}_m = \nabla' \times \mathbf{M} = \frac{1}{\rho}\frac{\partial}{\partial \rho}(\rho M_\phi)\mathbf{u}_z = 3K\rho\,\mathbf{u}_z. \quad (6.249)$$

and for the surface density current,

$$\mathbf{j}_{ms} = \mathbf{M} \times \mathbf{n}' = (0, M_\phi, 0) \times (1, 0, 0) = -K R^2\,\mathbf{u}_z. \quad (6.250)$$

These results show that the cylinder not only has $\mathbf{j}_{ms}$ like in the Problems 6.1 and 6.2, but also volumetric magnetization density current. The reason for this is found in the existence of a non-homogeneous magnetization of the body (Fig. 6.60).

First method

(b) ($\rho < R$) In Chap. 5 we devoted some time to explain in detail the utility of studying the symmetries of a problem (see for example Problem 5.1). With the same idea, we first analyze the symmetries of our system and then we try to solve it.

In the present problem the bar is very large, thus following what was explained previously we conclude that the magnetized rod has rotational and translational symmetry. It means we may employ the Ampère theorem for solving the question. However an important question arises, namely which of the Ampère laws interests us. In effect, in the preceding Chapter we studied Ampère's integral theorem for the magnetic field **B** (5.7). We saw that the circulation of **B** along a closed curve Γ only depends of the net current crossing the open surface whose delimitation is Γ. At that

Solved Problems

time we knew neither about magnetization nor about the existence of volumetric and surface magnetization currents, so we identified the net current I with the current of free charge I_f. Now we know that this point of view does not correspond with the general case B (see (6.15)), in which $\oint_\Gamma \mathbf{B} \cdot d\mathbf{l}$ depends not only on I_f but on I_m and I_{ms} as well. On the other hand in this chapter we have defined the magnetic field $\mathbf{H}$, whose sources are either currents of free charge, or regions in which $\nabla \cdot \mathbf{M} \neq 0$ (for the magnetostatic field). Hence, why should we use (5.7) or (6.15)? In principle we can employ either of the two equations, but the exact procedure depends on the specific problem. So, if in addition to magnetic materials we have currents of free charge, and we can neglect the demagnetizing fields, if any, it is usually easier to begin with Ampère's theorem for the magnetic field H. Later, as this field is known we get B by means of (6.12). The reason to proceed in this way is that the theorem (6.14) is directly related with I_f. On the contrary, if we start with $\oint_\Gamma \mathbf{B} \cdot d\mathbf{l}$ we must work with the magnetization currents, which are not known a priori.

In our case we have neither conduction currents, nor demagnetizing fields (since the cylinder is very long). On the other hand, we have calculated the magnetization currents. Hence we can apply both equations and, as we will demonstrate, we obtain the same result.

In fact, for obtaining the magnetic field $\mathbf{B}$ in the interior of the cylinder, due to its rotational symmetry, we choose a circular curve of radius $\rho < R$ and directly apply (6.15). Setting $d\mathbf{l} = (0, dl, 0)$ so that the surface S is oriented in the positive direction of the OZ axis, we have

$$\oint_\Gamma \mathbf{B} \cdot d\mathbf{l} = \mu_0 \left\{ \int_S \mathbf{j}_f \cdot d\mathbf{S} + \int_S \mathbf{j}_m \cdot d\mathbf{S} + \oint_\Gamma \mathbf{j}_{ms} \cdot \mathbf{n} dl \right\}. \tag{6.251}$$

Integrating the first member leads to

$$\oint_\Gamma (B_\rho, B_\phi, B_z) \cdot (0, dl, 0) = \oint_\Gamma B_\phi \, dl = \int_0^{2\pi} B_\phi \rho \, d\phi = 2\pi \rho B_\phi. \tag{6.252}$$

For the second part of (6.251), as the curve Γ does not reach the surface magnetization density current $\mathbf{j}_{ms}$ does not appear, then only $\mathbf{j}_m$ matters because we do not have $\mathbf{j}_f$.

$$\int_{S \cap V_j} \mathbf{j}_m \cdot d\mathbf{S} = \int_{S_\rho} 3K\rho \mathbf{u}_z \cdot d\mathbf{S} \, \mathbf{u}_z = 3K \int_0^{\rho<R} \int_0^{2\pi} \rho^2 \, d\rho \, d\phi =$$

$$3K \int_0^{2\pi} d\phi \int_0^{\rho<R} \rho^2 \, d\rho = 2\pi K \rho^3 \tag{6.253}$$

$$B_\phi = \mu_0 K \rho^2. \tag{6.254}$$

Once we have the magnetic field $\mathbf{B}$ we may determine $\mathbf{H}$ through (6.12) because we know the value of the magnetization $\mathbf{M}$,

$$\mathbf{H} = \frac{\mathbf{B}}{\mu_0} - \mathbf{M} = \frac{\mu_0 K \rho^2}{\mu_0}\mathbf{u}_\phi - K\rho^2 \mathbf{u}_\phi = 0. \quad (6.255)$$

Second method

(b*) ($\rho < R$) Now we try to obtain the same results but from another point of view. Let us write (6.15) as a function of the magnetization. In this case we have

$$\oint_{\partial S} \mathbf{B} \cdot d\mathbf{l} = \mu_0 \left\{ \int_S \mathbf{j}_m \cdot d\mathbf{S} + \oint_\Gamma \mathbf{j}_{ms} \cdot \mathbf{n} dl \right\} = \mu_0 \int_S (\nabla \times \mathbf{M}) \cdot d\mathbf{S} + \mu_0 \oint_\Gamma (\mathbf{M} \times \mathbf{n}') \times \mathbf{n} dl,$$

in which we related the magnetic field B with the magnetization M. As we have previously seen for $\rho < R$, only the volumetric magnetization currents are in place, thus we can rewrite

$$\oint_{\partial S} \mathbf{B} \cdot d\mathbf{l} = \mu_0 \int_S (\nabla \times \mathbf{M}) \cdot d\mathbf{S}. \quad (6.256)$$

By applying Stokes's theorem we convert this double integration into a curvilinear integral along the delimiting curve of the surface S

$$\oint_\Gamma \mathbf{B} \cdot d\mathbf{l} = \mu_0 \oint_\Gamma \mathbf{M} \cdot d\mathbf{l}. \quad (6.257)$$

By introducing $\mathbf{M} = \mathbf{M}_\phi = K\rho^2 \mathbf{u}_\phi$ into (6.256), we have

$$\oint_\Gamma (B_\rho, B_\phi, B_z) \cdot d\mathbf{l} = \mu_0 \int_\Gamma (0, M_\phi, 0) \cdot d\mathbf{l} \Rightarrow \oint_\Gamma (B_\rho, B_\phi, B_z)(0, dl, 0) = \mu_0 \oint_\Gamma (0, K\rho^2, 0)(0, dl, 0) \Rightarrow \quad (6.258)$$

$$\oint_\Gamma B_\phi dl = \mu_0 \oint_\Gamma K\rho^2 dl \Rightarrow \int_0^{2\pi} B_\phi \rho d\phi = \mu_0 \int_0^{2\pi} K\rho^2 \rho d\phi \Rightarrow \quad (6.259)$$

$$B_\phi \rho \int_0^{2\pi} d\phi = \mu_0 K \rho^3 \int_0^{2\pi} d\phi \Rightarrow$$

$$B_\phi 2\pi \rho = 2\pi \mu_0 K \rho^3 \Rightarrow B_\phi = \mu_0 K \rho^2,$$

and in vectorial form

$$\mathbf{B}_\phi = \mu_0 K \rho^2 \mathbf{u}_\phi. \quad (6.260)$$

obtaining the same result. Referring to field H, there are no differences in the procedure for the calculation with respect to those shown in section (b).

First method

($\rho > R$)

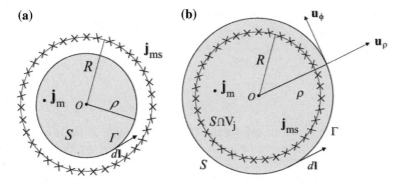

Fig. 6.61 a Non-homogeneous magnetized cylinder. Notice that M grows with distance from the axis of symmetry (OZ). **b** Magnetization density currents

Returning to the first procedure seen in (b), we calculate the fields for $\rho > R$. In this case the curve chosen is also circular but the radius ρ is greater than R. As we show in Fig. 6.61b, the surface S is crossed by $\mathbf{j}_m$ and $\mathbf{j}_{ms}$, thus we can write

$$\oint_\Gamma \mathbf{B} \cdot d\mathbf{l} = \mu_0 \left\{ \int_S \mathbf{j}_m \cdot d\mathbf{S} + \oint_\Gamma \mathbf{j}_{ms} \cdot \mathbf{n} dl \right\} \quad (6.261)$$

Taking into consideration that outside the cross section of the cylinder isn't any magnetization currents (Fig. 6.61b), the effective surface of integration is not $S = \pi \rho^2$ but $S \cap V_j = S_R$, then it gives

$$\int_{S \cap V_j} \mathbf{j}_m \cdot d\mathbf{S} = \int_{S_R} 3K\rho \, \mathbf{u}_z \cdot d\mathbf{S} \, \mathbf{u}_z = 3K \int_0^R \int_0^{2\pi} \rho^2 \, d\rho \, d\phi = 2\pi K R^3 \quad (6.262)$$

Due to the surface magnetization density current flowing through the surface of the cylinder, it seems to be adequate employing the Dirac's delta distribution (if we want to work with dS, and the Dirac delta, instead of dl). If we put $\mathbf{j}_{ms} = -K R^2 \delta(\rho - R) \, \mathbf{u}_z$ into (6.15), we obtain

$$\int_{S \cap V_j} \mathbf{j}_{ms} \cdot d\mathbf{S} = \int_0^{\rho > R} \int_0^{2\pi} -KR^2 \, \delta(\rho - R) \, \mathbf{u}_z \cdot d\mathbf{S} \, \mathbf{u}_z = \quad (6.263)$$

$$= -K \lim_{a \to \infty} \int_0^{a(\rho > R)} \int_0^{2\pi} R^2 \, \delta(\rho - R) \, \rho \, d\rho \, d\phi = -2\pi K R^3,$$

and all together[15]

$$2\pi B_\phi = 2\pi R^3 - 2\pi R^3 = 0 \Rightarrow B_\phi = 0. \quad (6.264)$$

---

[15] In the former calculation appears $\lim_{a \to \infty} \int_0^a \ldots$ This is only in order to be mathematically correct by definition of the Dirac delta distribution.

The magnetic field H once again by means of (6.12),

$$\mathbf{H} = \frac{\mathbf{B}}{\mu_0} - \mathbf{M} = 0 - 0 = 0. \tag{6.265}$$

Second method

(b*) ($\rho > R$)

As the region in which we are going to obtain the magnetic fields is outside of the bar, we again choose a circular curve Γ with radius $\rho > R$ together with a plane surface S for applying Ampère's theorem (see Fig. 6.61a), which may be now written in the following form

$$\oint_\Gamma \mathbf{B} \cdot d\mathbf{l} = \mu_0 \oint_\Gamma \mathbf{M} \cdot d\mathbf{l} + \oint_\Gamma (\mathbf{M} \times \mathbf{n}') \times \mathbf{n} dl. \tag{6.266}$$

Upon calculating the integrals, a doubt arises. In fact, we have seen that the integration of the magnetic field $\mathbf{B}$ is carried out over the curve with radius ρ, independent of the fact that at points belonging up Γ there is current density. On the other hand, the calculation of the surface integrals (6.15) are taken through the corresponding effective surface $S_{ef} = S \cap V_{j_m}$, but now we have converted $\int_S (\nabla \times \mathbf{M}) \cdot d\mathbf{S}$ into a line integral, and the question is whether or not the integration curve is also Γ as for the magnetic field $\mathbf{B}$. To answer this question we must remember that before performing Stokes' theorem over $\nabla \times \mathbf{M}$, its double integral extends over the effective surface $S_{ef} = S \cap V_{j_m}$, the integration curve for $\mathbf{M}$ in (6.266) must be the corresponding boundary of $S_{ef} = \pi R^2 = S_R$. The same reasoning holds for $\int_S (\mathbf{M} \times \mathbf{n}) \cdot d\mathbf{S}$ (see 6.262). The density $\mathbf{j}_{ms}$ flows on the lateral surface of the magnetized rod, then the product $(\mathbf{M} \times \mathbf{n}) \cdot d\mathbf{S}$ (actually $(\mathbf{M} \times \mathbf{n}') \times \mathbf{n} dl$. Look at the physical units) is distinct from zero where $\mathbf{j}_{ms}$ and $d\mathbf{S}$ intersect, which in this case is not actually a surface but the curve $V_{j_{ms}} \cap S = S_\ell = \ell = 2\pi R = \Gamma_\ell$.

Adequately setting the integration limits in (6.266) we can write

$$\oint_\Gamma \mathbf{B} \cdot d\mathbf{l} = \mu_0 \left\{ \oint_{\Gamma_R} \mathbf{M} \cdot d\mathbf{l} + \int_{S_\ell} (\mathbf{M} \times \mathbf{n}) \cdot d\mathbf{S} \right\},$$

Γ_R being the boundary of S_R

$$\oint_\Gamma (B_\rho, B_\phi, B_z)(0, dl, 0) = \mu_0 \left\{ \oint_{\Gamma_R} (0, K\rho^2, 0)(0, dl, 0) + \int_{S_\ell} (0, 0, -K\rho^2)(0, 0, dS) \right\}.$$

As we have already commented, actually the last integral does not extend over a surface but a line (otherwise the units have no sense). For this reason we can express $\int_S (-K\rho^2) \mathbf{u}_z \, dS \, \mathbf{u}_z$ by means of Dirac's delta distributions. In effect, we may write $\mathbf{j}_{ms} = -K\rho^2 \delta(\rho - R) \mathbf{u}_z$, and introducing it in the expression above

$$\int_{S_\ell} (\mathbf{M} \times \mathbf{n}) \cdot d\mathbf{S} = \int_S (-K\rho^2) \delta(\rho - R) \mathbf{u}_z \, dS \, \mathbf{u}_z = -\int_S K\rho^2 \delta(\rho - R) \, dS. \tag{6.267}$$

Fig. 6.62 Tangential magnetic field B versus distance. Observe that no field exist out of the cylinder

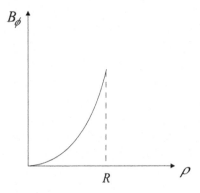

Changing to polar coordinates and taking into account that to do the integration along the curve Γ_R, the magnetization must be particularized for $\rho = R$, thus we have

$$\int_0^{2\pi} B_\phi \rho d\phi = \mu_0 \left\{ \int_0^{2\pi} K R^2 R d\phi - \int_0^\infty K \rho^2 \delta(\rho - R) \rho d\rho \int_0^{2\pi} d\phi \right\} \Rightarrow$$

$$B_\phi \rho \int_0^{2\pi} d\phi = \mu_0 \left\{ K R^3 \int_0^{2\pi} d\phi - K R^3 \int_0^{2\pi} d\phi \right\} = 0 \Rightarrow$$

$$B_\phi = 0,$$

which agrees with (6.254). Referred to H the procedure is the same as shown in the first method.

(c) Taking into consideration all results of the preceding sections, the magnetic field B_ϕ as a function of distance ρ has the form shown in Fig. 6.62.

6.24 Let us suppose we will construct a permanent magnet whose iron core length and cross-section are l_i and S_i, respectively, whereas for the air gap those magnitudes are l_g and S_g. Find: (a) The operating straight length of the iron core. (b) The minimum volume of ferromagnetic material we need for reaching a magnetic field B_i in the air gap. (c) Under the conditions of section (b), how does the iron volume increase if we wish to increase the magnetic field H_g?, what implications does that have?

Solution

(a) By applying the Ampère theorem over the mean length of the electromagnet it yields

$$H_h l_h + H_g l_g = 0 \Rightarrow H_h l_h + \frac{B_g l_g}{\mu_0} = 0. \qquad (6.268)$$

On the other hand, from the flux of the magnetic field B throughout a closed surface containing both an iron core and an air-gap we obtain $B_i S_i = B_g S_g$, and then

$$H_i l_i + \frac{B_i S_i}{\mu_0 S_g} l_g = 0, \tag{6.269}$$

that is to say

$$B_i = -\frac{\mu_0 l_i S_g}{S_i l_g} H_i. \tag{6.270}$$

This result is the same as we saw in (6.31).

(b) Starting from the equality, $B_i S_i = B_g S_g$, and considering the material equation for the air gap $B_g = \mu_0 H_g$, we can write

$$B_i S_i = B_g S_g = \mu_0 H_g S_g. \tag{6.271}$$

From Ampère's law (6.268) we know

$$H_i l_i = -H_g l_g. \tag{6.272}$$

By multiplying (6.271) and (6.272) we have

$$B_i S_i H_i l_i = \mu_0 H_g S_g H_g l_g. \tag{6.273}$$

But $S_i l_i$ and $S_g l_g$ are the volume of the iron core V_i and the gap v_g, respectively, therefore (6.273) leads to

$$B_i H_i V_i = -\mu_0 H_g^2 v_g, \tag{6.274}$$

and the volume V_i

$$V_i = \frac{\mu_0 H_g^2 v_g}{B_i H_i}. \tag{6.275}$$

Equation (6.275) provides a reasonable value for the volume of the permanent magnet supposing the leakage of the magnetic flux negligible. On the other hand, (6.275) shows that the minimum volume material corresponds to the maximum of the product $B_i H_i$. Let us find the operating point corresponding to this minimum volume of material.

As $B_i H_i$ must be a maximum, we differentiate such a product and we impose the condition to be an extreme,

$$d(B_i H_i) = B_i dH_i + H_i dB_i = 0. \tag{6.276}$$

But we know that $dB_i = (\frac{dB_i}{dH_i}) dH_i$, so introducing this expression into (6.276) it yields

$$B_i dH_i + H_i \left(\frac{dB_i}{dH_i}\right) dH_i \Rightarrow \left(B_i + H_i \left(\frac{dB_i}{dH_i}\right)\right) dH_i = 0. \tag{6.277}$$

To fulfill this result the term in the bracket must be zero

Fig. 6.63 First and a part of the second hysteresis curve. Observe that, due to the air-gap the operating point of the permanent magnet is located in the second quadrant of the loop

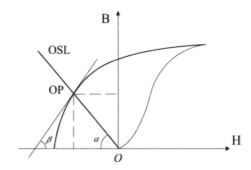

$$B_i + H_i \left(\frac{dB_i}{dH_i}\right) = 0 \Rightarrow \frac{dB_i}{dH_i} = -\frac{B_i}{H_i}. \tag{6.278}$$

where H_i and B_i are the coordinates of the operating point OP. From the Fig. 6.63 we see that

$$\frac{B_i}{H_i} = \tan \alpha.$$

On the other hand the derivative $\frac{dB_i}{dH_i}$ represents the straight line tangent at OP, whose slope is $\tan \beta$ (see Fig. 6.63), hence it holds

$$\frac{dB_i}{dH_i} = \tan \beta = -\frac{B_i}{H_i} = -\tan \alpha \Rightarrow \tan \beta = -\tan \alpha \Rightarrow \alpha = \beta. \tag{6.279}$$

This result means that when both angles are equal, the volume needed to construct the electromagnet is minimum. This angle α depends only of the geometrical characteristics of the magnetic circuit. In fact, employing (6.270) we have

$$\frac{B_i}{H_i} = -\frac{\mu_0 \, l_i \, S_g}{S_i l_g} = -\tan \alpha, \tag{6.280}$$

which is a function of the lengths and cross sections of both the iron core and the air-gap.

(c) The basic characteristics in the design of a permanent magnet depend on the applications for which it must be used. However, in general, for a specific volume of the air-gap v_g, from (6.275) we see that the volume of ferromagnetic material V_i needed for the construction of a permanent magnet increases proportional to the square of the magnetic field H_i in the air-gap. It means that the price of the device grows very much with the field we must reach in this gap.

6.25 The electromagnet of the attached Fig. 6.64 was constructed with a material for which the first magnetization curve is $H = KB^2$. The cross section of the system is circular, with area of constant value S_1 along the length l_1, and it narrows to $S_1/2$ in the polar pieces of lengths l_2. The values of the different parts of

Fig. 6.64 Electromagnet of variable cross section

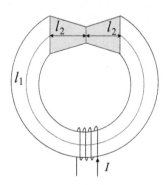

Fig. 6.65 Electromagnet of variable section

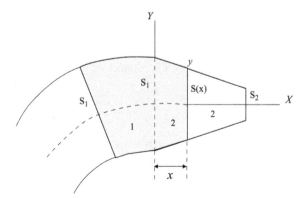

the electromagnet are: $l_1 = 0.60$ m, $l_2 = 0.10$ m; $K = 10{,}000\,\text{Am}^{-1}\,\text{T}^{-2}$ and $NI = 1000$ A. Obtain the magnetic field B_1 in the longer part of the material. There is no flux leakage.

Solution

As we have seen in other problems, we proceed in three steps. Firstly, we apply the Ampère law along the mean length of the material,

$$\oint_\Gamma \mathbf{H} d\mathbf{l} = I_c \Rightarrow H_1 l_1 + 2 \int_0^{l_2} H_2\, dl = NI. \tag{6.281}$$

Although the material is the same, we cannot write $H_2 l_2$ for the polar pieces, since the sections vary. As a result, the field H_2 must be different from H_1, in principle. The second step is to use the equality of fluxes over a closed surface. As we need to relate the field in the two parts of the system, we choose a surface which encompasses them (Fig. 6.65). Observe that this closed surface does not reach the right side S_2. The reason for this is that the field H in region 2 depends on x, thus we must leave the flux equation as a function of a generic point within region 2. Applying the integral equation for the flux of B over this surface, we have

Fig. 6.66 Electromagnet of variable section

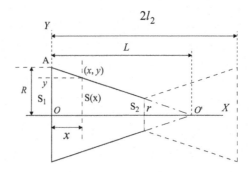

$$\oint_S \mathbf{B} d\mathbf{S} = 0 \Rightarrow B_1 S_1 = \int\int_{S(x)} B(x) \, dS = B(x) S(x) = B(x) \pi y^2, \quad (6.282)$$

and then, the magnetic field on the right side at a distance x is

$$B_1 = B(x) \frac{\pi y^2}{\pi R^2} = B(x) \frac{y^2}{R^2} \Rightarrow B(x) = B_1 \frac{R^2}{y^2}. \quad (6.283)$$

This equation depends on the height y at point x inside the polar piece of length l_2, but in (6.283) x does not appear, then we need to find the magnetic field B as function of x. With this aim, let us look at Fig. 6.66. On the plane of this figure we can identify two similar triangles, namely AOO' and yxO'. For them it holds

$$\frac{R}{L} = \frac{y}{L - x} \Rightarrow y = \frac{R(L - x)}{L}. \quad (6.284)$$

The difficulty of this result is that we do not know the value of L, however we can deduce it from the initial data of the problem. In fact, as the surface $S_2 = \frac{S_1}{2}$, we may write

$$\pi r^2 = \frac{1}{2} \pi R^2 \Rightarrow R = \sqrt{2} \, r, \quad (6.285)$$

which represents the relation between the radii of the polar piece surfaces. On the other hand, as equation (6.284) is valid in the interval $(0 < x < L)$, we particularize it for $y = r$ obtaining

$$r = \frac{R(L - l_2)}{L} \Rightarrow r = \frac{\sqrt{2}\, r(L - l_2)}{L} \Rightarrow L = \frac{\sqrt{2}}{\sqrt{2} - 1} l_2 = 3.4 \, l_2 = 0.34 \, \text{m}, \quad (6.286)$$

and then

$$y = \frac{R(0.34 - x)}{0.34}. \quad (6.287)$$

Introducing (6.287) in (6.283) gives

Fig. 6.67 Electromagnet of variable section

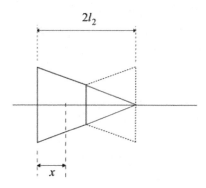

$$B(x) = B_1 \frac{R^2}{\left(\frac{R(L-x)}{L}\right)^2} = B_1 \left(\frac{0.34}{(0.34-x)}\right)^2, \quad (6.288)$$

which corresponds to the magnetic field in the polar piece (denoted by 2 in Fig. 6.65). Now, taking into account that we know the the first magnetization curve $B = B(H)$, we introduce it into (6.281) resulting in (third step)

$$H_1 l_1 + 2 \int_0^{l_2} K B_2^2(x) \, dl = NI, \quad (6.289)$$

and through (6.283), we have

$$H_1 l_1 + 2 \int_0^{l_2} K B_1^2 \left(\frac{L}{L-x}\right)^4 dl = NI \Rightarrow H_1 l_1 + 2 \int_0^{l_2} K B_1^2 \left(\frac{L}{L-x}\right)^4 dl = NI. \quad (6.290)$$

Under the basic approximation made for solving magnetic circuits, the average magnetic field B_1 along the length l_1 is the same, hence we can write (Fig. 6.67)

$$H_1 l_1 + 2K B_1^2 L^4 \int_0^{l_2} \left(\frac{1}{L-x}\right)^4 dl = NI \Rightarrow H_1 l_1 + 2K B_1^2 L^4 \frac{1}{3}\left(\frac{1}{(L-l_2)^3} - \frac{1}{L^3}\right) = NI. \quad (6.291)$$

This expression represents the operating straight length (OSL) of the electromagnet. For finding the operating point we must calculate the intersection of (6.290) with the magnetization curve $H = K B^2$. To accomplish this we substitute H_1 by $K B_1^2$, getting

$$K B_1^2 l_1 + 2K B_1^2 L^4 \frac{1}{3}\left(\frac{1}{(L-l_2)^3} - \frac{1}{L^3}\right) = NI, \quad (6.292)$$

and then we have

$$B_1 = \sqrt{\frac{NI}{K\left[l_1 + \frac{2L^4}{3}\left(\frac{1}{(L-l_2)^3} - \frac{1}{L^3}\right)\right]}} = 0.31 \, \text{T}. \quad (6.293)$$

6.26 The system shown in the figure below is constructed of the same ferromagnetic material and comprises a hollow cylinder of radius $R_1 = 4$ cm and $R_2 = 6$ cm, height $h = 10$ cm, and base thickness of $d = 1$ cm. This cylinder is traversed in the direction of its symmetry axis by a slender bar of radius $R = 1$ cm, on which a coil of $N = 100$ turns is wound. Assuming there is no magnetization saturation, the first magnetization curve, excluding the origin and its near neighboring points, is

$$B = aH^2, \quad 0 < H \leq H_{max},$$

where $a = 1 \times 10^{-5}$ Tm^{2}A^{-2}. If the current circulating through the wires is $I = 0.5$ A, and the leakage of flux is negligible, obtain the magnetic field B at point P in the bar (region 1) and in the external region delimited between R_1 and R_2 (region 3).

Solution

In the same manner that we have seen in other problems, we begin with the application of Ampère's theorem. To do so we choose a curve Γ as shown in Fig. 6.68,

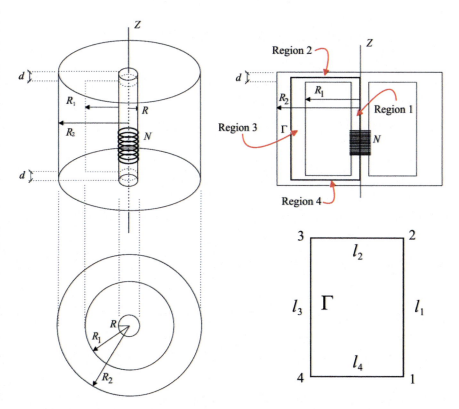

Fig. 6.68 Front and plane view of the system. On the right the curve Γ. Observe that l_1 and l_3 are equal and the same occurs between l_2 and l_4

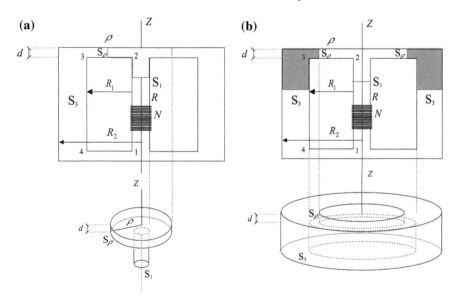

Fig. 6.69 Surfaces for calculating the magnetic flux. **a** This closed surface relates the magnetic flux across the surfaces S_1 and S_ρ. **b** Idem for the surfaces S_ρ and S_3

$$\oint_\Gamma \mathbf{H} d\mathbf{l} = I_f \Rightarrow H_1 l_1 + \int_2^3 H\, dl + H_3 l_3 + \int_4^1 H\, dl = NI. \quad (6.294)$$

To find the flux we first choose a closed curve as depicted in Fig. 6.69. Supposing that there is no flux leakage we can write

$$\oint_S \mathbf{B} d\mathbf{S} = 0 \Rightarrow B_1 S_1 = B_\rho S_\rho \Rightarrow B_1 \pi R^2 = B_\rho 2\pi \rho d \Rightarrow B_r = B_1 \frac{R^2}{2\rho d}. \quad (6.295)$$

At the same time we must relate the magnetic field B_1 with B_3. With this end, we can again use the flux law for relating region 3 (surface S_3) with the intermediate zone between 1 and 3 (surface S_ρ) (see Fig. 6.69), i.e. the part ρ. Thus, it results that

$$\oint_S \mathbf{B} d\mathbf{S} = 0 \Rightarrow B_\rho S_\rho = B_3 S_3 \Rightarrow B_\rho 2\pi\rho d = B_3 \pi (R_2^2 - R_1^2) \Rightarrow B_\rho = B_3 \frac{(R_2^2 - R_1^2)}{2\rho d}, \quad (6.296)$$

which gives B_r as a function of B_3. Once we know (6.296), we can obtain the relation between B_1 and B_3 by means of (6.295)

$$B_3 \frac{(R_2^2 - R_1^2)}{2\rho d} = B_\rho = B_1 \frac{R^2}{2\rho d} \Rightarrow B_3 = B_1 \frac{R^2}{(R_2^2 - R_1^2)}. \quad (6.297)$$

Now, setting this latter equation into (6.295) it yields

$$H_1 l_1 + 2 \int_2^3 H \, dl + H_3 l_3 = NI$$

$$\Rightarrow \sqrt{\frac{B_1}{a}} l_1 + 2 \int_0^{\frac{(R_1+R_2)}{2}} R \sqrt{\frac{B_1}{2a\rho d}} \, d\rho + R \sqrt{\frac{B_1}{a(R_2^2 - R_1^2)}} l_3 = NI$$

$$\Rightarrow \sqrt{\frac{B_1}{a}} l_1 + 2R \sqrt{\frac{B_1}{2ad}} \int_0^{\frac{(R_1+R_2)}{2}} \frac{d\rho}{\sqrt{\rho}} + R \sqrt{\frac{B_1}{a(R_2^2 - R_1^2)}} l_3 = NI \qquad (6.298)$$

Observe that the second term of the addition in (6.298) is an integral instead a single product. This is due to the fact that over the segment 2–3 (l_2) and 4–1 (l_4) the magnetic field H varies with distance ρ to OZ axis. Along the segments 1–2 (l_1) and 3–4 (l_3) the field H remains constant (mean average line-see Fig. 6.68).

By integrating,

$$\sqrt{\frac{B_1}{a}} l_1 + 4R \sqrt{\frac{B_1}{2ad}} [\sqrt{r}]_0^{\frac{(R_1+R_2)}{2}} + R \sqrt{\frac{B_1}{a(R_2^2 - R_1^2)}} l_3 = NI \Rightarrow$$

$$\sqrt{\frac{B_1}{a}} l_1 + 4R \sqrt{\frac{B_1}{2ad}} \sqrt{\frac{(R_1 + R_2)}{2}} + R \sqrt{\frac{B_1}{a(R_2^2 - R_1^2)}} l_3 = NI \Rightarrow$$

$$\sqrt{\frac{B_1}{a}} \left\{ l_1 + \frac{2R}{\sqrt{d}} \sqrt{(R_1 + R_2)} + \frac{R}{\sqrt{(R_2^2 - R_1^2)}} l_3 \right\} = NI \qquad (6.299)$$

$$B_1 = \frac{a(NI)^2}{\left\{ l_1 + \frac{2R}{\sqrt{d}} \sqrt{(R_1 + R_2)} + \frac{R}{\sqrt{(R_2^2 - R_1^2)}} l_3 \right\}^2} = 0.83 \, \text{T} . \qquad (6.300)$$

To calculate the magnetic field B_3 (segment 2–3) corresponding to the exterior region ($R_1 < r < R_2$), we employ (6.297)

$$B_3 = \left(\frac{a(NI)^2}{\left\{ l_1 + \frac{2R}{\sqrt{d}} \sqrt{(R_1 + R_2)} + \frac{R}{\sqrt{(R_2^2 - R_1^2)}} l_3 \right\}^2} \right) \frac{R^2}{(R_2^2 - R_1^2)} = 0.04 \, \text{T} . \qquad (6.301)$$

6.27 A cube of side L has magnetization **M** uniform and parallel to an edge. Calculate the magnetic field **B** at an exterior point P, situated at a distance d ($d \ll L$) from one of the faces for which **B** is tangent, and is separated from the edges.

Solution

There are no conduction currents, and therefore $\mathbf{j}_c = 0$.

The magnetization **M** is uniform inside the cube, and hence does not depend on the coordinates of the point. Its curl is therefore zero and the current of volumetric

Fig. 6.70 Magnetized cube

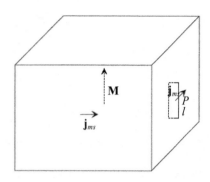

magnetization is

$$\mathbf{j}_m = \nabla \times \mathbf{M} = 0$$

The surface current density of magnetization is

$$\mathbf{j}_{ms} = \mathbf{M} \times \mathbf{n}$$

Hence $\mathbf{j}_{ms}$ is null on the faces where $\mathbf{M}$ is perpendicular to the face, whereas $j_{ms} = M$ on those faces where $\mathbf{M}$ is parallel to the face. The problem is thereby reduced to the calculation of field $\mathbf{B}$ originated by the surface current density $\mathbf{j}_{ms}$ which flows on four of the vertical faces (Fig. 6.70).

Since the distance d is significantly less than L, then d is also significantly less than any typical distance of the cube. In other words: from point P only the presence of the closest face is detectable, and the problem is then comparable to a single, infinitely large face transporting the uniform current $\mathbf{j}_{ms}$. The relationship is

$$\nabla \times \mathbf{B} = \mu_0 \left(\mathbf{j}_c + \mathbf{j}_m \right) = \mu_0 \mathbf{j}_m \quad \Rightarrow \quad \oint \mathbf{B} \cdot d\mathbf{l} = \mu_0 I_m$$

By drawing a rectangle of base $2d$, with a side of length l inside the cube and another side containing point P, and by imagining a vertical axis, OZ, we can apply the preceding equation on this rectangle to obtain:

$$\oint \mathbf{B} \cdot d\mathbf{l} = B_z l - B'_z l = \mu_0 I_m = \mu_0 j_{ms} l$$

Since, through symmetry, the inner and outer vertical components of $\mathbf{B}$ are of equal magnitude and of opposite directions, we finally obtain:

$$2B_z = \mu_0 j_{ms} \quad \Rightarrow \quad B_z = \mu_0 j_{ms}/2 = \mu_0 M/2$$

6.28 A ferromagnetic material has a hysteresis loop, which can be approximated as a square in the diagram $H - B/\mu_0$, as shown in the Fig. 6.71b.

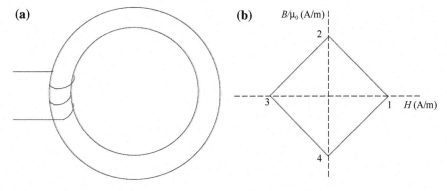

Fig. 6.71 a Magnetic material. b Hysteresis loop

With such material, a torus of average length l, and cross sectional area S, is created, onto which N turns of a conducting wire are wound. At point 1, $H = b$. Calculate the flow of **B** through the induction coil in each segment of the loop.

Solution

In order to calculate the flux, it is necessary to express **B** within the coil in terms of current I.

Ampère's law, over an inner circumference of the torus and of the mean radius, gives

$$\oint_c \mathbf{H} \cdot d\mathbf{l} = \oint_c H \, dl = Hl = I_c = NI \quad \Rightarrow \quad H = \frac{NI}{l}$$

Segment 1–2. The relationship $B(H)/\mu_0$ is, see figure,

$$\frac{B}{\mu_0} = -H + b \quad \Rightarrow \quad B = -\mu_0 \frac{N}{l} I + \mu_0 b$$

The flow of **B** is

$$\Phi = \mu_0 \left(-\frac{N}{l} I + b \right) NS$$

In the same way, it is deduced that:

Segment 2–3.

$$B = \mu_0 \frac{N}{l} I + \mu_0 b \quad \Rightarrow \quad \Phi = \mu_0 \left(\frac{N}{l} I + b \right) NS$$

Segment 3–4.

$$B = -\mu_0 \frac{N}{l} I - \mu_0 b \quad \Rightarrow \quad \Phi = \mu_0 \left(-\frac{N}{l} I - b \right) NS$$

Fig. 6.72 Electromagnet

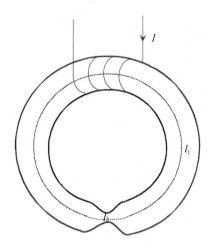

Segment 4–1.

$$B = \mu_0 \frac{N}{l} I - \mu_0 b_3 \quad \Rightarrow \quad \Phi = \mu_0 \left(\frac{N}{l} I - b\right) NS$$

6.29 The electromagnet depicted in the Fig. 6.72 is constructed of a material whose magnetization curve is $H = kB^2$. The cross section is circular, of constant area S_1 all along the length l_1 and narrows to $S_2 = S_1/2$ at the poles, which are of length l_2. The expression is required of B_1 in the wide part that supposedly has negligible flux leakage. Calculate B_1 for: $l_1 = 0.60$ m, $l_2 = 0.10$ m, $k = 10{,}000$ AT^{-2} m^{-1}, $NI = 1000$ A.

Solution

The flux theorem of the magnetic field **B** applied to one of the conical zones yields:

$$B_1 S_1 = B_2 S_2 = B_2 \frac{S_1}{2} \quad \Rightarrow \quad B_2 = 2B_1$$

where $\mathbf{B}_2$ is the field in the narrowest part.

Ampere's law along the length of the mean circumference and in an anti-clockwise direction takes the form

$$\oint \mathbf{H} \cdot d\mathbf{l} = H_1 l_1 + H_2 l_2 = I_c = NI$$

where the substitution of the material property:

$$H_1 = kB_1^2, \quad H_2 = kB_2^2$$

leads to the following relationship

$$NI = kB_1^2 l_1 + kB_2^2 l_2 = kB_1^2 l_1 + 4kB_1^2 l_2 \Rightarrow B_1 = \left(\frac{NI}{k(l_1 + 4l_2)}\right)^{1/2}$$

For the specific case of data supplied, this becomes

$$B_1 = \left(\frac{NI}{k(l_1 + 4l_2)}\right)^{1/2} = \left(\frac{1000}{10{,}000(0.60 + 4 \times 0.10)}\right)^{1/2} \quad T = 0.32 \text{ T}$$

6.30 A cylinder of radius R_1 and of relative permeability μ_r is given. Concentric to this cylinder there is a hollow cylinder of the same material with an inner radius $R_2 > R_1$ and exterior radius of $R_3 > R_2$, through which a steady current of uniform intensity I flows. Determine $\mathbf{B}$ and $\mathbf{M}$ at all points inside and outside the cylinders. Both cylinders are very long.

Solution

The system presents an axis of symmetry, and hence we expect the magnetic field to be tangential to a concentric circumference. The current is specified by the value of the conduction current, and therefore the magnetic field $\mathbf{H}$ that only depends on this current should be calculated first, and then the relationship $\mathbf{B} = \mu_r \mu_0 \mathbf{H}$ should be applied. We apply Ampere's law to a circumference c that is concentric with the system, located successively (Fig. 6.73):

(a) Inside the inner cylinder

$$\oint_{c_{in}} \mathbf{H} \cdot d\mathbf{l} = \oint_{c_{in}} H.dl = H_{in} 2\pi r = I_{cin} = 0 = \Rightarrow H_{in} = 0 \Rightarrow B_{in} = 0$$

(b) In the empty space between the two cylinders

$$\oint_{c_e} \mathbf{H} \cdot d\mathbf{l} = \oint_{c_e} H.dl = H_e 2\pi r = I_{ce} = 0 = \Rightarrow H_e = 0 \Rightarrow B_e = 0$$

Fig. 6.73 Cylindrical region

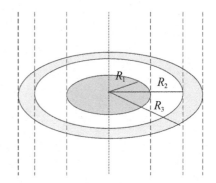

(c) Inside the outer cylinder

$$\oint_{cio} \mathbf{H} \cdot d\mathbf{l} = \oint_{cio} H.dl = H_{io} 2\pi r = I_c$$

Since the current density in the hollow cylinder is

$$j = \frac{I}{\pi(R_3^2 - R_2^2)}$$

then the intensity I_c across the circular corona of the radii r and R_2 is

$$I_c = jS = \frac{I}{\pi(R_3^2 - R_2^2)} \pi(r^2 - R_2^2) = \frac{(r^2 - R_2^2)}{(R_3^2 - R_2^2)} I$$

Hence

$$H_{io} 2\pi r = \frac{(r^2 - R_2^2)}{(R_3^2 - R_2^2)} I \quad \Rightarrow \quad H_{io} = \frac{(r^2 - R_2^2)}{2\pi r (R_3^2 - R_2^2)} I$$

(d) Outside the cylinders

$$\oint_{co} \mathbf{H} \cdot d\mathbf{l} = \oint_{co} H_o.dl = H_o 2\pi r = I_c = I \quad \Rightarrow \quad H_o = \frac{I}{2\pi r}$$

The relations $\mathbf{B} = \mu_r \mu_0 \mathbf{H}$ and $\mathbf{M} = \mathbf{B}/\mu_0 - \mathbf{H}$ for (a)–(c) determine $\mathbf{B}$ and $\mathbf{M}$ respectively.

6.31 A material is dielectric, isotropic and its magnetic properties are governed by the equation $\mathbf{B} = c\mathbf{H}^2$. A cone is constructed with that material. The cone has a height h and radius R_2. An axial hole of radius R_1 is drilled through which a non-magnetic conductor of radio R_1 is inserted. An electric current of intensity I is made to flow along the conductor (Fig. 6.74).

Fig. 6.74 Cone with cylindrical hole

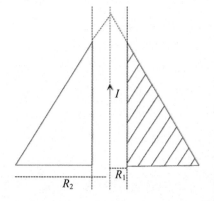

Solved Problems

Fig. 6.75 Curve for applying Ampère's theorem

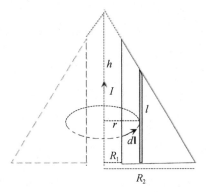

(1) Obtain the flux of **B** across the striped triangle.
(2) Apply the result to the particular case of $R_2 = 24$ cm, $h = 50$ cm, $R_1 = 4.0$ cm, $I = 6.7$ A, and $c = 0.055$ SI units.

Solution

(1) In order to calculate the flux of **B** across the triangle, it is necessary to determine **B** at every point of the triangle. However, since the total currents remain unknown but data on the conduction currents is available, the solution should start with the calculation of **H**.

Ampere's law along a concentric circumference within the material is (see Fig. 6.75):

$$\oint_c \mathbf{H} \cdot d\mathbf{l} = \oint_c H_\varphi . dl = H_\varphi 2\pi r = I_c = I \quad \Rightarrow \quad H_\varphi = \frac{I}{2\pi r} \quad \Rightarrow \quad B_\varphi = cH_\varphi^2 = c\left(\frac{I}{2\pi r}\right)^2$$

As B_ϕ depends on r, a vertical strip, as shown in the diagram, should be considered whose points are at the same distance r from the axis of symmetry. The flux is:

$$\Phi = \int \mathbf{B} \cdot d\mathbf{S} = \int B_\varphi .dS = \int c\left(\frac{I}{2\pi r}\right)^2 l dr$$

From the figure, it is deduced due to the similarity of the triangles that

$$\frac{h}{R_2} = \frac{l}{R_2 - r} \quad \Rightarrow \quad l = \frac{h}{R_2}(R_2 - r)$$

and hence

$$\Phi = \int \mathbf{B} \cdot d\mathbf{S} = \int B_\varphi .dS = \int_{R_1}^{R_2} c\left(\frac{I}{2\pi r}\right)^2 \frac{h}{R_2}(R_2 - r) dr = \frac{cI^2 h}{4\pi^2 R_2}\left(-1 + \frac{R_2}{R_1} - \ln\frac{R_2}{R_1}\right)$$

(2) In the particular case specified above, this becomes

Fig. 6.76 Piece of magnetic material

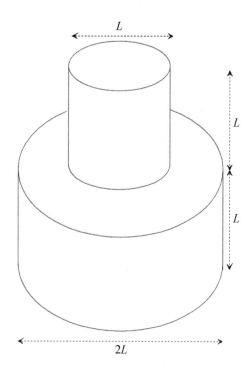

$$\Phi = \frac{0.055 \times 6.7^2 \times 0.5}{4\pi^2 \times 0.24}\left(-1 + \frac{24}{4} - \ln\frac{24}{4}\right) \text{Wb} = 0.42 \text{ Wb}$$

6.32 A material in the form of two coaxial cylinders has uniform magnetization **M** which is axial and in an upwards direction (Fig. 6.76).

(a) Obtain **B** in the centre of the base.
(b) Calculate **B** for M = 822 A/m.
Solution

(a) The conduction current density is null across the whole space. The magnetization current density is obtainable from **M**, thereby implying that **B** is directly attainable. Inside the piece, the magnetization **M** is independent from the coordinates and therefore its curl is null and hence

$$\mathbf{j}_m = \nabla \times \mathbf{M} = 0$$

On the surface of the piece, the current of magnetization is

$$\mathbf{j}_{ms} = \mathbf{M} \times \mathbf{n}$$

In the three bases, the unit normal vector is vertical, as is **M**, and hence the surface magnetization currents are null.

Solved Problems

Fig. 6.77 Surface current density

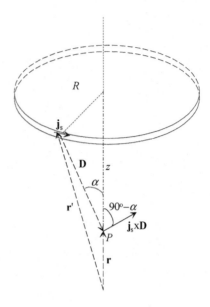

In the cylindrical surface, **n** and **M** are perpendicular, and therefore

$$j_{ms} = M$$

and the direction of $\mathbf{j}_{ms}$ is horizontal and towards the right in the strip in the diagram, as shown in Fig. 6.77.

A cylindrical surface of radius R carrying an electric current of magnetization can be considered as formed by a set of cylindrical strips of radius R and width dz, as in Fig. 6.77. The magnetic field **B**, created by the strip at a point P of its axis, must, according to the Biot–Savart law, be

$$d\mathbf{B} = \frac{\mu_0}{4\pi} \int_{strip} \frac{\mathbf{j}_{ms} \times (\mathbf{r} - \mathbf{r}')}{|\mathbf{r} - \mathbf{r}'|^3} dS' = \frac{\mu_0}{4\pi} \int_{strip} \frac{\mathbf{j}_{ms} \times \mathbf{D}}{D^3} dS'$$

where **D** denotes the vector $\mathbf{r} - \mathbf{r}'$. The magnetic field is infinitely small due to the tiny current intensity carried by the strip.

The projection of $d\mathbf{B}$ over the axis of symmetry is

$$dB = \frac{\mu_0}{4\pi} \int_{strip} \frac{j_{ms} D}{D^3} dS' \cos(90° - \alpha)$$

$$= \frac{\mu_0}{4\pi} \int_{strip} \frac{M}{D^2} 2\pi R dz \sin\alpha = \frac{\mu_0}{2} M \int_{strip} \frac{R}{D^2} dz \sin\alpha$$

Fig. 6.78 Lateral cross-section

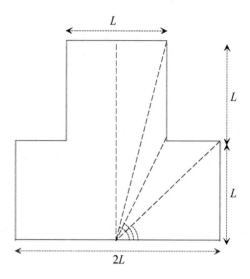

The trigonometric relationship

$$\tan\alpha = \frac{R}{z} \Rightarrow dz = -\frac{z^2}{R\cos^2\alpha}d\alpha$$

yields

$$dB = -\frac{\mu_0}{2}M\int_{strip}\sin\alpha\, d\alpha$$

Each cylindrical surface is a set of strips, and hence the resulting magnetic field in the centre of the base of the piece is (Fig. 6.78)

$$B = -\frac{\mu_0}{2}M\int_{all}\sin\alpha\, d\alpha = -\frac{\mu_0}{2}M\int_{large}\sin\alpha\, d\alpha - \frac{\mu_0}{2}M\int_{small}\sin\alpha\, d\alpha$$

$$= -\frac{\mu_0}{2}M\int_{\pi/2}^{\arctan(L/L)}\sin\alpha\, d\alpha - \frac{\mu_0}{2}M\int_{\arctan(L/2/l)}^{\arctan(L/2/(2L))}\sin\alpha\, d\alpha$$

$$= -\frac{\mu_0}{2}M\int_{\pi/2}^{\arctan 1}\sin\alpha\, d\alpha - \frac{\mu_0}{2}M\int_{\arctan 0.5}^{\arctan 0.25}\sin\alpha\, d\alpha = \frac{\mu_0 M}{2}\left\langle [\cos\alpha]_{\pi/2}^{\arctan 1} + [\cos\alpha]_{\arctan 0.5}^{\arctan 0.25}\right\rangle$$

$$= 3.96 \times 10^{-7}M$$

(b) With the numerical data supplied, this becomes

$$B = 3.96 \times 10^{-7}M = 3.96 \times 10^{-7} \times 822\text{ T} = 3.26 \times 10^{-4}\text{ T}$$

6.33 The magnetic circuit of the Fig. 6.79 is of constant cross section, without flux leakage, and contains 10,000 turns. The left column has a mean length $L = 20$ cm and is of linear magnetic material. The remaining material is fer-

Fig. 6.79 Electromagnet

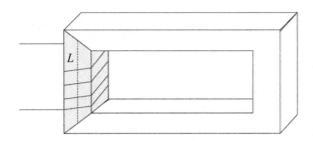

Fig. 6.80 A part of the hysteresis loop

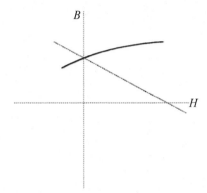

romagnetic. The mean length of each horizontal beam is $2L$. When the wire carries a given current, the circuit is magnetized. Then the intensity of the current is decreased to the value $I = 10\,\text{A}$. In this stage of the reduction of I, it is known that the equation $B_f^2 = k_1 H_f + k_2$ with $k_2 = 0.4\,T^2$ in the ferromagnetic material is satisfied. Calculate the magnetic susceptibility of the linear material, indicating whether it is paramagnetic or diamagnetic.

Solution

In the linear material
$$B_l = \mu_r \mu_0 H_l$$

In the ferromagnetic material for the given current I (Fig. 6.80)
$$B_f^2 = k_1 H_f + k_2$$

Since there is no flux leakage, that is to say, the magnetic field outside the core is null, hence the fluxes of **B** in the linear section and in the non-linear section are equal, and therefore we can write

$$\Phi = \Phi_l \;\Rightarrow\; B_f S = B_l S \;\Rightarrow\; B_f = B_l.$$

Ampère's law relating to H, along the mean line and denoting H_l as the longitudinal field in the linear material, yields

$$\oint \mathbf{H} \cdot d\mathbf{l} = H_l L + 2H_f L + H_f L + 2H_f L = (H_l + 5H_f)L = I_c = NI \Rightarrow H_l + 5H_f = \frac{NI}{L}$$

When, in the second equation, $H_f = 0$, then

$$B_{f,H=0} = \sqrt{k_2}$$

The first and third equations give

$$B_f = \mu_r \mu_0 H_l \Rightarrow H_l = \frac{1}{\mu_r \mu_0} B_f$$

which substituted into the fourth equation yields

$$\frac{1}{\mu_r \mu_0} B_f + 5H_f = \frac{NI}{L} \Rightarrow B_f = -5\mu_r \mu_0 H_f + \frac{\mu_r \mu_0 NI}{L}$$

This line is called the operation line. If $H_f = 0$ is substituted into this line and the fifth equation is taken into consideration, then

$$\sqrt{k_2} = -0 + \frac{\mu_r \mu_0 NI}{L} \Rightarrow \mu_r = \frac{L\sqrt{k_2}}{\mu_0 NI}$$

With the numerical data provided, we obtain

$$\mu_r = \frac{L\sqrt{k_2}}{\mu_0 NI} \Rightarrow \chi = \mu_r - 1 = \frac{0.20 \times \sqrt{0.4}}{4\pi \times 10^{-7} \times 10{,}000 \times 10} - 1 = 0.007$$

The linear material is therefore paramagnetic. With such a small value of susceptibility it is difficult to justify that there is no flux leakage.

6.34 A ring of mean length L and constant cross-section is made up of two ferromagnetic materials, each of mean length $L/2$. Material a has a hysteresis loop which, owing to saturation, is similar to a line passing through the point $H_a = 0\,\text{A/m}$, $B_a = 1\,\text{T}$ and through the point $H_{a'} = -200\,\text{A/m}$, $B_{a'} = 0\,\text{T}$. The similar line of material b is parallel to that of material a and passes through $H_b = 0\,\text{A/m}$, $B_b = 0.5\,\text{T}$. A conductor is wound around the ring, which is then subjected to a current sufficient for the saturation of both materials. The current is then switched off. Find the values of H and B acquired by the two materials.

Solution

Assuming that there is no flux leakage, then the flux theorem of **B** yields

$$\oint_S \mathbf{B} \cdot d\mathbf{S} = 0 \quad \Rightarrow \quad B_a = B_b$$

Ampere's law along the circumference of the mean length, applied to **H**, gives

$$\oint \mathbf{H} \cdot d\mathbf{l} = H_a L/2 + H_b L/2 = I_c = NI = 0 \quad \Rightarrow \quad H_a + H_b = 0$$

where it has been taken into account that at the moment when the calculations are made, the current has already been switched off.
From the Fig. 6.81, it can be deduced that both lines have a slope of 1/200 T/(A/m) and their equations are, respectively:

$$B_a = \frac{1}{200} H_a + 1$$

$$B_b = \frac{1}{200} H_b + 0.5$$

Substitution of these two relations into the first equation yields

$$\frac{1}{200} H_a + 1 = \frac{1}{200} H_b + 0.5 \quad \Rightarrow \quad H_a - H_b - = -100$$

This equation together with the second equation give:

$$H_a = -50 \text{ A/m}$$
$$H_b = 50 \text{ A/m}$$

Fig. 6.81 Hysteresis curve

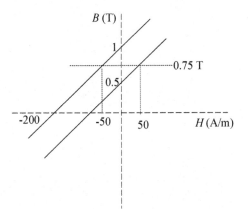

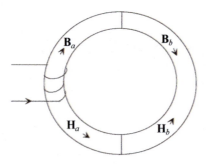

Fig. 6.82 Magnetic materials

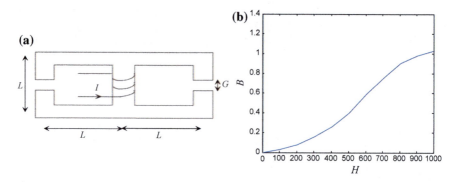

Fig. 6.83 a Electromagnet. b First magnetization curve

By substituting one of these values into the equation which relates B_a with H_a, we finally obtain

$$B_b = B_a = \frac{1}{200}(-50) + 1 \text{ T} = 0.75 \text{ T}$$

Observe that if the direction of **B** is clockwise, then the direction of **H** changes from one material to the other (Fig. 6.82).

6.35 Figure 6.83a depicts a double and symmetric electromagnet. It is desired that a magnetic field B be reached within the air gaps in its first magnetization process. The curve of the first magnetization is drawn in the Fig. 6.83b. Calculate the length G of the air gaps. Data: $L = 45$ cm, area of cross section $S = 486$ cm^2, number of turns $N = 600$, $I = 40$ A, $B = 0.40$ T.

Solution

Let 1 be the left branch, 2 be the central branch, and 3 be the right-hand branch.

Assuming that there is no flux leakage, the application of the conservation of the flux to the highest node gives

$$0 = \oint_S \mathbf{B} \cdot d\mathbf{S} = B_1 S - B_2 S + B_3 S \quad \Rightarrow \quad B_1 - B_2 + B_3 = 0$$

The application of Ampere's law to the closed line 1–2, running in an anti-clockwise direction, yields

$$\oint \mathbf{H} \cdot d\mathbf{l} = H_2 L + H_1 L + H_1 (L - G) + H_G G + H_1 L = (H_2 + 3H_1)L + (H_G - H_1) G = I_c = NI$$
$$\Rightarrow \quad G = \frac{NI - (H_2 + 3H_1)L}{H_G - H_1}$$

The application of Ampere's law to the closed line 2–3, running in a clockwise direction, yields

$$\oint \mathbf{H} \cdot d\mathbf{l} = H_2 L + H_3 L + H_3 (L - G) + H_G G + H_3 L = H_2 L + H_3 (3L - G) + H_G G = I_c = NI$$

A comparison of these two latter equations brings us to the conclusion that

$$H_1 = H_3 \quad \Rightarrow \quad B_1 = B_3$$

where it is taken into account that materials 1 and 3 are equal, the equality of H corresponds to the equality of B. This result could have been reached through the symmetry of the figure.

The application of the conservation of the flux to the left-hand air gap yields

$$B_1 = B$$

where B is a data.

Hence the first equation can be written:

$$B_1 - B_2 + B_1 = 0 \quad \Rightarrow \quad B_2 = 2B_1 = 2B$$

The air gap is linear and with $\mu_r = 1$, therefore

$$B = \mu_0 H_G \quad \Rightarrow \quad H_G = B/\mu_0$$

The values obtained for B_1 and B_2 permit the values of H_1 and H_2 to be deduced from the graph provided.

With the data $B_1 = B = 0.40\,\text{T}$, the graph gives $H_1 = 500\,\text{A/m}$, and with the data $B_2 = 2B = 0.80\,\text{T}$, the graph gives 740 A/m. Substitution in the equation that provides the length of the air gap finally results in:

$$G = \frac{600 \times 40 - (740 + 3 \times 500)0.45}{40/(4\pi \times 10^{-7}) - 500}\,\text{m} = 0.072\,\text{m}$$

Fig. 6.84 Electromagnet

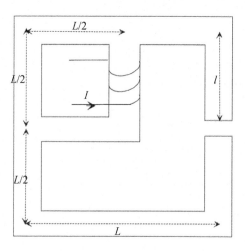

6.36 A certain company wants to manufacture a device as indicated in the Fig. 6.84, where each branch has a circular cross section. A magnetic field in the air gap of $H_g = 200$ A/m is required from this device when a current intensity of $I = 1$ A flows through the coil. For technical reasons, no more than 203 m of conductor cable can be employed. The material of the device is paramagnetic. What should the value of its μ_r be? Apply values of $L = 1$ m, $l = 0.04$ m, and radius of cross section $r = 5$ cm.

Solution

Let 1 be the upper left-hand branch of the magnetic circuit, 2 be the central branch, and 3 be the lower right-hand branch, as shown in Fig. 6.84. Suppose that there is no flux leakage dispersion (this approximation is imprecise since it is not a ferromagnetic material). The flux theorem for **B** applied to a small closed surface that surrounds the upper node relates the tangential components of **B** (Fig. 6.85):

$$0 = \oint_S \mathbf{B} \cdot d\mathbf{S} = B_1 A - B_2 A + B_3 A \quad \Rightarrow \quad B_1 - B_2 + B_3 = 0$$

and applied to a small cylinder that surrounds the upper magnetic pole yields:

$$0 = \oint_S \mathbf{B} \cdot d\mathbf{S} = B_g A - B_3 A \quad \Rightarrow \quad B_3 = B_g$$

In the three branches of paramagnetic material, the following equations hold, respectively:

$$B_1 = \mu_r \mu_0 H_1, \quad B_2 = \mu_r \mu_0 H_2, \quad B_3 = \mu_r \mu_0 H_3$$

Fig. 6.85 Magnetic circuit

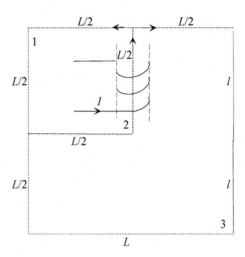

and in the air gap, this is

$$B_g = \mu_0 H_g$$

Substitution of these expressions into the first two equations yields:

$$\mu_r \mu_0 H_1 - \mu_r \mu_0 H_2 + \mu_r \mu_0 H_3 = 0 \quad \Rightarrow \quad H_1 - H_2 + H_3 = 0$$

$$\mu_r \mu_0 H_3 = \mu_0 H_g \quad \Rightarrow \quad \mu_r H_3 = H_g$$

Through the application of Ampere's law to **H** along the closed line 1–2, running in an anti-clockwise direction, another relationship between the tangential components of **H** is obtained:

$$\oint \mathbf{H} \cdot d\mathbf{l} = H_2 L + H_1 L = I_c = NI$$

The same law applied to the closed line 2–3, running in a clockwise direction, gives

$$\oint \mathbf{H} \cdot d\mathbf{l} = H_2 L + 2H_3(L + l) + H_g (L - 2l) = I_c = NI$$

The system of the last four equations enables the unknown values H_1, H_2, H_3, and μ_r to be solved. The solution is

$$\mu_r = \frac{(5L + 4l) H_g}{NI - 2(L - 2l) H_g} = \frac{(5 \times 1 + 4 \times 0.04) 200}{\frac{203}{2\pi \times 0.05} \times 1 - 2(1 - 2 \times 0.04) 200} = 3.94$$

Fig. 6.86 Ferromagnetic system

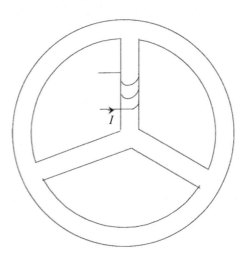

6.37 A bar of cross section $S = 4.0\,\text{cm}^2$ is given from which a circumference and three radii are formed. Each radius is of length $r = 20\,\text{cm}$ and forms an angle of $120°$ with the other radii. The curve of the first magnetization of the material used can be expressed as $B = CH^3$. From the virgin state, $N = 300$ turns are wound and subjected to a current of $I = 30\,\text{A}$. Calculate **B** in the left-hand radius, disregarding the flux leakage (Fig. 6.86).

Solution

Although this problem could be solved with the systematic methodology of applying the flux theorem to the three nodes and Ampere's theorem to the two loops, this problem will be solved here by taking advantage of the symmetry of the figure.

Due to the symmetry, if **B** in the lower branch were, for any reason, on the right-hand direction, then the same reason would exist for it to be on the left-hand direction. Therefore **B** must be null in the lower branch and this branch can be disregarded for the solution of this problem, see Fig. 6.87.

Again owing to symmetry, if **B** in the upper left-hand branch has anti-clockwise direction and its module is B_1, then the upper right-hand branch has clockwise direction, and **B** has the same module. Let B_2 be the projection towards the top of the field in the vertical radial branch.

The flux theorem in the upper node yields:

$$0 = \oint_S \mathbf{B} \cdot d\mathbf{S} = B_1 S - B_2 S + B_3 S = 2B_1 S - B_2 S \quad \Rightarrow \quad B_2 = 2B_1$$

Ampere's law applied to the line formed by the vertical radial branch, the upper left-hand branch and the left-hand radial branch gives:

$$\oint \mathbf{H} \cdot d\mathbf{l} = H_2 r + H_1 \left(\frac{1}{3}2\pi r + r\right)_4 = H_2 r + \frac{2\pi + 3}{3} H_1 r = I_c = NI$$

Fig. 6.87 Magnetic circuit

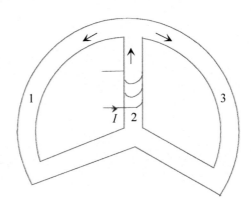

Substitution of the relation $B = CH^3$ into the first equation gives:

$$CH_2^3 = 2CH_1^3 \quad \Rightarrow \quad H_2 = \sqrt[3]{2}H_1$$

And the substitution of this expression into that above yields:

$$\sqrt[3]{2}rH_1 + \frac{2\pi + 3}{3}H_1 r = NI \quad \Rightarrow \quad H_1 = \frac{NI}{\left(\sqrt[3]{2} + \frac{2\pi+3}{3}\right)r}$$

Hence

$$B_1 = CH_1^3 = C\left[\frac{NI}{\left(\sqrt[3]{2} + \frac{2\pi+3}{3}\right)r}\right]^3$$

With the numerical data provided, this finally results in

$$B_1 = 1.0 \times 10^{-12}\left[\frac{300 \times 30}{\left(\sqrt[3]{2} + \frac{2\pi+3}{3}\right)0.20}\right]^3 \text{T} = 1.1 \text{ T}$$

Chapter 7
Methods for Solving Electrostatic and Magnetostatic Problems

Abstract In the preceding chapters we have studied properties of the electrostatic and magnetostatic fields in vacuum and in the presence of matter. We have also seen how we can generate them and some techniques for calculating **E** and **B**. However, it is not always possible to obtain a solution by means of these techniques for different reasons. For instance, when we wanted to calculate the electric field produced by a system of charges, we needed to know exactly its distribution and then apply Coulomb's law. This method is very clear, but sometimes does not easily work. In fact, when the configuration of charges has a low symmetry the integral to be computed cannot be resolved in terms of elementary functions. The same occurs with the magnetic field. Even in case of symmetry its calculation may be of hight difficulty if the region where the field must be calculated is off the axis of revolution of the system (see Chap. 5). To solve these drawbacks different methods of analysis have been developed. In this chapter we try to introduce the reader to some of the most important analytical and numerical techniques for solving non-time dependent electromagnetic problems. We present a set of exercises with the aim to show the basic ideas so that the student can understand a further reading in specialized books about this subject. Because of the characteristics of the fields to be studied in this chapter, we focus our attention on the Laplace equation. As we will see many problems appearing in electromagnetics may be posed on the basis of this equation, or on the Poisson equality, depending on the existence of charges in the place where the potential must be obtained. The question is then to seek a solution subjected to some boundary conditions, which depend on the specific problem to be investigated.

7.1 The Laplace Equation

In a charge or current free region of the space where there does not exist time varying fields it is in principle possible to obtain a potential function, because $\nabla \times \mathbf{E} = 0$ and $\nabla \times \mathbf{H} = 0$. Both cases bring to differential equation of Laplace type for the respective potentials, that is,

$$\nabla^2 V(\mathbf{r}) = 0 \tag{7.1}$$

for the electric potential and

$$\nabla^2 V_m(\mathbf{r}) = 0 \qquad (7.2)$$

for the magnetic scalar potential. Scalar functions that verify the Laplace equation are called harmonic functions, and they are very important in the potential theory.

From a mathematical viewpoint Laplace's equation is a partial differential equation of elliptic type whose specific solutions depend on the conditions imposed on the boundary of the region where the problem is defined. Physically, the three most important cases we have are known as the Dirichlet, Neumann and Robin problems, which differ to each other in the boundary conditions.

Let us suppose a region D delimited by a surface S. Under these schema we have different possibilities for the solution of the problem in the interior and exterior of the boundary.

(a) The Dirichlet problem consist of finding a function V that accomplishes (7.1) (or (7.2), for the magnetostatic potential) in the interior of D and it is subjected to the condition $V(\mathbf{r}) = f(\mathbf{r})$ on the surface S. Physically it would represent the problem of calculating the potential at any interior/exterior point of the system whose potential at the surface is prescribed by the function f.

(b) The Neumann problem is posed in the same way but giving the boundary conditions on the normal derivatives. In this case we impose the condition $\frac{\partial V(\mathbf{r})}{\partial n} = h(\mathbf{r})$ on S. From a physical point of view it represent the problem of finding the potential at any interior/exterior point of the system whose potential has a specific normal derivative at the surface.

(c) Mixed problems. In this case, Dirichlet and Neumann conditions are satisfied on the boundary in such a way that on some parts of it we impose the value of the potential, and on other parts we know the value of the normal derivative, that is

$$\partial S = D_i \bigcup_1^n N_i, \qquad (7.3)$$

where D_i and N_i represent the regions on the boundary where the Dirichlet and Neumann conditions apply, respectively.

(d) The Robin problem is the most generally posed of the study, and uses boundary conditions of the two aforementioned problems but not disjointed. In fact, we seek solutions for the conditions on the surface

$$V(\mathbf{r}) + f(\mathbf{r}) \frac{\partial V(\mathbf{r})}{\partial \mathbf{n}} = h(\mathbf{r}). \qquad (7.4)$$

For solving the differential equations it is useful to sum up the frame in which we must locate an electrostatic or magnetostatic problem. In this context, we are going to give characteristics of both fields.

In the case of electrostatics in a vacuum, the potential verifies the following properties

7.1 The Laplace Equation

(1) $\nabla^2 V = 0$ for every point of the space where there is no charge.
(2) $V(x, y, z)$ is constant on any conductor system.
(3) $\frac{\partial V}{\partial \mathbf{n}} = -\sigma$, σ being the conductivity.
(4) The total charge on the conductor surface may be calculated by means of the following integral

$$-\int\int_S \frac{\partial V}{\partial \mathbf{n}} dS \qquad (7.5)$$

(5) The potential $V(x, y, z) \longrightarrow 0$ at infinity, if the system of charges is confined in a finite region of space.
(6) If there exist discontinuities in the potential they must be placed at the charges, dipoles or similar structures.

In case we have dielectrics, the basic information to be taken into consideration is

(1) $\nabla.(\varepsilon \nabla V(x, y, z)) = -\rho$.
(2) If there are two different contacting materials of permittivities ε_1 and ε_2, on the surface of separation of both media, the potential and its normal derivative realize the following relations

$$V_1(x, y, z) = V_2(x, y, z), \qquad (7.6)$$

and

$$\varepsilon_1 \frac{\partial V_1}{\partial \mathbf{n}} = \varepsilon_2 \frac{\partial V_2}{\partial \mathbf{n}}. \qquad (7.7)$$

With respect to Magnetostatics, we also have additional information to apply in solving problems.

(1) In a current-free region of space a magnetic potential $V_m(x, y, z)$ may be found so that the magnetic field $\mathbf{H}$ fulfills $\mathbf{H} = -\nabla V_m(x, y, z)$.
(2) $\nabla.(\mu \nabla V_m(x, y, z)) = 0$.
(3) If there are two different contacting materials of susceptibilities μ_1 and μ_2, on the surface of separation of both media the potential and its normal derivative accomplish the following relations

$$V_{m1}(x, y, z) = V_{m2}(x, y, z), \qquad (7.8)$$

and

$$\mu_1 \frac{\partial V_1}{\partial \mathbf{n}} = \mu_2 \frac{\partial V_2}{\partial \mathbf{n}}. \qquad (7.9)$$

7.2 The Method of Separation of Variables

There are different techniques for solving the Laplace equation and, in general, partial differential equations (PDF). One of them is the method of separation of variables (MSV). This technique consists of finding solutions in the form of products of functions; each of them depending only of one of the variables that appears in the unknown function. This method is a very powerful tool if the problem has some symmetries, that is, problems that can be expressed easily in a coordinate system, such as cylindrical, spherical, or rectangular among others. In others cases where there does not exist the possibility of identifying the geometrical characteristics of physical system with a specific coordinate frame, the MSV is not a useful way to calculate the solution, if not impossible. However, there are many actual problems were the geometries involved have some symmetries (translational or rotational symmetry, for instance). It is not the aim of this chapter to demonstrate all the possibilities we can have, but only to show the final results for the most important cases.

When analyzing a problem in cartesian coordinates the Laplace equation for a scalar potential $V(x, y, z)$ has the form

$$\Delta V(x, y, z) \equiv \frac{\partial^2 V}{\partial x^2} + \frac{\partial^2 V}{\partial y^2} + \frac{\partial^2 V}{\partial z^2} = 0. \tag{7.10}$$

The technique of separation of variables supposes that the general solution may represented by means of the following product

$$V(x, y, z) = X(x) Y(y) Z(z). \tag{7.11}$$

As we can observe the functions appearing above depend only on one of the three variables involved. By introducing (7.11) into (7.10), and manipulating this expression we find

$$\frac{1}{X(x)} \frac{d^2 X(x)}{dx^2} = k_x^2, \tag{7.12}$$

$$\frac{1}{Y(y)} \frac{d^2 Y(y)}{dy^2} = k_y^2, \tag{7.13}$$

and

$$\frac{1}{Z(z)} \frac{d^2 Z(z)}{dz^2} = k_z^2, \tag{7.14}$$

where k_x, k_y, and k_z are constants which satisfy the relation

$$k_x^2 + k_y^2 + k_z^2 = 0. \tag{7.15}$$

7.2 The Method of Separation of Variables

We have different solutions, but we can express in a general form all of them as a combination of the following possibilities[1]

$$X(x) = A\exp(k_x x) + B\exp(-k_x x), \quad (7.16)$$

$$Y(y) = C\exp(k_y y) + D\exp(-k_y y), \quad (7.17)$$

and

$$Z(z) = E\exp(k_z z) + F\exp(-k_z z). \quad (7.18)$$

In case that $k_x^2 < 0$, $k_y^2 < 0$, and $k_z^2 < 0$ the solutions are

$$X(x) = A\sin(k_x x) + B\cos(k_x x), \quad (7.19)$$

$$Y(y) = C\sin(k_y y) + D\cos(k_y y), \quad (7.20)$$

and

$$Z(z) = E\sin(k_z z) + F\cos(k_z z) \quad (7.21)$$

A particular solution occurs when $k_x = k_y = k_z = 0$. In this case we can construct a polynomial function containing a combination of all variables

$$V(x, y, z) = X(x)Y(y)Z(z) = (A + Bx)(C + Dy)(E + Fz). \quad (7.22)$$

In cylindrical coordinates, by making the substitutions $x = \rho\cos\phi$, $y = \rho\sin\phi$, and $z = z$ we have

$$\Delta \equiv \frac{1}{\rho}\frac{\partial}{\partial \rho}\left(\rho\frac{\partial V}{\partial \rho}\right) + \frac{1}{\rho^2}\frac{\partial^2 V}{\partial \phi^2} + \frac{\partial^2 V}{\partial z^2} = \frac{\partial^2 V}{\partial \rho^2} + \frac{1}{\rho}\frac{\partial V}{\partial \rho} + \frac{1}{\rho^2}\frac{\partial^2 V}{\partial \phi^2} + \frac{\partial^2 V}{\partial z^2} = 0. \quad (7.23)$$

Following the same method as explained before, we try to find solutions in the form of products of functions of one variable

$$V(\rho, \phi, z) = R(\rho)\,\Phi(\phi)\,Z(z). \quad (7.24)$$

Taking into account that $V(\phi) = V(\phi + 2\pi)$, the different combinations we can have are of the form

$$V(\rho, \phi, z) = \exp(\pm kz)Z_m(k\rho)(A\cos(m\phi) + B\sin(m\phi)), \quad (7.25)$$

$$V(\rho, \phi, z) = \exp(\pm kz)Z_0(k\rho)(A + B\phi), \quad (7.26)$$

---

[1] In these equations the parameters A, B, etc. depend on a subindex related to its corresponding k_i ($i = x, y, z$). We do not choose here A_n, B_n, etc., for simplicity of the notation. Otherwise we should write k_{x_n}, and so on.

$$V(\rho, \phi, z) = (C + Dz) \sum_{1}^{\infty} \left(a_m \rho^m + \frac{b_m}{\rho^m} \right) (A_m \cos(m\phi) + B_m \sin(m\phi)), \quad (7.27)$$

and

$$V(\rho, \phi, z) = (C + Dz)(a + b \ln \rho)(A + B\phi), \quad (7.28)$$

where $Z_m(k\rho)$ are the cylinder functions obtained from the Bessel differential equation, and A, B, C, D, a, and b are constants which depend on the specific conditions of the problem, and $m = 0, \pm 1, \pm 2, \pm 3, \ldots$.

In spherical coordinates, using the substitutions $x = r \cos \phi \sin \theta$, $y = r \sin \phi \sin \theta$, and $z = r \cos \theta$ we have

$$\Delta \equiv \frac{1}{r^2} \frac{\partial}{\partial r} \left(r^2 \frac{\partial V}{\partial r} \right) + \frac{1}{r^2 \sin \theta} \frac{\partial}{\partial \theta} \left(\sin \theta \frac{\partial V}{\partial \theta} \right) + \frac{1}{r^2 \sin \theta} \frac{\partial^2 V}{\partial \phi^2} = 0. \quad (7.29)$$

Introduction a function $V(r, \phi, \theta) = R(r) \Phi(\phi) \Theta(\cos \theta)$ into (7.29) leads to

$$V(r, \phi, \theta) = \sum_{1}^{\infty} \left(a_m r^m + \frac{b_m}{r^{m+1}} \right) Y_m(\phi, \theta) \quad (7.30)$$

$Y_i(\phi, \theta)$ being the spherical functions,

$$Y_m(\phi, \theta) = \sum_{j=0}^{m} P_m^j(\cos \theta)(a_j \cos(j\phi) + b_j \sin(j\phi)), \quad (7.31)$$

where $P_m^j(\cos \theta)$ are the associated Legendre functions of the first kind ($j = 1, 2, 3, \ldots$). Sometimes, because of the symmetries of the problem we can reduce our study to two variables. In such a case the solutions we can find working with cartesian coordinates is

$$V(x, y) = (A + Bx)(C + Dy) + \sum_{n=1}^{\infty} (E_n \sin(k_x x) + F_n \cos(k_x x)) + \sum_{n=1}^{\infty} (G_n \sinh(k_x y) + H_n \cosh(k_x y)).$$
$$(7.32)$$

If we have rotational symmetry, the solution may be expressed as a function of ρ and ϕ, i.e.

$$V(\rho, \phi) = (C + D \ln \rho)(E + F\phi) + \sum_{m=1}^{\infty} \left(a_m \rho^m + \frac{b_m}{\rho^m} \right) (A_m \cos(m\phi) + B_m \sin(m\phi)).$$
$$(7.33)$$

These two last results are the most general expression we can construct in two dimensions.

7.3 Green's Function Method

The method of separation of variables is very useful when there is a coordinate system in which the intervening variables of the differential equation can be separated. This is not always possible. In actual problems the domains in which differential equations are solved do not always have the desired symmetry, then the separation of variables does not work. On the other hand, solutions using this technique appear as sums of functions, which are required to verify the boundary conditions of the specific problem. Therefore, obtaining the numerical value of the potential at a point in space (or plane) for a problem is a laborious task, since a sufficient number of terms in the series are needed.

In this context, Green functions give the solution of electrostatic and magnetostatic problems by means of integral representations.

The Green function is named as G and is defined as follows. Let us suppose a domain D bounded by a surface S and two points P, Q belonging to D, then Green's function satisfies

$$\nabla^2 G(\mathbf{r}, \mathbf{r}') = \delta(\mathbf{r} - \mathbf{r}') \tag{7.34}$$

where $\delta(\mathbf{r} - \mathbf{r}')$ is the Dirac's delta distribution. Green's function has some properties, which may be summing up as follows:

(a) This function is of the form

$$G(\mathbf{r}, \mathbf{r}') = \frac{1}{|\mathbf{r} - \mathbf{r}'|} + g(\mathbf{r}, \mathbf{r}') \tag{7.35}$$

where $g(\mathbf{r}, \mathbf{r}')$ is a continuous function for every point on S which also verifies Laplace equation.

(b) $G(\mathbf{r}, \mathbf{r}') = 0$ for all P on S. For unbounded domains $g(\mathbf{r} - \mathbf{r}') \longrightarrow 0$ when $r \longrightarrow \infty$.

(c) Green's function is symmetric, i.e. $G(P, Q) = G(Q, P)$, which has an easy physical interpretation. It means that the potential at a point P (in the space or the plane) due to an electric charge located at point Q, is the same as the electrostatic potential at Q if the charge is now placed at P.

(d) $G(\mathbf{r}, \mathbf{r}')$ is unique for the domain where it is defined.

As we have commented in Sect. 7.1, in many problems the boundary conditions are not given on the value of the function (potential) on the surface but on its derivative in the direction normal to S. For this reason, in order to distinguish the two possibilities, we will define the Neumann function.

Let D be a region delimited by the surface S. Neumann's function denoted by $N(\mathbf{r}, \mathbf{r}')$ verifies

$$\nabla^2 N(\mathbf{r}, \mathbf{r}') = \delta(\mathbf{r} - \mathbf{r}'), \tag{7.36}$$

in D, with boundary conditions

$$\left(\frac{\partial N(\mathbf{r}, \mathbf{r}')}{\partial n}\right)_S = C, \tag{7.37}$$

C being a constant. In short, the Neumann function is the Green function for boundary conditions on the normal derivative. As we have seen for G, Neumann's function is symmetric too, i.e. $N(P, Q) = N(Q, P)$. On the other hand, it may be observed that the function N is not unique. This means that if we find a function N that solves the problem (7.36), any other function that differs from N by a constant is also a solution. In order to eliminate this difficulty a normalizing condition for N is often chosen as follows

$$\int\int_{S'} N(P, Q) dS' = 0. \tag{7.38}$$

Once we have defined the functions $G(\mathbf{r}, \mathbf{r}')$ and $N(\mathbf{r}, \mathbf{r}')$, we can obtain the general solutions for Dirichlet and Neumann boundary conditions. In fact, when taking Dirichlet's problems into account, the problem is posed in the following form

$$\begin{cases} \nabla^2 V = 0 & \text{in D,} \\ V = f & \text{on S.} \end{cases} \tag{7.39}$$

where $V \to 0$ at infinity. Now, if we know the Green function, the result may be expressed in integral form as follows

$$V(\mathbf{r}) = \int\int_{S'} f \left(\frac{\partial G(\mathbf{r}, \mathbf{r}')}{\partial n'}\right) dS'. \tag{7.40}$$

Formula (7.40) gives the value of the function $V(\mathbf{r})$ at any interior point of D if the derivative of G and the value of f on the surface S' are known. Equation (7.40) is normally called Poisson's formula.

In the case of the non-homogeneous problem, i.e. for the Poisson equation for a bounded domain D where the normal derivative on the smooth surface S' is regular, we have

$$\begin{cases} \nabla^2 V = u & \text{in D,} \\ V = f & \text{on S.} \end{cases} \tag{7.41}$$

The solution of this differential equation is given by the following formula

$$V(\mathbf{r}) = -\int\int\int_{V'} G(\mathbf{r}, \mathbf{r}') u(\mathbf{r}') dV' + \int\int_{S'} f \left(\frac{\partial G(\mathbf{r}, \mathbf{r}')}{\partial n'}\right) dS'. \tag{7.42}$$

Observe that if we put $u = 0$, we have (7.40) again.

In the same way expressed before, when analyzing Neumann's boundary conditions the problem is posed as

7.3 Green's Function Method

$$\begin{cases} \nabla^2 V = 0 & \text{in D,} \\ \frac{\partial V}{\partial n'} = h & \text{on S.} \end{cases}$$

The function $h(\mathbf{r}')$ must be correctly chosen because of the Gauss integral theorem which says that if there are no sources enclosed by the surface S', the integral over this surface of the derivative of the potential V in the direction of $\mathbf{n}'$ must be zero, i.e.

$$\int\int_{S'} \left(\frac{\partial V}{\partial \mathbf{n}'}\right) dS' = 0, \tag{7.43}$$

the result of which is valid for any harmonic function in a finite domain. With the same idea shown in (7.38), to avoid the non-unicity of the solution, a normalizing condition is needed. As a result, a value for V is often chosen so that

$$\int\int_{S'} V \, dS' = 0, \tag{7.44}$$

over the surface S'.

Taking into account (7.43), and applying the Green formula, the solution may be expressed as

$$V(\mathbf{r}) = -\int\int_{S'} h(\mathbf{r}')N(\mathbf{r},\mathbf{r}')dS', \tag{7.45}$$

where $N(\mathbf{r},\mathbf{r}')$ is the Neumann function defined before. Similarly, for the Poisson equation with Neumann boundary conditions it holds that

$$V(\mathbf{r}) = -\int\int\int_{V'} N(\mathbf{r},\mathbf{r}')u(\mathbf{r}')dV' - \int\int_{S'} N(\mathbf{r},\mathbf{r}')\left(\frac{\partial V(\mathbf{r},\mathbf{r}')}{\partial \mathbf{n}'}\right)dS' + C, \tag{7.46}$$

where C is a constant unknown. This constant does not represents a problem because the scalar potential is not uniquely determined (see Chap. 2).

The most important thing about this method for solving the Laplace equation is, perhaps, that the way for obtaining a solution (see (7.40) and (7.45)) is quite different from those presented in Sect. 7.2. In effect, there the solution was sought in the form of a product of functions which had to verify their corresponding boundary conditions. In the present case, by means of the Green and Neumann functions, the calculation is presented in the form of integral equations. This approach does not need additional boundary conditions because this information is included explicitly in the integrands of the integral. On the other hand, another advantage of this technique is that the region where the integrals must be performed is not the volume, but only the surfaces delimiting the region D (7.40) and (7.45), even in cases of non-linear materials. Therefore, if we know the potential V and its normal derivative on S, we can find a solution for the problem everywhere.

7.4 Method of Images

The method of images is a procedure to solve the Laplace equation without analyzing the differential equation. This procedure is not applicable in all cases, however it is a useful tool under some circumstances where symmetries exist. The basic idea is to seek a charge distribution that creates the same potential like the actual problem to be solved. To make this possible it is necessary that the surfaces of the conductors of our problem coincide with the equipotential surfaces for the charge distribution that produce the same effect. Mathematically this is justified by the existence and uniqueness theorem for the equation of Laplace. Indeed, if a function verifies the equation and boundary conditions of a specific problem, the function obtained is unique. Therefore, if we reach a charge disposition that produces the same potential, the solution can be calculated using the fictitious load distribution, since the solution is the same. In order to understand the procedure we will begin with a simple example, which also is the basis for solving other problems.

We will deal with the potential created by two charges of different sign. Let us consider the picture appearing in Fig. 7.1. Two charges q_1 and $-q_2$ separated by a distance a over the OY axis. If we calculate the potential created at an arbitrary point $P(x, y, z)$ of space, we have

$$V(x, y, z) = \frac{1}{4\pi\varepsilon_0} \frac{q_1}{\sqrt{x^2 + y^2 + z^2}} + \frac{1}{4\pi\varepsilon_0} \frac{-q_2}{\sqrt{x^2 + (y-a)^2 + z^2}}. \qquad (7.47)$$

The surfaces of constant potential may be determined by matching $V(x, y, z)$ to be a constant C. The result is shown in Fig. 7.2. In this graphic different curves for the distinct values of C are obtained. If we examine the equipotential surfaces for $C = 0$, (7.47) yields

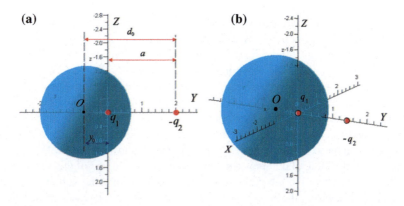

Fig. 7.1 **a** Plane view of the surface of constant potential for $C = 0$ when $|q_1| < |q_2|$. **b** The same in a three-dimensional perspective. Observe that, for this case, the centre is displaced on the *left*

7.4 Method of Images

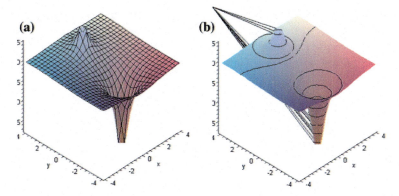

Fig. 7.2 **a** Two-dimensional potential of two charges $q_1 = q$ and $q_2 = -3q$. **b** Two-dimensional curves of constant potential for different values of the constant C

$$V(x, y, z) = \frac{1}{4\pi\varepsilon_0} \frac{q_1}{\sqrt{x^2 + y^2 + z^2}} + \frac{1}{4\pi\varepsilon_0} \frac{-q_2}{\sqrt{x^2 + (y-a)^2 + z^2}} = C = 0 \quad (7.48)$$

or

$$x^2(1 - \frac{q_2^2}{q_1^2}) + y^2(1 - \frac{q_2^2}{q_1^2}) + z^2(1 - \frac{q_2^2}{q_1^2}) - 2ay + a^2 = x^2 + (y - \frac{aq_1^2}{(q_1^2 - q_2^2)})^2 + z^2 = \frac{a^2 q_1^2 q_2^2}{(q_1^2 - q_2^2)^2}. \quad (7.49)$$

This is a sphere of radius

$$R = \left| \frac{aq_1 q_2}{(q_1^2 - q_2^2)} \right| \quad (7.50)$$

whose center is displaced from the origin of coordinates by a quantity $y_0 = \frac{aq_1^2}{(q_1^2 - q_2^2)}$ over the OY axis. Depending on the relation between the value of the charges q_1 and q_2, the location of y_0 may be on the left or on the right with respect the location of the charge q_1. In fact, if $|q_1| < |q_2|$, then $(q_1^2 - q_2^2) < 0$, which displace the center of the sphere on the left side (Fig. 7.1). The contrary occurs when $|q_1| > |q_2|$. By this possibility $(q_1^2 - q_2^2) > 0$, and $y_0 > 0$, locating the centre on the right.

For simplicity, if we rewrite the expression of R^2 which appears in (7.49) by using only, for instance, the coordinate y, we obtain

$$R^2 = \frac{a^2 q_1^2}{(q_1^2 - q_2^2)} \left(\frac{q_2^2}{q_1^2 - q_2^2} \right) = \frac{a^2 q_1^2}{(q_1^2 - q_2^2)} \left(\frac{q_1^2}{q_1^2 - q_2^2} - 1 \right) = ay_0 \left(\frac{y_0}{a} - 1 \right) = y_0^2 - ay_0 = y_0(y_0 - a), \quad (7.51)$$

where the sign of y_0 must be taking into account. For example, if $y_0 < 0$ (see Fig. 7.1) (7.51) has the form $R^2 = y_0(y_0 + a)$. In general we can write

$$R^2 = y_0(y_0 - a) = y_0 d_0. \quad (7.52)$$

Equation (7.52) means that the product of the distance from the center of the sphere to the image charge q_1 times the distance between O and the position of the actual charge q_2 is a constant of value R^2. On the other hand, considering the value of a as a function of R and introducing it into y_0, we have

$$\frac{R(q_1^2 - q_2^2)}{(q_1 q_2)} \left(\frac{q_1^2}{q_1^2 - q_2^2} \right) = y_0, \tag{7.53}$$

that is,

$$q_1 = \frac{y_0}{R} q_2. \tag{7.54}$$

By using (7.53) and taking into consideration that the values of y_0 for the system of coordinates chosen is negative, (7.54) takes the form

$$q_1 = \frac{R}{d_0} q_2, \tag{7.55}$$

where $d_0 = (y_0 + a)$. Note that in these equations the signs of the charges are opposite to each other.[2]

From this result we can see that, for instance, if we have a metallic spherical surface of radius R held to zero potential in front of a charge q_2, adjusting with an adequate image charge q_1, we could obtain an equipotential surface ($V = 0$) coinciding with the sphere. Therefore, as the solution is unique, the fictitious disposition for the charge q_1 makes the problem equivalent to the actual problem (it produces the same effect), thus questions about the potential at any point of space, or the induced distribution of charge on the conducting sphere, may be answered by referring to calculations based on the aforementioned *fictitious* system of charges. This idea is the basis of the method for any geometry where this technique can be applied.

From a more mathematical way the problem may be posed in a different way.

In the foregoing section we saw that the solution of the Dirichlet and Neumann problems may be reduced to the calculation of the corresponding Green's function which has the form (see (7.35) in Sect. 7.3)

$$G(\mathbf{r}, \mathbf{r}') = \frac{1}{|\mathbf{r} - \mathbf{r}'|} + g(\mathbf{r}, \mathbf{r}'). \tag{7.56}$$

This equation tells us that, in principle, we can find a charge distribution $g(\mathbf{r}, \mathbf{r}')$ which is capable of exactly compensating the potential $\frac{1}{|\mathbf{r} - \mathbf{r}'|}$ created by a point unit charge, so that on the surface $G(\mathbf{r}, \mathbf{r}') = 0$ (see properties again), or $\frac{\partial G(\mathbf{r}, \mathbf{r}')}{\partial n} = \frac{1}{S}$, S being the surface area.[3] As it can be examined, this result is the same as exposed before (Fig. 7.3).

---

[2] If we have considered the sign of the charges at the principle of the problem, then both members of (7.55) appear positive. If we do not do it (7.55) would appear as $q_1 = -\frac{R}{d_0} q_2$.

[3] See Problem 7.14.

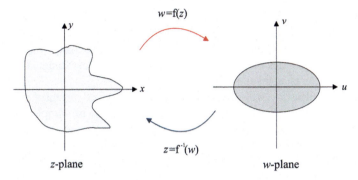

Fig. 7.3 The mapping $w = f(z)$ transforms the domain D into another D^* where the solution may be calculated easier

7.5 Application of Complex Analysis to Electromagnetism

In this section we will briefly explain the fundamentals of another technique based on complex analysis for determining the solution of electromagnetic problems under certain conditions. Because of the specific characteristics of the many different problems, sometimes they should be simplified so that the solution is easier to find. For instance, as we have seen in foregoing chapters, the calculation of the electric or magnetic fields with symmetries allow us to consider reducing the number of variables necessary for calculations. Specifically, when the system to be studied has translational symmetry in one of the axes chosen for its representation, we can analyze the problem like a two dimensional problem, considering only two variables for the computations. The final result in $\Re^3$ is then obtained by translating the solution through the symmetry axis. In this regard, methods of complex analysis play an important role in these kind of systems due to their versatility to transform plane regions difficult to study to domains where the problem may be posed in a simpler and comprehensive way. Basically, the application of complex analysis to solve the problem is based on conformal mapping.[4] By this method we transform the region of the actual two-dimensional electromagnetic problem defined on the z-plane (domain D), onto another in the w-plane (domain D^*) in which either we can calculate the solution easily or we already *know* the solution of the transformed problem. Once we have solved the solution in w, the final result of our original problem is obtained going back to the z-plane.

---

[4]The existence of the symmetry along one axis provides the study a system in an easier form. In the way we present this method based on complex variables, it could seem that this technique is only useful when symmetries exist, however that is not complectly true. In fact, recently it has been demonstrated that conformal mapping may be also employed for systems where the symmetry is partially broken. Specifically, it is possible to find the electrostatic potential in a space limited axisymmetric geometry by mapping the plane parallel to the axis and rotating it (see, for example, [116]).

When transforming the domain D on D^* in the w-plane we must examine the boundary conditions given. If the boundary conditions corresponds to the Dirichlet problem, the boundary values of the region D get mapped unchanged to the corresponding values of ∂D^*. However, if we work under Neumann conditions the transformation changes the data.

To fix these ideas we will denote a function of complex variables as usual, i.e. $f(z) = u(x, y) + iv(x, y)$ where $z = x + iy$. In this context of complex analysis, a function assigns at every point z of the complex plane in the domain D a one-to-one value $f(z)$ belonging to D^*. As we will succinctly comment, sometimes a function is multiple-valued, i.e. points of D can have more than one image in D^*. To solve this drawback the concept of a branch of the complex function $f(z)$ is introduced. It consists of delimiting the region on D where, for sure, at every point $z \in D$ corresponds only one value in D^*. To do so, a cut line in D is chosen in such a manner that no value of z may cross through it, then guarantying the one-to-one correspondence between z and $f(z)$.

There is another more general way to obtain a similar result, but in a more complicated way. In effect, a multiply-valued function may be regarded as a Riemann surface. From a geometrical viewpoint we can imagine a non single-valued function of a complex variable z as a construction containing different sheets of the same $f(z)$.

In effect, let us suppose that two functions $f_1(z)$ and $f_2(z)$, each one defined in its respective domain D_1 and D_2, have a part of themselves where they coincide ($D_1 \cap D_2 = \varpi \neq 0$). In this case we can obtain a single-valued function by eliminating the sheet corresponding to the overlapping regions (ϖ), and joining what remains of D_1 and D_2 along (ϖ).[5]

7.5.1 Transforming Boundary Conditions

Taking into account the key idea expressed of using methods of conformal representation, an important question that must be taken into account refers to how do boundary conditions change when mapped. In fact, let us suppose that the 2D-curve delimiting D is parameterizable. Then we can write $x = x(t)$ and $y = y(t)$, t being the parameter. By introducing these functions into the functions $u(x, y)$ and $v(x, y)$, we have

$$u = u(x, y) = u(x(t), y(t)) \equiv \alpha(t), \quad (7.57)$$

and similar to $v(x, y)$

$$v = u(x, y) = v(x(t), y(t)) \equiv \beta(t). \quad (7.58)$$

---

[5]In some extent a Riemann surface may be regarded as an analytic continuation of the function $f_1(z)$.

7.5 Application of Complex Analysis to Electromagnetism

By using Neumann's boundary conditions, the form of the equation, considering the new variables is

$$b(t)\frac{\partial V(x(t), y(t))}{\partial n} = d(t), \qquad (7.59)$$

where $b(t)$ and $d(t)$ are prescribed in the specific problem.

$$b(t)\frac{\partial V(x(t), y(t))}{\partial n} = b(t)\left(\frac{\partial V(x, y)}{\partial x}\frac{\dot{y}}{(\dot{x}^2 + \dot{y}^2)} - \frac{\partial V(x, y)}{\partial y}\frac{\dot{x}}{(\dot{x}^2 + \dot{y}^2)}\right) = d(t), \qquad (7.60)$$

where b(t) and d(t) denote functions well defined on the boundary. The values of the boundary condition are transformed by means of the following relation

$$b(t)\sqrt{\left(\frac{\dot{\alpha}^2 + \dot{\beta}^2}{\dot{x(t)}^2 + \dot{y(t)}^2}\right)}\left(\frac{\partial V(u, v)}{\partial u}\frac{\dot{\beta}}{(\dot{\alpha}^2 + \dot{\beta}^2)} - \frac{\partial V(u, v)}{\partial v}\frac{\dot{\beta}}{(\dot{\alpha}^2 + \dot{\beta}^2)}\right) = d(t). \qquad (7.61)$$

As we can see, in this case the boundary conditions change in the mapping.

7.5.2 Conformal Mapping

A single valued transformation $w = f(z)$ between the domains D and D^* is said to be conformal if at all points z belonging to D the mapping conserves the angles and invariance of stretching. Conceptually it means that, when transforming points $z \in D$ into points $w \in D^*$ by means of $w = f(z)$, supposing $f'(z) \neq 0$, all curves Γ that cross a point z_0 are transformed into curves Γ^* that go through $w_0 = f(z_0)$ in such a manner that the angle between tangents to any two curves Γ_i at z_0 is the same as the angle at w_0 of the tangents corresponding to the transformed curves Γ_i^* (Fig. 7.4). When the transformation preserves the signs of the angles the conformal mapping is called of the first kind. If in the transformation the angles between tangents change sign (conserving the absolute value) we speak of conformal mapping of the second kind. This class of transformation is fulfilled by complex conjugates of analytic functions which have derivatives distinct from zero. Starting with the aforementioned definitions, two important characteristics of this transformation can be demonstrated,

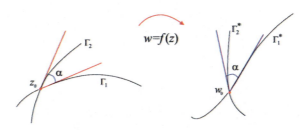

Fig. 7.4 Conformal mapping. Observe that the angles in the transformation are conserved

namely: (a) A conformal mapping transforms infinitesimal circumferences belonging to D onto infinitesimal circumferences in D^*. (b) It conserves the angle between curves at its intersection point. From a practical standpoint, when solving problems it is interesting to establish mathematically how can we perform calculations in order to know whether a transformation is conformal or not. In this regard, the necessary and sufficient condition for a mapping to be conformal is that $w = f(z)$ be univalued and analytic.[6]

7.5.3 Some Conformal Transformations

In order to show examples of conformal mappings, we present some of the most important elementary functions.
- Linear function

Let $f(z) = u(x, y) + iv(x, y)$ be the function of complex variable z defined as

$$f(z) = a + bz, \tag{7.62}$$

where a and b are complex constants. The geometrical significance of (7.62) consist of a translation and a stretching. For this reason this mapping may be used in problems where we want to transform figures so that their characteristics do not change, but only the locations or/and the magnitudes of them.

For analyzing whether this function transforms points $z \in D$ into points $w \in D^*$ conformally, we must examine its characteristics. If the mapping is conformal, $f(z)$ must be univalued and analytic. In fact, given generic points z_1 and z_2, it is verified that $f(z_1) \neq f(z_2)$, which means that this function is univalued. On the other hand it is easy to see that $f'(z) = b \neq 0 \ \forall z \in D$, hence the mapping is conformal.[7]
- Bilinear transformation

A bilinear or fractional transformation is defined as follows

$$f(z) = \frac{a + bz}{c + dz}, \tag{7.63}$$

a, b, c, and d being complex constants and $bc - ad \neq 0$. This transformation also maps points $z \in D$ onto points $w \in D^*$ conformally, because

---

[6]There are four equivalent ways to define when $w = f(z)$ is analytic. The function $f(z) = u(x, y) + iv(x, y)$ is analytic in D if it can be expanded by polynomials of the form $f(z) = \Sigma_{n=0}^{\infty} a_n (z - z_0)^n$, where $z_0 \in D$. If there exist $f'(z) \neq 0$ for any point of D. If its real and imaginary parts verify the Cauchy–Riemann conditions, i.e. $\frac{\partial u(x,y)}{\partial x} = \frac{\partial v(x,y)}{\partial y}$, and $\frac{\partial u(x,y)}{\partial y} = -\frac{\partial v(x,y)}{\partial x}$. This last equality is the same as to say that $u(x, y)$ and $v(x, y)$ are harmonics, and they verify Laplace equation. If $\oint_\Gamma f(z) = 0$, for any closed curve $\Gamma \in D$.

[7]Actually, for proving that the transformation is conformal we also must study the behavior of $f(z)$ in the neighbor of $z = \infty$. To do this it is enough to change the variable z by $\frac{1}{q}$, and by introducing it into (7.62) we can analyze $f'(q)$ for $q \to 0$.

7.5 Application of Complex Analysis to Electromagnetism

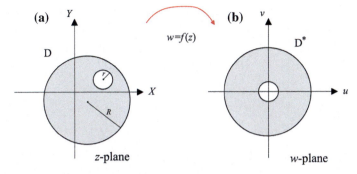

Fig. 7.5 a Two cylindrical conductors whose symmetry axes are parallel to each other but not coincident. **b** After transforming conformally both have their centers at the same point in the w-plane

$$f'(z) = \frac{bc - ad}{(c + dz)^2} \neq 0, \tag{7.64}$$

and transforms the extended z-plane onto the extended w-plane in an infinite number of different ways. However, for solving problems it is necessary to choose the adequate constants that appear in (7.64). In this sense it is possible to find the exact form of the function if we can specify three points $z \in D$ and their corresponding w in D^*. The expression that allows us to find these constants is the following,

$$\frac{(w_1 - w)(w_2 - w_3)}{(w_1 - w_3)(w_2 - w)} = \frac{(z_1 - z)(z_2 - z_3)}{(z_2 - z)(z_1 - z_3)}. \tag{7.65}$$

In electromagnetics, often the systems' object of study are formed by cylinders (capacitors, wires, etc.). In some cases the spacial disposition of each one can break the symmetry of the system, making the calculations difficult. For analyzing such cases it would be useful to have a transformation so that the mapping changes the geometry of the problem in D^* in such a way that the study is easier. For example, Fig. 7.5a depicts a system formed by two parallel metallic cylinder whose centers are not coincident. The study of the electric field and charge distribution on the surfaces is not easy to compute. However, if we would have the picture shown in (b) the problem could be solved immediately, because it corresponds to a cylindrical capacitor whose solution is known. Then working with this kind of system the bilinear transformation plays an important role because of its basic characteristics, namely: (a) Points z symmetric to a circle[8] will be mapped into points symmetric with respect to the transformed circle in D^*. (b) Every circle belonging to the z-plane is transformed into a circumference of the w-plane. As we will see in the worked problems, we can find an adequate procedure for mapping these types of regions. In this context

---

[8] Two points z_1 and z_2 are said to be symmetric with respect to a circle Γ defined as $|z_1 - z_0| = R$ if they lie on the same segment containing z_0 and $|z_1 - z_0| |z_2 - z_0| = R^2$.

Fig. 7.6 Mapping of a region D into the unit *circle*

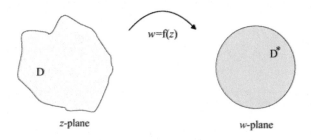

of transforming regions for obtaining other more simpler domains, a very important theorem due to Riemann, must be commented on. Let us suppose a singly connected domain D of the complex z-plane whose boundary is ∂D. It can be demonstrated, that this region D can be conformally mapped onto the interior of the unit circle $|w| < 1$ of the w-plane. The problem of this theorem is that, even if a mapping can convert (see Fig. 7.6) a difficult domain D into the unit circle, it does not say how we can construct the transformation. To seek this mapping, intuition and experience come into play, and in no case is it easy when working with very irregular domains, where numerical techniques must be employed.

7.5.4 Complex Potential

In this section we will present a procedure that does allow geometrical characteristics of the fields. With this aim, let us suppose a transformation $(x, y) \in \Re^2 \longrightarrow \mathcal{C}$ so that at every point of the plane OXY where an electric field $\mathbf{E}(x, y)$ is defined, we map it into the following complex variable function

$$E(z) = E_x(x, y) + iE_y(x, y). \tag{7.66}$$

If in the region of interest we have neither charges nor currents, we can derive the electric field from a scalar potential, i.e. $\mathbf{E}(x, y) = -\nabla V(x, y)$, then

$$E_x(x, y) = -\frac{\partial V(x, y)}{\partial x} \quad E_y(x, y) = -\frac{\partial V(x, y)}{\partial y}, \tag{7.67}$$

and

$$E(z) = -\frac{\partial V(x, y)}{\partial x} - i\frac{\partial V(x, y)}{\partial y}. \tag{7.68}$$

Since the field is irrotational and solenoidal, the potential $V(x, y)$ verifies the Laplace equation, thus it is represented by a harmonic function. For this reason we can form an analytic function in such a way that one its parts (real or imaginary) coincides with $V(x, y)$,

7.5 Application of Complex Analysis to Electromagnetism

$$f(z) = V(x, y) + i\Pi(x, y), \qquad (7.69)$$

$\Pi(x, y)$ being the imaginary part of the analytic function $f(z)$, and at the moment unknown. However, with $V(x, y)$ being harmonic in a simple connected domain, we can construct $\Pi(x, y)$ by means of the following formula

$$\Pi(x, y) = \int_{z_0}^{z} -\frac{\partial V(x, y)}{\partial y} dx + \frac{\partial V(x, y)}{\partial x} dy + C \qquad (7.70)$$

where C is a constant. The election of $V(x, y)$[9] for the real part of the complex potential $f(z)$ is complectly arbitrary. We could also choose the imaginary part (see Problem 15).

The two parts of (7.69) have an important interpretation. In order to understand its significance, let us consider the curves $V(x, y) = C_1$ and $\Pi(x, y) = C_2$, where C_1 and C_2 are constants. As we know such equations represent an independent family of level curves for both parts of $f(z)$. Calculating at any point the scalar product of the gradients for $V(x, y)$ and $\Pi(x, y)$, we could investigate the geometrical relation between both level lines, i.e.,

$$\nabla V(x, y) \cdot \nabla \Pi(x, y) = \left(\frac{\partial V}{\partial x}, \frac{\partial V}{\partial y}\right) \cdot \left(\frac{\partial \Pi}{\partial x}, \frac{\partial \Pi}{\partial y}\right) = \frac{\partial V}{\partial x}\frac{\partial \Pi}{\partial x} + \frac{\partial V}{\partial y}\frac{\partial \Pi}{\partial y}. \qquad (7.71)$$

On the other hand, as $f(z)$ is analytic it must fulfill the Cauchy–Riemann conditions,

$$\frac{\partial V}{\partial x} = \frac{\partial \Pi}{\partial y}, \quad \frac{\partial V}{\partial y} = -\frac{\partial \Pi_z}{\partial x}, \qquad (7.72)$$

and therefore setting (7.72) into (7.71) leads to

$$\nabla V(x, y) \cdot \nabla \Pi(x, y) = -\frac{\partial V}{\partial x}\frac{\partial V}{\partial y} + \frac{\partial V}{\partial y}\frac{\partial V}{\partial x} = 0. \qquad (7.73)$$

From this result we can interpret that, as the gradients at any point are orthogonal and at the same time each gradient is perpendicular to its respective level line, the family of curves $V(x, y) = C_1$ and $\Pi(x, y) = C_2$ must be also perpendicular to each other (see Fig. 7.7). As we can see, owing to the foregoing perpendicularity, $\nabla V(x, y)$ is parallel to the level curves $\Pi(x, y) = C$, then the electric field $\mathbf{E}$ is tangent to $\Pi(x, y)$ at each point. For this reason we can identify the imaginary part of the complex potential (7.69) as the line forces of the field.

---

[9] It is necessary to be careful with the notation presented. For us a complex function has the form $f(z) = u(x, y) + iv(x, y)$. By the way we have written the complex potential $f(z)$, $V(x, y)$ corresponds to $u(x, y)$ but not with $v(x, y)$. We have selected V instead of $U(x, y)$, for instance, because we have chosen V for the scalar potential in all the book.

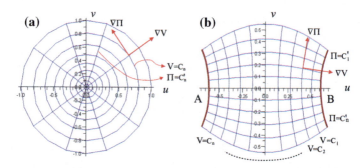

Fig. 7.7 **a** Lines $V(x, y) = C_i$ and $\Pi(x, y) = C_i$ corresponding to the complex potential of a infinite conducting wire of constant charge per unit length. **b** Idem but corresponding to the potential between two curved metallic plates of a capacitor. Note that $\nabla \Pi(x, y)$ is tangent to the curves of constant potential $V(x, y) = C$, and $\nabla V(x, y)$ also to $\Pi(x, y) = C$

From the above definition of the complex potential we see its utility for studying electric or magnetic potentials and fields but if $f(z)$ it is known, but the knowledge of the conformal representation for a specific problem is sometimes hard to come by.

Once the complex potential is known, it is possible to set up its relationship with the electric field. Taking into account the Cauchy–Riemann relations and our definition (7.66), we have

$$E(z) = -\frac{\partial V(x,y)}{\partial x} - i\frac{\partial V(x,y)}{\partial y} = -\frac{\partial V(x,y)}{\partial x} + i\frac{\partial \Pi(x,y)}{\partial x} = -\left(\frac{\partial V(x,y)}{\partial x} - i\frac{\partial \Pi(x,y)}{\partial x}\right) = -\overline{f'(z)}, \tag{7.74}$$

where $\overline{f'(z)}$ is the conjugate of $f'(z)$. From (7.74) we can obtain the expression of the potential as a function of the complex electric field

$$f(z) = -\int \overline{E(z)}\,dz + a, \tag{7.75}$$

a being a constant.

Now we will calculate the integral of $-f'(z) = \overline{E(z)}$, because, as we are going to demonstrate it has an important physical significance. Observe that we focus our attention on the conjugate of $E(z)$ and not on the electric field self. To clarify the ideas, the vector field $\overline{E}(z)$ is called the Pólya vector field of $E(z)$. In general (not only for the electric field), the Pólya vector field of a vector $P(z)$ is simply its conjugate $\overline{P}(z)$.

By means of the derivative and using the Cauchy–Riemann conditions it holds that

$$-\oint_\Gamma f'(z)\,dz = \oint_\Gamma \overline{E}(z)\,dz = -\oint_\Gamma \left(\frac{\partial V(x,y)}{\partial x} + i\frac{\partial \Pi(x,y)}{\partial x}\right)(dx + i\,dy) =$$

$$-\oint_\Gamma \left(\frac{\partial V(x,y)}{\partial x}dx - \frac{\partial \Pi(x,y)}{\partial x}dy\right) - i\oint_\Gamma \left(\frac{\partial V(x,y)}{\partial x}dy + \frac{\partial \Pi(x,y)}{\partial x}dx\right) =$$

7.5 Application of Complex Analysis to Electromagnetism

$$-\oint_\Gamma \left(\frac{\partial V(x,y)}{\partial x}dx + \frac{\partial V(x,y)}{\partial y}dy\right) - i\oint_\Gamma \left(\frac{\partial V(x,y)}{\partial x}dy - \frac{\partial V(x,y)}{\partial y}dx\right) =$$

$$\oint_\Gamma \langle \mathbf{E}\cdot\mathbf{u}_t\rangle\, dl + i\oint_\Gamma \langle \mathbf{E}\cdot\mathbf{n}\rangle\, dl = \oint_\Gamma E_t\, dl + i\oint_\Gamma E_n\, dl, \qquad (7.76)$$

where E_t and E_n are the tangential and normal bidimensional components of the electric field, and $\mathbf{u}_t$ and $\mathbf{u}_n$ are unitary vectors. From a physical point of view, the first integral of the last result shown in (7.76) represents the circulation of the electric field over the closed curve chosen, and the second one the flux of E throughout Γ. Labelling the circulation Ω and the flux Φ, we obtain

$$\oint_\Gamma \overline{E}(z)\, dz = -\oint_\Gamma f'(z)\, dz = \Omega + i\,\Phi. \qquad (7.77)$$

This result shows that we can compute the circulation (work) and flux of the electric field along the integration contour Γ by using its corresponding Pólya field.[10]

Observe that by applying Green's theorem we also can express (7.76) and (7.77) by means of the rotational and divergence in a similar way to the three-dimensional case. To do this, let us first consider Ω as a function of the field projections instead the potentials by substituting $E_x = -\frac{\partial V(x,y)}{\partial x}$ and $E_y = -\frac{\partial V(x,y)}{\partial y}$

$$\Omega = \oint_\Gamma \left(-\frac{\partial V(x,y)}{\partial x}dx - \frac{\partial V(x,y)}{\partial y}dy\right) = \oint_\Gamma E_x\, dx + E_y\, dy = \iint_S \left(\frac{\partial E_y(x,y)}{\partial x} - \frac{\partial E_x(x,y)}{\partial y}\right) dS, \qquad (7.78)$$

which represents the rotational in two-dimensions. In the same manner, we obtain for Φ the following formula

$$\Phi = \oint_\Gamma \left(-\frac{\partial V(x,y)}{\partial x}dy + \frac{\partial V(x,y)}{\partial y}dx\right) = \oint_\Gamma E_x\, dy - E_y\, dx = \iint_S \left(\frac{\partial E_x(x,y)}{\partial x} + \frac{\partial E_y(x,y)}{\partial y}\right) dS, \qquad (7.79)$$

whose right side is the divergence, hence we can write

$$\oint_\Gamma \overline{E}(z)\, dz = \iint_S \nabla \times E\, dS + i \iint_S \nabla \cdot E\, dS. \qquad (7.80)$$

This equation gives us a complementary interpretation of (7.77). It tells that the circulation of the Pólya field of $E(z)$ may be also understood in terms of the rotational and the divergence of electric field. It is very important because it sheds light on important theorems in electromagnetism.

In fact, in the case we are going to study where we do not have fields depending on time ($\nabla \times \mathbf{E} = 0$), then there is no circulation (irrotational) and, as a result, Ω must be zero. Rewriting (7.77) yields,

---

[10] In this moment we deal with $\mathbf{E}$, but the idea is also applicable to any vector field.

$$-\oint_\Gamma f'(z)\,dz = i\oint_\Gamma E_n\,dl = i\int\int_S \nabla \cdot E\,dS = i\,\Phi, \qquad (7.81)$$

thus

$$\oint_\Gamma E_n\,dl = \int\int_S \nabla \cdot E\,dS = \frac{\lambda}{\varepsilon_0}, \qquad (7.82)$$

λ being the charge per unit length (remember the symmetry). Equation (7.82) represents Gauss's law when working with complex potentials and shows that the flux of the electric field across Γ depends on the charge enclosed by the curve (see Chap. 2). Therefore, we can conclude that the Gauss theorem for the electric field **E** is connected with the imaginary part of the circulation of its Pólya field along the curve Γ.

If instead of **E** we investigate the magnetic field **H**, it is also possible to define a complex potential in a similar way, however some differences appear. In fact, let us suppose a region free of currents where our system is symmetric with respect to a specific direction in space. Under these condition we can find a scalar function V_m so that $\mathbf{H} = -\nabla V_m$, and therefore a complex potential can be defined in such a manner that its real part coincides with V_m, i.e.,[11]

$$F(z) = V_m(x,y) + i\Pi_m(x,y), \qquad (7.83)$$

where $V_m(x,y)$ represents the potential and $\Pi_m(x,y)$ the function of force lines.[12] The process to determine the expressions for H is the same as shown for the electric field, however some results differ. For instance, (7.77) is modified because of the own characteristics of the magnetic field. As we saw in Chap. 5, for non-varying fields we worked with Ampère's law, which established a relation between the circulation of **H** (or **B**) and the intensity pierced by the closed curve chosen for performing the integration. On the other hand we know that $\nabla \cdot \mathbf{H} = -\nabla \cdot \mathbf{M}$ because $\nabla \cdot \mathbf{B} = 0$, then in a region free of magnetized matter $\nabla \cdot \mathbf{M} = 0$, and therefore $\nabla \cdot \mathbf{H} = 0$. This means that we cannot have a flux of H across a closed curve (in the context spoken before), but only a circulation because of the tangential component. As a consequence, the equivalent equation to (7.77) for H is the following

$$\oint_\Gamma \overline{H}(z)\,dz = -\oint_\Gamma f'(z)\,dz = \oint_\Gamma H_t\,dl = \int\int_S \nabla \times H\,dS = \Omega, \qquad (7.84)$$

---

[11] We have supposed that the physical system has the direction OZ as the axis of symmetry.

[12] The definition of complex potential is also very useful in other branches of physics such as Fluid Mechanics or Thermodynamics. By working with incompressible fluids a potential is defined as $f(z) = \phi(x,y) + i\psi(x,y)$ for the velocities, where the significance of the real and imaginary parts are changed with respect to our presentation of the electric and magnetic fields. Indeed, $\phi(x,y) = C$, C being constant, corresponds to stream lines, and $\psi(x,y) = C$ are the equipotential curves. Besides, some signs are also modified.

where $\overline{H}(z)$ is the Pólya field of $H(z)$. Observe that this new expression depends on Ω and not on Φ, which is related with the rotational. In the same manner we wrote for the Gauss theorem in (7.82), remembering that for the magnetic field $\nabla \times \mathbf{H} = \mathbf{j}$, (7.84) may be expressed as

$$\oint_\Gamma H_t \, dl = \int\!\!\int_S \nabla \times H \, dS = I, \tag{7.85}$$

I being the current crossing the surface delimited by the curve Γ. Equation (7.85) is the form of Ampère's theorem with complex variables, and as we can see it is linked with the real part of the line integral of the Pólya field of H. The other formulae we can obtain are

$$H(z) = -\frac{\partial V_m(x,y)}{\partial x} - i\frac{\partial V_m(x,y)}{\partial y} = -\overline{F'}(z), \tag{7.86}$$

where $\overline{F'}(z)$ is the conjugate of $F'(z)$. From (7.74) we can obtain the expression of the potential as a function of the complex electric field

$$F(z) = -\int \overline{H}(z) \, dz + b, \tag{7.87}$$

where b is again a constant.

7.6 Numerical Techniques

There are distinct important numerical methods that may be employed for solving electromagnetic problems. The election of one of them in particular may depend on the kind and characteristics of problem to be solved. So, the geometry, boundary conditions, precision, and time of calculation are some of the most basic questions that usually are posed before choosing a specific technique. In this section we are first dealing with the method of finite differences which, as we will see, has some advantages. Additionally, some basic ideas of other techniques will be also presented but not in detail, because they are beyond the scope of this book.

7.6.1 The Finite Difference Method

The method of finite differences (FDM) is one of the most successful numerical techniques used, not only in electromagnetics but also in other scientific and technical branches. This technique consists of dividing the region of interest into a mesh formed by parallel lines, which leads to a set of nodes spaced a length a from each other. A numerical value is given to each node and, under some approximations, the potential

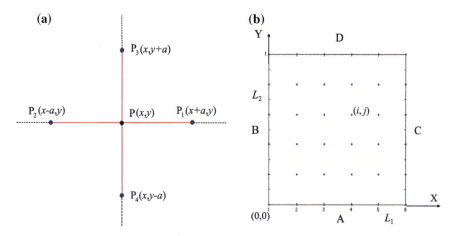

Fig. 7.8 **a** Grid and nodes corresponding to the nearest neighbors. **b** Mesh in a rectangular domain

value at each point may be related with the potentials of the nearest neighbors by linear equations (Fig. 7.8).

Let us suppose we want to know the potential V at point P of coordinates (x, y), approximately. For relating the potential at this node with the potentials at P_1, P_2, P_3, and P_4, if the function representing V is continuous, we can expand the potential in Taylor series as follows

$$V(x+a, y) \approx V(x, y) + \frac{\partial V(x, y)}{\partial x}a + \frac{1}{2}\frac{\partial^2 V(x, y)}{\partial x^2}a^2 + \vartheta(x^3), \qquad (7.88)$$

and for the point on the left

$$V(x-a, y) \approx V(x, y) - \frac{\partial V(x, y)}{\partial x}a + \frac{1}{2}\frac{\partial^2 V(x, y)}{\partial x^2}a^2 + \vartheta(x^3). \qquad (7.89)$$

In the same manner, for the neighbors above and below

$$V(x, y+a) \approx V(x, y) + \frac{\partial V(x, y)}{\partial y}a + \frac{1}{2}\frac{\partial^2 V(x, y)}{\partial y^2}a^2 + \vartheta(y^3), \qquad (7.90)$$

and

$$V(x, y-a) \approx V(x, y) - \frac{\partial V(x, y)}{\partial y}a + \frac{1}{2}\frac{\partial^2 V(x, y)}{\partial y^2}a^2 + \vartheta(y^3). \qquad (7.91)$$

7.6 Numerical Techniques

The addition of the two first equations leads to

$$\frac{\partial^2 V(x,y)}{\partial x^2} \approx \frac{V(x+a,y) + V(x-a,y) - 2V(x,y)}{a^2}, \quad (7.92)$$

and the same for (7.90) and (7.91)

$$\frac{\partial^2 V(x,y)}{\partial y^2} \approx \frac{V(x,y+a) + V(x,y-a) - 2V(x,y)}{a^2}. \quad (7.93)$$

From these results the approximate expression of the Laplace equation at point P is

$$\nabla^2 V(x,y) = \frac{\partial^2 V(x,y)}{\partial x^2} + \frac{\partial^2 V(x,y)}{\partial y^2} \approx$$
$$\approx \frac{V(x+a,y) + V(x-a,y) + V(x,y+a) + V(x,y-a) - 4V(x,y)}{a^2} = 0, \quad (7.94)$$

and then the potential

$$V(x,y) \approx \frac{V(x+a,y) + V(x-a,y) + V(x,y+a) + V(x,y-a)}{4}. \quad (7.95)$$

By this procedure we have reduced the solution of the Laplace differential equation to a set of algebraic equations that allow us to estimate the potential at each point of the net. Relation (7.95) is then the basis of the method and, according to that, applying it at the nodes of the mesh we should obtain a system of N linear algebraic equations with N unknowns in order to have a solution. These results may be expressed in a matrix form as

$$\tilde{S}V = B, \quad (7.96)$$

where $\tilde{S}$ represents a matrix with the respective coefficients of the potentials at point (x_i, y_j), V is a vector containing all unknown V_k with $k = 1, \ldots N$, and B is another vector with known values. Thus the problem solution consists of determining the inverse of $\tilde{S}$ and multiplying it by B.

The method of finite differences explained has the advantage that its mathematical procedure is very easy and simple for programming, however it has some drawbacks. For example, when studying complex irregular boundary shapes it is necessary, but not always easy, to match with enough accuracy the boundary with a suitable grid and density nodes. Another difficulty arises when, because of the characteristics of the problem, a higher terms of the Taylor's series must be introduced in order to augment the precision of the results. In this case the degree of complexity makes the FDM not be so flexible and easy to apply, and then not very useful if the precision, programming time and truncation errors are important conditions of the potential solution.

If the distances of the neighbors to the point $P(x, y)$ are not equal, the expressions that relate the potential $V(x, y)$ with the nearest potentials to P is

$$V(x, y) \approx \frac{h_3 h_4 (h_2 V_1 + h_1 V_2)}{(h_1 + h_2)(h_1 h_2 + h_3 h_4)} + \frac{h_1 h_2 (h_4 V_3 + h_3 V_4)}{(h_3 + h_4)(h_1 h_2 + h_3 h_4)}, \quad (7.97)$$

where h_1, h_2, h_3 and h_4 are the unequal distances between point P and its surrounding points.

The formulae shown above started on conditions for the potential on the boundary. However, as we saw with the Laplace equation, sometimes we do not have knowledge of the potential but on its normal partial derivatives (Neumann). In such a case additional equation are needed to pose the problem. The basic idea consists of adding new virtual points outside the region of interest, but near to it, and work with them in a similar way. By this procedure, with the new equations generated it is possible to rearrange the variables in such a manner that the *virtual points* do not appear in the final set of equations.

For simplicity let us suppose we have a rectangular geometry with Neumann's boundary conditions, i.e. $\frac{\partial V}{\partial n} = h$ on its sides. If the value of h on the edges are, respectively, h_A, h_B, h_C, and h_D, the equations that must be considered at each point of the boundary together with (7.95) are the following

$$V(i + a, 0) + V(i - a, 0) - 2ah(i, 0) = 0 \Rightarrow V(i + a, 0) + V(i - a, 0) = 2ah_A, \quad (7.98)$$

$$V(i + a, L_1) + V(i - a, L_1) - 2ah(i, L_1) = 0 \Rightarrow V(i + a, L_1) + V(i - a, L_1) = 2ah_D, \quad (7.99)$$

$$V(0, j + a) + V(0, j - a) - 2ah(0, j) = 0 \Rightarrow V(0, j + a) + V(0, j - a) = 2ah_B, \quad (7.100)$$

and

$$V(L_2, j + a) + V(L_2, j - a) - 2ah(L_2, j) = 0 \Rightarrow V(L_2, j + a) + V(L_2, j - a) = 2ah_C. \quad (7.101)$$

At points on the corners it is usual to take values obtained by averaging the two nearest neighbors. For instance, at point $(0, 0)$ we have

$$V(1, 0) + V(0, 1) - 2ah(0, 0) = 0, \quad (7.102)$$

and similar equations hold for the other three.

There exist more possibilities to construct a solution. So, we could choose the nearest four neighbors points located on the diagonal (see Fig. 7.9). Proceeding in the same way as at the beginning of this subsection we get

$$V(x, y) \approx \frac{V(x + a, y + a) + V(x - a, y + a) + V(x - a, y - a) + V(x + a, y - a)}{4}. \quad (7.103)$$

7.6 Numerical Techniques

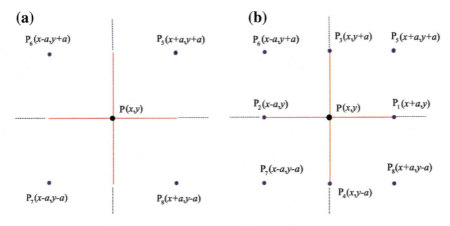

Fig. 7.9 a Calculation by considering the four nearest point over the diagonal. **b** Idem but using eight

Another way can be selecting all the points of the foregoing schemes presented, i.e. the eight neighboring points. In this case the approximate solutions adopts the following form

$$V(x, y) \approx \frac{V(x+a, y) + V(x-a, y) + V(x, y+a) + V(x, y-a)}{20} + \frac{V(x+a, y+a) + V(x-a, y+a) + V(x-a, y-a) + V(x-a, y+a)}{20}.$$
(7.104)

7.6.2 Other Important Techniques

There exist more ways for finding an approximate solution of a well behaved boundary value problem. One of the most used techniques is the Finite element method (FEM). This technique is a very powerful tool for solving all types of problems, not only in electromagnetics but also in many other branches of knowledge such as fluid mechanics, elasticity, building structures, thermal problems and vibration analysis among others. This technique is a variational procedure which, starting on a set of subregions D_i belonging to a whole region D, it approximates the values of the unknowns by functions inside of each subdomain. Such a subregion of D constitutes the *finite element* (Fig. 7.10). The functions chosen are usually algebraic polynomials generated by interpolating the unknown function by means of its values at the nodes of the geometrical element. The way we approximate does not depend

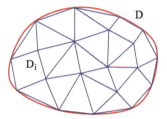

Fig. 7.10 Domain D and subdomains D_i. Observe the different geometries of the finite elements

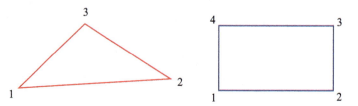

Fig. 7.11 A *triangle* and a *rectangle* as finite elements

on the boundary conditions of the problem. For instance, for a triangle, only three neighboring nodes are chosen and the following function may be employed[13]

$$V(x, y) = a_1 + a_2 x + a_2 y, \tag{7.105}$$

and for a rectangle

$$V(x, y) = a_1 + a_2 x + a_2 y + a_3 xy. \tag{7.106}$$

Once we know the functions to be employed, a variational appropriate model is chosen in order to develop a system of algebraic equations in which a relationship among the values at each point of the element is obtained. This process holds for any of the elements used for constructing the mesh, thus repeating the procedure for each of them we get more equations, making the connections among points shared by different elements. In the same manner we have seen for the method of finite differences, in some of the steps of the procedure the boundary conditions must be imposed, with which a whole system of equations is created for solving the unknowns (Fig. 7.11).

As we can see this method is a very powerful method for any kind of geometry, but has the inconvenience of employing a huge number of steps to obtain the results even for only a few elements, which makes it not practicable by hand.

In the same context of the variational procedures, other important techniques should be mentioned. Of special relevance are the methods of Ritz–Rayleigh, Galerkin (GM), Kantorovich (KM), Trefftz (TM), and least squares (LSM).

---

[13] In each element we can choose more nodes, but in this case the algebraic function must be changed by adding more terms.

7.6 Numerical Techniques

The Ritz–Rayleigh technique allows finding solutions to the variational problem posed by approximating $V(x, y)$ by a linear combinations of basis functions

$$V(x, y) = \sum_{i=1}^{N} \sum_{j=1}^{N} c_{ij} V_{ij}, \qquad (7.107)$$

which must also verify the corresponding boundary conditions. Coefficients c_{ij} are parameters to be determined. The election of the functions in (7.107) should be made with the aim of obtaining matrices easy to deal with, so that an analytical evaluation of the integrals that appear is possible. Common basis functions employed are of the Chebishev, Legendre polynomials, and functions of the form $V_i(x, y) = a_{lm} x^l y^m$ as well. Of course, the functions chosen depends on the geometrical characteristics of the specific problem. With all this information a functional $F(c_{11}, c_{12a}, \ldots, c_{nn})$ can be formed and a variational principle applied to calcule the terms c_{ij}. Ritz–Rayleigh method (RRM) has some advantages to some problems when compared with FEM and other techniques. For instance using global functions is more accurate per degree of freedom of the problem. Besides, from another viewpoint, RRM is like breaking the global problem into smaller problems which it is not possible to do with the FEM.

The Galerking method also starts with a linear combination of basis functions, but it calculates the coefficients c_{ij} imposing that the residual R, i.e. the difference between the exact and the approximate solution, is orthogonal to the functions V_{ij}. When this technique is used in variational problems, including quadratic functionals, it is essentially the same as the Ritz–Rayleigh method.

The Kantorovich method basically consists in reducing the integration of the partial differential equation to the integration of a system of ordinary differential equations in terms of unknown functions. For this reason it is very useful when resolving problems in which the variables are independent.

By the Trefftz technique (TM), different from the other methods we have seen, the basis functions are chosen for satisfying the differential equation but not necessarily the boundary conditions. In this approach the method starts from a variational principle as before, seeking the solution parameters c_{ij} in such a way to obtain the boundary conditions required. In other words; this procedure focuses on obtaining the adjustment of the boundary values. In some extent the idea is complementary to the other methods presented, in which the boundary conditions are imposed, known data.

In relation to the least square method (LSM) the only difference with the Galerkin method is in the conditions imposed to the residual when minimizing the functionals. In this new case the constrain holds on R in the following form

$$\int \int_D R^2 \, dS, \qquad (7.108)$$

which must be a minimum.

The differences among these methods are in the form of the variational statement used and on the functions selected. However, compared with the method of finite differences these last techniques shown may be somehow considered as *semi-analytical* procedures. In fact, though they give numerical approximate values in the region of study, they use analytical functions for constructing an approach to the problem.

Solved Problems

Problems C

7.1 Consider two parallel conducting square plates located at $z_1 = -d$ and $z_2 = d$. If the potentials of the plates are V_1 and V_2, respectively, find: (a) The expression of the potential between plates. (b) The capacity of the capacitor.

Solution

This dispositive formed by two parallel thin metallic plates is known as a capacitor (see Chap. 2). This set-up is one of the most simple geometries we can have for a condenser. For solving this problem we suppose the distance between plates is much smaller than the length of the side, that is, $|z_2 - z_1| \ll L$, L being the length of the square sheet. By this way we can neglect end effects that capacitors have. Starting from this viewpoint, we can directly use the Laplace equation in cartesian coordinates. However, due to the symmetries of the device we may reduce the number of differential equations to be solved. In fact, as the ratio $\frac{2d}{L} \ll 1$, regardless of end effects, we can consider the plates as almost infinite, and then along the directions OX, and OY, translational symmetry holds. In this case, we do not need to solve the differential equations for all the variables. On the contrary, we must focus our attention only to z, where we have no symmetry. Employing (7.10) we can write

$$\frac{\partial^2 V}{\partial z^2} = \frac{d^2 V}{dz^2} = 0, \qquad (7.109)$$

which is a differential equation of only one variable. Integrating it gives

$$\frac{dV}{dz} = C_1, \qquad (7.110)$$

where C_1 is a constant. Integrating again

$$V = C_1 z + C_2, \qquad (7.111)$$

C_2 being another constant. The problem is now to determine both constants. To this aim we use the boundary conditions, that is, the value of the potential on each plate of the capacitor. We impose the following conditions

$$V(z_1) = V(-d) = V_1, \qquad (7.112)$$

and

$$V(z_2) = V(d) = V_2, \qquad (7.113)$$

then we get

$$V(-d) = -C_1 d + C_2 = V_1, \qquad (7.114)$$

and

$$V(d) = C_1 d + C_2 = V_2. \qquad (7.115)$$

We have obtained a system of two linear equations with two unknowns, whose solutions is

$$C_1 = \frac{V_2 - V_1}{2d}, \qquad (7.116)$$

and

$$C_2 = \frac{V_2 + V_1}{2}, \qquad (7.117)$$

Introduction of these constants into (7.111) leads to

$$V(z) = \frac{V_2 - V_1}{2d} z + \frac{V_2 + V_1}{2} = \frac{1}{2}\left(\frac{V_2 - V_1}{d} z + (V_2 + V_1)\right). \qquad (7.118)$$

As a particular case if $V_1 = 0$, and making $V_2 = V$, we have

$$V(z) = \frac{V}{2}\left(\frac{z}{d} + 1\right). \qquad (7.119)$$

7.2 The capacitor of the figure consists of two metallic identical non-parallel plane plates a and b. The edge perpendicular to the plane of the Fig. 7.12 is L_1, and the other one L_2. The side L_1 is very large and L_2 is large enough so that we can neglect end effects. The angle between these plates is θ and the distance from the origin O to both edges parallel to OZ is c. The potential of the first one is V_1 [V], and of the upper plate V_2 [V]. Obtain: (a) The potential in the interior region of the plates. (b) The electric field. (c) The density of charge on the plates. (d) The total charge on a plate per unit length L_1.

Solution

As we can observe, both plates are located in such a manner that two of their edges are parallel to the OZ axis, then considering both sheets as very large in this direction the problem has translational symmetry along OZ. Due to this fact, we can suppress the z coordinate in (7.23) simplifying the problem, thus we may write

$$\Delta \equiv \frac{1}{\rho}\frac{\partial}{\partial \rho}\left(\rho \frac{\partial V(\rho, \phi)}{\partial \rho}\right) + \frac{1}{\rho^2}\frac{\partial^2 V(\rho, \phi)}{\partial \phi^2} = 0. \qquad (7.120)$$

Fig. 7.12 Two thin metallic plates forming an angle α. The surface of each plate is $L_1.L_2$

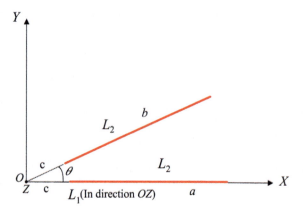

Using the method of separation of variables we can try solutions of the form $V(\rho, \phi) = R(\rho)\Phi(\phi)$. Introduction of this function into (7.120) leads to

$$\frac{1}{\rho}\frac{\partial}{\partial \rho}\left(\rho\frac{\partial(R(\rho)\Phi(\phi))}{\partial \rho}\right) + \frac{1}{\rho^2}\frac{\partial^2(R(\rho)\Phi(\phi))}{\partial \phi^2} = 0, \quad (7.121)$$

and computing the derivatives this equation may be expressed as

$$\frac{\rho}{R(\rho)}\frac{\partial}{\partial \rho}\left(\rho\frac{\partial R(\rho)}{\partial \rho}\right) + \frac{1}{\Phi(\phi)}\frac{\partial^2(R(\rho)\Phi(\phi))}{\partial \phi^2} = 0. \quad (7.122)$$

The first part of the last equation depends only on ρ and the second only on ϕ, therefore this equality holds if

$$\frac{\rho}{R(\rho)}\frac{\partial}{\partial \rho}\left(\rho\frac{\partial R(\rho)}{\partial \rho}\right) = -\frac{1}{\Phi(\phi)}\frac{\partial^2(R(\rho)\Phi(\phi))}{\partial \phi^2} = k^2, \quad (7.123)$$

k being a constant. Resolving separately both equations we have

$$\frac{\rho}{R(\rho)}\frac{\partial}{\partial \rho}\left(\rho\frac{\partial R(\rho)}{\partial \rho}\right) = k^2. \quad (7.124)$$

Here we have two possibilities, namely $k = 0$ and $k \neq 0$. For the first case we get

$$\frac{\rho}{R(\rho)}\frac{\partial}{\partial \rho}\left(\rho\frac{\partial R(\rho)}{\partial \rho}\right) = 0 \Rightarrow \rho\frac{\partial R(\rho)}{\partial \rho} = A_1, \quad (7.125)$$

where A_1 is a constant. The solution is

$$R(\rho) = A_1 \ln \rho + A_2, \quad (7.126)$$

for A_2 constant, too. If $k \neq 0$ the solution of (7.124) is

$$R(\rho) = B_1 \rho^k + B_2 \rho^{-k}. \tag{7.127}$$

The equation for $\Phi(\phi)$ is

$$\frac{1}{\Phi(\phi)} \frac{\partial^2 (R(\rho)\Phi(\phi))}{\partial \phi^2} = -k^2 \Rightarrow \frac{\partial^2 \Phi(\phi)}{\partial \phi^2} + k^2 \Phi(\phi) = 0, \tag{7.128}$$

which consist of two solutions as before for $R(\rho)$. In effect, when $k = 0$

$$\frac{1}{\rho^2} \frac{\partial^2 V(\rho, \phi)}{\partial \phi^2} = 0 \Rightarrow \frac{\partial V(\rho, \phi)}{\partial \phi} = C_1, \tag{7.129}$$

and integrating for C_1 constant

$$V(\rho, \phi) = C_1 \phi + C_2 \tag{7.130}$$

If $k \neq 0$ the solution is

$$V(\rho, \phi) = D_1 \cos(k\phi) + D_2 \sin(k\phi). \tag{7.131}$$

With all these results we must find what solution is valid for our case. Owing to the boundary conditions of the problem, that is, each of the plates are held a constant potential, the solution must not depend on the distance ρ. For this reason we, in principle, can exclude (7.126) and (7.127). From the other two referring to the angle, (7.131) does not work, because it cannot fulfil the boundary conditions of constant potential on each metallic plate of the capacitor. Thus, we have only (7.130). Imposing that $V(\phi = 0) = V_1$ and $V(\phi = \theta) = V_2$, we easily get

$$V(\rho, \phi) = \frac{(V_2 - V_1)}{\theta} \phi + V_1. \tag{7.132}$$

The electric field inside, neglecting end effects, may be calculated be means of the gradient, i.e.,

$$\nabla V = \frac{\partial V}{\partial \rho} + \frac{1}{\rho} \frac{\partial V}{\partial \phi} + \frac{\partial V}{\partial z}. \tag{7.133}$$

As the potential only depends on ϕ, we can write for the electric field

$$\mathbf{E} = -\nabla V = -\frac{1}{\rho} \frac{\partial V(\phi)}{\partial \phi} \mathbf{u}_\phi, \tag{7.134}$$

which leads to

$$\mathbf{E} = -\nabla V = -\frac{1}{\rho} \frac{(V_2 - V_1)}{\theta} \mathbf{u}_\phi. \tag{7.135}$$

The surface charge density may be calculated by means of the following relation (see Chap. 4)

$$\sigma(\phi = 0) = \varepsilon E = -\frac{1}{\rho}\frac{(V_2 - V_1)}{\theta}\varepsilon, \qquad (7.136)$$

and for the upper plate

$$\sigma(\phi = \theta) = -\varepsilon E = \frac{1}{\rho}\frac{(V_2 - V_1)}{\theta}\varepsilon. \qquad (7.137)$$

(d) For calculating the total charge per unit length L_1 on a plate we can employ the definition of surface charge density, i.e.

$$\sigma(x, y) = \frac{dQ(x, y)}{dS}. \qquad (7.138)$$

By integrating this equation we have

$$Q(x, y) = \int\int_S \sigma(x, y)\, dS. \qquad (7.139)$$

Let us suppose we choose the lower plate. Even though we have used at the beginning of the problem polar coordinates, since this conducting plate lies on the plane OXZ, for simplicity we can work with x and z coordinates. In this case the differential element of surface is $dS = dx\, dz$, then

$$Q(x, y) = \int\int_S \sigma(x, y)dxdz = -\int\int_S \frac{\varepsilon}{x}\frac{(V_2 - V_1)}{\theta}dxdz = -\int_0^{L_1} dz \int_c^{c+L_2} \frac{\varepsilon}{x}\frac{(V_2 - V_1)}{\theta}dx = \\ \frac{(V_2 - V_1)L_1}{\theta}\ln\left(\frac{c + L_2}{c}\right), \qquad (7.140)$$

thus

$$\frac{Q(x, y)}{L_1} = \frac{(V_2 - V_1)}{\theta}\ln\left(\frac{c + L_2}{c}\right), \qquad (7.141)$$

where we suppose L_2 large in order to avoid end effects, but not infinite.[14]

7.3 Let us consider four metallic thin rectangular plates with sides $a \times L$ and $b \times L$ forming a prismatic system as shown in Fig. 7.13a. The edges of length L are much longer than the other. The side AD is held to a potential $V_0 = 200$ V, and the other parts are grounded. (a) Calculate the potential in the region $0 < x < a$ and $0 < y < b$, if $a = 20$ cm and $b = 10$ cm. (b) Obtain the electric field $\mathbf{E}(x, y)$.

---

[14]Equation (7.141) could also be valid for theoretical cases in which the edge L_1 is mathematically infinite, because the result is expressed per unit length. This procedure is very useful to prevent us from having difficulties with infinite quantities. Observe that in this problem the length L_1 must be large in order we can apply reasoning of symmetries which lead to a simplification of the differential equation.

Solved Problems

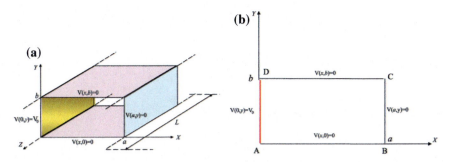

Fig. 7.13 **a** Prismatic system formed by four metallic plates one of them is grounded. **b** Cross-sectional view of the system

Solution

As the edge of length L verifies that $L \gg a$ and $L \gg b$, we can consider that the system has translational symmetry with respect to the OZ axis. It then means that $V(x, y, z) = V(x, y, z+h)$ and $\mathbf{E}(x, y, z) = \mathbf{E}(x, y, z+h)$, h being a constant and neglecting end effects. For this reason we can work with the two-dimensional system depicted in Fig. 7.13b, which is easier. Bearing in mind the geometrical characteristics of the prismatic set of metallic plates, we can find the solution of the problem by employing the two dimensional Laplace equation in cartesian coordinates, that is,

$$\Delta V(x, y, z) \equiv \frac{\partial V}{\partial x} + \frac{\partial V}{\partial y} = 0, \qquad (7.142)$$

constrained by the boundary conditions

$$V(0, y) = V_0 = 200, \qquad (7.143)$$

$$V(x, 0) = 0, \qquad (7.144)$$

$$V(a, y) = 0, \qquad (7.145)$$

and

$$V(x, b) = 0. \qquad (7.146)$$

The first difficulty to account for is that we do not have homogeneous boundary conditions, and if we would like to use the method of separation of variables we must have them. To solve this difficulty we first try solutions of the form

$$V(x, y, z) = X(x) Y(y), \qquad (7.147)$$

and later, we pursue to adjust some coefficients that will appear. Let us introduce (7.147) into (7.142)

$$Y(y)\frac{\partial^2 X(x)}{\partial x^2} + X(x)\frac{\partial^2 Y(y)}{\partial y^2} = 0. \tag{7.148}$$

Now, dividing by the product $X(x)Y(y)$ we have

$$\frac{1}{X(x)}\frac{\partial^2 X(x)}{\partial x^2} + \frac{1}{Y(y)}\frac{\partial^2 Y(y)}{\partial y^2} = 0. \tag{7.149}$$

The only possibility for a non-trivial solution is

$$\frac{1}{X(x)}\frac{d^2 X(x)}{dx^2} = -\frac{1}{Y(y)}\frac{d^2 Y(y)}{dy^2} = \pm k^2. \tag{7.150}$$

k being a constant. The election of plus or minus in this equation depends on the conditions of the problem. So, in our case we know that the potential must be zero on sides AB, BC, and CD, then $Y(y)$ must be zero at $y = 0$ and $y = b$, and $X(x) = 0$ at $x = a$, exclusively, because at $x = 0$ it is non-homogeneous. Thus, one of these two possibilities does not work.

As we know, the general solution of the one dimensional differential equation $\frac{d^2 Q(s)}{ds^2} = k^2 Q(s)$ is $Q(s) = A\exp(ks) + A\exp(-ks)$, which may also be written as an addition of hyperbolic sines and cosines. These kind of functions cannot be zero at two points, namely $y = 0$ and $y = b$ at the same time. On the contrary, the solution of $\frac{d^2 Q(s)}{ds^2} = -k^2 Q(s)$ is the combination of sinusoidal functions, $Q(s) = A\cos(ks) + A\sin(ks)$, which reach zero periodically. From these solutions we see that the only way for this problem to be solved is

$$\frac{1}{Y(y)}\frac{d^2 Y(y)}{dy^2} = -k^2, \quad \text{with} \quad Y(0) = Y(b) = 0, \tag{7.151}$$

and

$$\frac{1}{X(x)}\frac{d^2 X(x)}{dx^2} = k^2, \quad \text{with} \quad X(a) = 0. \tag{7.152}$$

The solution of the first equation must be zero at the first and last points of the interval, which may be accomplished by periodic functions. The second one (7.152) will be zero only at a point, and therefore a linear combination of hyperbolic functions holds.

The general solution of (7.151) is

$$Y(y) = A\cos ky + B\sin ky. \tag{7.153}$$

Imposing the boundary conditions it yields

$$Y(0) = A = 0, \tag{7.154}$$

and for the extreme of the rectangle at $y = b$

$$Y(b) = B \sin kb = 0 \Rightarrow kb = n\pi \Rightarrow k = \frac{n\pi}{b}. \tag{7.155}$$

$$Y(y) = B \sin\left(\frac{n\pi y}{b}\right). \tag{7.156}$$

Introduction of the values of k obtained in (7.155) into (7.153) gives

$$\frac{d^2 X(x)}{dx^2} = \left(\frac{n\pi}{b}\right)^2 X(x). \tag{7.157}$$

This is the differential equation for the function $X(x)$ to be solved, with value $X(a) = 0$. As we can observe, we only have homogeneous conditions at $x = a$, then this problem is not properly a boundary value problem as in the preceding case for $Y(y)$. However, we will avoid this drawback later.

The general solution of this equation may be written in the form of real exponentials or as a combination of hyperbolic functions, that is,

$$X(x) = C \cosh\left(\left(\frac{n\pi}{b}\right)(x-L)\right) + D \sinh\left(\left(\frac{n\pi}{b}\right)(x-a)\right). \tag{7.158}$$

This solution must satisfy $X(a) = 0$, thus

$$X(a) = C + 0 = 0 \Rightarrow C = 0, \tag{7.159}$$

and therefore

$$X(x) = D \sinh\left(\left(\frac{n\pi}{b}\right)(x-a)\right). \tag{7.160}$$

Knowing the expressions for $X(x)$ and $Y(y)$, we can put the solution as the product of (7.156) and (7.160)

$$V(x, y) = X(x)Y(y) = BD \sin\left(\frac{n\pi y}{b}\right) \sinh\left(\left(\frac{n\pi}{b}\right)(x-a)\right). \tag{7.161}$$

This function $V(x, y)$ verify $V(0, y) = V_0 = 200$, but (7.161) does not fulfill this non-homogeneous boundary condition at all (it is not a boundary value problem). In order to avoid this trouble, we try with linear combinations of such functions (7.161). Denoting the product $BD = \alpha$, we can put

$$V(x, y) = X(x) Y(y) = \sum_{1}^{\infty} \alpha_n \sin\left(\frac{n\pi y}{b}\right) \sinh\left(\left(\frac{n\pi}{b}\right)(x-a)\right). \tag{7.162}$$

The objective is now to obtain the coefficients α_n of the series so that the solution meets the aforementioned condition at $x = 0$. For calculating them we will employ the value of the potential on the side AD, that is,

$$V(0, y) = \sum_{1}^{\infty} \alpha_n \sin\left(\frac{n\pi y}{b}\right) \sinh\left(\frac{-n\pi a}{b}\right) = V_0. \quad (7.163)$$

For obtaining the values of α_n we multiply both members by $\sin\left(\frac{m\pi y}{b}\right)$, m being an integer and we integrate in the interval from 0 to b for the variable y

$$\sum_{1}^{\infty} \int_0^b \alpha_n \sin\left(\frac{n\pi y}{b}\right) \sin\left(\frac{m\pi y}{b}\right) \sinh\left(\frac{-n\pi a}{b}\right) dy = \int_0^b V_0 \sin\left(\frac{m\pi y}{b}\right) dy. \quad (7.164)$$

As $\sinh(-x) = -\sinh(x)$, we have

$$-\sum_{1}^{\infty} \alpha_n \sinh\left(\frac{-n\pi a}{b}\right) \int_0^b \sin\left(\frac{n\pi y}{b}\right) \sin\left(\frac{m\pi y}{b}\right) dy = \int_0^b V_0 \sin\left(\frac{m\pi y}{b}\right) dy. \quad (7.165)$$

We must now compute the integral for all n and m. Taking into consideration that

$$\int \sin(py) \sin(qy) dy = \frac{\sin(p-q)y}{2(p-q)} - \frac{\sin(p+q)y}{2(p+q)}, \quad (7.166)$$

we get

$$\int_0^b \sin\left(\frac{n\pi y}{b}\right) \sin\left(\frac{m\pi y}{b}\right) dy = \begin{cases} 0, & \text{if } n \neq m, \\ \frac{b}{2}, & \text{if } n = m. \end{cases} \quad (7.167)$$

$$-\alpha_n \frac{b}{2} \sinh\left(\frac{n\pi a}{b}\right) = \int_0^b V_0 \sin\left(\frac{m\pi y}{b}\right) dy, \quad (7.168)$$

thus

$$\alpha_n = -\frac{2}{b \sinh\left(\frac{n\pi a}{b}\right)} \int_0^b V_0 \sin\left(\frac{n\pi y}{b}\right) dy = -\frac{2V_0}{b \sinh\left(\frac{n\pi a}{b}\right)} \frac{b}{n\pi} (-\cos(n\pi) + 1), \quad (7.169)$$

which depends on the values of n. In a general form it may be written[15]

$$\alpha_n = \begin{cases} 0 & \text{if } n \text{ even}, \\ -\frac{4V_0}{n\pi \sinh\left(\frac{n\pi a}{b}\right)} & n \text{ odd}. \end{cases} \quad (7.170)$$

To be clear, right now we will denote these odd numbers by l instead of n. With all the data the final expression for the potential is

---

[15] If instead of constant the potential on the side AD would be a function of y, namely $V(0, y) = f(y)$, the expression we get for the coefficients is $\alpha_n = \frac{2}{b \sinh\left(\frac{-n\pi a}{b}\right)} \int_0^b f(y) \sin\left(\frac{m\pi y}{b}\right) dy$.

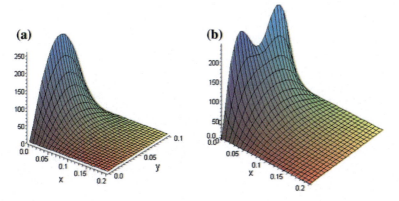

Fig. 7.14 a Graphic of the first element of the expansion of (7.171). **b** This corresponds to the potential $V(x, y)$ considering $n = 1$ and $n = 3$

$$V(x, y) = \sum_{l=odd}^{\infty} -\frac{4V_0}{\sinh\left(\frac{l\pi a}{b}\right)} \sin\left(\frac{l\pi y}{b}\right) \sinh\left(\left(\frac{l\pi}{b}\right)(x-a)\right). \quad (7.171)$$

In Figs. 7.14, 7.15, and 7.16 the solution considering up to six terms of the expansion (7.171) is plotted. As we can see the potential reaches its maximum at $x = 0$ and decrees progressively when x grows, being zero at $x = a$ and on sides $y = 0$ and $y = b$. Due to the finite number of terms chosen (six), over the side $(0, y)$ the solution oscillates not being V_0 exactly. To reproduce an almost constant potential on this side in this graphic, we would need more terms of the series. Figure 7.17a represents curves of constant potential for $n = 1$ (Fig. 7.14a), and Fig. 7.17b the same in the case of $n = 11$. As we can observe, both family of curves are similar to each other from $x = 0.04$ m, being the major difference in the vicinity of the side AD. Therefore, in this problem the election of the number of terms of the series mostly affects the result near the edge not grounded.

(b) Once we have calculated the potential for any point interior to the region delimited by the four sides of the rectangle, the electric field may be obtained by means of the gradient. In fact,

$$\mathbf{E} = -\nabla V(x, y) = \sum_{l=odd}^{\infty} \frac{4V_0 l}{nb \sinh\left(\frac{l\pi a}{b}\right)} \sin\left(\frac{l\pi y}{b}\right) \cosh\left(\left(\frac{l\pi}{b}\right)(x-a)\right) \mathbf{u}_x +$$
$$\sum_{l=odd}^{\infty} \frac{4V_0 l}{nb \sinh\left(\frac{l\pi a}{b}\right)} \cos\left(\frac{l\pi y}{b}\right) \sinh\left(\left(\frac{l\pi}{b}\right)(x-a)\right) \mathbf{u}_y. \quad (7.172)$$

Figures 7.18 and 7.19 show the electric field (red vectors) at every interior point of the cavity $ABCD$. Note that from 0.1 m, approximately, the electric field is very small. We can easily understand this result by looking at the graphics of Fig. 7.16 (the other

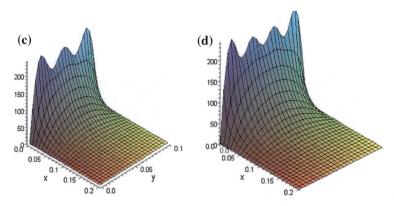

Fig. 7.15 **c** Representation considering the three first terms. **d** Idem for four terms of the series.

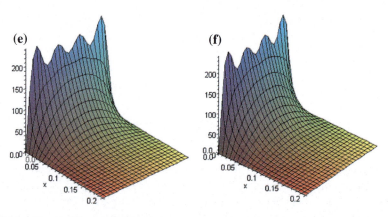

Fig. 7.16 **e** Graphic with five terms. **f** Idem for six terms

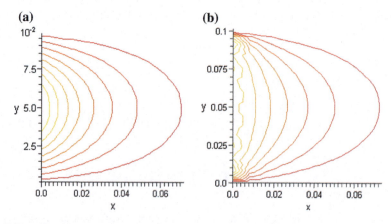

Fig. 7.17 **a** Graphic of the first element of the expansion of (7.171). **b** This corresponds to the potential $V(x, y)$ considering $n = 1$ and $n = 3$

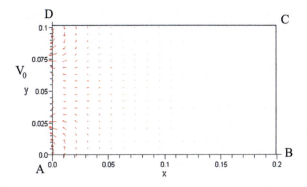

Fig. 7.18 Graphic of the electric field with only the first element of the expansion of (7.172)

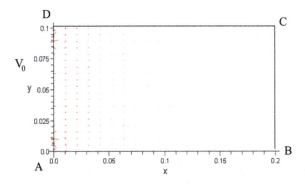

Fig. 7.19 This corresponds to the field $\mathbf{E}(x, y)$ with $n = 1$ and $n = 3$

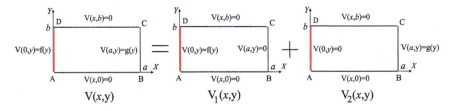

Fig. 7.20 The calculation procedure is studied as two independent problems

pictures of Figs. 7.14 and 7.15 are also valid, but Fig. 7.16 is more precise). In fact, the function $V(x, y)$ is very flat up $x = 0.1$ m, even in the solution regarding only the first term. As a result, as the electrostatic field is related with the potential by a gradient, the electric field must be smaller on this part than in regions where the slope of the potential is larger (the region close to the side AD). Note: As we have commented in this problem, for correctly applying the method of separation of variables the partial differential equation and the boundary conditions must be linear and homogeneous. The importance of this exercise lies on the fact that other problems with the same symmetry, but with other boundary conditions, may be solved taking this result as a basis. For instance if we would have the same prismatic system but besides the side AD, the side AB held to a potential $V(a, y) = g(y)$, we can pose the problem as two independent problems, each of one has its boundary conditions, and at the end apply the principle of superposition. This is schematically represented in the picture (Fig. 7.20). Therefore the final solution $V(x, y)$ of the problem may be obtained as the sum two problems in which only one side is held to a potential. We mean that

$$V(x, y) = V_1(x, y) + V_2(x, y). \tag{7.173}$$

The same principle applies in the most general case if every side has a nonzero potential. For this case we can write

$$V(x, y) = V_1(x, y) + V_2(x, y) + V_3(x, y) + V_4(x, y). \tag{7.174}$$

7.4 A dielectric sphere of constant ε_1 and radius R is placed coinciding its centre with the origin of the coordinates system $OXYZ$. The sphere is surrounded by a system of permittivity ε_2. If before locating the sphere in the medium there was a homogeneous electric field $\mathbf{E} = E\mathbf{u}_z$, calculate the electric field inside and outside of the dielectric ball.

Solution

Due to the rotational symmetry of this problem we can employ spherical coordinates (see Fig. 7.21). Taking into consideration that we do not have free charges, the differential equation to be solved is that of the Laplace (7.29), thus

$$\Delta V \equiv \frac{1}{r^2}\frac{\partial}{\partial r}\left(r^2\frac{\partial V}{\partial r}\right) + \frac{1}{r^2 \sin\theta}\frac{\partial}{\partial \theta}\left(\sin\theta \frac{\partial V}{\partial \theta}\right) + \frac{1}{r^2 \sin\theta}\frac{\partial^2 V}{\partial \phi^2} = 0 \tag{7.175}$$

In this problem the solution must not depend on the angle ϕ, because of the revolution symmetry, that is, if the field at a point $P(r, \phi, \theta)$ is $\mathbf{E}$, the components of $\mathbf{E}$ over the spherical unitary vectors basis $\mathbf{u}_r$, $\mathbf{u}_\phi$, and $\mathbf{u}_\theta$ displaced at any other point $P'(r, \phi +$

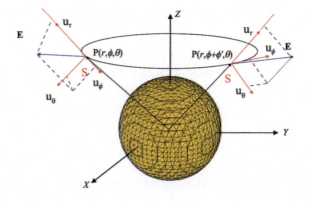

Fig. 7.21 Electric field $\mathbf{E}$ at two points on the same plane but with different coordinate ϕ. Because of the rotational symmetry, its projections over the basis $\{\mathbf{u}_r, \mathbf{u}_\phi, \mathbf{u}_\theta\}$ are the same when maintaining r and θ constant and ϕ varies to $(\phi + \phi')$

ϕ', θ) must be the same. For this reason we can eliminate the derivative with respect to ϕ in the above equation, i.e.,

$$\frac{1}{r^2}\frac{\partial}{\partial r}\left(r^2\frac{\partial V}{\partial r}\right) + \frac{1}{r^2 \sin\theta}\frac{\partial}{\partial \theta}\left(\sin\theta \frac{\partial V}{\partial \theta}\right) = 0. \tag{7.176}$$

Considering the symmetry of the problem seem to be adequate, looking for solutions of the form shown in (7.30), that is, the spherical harmonics

$$V(r, \phi, \theta) = \sum_0^\infty \left(a_m r^m + \frac{b_m}{r^{m+1}}\right) P_m(\cos\theta), \tag{7.177}$$

where $P_i^m(\cos\theta)$ are the Legendre polynomials. However, the solution will be only valid if it satisfies the following boundary conditions:

$$V_1(R, \phi, \theta) = V_2(R, \phi, \theta), \tag{7.178}$$

$$\varepsilon_1 \left(\frac{\partial V_1}{\partial r}\right) = \varepsilon_2 \left(\frac{\partial V_2}{\partial r}\right), \tag{7.179}$$

and

$$V_2(R, \phi, \theta) \longrightarrow V, \quad r \longrightarrow \infty. \tag{7.180}$$

The first one means that the potentials V_1 and V_2 on the sphere surface must be the same. The second condition depicts that the normal components of the displacement vector **D** on the boundary are equal. The third shows that at large distances the potential takes the same values as before the dielectric sphere was placed into the electric field. Physically it is the same to say that the sphere has no influence on the potential far away of its loci; if an effect exists, it is in proximity to the sphere (Fig. 7.22).

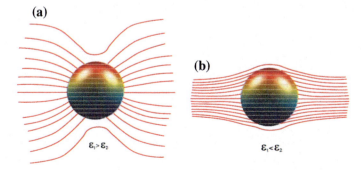

Fig. 7.22 **a** Lines of electric field for $\varepsilon_1 > \varepsilon_2$. **b** Field lines for $\varepsilon_1 < \varepsilon_2$

To solve the problem we first study the potential inside of the dielectric ball. There holds

$$V_1(r,\theta) = a_0 P_0(\cos\theta) + \frac{b_0}{r} P_0(\cos\theta) + a_1 r P_1(\cos\theta) + \frac{b_1}{r^2} P_1(\cos\theta) + a_2 r^2 P_2(\cos\theta) + \cdots \quad (7.181)$$

The Legendre polynomials may be obtained by means of the following relations

$$P_m(\cos\theta) = \frac{1}{2^m m!} \frac{d^m}{d(\cos\theta)^m}(\cos\theta^2 - 1), \quad (7.182)$$

or through

$$P_{m+1}(\cos\theta) = \frac{(2m+1)\cos\theta \, P_m(\cos\theta) + m P_{m-1}(\cos\theta)}{(m+1)}. \quad (7.183)$$

For the calculation it is not necessary to introduce the expanded expression of $P_m(\cos\theta)$. As we will see, we can work only with P_0 and P_1 at hand. These first two polynomials are $P_0(\cos\theta) = 1$ and $P_1(\cos\theta) = \cos\theta$. Introduction of both terms into (7.181) gives

$$V_1(r,\theta) = a_0 + \frac{b_0}{r} + a_1 r \cos\theta + \frac{b_1}{r^2}\cos\theta + a_2 r^2 P_2(\cos\theta) + \frac{b_2}{r^3} P_2(\cos\theta) + \cdots \quad (7.184)$$

This expression may be simplified if we take into account some properties that the potential inside of the sphere must have. So, the potential at the center of the ball must be well valued, but if we examine equality (7.181) we see that for $r=0$ it tends to infinity. To avoid this difficulty we make all $b_i = 0$, for $i = 0, 1, 2, \ldots$, and then we have

$$V_1(r,\theta) = a_0 + a_1 r \cos\theta + a_2 r^2 P_2(\cos\theta) + \cdots = \sum_{n=0}^{\infty} a_n r^n P_n(\cos\theta) \quad (7.185)$$

Outside of the sphere we have an expansion identical to (7.181) but with other coefficients (now with primes'), i.e.,

$$V_2(r,\theta) = a'_0 + \frac{b'_0}{r} + a'_1 r \cos\theta + \frac{b'_1}{r^2}\cos\theta + a'_2 r^2 P_2(\cos\theta) + \frac{b'_2}{r^3} P_2(\cos\theta) + \cdots = \quad (7.186)$$

$$= a'_0 + \frac{b'_0}{r} + a'_1 r \cos\theta + \frac{b'_1}{r^2}\cos\theta + \sum_{n>1} \frac{b'_n}{r^{n+1}} P_n(\cos\theta) + \sum_{n>1} a'_n r^n P_n(\cos\theta).$$

This equation must satisfy the boundary conditions too. As we commented before, far away from the ball the potential and the electric field must not depend on the presence of the dielectric sphere. It means that if the field is $\mathbf{E} = E\mathbf{u}_z$, then it holds for the potential

$$\mathbf{E} = -\nabla V \Rightarrow V(r, \theta) = -\int \mathbf{E} \cdot d\mathbf{l} = -Ez + V_0, \qquad (7.187)$$

V_0 being a constant. For an arbitrary point of space r, as $z = r \cos \theta$, (7.187) may be written as

$$V_2(r, \theta) = -Er \cos \theta + V_0. \qquad (7.188)$$

This is the form of that the potential has at infinity and this condition must be valid for (7.184), and therefore

$$\lim_{r \to \infty} V_2(r, \theta) = -Er \cos \theta + V_0 = \qquad (7.189)$$

$$= \lim_{r \to \infty} \left(a_0' + \frac{b_0'}{r} + a_1' r \cos \theta + \frac{b_1'}{r^2} \cos \theta + \sum_{n>1}^{\infty} \frac{b_n'}{r^{n+1}} P_n(\cos \theta) + \sum_{n>1}^{\infty} a_n' r^n P_n(\cos \theta) \right).$$

From this equality it follows

$$-Er \cos \theta + V_0 = a_0' + a_1' r \cos \theta, \qquad (7.190)$$

thus

$$a_0' = V_0, \qquad (7.191)$$

$$a_1' = -E, \qquad (7.192)$$

and for $n > 1$

$$a_n' = 0, \qquad (7.193)$$

otherwise the potential at large distances would be divergent. Substitution of these constants into (7.184) leads to

$$V_2(r, \theta) = V_0 + -Er \cos \theta + \frac{b_1'}{r^2} \cos \theta + \sum_{n \neq 1}^{\infty} \frac{b_n'}{r^{n+1}} P_n(\cos \theta). \qquad (7.194)$$

Once we have the general expressions for the potential inside and outside of the sphere we can apply the other boundary conditions. The first one equates the potential on the ball surface, that is, $V_1(r, \phi, \theta) = V_2(r, \phi, \theta)$, then

$$\sum_{n=0}^{\infty} a_n r^n P_n(\cos \theta)|_{r=R} = V_0 + \left(-Er + \frac{b_1'}{R^2} \right) \cos \theta + \sum_{n \neq 1}^{\infty} \frac{b_n'}{r^{n+1}} P_n(\cos \theta)|_{r=R}.$$

$$(7.195)$$

Since the last equation must hold for any value of θ (remember that it does not depend on ϕ), the only possibility we have is

$$a_0 = V_0, \tag{7.196}$$

$$\left(-ER + \frac{b'_1}{R^2}\right) = R a_1, \tag{7.197}$$

and

$$\frac{b'_n}{R^{n+1}} = R^n a_n, \quad n > 1. \tag{7.198}$$

In order to apply (7.179) we take the derivative of (7.184) and (7.186), obtaining

$$\left(\frac{\partial V_1}{\partial r}\right)_{r=R} = \sum_{n>1}^{\infty} a_n r^{n+1} P_n(\cos\theta), \tag{7.199}$$

and

$$\left(\frac{\partial V_2}{\partial r}\right)_{r=R} = -\left(E + \frac{2b'_1}{R^3}\right)\cos\theta - \sum_{n\neq 1}^{\infty}(n+1)\frac{b'_n}{r^{n+2}}P_n(\cos\theta), \tag{7.200}$$

then using (7.179)

$$\varepsilon_1 \sum_{n>1}^{\infty} a_n r^{n+1} P_n(\cos\theta) = -\varepsilon_2\left(E + \frac{2b'_1}{R^3}\right)\cos\theta - \varepsilon_2 \sum_{n\neq 1}^{\infty}(n+1)\frac{b'_n}{r^{n+2}}P_n(\cos\theta), \tag{7.201}$$

which gives

$$\varepsilon_1 a_1 = -\varepsilon_2\left(E + \frac{2b'_1}{R^3}\right), \tag{7.202}$$

and for $n \neq 1$

$$\varepsilon_1 n a_n R^{n-1} = -\varepsilon_2(n+1)\frac{b'_n}{R^{n+2}}. \tag{7.203}$$

Combining (7.197) and (7.202) we get

$$a_1 = -\frac{3\varepsilon_2}{\varepsilon_1 + 2\varepsilon_2}E, \tag{7.204}$$

and

$$b'_1 = \frac{\varepsilon_1 - \varepsilon_2}{\varepsilon_1 + 2\varepsilon_2} R^3 E. \tag{7.205}$$

On the other hand (7.198) and (7.203) are only possible if $a_0 = b'_0 = 0$ and $a_n = b'_n = 0$ for $n > 1$. By inspection of (7.196) we observe that as $a_0 = 0$, the potential $V_0 = 0$, that is, the potential at the center of the dielectric sphere vanishes (see (7.184) and note that $V_1(0, \theta) = 0$). Once we have the value of the constants the potential and field inside and outside of the ball are, respectively

$$V_1(r, \theta) = -\frac{3\varepsilon_2}{\varepsilon_1 + 2\varepsilon_2} E r \cos\theta, \quad r < R, \tag{7.206}$$

and

$$V_2(r, \theta) = -\left(1 - \frac{\varepsilon_1 - \varepsilon_2}{\varepsilon_1 + 2\varepsilon_2} \frac{R^3}{r^3}\right) r E \cos\theta, \quad r > R. \tag{7.207}$$

The electric field may be determined through the gradient. For $r < R$, we have

$$\mathbf{E}_1 = -\nabla V = -\frac{\partial V_1}{\partial z} = -\frac{\partial V_2}{\partial (r \cos\theta)} \mathbf{u}_z = \frac{3\varepsilon_2}{\varepsilon_1 + 2\varepsilon_2} E \mathbf{u}_z \quad r < R. \tag{7.208}$$

As it is known, when a dielectric material remains in the presence of an electric field, a local redistribution of its charges is produced and the material will polarize. As a consequence, in general, two kinds of bounded charges appear: a volume distribution of polarization charge $\rho_p = -\nabla \cdot \mathbf{P}$, and a surface polarization charge $\sigma_p = -\mathbf{P} \cdot \mathbf{n}$, where $\mathbf{P}$ and $\mathbf{n}$ are the polarization field and the unitary normal vector to the surface, respectively (see Chap. 2). In the case studied we see from (7.208) that inside the ball the electric field is uniform, hence the polarization $\mathbf{P} = \varepsilon_0 \chi_e \mathbf{E}$ too, which means that we do not have ρ_p but only σ_p. The apparition of these bounded surface charges behaves as a source of electric field modifying the electric properties inside of the sphere and in its outside surroundings. Physically it means that the field inside of the ball is the result of the addition of the exterior electric field $\mathbf{E}$ (generated before the ball was placed there) and the field created in the interior by the bound charges that appear on the surface of the dielectric sphere. Calling $\mathbf{E}'_{\sigma_p}$ the electric field created by σ_p inside, it yields

$$\mathbf{E}_1 = \mathbf{E}_{exterior} + \mathbf{E}'_{\sigma_p} \Rightarrow \mathbf{E}'_{\sigma_p} = \mathbf{E}_1 - \mathbf{E}_{exterior} = \frac{3\varepsilon_2}{\varepsilon_1 + 2\varepsilon_2} E \mathbf{u}_z - E \mathbf{u}_z = \left(\frac{\varepsilon_2 - \varepsilon_1}{\varepsilon_1 + 2\varepsilon_2}\right) E \mathbf{u}_z. \tag{7.209}$$

Outside we obtain the field in the same way

$$\mathbf{E}_2(r, \theta) = -\nabla V = -\frac{\partial V_2}{\partial r} = -\frac{\partial V_2}{\partial r} \mathbf{u}_r - \frac{1}{r}\frac{\partial V_2}{\partial \theta} \mathbf{u}_\theta = \left(1 + 2\frac{\varepsilon_1 - \varepsilon_2}{\varepsilon_1 + 2\varepsilon_2} \frac{R^3}{r^3}\right) E \cos\theta \, \mathbf{u}_r \tag{7.210}$$

$$-\left(1 - \frac{\varepsilon_1 - \varepsilon_2}{\varepsilon_1 + 2\varepsilon_2} \frac{R^3}{r^3}\right) E \sin\theta \, \mathbf{u}_\theta, \quad r > R.$$

As we can prove, contrary to what occurs inside, the field is not homogeneous and depends on the point examined. Equation (7.210) shows that in the surrounding material of the sphere we will observe the major modification of the electric field.

However, a very different behavior of the field lines happens depending on the values of ε_1 and ε_2. So, from (7.208) it follows that

$$\mathbf{E}_1 = \frac{3}{\left(2 + \frac{\varepsilon_1}{\varepsilon_2}\right)} E\,\mathbf{u}_z. \tag{7.211}$$

and then, if $\varepsilon_1 > \varepsilon_2$ the electric field inside is smaller than $\mathbf{E}$. On the contrary, when $\varepsilon_1 < \varepsilon_2$ holds $\mathbf{E}_1 > \mathbf{E}$ (the contrary occurs with the displacement vector $\mathbf{D}$). This case is very important in industrial production processes. In fact, the existence of air bubbles in a dielectric matrix creates an increment of the electric field inside of the microspheres. As the gas bubbles have a dielectric constant $\varepsilon_1 \approx 1$, if $\varepsilon_1 < \varepsilon_2$ (or $<<$), the electric field inside them can be very high, which may produce discharges into the system. For this reason, in most applications where it matters, a control in the material fabrication is needed in order to avoid this problem. As a particular case, if the permittivities are the same, that is $\varepsilon_1 = \varepsilon_2$, then $\mathbf{E}_1 = \mathbf{E}$ and the field are not affected.

The understanding of this study shown is very important in some fields of applied physics and technology. For example, sometimes researchers try to modify dielectric properties of materials by means of the inclusion of dielectric (and metallic too) balls into a dielectric matrix. In those cases the problem is to calculate the effective permittivity ε_{eff} of the new system. To solve this question some formulae may be used. The two most important are the relationship of Maxwell-Wagner and the Rayleigh formula. The first one is

$$\varepsilon_{eff}(\alpha) = \left(\frac{2\varepsilon_2 + \varepsilon_1 + 2\alpha(\varepsilon_1 - \varepsilon_2)}{2\varepsilon_2 + \varepsilon_1 - \alpha(\varepsilon_1 - \varepsilon_2)}\right)\varepsilon_2, \tag{7.212}$$

where ε_1 and ε_2 are the permittivities of the spheres and of the medium (matrix), respectively. For the simplified relation of Rayleigh ($\alpha << 1$) we have

$$\varepsilon_{eff}(\alpha) = \left(1 + 3\alpha\frac{(\varepsilon_1 - \varepsilon_2)}{(2\varepsilon_1 + \varepsilon_2)}\right)\varepsilon_2. \tag{7.213}$$

Recently, however, recent attention has shifted to the employment of inhomogeneous materials containing inclusions for modifying optical properties.[16] Specifically, it is possible to induce surface anisotropy by different mechanisms by introducing inclusions adequately (dielectric or metallic-see next problem), leading to a change in the refractive index of the material (see Chapter 14). In fact, it can be shown that the anisotropic induced properties may be controlled by adjusting the concentration of the spheres and the thickness of the corresponding layer, which opens a large number of scientific and technical applications.

---

[16]See, for example, [111].

7.5

A metallic sphere of radius R is located with its center coinciding with the origin of the coordinates system $OXYZ$. The sphere is immersed into a dielectric substance of permittivity ε_2. If before locating the sphere in the medium there was a uniform electric field $\mathbf{E} = E\mathbf{u}_z$, obtain the electric field inside and outside of the sphere.

Solution

This problem is similar to the one explained previously, but now the ball is made of a metallic material instead of a dielectric. At first sight we can proceed in the same manner, however due to the characteristics of the conductors, some new information about the boundary conditions appears. In fact, the potential must fulfill the following conditions

$$V(R, \phi, \theta) = C, \tag{7.214}$$

C being a constant, and

$$V(R, \phi, \theta) \longrightarrow V, \quad r \longrightarrow \infty. \tag{7.215}$$

The first condition shows that the potential at $r = R$ is constant. Actually this potential is constant in all the ball, because it is a conductor, hence the electric field inside vanishes. The second BC is the same as we studied before for the dielectric sphere and physically means that the effect of the metallic ball on the field is only important in its owing vicinity, and not far away from it.

The resolution of this problem again requires using the Laplace equation in spherical coordinates where the variable ϕ does not appear due to the symmetry of the system. However, instead of solving this equation together with the boundary conditions, we can proceed in an easier way if we consider the results of the preceding exercise. In effect, as the ball is made of a metallic material, we can consider that the constant ε_1 of the ball tends to infinity. If that is true, we could directly make the substitution $\varepsilon_1 \to \infty$ in (7.206), (7.207), (7.208) and (7.209) for obtaining the solution. According to this idea we get

$$V_1(r, \theta) = 0 + C, \quad r < R, \tag{7.216}$$

and

$$V_2(r, \theta) = -\left(1 - \frac{R^3}{r^3}\right) rE \cos\theta, \quad r > R. \tag{7.217}$$

For the electric inside of the ball ($r < R$), we have

$$\mathbf{E}_1 = 0 \quad r < R, \tag{7.218}$$

and outside ($r > R$)

$$\mathbf{E}_2(r, \theta) = \left(1 + 2\frac{R^3}{r^3}\right) E \cos\theta\, \mathbf{u}_r - \left(1 - \frac{R^3}{r^3}\right) E \sin\theta\, \mathbf{u}_\theta, \quad r > R. \quad (7.219)$$

In the same manner as we commented in the preceding problem, the inclusion of metallic balls in a dielectric matrix is of great practical interest. In this way we can modify the physical properties of the matrix varying the concentrations of the metallic balls in the dielectric. A simple formula that gives the modified permittivity is due to Bruggeman

$$\epsilon_{eff}(\alpha) = \frac{\epsilon_2}{(1-\alpha)^3}, \quad (7.220)$$

where α is the volume concentrations of the spheres and ϵ_2 is the permittivity of dielectric matrix.

7.6 A long wire of diameter R carries a current I. Find (a): The vector potential $\mathbf{A}$ for $\rho < R$. (b) The vector potential $\mathbf{A}$ for $\rho > R$. (c) The magnetic field $\mathbf{B}$.

Solution

(a) This problem could be solved using techniques we have seen in Chap. 5. In fact, by means of Ampère's circuital law we can obtain the magnetic field everywhere and, once we know it, the relation between both fields, i.e., the potential vector and $\mathbf{B}$ is determined by the following relation

$$\oint_\Gamma \mathbf{A} d\mathbf{l} = \int\int_S \mathbf{B} \cdot d\mathbf{S}. \quad (7.221)$$

However, taking into account the methods exposed in the introduction, we will try to seek a solution directly solving the corresponding differential equation.

For non-varying electromagnetic fields (see Chap. 5) it holds that

$$\nabla^2 \mathbf{A} = -\mu_0 \mathbf{j}. \quad (7.222)$$

Let us suppose we place the wire coinciding its length with the OZ axis. As the wire is very long and its cross section is constant, we have translational symmetry along OZ in the direction of the current and rotational symmetry, too. Starting from a cylindrical coordinate frame it means that the solution will depend neither on the coordinate z, nor the angle ϕ. For this reason we can study the problem by eliminating the partial derivatives of these variables. Doing so, we have

$$\frac{1}{\rho}\frac{\partial}{\partial \rho}\left(\rho \frac{\partial A_z}{\partial \rho}\right) = \mu_0 j, \quad (7.223)$$

whose solution may be obtained by direct integration

$$\rho \frac{\partial A_z}{\partial \rho} = -\mu_0 \int j\rho\, d\rho = -\mu_0 j \frac{\rho^2}{2} + C_1, \quad (7.224)$$

and then

$$\frac{\partial A_z}{\partial \rho} = \frac{C_1}{\rho} - \mu_0 j \frac{\rho}{2} \Rightarrow A_z(\rho) = C_1 \ln \rho - \mu_0 j \frac{\rho^2}{4} + C_2, \quad (7.225)$$

where C_1 and C_2 are two constants. As the vector potential must be finite in the region examined, to accomplish this condition, the only possibility we have is that $C_1 = 0$, otherwise the logarithm becomes infinity for $\rho = 0$, with then yields

$$A_z(\rho) = -\mu_0 j \frac{\rho^2}{4} + C_2. \quad (7.226)$$

(b) In this case the calculation refers to exterior region of the wire, where there is no current. As a result, (7.222) adopts the form

$$\nabla^2 \mathbf{A} = 0. \quad (7.227)$$

Integrating again, we get

$$\frac{1}{\rho}\frac{\partial}{\partial \rho}\left(\rho \frac{\partial A_z}{\partial \rho}\right) = 0 \Rightarrow \frac{\partial}{\partial \rho}\left(\rho \frac{\partial A_z}{\partial \rho}\right) = 0 \Rightarrow \left(\rho \frac{\partial A_z}{\partial \rho}\right) = D_1, \quad (7.228)$$

D_1 being a constant to be determined. An integration more gives

$$A_z = \int \frac{D_1}{\rho} d\rho + D_2 = D_1 \ln \rho + D_2. \quad (7.229)$$

At this point we are tempted to put $D_1 = 0$ in (7.229), because for $\rho \longrightarrow \infty$ the $\ln \rho \longrightarrow \infty$, and making D_1 this constat zero this problem is avoided; however it would be wrong. The reason for this is that, contrary to what happens with the magnetic field, the vector potential can be unbounded if the region where the current extends *is not confined in a finite domain* of space, as in our case (the wire is infinite (mathematically) see Chap. 5). Therefore, the expression (7.229) is valid, in principle, it being necessary to seek other possibilities to determine the constants.

(c) The magnetic field can be computed by using the relation between B and A (5.11), i.e.

$$\mathbf{B}(\mathbf{r}) = \nabla \times \mathbf{A}(\mathbf{r}). \quad (7.230)$$

Considering that the only component of the vector potential we have is the z component, we can write

$$B_\phi = -\frac{\partial A_z}{\partial \rho} = \frac{1}{2}\mu_0 j \rho, \quad (7.231)$$

which represents the general formula for B inside of the wire. In the same manner, calculating the rotational in a exterior region (5.11) leads to

$$B_\phi = -\frac{\partial A_z}{\partial \rho} = -\frac{D_1}{\rho}. \tag{7.232}$$

Now, for obtaining D_1 and D_2, we impose the continuity of the magnetic field on the surface of the wire

$$B_\phi(1)_{\rho=R} = B_\phi(2)_{\rho=R}, \tag{7.233}$$

yielding

$$\mu_0 j \frac{R}{2} = -\frac{D_1}{R} \Rightarrow D_1 = -\mu_0 j \frac{R^2}{2}, \tag{7.234}$$

thus

$$B_\phi = \frac{\mu_0 j R^2}{2\rho}. \tag{7.235}$$

In relation with B, we do not need anything else. Observe that, even though we did not obtain constants C_1 and D_2 in (7.224) and (7.229), the magnetic field has been determined without difficulty. The reason of that is due to the fact that for computing B we must calculate a partial derivative, then eliminating C_1 and D_2, indirectly. Nevertheless, we can determine such constants by imposing the boundary conditions for the vector potential

$$A_z(1)_{\rho=R} = A_z(2)_{\rho=R}, \tag{7.236}$$

which gives

$$-\mu_0 j \frac{R^2}{4} + C_2 = -\mu_0 j \frac{R^2}{2} \ln R + D_2. \tag{7.237}$$

This result has no unique solution, because we have two constants and only one equation. However, we can impose a normalization condition at the origin of coordinates. For instance, we can choose $A_z(1)_{\rho=0} = 0$ which gives $C_2 = 0$ (or any other value), and then we get

$$-\mu_0 j \frac{R^2}{4} = -\mu_0 j \frac{R^2}{2} \ln R + D_2 \Rightarrow D_2 = \frac{1}{4}\mu_0 j R^2 (\ln R^2 - 1). \tag{7.238}$$

Introducing these values into (7.226) and (7.229), holds

$$A_z(\rho) = -\frac{1}{4}\mu_0 j \rho^2, \tag{7.239}$$

and

$$A_z(\rho) = -\frac{1}{2}\mu_0 j R^2 \ln \rho + \frac{1}{4}\mu_0 j R^2 (\ln R^2 - 1). \tag{7.240}$$

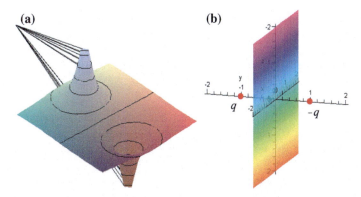

Fig. 7.23 Potential due to two identical charges of opposite signs. **a** The *straight line* in the middle of this figure represents the points in two dimensions where the potential is zero. **b** In a three dimensional representation we have a plane

7.7 Consider two charges of different signs separated by distance $2a$ over the OY axis. Show that if the absolute value of them is the same, the geometric points of zero potential are located on a plane.

Solution

As we explained in Sect. 7.4, when two charges of opposite signs, namely q_1 and q_2, are placed at a distance a, a family of different surfaces are obtained, each one corresponds to a constant C (7.48). In the example seen there, for $C = 0$ we obtained a spherical surface whose centre was displaced with respect to the location of the charge q_1 (Fig. 7.23a). In the present problem the charges have the same absolute values, i.e. $|q_1| = |q_2| = q$, thus probably the surface of constant potential changes (see the example in Sect. 7.4).

To demonstrate it, let us first suppose that the charges are located at points $P_1(0, -a, 0)$ and $P_2(0, a, 0)$. To see what are the loci of constant potential for this configuration we introduce $|q_1| = |q_2| = q$ into (7.48), and the distance between charges $2a$, which leads to

$$(x^2 + (y-a)^2 + z^2) = (x^2 + (y+a) + z^2). \tag{7.241}$$

This equation is verified only in the case that $y = 0$, which represents the equation of a plane (see Fig. 7.23b). This result is important and provides a means to other questions that may be solved by the method of images. In fact, it is tied to a problem that we could pose as follows. Let us imagine a point charge q placed at the point of coordinates $P(0, a, 0)$ in front of a semi-infinite conducting block whose face coincides with the OXZ plane. If the metallic block is held to zero potential, calculate the potential at a generic point of space $P(x, y, z)$, $y > 0$. To stress this issue, we must find a charge disposition in such a way that the potential on the plane $y = 0$ disappears. For this to occur we can place a *fictitious* charge $-q$ in front of the charge q, at a distance a with respect the origin of the coordinate frame. By this procedure we immediately see that if we add the potentials corresponding to both charges we again obtain

$$V(x, y, z) = V_1(x, y, z) + V_2(x, y, z) = \frac{1}{4\pi\varepsilon_0} \frac{q}{\sqrt{x^2 + (y-a)^2 + z^2}} + \frac{1}{4\pi\varepsilon_0} \frac{-q}{\sqrt{x^2 + (y+a)^2 + z^2}}.$$
(7.242)

As we can prove (7.242) fulfills the Laplace equation and the boundary conditions, namely $V(x, 0, z) = 0$. It means that, following the uniqueness theorem, expression (7.242) must be the solution for the potential created by a charge q in front of a metallic plane.

Once we know the potential, other questions as how does the density charge distribute on the metallic surface, or the force between the charge and the plane, can be answered easily.

For the new density of charge on the metallic plane as a result of the presence of the charge q at $P(0, a, 0)$ we have

$$\sigma(x, 0, z) = \epsilon_0 \left(\frac{\partial V(x, y, z)}{\partial n} \right)_{y=0} = \epsilon_0 \mathbf{n} \mathbf{E},$$
(7.243)

n being the normal unitary vector to the surface. The force is $\mathbf{F} = -q\nabla V_2(x, y, z)$, which coincides with the force that the image (fictitious) charge creates on the actual charge q.

7.8 A grounded conducting sphere of radius R has its center coinciding with the origin of coordinates $O(0, 0, 0)$. A point charge q_2 is located at the point $P(0, d_0, 0)$ with $d_0 > R$. Find: (a) The potential at any point of space. (b) The induced surface charge density on the sphere (Fig. 7.24).

Solution

(a) Because the sphere is grounded, this problem is easily solved using the result seen in the theoretical introduction. In this introduction we have demonstrated that

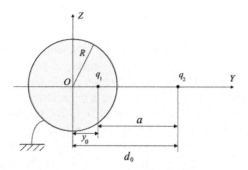

Fig. 7.24 Metallic sphere in presence of a charge q_2. The charge q_1 is the corresponding image charge

Solved Problems

when two electric charges of different values are placed at a distance a from each other, the equipotential surface corresponding to the zero potential is a sphere whose center lies on the line connecting the charges. Thus, the problem can be understood as follows: given a metal sphere S and an external charge $q = q_2$, where an additional charge should be placed, such that the surface of zero potential obtained for the two charges does coincide with the spherical surface S?

To answer this question we can employ the equations seen in the theory (see Fig. 7.1). In particular those of most interest are

$$q_1 = -\frac{R}{y_0 + a} q_2 = -\frac{R}{d_0} q_2, \qquad (7.244)$$

and

$$y_0 = \frac{R^2}{d_0}, \qquad (7.245)$$

which represent the fictitious charge that together with q_2 produces zero potential on the surface of radius R, and the distance between q_1 and the center of the sphere, respectively.[17] Physically it allows us to substitute the metallic sphere by a charge of magnitude given by (7.244), at y_0, because the effect is the same (it fulfills the boundary conditions). The potential created by this electric system outside of the sphere is

$$V(x, y, z) = \frac{1}{4\pi\varepsilon_0} \frac{R}{d_0} \frac{-q_2}{\sqrt{x^2 + (y - \frac{R^2}{d_0})^2 + z^2}} + \frac{1}{4\pi\varepsilon_0} \frac{q_2}{\sqrt{x^2 + (y - d_0)^2 + z^2}}. \qquad (7.246)$$

The potential for points in the interior may be deduced by using the Gauss theorem shown in Chap. 2. As it was demonstrated the electric field in the interior of a conductor in equilibrium is zero. As a result, the potential is, in principle, a constant. In our case the sphere was grounded which means that the potential on the surface must be zero, and in the interior of the surface too, because the potential inside has to take the value zero when approaching the surface.

(b) The surface density charge may be calculated by means of the boundary condition in the proximity of the conductor

$$\sigma(x, y, z) = \epsilon_0 \frac{\partial V(x, y, z)}{\partial n} = \epsilon_0 \mathbf{nE}. \qquad (7.247)$$

This means we have to obtain firstly the electric field, and then σ. The electric field is easy to calculate through the potential $V(x, y, z)$, by using gradient, i.e. $\mathbf{E} = -\nabla V(x, y, z)$, obtaining

---

[17]Observe that here we have written the equation keeping the signs. In the theory (7.55) appears as $q_1 = \frac{R}{d_0} q_2$ because we knew q_2 was negative.

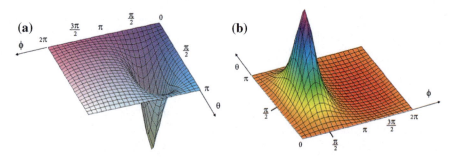

Fig. 7.25 Density of charge versus the angles ϕ and θ corresponding to spherical coordinates. The maximum absolute value of $\sigma(x, y)$ is reached at $(\phi, \theta) = (\pi/2, \pi/2)$. The difference between the graphics in (*a*) and (*b*) corresponds only to the sign of the charge induced

$$E(x, y, z) = \frac{q_2}{4\pi\varepsilon_0}\left(\frac{1}{|\mathbf{r} - \mathbf{r}_2|^3}(\mathbf{r} - \mathbf{r}_2) - \frac{R}{d_0|\mathbf{r} - \mathbf{r}_1|^3}(\mathbf{r} - \mathbf{r}_1)\right) \quad (7.248)$$

where $\mathbf{r}_2$ and $\mathbf{r}_1$ refer to the position vectors of the charge $q = q_2$ and the image $q_1 = -\frac{R}{d_0}q_2$, respectively.

Due to the symmetry of the problem, it seems to be adequate to choose the normal **n** to the surface in spherical coordinates, which coincides with $\mathbf{u}_\rho$.

$$E(x, y, z) = \frac{q_2}{4\pi\varepsilon_0}\left(\frac{1}{|\mathbf{r} - \mathbf{r}_2|^3}(r\mathbf{u}_r - d_0\mathbf{u}_y) - \frac{R}{d_0|\mathbf{r} - \mathbf{r}_1|^3}\left(r\mathbf{u}_r - \frac{R^2}{d_0}\mathbf{u}_y\right)\right). \quad (7.249)$$

In order to make the scalar product of the electric field with the normal vector we set $\mathbf{u}_y = \sin\theta\sin\phi\,\mathbf{u}_\rho + \cos\theta\sin\phi\,\mathbf{u}_\theta + \cos\theta\,\mathbf{u}_\phi$, obtaining,

$$\sigma(x, y, z) = \epsilon_0 \mathbf{n}\mathbf{E} = \frac{q_2}{4\pi}\frac{1}{(x^2 + (y - d_0)^2 + z^2)^{3/2}}(R - d_0\sin\theta\sin\phi). \quad (7.250)$$

$$-\frac{q_2}{4\pi}\frac{R}{d_0(x^2 + (y - \frac{R^2}{d_0})^2 + z^2)^{3/2}}\left(R - \frac{R^2}{d_0}\sin\theta\sin\phi\right)$$

Obviously, the induced surface charge density is not homogeneous (Figs. 7.25 and 7.26). This is logical because the presence of the charge q_2 in front of the metallic sphere modifies the charge distribution on its surface. Observe that, for a fixed distance between the charge and the sphere, the accumulation of induced charge is a maximum in the direction $(\phi, \theta) = (\pi/2, \pi/2)$ with respect to the coordinate frame chosen.

7.9 A metallic sphere of radius R has its center coinciding with the origin of coordinates $O(0, 0, 0)$. A point charge q_2 is located at the point $P(0, d_0, 0)$ with $d_0 > R$.

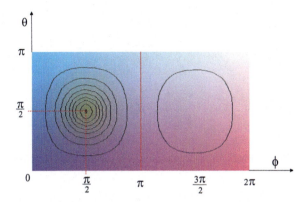

Fig. 7.26 Curves of constant density charge in the space of the angles ϕ and θ

If the sphere is connected to a constant potential V_0, find: (a) The potential for all points of $\Re^3$. (b) The electric field. (c) The surface charge density on the surface. (d) The force exerted on the charge (Fig. 7.27).

Solution

If we wish to solve this problem directly as in the preceding case we have some difficulties, because the results obtained in the introduction were developed for the case of zero potential. At first glance we could think in solving the (7.48) for an arbitrary value of C, and proceed like in the previous exercise, that is, adjusting an image charge and its location so that the spherical surface of constant potential obtained coincides with the potential V. However this is not possible. In fact, if we solve the (7.48) for an arbitrary value of the constants C, it can be demonstrated that the surface of constant potential is not a sphere, which means that it is not possible to solve this problem by only using one image charge (Fig. 7.28).

To face this problem we can divide the system into more subproblems, but being careful that the boundary conditions of the initial system are maintained. To do so, we will employ the principle of superpositions of fields. This procedure is justified because of the linearity of the Laplace equation.[18] In fact, if $V_1(\mathbf{r})$ and $V_2(\mathbf{r})$ separately verify this equation, the function constructed by the sum, i.e. $V(\mathbf{r}) = V_1(\mathbf{r}) + V_2(\mathbf{r})$ is also a solution. On the other hand, if we obtain a solution by means of a configuration of fictitious charges, due to the uniqueness theorem, the solution sought is the same as the actual problem (Fig. 7.29).

If we remember the preceding problem, we obtained a spherical surface of zero potential. As in our case the problem is to simulate the metallic sphere that has a potential V_0, we must only add a configuration of charges in such a way that its corresponding potential coincides with V_0. Usually it would be a very, if not impossibly, difficult task. However, due to the spherical symmetry of the problem, it seems to be plausible to seek a geometrical disposition of charges that fulfills the requirements. In fact, as the potential generated by an electric charge has spherical

---

[18]If the partial differential equation is non-linear, the principle of superposition does not apply.

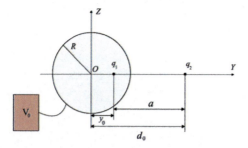

Fig. 7.27 Metallic sphere held to a constant potential V_0 in front of a charge q_2

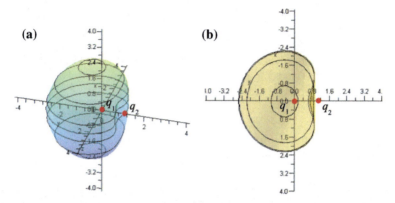

Fig. 7.28 **a** Surface of constant potential $V(x, y, z) = C = 0.25$ of two charges q_1 and $q_2 = -\frac{R}{d_0}q_1$, for $\frac{R}{d_0} = 0.5$. **b** Two dimensional cross-section. Observe that only changing the potential from $V = C = 0$ to $V = C = 0.25$, the equipotential surface obtained does not correspond to a sphere

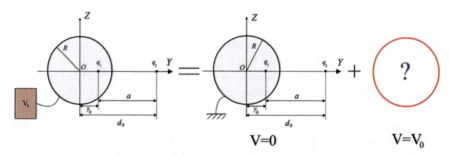

Fig. 7.29 Considering the principle of superposition the problem is equivalent to the sum of the two problems. If we find the charge distribution that creates a potential V_0 on a surface of radius R, the solution may be expressed as sum of the potential due to two charges q_1 and q_2, so that $q_1 = -\frac{R}{d_0}q_2$ (we obtain zero potential), and the potential produced by the other system of charges

symmetry, the easiest solution one may think is to locate a charge Q at the center of the sphere, whose exact value is at the beginning unknown. In order to obtain the same behavior for the system, the value and sign of Q must be adjusted so that the potential on the sphere surface is V_0. To calculate its magnitude we need only employ the expression for the potential of an electric charge Q, i.e.

$$V(r) = \frac{1}{4\pi\varepsilon_0}\frac{Q}{r}, \qquad (7.251)$$

and then the potential V_0 may be obtained from this latter equation as follows

$$Q = 4\pi\varepsilon_0 R V(R) = 4\pi\varepsilon_0 R V_0. \qquad (7.252)$$

Applying the commented principle of superposition, the potential at any point of space may be regarded as the sum of the potential generated by two charges q_1 and q_2 (see the former problem) and the potential due to a point charge at the origin of the coordinate frame

$$V(x,y,z) = \frac{1}{4\pi\varepsilon_0}\frac{R}{d_0}\frac{-q_2}{\sqrt{x^2+(y-\frac{R^2}{d_0})^2+z^2}} + \frac{1}{4\pi\varepsilon_0}\frac{q_2}{\sqrt{x^2+(y-d_0)^2+z^2}} + \frac{RV_0}{\sqrt{x^2+y^2+z^2}}.$$
(7.253)

For points $r < R$ we proceed using Gauss' theorem. In effect, we know that the electric field inside of the conducting sphere is zero, and the potential is a constant magnitude. However, contrary to the preceding exercise, the potential now is V_0 if $r < R$, which fulfils the boundary condition on the surface.

(b) The electric field may be directly calculated by means of the gradient of the potential,

$$\mathbf{E} = -\nabla V(x,y,z) = \frac{q_2}{4\pi\varepsilon_0}\left(\frac{1}{|\mathbf{r}-\mathbf{r}_2|^3}(\mathbf{r}-\mathbf{r}_2) - \frac{R}{d_0|\mathbf{r}-\mathbf{r}_1|^3}(\mathbf{r}-\mathbf{r}_1)\right) + \frac{RV_0}{r^3}\mathbf{r},$$
(7.254)

where $\mathbf{r}_2 = d_0\,\mathbf{u}_y$, and $\mathbf{r}_1 = \frac{R^2}{d_0}\mathbf{u}_y$. Introduction of these values in the latter equality leads to

$$E(x,y,z) = \frac{q_2}{4\pi\varepsilon_0}\left(\frac{1}{|\mathbf{r}-\mathbf{r}_2|^3}(r\mathbf{u}_r - d_0\mathbf{u}_y) - \frac{R}{d_0|\mathbf{r}-\mathbf{r}_1|^3}(r\mathbf{u}_r - \frac{R^2}{d_0}\mathbf{u}_y)\right) + \frac{RV_0}{r^2}\mathbf{u}_r. \qquad (7.255)$$

In the interior of the sphere the electric field is zero.

(c) Using the formula (7.247), the surface charge density is

$$\sigma(x,y,z) = \epsilon_0\frac{\partial V(x,y,z)}{\partial n}, \qquad (7.256)$$

and then

$$\sigma(x, y, z) = \epsilon_0 \mathbf{nE} = \frac{1}{4\pi} \left(\frac{1}{(x^2 + (y - d_0)^2 + z^2)^{3/2}} (R - d_0 \sin\theta \sin\phi) \right)$$

$$- \frac{1}{4\pi} \left(\frac{R}{d_0(x^2 + (y - \frac{R^2}{d_0})^2 + z^2)^{3/2}} (R - \frac{R^2}{d_0} \sin\theta \sin\phi) \right) + \frac{\epsilon_0 R V_0}{r^3}. \quad (7.257)$$

(d) The force on the charge by the conducting sphere is very hard to compute. However we have found an electrical configuration which, in some extent (see next problem), is equivalent to the system in Fig. 7.27. For this reason the calculus may be simplified if we take into consideration that this force is the same as that which would be produced by the two image charges. The force is

$$\mathbf{F} = \frac{-1}{4\pi\varepsilon_0} \frac{R d_0 q_2^2}{(d_0^2 - R^2)^2} \mathbf{u}_r + \frac{R V_0 q_2}{d_0^2} \mathbf{u}_r. \quad (7.258)$$

7.10 For the geometrical disposition of the preceding problem, let us suppose that initially the charge q_2 was far away of the sphere. If the sphere is maintained at constant potential V_0, calculate the work necessary to bring the charge q_2 from its initial position to the point d_0.

Solution

Applying the definition of work we will calculate it. Starting from $P_1 = \infty$ and finishing at $P_2 = d_0$, we have

$$W = \int_{P_1}^{P_2} \mathbf{F}(\mathbf{r}).d\mathbf{r} = \lim_{P_1 \to \infty} \int_{P_1}^{P_2} q_2 \mathbf{E}(\mathbf{r}).d\mathbf{r} = \lim_{P_1 \to \infty} \int_{P_1}^{P_2} -q_2 \nabla V(x, y, z).d\mathbf{r} =$$

$$= -q_2 (V(P_2) - V(\infty)) = -q_2 V(P_2), \quad (7.259)$$

that is,

$$W = \frac{-1}{8\pi\varepsilon_0} \frac{R q_2^2}{(d_0^2 - R^2)} + \frac{R V_0 q_2}{d_0}. \quad (7.260)$$

7.11 On the same metallic sphere of the preceding exercises a charge Q is placed on its surface. A point charge q is located at the point $P(0, b, 0)$ where $b > R$. Determine: (a) The potential for all points of space. (b) The electric field. (c) The surface charge density of the conducting sphere.

Solution

This problem can be resolved with the help of the results obtained in the preceding exercises. As we have shown in the problem (7.8), by means of an image charge q_1 with respect to an external charge q_2, we can only obtain a spherical surface if the potential is zero ($V = C = 0$), otherwise it is necessary to introduce changes in the system in order to reach the desired potential. In a similar manner, when a charge Q

is placed on a metallic sphere, the only thing we know is that the conducting surface will be a sphere of constant potential, but not necessarily zero. The condition we have is that the total charge enclosed in the surface is Q, then employing the Gauss theorem it may be written

$$\oint_S \mathbf{E} \cdot d\mathbf{S} = \frac{Q}{\varepsilon_0}. \tag{7.261}$$

Due to the spherical symmetry of the problem, the idea consists of introducing another charge q_3 at the centre of the sphere whose value must be determined, and adjusting it so that we obtain the constant potential desired. To find out q_3, we can apply (7.261) as follows. Let us first thing of a charge q_2 in the exterior and a grounded spherical surface. For understanding this system we can put an fictitious image charge q_1 inside of the sphere whose effect together with q_2 is to bring the metallic sphere to zero potential. If now we have a actual total charge Q on the surface the potential changes. In this situation, applying (7.261) holds

$$\oint_S \mathbf{E} \cdot d\mathbf{S} = \frac{Q}{\varepsilon_0} = \frac{q_1 + q_3}{\varepsilon_0}. \tag{7.262}$$

Physically this latter result means that the final effect of having a charge Q on the sphere is equivalent to those generated by two fictitious charges at points $y_0 = \frac{R^2}{b}$, and $y = 0$. From (7.262) we have

$$Q = q_1 + q_3 \Rightarrow q_3 = Q - q_1,$$

thus introducing in this expression the value of Q_1 (7.244) calculated previously it yields

$$q_3 = Q + q_2 \frac{R}{b}, \tag{7.263}$$

and then the potential is

$$V(x, y, z) = \frac{1}{4\pi\varepsilon_0} \left(\frac{R}{b} \frac{-q_2}{\sqrt{x^2 + (y - \frac{R^2}{b})^2 + z^2}} + \frac{q_2}{\sqrt{x^2 + (y - b)^2 + z^2}} + \frac{Q + q_2 \frac{R}{b}}{\sqrt{x^2 + y^2 + z^2}} \right) \tag{7.264}$$

For the interior of the metallic sphere, as we know, the potential is uniform, but due to the boundary conditions imposed it must not be necessarily the same as we saw in latter problems. In fact, the potential depends only on the net charge enclosed by the surface, and this is the charge Q located externally and the image charge that appears as a consequence of q_2. Thus, we have

$$V(x, y, z) = \frac{1}{4\pi\varepsilon_0} \frac{Q + q_2 \frac{R}{b}}{\sqrt{x^2 + y^2 + z^2}} = \frac{1}{4\pi\varepsilon_0} \frac{Q + q_2 \frac{R}{b}}{R} \tag{7.265}$$

(b) The electric field may be directly calculated by means of the gradient

$$\mathbf{E}(x, y, z) = \frac{1}{4\pi\varepsilon_0} \left(\frac{q_2}{|\mathbf{r} - \mathbf{r}_2|^3} (r\mathbf{u}_r - b\mathbf{u}_y) - \frac{q_2 R}{b|\mathbf{r} - \mathbf{r}_1|^3} (r\mathbf{u}_r - \frac{R^2}{b}\mathbf{u}_y) + \frac{Q + q_2 \frac{R}{b}}{R^2} \mathbf{u}_r \right) \quad (7.266)$$

(c) The surface charge density is (7.247)

$$\sigma(x, y, z) = \epsilon_0 \frac{\partial V(x, y, z)}{\partial n}, \quad (7.267)$$

thus

$$\sigma(x, y, z) = \epsilon_0 \mathbf{nE} = \frac{1}{4\pi} \left(\frac{q_2}{(x^2 + (y - b)^2 + z^2)^{3/2}} (R - b\sin\theta \sin\phi) - \quad (7.268) \right.$$

$$\left. \frac{1}{4\pi} \left(\frac{q_2 R}{b(x^2 + (y - \frac{R^2}{b})^2 + z^2)^{3/2}} (R - \frac{R^2}{b} \sin\theta \sin\phi) + \frac{Q + q_2 \frac{R}{b}}{R^2} \right). \right.$$

7.12 The system of the figure is composed by a metallic sphere of radius R_1 and a circular conducting wire of radius R_2. The coil is charged with a total charge Q_2 (Fig. 7.30). Find the electric field at any point P on the OZ axis of a cartesian coordinate frame.

Solution

To face this problem we can use the results obtained in the Exercise 7.8, to help us. In fact, we will first focus our attention on the potential generated by a system composed of a metallic sphere and a charge q_2. As we are going to see, the solution found is directly applicable to our present problem. To explain it, look at the Fig. 7.31a. Let us suppose we examine only one of the charges located over the conducting ring of radius R_2, namely q_2. Setting aside the other charges, to q_2 corresponds an image charge q_1 located inside of the sphere. Now, if we fix our attention only to another charge q_2' infinitely close to q_2, as it would be alone, we can again find a fictitious charge q_1'. Both image charges obtained are placed at distinct points in the space,

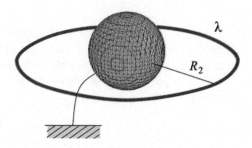

Fig. 7.30 A metallic sphere with a circular conducting ring

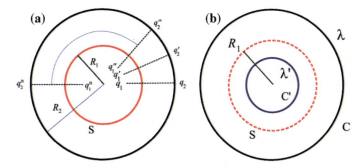

Fig. 7.31 Planar view of the system. **a** The charges $q'_1, q''_1, q'''_1, \ldots, q'^n_1$ are the respective images of the charges $q'_2, q''_2, q'''_2, \ldots, q'^n_2$ located in the conducting wire. **b** The *circle* in *red* corresponds to the metallic sphere of radius R_1

but at the same distance to the surface S. It seems to be logical that, if we repeat this reasoning with all charges composing the metallic ring we will obtain a set of image charges q'^n_1 whose location forms another circle of radius $r = y_0 = \frac{R_1^2}{R_2}$ (7.52). This new circle represents the image of the external circular conducting wire, and having a total charge $Q_1 = \sum_{i=1}^{n} q'^n_1$. From the viewpoint of the method of images, this reasoning shown leads to an equivalent pose of the problem in the following terms: calculate the electric field on the axis OZ, generated by two concentric and parallel metallic rings of radius[19] R_2 and R_1, charged with Q_2 and Q_1 coulombs, respectively.

To respond to this question we can directly use (7.244). Introducing the notation chosen it yields

$$Q_1 = -\frac{R_1}{R_2} Q_2. \qquad (7.269)$$

In Chap. 2 we studied the electric field produced by a circular conducting wire charged by a homogeneous lineal charge density $\lambda = 2\pi R_2 Q_2$. The formula is

$$\mathbf{E} = \frac{1}{4\pi\varepsilon_0} \frac{zq}{(R^2 + z^2)^{\frac{3}{2}}} \mathbf{u}_z \qquad (7.270)$$

Taking into account this result, substituting $R = R_2$ and $q = Q_2$, we get the field $\mathbf{E}_w$ produced by the metallic wire

$$\mathbf{E}_w = \frac{1}{4\pi\varepsilon_0} \frac{zQ_2}{(R_2^2 + z^2)^{\frac{3}{2}}} \mathbf{u}_z. \qquad (7.271)$$

---

[19] Remember that, for a point charge located at a distance d_0 from the origin of a conducting sphere of radius R, the following formula holds $R^2 = y_0 d_0$, where y_0 is the distance of its center to the image charge.

In the same way, setting $r = \frac{R_1^2}{R_2}$ and (7.269) into (7.270) we get the electric field generated by the image ring

$$\mathbf{E}_R = \frac{1}{4\pi\varepsilon_0} \frac{zQ_1}{(r^2+z^2)^{\frac{3}{2}}}\mathbf{u}_z = \frac{-1}{4\pi\varepsilon_0} \frac{zR_1 Q_2}{R_2((\frac{R_1^2}{R_2})^2+z^2)^{\frac{3}{2}}}\mathbf{u}_z. \qquad (7.272)$$

Applying the principle of superposition

$$\mathbf{E} = \mathbf{E}_w + \mathbf{E}_R = \frac{zQ_2}{4\pi\varepsilon_0}\left(\frac{1}{(R_2^2+z^2)^{\frac{3}{2}}} - \frac{R_1}{R_2((\frac{R_1^2}{R_2})^2+z^2)^{\frac{3}{2}}}\right)\mathbf{u}_z, \qquad (7.273)$$

which represents the field at any point on the revolution axis of the system.

7.13 Let us suppose that the Neumann boundary conditions apply. Demonstrate that the election of $\frac{\partial G}{\partial n} = 0$ for the Green function on the surface yields to a contradiction.

Solution

In the development of the theory we have seen for the definition of Neumann's function imposed that the normal derivative on the surface S must be a constant (7.37). Taking into consideration the definition of Green's function it seems more logical to choose this derivative to be zero because it makes the solution easier. However, if we analyze the problem in more detail we conclude that this choice leads to a contradiction. In fact, by applying Gauss' theorem on the left side of (7.36) we obtain

$$\iiint_D \Delta N(\mathbf{r},\mathbf{r}')dV' = \iiint_D \nabla\cdot\nabla N(\mathbf{r},\mathbf{r}')dV' = \iint_{S'}\left(\frac{\partial N(\mathbf{r},\mathbf{r}')}{\partial n'}\right)dS' = 0. \qquad (7.274)$$

Taking volume integrals on the right side of (7.36), it results

$$\iiint_D \Delta N(\mathbf{r},\mathbf{r}')dV' = \iiint_D \delta(r,r')dV' = 1, \qquad (7.275)$$

that is $1 = 0$, which is a contradiction.

7.14 Show that in the case of the Neumann problem, the constant value chosen for the normal derivative on the surface S' is not arbitrary, and calculate it.

Solution

By applying Gauss's theorem to (7.36), and taking into consideration (7.37), we have

$$\iiint_D \Delta N(\mathbf{r},\mathbf{r}')dV' = \iiint_D \nabla\cdot\nabla N(\mathbf{r},\mathbf{r}')dV' = \iint_{S'}\left(\frac{\partial N(\mathbf{r},\mathbf{r}')}{\partial n'}\right)dS' = C\iint_{S'} = CS. \qquad (7.276)$$

On the other hand, the first integral on the left may be calculated

$$\int\int\int_D \Delta N(\mathbf{r},\mathbf{r}')dV' = \int\int\int_D \delta(r,r')dV' = 1, \qquad (7.277)$$

then,

$$C = \frac{1}{S}, \qquad (7.278)$$

where S is the boundary surface of the region D.

7.15 When studying the conformal mapping in the theoretical introduction we proposed a formulation for defining a complex potential under some circumstances. By using the Maxwell equations show another procedure to introduce the complex potential.

Solution

As we saw in the introduction of this chapter, sometimes a problem in three dimensions can be reduced to two variables because of symmetries. This is the case, for example, of problems in which the fields are the same on every section perpendicular to an axis. As we shall show in the next problems, for these specials problems it is possible to use methods that are based on complex variables to find the solution. To show this, consider a region of space free of charge, then

$$\nabla \cdot \mathbf{E} = 0, \qquad (7.279)$$

$$\nabla \times \mathbf{E} = 0. \qquad (7.280)$$

Equation (7.279) means that a scalar potential $V(\mathbf{r})$ for $\mathbf{E}$ may be found, i.e.

$$\mathbf{E} = -\nabla V(\mathbf{r}). \qquad (7.281)$$

On the other hand, as in the case of the magnetic field ($\nabla \cdot \mathbf{B} = 0$), (7.280) allows us to introduce a vector potential for $\mathbf{E}$. In effect, let this potential be Π, then

$$\mathbf{E} = \nabla \times \Pi. \qquad (7.282)$$

To study a plane electric (or magnetic) field we only need two variables, namely (x, y). In this case, the third component of $\mathbf{E}$ does not appear, and then component z of the rotational (7.282) must be zero, i.e.

$$E_z = \frac{\partial \Pi_y}{\partial x} - \frac{\partial \Pi_x}{\partial y} = 0. \qquad (7.283)$$

The simplest possibility to fulfill (7.283) is to choose $\Pi_x = \Pi_y = 0$, hence the vector potential Π has only the component along the OZ axis,

$$\mathbf{E} = \frac{\partial \Pi_z}{\partial y}\mathbf{u}_x - \frac{\partial \Pi_z}{\partial x}\mathbf{u}_y. \tag{7.284}$$

Simultaneously, from (7.279) we have for the x and y components of the electric field $\mathbf{E}$,

$$\mathbf{E} = -\frac{\partial V}{\partial x}\mathbf{u}_x - \frac{\partial V}{\partial y}\mathbf{u}_y, \tag{7.285}$$

and comparing (7.284) with (7.285), we have

$$E_x = -\frac{\partial V}{\partial x} = \frac{\partial \Pi_z}{\partial y}, \quad E_y = -\frac{\partial V}{\partial y} = -\frac{\partial \Pi_z}{\partial x}. \tag{7.286}$$

These (7.286) have the same structure as the Cauchy–Riemann relations of complex analysis, and as we have commented in Sect. 7.5.4, we can form a function of complex variable z which physically represents the potential. In fact, we can write

$$f(z) = \Pi(x, y) + iV(x, y). \tag{7.287}$$

This equation is similar to (7.69), but its real and imaginary parts have been interchanged. Now the curves of constant potential $V(x, y) = C_1$ correspond to the imaginary part of (7.287), and the lines of force $\Pi(x, y) = C_2$ take the place of the imaginary part of $f(z)$. This result affects neither the key idea of the method nor the basic procedure for solving problems. Actually, the choice of $V(x, y)$ and $\Pi(x, y)$ as real or imaginary parts of the potential is completely arbitrary. We could have develop in the same manner and with the same validity the technique exposed in the theory with an expression like (7.287), but the formulae seen there changed some signs and constants. For instance, considering the potential in the form of (7.287) for the electric field holds

$$E(z) = -\frac{\partial V(x, y)}{\partial x} - i\frac{\partial V(x, y)}{\partial y} = -\frac{\partial V(x, y)}{\partial x} - i\frac{\partial \Pi(x, y)}{\partial x} = -\left(\frac{\partial V(x, y)}{\partial x} + i\frac{\partial \Pi(x, y)}{\partial x}\right) = \tag{7.288}$$

$$-i\left(\frac{\partial \Pi(x, y)}{\partial x} - i\frac{\partial V(x, y)}{\partial x}\right) = -i\overline{f'(z)}.$$

As we can see the only difference with (7.74) is the factor i.

7.16 In Chap. 2 we have seen the electric field created by a very long conducting wire charged with a homogeneous density of charge λ. Find its complex potential and show the potential level curves and the force lines.

Solution

Let us suppose, for simplicity, that the wire is located along the OZ axis coinciding with the origin of coordinates of the cartesian system. As the wire is very large, we have translational symmetry along OZ. This means that we can reduce the calculation to a plane perpendicular to the OZ direction, with the result we will obtain the same for every plane parallel to the OXY plane (see theory).

For finding the complex potential we start with the formula of the electric field generated by a large metallic wire holding a density of charge per unit length λ. As it was demonstrated in Chap. 2, the expression of $\mathbf{E}$ is

$$\mathbf{E} = \frac{\lambda}{2\pi\varepsilon_0 \rho}\mathbf{u}_\rho, \tag{7.289}$$

where ρ is the distance from the point where the field is calculated to the wire, and ε_0 is the permittivity. Setting $\mathbf{u}_\rho$ as functions of its cartesian components we obtain

$$\mathbf{E} = \frac{\lambda}{2\pi\varepsilon_0\sqrt{x^2+y^2}}(\cos\phi\,\mathbf{u}_x + \sin\phi\,\mathbf{u}_y) = \frac{\lambda}{2\pi\varepsilon_0\sqrt{x^2+y^2}}(\frac{x}{\sqrt{x^2+y^2}}\mathbf{u}_x + \frac{y}{\sqrt{x^2+y^2}}\mathbf{u}_y) = \\ \frac{\lambda}{2\pi\varepsilon_0}\left(\frac{x}{(x^2+y^2)}\mathbf{u}_x + \frac{y}{(x^2+y^2)}\mathbf{u}_y\right). \tag{7.290}$$

Now, we construct a complex number in such a way that its real and imaginary parts coincide with the components of (7.290)

$$E = E_x + iE_y = \frac{\lambda}{2\pi\varepsilon_0}\frac{x+iy}{(x^2+y^2)}. \tag{7.291}$$

Transforming this equation to z coordinates, we have

$$E = \frac{\lambda}{2\pi\varepsilon_0}\frac{z}{z\bar{z}} = \frac{\lambda}{2\pi\varepsilon_0}\frac{1}{\bar{z}}. \tag{7.292}$$

Introduction of this last result into (7.75) gives

$$f(z) = -\int \overline{E}(z)\,dz + a = -\int \frac{\lambda}{2\pi\varepsilon_0}\frac{1}{z}dz + a = -\frac{\lambda}{2\pi\varepsilon_0}\int \frac{1}{z}dz + a = \frac{\lambda}{2\pi\varepsilon_0}\ln\frac{1}{z} + a, \tag{7.293}$$

a being a constant. This result represents the complex potential corresponding to the electric field (7.289). Taking into consideration that $z = \rho\exp i\theta$ (7.293) yields[20]

$$f(z) = \frac{\lambda}{2\pi\varepsilon_0}(\ln\rho + i\theta), \tag{7.294}$$

then

$$u(\rho,\theta) = \frac{\lambda}{2\pi\varepsilon_0}\ln\rho, \tag{7.295}$$

---

[20] For simplicity we will consider in what follows $a = 0$.

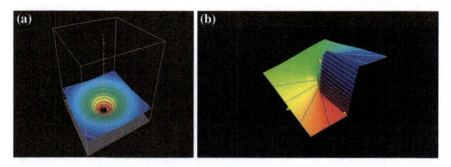

Fig. 7.32 a Real part of the complex potential. Observe that the *circles* represent the curves of constant potential. **b** Imaginary part of $f(z)$. The level *curves* are *straight lines* starting from the origin of coordinates, where the wire is located

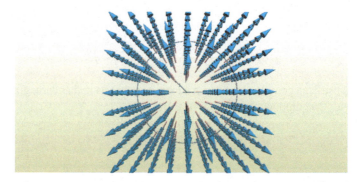

Fig. 7.33 Electric field created by the charged wire. The field lines are perpendicular to the level curves of the potential (real part). The direction coincides with the *force lines* shown in Fig. 7.32a

and
$$v(\rho, \theta) = \frac{\lambda}{2\pi\varepsilon_0} \theta. \tag{7.296}$$

As we commented in the theory, the level lines $u = C_1$ (Fig. 7.32a) represent the equipotential curves, and $v = C_2$ (Fig. 7.32b) the constant lines of force. If $\lambda > 0$ the lines of force are directed away of the singularity, and for $\lambda < 0$ the field lines are pointed toward $z = 0$. The electric field is depicted in Fig. 7.33. Note that these vectors are perpendicular to the curves $u = C_1$.

7.17 Let us suppose a very large metallic wire carries a homogeneous density current j. Give the corresponding complex potential and study its real and imaginary parts.

Solution

As we explained in Chap. 5 (see Problem 5.1), the magnetostatic field produced by a very large metallic wire only has a tangential component, i.e.

$$\mathbf{H}(\rho) = \frac{I}{2\pi\rho}\mathbf{u}_\phi. \qquad (7.297)$$

By introducing the expression of the unitary vector $\mathbf{u}_\phi = (-\sin\phi\,\mathbf{u}_x + \cos\phi\,\mathbf{u}_y)$ (7.297) leads to

$$\mathbf{H} = \frac{I}{2\pi\sqrt{x^2+y^2}}(-\sin\phi\,\mathbf{u}_x + \cos\phi\,\mathbf{u}_y) = \frac{I}{2\pi\sqrt{x^2+y^2}}\left(\frac{-y}{\sqrt{x^2+y^2}}\mathbf{u}_x + \frac{x}{\sqrt{x^2+y^2}}\mathbf{u}_y\right) =$$
$$= \frac{I}{2\pi}\left(\frac{-y}{x^2+y^2}\mathbf{u}_x + \frac{x}{x^2+y^2}\mathbf{u}_y\right). \qquad (7.298)$$

With this last formula the magnetic field expressed as a complex number adopts the following form

$$H = H_x + iH_y = \frac{I}{2\pi}\frac{-y+ix}{(x^2+y^2)} = \frac{I}{2\pi}\frac{i(x+iy)}{(x^2+y^2)} = \frac{I\,i}{2\pi}\frac{z}{z\bar{z}} = \frac{I\,i}{2\pi\bar{z}}. \qquad (7.299)$$

Once the field is known, the complex potential may be obtained by means of (7.87)

$$F(z) = -\int \overline{H}(z)\,dz + b = -\int \frac{-I\,i}{2\pi z}dz + b = \frac{I\,i}{2\pi}\int \frac{1}{z}dz + b = \frac{I\,i}{2\pi}\ln z = \frac{I\,i}{2\pi}(\ln\rho + i\theta) + b, \qquad (7.300)$$

where b is a constant which for simplicity we consider zero. Separating in real and imaginary parts $F(z) = \frac{I}{2\pi}(-\theta + i\ln\rho) = u + iv$, we have

$$u(\rho,\theta) = -\frac{I}{2\pi}\theta, \qquad (7.301)$$

and

$$v(\rho,\theta) = \frac{I}{2\pi}\ln\rho. \qquad (7.302)$$

As we can observe the result is similar to the former problem, but the value of the real and imaginary parts are switched. In fact, the curves of constant potential in a reference frame of polar coordinates are now $-\frac{I}{2\pi}\theta = C_1$, which represent straight lines as depicted in Fig. 7.32b for the electric field of a wire. On the contrary, $\frac{I}{2\pi}\ln\rho = C_2$ are the line forces and as the reader can probe there are perpendicular to the isolines $u(\rho,\theta) = Constant$ at each point. The field lines, i.e. the magnetostatic field generated by the wire currying an intensity I has the form shown in Fig. 7.34.

7.18 The attached figure shows a system composed by two very long conducting cylinders of radii R and r, respectively ($r < R$), whose revolution axes are parallel to each other, but they do not coincide. The inner cylinder is held to zero potential and the other one to V_0 volts. Determine the potential in the region between cylinders and the capacitance per unit length of the system.

Fig. 7.34 Magnetic field generated by a very large wire. The direction of the field coincides with the lines of force represented by the imaginary part of the complex magnetic potential $F(z)$

Solution

The problem in its presented form is very difficult to solve. It would be easier if we had two exact concentric cylinders as depicted in Fig. 7.35, because in such a case the solution is well known. In order to obtain these new geometric configuration, we must find a mapping for which our problem is transformed. With this aim, we will investigate the characteristics we must have. In the introductory theory given about this subject (Sect. 7.5.3), we saw that the fractional transformation (Möbius mapping) has properties very adequate for this case. For instance, circles in the z-plane are mapped into circles in the w-plane, then it seems to be logical to try using these kind of functions for solving the question. Actually, we must find a transformation that maps the exterior circle onto itself, and the center of the small cylinder transforms into the origin of coordinates in the w-plane. Thus, we would have two concentric circles in w, as we wanted. In this way, by means of the bilinear transformation, points symmetric in the z-plane with respect to a circle are mapped into points symmetric with respect to the transformed circle in the plane w. Taking it into account we see that the circles will be concentric if we find two points in z-plane that are symmetric with respect to both circles, *simultanously*. These two symmetric points will be transformed into two symmetric points in w with respect to the new geometry. Considering the disposition shown in Fig. 7.35, and labelling x_1 and x_2 the coordinates over the OX axis in the z-plane (because of the disposition), the equations that must be fulfilled in order to find a transformation are the following

$$x_1 \cdot x_2 = R^2, \tag{7.303}$$

and

$$(x_1 - d) \cdot (x_1 - d) = r^2. \tag{7.304}$$

The first one gives the symmetric point with respect to the big cylinder, and (7.304) allows us to compute the same for the small circle. Observe that if we would only have a centered circle (in this case of radius R), the symmetric point to $(0, 0)$ is the point $z = \infty$. However, because of the two conditions (7.303), and (7.304), the

Fig. 7.35 Two metallic cylinders whose revolution axes are parallel to each other

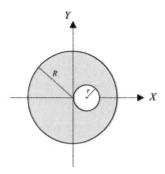

solution is different. Combining both equations, we obtain a new expression where only one variable appears,

$$d x_1^2 + (r^2 - R^2 - d^2) x_1 + d R^2 = 0. \tag{7.305}$$

The solution of (7.305) gives two roots that solve the problem. The bilinear transformation we find has the following form

$$w = f(z) = \alpha \frac{z - x_1}{z - x_2}, \tag{7.306}$$

where x_1 and x_2 are yet unknown. The parameter α is, in general, a complex number that does not affect the final result. Introductions of the values $r = 0.2$, $d = 0.2$, and $R = 1$ into (7.305) $0.2 x_1^2 - x_1 + 0.2 = 0$ yields $x_1 = 0.21$ and $x_2 = 4.79$, therefore

$$f(z) = \frac{z - 0.21}{z - 4.79} = w. \tag{7.307}$$

The mapping (7.307) transforms the two eccentrical circles of the z-plane into two centered circles in the w-plane. The new radii of the cylinders can be calculated by means of the (7.307). In fact, $w_1 = |f(0)| \approx 0.04$ and $w_2 = |f(R = 1)| \approx 0.2$. The expression of the potential between two concentric metallic cylinders held to potentials $V_1 = 0$ and $V_2 = v_0$ in the coordinates of the plane w is[21]

$$\varphi(w) = \frac{V_0}{\ln\left(\frac{w_2}{w_1}\right)} \ln\left(\frac{|w|}{w_1}\right), \tag{7.308}$$

where $|w| = \sqrt{u^2 + v^2}$ represents in that plane the same as ρ in the z-plane. The solution is determined by mapping (7.308) back to the z-plane. It can be immediately performed by substituting (7.307) into (7.308)

---

[21] The calculation of the potential in a OXY coordinate frame, may be directly obtained from the differential equation $\frac{1}{\rho} \frac{\partial}{\partial \rho} \left(\rho \frac{\partial V}{\partial \rho} \right) = 0$, with boundary conditions $V(r) = V_1$ and $V(R) = V_2$ ($r < R$). The solution is $V(\rho) = \frac{V_2 - V_1}{\ln\left(\frac{R}{r}\right)} \ln\left(\frac{\rho}{r}\right) + V_1$.

$$V(x,y) = \frac{V_0}{\ln(5)} \ln\left|\frac{25(z-0.21)}{(z-4.79)}\right| = \frac{V_0}{\ln(5)}\left[\ln(25) + \frac{1}{2}((x-0.21)^2 + y^2) - \frac{1}{2}((x-4.79)^2 + y^2))\right].$$
(7.309)

This result is represented in Fig. 7.36a. Figure 7.36b shows the curves of constant potential between the two cylinders. The imaginary part of the complex potential which gives information of the line forces can be computed with the (7.70) by means of a simple integration. The result is depicted in Figs. 7.37 and 7.38.

7.19 The cross section of a square prismatic metallic system (see Problem 7.3) is formed by four sides of length L. The depth h of the system is larger than L ($L \ll h$) and each of its sides is held to potentials A, B, C, and D, respectively. Calculate the potential at any interior point of the system in the following cases: (a) $A = 200$, $B = 0$, $C = 0$, and $D = 0$. (b) $A = 500$, $B = 0$, $C = 0$, and $D = 0$. (c) $A = 500$, $B = 100$, $C = 0$, and $D = 0$. (d) $A = 500$, $B = 100$, $C = 300$, and $D = 400$.

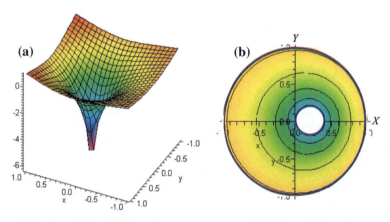

Fig. 7.36 **a** Graphic of the potential $V(x, y)$. **b** Curves of constant potential on the OXY plane

Fig. 7.37 The geometry of the problem in the w-plane corresponds to two concentric cylinders of radii w_1 and w_2

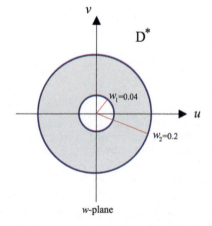

Fig. 7.38 Line forces between the *two circles* in the z-plane. Observe that this vector field is perpendicular to the level potential curves

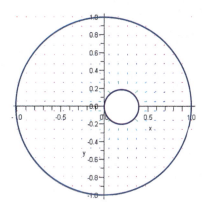

Solution

Due to the side of the square section being much smaller than the length of the prismatic hole ($L << h$), we can study the system as a two dimensional problem (Fig. 7.39). The only solution we are going to find differs from the actual one in the vicinity of the top and bottom of the prismatic geometry, where the end effects are present. Accepting this viewpoint, we will work with a cross-section of the four metallic plates as depicted in Fig. 7.40. For obtaining the electric field inside of the system we have previously successfully employed the technique of separation of variables. However, now we want to apply the method of finite differences with the aim to explain the procedure. As we will see, by examining only a few points we are able to know roughly the electric field between the metallic sheets. By this method we want to obtain a numerical solution of the laplace equation inside of the region of interest. Because of its characteristics, it follows that we can perform the calculation only at a finite number of points, and therefore we must first construct a discretized model adapted to the specifications of the problem to be solved. The points selection over the domain of interest depends on the geometry and resolution we want to reach.

We begin to construct a grid of points with the intersections of crossing parallel lines to the coordinate frame. The more resolution, the more points, but also more computation steps and time consuming. As we have seen in Sect. 7.6.1, the basic idea consists in determining an approximate solution at any node of the mesh by averaging of its nearest four neighbors. Boundary points have information of the

Fig. 7.39 *Square* metallic guide

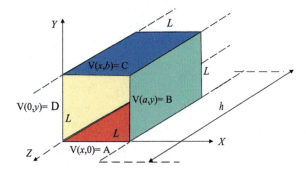

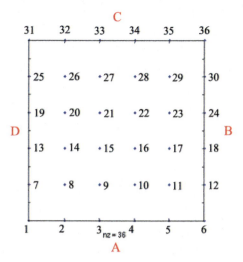

Fig. 7.40 *Square* sections with the 36 points to be studied

boundary conditions given, and their values and the mesh size must be introduced in the equations correctly. Otherwise the solution propagates with errors which are amplified at every step.

In order to have all possibilities for the different Dirichlet boundary conditions on all the sides, we suppose that the potential on the edges of the square section are A, B, C and D, respectively. It means that the potential for the points over these lines are known. We start at point 8 of the Fig. 7.40, which depends on V_2, V_7, V_9, and V_{14}, one of them is located on the boundary (V_2), and continue in order to the point 29. The general system of equations for the 16 points inside of the square is the following

$$4V_8 - V_9 - V_{14} = A + D \tag{7.310}$$
$$4V_9 - V_8 - V_{10} - V_{15} = A \tag{7.311}$$
$$4V_{10} - V_9 - V_{11} - V_{16} = A \tag{7.312}$$
$$4V_{11} - V_{10} - V_{17} = A + B \tag{7.313}$$
$$4V_{14} - V_8 - V_{20} - V_{15} = D \tag{7.314}$$
$$4V_{15} - V_9 - V_{14} - V_{16} - V_{21} = 0 \tag{7.315}$$
$$4V_{16} - V_{10} - V_{15} - V_{17} - V_{22} = 0 \tag{7.316}$$
$$4V_{17} - V_{11} - V_{16} - V_{23} = B \tag{7.317}$$
$$4V_{20} - V_{14} - V_{21} - V_{26} = D \tag{7.318}$$
$$4V_{21} - V_{15} - V_{20} - V_{22} - V_{27} = 0 \tag{7.319}$$
$$4V_{22} - V_{16} - V_{21} - V_{23} - V_{28} = 0 \tag{7.320}$$
$$4V_{23} - V_{17} - V_{22} - V_{29} = B \tag{7.321}$$
$$4V_{26} - V_{20} - V_{27} = C + D \tag{7.322}$$
$$4V_{27} - V_{21} - V_{26} - V_{28} = C \tag{7.323}$$
$$4V_{28} - V_{22} - V_{27} - V_{29} = C \tag{7.324}$$

Solved Problems   493

$$4V_{29} - V_{23} - V_{28} = C + B. \tag{7.325}$$

It is interesting to note that for Dirichlet boundary conditions, the procedure does not begin at points over the boundary. We start with points placed over the next line. This situation differs from that corresponding to Neumann. In that case the conditions are over the normal derivative on the surface, which leads to one additional system of equations taking the points over the boundary.

Setting the (7.310)–(7.325) in matrix form we get

$$\begin{bmatrix} 4 & -1 & 0 & 0 & -1 & 0 & 0 & 0 & 0 & 0 & 0 & 0 & 0 & 0 & 0 & 0 \\ -1 & 4 & -1 & 0 & 0 & -1 & 0 & 0 & 0 & 0 & 0 & 0 & 0 & 0 & 0 & 0 \\ 0 & -1 & 4 & -1 & 0 & 0 & -1 & 0 & 0 & 0 & 0 & 0 & 0 & 0 & 0 & 0 \\ 0 & 0 & -1 & 4 & 0 & 0 & 0 & -1 & 0 & 0 & 0 & 0 & 0 & 0 & 0 & 0 \\ -1 & 0 & 0 & 0 & 4 & -1 & 0 & 0 & -1 & 0 & 0 & 0 & 0 & 0 & 0 & 0 \\ 0 & -1 & 0 & 0 & -1 & 4 & -1 & 0 & 0 & -1 & 0 & 0 & 0 & 0 & 0 & 0 \\ 0 & 0 & -1 & 0 & 0 & -1 & 4 & -1 & 0 & 0 & -1 & 0 & 0 & 0 & 0 & 0 \\ 0 & 0 & 0 & -1 & 0 & 0 & -1 & 4 & 0 & 0 & 0 & -1 & 0 & 0 & 0 & 0 \\ 0 & 0 & 0 & 0 & -1 & 0 & 0 & 0 & 4 & -1 & 0 & 0 & -1 & 0 & 0 & 0 \\ 0 & 0 & 0 & 0 & 0 & -1 & 0 & 0 & -1 & 4 & -1 & 0 & 0 & -1 & 0 & 0 \\ 0 & 0 & 0 & 0 & 0 & 0 & -1 & 0 & 0 & -1 & 4 & -1 & 0 & 0 & -1 & 0 \\ 0 & 0 & 0 & 0 & 0 & 0 & 0 & -1 & 0 & 0 & -1 & 4 & 0 & 0 & 0 & -1 \\ 0 & 0 & 0 & 0 & 0 & 0 & 0 & 0 & -1 & 0 & 0 & 0 & 4 & -1 & 0 & 0 \\ 0 & 0 & 0 & 0 & 0 & 0 & 0 & 0 & 0 & -1 & 0 & 0 & -1 & 4 & -1 & 0 \\ 0 & 0 & 0 & 0 & 0 & 0 & 0 & 0 & 0 & 0 & -1 & 0 & 0 & -1 & 4 & -1 \\ 0 & 0 & 0 & 0 & 0 & 0 & 0 & 0 & 0 & 0 & 0 & -1 & 0 & 0 & -1 & 4 \end{bmatrix} \begin{bmatrix} V_8 \\ V_9 \\ V_{10} \\ V_{11} \\ V_{14} \\ V_{15} \\ V_{16} \\ V_{17} \\ V_{20} \\ V_{21} \\ V_{22} \\ V_{23} \\ V_{26} \\ V_{27} \\ V_{28} \\ V_{29} \end{bmatrix} = \begin{bmatrix} A+D \\ A \\ A \\ A+B \\ D \\ 0 \\ 0 \\ B \\ D \\ 0 \\ 0 \\ B \\ C+D \\ C \\ C \\ C+B \end{bmatrix}$$

(7.326)

The problem reduces to calculating the inverse matrix and multiplying it by the vector containing a combination of values of boundary conditions. Specifically for the first case (case (a)) the side A is held to potential $V = 200\,\text{v}$, then introducing in (7.326) $A = 200$, we obtain

$$\begin{bmatrix} V_8 \\ V_9 \\ V_{10} \\ V_{11} \\ V_{14} \\ V_{15} \\ V_{16} \\ V_{17} \\ V_{20} \\ V_{21} \\ V_{22} \\ V_{23} \\ V_{26} \\ V_{27} \\ V_{28} \\ V_{29} \end{bmatrix} = \begin{bmatrix} 90.9 \\ 118.9 \\ 118.9 \\ 90.9 \\ 44.7 \\ 65.9 \\ 65.9 \\ 44.7 \\ 22.0 \\ 34.1 \\ 34.1 \\ 22.0 \\ 9.1 \\ 14.4 \\ 14.4 \\ 9.1 \end{bmatrix}. \tag{7.327}$$

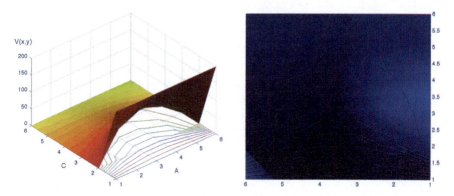

Fig. 7.41 Surface representing the potential inside of the prismatic system for $A = 200$, $B = 0$, $C = 0$, and $D = 0$. Observe the level curves in both graphics

Fig. 7.42 Three dimensional representation of $V(x, y)$ for $A = 500$, $B = 0$, $C = 0$, and $D = 0$

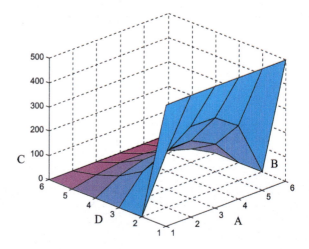

Figure 7.41 represents the scalar surface corresponding to potential. Observe that even if the mesh resolution chosen is bad, the result is enough to see the behavior of the potential (Fig. 7.42).

(b) Starting with the same scheme we substitute $A = 500$, $B = 0$, $C = 0$, and $D = 0$ into (7.326), obtaining the following vector with values of the potential for all points studied, i.e.

$$\begin{bmatrix} V_8 \\ V_9 \\ V_{10} \\ V_{11} \\ V_{14} \\ V_{15} \\ V_{16} \\ V_{17} \\ V_{20} \\ V_{21} \\ V_{22} \\ V_{23} \\ V_{26} \\ V_{27} \\ V_{28} \\ V_{29} \end{bmatrix} = \begin{bmatrix} 227.3 \\ 297.3 \\ 297.3 \\ 227.3 \\ 111.7 \\ 164.8 \\ 164.8 \\ 111.7 \\ 54.9 \\ 85.2 \\ 85.2 \\ 54.9 \\ 22.7 \\ 36.0 \\ 36.0 \\ 22.7 \end{bmatrix} \qquad (7.328)$$

The diagram for this case is exhibited in Sect. 7.6.1. The figure has the same basic characteristics like the picture presented in Fig. 7.41. The most important difference appears in the potential on the side A, because of the initial condition here. The rest of the interior points take also other values, but the form of the surface has the same structure.

(c) In the same manner that we have seen in the foregoing sections, if two sides of the system are held to potentials $A = 500$, $B = 100$, and on the edges C and D are zero, we put such a values in (7.326) again yielding

$$\begin{bmatrix} V_8 \\ V_9 \\ V_{10} \\ V_{11} \\ V_{14} \\ V_{15} \\ V_{16} \\ V_{17} \\ V_{20} \\ V_{21} \\ V_{22} \\ V_{23} \\ V_{26} \\ V_{27} \\ V_{28} \\ V_{29} \end{bmatrix} = \begin{bmatrix} 231.8 \\ 308.3 \\ 319.7 \\ 272.7 \\ 118.9 \\ 181.8 \\ 197.7 \\ 171.2 \\ 62.1 \\ 102.2 \\ 118.2 \\ 114.4 \\ 27.3 \\ 47.0 \\ 58.3 \\ 68.2 \end{bmatrix} \qquad (7.329)$$

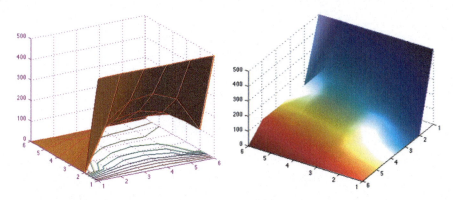

Fig. 7.43 Two perspectives of the surface representing the potential inside of the prismatic system for $A = 500, B = 100, C = 0$, and $D = 0$. Note that in this case the curves of constant potential are not symmetric like Fig. 7.17

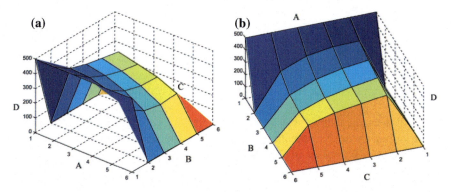

Fig. 7.44 Graphic when $A = 500, B = 100, C = 300$, and $D = 400$. **a** Lateral view of the function $V(x, y)$. **b** Same representation but viewed from the *right*

The potential versus the coordinates x and y is shown in Fig. 7.43. Now, the level curves[22] are not similar to that which we saw in Fig. 7.17 for the rectangular plate. This follows from the introduction of the potential on side B, which breaks the symmetry with respect to an axis passing across the geometrical center and parallel to B (Fig. 7.44).

(d) Introduction of $A = 500$, $B = 100$, $C = 300$, and $D = 400$ into (7.326) leads to

---

[22]Remember what we saw in Chap. 1. The level curve (or level surface) is in the domain of the scalar function considered.

$$\begin{bmatrix} V_8 \\ V_9 \\ V_{10} \\ V_{11} \\ V_{14} \\ V_{15} \\ V_{16} \\ V_{17} \\ V_{20} \\ V_{21} \\ V_{22} \\ V_{23} \\ V_{26} \\ V_{27} \\ V_{28} \\ V_{29} \end{bmatrix} = \begin{bmatrix} 427.3 \\ 419.3 \\ 385.2 \\ 304.5 \\ 389.8 \\ 364.8 \\ 317.0 \\ 233.0 \\ 367.0 \\ 333.0 \\ 285.2 \\ 210.2 \\ 345.5 \\ 314.8 \\ 280.7 \\ 222.7 \end{bmatrix}. \qquad (7.330)$$

The picture corresponding to this vector is represented below.

7.20 A big ferromagnetic material has a hole inside with the shape of a square prism. In the centre of it a very large metallic wire carrying a current $I = 1$ A is located (see figure attached). Supposing that μ is very high and the edge of the hole has length L, using the finite difference method (FDM), determine the magnetostatic potential in the square region showed in the figure and sketch the magnetic field lines.

Solution

Before beginning the calculation we are going to analyze the symmetries, because it shows us how to simplify the problem.

As we can observe, the system of Fig. 7.45 has translational symmetry in the direction of the axis perpendicular to the plane view. Let us choose this direction as the OZ axis. Because of this characteristic we can consider the system as a two dimensional problem, and therefore the solution we will find to be the same for all parallel planes to that figure. On the other hand, according to its square cross section, we also have a four order rotational symmetry with respect to OZ, which means that we do not need to calculate the potential directly at every point of the grid, but only at those locations inside a triangle of side L, because the solution for the other parts are connected together. Now, let us follow with the scalar potential.

In a region of the space where there are no currents $\nabla \times \mathbf{H} = 0$, so in a way similar to what we have seen for the electric field, we can define a magnetostatic potential V_m so that

$$\mathbf{H} = -\nabla V_m. \qquad (7.331)$$

By virtue of that equation the linear integral between two points a and b not passing throughout the current depends on such endpoints, i.e.,

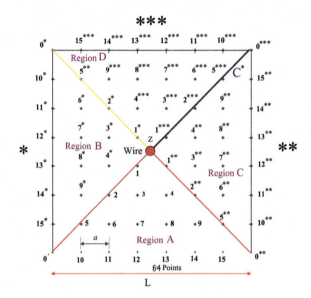

Fig. 7.45 *Rectangular* grid. The first region corresponds to the *triangle* without an asterisk. The other regions on the *left*, on the *right* and on the *upper triangle* are depicted with one asterisk ∗, two ∗∗ and three ∗ ∗ ∗, respectively

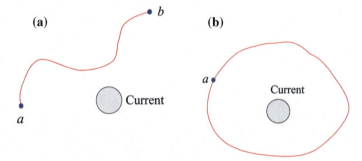

Fig. 7.46 a Open curve with extremes a and b. **b** Closed curve. The initial point a on the plane OXY is the same as the final point in the integration

$$\int_a^b \mathbf{H}\, d\mathbf{l} = -\int_a^b \nabla V_m\, d\mathbf{l} = V_m(a) - V_m(b). \quad (7.332)$$

In principle, it seems to be the same as we studied for the electric potential, however there are some differences. The most important difference between both potentials deals with the region where the magnetostatic potential is defined. To explain them, let us suppose a point a as shown in Fig. 7.46, from which we begin to construct a closed curve. If we go step by step we have an open curve, but at the end of the trace we reach the same point a, then we draw a closed line. Applying Ampère's law we have for the magnetic field **H**

$$\oint_\Gamma \mathbf{H}\, d\mathbf{l} = I, \tag{7.333}$$

which means that the integral over a closed curve Γ is not zero. Now, if we combine this result with (7.332) it yields

$$V_m(a) - V_m(a) = I \ (!!!), \tag{7.334}$$

a result that seems to be illogical because $V_m(a) - V_m(a)$ should be zero, and not I. The only possibility to make (7.334) feasible is that the value of the function V_m at point a when starting to follow the curve Γ is not the same as the value it reaches when arriving again at a after going around the curve; in other words it means that the potential is a multivalued function of position, that is to say, that V_m maps different images at the same point when closing the curve. In order to be clear and distinguish the two values, we will label with an asterisk $*$ the second potential, then the (7.333) converts to

$$V_m(a) - V_m^*(a) = I. \tag{7.335}$$

Mathematically this equation means that a discontinuity exists when completing a whole closed circulation, and it is due to the fact that the region where we have the potential is multiply connected (see Chap. 1).[23] In such a case we can understand this behavior for V_m as if it displaces on another sheet when passing across the discontinuity (see Fig. 1.30 from Chap. 1).[24] However we can prevent this problem by introducing a cut barrier which avoids having a closed curve in such a way to link currents. Once we have located the cut (there are infinite possibilities for that), the potential in this constrained region is single-valued, then the curvilinear integral of H over any closed curve will be zero.

To determine the scalar potential of this problem we must use the (5.60) Problem 5.1 of the magnetic field created by a very large metallic wire currying a current I that we have seen in Chap. 5, i.e.,

$$\mathbf{B} = \frac{\mu_0 I}{2\pi\rho}\mathbf{u}_\phi \Rightarrow \mathbf{H} = \frac{I}{2\pi\rho}\mathbf{u}_\phi. \tag{7.336}$$

Taking into account the symmetry of the system, by virtue of (7.331) and using cylindrical coordinates we can write

$$\mathbf{H} = -\nabla V_m \mathbf{u}_\phi = -\frac{1}{\rho}\frac{\partial V_m}{\partial \phi}\mathbf{u}_\phi = \frac{I}{2\pi\rho}\mathbf{u}_\phi, \tag{7.337}$$

thus

$$\frac{1}{\rho}\frac{\partial V_m}{\partial \phi} = -\frac{I}{2\pi\rho} \Rightarrow V_m(\phi) = -\int \frac{I}{2\pi}d\phi + D = -\frac{I}{2\pi}\phi + D, \tag{7.338}$$

---

[23] If we complete n circulations around a current we get the expression $V_m(a) - V_m^*(a) = nI$.
[24] Observe that this function seen in Chap. 1 ($\tan^{-1}\left(\frac{y}{x}\right)$) is the same as shown in Fig. 7.32.

D being an arbitrary constant. For this problem we choose $D = 0$, which does not affect to the final result in any way.

Once we know the function $V_m(\phi)$ for the wire alone, we can employ the finite difference method for calculating the scalar potential of the system represented in Fig. 7.45. As we have seen in the previous problem, for applying the FDM we first divide the region to be studied in subdomains by means of a mesh formed by parallel lines, which intersect at points where we will calculate the solution of the differential equation. However, in the present problem, we must not compute the calculations for all the points we see in Fig. 7.45. On the contrary, due to the symmetry of the rectangular cavity, we can perform the calculations in one of the triangles depicted, and then the results for the other points may be computed by using the relation between triangles which, as we are going to see, differ in quantities determined by (7.338).

At the starting point let us divide the square region into four identical triangles (Fig. 7.45). We will denote with asterisks the triangles on the left, on the right and on the upper part of the square, and the first one on the lower region without signs. For applying the finite difference method we mesh the region of interest by parallel lines and the intersections of them are labelled with numbers. In this problem, with the aim to be didactic and show the procedure, we have drawn in the square only 64 points. To explain the calculations it is enough and it does not lose generality. It only has influence on the resolution of the final result, but it does not affect the idea of the technique. Once we have identified the points, we choose one of the triangular regions of the square. We begin with the lower triangle where every point has a number from 1 to 15. Observe that at the corner we have written 0, then we have 16 points to be studied in each triangle. All points in this subregion have their corresponding point in the other parts, which corresponds to the same number but labelled with asterisks (see Fig. 7.45). Now, we will apply (7.95), however as it can be seen some problems occur. As we have explained in Sect. 7.6.1, using FDM requires the knowing of the potential values at the neighbors of the points where the potential will be computed. As a result in our case we see that points such as 1, 2 or 5 need data from region $*$, which are not known *a priori*. This difficulty may be solved by employing (7.338). In fact, as the magnetic potential grows linearly with angle ϕ, then every point i^* of the region b will differ by a quantity $\frac{l}{4}$, that is to say, if the potential at point 4 is V_4 its equivalent in the region b, $V_4^* = V_4 + \frac{l}{4}$. The same occurs for domains c and d. For these subregions it holds that $V_j^{**} = V_j - \frac{l}{4}$ and $V_j^{***} = V_j + \frac{l}{2}$ for $j = 1, \ldots N$, N being the number of points. With these relations, the values of the potential at set points of a domain are connected with the other ones. The second important thing refers to the conditions on the boundaries. Let us suppose two media as shown in Fig. 7.47, of permeabilities μ_1 and μ_2, respectively. If the angles of incidence with respect the normal to the surface are α_1 and α_2, from the figure it yields

$$\frac{\tan \alpha_1}{\tan \alpha_2} = \frac{\left(\frac{B_{1t}}{B_{1n}}\right)}{\left(\frac{B_{2t}}{B_{2n}}\right)}, \tag{7.339}$$

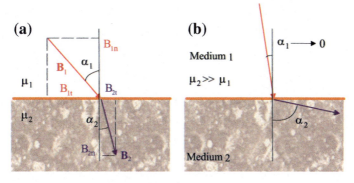

Fig. 7.47 a Magnetic field in the proximity of a boundary surface. **b** If $\mu_2 \gg \mu_2$ the *field lines* almost do not penetrate into the *medium 2*

but as $B_{1n} = B_{2n}$, $H_{1t} = H_{2t}$ and $B_{it} = \mu_i H_{it}$ ($i = 1, 2$), this equation becomes

$$\frac{\tan \alpha_1}{\tan \alpha_2} = \frac{B_{1t}}{B_{2t}} = \frac{\left(\frac{H_{1t}}{\mu_1}\right)}{\left(\frac{H_{2t}}{\mu_2}\right)} = \frac{\mu_1}{\mu_2}. \qquad (7.340)$$

This expression gives the behavior of the magnetic lines in the proximity of the boundary of two media. If one of them has a high permeability (as in our case), for instance material 2, $\mu_2 \gg \mu_1$ separation, and therefore from (7.340) holds

$$\frac{\tan \alpha_1}{\tan \alpha_2} = \frac{\mu_1}{\mu_2} \approx 0 \Rightarrow \alpha_1 \approx 0. \qquad (7.341)$$

This important result means that the magnetic lines in the region of lower permeability are practically perpendicular to the boundary surface (Fig. 7.47). As a consequence the separation surface corresponding to the domain of high μ may be approximately considered as an equipotential surface. Taking into account that the material of the problem is ferromagnetic, we cannot exactly speak about the magnetic permittivity, because of its non-linear behavior. We mean that μ is a function of the magnetic field and not a constant (see Chap. 6). However, even though in case of low **H** fields, the corresponding $\mu(H)$ is usually larger than μ_0 (vacuum or air), hence we can employ the aforementioned conclusions. Thus, choosing for instance that the potential at all points on the surface domain A are zero (at 0, 10, 11, 12, 13, 14 and 15),[25] and applying (7.94) to the points of the first triangle, we have

---

[25] Due to the potential at the boundary being constant, we could take another constant value for these points, but in order to perform the calculations as simply as possible we have set it to zero.

$$4V_1 - V_4^* - V_1^* - V_3 - V_1^{**} = 0 \Rightarrow 4V_1 - \left(V_4 + \frac{I}{4}\right) - \left(V_1 + \frac{I}{4}\right) - V_3 - \left(V_1 - \frac{I}{4}\right) = 0 \quad (7.342)$$

$$4V_2 - V_3 - V_6 - V_4^* - V_9^* = 0 \Rightarrow 4V_2 - V_3 - V_6 - \left(V_4 + \frac{I}{4}\right) - \left(V_9 + \frac{I}{4}\right) = 0 \quad (7.343)$$

$$4V_3 - V_1 - V_2 - V_4 - V_7 = 0 \quad (7.344)$$

$$4V_4 - V_3 - V_2^{**} - V_8 - V_1^{**} = 0 \Rightarrow 4V_4 - V_3 - \left(V_2 - \frac{I}{4}\right) - V_8 - \left(V_1 - \frac{I}{4}\right) = 0 \quad (7.345)$$

$$4V_5 - V_6 - V_{15}^* - V_{10} - V_9^* = 0 \Rightarrow 4V_5 - V_6 - \left(0 + \frac{I}{4}\right) - V_{10} - \left(V_9 + \frac{I}{4}\right) = 0 \quad (7.346)$$

$$4V_6 - V_2 - V_5 - V_7 - V_{11} = 0 \Rightarrow 4V_6 - V_2 - V_5 - V_7 - 0 = 0 \quad (7.347)$$

$$4V_7 - V_3 - V_6 - V_8 - V_{12} = 0 \Rightarrow 4V_7 - V_3 - V_6 - V_8 - 0 = 0 \quad (7.348)$$

$$4V_8 - V_4 - V_7 - V_9 - V_{13} = 0 \Rightarrow 4V_8 - V_4 - V_7 - V_9 - 0 = 0 \quad (7.349)$$

$$4V_9 - V_8 - V_2^{**} - V_5^{**} - V_{14} = 0 \Rightarrow 4V_9 - V_8 - \left(V_2 - \frac{I}{4}\right) - \left(V_5 - \frac{I}{4}\right) - 0 = 0. \quad (7.350)$$

Simplifying these equations we obtain

$$2V_1 - V_3 - V_4 = \frac{I}{4} \quad (7.351)$$

$$4V_2 - V_3 - V_4 - V_6 - V_9 = \frac{I}{2} \quad (7.352)$$

$$4V_3 - V_1 - V_2 - V_4 - V_7 = 0 \quad (7.353)$$

$$4V_4 - V_1 - V_2 - V_3 - V_8 = -\frac{I}{2} \quad (7.354)$$

$$4V_5 - V_6 - V_9 = \frac{I}{2} \quad (7.355)$$

$$4V_6 - V_2 - V_5 - V_7 = 0 \quad (7.356)$$

$$4V_7 - V_3 - V_6 - V_8 = 0 \quad (7.357)$$

$$4V_8 - V_4 - V_7 - V_9 = 0 \quad (7.358)$$

$$4V_9 - V_2 - V_5 - V_8 = -\frac{I}{2}, \quad (7.359)$$

which form a system of 9 linear equation with 9 unknowns.

$$\begin{bmatrix} 2 & 0 & -1 & -1 & 0 & 0 & 0 & 0 & 0 \\ 0 & 4 & -1 & -1 & 0 & -1 & 0 & 0 & -1 \\ -1 & -1 & 4 & -1 & 0 & 0 & -1 & 0 & 0 \\ -1 & -1 & -1 & 4 & 0 & 0 & 0 & -1 & 0 \\ 0 & 0 & 0 & 0 & 4 & -1 & 0 & 0 & -1 \\ 0 & -1 & 0 & 0 & -1 & 4 & -1 & 0 & 0 \\ 0 & 0 & -1 & 0 & 0 & -1 & 4 & -1 & 0 \\ 0 & 0 & 0 & -1 & 0 & 0 & -1 & 4 & -1 \\ 0 & -1 & 0 & 0 & -1 & 0 & 0 & -1 & 4 \end{bmatrix} \begin{bmatrix} V_1 \\ V_2 \\ V_3 \\ V_4 \\ V_5 \\ V_6 \\ V_7 \\ V_8 \\ V_9 \end{bmatrix} = \begin{bmatrix} \frac{I}{4} \\ \frac{I}{2} \\ 0 \\ -\frac{I}{2} \\ \frac{I}{2} \\ 0 \\ 0 \\ 0 \\ -\frac{I}{2} \end{bmatrix} \quad (7.360)$$

Calculating the inverse and multiplication by the column on the right side introducing $I = 1$ we get

$$\begin{bmatrix} V_1 \\ V_2 \\ V_3 \\ V_4 \\ V_5 \\ V_6 \\ V_7 \\ V_8 \\ V_9 \end{bmatrix} = \begin{bmatrix} 0.125 \\ 0.125 \\ 0.055 \\ -0.055 \\ 0.125 \\ 0.069 \\ 0.025 \\ -0.025 \\ -0.069 \end{bmatrix}. \qquad (7.361)$$

Knowing the potential for all points in the subdomain a, we can determine the other values by using the relations between homologous points $V_j^{**} = V_j - \frac{I}{4}$, $V_j^* = V_j + \frac{I}{4}$ and $V_j^{***} = V_j + \frac{I}{4}$, the potential for the other points in b, c and d may be determined. The results for these potentials are

$$\begin{bmatrix} V_1^* \\ V_2^* \\ V_3^* \\ V_4^* \\ V_5^* \\ V_6^* \\ V_7^* \\ V_8^* \\ V_9^* \end{bmatrix} = \begin{bmatrix} 0.375 \\ 0.375 \\ 0.305 \\ 0.195 \\ 0.375 \\ 0.319 \\ 0.275 \\ 0.225 \\ 0.181 \end{bmatrix}, \qquad (7.362)$$

$$\begin{bmatrix} V_1^{**} \\ V_2^{**} \\ V_3^{**} \\ V_4^{**} \\ V_5^{**} \\ V_6^{**} \\ V_7^{**} \\ V_8^{**} \\ V_9^{**} \end{bmatrix} = \begin{bmatrix} -0.125 \\ -0.125 \\ -0.195 \\ -0.305 \\ -0.125 \\ -0.181 \\ -0.225 \\ -0.275 \\ -0.319 \end{bmatrix}, \qquad (7.363)$$

and

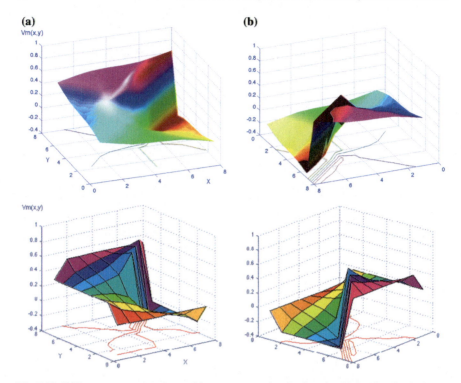

Fig. 7.48 Different perspective views of the reconstructed scalar function $V_m(x, y)$. **a** On the lower graphic we can identify the points where the potential has been calculated by means of the FDM. The *upper side* represents the same function after smoothing. **b** Idem from a viewpoint parallel to the cut line C^*. Observe the curves of constant potential on the plane OXY

$$\begin{bmatrix} V_1^{***} \\ V_2^{***} \\ V_3^{***} \\ V_4^{***} \\ V_5^{***} \\ V_6^{***} \\ V_7^{***} \\ V_8^{***} \\ V_9^{***} \end{bmatrix} = \begin{bmatrix} 0.625 \\ 0.625 \\ 0.555 \\ 0.445 \\ 0.625 \\ 0.569 \\ 0.525 \\ 0.475 \\ 0.431 \end{bmatrix} . \qquad (7.364)$$

These data allow reconstruction of the function $V_m(x, y)$ in the region LxL. Figure 7.48 show the mathematical surface corresponding to the aforementioned values. Observe that this function is similar to the graphic (1.30) studied in Chap. 1 (this similitude is not fortuitous-see Problem 1.17).

The magnetic field may be sketched approximating the partial derivatives that appear in $\mathbf{H} = -\nabla V_m(x, y)$ by means of the following formulae

Fig. 7.49 Magnetic field at points studied. Note that on the *right side* and on the upper row do not appear vectors because the derivative needs two points to be performed. Over the edge of *triangle C* we can calculate the partial derivative with respect the coordinate y, but not x. The contrary occurs for the points over the *upper side*

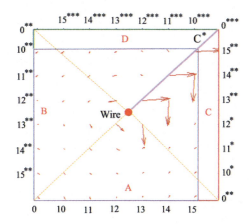

$$\frac{\partial V_m(x,y)}{\partial x} \approx \frac{\Delta V_{\text{x-direction}}}{\Delta x}, \tag{7.365}$$

and

$$\frac{\partial V_m(x,y)}{\partial y} \approx \frac{\Delta V_{\text{y-direction}}}{\Delta y}, \tag{7.366}$$

where $\Delta V_{\text{x-direction}}$ represents the difference between two contiguous potentials in the direction of the x coordinate, and $\Delta V_{\text{y-direction}}$ the same for y. For example, the gradient at point 3 in the domain A (first triangle) is

$$\mathbf{H} = -\nabla V_m(x,y) \approx -\frac{V_4 - V_3}{a}\mathbf{u}_x - \frac{V_1 - V_3}{a}\mathbf{u}_y. \tag{7.367}$$

Computing all the differences, the magnetic field H in the hole has the form represented in Fig. 7.49. Obviously a part of the vector field depicted is wrong. As we commented at the beginning of this problem, we have a symmetry of fourth order, but this cannot be seen in all regions of this figure. The result seems to be correct on the domains A, B and D, and fails for some points of C. Specifically, at points 1**, 4**, 9**, 1***, 2***, and 5***, the magnetic field does not correspond to the actual H. Because of the symmetry, at each of these points the field should theoretically have the same modulus as its complementary point of the other parts by changing the sign of the components adequately. For instance, at point 4 the field is $\mathbf{H}_4 = \frac{0.07}{a}\mathbf{u}_x + \frac{0.07}{a}\mathbf{u}_y$, and at 4** it would be $\mathbf{H}_{4**} = -\frac{0.07}{a}\mathbf{u}_x + \frac{0.07}{a}\mathbf{u}_y$. The reason for this lies in the fact that for calculating the partial derivatives at these points we need points over the cut line C^*(see Fig. 7.45).[26] It leads to a great error because the points on this line are multivalued, and then the result must be wrong. On the other hand, if we analyze the values of the well behaved points in domain A, we observe that they do not exactly maintain the foreign principle of symmetry stated. At point

---

[26]The election of the cut C^* along this edge is completely arbitrary.

Fig. 7.50 Sketch of the lines of magnetic field **H**

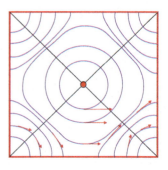

3 the field is $\mathbf{H}_3 = \frac{0.11}{a}\mathbf{u}_x - \frac{0.07}{a}\mathbf{u}_y$, and at point 4 it reaches $\mathbf{H}_4 = \frac{0.07}{a}\mathbf{u}_x + \frac{0.07}{a}\mathbf{u}_y$ instead of $\mathbf{H}'_4 = \frac{0.11}{a}\mathbf{u}_x + \frac{0.07}{a}\mathbf{u}_y$. The reason may be found in the precision of the grid elements we have chosen. Our election was very rough with the only objective to present the method didactically without addressing other concerns. If much smaller elements would be selected, the difference between the field components at such points would be negligent. The easiest way to generate the magnetic field H the rest of the points in regions B, C and D, is to take as a basis the results obtained in the triangle A, and then determine their field by means of a rotation around one axis coinciding with the wire axis. So, in this sense, for calculating **H** at points belonging to the domain C, we apply the following transformation

$$R_z(\phi) = \begin{bmatrix} \cos\phi & -\sin\phi \\ \sin\phi & \cos\phi \end{bmatrix}_{\phi=\frac{\pi}{2}} = \begin{bmatrix} 0 & -1 \\ 1 & 0 \end{bmatrix}. \tag{7.368}$$

For instance, let us suppose we want to compute the field at point 3**. Applying this matrix we have

$$\mathbf{H}^3_{**} = R_z(\frac{\pi}{2})\mathbf{H}_3 = \begin{bmatrix} 0 & -1 \\ 1 & 0 \end{bmatrix} \begin{bmatrix} \frac{0.11}{a} \\ -\frac{0.07}{a} \end{bmatrix} = \begin{bmatrix} \frac{0.07}{a} \\ \frac{0.11}{a} \end{bmatrix}. \tag{7.369}$$

For the domain B, the following transformations holds

$$\begin{bmatrix} -1 & 0 \\ 0 & -1 \end{bmatrix}, \tag{7.370}$$

and for D

$$\begin{bmatrix} 0 & 1 \\ -1 & 0 \end{bmatrix}. \tag{7.371}$$

A hand sketch of the magnetic lines has been represented in Fig. 7.50.

Solved Problems

7.21 Let a closed curve be the cross section of a conducting cylindrical shell, whose cross section has one axis of symmetry (Fig. 7.51). Let this shell be divided into four parts by two planes at right angles, the line of intersection of the planes being parallel to the generator and one of the planes containing the symmetry axis. Show that the direct capacitance, per unit length of the cylinder, between opposing parts of the shell due to the field inside is a constant of value $C_0 = 2, 8$ pFm^{-1}. This result is known as the Thompson-Lampard theorem.

Solution

To demonstrate this theorem we calculate the cross capacitance between two segments of the Fig. 7.51. Let us consider, for example, the arcs (a, b) and (c, d). The cross capacitance C_{13} is the ratio of the negative of the charge Q_3 to the potential V on the surface S_1. For simplicity we suppose that the surface S_1 is held at unity potential and the rest of the system is held at zero. The calculation of charge for a generic geometry with an axis of symmetry like that shown in Fig. 7.51 may be very difficult. In order to make it easier, we can use the Riemann theorem. According to this the-

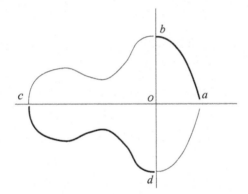

Fig. 7.51 Plane view of a cross section of a system with an axis of symmetry

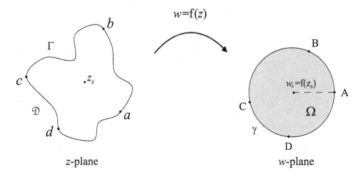

Fig. 7.52 Transformation of the region delimited by the curve Γ in the z-plane onto the *circle* (region Ω) of the w-plane

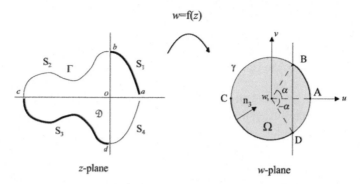

Fig. 7.53 Transformation of the region D belonging to the z-plane into the w-plane

orem, any simply connected region D of the plane can be transformed conformally into a unit circle (Fig. 7.52). Thus in our case, we can map any cylinder cross section onto a circle in which each segment of the original geometry ((a, b),(b, c), etc.) over the curve Γ has its corresponding arc over γ (circle) in the w-plane (Fig. 7.52). Using an adequate phase in the conformal mapping we can get the segment (a, c) over the OX-axis to be transformed into the diameter of the circle (A, C) (Fig. 7.53). The problem reduces to calculating the cross capacitance of the arcs (A, B) and (C, D) in the w-plane. As we have seen in the theory, if the boundary conditions are those of Dirichlet, the values of the potential in the boundary Γ remain unaltered on γ when the region D is transformed into Ω. The potential $\Phi(\rho, \phi)$ inside the circle is well known and may be found by using the following series

$$\Phi(\rho, \phi) = \sum_{n=1}^{\infty} \rho^n [a_n \cos(n\phi) + b_n \sin(n\phi)], \qquad (7.372)$$

where the coefficients of (7.372) for the unit circle ($\rho = 1$) are

$$a_n = \frac{1}{\pi} \int_0^{2\pi} \Phi(1, \phi) \cos(n\phi) d\phi, \qquad (7.373)$$

$$b_n = \frac{1}{\pi} \int_0^{2\pi} \Phi(1, \phi) \sin(n\phi) d\phi, \qquad (7.374)$$

respectively. In our case

$$a_n = \frac{1}{\pi} \int_0^{2\pi} \Phi(1, \phi) \cos(n\phi) d\phi = \frac{1}{\pi} \int_0^{\alpha} \cos(n\phi) d\phi = \frac{1}{n\pi} \sin n\alpha, \qquad (7.375)$$

and

$$b_n = \frac{1}{\pi}\int_0^{2\pi} \Phi(1,\phi)\sin(n\phi)d\phi = \frac{1}{\pi}\int_0^{\alpha} \cos(n\phi)d\phi = \frac{1}{n\pi}[1 - \cos n\alpha].$$
(7.376)

As we have commented, for calculating the cross capacitance C_{13} it is necessary to know the charge on the arc (C, D), which may be obtained by means of the following expression

$$Q_3 = \int_{S_3} \sigma \, dS = \varepsilon_0 \int_{S_3} \frac{\partial \Phi}{\partial n_3} \, dS.$$
(7.377)

Introducing (7.372) into (7.377), and knowing that due to the circular symmetry $\frac{\partial \Phi}{\partial n_3} = \frac{\partial \Phi}{\partial \rho}$, it yields for $V = 1$

$$C_{13} = \frac{Q_3}{V} = \frac{Q_3}{\varepsilon_0} = \sum_{n=1}^{\infty}\int_{\pi}^{2\pi-\alpha} n[a_n \cos(n\phi) + b_n \sin(n\phi)] =$$

$$\sum_{n=1}^{\infty}\left\{ n\frac{1}{n\pi}\sin n\alpha \int_{\pi}^{2\pi-\alpha}\cos(n\phi)d\phi + n\frac{1}{n\pi}[1-\cos n\alpha]\int_{\pi}^{2\pi-\alpha}\sin(n\phi)d\phi \right\} =$$

$$\sum_{n=1}^{\infty}\left\{ \frac{1}{n\pi}\sin n\alpha \sin[(2\pi-\alpha)n] + \frac{1}{n\pi}[1-\cos n\alpha][-\cos(2\pi-\alpha)n + \cos n\pi] \right\} =$$

$$\sum_{n=1}^{\infty}\left\{ -\frac{1}{n\pi}\sin^2 n\alpha - \frac{1}{n\pi}[1-\cos n\alpha][\cos n\alpha - (-1)^n] \right\}. \quad (7.378)$$

However, taking into account that $-(-1)^n = (-1)^{n-1}$ we have

$$\frac{-Q}{\varepsilon_0} = \sum_{n=1}^{\infty}\left\{ -\frac{1}{n\pi}\sin^2 n\alpha - \frac{1}{n\pi}[\cos n\alpha - \cos^2 n\alpha - (-1)^{n-1}\cos n\alpha + (-1)^{n-1}] \right\} =$$

$$-\sum_{n=1}^{\infty}\frac{1}{n\pi}(-1)^{n-1} - \sum_{n=1}^{\infty}\frac{1}{n\pi}\left\{ \sin^2 n\alpha - \cos^2 n\alpha + \cos n\alpha[1-(-1)^{n-1}] \right\} =$$

$$-\sum_{n=1}^{\infty}\frac{1}{n\pi}(-1)^{n-1} - \sum_{n=1}^{\infty}\frac{1}{n\pi}\left\{ -\cos 2n\alpha + \cos n\alpha[1-(-1)^{n-1}] \right\}. \quad (7.379)$$

$$\sum_{n=1}^{\infty} \frac{(-1)^{n-1}}{n} = 1 - \frac{1}{2} + \frac{1}{3} - \frac{1}{4} + \frac{1}{5}\cdots. \quad (7.380)$$

This is an alternating series whose sum may be calculated by means of different procedures. Perhaps the easiest way is to compare it with the Taylor series of a known function whose coefficients coincide with those appearing in (7.380) (if possible). In this case we know that the terms of the logarithmic function are alternately positive and negative, and the denominator of the fractions grow with the number of terms n, then we try the expansion

$$\ln(1+x) = x - \frac{x^2}{2} + \frac{x^3}{3} - \frac{x^4}{4} + \frac{x^5}{5} \cdots, \tag{7.381}$$

which is convergent in the interval $-1 < x \leq 1$. If we put $x = 1$ in (7.381), we have

$$\ln(2) = 1 - \frac{1}{2} + \frac{1}{3} - \frac{1}{4} + \frac{1}{5} \cdots, \tag{7.382}$$

which is the same as (7.380). Thus we can write

$$\sum_{n=1}^{\infty} \frac{(-1)^{n-1}}{n} = \ln 2 \tag{7.383}$$

$$\frac{-Q}{\varepsilon_0} = -\ln 2 + \frac{1}{\pi} \sum_{n=1}^{\infty} \frac{1}{n} \cos 2n\alpha - \frac{1}{\pi} \sum_{n \, even}^{\infty} \frac{2}{n} \cos n\alpha =$$
$$-\ln 2 + \frac{1}{\pi} \left\{ \cos 2\alpha + \frac{1}{2} \cos 4\alpha + \frac{1}{3} \cos 6\alpha + \cdots - \cos 2\alpha - \frac{1}{2} \cos 4\alpha - \frac{1}{4} \cos 6\alpha - \cdots \right\}. \tag{7.384}$$

If we observe this expression for the charge on the arc (γ, δ) we see that the terms in the brackets cancel each other out and result in a constant, hence we can write

$$\frac{Q}{\varepsilon_0} = \ln 2. \tag{7.385}$$

Chapter 8
Electromagnetic Induction

Abstract Faraday (1831) discovered experimentally that an electric current flows in a circuit when the circuit moves in presence of a magnet, or the circuit is fixed and the position of the magnet changes. Furthermore, the electric current is also induced in a circuit when it is close to another circuit carrying a varying current. Non-steady currents give rise to non-steady magnetic fields, which also give rise to electric fields. In all these cases, generation of the induced currents is due to either the circuit moving in a magnetic field or the magnetic field changes with time; i.e. whenever the magnetic flux through the circuit varies. This chapter is concerned with electromagnetic induction. First, we consider Faraday's law of electromagnetic induction for a stationary closed loop. Secondly, we will discuss the electromotive force induced in conductors moving in a static magnetic field. Then, the general case of a circuit moving in a time-varying magnetic field is considered. Finally, we will study the phenomena of self-inductance and mutual inductance of closed loops along with some applications of Faraday's law.

8.1 Electromotive Force

Figure 8.1 shows a closed curve Γ, at point P the resulting electric force on a charge q is **F**. The effective electric field $\mathbf{E}_e$, equal to the total electric force per unit of charge q, can be written as

$$\mathbf{E}_e = \frac{\mathbf{F}}{q}. \tag{8.1}$$

The electromotive force (e.m.f.) $\mathcal{E}$ along the curve Γ is defined as the line integral around the closed curve of the effective electric field $\mathbf{E}_e$. Hence, the e.m.f. calculated, following the direction given by $d\mathbf{l}$, can be expressed as

$$\mathcal{E} = \oint_\Gamma \mathbf{E}_e \cdot d\mathbf{l}. \tag{8.2}$$

It should be noted that the curve along which the e.m.f. is calculated must be closed. In many applications, either part or the whole curve will be a conducting wire. In the

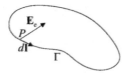

Fig. 8.1 A closed loop Γ around which the electromotive force is calculated. The effective electric field at the elemental longitudinal element $d\mathbf{l}$ is $\mathbf{E}_e$

latter case, a current around the conducting wire will flow whenever the flux linked by the circuit changes. Electromotive force is measured in units of volts (V).

Since $\mathbf{E}_e = \mathbf{F}/q$, the tangential component of the force per unit charge, integrated around the closed curve, will only contribute to the resulting electromotive force. The result of integration will be equal to the work done on a single charge that travels once around the circuit. In general the resulting effective electric field can be expressed as the sum of two fields: one being a conservative electric field, $\mathbf{E}_{cs}$, and the other non-conservative, $\mathbf{E}_{ncs}$. When the line integration is done over the closed curve, the conservative electric field will not contribute to generation of e.m.f., whereas the closed line integral of the non-conservative electric field will produce the net electromotive force around such a curve. Non-conservative electric fields induced by changing magnetic fields or associated with physical circuits moving in the presence of a magnetic field will cause e.m.f. to be induced. According to Lorentz's law, the effective electric field can be expressed as

$$\mathbf{E}_e = \mathbf{E} + \mathbf{v} \times \mathbf{B}, \qquad (8.3)$$

the first term on the right side will give rise to e.m.f., when time-dependent magnetic fields are present, and the second one will produce e.m.f., when there are moving charges in the presence of magnetic fields.

8.2 Faraday's law

Let us consider a stationary closed loop situated in a region where there is a time-varying magnetic field, as shown in Fig. 8.2. Part of or the whole contour of the loop can be a physical circuit. However, the loop can also be a mathematical curve. According to Faraday's law, the point relationship between the varying magnetic field and the induced electric field is given by

$$\nabla \times \mathbf{E} = -\frac{\partial \mathbf{B}}{\partial t}. \qquad (8.4)$$

This equation shows that a time-varying magnetic field produces an electric field, which is non-conservative and, therefore, cannot be expressed as the negative gradient

8.2 Faraday's law

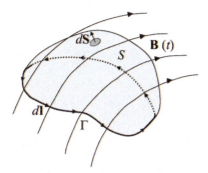

Fig. 8.2 The electromotive force is evaluated around the fixed closed curve Γ, boundary of the open surface S. The orientation of $d\mathbf{S}$ is determined by the direction of $d\mathbf{l}$, set by the right-hand rule

of a scalar potential. The electromotive force along the loop is given by the closed line integral around the contour of the effective electric field, (8.2). As the loop is considered to be stationary, the electromotive force is only induced by the electric field. Taking the surface integral of both sides of (8.4) over an open surface S with contour Γ, and applying Stoke's theorem, it is found that

$$\mathcal{E} = \oint_\Gamma \mathbf{E}_e \cdot d\mathbf{l} = \oint_\Gamma \mathbf{E} \cdot d\mathbf{l} = -\int_S \frac{\partial \mathbf{B}}{\partial t} \cdot d\mathbf{S} = -\frac{d}{dt}\int_S \mathbf{B} \cdot d\mathbf{S} = -\frac{d\Phi}{dt}. \quad (8.5)$$

This result shows that the electromotive force induced in a closed stationary loop is equal to the negative rate of change of the magnetic flux through any surface bounded by the loop. If the e.m.f is evaluated following the direction $d\mathbf{l}$, then the surface element $d\mathbf{S}$ used in calculating the flux through the open surface, whose perimeter coincides with Γ, is in the direction given by the right-hand rule; i.e. if the fingers of the right hand follow the direction of $d\mathbf{l}$, the thumb points in the direction of $d\mathbf{S}$, as shown in Fig. 8.2. The electric field can be expressed in terms of the vector potential $\mathbf{A}$ and the scalar potential V as

$$\mathbf{E} = -\nabla V - \frac{\partial \mathbf{A}}{\partial t}. \quad (8.6)$$

The line integral of the first term on the right side of (8.6) around a closed loop is zero and, therefore, the vector potential $\mathbf{A}$, related to the magnetic field by $\mathbf{B} = \nabla \times \mathbf{A}$, is the term that will create an induced electric field. Thus, the electromotive force induced in a closed stationary loop can be expressed in terms of the vector potential as

$$\mathcal{E} = \oint_\Gamma \mathbf{E}_e \cdot d\mathbf{l} = \oint_\Gamma \mathbf{E} \cdot d\mathbf{l} = \oint_\Gamma -\frac{\partial \mathbf{A}}{\partial t} \cdot d\mathbf{l} = -\frac{d}{dt}\oint_\Gamma \mathbf{A} \cdot d\mathbf{l}. \quad (8.7)$$

According to Faraday's law, an electric field is induced whenever the magnetic flux through a loop varies. This field will induce, in turn, a current if the closed loop is a conducting circuit. The direction of the current is given by Lenz's law. This law states that the current induced is in a direction such that it produces a magnetic flux

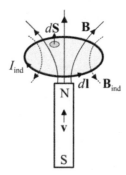

Fig. 8.3 A magnet approaching with velocity **v** a stationary closed conducting wire. As a result of the time-dependent magnetic field, a current I is induced, whose field is indicated in *dashed line*

tending to maintain the total flux through the circuit. The minus sign means that the current induced produces a magnetic flux through the circuit that tends to oppose the original change of flux.

As one example of the application of Lenz's law, let us consider a permanent magnet approaching a stationary closed conducting loop, as shown in Fig. 8.3. The e.m.f. is calculated around the contour following the direction $d\mathbf{l}$ (counterclockwise), and the direction of the corresponding surface element $d\mathbf{S}$ is given by the right-hand rule, thus the resulting flux through the open surface is positive. As the magnet moves towards the loop, the magnetic flux increases, and, consequently, $d\Phi/dt > 0$. The line integral of the induced electric field, the e.m.f., is negative and, therefore, the induced current will flow in the opposite direction to that of $d\mathbf{l}$. The induced current is thus in a clockwise direction and produces a magnetic field that is shown by the dashed lines. This field tends to stop the flux through the loop from increasing, reducing the change of the magnetic flux, caused initially by the magnet motion.

Figure 8.4 shows another example of an application of Lenz's law. In (a) the switch in circuit 1 has just been closed and a current I_1 flows in the direction indicated that produces the magnetic field $\mathbf{B}_1$. In evaluating the e.m.f. around the upper coils,

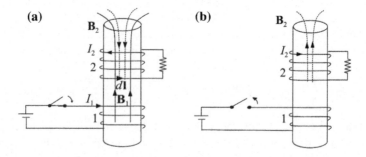

Fig. 8.4 a In the coil denoted 1, current I_1 flows when the switch is closed, resulting in an induced clockwise current I_2 in coil 2, according to Lenz's law. **b** When the switch is open, the induced current in coil 2 is counterclockwise and tries to maintain a constant flux through the coils

8.2 Faraday's law

circuit 2, the line integral in the direction $d\mathbf{l}$ produces a negative result since the magnetic flux through the coils increases. The induced current I_2 will tend to lessen the change of the magnetic flux, the corresponding field being $\mathbf{B}_2$, in the opposite direction to $\mathbf{B}_1$. When the equilibrium is reached in 1, no more induced current flows in coil 2. In Fig. 8.4b the switch is open and the magnetic flux decreases as a result of the decay of the current I_1. Then, the direction of the current is such that the corresponding magnetic field $\mathbf{B}_2$ tends to stop the flux through the coils from falling.

8.3 Motional Electromotive Force

Let us consider a physical circuit that either moves or is deformed (or both), in the presence of a stationary magnetic field. According to Lorentz's law, the effective electric field at a line element $d\mathbf{l}$ will be $\mathbf{E}_e = \mathbf{v} \times \mathbf{B}$, where $\mathbf{v}$ is the velocity of the element and $\mathbf{B}$ the magnetic field (where the element is located). Firstly, we consider an open circuit with terminals ab, as shown in Fig. 8.5. The e.m.f. around the closed path aba, including the physical circuit ab and a mathematical dashed line ba, results in,

$$\mathcal{E} = \oint_{\Gamma(aba)} \mathbf{E}_e \cdot d\mathbf{l} = \int_{ab(\text{circuit})} (\mathbf{v} \times \mathbf{B}) \cdot d\mathbf{l} + \int_{ba(\text{exterior})} (\mathbf{v} \times \mathbf{B}) \cdot d\mathbf{l} = \int_{a(\text{circuit})}^{b} (\mathbf{v} \times \mathbf{B}) \cdot d\mathbf{l}, \tag{8.8}$$

which is generated by the magnetic force acting on the moving circuit.

It should be noted that no matter how the loop is closed, only the physical circuit ab contributes to the generation of e.m.f. However, in defining the electromotive force the line integral must be closed. The conservative field will give a resulting integral equal to zero and only the non-conservative field $(\mathbf{v} \times \mathbf{B})$ will cause e.m.f. to be generated.

As a simple example of the generation of e.m.f. in an open circuit, consider a conducting rod with length ℓ, moving with velocity $\mathbf{v}$, perpendicular to a uniform magnetic field $\mathbf{B}$, as shown in Fig. 8.6. The e.m.f. calculated following the path $ab(\text{rod})$-$ba(\text{exterior})$ is given by

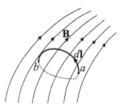

Fig. 8.5 Closed path formed by the open circuit with terminals ab (*solid line*) and the mathematical line ba (*dashed line*)

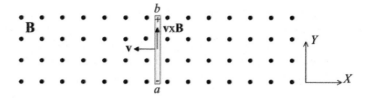

Fig. 8.6 Conducting rod with terminal ab moving in a stationary magnetic field directed out of the page

$$\mathcal{E} = \int_{aba} (\mathbf{v} \times \mathbf{B}) \cdot d\mathbf{l} = \int_{a(\text{rod})}^{b} (\mathbf{v} \times \mathbf{B}) \cdot d\mathbf{l} = \int_{a}^{b} [(-v, 0, 0) \times (0, 0, B)] \cdot (0, dy, 0) \quad (8.9)$$

$$= \int_{a}^{b} (0, vB, 0) \cdot (0, dy, 0) = \int_{0}^{\ell} vB \, dy = vB\ell. \quad (8.10)$$

There is a magnetic force $\mathbf{F} = q(\mathbf{v} \times \mathbf{B})$ on each charge in the rod. The free electrons in the rod will move towards the end a and terminal b will be then positively charged. Then, this new charge distribution will give rise to an electric field opposite to $\mathbf{v} \times \mathbf{B}$. Migration of electrons halts when electric and magnetic forces balance each other and a state of equilibrium is reached. As the conducting circuit is not closed, no resulting current flows through the rod.

For a closed circuit moving in a static magnetic field, as shown in Fig. 8.7, the electromotive force around the closed contour Γ will be

$$\mathcal{E} = \oint_{\Gamma} \mathbf{E}_e \cdot d\mathbf{l} = \oint_{\Gamma} (\mathbf{v} \times \mathbf{B}) \cdot d\mathbf{l}. \quad (8.11)$$

If the closed circuit is a conducting one, the e.m.f. results in motion of electrons and, consequently, an electric current will flow in the circuit. By manipulation of (8.11), it can also be expressed in terms of the change of magnetic flux. Then, the electromotive force around a closed circuit moving in a stationary magnetic field can be expressed as,

$$\mathcal{E} = \oint_{\Gamma} \mathbf{E}_e \cdot d\mathbf{l} = \oint_{\Gamma} (\mathbf{v} \times \mathbf{B}) \cdot d\mathbf{l} = -\frac{d\Phi}{dt} = -\frac{d}{dt} \int_{S} \mathbf{B} \cdot d\mathbf{S}, \quad (8.12)$$

where the variation of magnetic flux is only due to the motion of the circuit. When the circuit is closed, the motional electromotive force given by (8.11) and that given by the rate of change of the magnetic flux (8.12) are equivalent expressions. However, for open circuits, (8.8) can always be applied but (8.12) can only be used when the path is closed in such a way that the change of magnetic flux is equal to the flux swept out by the open circuit in its motion.

8.3 Motional Electromotive Force

Fig. 8.7 Closed circuit moving in a stationary magnetic field

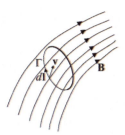

A simple example of e.m.f. induced in a closed circuit follows. Figure 8.8 shows a conducting rod with length ℓ sliding over parallel rails in a stationary magnetic field **B**.

Let us calculate the e.m.f. around the closed circuit formed by the bar-rails-resistance, following the path *abcda* (clockwise direction). By applying (8.11), it follows that

$$\mathcal{E} = \oint_\Gamma \mathbf{E}_e \cdot d\mathbf{l} = \oint_{abcda} (\mathbf{v} \times \mathbf{B}) \cdot d\mathbf{l} = \int_a^b (\mathbf{v} \times \mathbf{B}) \cdot d\mathbf{l} = vB\ell. \qquad (8.13)$$

The net contribution to the line integral comes from the portion *ab* of the loop.

If the flux rule is applied to calculate the e.m.f. at an instant t, as shown in Fig. 8.8, the magnetic flux through the circuit will be $\Phi = \int_S \mathbf{B} \cdot d\mathbf{S}$, with $d\mathbf{S}$ opposite to **B**. As **B** is uniform, $\Phi = \mathbf{B} \cdot \mathbf{S} = -BS = -B\ell x = -B\ell vt$, where $S = \ell x$ and $x = vt$. Therefore,

$$\mathcal{E} = -\frac{d\Phi}{dt} = B\ell v, \qquad (8.14)$$

which agrees with the result found in (8.13).

As both rod and rails are conducting media, an electric current will be induced. The term $(\mathbf{v} \times \mathbf{B})$ causes electrons to move around the contour defined by the circuit, which, in turn, results in a stationary electric current $\mathbf{j} = nQ\mathbf{v}$, in the direction shown in Fig. 8.8. The current is such that the induced magnetic field points into the page and tends to keep the total flux through the circuit constant, in agreement with Lenz's law. If the self-inductance of the loop and the resistance of the bar and the rails are negligible, the induced current is given by

$$I = \frac{\mathcal{E}}{R} = \frac{B\ell v}{R}, \qquad (8.15)$$

in a clockwise direction.

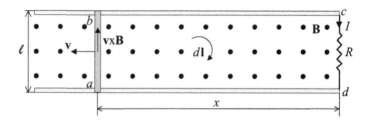

Fig. 8.8 A conducting rod moving with velocity **v** on two conducting rails in an uniform and stationary magnetic field directed out of the page. The rails are terminated in a resistance R

8.4 The General Law of Electromagnetic Induction

Let us now consider a closed moving circuit in a time-varying magnetic field. The total electromotive force induced in the closed contour is due to both the time variation of the magnetic field and to the motion of the closed circuit in the magnetic field. The resulting effective electric field is the sum of the electric induced field, as a result of the change of the magnetic field, plus the term $(\mathbf{v} \times \mathbf{B})$ given by Lorenz's law. Therefore, the electromotive force around the contour is given by

$$\mathcal{E} = \oint_\Gamma \mathbf{E}_e \cdot d\mathbf{l} = \oint_\Gamma \mathbf{E} \cdot d\mathbf{l} + \oint_\Gamma (\mathbf{v} \times \mathbf{B}) \cdot d\mathbf{l}. \tag{8.16}$$

The first term on the right side can be expressed in terms of the variation of **B** with time,

$$\mathcal{E} = \oint_\Gamma \mathbf{E}_e \cdot d\mathbf{l} = \int_S -\frac{\partial \mathbf{B}}{\partial t} \cdot d\mathbf{S} + \oint_\Gamma (\mathbf{v} \times \mathbf{B}) \cdot d\mathbf{l}. \tag{8.17}$$

As shown above, the terms on the right-hand side of (8.17) can be written as the variation of magnetic flux,

$$\mathcal{E} = \left(-\frac{\partial \Phi}{\partial t}\right)_{\mathbf{B}(t)} + \left(-\frac{\partial \Phi}{\partial t}\right)_{\text{motion}}, \tag{8.18}$$

where the change of flux through the contour is due to the time variation of **B** (first term) and to the motion of the circuit in **B** (second term).

In general, (8.18) is equivalent to

$$\mathcal{E} = -\frac{d\Phi}{dt}, \tag{8.19}$$

which is of the same form as (8.5), but the change of the magnetic flux linked by the circuit can be due to one or both of the aforementioned causes. Equation (8.19) is another form of Faraday's law and either (8.17) or (8.19) can be used to evaluate the induced electromotive force in the general case.

8.4 The General Law of Electromagnetic Induction

As an example of application of the general form of Faraday's law, let us consider the example included in Sect. 8.3 (see Fig. 8.8), but with the magnetic field varying with the time in the form $B(t) = B_0 \sin(\omega t)$. Then, the resulting flux at a given instant t, $d\mathbf{S}$ used in calculating the flux according to the right-hand screw rule, is given by

$$\Phi = \int_S \mathbf{B} \cdot d\mathbf{S} = \mathbf{B} \cdot \mathbf{S} = -BS = -B_0 \sin(\omega t)\ell x = -B_0 \sin(\omega t)\ell vt, \qquad (8.20)$$

where the surface determined by the circuit at the instant t is $S = \ell x = \ell vt$ and since $\mathbf{B}$ is a uniform field, it can be taken outside the surface integral. The resulting e.m.f. is given by

$$\mathcal{E} = -\frac{d\Phi}{dt} = B_0 \omega \cos(\omega t)\ell vt + B_0 \sin(\omega t)\ell v, \qquad (8.21)$$

the first term on the right side is due to the variation of $\mathbf{B}$ with the time, while the second one is due to the motion of the circuit, in this case part of the circuit, in the presence of a magnetic field.

8.5 Self-inductance and Mutual Inductance

8.5.1 Self-inductance

Figure 8.9 shows a closed quasi-filamentary stationary loop with contour Γ and bounding surface S. If a current flows in Γ, a magnetic field will be created. The magnetic flux through the loop will be produced by the current in the circuit itself. A time-varying current in the circuit will induce at all points an electric field. The self-induced electric field along the loop will give rise to an electromotive force in the loop. Therefore, any change of the current in the loop itself induces an e.m.f. in the loop that can be written in the form

$$\mathcal{E} = -\frac{d\Phi}{dt} = -\frac{d\Phi}{dI}\frac{dI}{dt} = -L\frac{dI}{dt}, \qquad (8.22)$$

$$L = \frac{d\Phi}{dI}. \qquad (8.23)$$

Fig. 8.9 A closed fixed filamentary circuit with a current intensity I. A change in its current produces an induced e.m.f. in the circuit itself

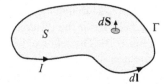

The coefficient L is known as the self-inductance of the circuit. The unit of self-inductance is the henry (H), which is equal to Wb/A.

If the flux Φ is proportional to the current I, the self-inductance coefficient can be written as

$$L = \frac{\Phi}{I}. \tag{8.24}$$

This equation can be used to calculate the self-inductance of a circuit in air or when it is situated in a linear magnetic medium. Then, the self-inductance of the circuit depends on its geometry and on the magnetic properties of the surrounding medium. In a linear medium, self-inductance does not depend on the current in the circuit. However, (8.23) must be used when the relation between flux and current is non-linear.

The e.m.f. induced in a circuit by any change of the current in the circuit itself calculated in accordance with (8.22) is in agreement with Lenz's law. Hence, it must also be included when other electromotive forces are induced due to other causes.

As an example of the calculation of L, let us consider a long straight solenoid of length ℓ, with radius a and with n turns per unit of length, which carries a current I. The end effect is considered to be negligible. The magnetic field $\mathbf{B}$ at all points within the solenoid is $B = \mu_0 n I$, and in the direction of solenoide axis. The magnetic flux through each turn is

$$\Phi_{one} = \int_S \mathbf{B} \cdot d\mathbf{S} = \mathbf{B} \cdot \mathbf{S} = \mu_0 n I \pi a^2, \tag{8.25}$$

where $S = \pi a^2$. The total magnetic flux through the $n\ell$ turns of the complete solenoid results as

$$\Phi = n\ell \Phi_{one} = \mu_0 n^2 I \pi a^2 \ell. \tag{8.26}$$

Then, the self-inductance of the solenoid is given by

$$L = \frac{d\Phi}{dI} = \frac{\Phi}{I} = \mu_0 n^2 \pi a^2 \ell, \tag{8.27}$$

which depends on the number of turns and on the size of the coil.

8.5.2 Mutual Inductance

Figure 8.10 shows two closed stationary loops, labelled 1 and 2, carrying currents I_1 and I_2, respectively. The current I_1 produces the field $\mathbf{B}_1$. The magnetic flux due to $\mathbf{B}_1$ will pass through the open surfaces S_1 and S_2 bounded by Γ_1 and Γ_2, respectively. The field $\mathbf{B}_1$ at the differential surface element $d\mathbf{S}_2$ is denoted by $\mathbf{B}_{21}$, as shown in Fig. 8.10. The relative directions of $d\mathbf{l}$ and $d\mathbf{S}$ follow the right-hand rule. As the current in loop 1 is varied, the magnetic flux will also change, and there will be an induced e.m.f. in loop 2. Analogously, if current I_2 varies, an e.m.f. will be induced in Circuit 1. Circuits 1 and 2 are magnetically coupled. The electromotive force induced

8.5 Self-inductance and Mutual Inductance

Fig. 8.10 Two magnetically coupled circuits labelled 1 and 2. $\mathbf{B}_1$ is the magnetic field produced by I_1, which is denoted $\mathbf{B}_{11}$ and $\mathbf{B}_{21}$ at the differential surface elements $d\mathbf{S}_1$ and $d\mathbf{S}_2$, respectively. A change in current in 1 induces an e.m.f. in 2, and vice versa

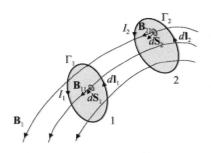

in a circuit as a result of the variation of the current in the magnetically coupled loop can be expressed in terms of the mutual inductance.

For Circuit 2, shown in Fig. 8.10, the total magnetic flux Φ_2 is equal to Φ_{21} due to $\mathbf{B}_{21}$, produced by I_1, plus Φ_{22} resulting from $\mathbf{B}_{22}$, caused by its own current I_2. The induced electromotive force in loop 2 is then given by

$$\mathcal{E}_2 = -\frac{d\Phi_2}{dt} = -\frac{\partial \Phi_{22}}{\partial I_2}\frac{dI_2}{dt} - \frac{\partial \Phi_{21}}{\partial I_1}\frac{dI_1}{dt} \tag{8.28}$$

$$= -L_2\frac{dI_2}{dt} - M_{21}\frac{dI_1}{dt}, \tag{8.29}$$

where the self-inductance of Circuit 2 is designated L_2 and the mutual inductance is denoted $M_{21} = \partial \Phi_{21}/\partial I_1$. In the same way, the e.m.f. induced in loop 1 is

$$\mathcal{E}_1 = -\frac{d\Phi_1}{dt} = -\frac{\partial \Phi_{11}}{\partial I_1}\frac{dI_1}{dt} - \frac{\partial \Phi_{12}}{\partial I_2}\frac{dI_2}{dt} \tag{8.30}$$

$$= -L_1\frac{dI_1}{dt} - M_{12}\frac{dI_2}{dt}, \tag{8.31}$$

where the mutual inductance is given by $M_{12} = \partial \Phi_{12}/\partial I_2$.

In general, mutual inductance between the circuit i and the circuit j, is defined as

$$M_{ij} = \frac{d\Phi_{ij}}{dI_j} \qquad i \neq j, \tag{8.32}$$

which can be used to calculate mutual inductance in linear and non-linear systems. The unit of magnetic inductance is the henry (H).

Let us consider two quasi-filamentary stationary loops situated in a linear magnetic medium. It is assumed that currents vary so slowly that the magnetic field created by them is equal to that produced by steady currents, having magnitudes equal to those of the varying currents at a given instant of time (quasi-stationary conditions). Then, it can be demonstrated that the general expressions for the coefficient M_{21} is given by

$$M_{21} = \frac{\mu_0}{4\pi}\oint_1\oint_2 \frac{d\mathbf{l}_1 \cdot d\mathbf{l}_2}{r_{21}}, \tag{8.33}$$

where r_{21} is the distance from the element $d\mathbf{l}_1$ in loop 1 to the element $d\mathbf{l}_2$ in 2, assuming that the loops are in the air. This equation is known as the Newmann formula for the mutual inductance between two quasi-filamentary loops. From (8.33) it follows that the quantities M_{21} and M_{12} have the same value and can be represented by the symbol M. The mutual inductance of the two loops depends on the circuit geometry and the magnetic properties of the surrounding medium.

For two rigid stationary circuits in a linear medium, with quasi-stationary currents, the magnetic flux Φ_{21} is proportional to I_1, and therefore $M_{21} = d\Phi_{21}/dI_1 = \Phi_{21}/I_1$. Taking into account that M_{21} between loops 2 and 1 is equal to M_{12} between loops 1 and 2, it follows that

$$M = M_{21} = M_{12} = \frac{\Phi_{21}}{I_1} = \frac{\Phi_{12}}{I_2}. \tag{8.34}$$

In general, the procedure for determining the mutual inductance between two circuits, labelled 1 and 2, is as follows:

- Assume current I_1 in loop Γ_1, and then determine the magnetic field due to this current.
- Determine the magnetic flux Φ_{21} created by I_1 through Γ_2.
- Determine the mutual inductance as

$$M = \frac{d\Phi_{21}}{dI_1} = \frac{\Phi_{21}}{I_1}.$$

It should be noted that the same result is obtained if a current is assumed to flow in Γ_2 and the effects are calculated on the circuit denoted 1.

As an example, let us calculate the mutual inductance of two long straight solenoids, as shown in Fig. 8.11. The length of both solenoids is ℓ and the radius is a. If the end effect is negligible, the magnetic field due to I_1 is $\mathbf{B}_1 = \mu_0 n_1 I_1 \mathbf{u}_z$, where n_1 is the number of turns per unit length of the solenoid 1. The flux through a coil of the second solenoid is

$$\Phi_{21_{\text{one}}} = \int_{S_2} \mathbf{B}_{21} \cdot d\mathbf{S}_2 = \mathbf{B}_{21} \cdot \mathbf{S}_2 = \mu_0 n_1 I_1 \pi a^2, \tag{8.35}$$

and the total magnetic flux through the second circuit with $n_2 \ell$ turns is

$$\Phi_{21} = n_2 \ell \Phi_{21_{\text{one}}} = \mu_0 n_1 I_1 \pi a^2 n_2 \ell. \tag{8.36}$$

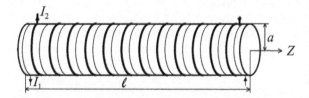

Fig. 8.11 Two mutually coupled straight solenoids: labelled 1 (*thin solid line*) and 2 (*thick solid line*)

8.5 Self-inductance and Mutual Inductance

The mutual inductance M is given by

$$M = \frac{\Phi_{21}}{I_1} = \mu_0 n_1 \pi a^2 n_2 \ell, \tag{8.37}$$

which depends on the geometry of the solenoids.

8.6 Voltage Between Two Points

In Fig. 8.12 a voltmeter at rest is connected to the two terminals of a circuit. Following the path between points a and b through the voltmeter, the voltage measured V_{ab} between these points can be expressed as the line integral of the electric field from point a to b,

$$V_{ab} = \int_a^b \mathbf{E} \cdot d\mathbf{l} = \int_a^b \left(-\nabla V - \frac{\partial \mathbf{A}}{\partial t} \right) \cdot d\mathbf{l}$$
$$= V_a - V_b - \frac{\partial}{\partial t} \int_a^b \mathbf{A} \cdot d\mathbf{l}.$$

This result shows that in the presence of a time-varying magnetic field, the voltage between two points differs from the potential difference. Therefore, the reading of the voltmeter depends on the path followed between a and b. On the other hand, in the static case, voltage and potential difference become identical, thus the reading of the voltmeter will only depend on the potential at points a and b.

In Fig. 8.12b the voltmeter is connected to terminals a and b following two different paths. The readings of the voltmeter V_{ab_1} and V_{ab_2} in the connections labelled 1 and 2, respectively, are different when time-varying magnetic fields are present. It can be easily shown that

$$V_{ab_1} = V_{ab_2} - \frac{d\Phi}{dt},$$

where Φ is the magnetic flux through the surface bounded by the contour $a1b2a$.

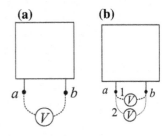

Fig. 8.12 a Voltmeter V connected between the two terminals a and b. b The voltmeter is connected following two different paths labelled 1 and 2

Solved Problems

Problems A

8.1 A square loop with a side of length a and resistance R moves, from the position shown in Fig. 8.13, at a constant velocity $\mathbf{v} = v\,\mathbf{u}_x$ in the presence of a uniform magnetic field $\mathbf{B} = B\,\mathbf{u}_z$, confined to the region shown in the figure. Find the induced current intensity in the square loop. Assume that self-inductance is negligible.

Solution

Electromotive force is induced by motion of the square loop in the magnetic field. The e.m.f. can be calculated by applying the "flux rule," (8.12), or the expression for motional e.m.f., (8.11). As an illustrative example, both procedures are used in calculating the e.m.f. in the square loop following the path 1–2–3–4–1 (see Fig. 8.14).

The following cases are studied:

- (a) For $0 < x < a$, at an instant t on the interval $0 < t < a/v$ (Fig. 8.14a).
  Magnetic field lines do not pass through the square loop. Then, the e.m.f. is zero as is the current intensity.
- (b) For $a < x < 2a$, $a/v < t < 2a/v$ (Fig. 8.14b).
  The motional e.m.f. calculated in the sides labelled 2–3 and 4–1 is zero since $\mathbf{v}$ is parallel to $d\mathbf{l}$. In the side labelled 1–2, the e.m.f. is also zero because $\mathbf{B} = 0$. Then, the e.m.f. in the loop is given by

$$\mathcal{E} = \oint_{1-2-3-4-1} (\mathbf{v} \times \mathbf{B}) \cdot d\mathbf{l} = \oint_{3-4} (\mathbf{v} \times \mathbf{B}) \cdot d\mathbf{l} = \int_{3}^{4} [(v,0,0) \times (0,0,B)] \cdot d\mathbf{l}$$

$$= \int_{3}^{4} (0, -vB, 0) \cdot (0, dy, 0) = \int_{y_1+a}^{y_1} -vB\,dy = vBa.$$

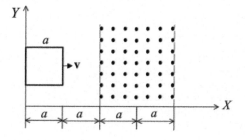

Fig. 8.13 The square loop begins to move with a constant velocity parallel to OX-axis

Fig. 8.14 The square loop at an instant on the interval:
a $0 < t < a/v$,
b $a/v < t < 2a/v$,
c $2a/v < t < 3a/v$,
d $3a/v < t < 4a/v$. The e.m.f. is calculated following a clockwise path

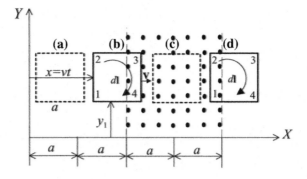

As a clockwise path is followed to calculate the e.m.f., the surface element used in evaluating the flux through the loop will be $d\mathbf{S} = dS\,(-\mathbf{u}_z) = -adx\,\mathbf{u}_z$. Calculating the flux through the loop first and then applying Faraday's law, it follows that

$$\Phi = \int_S \mathbf{B} \cdot d\mathbf{S} = \int_S (0,0,B) \cdot (0,0,-dS) = \int_{2a}^{vt+a} -Ba\,dx$$

$$= -Ba(vt - a) \Rightarrow \mathcal{E} = -\frac{d\Phi}{dt} = Bav.$$

Therefore, the current intensity induced in the square loop, calculated in a clockwise direction, is

$$I = \frac{\mathcal{E}}{R} = \frac{vBa}{R}.$$

As a positive sign is obtained for I, the induced current will have the same direction as that assumed in the calculation, i.e. that of $d\mathbf{l}$ (clockwise direction).

- (c) For $2a < x < 3a$, $2a/v < t < 3a/v$ (Fig. 8.14c).
  The resulting e.m.f. is zero since the value obtained for the line integral in side 3–4 is equal in magnitude to that obtained in 1–2, however, opposite in sign. The magnetic flux is constant, and hence no e.m.f. is induced. Then, no current intensity flows in the circuit.
- (d) For $3a < x < 4a$, $3a/v < t < 4a/v$ (Fig. 8.14d).
  In the side labelled 3–4, the e.m.f. is zero because $\mathbf{B} = 0$. The e.m.f. in the loop is given by

$$\mathcal{E} = \oint_{1-2-3-4-1} (\mathbf{v} \times \mathbf{B}) \cdot d\mathbf{l} = \oint_{1-2} (\mathbf{v} \times \mathbf{B}) \cdot d\mathbf{l} = \int_1^2 [(v,0,0) \times (0,0,B)] \cdot d\mathbf{l}$$

$$= \int_1^2 (0,-vB,0) \cdot (0,dy,0) = \int_{y_1}^{y_1+a} -vB\,dy = -vBa.$$

Fig. 8.15 The intensity induced in the square loop in terms of its position in the magnetic field

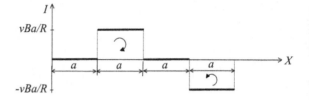

Calculating the flux through the loop first and then applying Faraday's law, it is found

$$\Phi = \int_S \mathbf{B} \cdot d\mathbf{S} = \int_S (0, 0, B) \cdot (0, 0, -dS) = \int_{vt}^{4a} -Ba\,dx$$
$$= -Ba(4a - vt) \Rightarrow \mathcal{E} = -\frac{d\Phi}{dt} = -Bva.$$

Therefore, following a clockwise path, the current intensity induced in the square loop is

$$I = \frac{\mathcal{E}}{R} = \frac{-vBa}{R}.$$

The minus sign means that the induced current flows in a counterclockwise direction; i.e. in opposite direction to that assumed in the calculation of the current intensity (Fig. 8.15).

8.2 Figure 8.16 shows a circular ring of radius r, a straight vertical wire and a bar of length r, electrically connected to the straight wire and to the ring. All these components are assumed to be perfect conductors. The bar rotates with a constant angular velocity $\boldsymbol{\omega} = \omega\,\mathbf{u}_z$ while the rest of components remain at rest. There is a uniform magnetic field of magnitude B, making an angle of α with OZ. A voltmeter is connected between the terminals a and b. What is the voltmeter reading V?

Solution

At an instant t, the e.m.f. is calculated following the closed path represented by the dotted lines shown in Fig. 8.17, i.e. the path $bcOdab$. The e.m.f. is caused by the motion of the bar in the presence of a magnetic field. Then,

$$\mathcal{E} = \oint_{bcOdab} \mathbf{E}_e \cdot d\mathbf{l} = \int_{Od} (\mathbf{v} \times \mathbf{B}) \cdot d\mathbf{l}.$$

The velocity of a point on Od is given by $\mathbf{v} = \omega\,\mathbf{u}_z \times \rho\,\mathbf{u}_\rho = \omega\rho\,\mathbf{u}_\phi$, where ρ is the distance from the point to the rotation axis (OZ-axis). The non-conservative field $\mathbf{v} \times \mathbf{B}$ can be calculated as

Solved Problems

Fig. 8.16 Horizontal ring and a rotating bar in a region where there is a uniform magnetic field

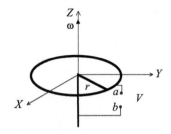

Fig. 8.17 Electromotive force is calculated following the closed path $bcOdab$ (*dotted lines*)

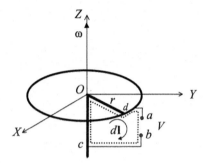

$$\mathbf{v} \times \mathbf{B} = \begin{vmatrix} \mathbf{u}_\rho & \mathbf{u}_\phi & \mathbf{u}_z \\ 0 & \omega\rho & 0 \\ B_\rho & B_\phi & B_z \end{vmatrix} = \omega\rho B_z \mathbf{u}_\rho - \omega\rho B_\rho \mathbf{u}_z.$$

The line integral of $\mathbf{E}_{ncs} = \mathbf{v} \times \mathbf{B}$ along the segment Od is

$$\mathcal{E} = \int_{Od} (\mathbf{v} \times \mathbf{B}) \cdot d\mathbf{l} = \int_{Od} (\omega\rho B_z, 0, -\omega\rho B_\rho) \cdot (d\rho, 0, 0) = \int_{Od} \omega\rho B_z d\rho$$
$$= \int_0^r \omega B \cos\alpha \, \rho d\rho = \frac{\omega B r^2}{2} \cos\alpha,$$

where it has been taken into account that $B_z = B\cos\alpha$.

The voltage measured by the voltmeter can be determined by calculating again the line integral of $\mathbf{E}_e$ along the path $bcOdab$ and taking into account that all components (ring, bar, vertical wire) are perfect conductors ($\mathbf{E}_e = 0$). Exterior to the physical system, the effective electric field can be expressed as the gradient of a scalar potential, $\mathbf{E}_e = -\nabla V$. Under these assumptions, the voltage between terminals a and b is

$$\mathcal{E} = \oint_{bcOdab} \mathbf{E}_e \cdot d\mathbf{l} = \int_{bcOda} \mathbf{E}_e \cdot d\mathbf{l} + \int_{ab} \mathbf{E}_e \cdot d\mathbf{l} = 0 + \int_{ab} -\nabla V \cdot d\mathbf{l} = V_a - V_b.$$

Hence, it is found that the voltage has the same value as that of the e.m.f.:

$$V_{ab} = V_a - V_b = \mathcal{E} = \frac{\omega B r^2}{2} \cos\alpha.$$

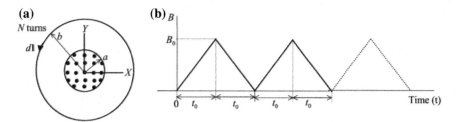

Fig. 8.18 A planar coil and the time-varying magnetic field confined to the region of radius a

8.3 Figure 8.18 shows a sketch of a planar coil with N turns with radius b located in the XY-plane. The magnetic field is confined to the region given by $\rho \leq a$, the field being in the OZ-direction, uniform within such a region but varying with time as shown in Fig. 8.18, where B_0 is a constant. (a) Determine the e.m.f. induced in the coil in terms of time. (b) If the resistance of the coil is R and its self-inductance L, not negligible, find the current flowing through the coil at an instant t on the interval $0 < t < t_0$.

Solution

(a) The electromotive force is produced by the time-varying magnetic field within the cylinder with radius a. Hence, e.m.f. is calculated by using Faraday's law (8.5). Let us study two different situations corresponding to the intervals:

- For $0 < t < t_0$, where the magnetic field is $B = (B_0/t_0)t$, pointing out of the plane of the page, as shown in Fig. 8.18. A counterclockwise path is followed in calculating the e.m.f. around the coil; then the corresponding surface element will be $d\mathbf{S} = dS\,\mathbf{u}_z$. The flux linking each turn, Φ_{one}, and the total flux, Φ, are

$$\Phi_{\text{one}} = \int_S \mathbf{B} \cdot d\mathbf{S} = \int_S (0, 0, B) \cdot (0, 0, dS) = \int_S B\,dS = BS_{\text{ef}}$$
$$= B\pi a^2 \Rightarrow \Phi = N\frac{B_0}{t_0}t\pi a^2.$$

Note that although the flux is calculated through the circle of radius b, only the magnetic lines of force crossing the surface $S_{\text{ef}} = \pi a^2$ contribute to the total flux; i.e. the area over which $\mathbf{B}$ exits. The resulting e.m.f. is

$$\mathcal{E} = -\frac{d\Phi}{dt} = -N\frac{B_0}{t_0}\pi a^2.$$

- For $t_0 < t < 2t_0$:
  By applying the same procedure and taking into account that $dB/dt = -B_0/t_0$, it is found that

$$\mathcal{E} = -\frac{d\Phi}{dt} = N\frac{B_0}{t_0}\pi a^2.$$

Fig. 8.19 The e.m.f. in the coil as a function of time, resulting from the time-varying magnetic field, where $\mathcal{E}_0 = N(B_0/t_0)\pi a^2$

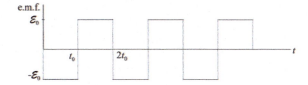

Figure 8.19 shows the result obtained for the e.m.f. in the coil versus time.

(b) By applying Kirchhoff's law: $\sum \mathcal{E} = \sum RI$ and denoting $\mathcal{E}_0 = N(B_0/t_0)\pi a^2$, and following a counterclockwise path around the loop, it is found that

$$\mathcal{E} - L\frac{dI}{dt} = RI \Rightarrow -\mathcal{E}_0 - L\frac{dI}{dt} = RI \Rightarrow \frac{dI}{RI + \mathcal{E}_0} = -\frac{dt}{L}.$$

By integrating this equation and taking into account that $t = 0$, the current intensity is $I = 0$, hence, it results

$$\int_0^I \frac{dI}{RI + \mathcal{E}_0} = \int_0^t -\frac{dt}{L} \Rightarrow \frac{1}{R}\ln(RI + \mathcal{E}_0) - \frac{1}{R}\ln(\mathcal{E}_0) = -\frac{t}{L},$$

$$I = \frac{\mathcal{E}_0}{R}\left(-1 + e^{-\frac{R}{L}t}\right) \Rightarrow I = \frac{NB_0}{t_0}\frac{\pi a^2}{R}\left(-1 + e^{-\frac{R}{L}t}\right).$$

8.4 Figure 8.20 shows a toroidal coil with N turns of conducting wire tightly wound on a toroidal frame with circular cross-section. The toroid has a mean radius R_m, and the radius of each turn is a. Find the self-inductance of the toroidal coil.

Solution

By assuming a current I in the conducting wire, and taking into account that due to the cylindrical symmetry, **B** has only the ϕ-component, which is constant along any circular path around the axis of the toroid, it is found by applying Ampère's law to a circular path with radius ρ ($R_m - a < \rho < R_m + a$):

$$\oint_C \mathbf{B} \cdot d\mathbf{l} = B_\phi 2\pi\rho = NI \Rightarrow B_\phi = \frac{\mu_0 NI}{2\pi\rho}.$$

It has been assumed that the toroid has an air core with permeability μ_0.

Fig. 8.20 A sketch of a closely wound toroidal coil

We assume that the dimensions of the cross-section of the core are very small in comparison to the mean radius, i.e. $a \ll R_m$, and the magnetic field inside the solenoid is approximately constant and equal to

$$B_\phi = \frac{\mu_0 NI}{2\pi R_m} = \frac{\mu_0 NI}{\ell},$$

where the toroidal mean length is $\ell = 2\pi R_m$.

Hence, the flux through one turn will be

$$\Phi_{one} \simeq B_\phi S = \frac{\mu_0 NIS}{\ell},$$

where the cross-sectional area $S = \pi a^2$. The total flux through the N turns,

$$\Phi = N\Phi_{one} \simeq \frac{\mu_0 N^2 IS}{\ell}.$$

Finally, the self-inductance of the toroidal coil will be,

$$L = \frac{d\Phi}{dI} = \frac{\Phi}{I} = \frac{\mu_0 N^2 S}{\ell}.$$

It should be noted that the self-inductance is not a function of I and is proportional to the square of the number of turns.

8.5 Figure 8.21 shows a sketch of two closed circuits labelled 1 and 2. Circuit 1 is a circular loop, with a small radius R_1, centered at the coordinate origin. Circuit 2 consists of "two almost circular loops" whose radii are R_2 and R_3, respectively, centered at the coordinate origin, and two straight wires, close to each other but there is no electrical contact between them. Find (a) self-inductance L_1 and (b) mutual inductance M.

Solution

It is assumed that $R_1 \ll R_2 < R_3$.
(a) The magnetic field produced by a counterclockwise current I_1 in Circuit 1 is that of a circular loop. Then, at the coordinate origin, the field will be

Fig. 8.21 Two closed loops labelled 1 and 2

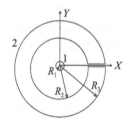

Fig. 8.22 A current intensity I_2 flows through Circuit 2 and the effects are calculated in 1

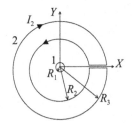

$$\mathbf{B}_{11} = \left.\frac{\mu_0 I_1 R_1^2}{2(R_1^2 + z^2)^{3/2}}\right|_{z=0} \mathbf{u}_z = \frac{\mu_0 I_1}{2R_1}\mathbf{u}_z.$$

It is assumed that as radius R_1 is small, the magnetic field produced by I_1 is approximately homogeneous over the area of the circular loop 1. Then, the magnetic flux through the closed loop with a radius of R_1 and the self-inductance are given, respectively, by:

$$\Phi_{11} = \int_{S_1} \mathbf{B}_{11} \cdot d\mathbf{S}_1 \simeq B_{11} S_1 = \frac{\mu_0 I_1}{2R_1}\pi R_1^2 = \frac{\mu_0 I_1 \pi R_1}{2},$$

$$L_1 = \frac{d\Phi_{11}}{dI_1} = \frac{\Phi_{11}}{I_1} = \frac{\mu_0 \pi R_1}{2}.$$

(b) Circuit 2 is assumed to carry a current I_2 as shown in Fig. 8.22. Let us calculate the effect of this current on Circuit 1. As R_1 is much smaller than R_2 and R_3, the magnetic field resulting from the current in 2, can be assumed to be homogeneous in the circular area defined by Circuit 1. The magnetic field at the coordinate origin produced by the "almost circular loops" with radii R_2 and R_3 are, respectively,

$$\mathbf{B}_{12(R_2)} = \frac{\mu_0 I_2}{2R_2}\mathbf{u}_z \qquad \mathbf{B}_{12(R_3)} = -\frac{\mu_0 I_2}{2R_3}\mathbf{u}_z.$$

The resulting field produced by the two straight wires at the coordinate origin is equal to zero, since the current intensities are in opposite directions. Then, the total flux is

$$\Phi_{12} = \int_{S_1} \mathbf{B}_{12} \cdot d\mathbf{S}_1 \simeq B_{12} S_1 = \left[\frac{\mu_0 I_2}{2R_2} - \frac{\mu_0 I_2}{2R_3}\right]\pi R_1^2.$$

The mutual inductance is therefore

$$M = \frac{d\Phi_{12}}{dI_2} = \frac{\Phi_{12}}{I_2} = \left[\frac{\mu_0}{2R_2} - \frac{\mu_0}{2R_3}\right]\pi R_1^2.$$

8.6 The transmission line shown in Fig. 8.23 consists of two long parallel wires with radius a carrying currents in opposite directions. The axes of the two wires are separated by a distance b such that $b \gg a$. Find the inductance per unit length of the line.

Fig. 8.23 Cross section of a transmission line with two long wires carrying currents in opposite directions

Fig. 8.24 The two-wire transmission line and the surface through which the magnetic flux is calculated

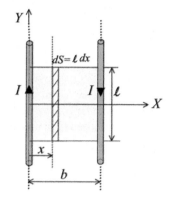

Solution

Since $b \gg a$, only the "external" inductance is considered which is calculated from the flux through the surface between the two wires due to their own currents. Assuming a current I in the wires, as shown in Fig. 8.24, the currents flowing through the left and right wires produce magnetic fields at a distance x from the left wire:

$$\left. \begin{array}{l} \mathbf{B}_l = \frac{\mu_0 I}{2\pi x}(-\mathbf{u}_z) \\ \mathbf{B}_r = \frac{\mu_0 I}{2\pi (b-x)}(-\mathbf{u}_z) \end{array} \right\} \mathbf{B} = \mathbf{B}_l + \mathbf{B}_r = -\left(\frac{\mu_0 I}{2\pi x} + \frac{\mu_0 I}{2\pi (b-x)} \right) \mathbf{u}_z.$$

By considering a length of wire ℓ, and a surface element $d\mathbf{S} = -\ell dx\, \mathbf{u}_z$, the flux through the rectangular surface between the two wires yields

$$\Phi = \int_S \mathbf{B} \cdot d\mathbf{S} = \int_S \left(0, 0, -\frac{\mu_0 I}{2\pi x} - \frac{\mu_0 I}{2\pi(b-x)} \right) \cdot (0, 0, -\ell dx) = \int_a^{b-a} \left(\frac{\mu_0 I}{2\pi x} + \frac{\mu_0 I}{2\pi(b-x)} \right) \ell dx$$

$$= \frac{\mu_0 I \ell}{2\pi} [\ln x - \ln(b-x)]_a^{b-a} = \frac{\mu_0 I \ell}{2\pi} [\ln(b-a) - \ln(a) - \ln(a) + \ln(b-a)]$$

$$= \frac{\mu_0 I \ell}{\pi} \ln\left(\frac{b-a}{a}\right).$$

The self-inductance is then obtained

$$L = \frac{d\Phi}{dI} = \frac{\mu_0 \ell}{\pi} \ln\left(\frac{b-a}{a}\right).$$

Finally, the self-inductance L' per unit length is

$$L' = \frac{L}{\ell} = \frac{\mu_0}{\pi} \ln\left(\frac{b-a}{a}\right) \simeq \frac{\mu_0}{\pi} \ln\left(\frac{b}{a}\right).$$

Problems B

8.7 Figure 8.25 shows three conducting straight wires labelled 1, 2, and 3, respectively. Wires 1 and 2 are connected to each other at O, making an angle θ. Wire 3 slides on the rails formed by wires 1 and 2; moving from the origin at constant velocity $\mathbf{v} = v\,\mathbf{u}_x$. There is a stationary magnetic field of magnitude B, confined in the regions shown in Fig. 8.25 and in the directions indicated. The resistance of the wires per unit length is R_u. Self-inductance is negligible. Numerical values: $B = 0.1$ T; $v = 3$ ms^{-1}; $\theta = 30°$; $R_u = 0.57$ Ωm^{-1}. Find the current induced in the closed circuit formed by the three wires.

Solution

Figure 8.26 shows the different positions of Wire 3. The position of the wire is determined by $x = vt$. The following situations are considered:
(a) For $0 < x < a$ ($0 < t < a/v$):
As the e.m.f. is induced due to the motion of Wire 3, the e.m.f. is calculated by applying (8.11). It should be noted that the "flux rule," (8.12), can also be applied. A counterclockwise direction is followed in evaluating the e.m.f. in the closed loop $OijO$, as shown in Fig. 8.26. Then, it is found

$$\mathcal{E} = \oint_{OijO} (\mathbf{v} \times \mathbf{B}) \cdot d\mathbf{l} = \int_{ij} (\mathbf{v} \times \mathbf{B}) \cdot d\mathbf{l} = \int_{ij} [(v, 0, 0) \times (0, 0, B)] \cdot (0, dy, 0)$$

$$= \int_i^j (0, -vB, 0) \cdot (0, dy, 0) = \int_0^{vt \tan \theta} -vB\,dy = -vBvt \tan \theta.$$

The total resistance will be $R = R_{Oi} + R_{ij} + R_{jO}$, which can be expressed as

$$R = R_u Oi + R_u ij + R_u jO = R_u vt + R_u vt \tan \theta + R_u \frac{vt}{\cos \theta} = R_u vt \left(1 + \tan \theta + \frac{1}{\cos \theta}\right).$$

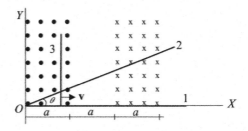

Fig. 8.25 A straight wire sliding over two conducting bars at constant velocity $\mathbf{v}$

Fig. 8.26 Wire 3 in the three different situations to be considered: **a** $0 < x < a$, **b** $a < x < 2a$, **c** $2a < x < 3a$

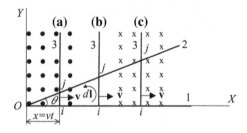

The current intensity, calculated following a counterclockwise direction, becomes

$$I = \frac{\mathcal{E}}{R} = -\frac{vB\tan\theta}{R_u\left(1 + \tan\theta + \frac{1}{\cos\theta}\right)} = -\frac{3 \times 0.1 \times \tan 30°}{0.57(1 + \tan 30° + 1/\cos 30°)} = -0.11 \text{ (A)},$$

the minus sign means that the current induced is in opposite direction to that used in the calculation of the current. Thus, a clockwise current intensity will be induced.
(b) For $a < x < 2a$ ($a/v < t < 2a/v$):

$$\mathcal{E} = \oint_{OijO} (\mathbf{v} \times \mathbf{B}) \cdot d\mathbf{l} = \int_{ij} (\mathbf{v} \times \mathbf{B}) \cdot d\mathbf{l} = 0.$$

The result obtained is due to the magnetic field being zero at all points in the bar.
(c) For $2a < x < 3a$ ($2a/v < t < 3a/v$):

$$\mathcal{E} = \oint_{OijO} (\mathbf{v} \times \mathbf{B}) \cdot d\mathbf{l} = \int_{ij} (\mathbf{v} \times \mathbf{B}) \cdot d\mathbf{l} = \int_{ij} [(v, 0, 0) \times (0, 0, -B)] \cdot (0, dy, 0)$$
$$= \int_i^j (0, vB, 0) \cdot (0, dy, 0) = \int_0^{vt\tan\theta} vB\, dy = vBvt\tan\theta.$$

In the same way as (a), it is found that $I = \mathcal{E}/R = 0.11$ A. In this case, a positive sign is obtained, hence, the induced current in the loop $OijO$ is counterclockwise (Fig. 8.27).

8.8 A conducting bar of length ℓ and mass m slides frictionlessly on two parallel conducting rails in the presence of a uniform magnetic field pointing out the page, as shown in Fig. 8.28. There is a resistance R connected across the rails. The bar is given an initial velocity $\mathbf{v}_0$ parallel to the rails. (a) Find the velocity and the current intensity in the bar as a function of time. (b) If $\ell = 20.0$ cm, $m = 20.0$ g,

Fig. 8.27 The induced intensity in the closed circuit formed by the three wires in the cases (a), (b), and (c) (see Fig. 8.26)

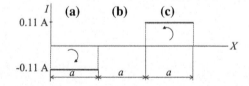

Fig. 8.28 The sliding bar at an instant t with velocity **v**

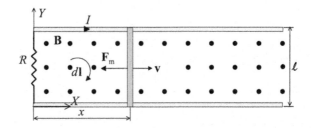

$R = 40.0\,\Omega$, $v_0 = 30.0$ m/s, and $B = 1$ T, find the magnitude of the velocity and the current intensity for $t = 3\tau$, $\tau = mR/\ell^2 B^2$. (c) In order for the bar to move at constant velocity $\mathbf{v}_0 = 30.0\,\mathbf{u}_x$ m/s, find the external force required and the power delivered to keep the bar in motion at constant velocity $\mathbf{v}_0$.

Solution

(a) It can be easily shown that at an instant t, when the velocity of the bar is **v**, the current intensity in the circuit shown in Fig. 8.28 is given by (8.15), $I = \mathcal{E}/R = vB\ell/R$, in the clockwise direction according to Lenz's law. The magnetic force $\mathbf{F}_m$ experienced by the bar as it moves to the right is given by

$$\mathbf{F}_m = I \int_{\text{bar}} d\mathbf{l} \times \mathbf{B} = I\left[-\ell\,\mathbf{u}_y \times B\mathbf{u}_z\right] = -I\ell B\,\mathbf{u}_x = -\frac{B^2\ell^2 v}{R}\,\mathbf{u}_x,$$

in the opposite direction to **v**. By applying Newton's second law to the sliding bar, rearranging the resulting equation, and then separating the variables v and t, we have

$$\mathbf{F} = m\mathbf{a} \Rightarrow ma_x = m\frac{dv}{dt} = -\frac{B^2\ell^2 v}{R} \Rightarrow \frac{dv}{v} = -\frac{B^2\ell^2}{mR}\,dt.$$

Integrating both sides of the last equation, taking into account that at $t = 0$ the velocity is v_0, it is found that

$$\int_{v_0}^{v} \frac{dv}{v} = \int_0^t -\frac{\ell^2 B^2}{mR}\,dt \Rightarrow \ln\left(\frac{v}{v_0}\right) = -\left(\frac{\ell^2 B^2}{mR}\right) t.$$

Therefore, if τ denotes the constant value $\tau = mR/(\ell^2 B^2)$, the velocity at any instant t can be expressed as

$$v = v_0\, e^{(-t/\tau)}.$$

As I is related to the velocity by (8.15), the current intensity can be expressed as

$$I = \frac{B\ell v_0}{R} e^{(-t/\tau)},$$

which decreases exponentially with time.

(b) For $t = 3\tau = 3(mR/B^2\ell^2) = 3(0.020 \times 40/1^2 \times 0.20^2) = 60$ s, substituting this value of t into the above equations for v and I, it is found that $v = 1.49$ m/s and $I = 7.5$ mA. These values correspond to approximately 5% the initial velocity and intensity, respectively.

(c) For the bar to move at a constant velocity v_0, the resultant force acting on the bar must be zero. Then, an external force $\mathbf{F}_{ext}$ should be applied to balance the magnetic one $\mathbf{F}_m$, i.e.

$$\mathbf{F}_{ext} = -\mathbf{F}_m = \frac{B^2\ell^2 v_0}{R}\mathbf{u}_x = \frac{1^2 \times 0.20^2 \times 30}{40}\mathbf{u}_x = 0.03\,\mathbf{u}_x \quad (N).$$

The power P delivered by $\mathbf{F}_{ext}$ is given by

$$P = \mathbf{F}_{ext} \cdot \mathbf{v}_0 = \left(\frac{B^2\ell^2 v_0}{R}, 0, 0\right) \cdot (v_0, 0, 0) = \frac{(B\ell v_0)^2}{R} = \frac{(1 \times 0.20 \times 30)^2}{40} = 0.9 \quad (W).$$

It should be noted that $P = (B\ell v_0)^2/R = \mathcal{E}^2/R = RI^2$, the power delivered to the circuit, $(F_{ext} v_0)$, is equal to the power transferred from the e.m.f. to the load resistance, as required by energy conservation.

8.9 Figure 8.29 shows a core of a linear magnetic material, with high permeability μ_r, mean length l_c, and with an air gap of length l_g. Around the core is wound a coil of wire, with N turns, carrying a current intensity I. The cross section of the core is a square with side length a. At the center of the air gap, there is a circular coil with diameter D ($D < a$), which initially lies in YZ-plane, as shown in Fig. 8.29. If from this position the circular coil starts to rotate, determine the e.m.f. induced in the coil as a function of time when the coil rotates around: (a) OX-axis with constant angular velocity $\boldsymbol{\omega}$ (ω_x, 0, 0). (b) OY-axis with $\boldsymbol{\omega}$ (0, ω_y, 0). (c) OZ-axis with $\boldsymbol{\omega}$ (0, 0, ω_z).

Numerical values: $\mu_r = 4000$; $N = 10,000$; $l_c = 1.5$ m; $l_g = 10$ cm, $I = 5$ A; $a = 6$ cm; $D = 1.5$ cm; $\omega_x = \omega_y = \omega_z = 50$ rad/s.

Fig. 8.29 An electromagnet with a rotating circular coil centered in the air gap

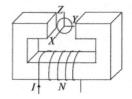

Solution

First of all, let us calculate the magnetic field B in the air gap by applying Ampère's law. The following assumptions are made: the magnetic field in the air gap is uniform, the magnetic field in a cross-section of the core is homogeneous, and flux leakage is negligible. Under these assumptions and since the cross-sectional area is constant, the magnetic field in the core B_c is constant and equal in magnitude to that in the air gap B_g. Then, $B = B_c = B_g$. In the core: $B = B_c = \mu_0 \mu_r H_c$ and in the air-gap: $B = B_g = \mu_0 H_g$. By applying Ampère's law, it follows

$$H_c l_c + H_g l_g = NI \Rightarrow \frac{B}{\mu_0 \mu_r} l_c + \frac{B}{\mu_0} l_g = NI,$$

$$B = \frac{\mu_0 \mu_r NI}{l_c + \mu_r l_g} = \frac{4\pi \times 10^{-7} \times 4000 \times 10^4 \times 5}{1.5 + 4000 \times 0.10} = 0.626 \text{ T}.$$

Therefore in the air gap, the magnetic field is uniform and can be expressed as

$$\mathbf{B} = 0.626 \, \mathbf{u}_y.$$

(a) In this case the surface of the coil $\mathbf{S}$ at any instant is parallel to $\mathbf{u}_x$ and, therefore, is perpendicular to the uniform magnetic field, as shown in Fig. 8.30a. Consequently, the flux through the coil $\Phi = 0$ and hence $\mathcal{E} = -d\Phi/dt = 0$.
(b) At instant t, the surface of the circular coil is $\mathbf{S}$ ($S \cos \omega_y t$, 0, $-S \sin \omega_y t$), with $S = \pi D^2/4$ (see Fig. 8.30b). Then, $\Phi = \int_S \mathbf{B} \cdot d\mathbf{S} = \mathbf{B} \cdot \mathbf{S} = (0, B, 0) \cdot (S \cos \omega_y t, 0, -S \sin \omega_y t) = 0 \Rightarrow \mathcal{E} = -d\Phi/dt = 0$.
(c) As shown in Fig. 8.30c, the surface of the circular coil at instant t is $\mathbf{S}$ ($S \cos \omega_z t$, $S \sin \omega_z t$, 0). Then, the magnetic flux and the e.m.f. are respectively,

$$\Phi = \int_S d\Phi = \int_S \mathbf{B} \cdot d\mathbf{S} = \mathbf{B} \cdot \mathbf{S} = (0, B, 0) \cdot (S \cos \omega_z t, S \sin \omega_z t, 0)$$

$$= BS \sin \omega_z t = B \frac{\pi D^2}{4} \sin \omega_z t,$$

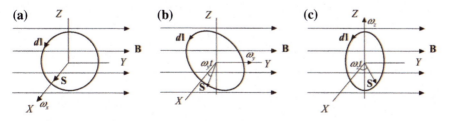

Fig. 8.30 The circular coil at an instant t when it is rotating about: **a** OX-axis, **b** OY-axis, and **c** OZ-axis

Fig. 8.31 A conducting disc rotating in an uniform magnetic field with a resistance between the connecting brushes

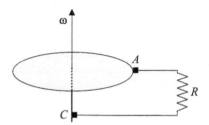

$$\mathcal{E} = -\frac{d\Phi}{dt} = -B\frac{\pi D^2}{4}\omega_z \cos\omega_z t = -0.626 \times \frac{\pi \times 0.015^2}{4} \times 50 \cos(50t)$$
$$= -5.53 \times 10^{-3} \cos(50t) \text{ (V)}.$$

8.10 A circular disc with diameter D rotates around its axis of revolution with a constant angular velocity ω in an uniform magnetic field of magnitude B. The direction of the magnetic field produces an angle of 30° with the axis of the disc. The terminals of a resistor R (always at rest) are electrically connected through the brushes to the rim of the disc and to the axis, as shown in Fig. 8.31. Both the disc and the axis are assumed to be perfect conductors. Find the current passing through the resistor: (a) if self-inductance is negligible, (b) if self-inductance is included in the analysis.

Solution

Motion of the conducting disc in the magnetic field causes e.m.f. to be induced and therefore a current intensity will flow through the wire with resistance R. Let us determine first the velocity of a point P on the disc at a distance ρ from the axis, then the e.m.f. is evaluated along the closed loop (dotted line), $OACO$, shown in Fig. 8.32. Finally, the induced current is calculated in the two cases: (a) negligible self-inductance; (b) including self-inductance in the analysis.
(a)

- Calculation of the velocity of P:

$$\mathbf{v} = \boldsymbol{\omega} \times \mathbf{r} = (0, 0, \omega) \times (\rho, \phi, 0) = \omega\rho\mathbf{u}_\phi.$$

The non-conservative field $\mathbf{v} \times \mathbf{B}$ can be easily obtained

$$\mathbf{v} \times \mathbf{B} = (0, \omega\rho, 0) \times (B_\rho, B_\phi, B_z) = (\omega\rho B_z, 0, -\omega\rho B_\rho).$$

- Calculation of the e.m.f. around the closed path:

$$\mathcal{E} = \oint_{OACO} (\mathbf{v} \times \mathbf{B}) \cdot d\mathbf{l} = \int_{OA} (\mathbf{v} \times \mathbf{B}) \cdot d\mathbf{l} = \int_{OA} (\omega\rho B_z, 0, -\omega\rho B_\rho) \cdot (d\rho, 0, 0)$$
$$= \int_0^{D/2} \omega\rho B_z \, d\rho = \omega B_z \frac{\rho^2}{2}\bigg|_0^{D/2} = \omega B \cos 30° \frac{D^2}{8} = \frac{\omega B D^2 \sqrt{3}}{16},$$

Fig. 8.32 The e.m.f. is calculated around the *closed dotted path*

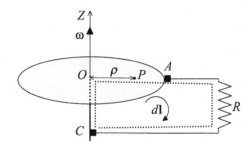

where it has been taken into account that only the segment OA on the disc contributes to the e.m.f. and that $B_z = B\cos 30°$.

- Calculation of the current induced:

$$I = \frac{\mathcal{E}}{R} = \frac{\omega B \sqrt{3} D^2}{16R}.$$

(b) If the self-inductance L of the closed path is included, we have by evaluating the e.m.f. around the aforesaid closed path, following a clockwise direction and applying Kirchhoff's second law,

$$\mathcal{E} - L\frac{dI}{dt} = RI \Rightarrow L\frac{dI}{dt} = \mathcal{E} - RI \Rightarrow \frac{dI}{\mathcal{E} - RI} = \frac{dt}{L},$$

$$\int_0^I \frac{dI}{\mathcal{E} - RI} = \int_0^t \frac{dt}{L} \Rightarrow -\frac{1}{R}\ln(\mathcal{E} - RI)\bigg|_0^I = \frac{t}{L}\bigg|_0^t,$$

$$-\frac{1}{R}[\ln(\mathcal{E} - RI) - \ln \mathcal{E}] = \frac{t}{L} \Rightarrow$$

$$I = \frac{\mathcal{E}}{R}\left(1 - e^{-\frac{R}{L}t}\right).$$

When a long time has passed, the current intensity tends to the value obtained in (a).

8.11 The semicircular loop of diameter D shown in Fig. 8.33 begins to rotate around its diameter with angular velocity $\boldsymbol{\omega} = \omega \mathbf{u}_y$ in the presence of a magnetic field $\mathbf{B} = B\mathbf{u}_z$, confined to the region $x \geq 0$. If the resistance of the loop is R and self-inductance is assumed to be negligible, find the current induced in the loop at any instant t. Numerical values: $D = 0.25\,\text{m}$; $\omega = 100\,\text{rad s}^{-1}$; $B = 0.8\,\text{T}$; $R = 2\,\Omega$.

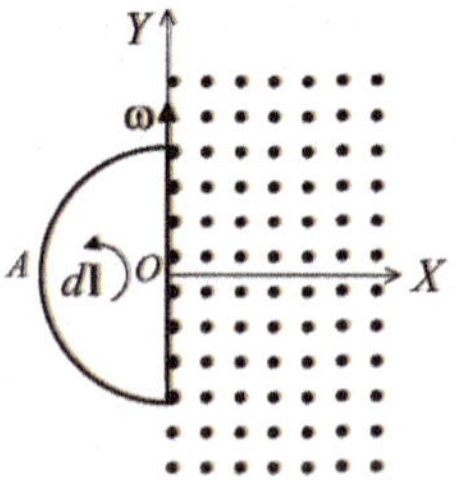

Fig. 8.33 Semicircular loop lying in XY-plane begins to rotates around OY-axis. Electromotive force around the loop is calculated following a counterclockwise direction

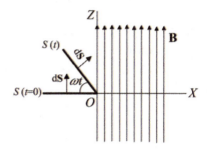

Fig. 8.34 The projection of the loop of Fig. 8.33 at an instant t on the XZ-plane

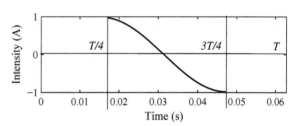

Fig. 8.35 The current intensity induced in terms of the time during one period

Solution

Let us calculate the e.m.f. and the induced intensity in the time period T.

- For $0 < \omega t < \frac{\pi}{2} \Rightarrow 0 < t < \frac{T}{4} = \frac{\pi}{2\omega}$:
  In this case, the flux through the semicircular loop is zero, as are the e.m.f. and the induced intensity (see Fig. 8.34).
- For $\frac{\pi}{2} < \omega t < \frac{3\pi}{2} \Rightarrow \frac{T}{4} = \frac{\pi}{2\omega} < t < \frac{3T}{4} = \frac{3\pi}{2\omega}$:
  The magnetic flux through the loop is given by

$$\Phi = \int_S \mathbf{B} \cdot d\mathbf{S} = \int_S BdS \cos(\omega t) = B\cos(\omega t) \int_S dS = B\cos(\omega t) S = B\frac{\pi D^2}{8} \cos(\omega t),$$

where the area of the loop is $S = \pi D^2/8$. Then, we find the e.m.f to be

$$\mathcal{E} = -\frac{d\Phi}{dt} = B\frac{\pi D^2}{8} \omega \sin(\omega t) = 0.8 \frac{\pi \times 0.25^2}{8} 100 \sin(100t) = 1.96 \sin(100t) \text{ (V)}.$$

Finally, the current intensity induced is

$$I = \frac{\mathcal{E}}{R} = \frac{1.96}{2}\sin(100t) = 0.98\sin(100t) \ (A).$$

- For $\frac{3\pi}{2} < \omega t < 2\pi \Rightarrow \frac{3T}{4} = \frac{3\pi}{2\omega} < t < T = \frac{2\pi}{\omega}$:
  In this interval the same conditions applied as in case (a). Magnetic lines of force do not pass through the semicircular loop. Hence, the e.m.f induced equals zero and the current is not induced in the loop.

  Figure 8.35 shows the result obtained for the intensity as a function of time.

8.12 A rectangular conducting loop with sides ℓ_1 and ℓ_2 is placed in a uniform magnetic field $\mathbf{B} = B_0\,\mathbf{u}_y$, confined to the region where the rectangle is situated. The coil rotates with an angular velocity ω about the X-axis, as shown in Fig. 8.36, and is initially placed on the XZ-plane with its center at the origin. The two ends of the loop are connected to two rings provided with sliding contacts. (a) Find the e.m.f. induced in the loop and the voltage between its terminals. (b) If a load resistance is connected to the terminals, and the coil has N turns, find the intensity flowing through the resistance and determine the external mechanical power required to maintain the rotation of the coil at an angular frequency ω.

Solution

(a) The e.m.f. in the loop is caused by the motion of the loop in the uniform magnetic field. Either (8.11) or the "flux rule" given by (8.12) can be used to calculate the e.m.f. around the coil. In order to calculate the e.m.f., we follow the counterclockwise closed path determined by the rectangular wire and a mathematical line between terminals ab: ba (wire) + ab (external line), see Fig. 8.37a. The flux of the magnetic field through the planar surface bounded by such a closed path is equal to the flux through the rectangle. At any time t, see Fig. 8.37b, the resulting flux is

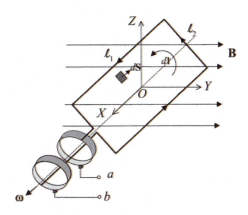

Fig. 8.36 A sketch of a simple generator consisting of a rectangular coil rotating about its own axis in a magnetic field

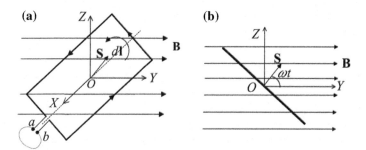

Fig. 8.37 a Sketch of the closed path followed to evaluate the e.m.f. **b** View of the rectangular loop from +OX direction at a certain instant in time t

$$\Phi = \int_S \mathbf{B} \cdot d\mathbf{S} = \mathbf{B} \cdot \mathbf{S} = BS\cos(\omega t) = B\ell_1\ell_2 \cos(\omega t),$$

where the area of the rectangular coil $S = \ell_1\ell_2$. Therefore, we have for the electromotive force of the simple generator

$$\mathcal{E} = -\frac{d\Phi}{dt} = -\frac{d(BS\cos(\omega t))}{dt} = BS\omega \sin(\omega t),$$

which varies in time according to a sinusoidal law. Assuming that the wire is made of a perfect conductor and following the above described counterclockwise closed path ba(wire)-ab(exterior), it is found

$$\mathcal{E} = \oint_{ba(\text{wire})-ab(\text{exterior})} \mathbf{E}_e \cdot d\mathbf{l} = BS\omega \sin(\omega t) = \int_{ba(\text{wire})} 0 \cdot d\mathbf{l} + \int_{ab(\text{exterior})} -\nabla V \cdot d\mathbf{l},$$

where it has been taken into account that in the wire $\mathbf{E}_e = 0$ and in the exterior $\mathbf{E}_e = -\nabla V$.

Hence, the voltage V_{ab} between the terminals can be easily obtained,

$$\mathcal{E} = BS\omega \sin(\omega t) = \int_{a(\text{exterior})}^{b} -\nabla V \cdot d\mathbf{l} = V_a - V_b = V_{ab}.$$

(b) If the circuit is completed through an external load (see Fig. 8.38), V_{ab} will produce a harmonic current. If the coil has N turns, the voltage between terminals will be $V_{ab} = NBS\omega \sin(\omega t)$. Assuming that the electrical resistance and self-inductance of the coil are negligible, the current in the load is

$$I = \frac{V_{ab}}{R} = \frac{NBS\omega}{R} \sin(\omega t).$$

Fig. 8.38 The coil has N turns and a load resistance is connected across the terminals of a generator

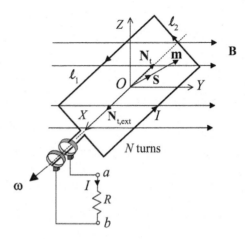

When the coil is open-circuited, no current flows and therefore no power is developed by the generator. On the other hand, if the circuit is closed through the load resistance, the power P dissipated in the resistance is

$$P = IV_{ab} = \frac{(NBS\omega)^2}{R} \sin^2(\omega t).$$

As the current flows, there is a torque $\mathbf{N}_t$ opposing the motion of the coil,

$$\mathbf{N}_t = \mathbf{m} \times \mathbf{B} = (NI\mathbf{S}) \times \mathbf{B} = NISB \sin(\omega t)(-\mathbf{u}_x),$$

where $\mathbf{m} = NI\mathbf{S}$ is the magnetic dipole moment of the coil and $\mathbf{S}$ is the vector perpendicular to the plane of the coil in a direction related to the current by the right-hand screw rule. Therefore, a external torque $\mathbf{N}_{t,ext}$, equal in magnitude but in opposite direction to $\mathbf{N}_t$, is required to maintain the rotation of the coil at a steady angular frequency ω. Thus, the external mechanical power supplied to rotate the loop is

$$P_{ext} = N_{t,ext} \,\omega = NISB \sin(\omega t) \,\omega = \frac{(NBS\omega)^2}{R} \sin^2(\omega t).$$

The power delivered from mechanical work is the same as the electric power generated and consumed by the resistance, as expected on energy conservation grounds.

8.13 Consider a very long solenoid with an air core, n turns of wire per unit length, and a circular cross-section of radius a. The cylindrical solenoid carries a varying current $I = I_0 \cos(\omega t)$. Determine the induced electric field both inside and outside the solenoid.

Fig. 8.39 Cross section of a very long cylindrical solenoid carrying a varying current $I = I_0 \cos(\omega t)$

Solution

As can be proved by the reader, the lines of the vector potential **A** created by a current I carried by a long cylindrical solenoid are circles centered at the solenoid axis. Assuming the current in our solenoid to be slowly varying, the instantaneous value of the vector potential is given by

- For $\rho < a$:

$$A_\phi = \frac{\mu_0 n \rho}{2} I_0 \cos(\omega t).$$

- For $\rho > a$:

$$A_\phi = \frac{\mu_0 n a^2}{2\rho} I_0 \cos(\omega t).$$

The varying solenoid current creates an electric field **E** at all points that can be calculated by applying Faraday's law (8.5). The induced electric field **E** can also be obtained from the vector potential **A** through (8.7). Cylindrical symmetry ensures that **E** has only the component E_ϕ, which is constant along any circular path around the axis of the solenoid. For a circular path with radius ρ, as shown in Fig. 8.39, (8.7) leads to

$$\oint_C \mathbf{E} \cdot d\mathbf{l} = E_\phi 2\pi\rho = -\frac{d}{dt}\int_0^{2\pi} A_\phi \rho d\phi \Rightarrow E_\phi = -\frac{dA_\phi}{dt}.$$

So, the induced electric field can be directly obtained from the vector potential, and the results obtained for the instantaneous value is

- For $\rho < a$:

$$E_\phi = -\frac{dA_\phi}{dt} = \frac{\mu_0 n I_0 \omega \rho}{2} \sin(\omega t).$$

- For $\rho > a$:

$$E_\phi = -\frac{dA_\phi}{dt} = \frac{\mu_0 n I_0 \omega a^2}{2\rho} \sin(\omega t).$$

Electric field lines are also circles centered on the solenoid axis. Finally, the electromotive force around circular paths also centered on the solenoid axis is given by

- For $\rho < a$:

$$\mathcal{E} = \oint_C \mathbf{E} \cdot d\mathbf{l} = \int_0^{2\pi} E_\phi \rho d\phi = E_\phi 2\pi\rho = \mu_0 n I_0 \omega \pi \rho^2 \sin(\omega t).$$

- For $\rho > a$:

$$\mathcal{E} = \oint_C \mathbf{E} \cdot d\mathbf{l} = \int_0^{2\pi} E_\phi \rho\, d\phi = E_\phi 2\pi\rho = \mu_0 n I_0 \omega \pi a^2 \sin(\omega t).$$

8.14 Four long conducting straight wires, parallel to the coordinate axes and situated in the XY-plane, move, from the coordinate origin, at constant velocities, as shown in Fig. 8.40. The resistance of the wires per unit length is r. There is a magnetic field $\mathbf{B} = B_0 \cos(\omega t) \left[\frac{\sqrt{3}}{2} \mathbf{u}_y + \frac{1}{2} \mathbf{u}_z \right]$. Find the current intensity in the circuit formed by the four conducting wires at any instant t. Self-inductance is negligible.

Solution

Let us determine, at an instant t, the e.m.f. around the closed loop formed by the four wires, the line element as shown in Fig. 8.40 (counterclockwise direction). Then, the surface element used in calculating the magnetic flux is $d\mathbf{S} = dS\, \mathbf{u}_z = dx dy\, \mathbf{u}_z$. At any instant, the resulting magnetic flux can then be written as

$$\Phi = \int_S \mathbf{B} \cdot d\mathbf{S} = \mathbf{B} \cdot \mathbf{S} = \left(0, B_0 \cos(\omega t) \frac{\sqrt{3}}{2}, B_0 \cos(\omega t) \frac{1}{2} \right) \cdot (0, 0, S) = \frac{B_0}{2} \cos(\omega t) S$$

$$= \frac{B_0}{2} \cos(\omega t) 4v^2 t^2 = 2B_0 \cos(\omega t) v^2 t^2,$$

where at the instant t, the surface defined by the four wires is $S = (2vt)^2 = 4v^2 t^2$. The electromotive force is then given by

$$\mathcal{E} = -\frac{d\Phi}{dt} = 2B_0 \omega \sin(\omega t) v^2 t^2 - 4B_0 \cos(\omega t) v^2 t.$$

The resistance of the closed circuit is $R = 4(2vt)r = 8vtr$. Then, we find the resulting current to be

$$I = \frac{B_0 \omega}{4r} \sin(\omega t) vt - \frac{1}{2r} B_0 \cos(\omega t) v.$$

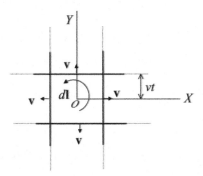

Fig. 8.40 The four moving conducting wires at an instant t; the e.m.f. is calculated following the path given by $d\mathbf{l}$

Fig. 8.41 A conducting bar sliding on two rails in presence of a time-variant magnetic field. At an instant t, the e.m.f. is calculated following the direction $d\mathbf{l}$

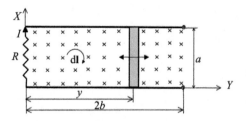

8.15 A bar slides frictionlessly on two parallel rails electrically connected by a resistor of resistance R, as shown in Fig. 8.41. Both the sliding bar and the rails can be considered perfect conductors. The bar moves parallel to the OY-axis, the distance from the bar to the OX-axis is given by $y = b[1 - \sin(\omega_1 t)]$, where b and ω_1 are constants, and t is the time. In the region shown in Fig. 8.41, there is a time-varying magnetic field $\mathbf{B} = B_0 \cos(\omega_2 t)\,\mathbf{u}_z$, where B_0 and ω_2 are also constants. (a) Find the current intensity flowing through the resistor as a function of the time. (b) Represent the result obtained for the intensity as a function of time in the case $\omega_1 = \omega_2$. Self-inductance is considered to be negligible.

(a) First of all, let us calculate the e.m.f. following the closed path: bar, rails, and the resistor (clockwise direction), as shown in Fig. 8.41. At a given instant t, the flux through the surface, whose boundary is the closed path, is given by

$$\Phi = \int_S \mathbf{B}\cdot d\mathbf{S} = \int_S (0,0,B)\cdot(0,0,dS) = \int_S B\,dS = BS = B_0\cos(\omega_2 t)\,ab\,[1-\sin(\omega_1 t)],$$

where $B = B_0 \cos(\omega_2 t)$ and $S = ay = ab[1 - \sin(\omega_1 t)]$.

Faraday's law becomes

$$\mathcal{E} = -\frac{d\Phi}{dt} = B_0 ab\,\omega_2 \sin(\omega_2 t)\,[1-\sin(\omega_1 t)] + B_0 ab\,\cos(\omega_2 t)\,\omega_1 \cos(\omega_1 t),$$

whereas the first term is produced by the time-variation of $\mathbf{B}$, and the second is due to the motion of the bar in the magnetic field.

Finally, the current intensity, calculated as indicated in Fig. 8.41, is given by,

$$I = \frac{\mathcal{E}}{R}.$$

(b) Substitution of $\omega_1 = \omega_2 = \omega$ into the value calculated of $\mathcal{E}$, yields

$$I = \frac{\mathcal{E}}{R} = \frac{B_0 ab\,\omega[\sin\omega t + \cos 2\omega t]}{R},$$

which is represented in Fig. 8.42. The induced intensity is periodic, the period being $2\pi/\omega$.

Fig. 8.42 The resulting induced current intensity as a function of time (Problem 8.15)

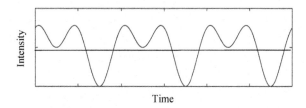

8.16 Figure 8.43 shows a sketch of a toroidal core along with its cross section. The core material is of high permeability μ_r. Coil labelled 1 has N_1 turns uniformly distributed and tightly wound on the entire toroid, while the N_2 turns of coil 2 are uniformly distributed along a portion of the toroid, between terminals c and d. Determine: (a) Self-inductance L_1 and mutual inductance M. If terminals c and d are connected to a generator and terminals a and b to a resistance R, (b) find the differential equation corresponding to the intensity through the resistance when the current intensity through Coil 2 is given by $I_{02}\sin(\omega t)$.

Solution

(a) Let us calculate the self-inductance L_1. Firstly, a current intensity I_1 is assumed to flow through Coil 1. Secondly, the magnetic field $\mathbf{B}_1$ is calculated from the $\mathbf{H}_1$-field by applying Ampère's law to a circular path Γ with radius ρ ($R < \rho < R+w$), see Fig. 8.43:

$$\oint_\Gamma \mathbf{H}_1 \cdot d\mathbf{l} = \int_S \mathbf{j}_c \cdot d\mathbf{S} \Rightarrow H_{\phi_1} 2\pi\rho = N_1 I_1 \Rightarrow H_{\phi_1} = \frac{N_1 I_1}{2\pi\rho},$$

where H_{ϕ_1} and ρ are constant around the circular path Γ that encircle a total current $N_1 I_1$. By taking into account the linear relation $B_1 = \mu_0 \mu_r H_1$, we have

$$B_{\phi_1} = \frac{\mu_0 \mu_r N_1 I_1}{2\pi\rho}.$$

Next we determine the flux through one loop in Coil 1:

$$\Phi_{11,\text{one}} = \int_{S_1} \mathbf{B}_1 \cdot d\mathbf{S} = \int_{S_1} \left(\frac{\mu_0 \mu_r N_1 I_1}{2\pi\rho} \mathbf{u}_\phi\right) \cdot \left(h d\rho\, \mathbf{u}_\phi\right) \qquad (8.38)$$

Fig. 8.43 A toroidal core and its cross section with size lengths w and h. The internal radius is denoted R. The two wire coils wound around the core are labelled Coil 1 (with terminals a and b) and Coil 2 (with terminals c and d)

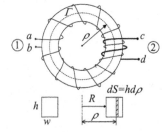

$$= \frac{\mu_0 \mu_r N_1 I_1 h}{2\pi} \int_R^{R+w} \frac{d\rho}{\rho} = \frac{\mu_0 \mu_r N_1 I_1 h}{2\pi} \ln \frac{(R+w)}{R} \quad (8.39)$$

Then, the total flux through Coil 1 is,

$$\Phi_{11} = N_1 \Phi_{11,\text{one}} = \frac{\mu_0 \mu_r N_1^2 I_1 h}{2\pi} \ln \frac{(R+w)}{R}.$$

Finally, we obtain for the self-inductance of Coil 1:

$$L_1 = \frac{d\Phi_{11}}{dI_1} = \frac{\Phi_{11}}{I_1} = \frac{\mu_0 \mu_r N_1^2 h}{2\pi} \ln \frac{(R+w)}{R}.$$

Note that the self-inductance of circuits in linear magnetic media is not a function of I and that it is proportional to the square number of turns.

Let us now calculate the mutual inductance M between Coil 1 and Coil 2. A current can be assumed in Coil 1 and then the effect on Coil 2 evaluated, or the current in Coil 2 assumed and the effect on Coil 1 determined. The former procedure seems much simpler. The flux linkage in Coil 2 due to the magnetic field produced by I_1 is

$$\Phi_{21} = N_2 \Phi_{21,\text{one}} = \frac{\mu_0 \mu_r N_1 N_2 I_1 h}{2\pi} \ln \frac{(R+w)}{R}.$$

Hence, the mutual inductance is

$$M = \frac{d\Phi_{21}}{dI_1} = \frac{\Phi_{21}}{I_1} = \frac{\mu_0 \mu_r N_1 N_2 h}{2\pi} \ln \frac{(R+w)}{R}.$$

(b) By applying Kirchhoff's second law to loop 1 in a clockwise direction (see Fig. 8.44), it is found:

$$\mathcal{E}_1 = -L_1 \frac{dI_1}{dt} - M \frac{dI_2}{dt} = RI_1,$$

$$L_1 \frac{dI_1}{dt} + RI_1 = -MI_{02}\omega \cos(\omega t).$$

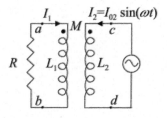

Fig. 8.44 Terminals a and b are connected to a resistance and c and d to a generator. Points represent the directions of current intensities in which the magnetic fields produced by I_1 and I_2 are in the same direction

Solved Problems

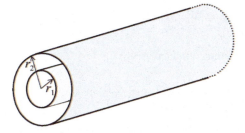

Fig. 8.45 A coaxial transmission line with a conductor tube of radius r_1 and a thin conductor with radius r_2

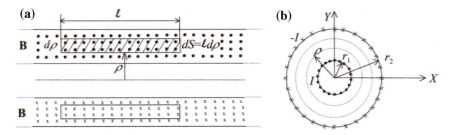

Fig. 8.46 Two sections of the transmission line

8.17 Determine the inductance per unit length of a coaxial transmission line with an inner hollow tube of radius r_1 and a thin hollow conductor of radius r_2 as shown in Fig. 8.45.

Solution

We assume that the inner conductor carries a current I and that identical current flows in the outer conductor but in opposite direction. By symmetry, the lines of **B** are circles, i.e. **B** has only the component B_ϕ. From Ampère's law applied to a circular path of radius ρ as shown in Fig. 8.46b, it is found that the values for **B** are

$$B_\phi(\rho) = \begin{cases} B_\phi = 0 & 0 < \rho < r_1, \\ B_\phi = \frac{\mu_0 I}{2\pi\rho} & r_1 < \rho < r_2, \\ B_\phi = 0 & \rho > r_2. \end{cases}$$

The resulting magnetic field is confined to the region between the two cylinders and B-field lines are concentric circles centered on the axis of the cylinders.

Consider a cylinder between the two conductors with an inner radius ρ, an outer radius $\rho + d\rho$, and a length ℓ, as shown in Fig. 8.46a. The flux of the magnetic field $B_\phi = \mu_0 I / 2\pi\rho$ through the surface $dS = \ell d\rho$, with cross hatches in Fig. 8.46a, and the total flux are, respectively,

$$d\Phi = \frac{\mu_0 I}{2\pi\rho}\ell d\rho \Rightarrow \Phi = \int_{r_1}^{r_2} \frac{\mu_0 I}{2\pi\rho}\ell d\rho = \frac{\mu_0 I \ell}{2\pi} \ln\rho\Big|_{r_1}^{r_2} = \frac{\mu_0 I \ell}{2\pi} \ln\frac{r_2}{r_1}.$$

The self-inductance is then obtained

$$L = \frac{d\Phi}{dI} = \frac{\mu_0 \ell}{2\pi} \ln\frac{r_2}{r_1}.$$

Finally, the inductance per unit length of the coaxial transmission line is therefore

$$L' = \frac{L}{\ell} = \frac{\mu_0}{2\pi} \ln\frac{r_2}{r_1}.$$

Problems C

8.18 A conducting rectangular loop, with sides of length ℓ_1 and ℓ_2, lies in the XY-plane and moves with a constant velocity $\mathbf{v} = v\,\mathbf{u}_x$, as shown in Fig. 8.47. At time $t = 0$, the vertex a is coincident with the origin of the coordinate system. Two different situations are considered:

- (a) The loop moves through a region where there is a magnetic field given by $\mathbf{B} = B_0 \cos(kx)\,\mathbf{u}_z$ (B_0 and k are constants). Calculate the induced e.m.f. in the loop as a function of time. Show the direction of the current induced in the loop at time $t = 1\,\mu s$.
- (b) If the magnetic field is now $\mathbf{B} = B_0 \cos(kx)\sin(\omega t)\,\mathbf{u}_z$, where ω is also a constant, find the e.m.f. induced in the loop.
  Numerical values: $\ell_1 = 5\,\text{cm}$; $\ell_2 = 10\,\text{cm}$; $v = 0.5\,\text{m/s}$; $B_0 = 1\,\text{T}$; $k = 40\,\text{m}^{-1}$; $\omega = 50\,\text{s}^{-1}$.

Solution

(a) The electromotive force in the rectangular loop is induced by the motion of the loop in the non-uniform magnetic field and can be calculated by applying (8.11) or (8.12). The latter will be used in the calculation of the e.m.f. At a certain instant t, when the square loop is located at the position shown in Fig. 8.48, the surface element being $d\mathbf{S} = \ell_2 dx\,\mathbf{u}_z$, thus the flux through the loop can be expressed as

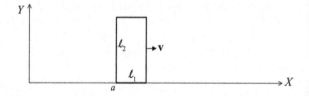

Fig. 8.47 A rectangular loop moving at a constant velocity **v** in a non-uniform magnetic field perpendicular to the XY-plane

Fig. 8.48 The rectangular loop at a certain instant t. Electromotive force is calculated following a counterclockwise path

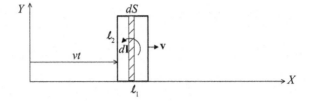

Fig. 8.49 The direction of the current induced in the loop at $t = 1\,\mu s$

$$\Phi = \int_S \mathbf{B}\cdot d\mathbf{S} = \int_S (0,0,B_0\cos(kx))\cdot(0,0,\ell_2 dx) = \int_{vt}^{vt+\ell_1} B_0\cos(kx)\ell_2 dx$$

$$= B_0\ell_2 \left.\frac{\sin(kx)}{k}\right|_{vt}^{vt+\ell_1} = \frac{B_0\ell_2}{k}\{\sin[k(vt+\ell_1)] - \sin[kvt]\}.$$

Then, the induced e.m.f can be calculated by applying (8.12), becoming

$$\mathcal{E} = -\frac{d\Phi}{dt} = -\frac{B_0\ell_2}{k} kv\{\cos[k(vt+\ell_1)] - \cos[kvt]\}$$
$$= B_0\ell_2 v\{\cos[kvt] - \cos[k(vt+\ell_1)]\}$$
$$= 0.05\,[\cos(20t) - \cos(20t+2)] \quad (V).$$

At $t = 1\,\mu s$, $\mathcal{E} = 0.07$ (V) > 0. Neglecting self-inductance, the current intensity calculated following a counterclockwise path around the loop becomes $I = \mathcal{E}/R > 0$, and the direction of the induced current is the same as that assumed in the calculation, shown in Fig. 8.49.

This result can also be easily obtained by applying the Leibniz rule for the differentiation of an integral. For the function $F(t)$ obtained by integration of the function $f(x,t)$,

$$F(t) = \int_{x_1(t)}^{x_2(t)} f(x,t)\,dx,$$

the derivative with respect to time can be expressed as

$$\frac{dF(t)}{dt} = \int_{x_1(t)}^{x_2(t)} \frac{\partial f(x,t)}{\partial t}\,dx + f[x_2(t),t]\frac{dx_2(t)}{dt} - f[x_1(t),t]\frac{dx_1(t)}{dt}.$$

Then for the flux obtained at t, by applying this rule it is found

$$\Phi(t) = \int_{x_1(t)=vt}^{x_2(t)=vt+\ell_1} B_0\cos(kx)\ell_2 dx \Rightarrow$$

$$\frac{d\Phi(t)}{dt} = 0 + B_0 \cos[k(vt + \ell_1)]\ell_2 v - B_0 \cos[kvt]\ell_2 v,$$

$$\mathcal{E} = -\frac{d\Phi(t)}{dt} = B_0 \ell_2 v \left\{\cos[kvt] - \cos[k(vt + \ell_1)]\right\}.$$

(b) In this case the change of the magnetic flux through the loop is due to both the circuit movement and the magnetic field variation with time. Then, the general law given in (8.19) is used to evaluate the induced e.m.f. in the rectangular loop. As in (a), the magnetic flux through the circuit at a certain instant t is first calculated. Then, the induced e.m.f. is obtained by applying the "flux rule" (8.19).

$$\Phi = \int_S \mathbf{B} \cdot d\mathbf{S} = \int_S (0, 0, B_0 \cos(kx)\sin(\omega t)) \cdot (0, 0, \ell_2 dx) = \int_{vt}^{vt+\ell_1} B_0 \cos(kx)\sin(\omega t)\ell_2 dx.$$

Applying the Leibniz rule, we have

$$\frac{d\Phi}{dt} = \int_{vt}^{vt+\ell_1} B_0 \cos(kx)\omega \cos(\omega t)\ell_2 \, dx + B_0 \cos[k(vt + \ell_1)]\sin(\omega t)\ell_2 v$$
$$- B_0 \cos[kvt]\sin(\omega t)\ell_2 v$$
$$= B_0 \omega \cos(\omega t)\ell_2 \left[\frac{\sin(kx)}{k}\right]_{vt}^{vt+\ell_1} + B_0 \cos[k(vt + \ell_1)]\sin(\omega t)\ell_2 v$$
$$- B_0 \cos[kvt]\sin(\omega t)\ell_2 v$$
$$= \frac{B_0 \omega \ell_2}{k} \cos(\omega t) \left\{\sin[k(vt + \ell_1)] - \sin[kvt]\right\}$$
$$+ B_0 \ell_2 v \sin(\omega t) \left\{\cos[k(vt + \ell_1)] - \cos[kvt]\right\}.$$

$$\mathcal{E} = -\frac{d\Phi}{dt}$$
$$= -\frac{B_0 \omega \ell_2}{k} \cos(\omega t) \left\{\sin[k(vt + \ell_1)] - \sin[kvt]\right\}$$
$$- B_0 \ell_2 v \sin(\omega t) \left\{\cos[k(vt + \ell_1)] - \cos[kvt]\right\}$$
$$= -0.125 \cos(50t) [\sin(20t + 2) - \sin(20t)]$$
$$- 0.05 \sin(50t) [\cos(20t + 2) - \cos(20t)] \quad (V).$$

The first term on the right side is produced by the time-varying magnetic field, while the second one results from the motion of the loop in the magnetic field.

8.19 Figure 8.50 shows a magnetic core and two conducting coils with N_1 and N_2 turns, respectively. The mean magnetic path length is also shown. The core is made of two linear materials with high permeabilities μ_{r1} and μ_{r2}, respectively.

Fig. 8.50 Core made of two magnetic materials around which two coils are wound

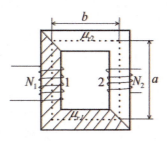

The cross-section is a square of side c. (a) Find the mutual inductance between the two coils. (b) If the coil labelled Coil 1 carries a slowly varying current $I_1 = 0.5\cos(100\pi t)$ (A), find the open-circuit voltage that appears between the terminals of the coil labelled Coil 2.
Numerical values: $a = 15\,\text{cm}; b = 10\,\text{cm}; c = 5\,\text{cm}; N_1 = 100; N_2 = 50; \mu_{r1} = 1000$, and $\mu_{r2} = 500$.

Solution

(a) It is assumed that the entire magnetic field is confined within the core. The magnetic flux through the cross-section of the core can be approximately expressed as $\Phi \approx BS$. As the core has the same cross sectional area throughout its length, the magnetic field B in the core is constant, i.e. $B = B_1 = B_2$. However, the H-field changes due to the two different constitutive equations of the core materials: $B_1 = \mu_0\mu_{r1}H_1$ and $B_2 = \mu_0\mu_{r2}H_2$. By assuming a current intensity I_1 in Coil 1 and applying Ampère's law:

$$H_1 l_1 + H_2 l_2 = N_1 I_1 \Rightarrow \frac{B}{\mu_0\mu_{r1}} l_1 + \frac{B}{\mu_0\mu_{r2}} l_2 = N_1 I_1,$$

where the total length of the path is $l_1 + l_2$, and $l_1 = l_2 = a + b$. Therefore, the magnetic field in the core becomes

$$B = \frac{\mu_0\mu_{r1}\mu_{r2}N_1 I_1}{(a+b)(\mu_{r1}+\mu_{r2})}.$$

For one turn in Coil 2, the magnetic flux is $\Phi_{21,\text{one}} = BS = Bc^2$, and the total flux trough Coil 2 due to the current intensity I_1 can be expressed as

$$\Phi_{21} = N_2 \Phi_{21,\text{one}} = \frac{N_1 N_2 \mu_0 \mu_{r1} \mu_{r2} c^2 I_1}{(a+b)(\mu_{r1}+\mu_{r2})}.$$

Fig. 8.51 Electromotive force is induced in Coil 2 as a result of a time-dependent current flowing through Coil 1

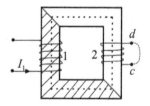

Finally, the mutual inductance is

$$M = \frac{d\Phi_{21}}{dI_1} = \frac{N_1 N_2 \mu_0 \mu_{r1} \mu_{r2} c^2}{(a+b)(\mu_{r1} + \mu_{r2})}$$
$$= \frac{100 \times 50 \times 4\pi \times 10^{-7} \times 1000 \times 500 \times 0.05^2}{(0.15 + 0.10)(1000 + 500)} = 0.021 \text{ (H)}.$$

(b) The current intensity flowing in Coil 1 is $I_1 = I_0 \cos(\omega t) = 0.5 \cos(100\pi t)$ (A), whereas in Coil 2 the current intensity is $I_2 = 0$. Then, the e.m.f. evaluated along the closed path formed by Coil 2 and the mathematical external line (dc) between its terminals, as shown in Fig. 8.51, is given by

$$\mathcal{E}_2 = \oint_\Gamma \mathbf{E}_e \cdot d\mathbf{l} = -\frac{d\Phi_{22}}{dt} = -M\frac{dI_1}{dt} = \frac{N_1 N_2 \mu_0 \mu_{r1} \mu_{r2} c^2 I_0 \omega \sin(\omega t)}{(a+b)(\mu_{r1} + \mu_{r2})}.$$

The e.m.f. can also be calculated as

$$\mathcal{E}_2 = \oint_\Gamma \mathbf{E}_e \cdot d\mathbf{l}_2 = \int_{\text{coil 2}} \mathbf{E}_e \cdot d\mathbf{l}_2 + \int_{\text{ext. line }(dc)} \mathbf{E}_e \cdot d\mathbf{l}_2 = 0 + \int_{\text{ext. line }(dc)} -\nabla V \cdot d\mathbf{l}_2$$
$$= \int_d^c -\nabla V \cdot d\mathbf{l}_2 = V_d - V_c = V_{dc},$$

where it is assumed that the electric field in Coil 2 is $\mathbf{E}_e = 0$ (the wire is a perfect conductor) and the magnetic field outside the core is negligible. Then, the voltage between the terminals of Coil 2 is equal

$$V_{dc} = \mathcal{E}_2 = \frac{N_1 N_2 \mu_0 \mu_{r1} \mu_{r2} c^2 I_0 \omega \sin(\omega t)}{(a+b)(\mu_{r1} + \mu_{r2})} = 3.29 \sin(100\pi t) \text{ (V)}.$$

8.20 The system shown in Fig. 8.52a consists of a long straight wire, w, and a closed square loop of wire with sides of length $b = 10$ cm, $N = 100$ turns, and a resistance of $R = 30\,\Omega$. The loop is placed in the XZ-plane at a distance a from the straight wire. If the straight wire carries a current intensity $I(t) = 3\left[1 - \exp(-0.1t)\right]$ (A) and $a = 5$ cm, find: (a) the e.m.f. induced in the square loop at the instant $t = 3$ s. (b) Mutual inductance between the straight wire and the square loop in this situation.

Fig. 8.52 a The straight wire and the square coil with N turns. b In this case, the square loop rotates with angular velocity Ω and the loop is located a long distance from the wire

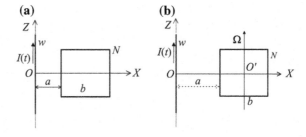

Fig. 8.53 The e.m.f. is calculated following the direction $d\mathbf{l}$, the surface element being $d\mathbf{S} = bdx\,\mathbf{u}_y$

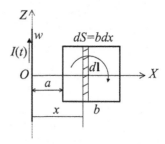

In a new situation, when a long time has elapsed and the distance between the loop and the wire is such that $a \gg b$, the loop begins to rotate with an angular velocity $\Omega = (0, 0, 200)\,\text{s}^{-1}$, around the symmetry axis parallel to the wire and passing through the centre O', as shown in Fig. 8.52b. (c) Calculate the current intensity induced in the square loop.

Solution

(a) In the plane of the Fig. 8.53, the magnetic field produced by the straight wire depends on the distance x to the wire and is given by:

$$\mathbf{B} = \frac{\mu_0 I}{2\pi x}\,\mathbf{u}_y.$$

The magnetic flux through one coil;

$$\Phi_{\text{one}} = \int_S \mathbf{B}\cdot d\mathbf{S} = \int_S (0, B, 0)\cdot(0, dS, 0) = \int_S B dS = \int_a^{a+b}\frac{\mu_0 I}{2\pi x}\,bdx = \frac{\mu_0 I b}{2\pi}\ln\frac{(a+b)}{a},$$

and the magnetic flux through the complete square loop becomes,

$$\Phi = N\Phi_{\text{one}} = \frac{\mu_0 NIb}{2\pi}\ln\frac{(a+b)}{a}.$$

Fig. 8.54 A view in the XY-plane of the rotating square loop at an instant t. As $a \gg b$, the distance from O' to O, $a + b/2$, is approximately equal to a

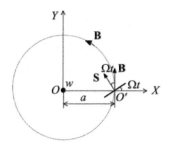

The e.m.f. obtained is therefore

$$\mathcal{E} = -\frac{d\Phi}{dt} = -\frac{\mu_0 Nb}{2\pi} \ln\frac{(a+b)}{a} \frac{dI}{dt} = -\frac{\mu_0 Nb}{2\pi} \ln\frac{(a+b)}{a}[-3(-0.1)\exp(-0.1t)]$$

$$= -\frac{4\pi \times 10^{-7} \times 100 \times 0.10}{2\pi} \ln\frac{0.05+0.10}{0.05} \times 0.3 \times \exp(-0.1 \times 3) = -4.88 \times 10^{-7} \quad (V).$$

(b) In order to calculate M, we evaluate the effect of the field created by the current in the straight wire on the coil. The magnetic flux through the coil has been already calculated. Then, we have for the mutual inductance

$$M = \frac{d\Phi_{21}}{dI_1} = \frac{\mu_0 Nb}{2\pi} \ln\frac{(a+b)}{a} = \frac{4\pi \times 10^{-7} \times 100 \times 0.10}{2\pi} \ln\frac{0.05+0.10}{0.05} = 2.20 \times 10^{-6} \quad (H).$$

(c) In this case, with $t \to \infty$, intensity tends to $I \to 3$ (A). As $a \gg b$, the current through the wire creates a magnetic field in the square coil that can be assumed to be approximately uniform. Under these assumptions, the magnetic field in the coil is $\mathbf{B} = (\mu_0 I)/(2\pi a)\, \mathbf{u}_y$, and at an instant t the magnetic flux through the coil results to be (see Fig. 8.54):

$$\Phi_{one} = \int_S \mathbf{B} \cdot d\mathbf{S} \simeq \mathbf{B} \cdot \mathbf{S} = BS\cos(\Omega t) = \frac{\mu_0 I}{2\pi a} b^2 \cos(\Omega t),$$

$$\Phi = N\Phi_{one} = \frac{\mu_0 NI}{2\pi a} b^2 \cos(\Omega t).$$

Finally, neglecting self-inductance, the e.m.f. and the induced intensity can be expressed as:

$$\mathcal{E} = -\frac{d\Phi}{dt} = \frac{\mu_0 INb^2}{2\pi a}\Omega\sin(\Omega t) \Rightarrow I_{ind} = \frac{\mathcal{E}}{R},$$

$$I_{ind} = \frac{\mu_0 INb^2}{2\pi aR}\Omega\sin(\Omega t) = \frac{4\pi \times 10^{-7} \times 3 \times 100 \times 0.10^2 \times 200}{2\pi \times a \times 30}\sin(200t)$$

$$= \frac{0.4 \times 10^{-5}}{a}\sin(200t) \quad (A).$$

Fig. 8.55 Two long straight wires in the XY-plane and a square loop lying in this plane

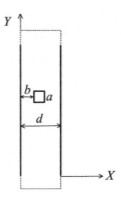

8.21 A rectangular loop is formed by two long straight wires and two short wire segments connected between their ends (see Fig. 8.55). The distance between the two long wires is d. A square loop with sides of length $a = d/4$ and a resistance R lies in the same plane as the two long wires, between them and at a distance $b = d/6$ from the left long wire, as shown in Fig. 8.55. (a) Find the mutual inductance between the two loops. (b) If the square loop is at rest, in the position shown in the figure, and the current carried by the long wires is $I = I_0 \cos(\omega t)$, find the e.m.f. induced in the square loop. (c) If the current flowing in the long straight wires is $I = 1$ A and the square loop begins to move at constant velocity $\mathbf{v} = v\,\mathbf{u}_x$, find the induced current in the square loop when its center is equidistant from the long parallel wires.

Solution

(a) A current, I_1, is made to flow through the long straight wires (see Fig. 8.56). The magnetic field of the two short straight wires connecting the ends of the long wires is considered to be negligible. Therefore, the flux through the square loop to be considered is only that due to the current flowing through the long wires. The magnetic fields produced by the left and right straight wires carrying the current I_1, at a point located between them, and at a distance x from the left wire are, respectively,

$$\mathbf{B}_l = \frac{\mu_0 I_1}{2\pi x}(-\mathbf{u}_z) \quad \text{and} \quad \mathbf{B}_r = \frac{\mu_0 I_1}{2\pi(d-x)}(-\mathbf{u}_z).$$

Following the square loop in a clockwise direction, the corresponding surface element is $d\mathbf{S} = -a\,dx\,\mathbf{u}_z$. Then, the magnetic fluxes through the square loop produced by $\mathbf{B}_l$ and $\mathbf{B}_r$ are given by

$$\Phi_l = \int_b^{b+a} \frac{\mu_0 I_1}{2\pi x} a\,dx = \frac{\mu_0 I_1 a}{2\pi} \ln x \big|_b^{b+a} = \frac{\mu_0 I_1 a}{2\pi} \ln\left(\frac{b+a}{b}\right) = \frac{\mu_0 I_1 a}{2\pi} \ln \frac{5}{2},$$

Fig. 8.56 The two long straight wires, circuit denoted 1, and the square loop, circuit denoted 2

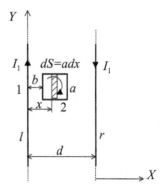

$$\Phi_r = \int_b^{b+a} \frac{\mu_0 I_1}{2\pi(d-x)} a dx = -\frac{\mu_0 I_1 a}{2\pi} \ln(d-x)\Big|_b^{b+a} = -\frac{\mu_0 I_1 a}{2\pi} \ln\left(\frac{d-(b+a)}{d-b}\right)$$
$$= -\frac{\mu_0 I_1 a}{2\pi} \ln\frac{7}{10}.$$

The resulting flux through the square loop results in

$$\Phi_{21} = \Phi_1 + \Phi_r = \frac{\mu_0 I_1 a}{2\pi} \ln\left(\frac{b+a}{b}\right) - \frac{\mu_0 I_1 a}{2\pi} \ln\left(\frac{d-(b+a)}{d-b}\right) = \frac{\mu_0 I_1 a}{2\pi} \ln\left[\frac{(b+a)(d-b)}{b(d-(b+a))}\right]$$
$$= \frac{\mu_0 I_1 a}{2\pi} \ln\frac{25}{7} = \frac{\mu_0 I_1 d}{8\pi} \ln\frac{25}{7}.$$

The mutual inductance is finally obtained as

$$M = \frac{\Phi_{21}}{dI_1} = \frac{\mu_0 d}{8\pi} \ln\left[\frac{(b+a)(d-b)}{b(d-(b+a))}\right] = \frac{\mu_0 d}{8\pi} \ln\frac{25}{7} \quad (H).$$

(b) The induced e.m.f. $\mathcal{E}$ in the square loop (see Fig. 8.57), produced by the current $I = I_0 \cos \omega t$ flowing through the long straight wires, can be expressed as

$$\mathcal{E} = -\frac{d\Phi}{dt} = -M\frac{dI}{dt} = -\frac{\mu_0 d}{8\pi} \ln\frac{25}{7} \frac{dI}{dt} = \frac{\mu_0 d}{8\pi} \ln\frac{25}{7} I_0 \omega \sin(\omega t) \quad (V).$$

(c) The induced electromotive force in the square loop caused by the motion of the loop in the non-uniform magnetic field produced by the current $I = 1$ (A) can be calculated by applying (8.11) or (8.12). By applying (8.11) when the square loop is located at the position shown in Fig. 8.58, it is found

$$\mathcal{E} = \oint_{ijkli} (\mathbf{v} \times \mathbf{B}) \cdot d\mathbf{l} = \int_{ij} (\mathbf{v} \times \mathbf{B}) \cdot d\mathbf{l} + \int_{jk} (\mathbf{v} \times \mathbf{B}) \cdot d\mathbf{l} + \int_{kl} (\mathbf{v} \times \mathbf{B}) \cdot d\mathbf{l} + \int_{li} (\mathbf{v} \times \mathbf{B}) \cdot d\mathbf{l}.$$

On sides labelled jk and li, the cross product $(\mathbf{v} \times \mathbf{B})$ is perpendicular to the corresponding longitudinal differential element. Thus, the line integrals along sides jk and li are equal to zero. Due to the symmetry conditions, on side ij the resulting

Fig. 8.57 The straight wires carry a current $I = I_0 \cos \omega t$ and the induced e.m.f. around the square loop is calculated following a counterclockwise direction

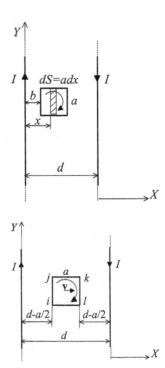

Fig. 8.58 The square loop with its center equidistant from the long parallel wires moves with constant velocity **v**. The current flowing through the long straight wires is $I = 1$ A

magnetic field $\mathbf{B}_{ij} = \mathbf{B}_{ij,1} + \mathbf{B}_{ij,r}$ is equal to $\mathbf{B}_{kl} = \mathbf{B}_{kl,1} + \mathbf{B}_{kl,r}$. As the path direction along side ij is opposite to that of side kl, the resulting line integral is zero. Then, the e.m.f. induced in the loop at the position shown in Fig. 8.58 is zero, as is the current induced in the square loop.

8.22 Two conducting rails with negligible resistance, separated by a distance $D = 1$ m, are connected through a fixed bar ab with a resistance of 6 Ω. There are two conducting bars cd and ef of resistance 8 Ω and 12 Ω, respectively, sliding frictionlessly on the parallel rails at constant velocities $v_{cd} = 6$ m/s and $v_{ef} = 8$ m/s (in direction parallel to the rails, and the bars moving away from ab). The angles that cd and ef make with the normal to the rails are $\alpha = 30°$ and $\beta = 45°$, respectively. Figure 8.59 shows the initial position of the two bars, the initial distances being $d_0 = 20$ cm and $f_0 = 15$ cm. There is a homogeneous and stationary magnetic field of magnitude $B = 0.2$ T, perpendicular to the plane of the figure and pointing inward. Find the current intensity flowing in bar ab at a given instant t.

Solution

Electromotive force is induced by the motion of bars cd and ef in the magnetic field, and hence the current in bar ab results. By calculating the e.m.f. around the loops labelled 1 and 2, see Fig. 8.60, following the clockwise and counterclockwise directions, respectively, it is found at a given instant t:

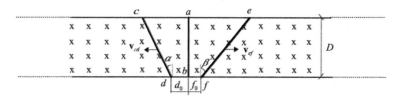

Fig. 8.59 Two parallel rails on which two bars cd and ef slide at constant velocity $\mathbf{v}_{cd}$ and $\mathbf{v}_{ef}$, respectively, in the presence of a uniform magnetic field pointing into the page

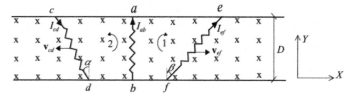

Fig. 8.60 Sketch showing, at an instant t, the direction assumed in calculating e.m.f. and the directions of the current intensities

$$\mathcal{E}_1 = \oint_{\Gamma_1} (\mathbf{v} \times \mathbf{B}) \cdot d\mathbf{l} = \int_{ef} [(v_{ef}, 0, 0) \times (0, 0, -B)] \cdot (dx, dy, 0)$$

$$= \int_e^f (0, v_{ef}B, 0) \cdot (dx, dy, 0) = \int_D^0 v_{ef}B\,dy = -v_{ef}BD,$$

$$\mathcal{E}_2 = \oint_{\Gamma_2} (\mathbf{v} \times \mathbf{B}) \cdot d\mathbf{l} = \int_{cd} [(-v_{cd}, 0, 0) \times (0, 0, -B)] \cdot (dx, dy, 0)$$

$$= \int_c^d (0, -v_{cd}B, 0) \cdot (dx, dy, 0) = \int_D^0 -v_{cd}B\,dy = v_{cd}BD.$$

Kirchhoff's laws, the first law applied to the node labelled a and the second law to loops 1 and 2, become:

$$\left. \begin{array}{l} I_{ab} + I_{ef} = I_{cd} \\ \mathcal{E}_1 = R_{ab}I_{ab} - R_{ef}I_{ef} \\ \mathcal{E}_2 = R_{ab}I_{ab} + R_{cd}I_{cd} \end{array} \right\}$$

By solving the equations set, it is found for I_{ab}:

$$I_{ab} = \frac{\mathcal{E}_1 R_{cd} + \mathcal{E}_2 R_{ef}}{R_{ab}R_{cd} + R_{ef}R_{cd} + R_{ab}R_{ef}} = \frac{(-v_{ef}BD)R_{cd} + (v_{cd}BD)R_{ef}}{R_{ab}R_{cd} + R_{ef}R_{cd} + R_{ab}R_{ef}}$$

$$= \frac{(-8 \times 0.2 \times 1)8 + (6 \times 0.2 \times 1)12}{6 \times 8 + 12 \times 8 + 6 \times 12} = 7.4 \times 10^{-3} \quad (A).$$

8.23 Figure 8.61 reminds us of Faraday's experiment to generate electric current from the flowing water in the River Thames. This figure shows a sketch of a river of width w and two rectangular electrodes, with side lengths a and b, which are

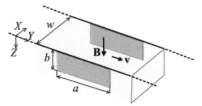

Fig. 8.61 A simple sketch of a river and the two electrodes used in the experiment to generate electric current

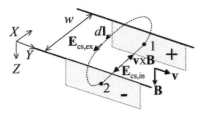

Fig. 8.62 The closed path around which the e.m.f. is calculated: 21 (between the electrodes) and 12 (exterior). As a result of the the term $\mathbf{v} \times \mathbf{B}$, positive and negative charges are accumulated on the electrodes, producing a electric field denoted $\mathbf{E}_{cs,in}$ (water) and $\mathbf{E}_{cs,ex}$ (exterior)

placed on both sides of the river. The velocity of the river is $\mathbf{v}$ and the component of the Earth's magnetic field in the region between the electrodes is $\mathbf{B}$, as shown in the figure. (a) Find the voltage between the two electrodes (open-circuit voltage). (b) If a resistance R is connected to the electrodes, find the current intensity I flowing through the resistance. The end effect of the electrodes can be neglected. (c) For the particular case: $R = 1\,\Omega$, $a = 190$ m, $b = 4.00$ m, $w = 18$ m, $v = 3$ m/s, $B = 0.50\,\mu\text{T}$, resistivity of water $\rho = 100\,\Omega\text{m}$, find the current intensity in the resistance.

Solution

(a) The term $\mathbf{v} \times \mathbf{B}$ causes charges to be accumulated on the electrodes until the field $\mathbf{E}_{cs}$ (electrostatic) produced by them balances the field $\mathbf{v} \times \mathbf{B}$. Then, between the electrodes $\mathbf{E}_{e,in} = \mathbf{v} \times \mathbf{B} + \mathbf{E}_{cs,in} = 0$. In the exterior, the resulting field is only due to the accumulation of charges, $\mathbf{E}_{e,ex} = \mathbf{E}_{cs,ex} = -\nabla V$ (see Fig. 8.62).

The e.m.f. around the closed path shown in Fig. 8.62, 21 (water)-12 (exterior), is

$$\mathcal{E} = \oint_{212} \mathbf{E}_e \cdot d\mathbf{l} = \oint_{212} (\mathbf{v} \times \mathbf{B}) \cdot d\mathbf{l} = \int_{21(\text{water})} (\mathbf{v} \times \mathbf{B}) \cdot d\mathbf{l} + \int_{12(\text{exterior})} (\mathbf{v} \times \mathbf{B}) \cdot d\mathbf{l}$$

$$= \int_{21(\text{water})} [(0, v, 0) \times (0, 0, B)] \cdot (dx, 0, 0) = \int_2^1 (vB, 0, 0) \cdot (dx, 0, 0) = \int_0^w vB\,dx = vBw.$$

Fig. 8.63 When the circuit is closed by a resistance R the current intensity I flows

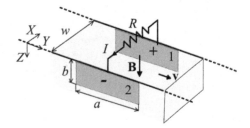

This equation can also be written as

$$\mathcal{E} = \oint_{212} \mathbf{E}_e \cdot d\mathbf{l} = vBw = \int_{21(\text{water})} \mathbf{E}_{e,\text{in}} \cdot d\mathbf{l} + \int_{12(\text{exterior})} \mathbf{E}_{e,\text{ex}} \cdot d\mathbf{l}$$

$$= 0 + \int_1^2 -\nabla V \cdot d\mathbf{l} = V_1 - V_2 = V_0 \Rightarrow V_0 = vBw,$$

where V_0 is the voltage between the electrodes (open-circuit).

(b) If the circuit is closed by a load resistance R connected to the electrodes as shown in Fig. 8.63, the charges escape from the electrodes producing a current in the resistance. Due to the unbalance, current flows between the electrodes, the current density in the water being $\mathbf{j} = \sigma(\mathbf{v} \times \mathbf{B} + \mathbf{E}_{cs,\text{in}})$, where σ is the conductivity of water.

By neglecting the end effects of the electrodes, the conservative electric field due to the accumulation of charges can be considered uniform, $\mathbf{E}_{cs,\text{in}} = E_{cs,\text{in}}(-\mathbf{u}_x)$. Then, the current density in the water can be expressed as

$$\mathbf{j} = \sigma[(\mathbf{v} \times \mathbf{B}) + \mathbf{E}_{cs,\text{in}}] = \sigma(vB - E_{cs,\text{in}})\mathbf{u}_x.$$

The new voltage V between the electrodes can be calculated as

$$V = \int_1^2 \mathbf{E}_{cs,\text{in}} \cdot d\mathbf{l} = E_{cs,\text{in}}w.$$

When a steady-state condition is reached, the current intensity I in the water is equal to the current flowing through the resistance. The current I through the water between the electrodes, with a section $S = ab$, is given by

$$I = \int_S \mathbf{j} \cdot d\mathbf{S} = \mathbf{j} \cdot \mathbf{S} = (\sigma(vB - E_{cs,\text{in}}), 0, 0) \cdot (ab, 0, 0) = \sigma(vB - E_{cs,\text{in}})ab,$$

where $\mathbf{j}$ is assumed to be uniform. The same current intensity flows through the resistance, which can be expressed in terms of the voltage between its terminals, $I = V/R$. The conservative field can be easily obtained by substituting $V = E_{cs,\text{in}}w$ into this equation for I,

Fig. 8.64 A simple sketch of an ideal transformer. Note that the secondary current is flowing in such a direction as to oppose the flux change due to I_1

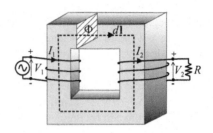

$$I = \frac{V}{R} = \frac{E_{cs,in} w}{R} \Rightarrow E_{cs,in} = \frac{RI}{w}.$$

Finally, substitution of $E_{cs,in}$ into the above expression of I in the river yields

$$I = \sigma \left(vB - \frac{RI}{w} \right) ab \Rightarrow I = \frac{\sigma abvBw}{w + \sigma abR}.$$

The problem can also be solved by evaluating the e.m.f. around the closed path 21(water)-12(resistance), see Fig. 8.63, and applying Kirchhoff's second law,

$$\mathcal{E} = vBw = (R + R_w)I \Rightarrow I = \frac{\sigma abvBw}{w + \sigma abR},$$

where the resistance of the water between the electrodes is $R_w = (1/\sigma)(w/ab)$.
(c) By substituting the values given in the expression obtained for I, it is found that $I = 8\,\mu A$, resulting in a very small value to be measured.

8.24 Figure 8.64 shows a sketch of an ideal transformer with two coils coupled magnetically through a ferromagnetic core with a mean path length ℓ and a uniform cross-section S. The primary coil, with N_1 turns, is connected to a source of e.m.f. with a voltage V_1 across its terminals. The secondary coil has N_2 turns and is connected to a load resistance R. Find: (a) the ratio of output voltage V_2 to the input voltage V_1, (b) the ratio of the currents in the primary and secondary coils, and (c) the value of the resistance seen by the source connected to the primary circuit.

Solution

(a) The varying current in the primary coil creates a varying magnetic flux confined to the core that produces a voltage in the secondary coil as a result of Faraday's law. If a load is connected to the secondary coil, electric current will flow through the load and therefore electric energy will be transferred from the primary circuit to the secondary circuit.

We assume that the transformer is ideal: the magnetic coupling between the coils is perfect, there is no dissipation of energy, and the core has a very high permeability (but it is not saturated). Then, the self-inductance of each coil is enormous, there is no flux leakage, and the resistance of the coils is negligible. Moreover, hysteresis and eddy-currents are also assumed to be negligible.

Under these assumptions, the same flux, Φ, passes through each turn of both coils, and an e.m.f., $-d\Phi/dt$, is induced around each turn. Since there is no resistance, the e.m.f. $\mathcal{E}_1$ produced by the generator is equal to the voltage V_1 across its terminals. According to Faraday's law, we obtain

$$V_1 - N_1 \frac{d\Phi}{dt} = 0, \quad V_2 - N_2 \frac{d\Phi}{dt} = 0.$$

Then, it follows that

$$\frac{V_2}{V_1} = \frac{N_2}{N_1} = n.$$

Therefore, the ratio of output voltage of the secondary to the primary voltage is the turns ratio $n = N_2/N_1$.

(b) By applying Ampère's law, the line integral of the **H**-field along the closed dashed path shown in Fig. 8.64 gives

$$\oint_\Gamma \mathbf{H} \cdot d\mathbf{l} = N_1 I_1 - N_2 I_2.$$

As the core has a uniform cross-section, the magnetic field is nearly uniform over any cross-section, and has the average magnitude $B = \Phi/S$. The magnetic field **B** is related to **H** by $\mathbf{B} = \mu\mathbf{H}$. Then,

$$N_1 I_1 - N_2 I_2 = \oint_\Gamma \mathbf{H} \cdot d\mathbf{l} = \frac{\oint_\Gamma \mathbf{B} \cdot d\mathbf{l}}{\mu} = \frac{B\ell}{\mu} = \frac{\Phi}{\frac{\mu S}{\ell}} \approx 0,$$

where it has been taken into account that the magnetic permeability is very large ($\mu \to \infty$). The quotient $\mu S/\ell$ is called *reluctance* of the magnetic circuit. From the last equation, we find

$$\frac{I_2}{I_1} = \frac{N_1}{N_2} = \frac{1}{n}.$$

This result shows that the ratio of the currents in the primary and secondary coils is equal to the inverse of the turns ratio n. The same result is obtained by equating the power in the primary and secondary circuits since there is no energy loss in an ideal transformer, i.e. $P_1 = P_2 = V_1 I_1 = V_2 I_2$.

(c) In the secondary circuit, in the case studied with a load resistance R,

$$V_2 = RI_2,$$

and from the equations for the ratio of voltages and currents, we have

$$V_1 = V_2 \frac{N_1}{N_2} = RI_2 \frac{N_1}{N_2} = RI_1 \left(\frac{N_1}{N_2}\right)^2 \Rightarrow (R)_1 = \frac{V_1}{I_1} = R \left(\frac{N_1}{N_2}\right)^2,$$

i.e. the load $(R)_1$ seen by the source connected to the primary coil is R/n^2.

Chapter 9
Energy of the Electromagnetic Field

Abstract This chapter deals with the energy associated with the electric and magnetic fields. In all branches of Physics, the concept of energy is present. The introduction of the idea of energy in electromagnetism provides a way of understanding many different phenomena, such as the properties of magnetic matter at different temperatures, circuits and networks, thermal electromagnetic radiation of bodies, and electromechanical machines. Due to the wide range of applications we can find, we have chosen for this chapter a viewpoint as simple as possible but, at the same time, accurate in order to explain the most important ideas involved.

9.1 The Electrostatic Energy of Charges

Consider a system formed exclusively by two electrically charged particles, one of which is fixed to a point P_1 in space and with charge q_1, and the other located at P_2 and with charge q_2. The distance between the two points is r_{12}. Suppose that both charges are positive. Since like charges repel, q_1 applies a repulsive force $\mathbf{F}$ on q_2 that causes it to accelerate and to attain the consequent kinetic energy. In order to perform the following demonstration, the device shown in Fig. 9.1 is formed. The charge q_1 is fixed. The second charge is attached to a thread which passes over a pulley and ends in a weight whose value is continually adjusted to maintain balance. The block is external to the electrical system and applies the force $\mathbf{F}_{ex}$. The balance implies that $\mathbf{F}_{ex} = -\mathbf{F}$ and that the magnitude of the force $F_b = F$. When lightening a small fraction of the block, it moves upwards by a small amount $d\mathbf{r}'$ such that $dr' = dr$: hence the charge q_2 applies work of $\mathbf{F}_b.d\mathbf{r}' = \mathbf{F}.d\mathbf{r}$ on the block. Therefore, the two-charge system has potential energy U_e measurable by the work that can be carried out on an external system: in this case, the block.

The work of the electrostatic force in the displacement $d\mathbf{r}$ is

$$dW = \mathbf{F}.d\mathbf{r} = Fdr = \frac{q_1 q_2}{4\pi\varepsilon_0 r_{12}^2} dr_{12}. \tag{9.1}$$

Fig. 9.1 Experimental device to evaluate the electrostatic energy

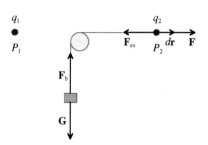

The total work to move q_2 away towards infinity is the electrostatic energy of the two-charge system:

$$U_e \equiv W = \int_{P_2}^{\infty} dW = \int_{r_{12}}^{\infty} \frac{q_1 q_2}{4\pi\varepsilon_0 r_{12}^2} dr_{12} = \frac{q_1 q_2}{4\pi\varepsilon_0} \int_{r_{12}}^{\infty} \frac{dr_{12}}{r_{12}^2}$$
$$= \frac{q_1 q_2}{4\pi\varepsilon_0}\left[-\frac{1}{r_{12}}\right]_{r_{12}}^{\infty} = \frac{q_1 q_2}{4\pi\varepsilon_0 r_{12}}. \qquad (9.2)$$

It can be seen that this work is the same as that which force $\mathbf{F}_{ex}$, originated by the block, would have to apply to q_2 for it to approach P_2 from infinity.

The (9.2) can be written thus

$$U_e = q_1 \frac{q_2}{4\pi\varepsilon_0 r_{12}} = q_1 V_1 = q_2 \frac{q_1}{4\pi\varepsilon_0 r_{12}} = q_2 V_2 = \frac{1}{2}(q_1 V_1 + q_2 V_2), \qquad (9.3)$$

that is, the potential energy of the pair of charges is the charge of one of them times the electrical potential that the other creates at the position that the first occupies, or the semi-sum of each charge times the potential at its point, due to the other charge.

Once the two-charge system is established, if it is wished to introduce a third charge, then it must be taken into account that the resulting force that is applied on it is the vectorial sum of those exerted by the first two, and it is therefore possible to reach the conclusion that the energy (electrostatic energy) of a system of N discrete point charges is

$$U_e = \sum_{i=1}^{N-1} \sum_{j=i+1}^{N} \frac{q_i q_j}{4\pi\varepsilon_0 r_{ij}} = \frac{1}{2}\sum_{i=1}^{N} q_i V_i, \qquad (9.4)$$

where V_k, the electrical potential at q_k, is caused by all the other charges.

9.2 The Energy of a Capacitor

Consider a capacitor during the charging process. In a certain intermediate state, the charge of the plate 1 is q, that of plate 2 is $-q$, and the difference of potential $V_1 - V_2$ is V. In order to increase the charge from q to $q + dq$, a charge dq is taken from

9.2 The Energy of a Capacitor

the second plate and given to the other. The required work is $dW = V\,dq = q\,dq/C$. The total energy that the capacitor acquires when its charge changes from zero to a certain final value q is

$$U_e = \int_{q=0}^{q=q} dW = \int_{q=0}^{q=q} \frac{q}{C} dq = \frac{1}{2}\frac{q^2}{C} = \frac{1}{2}qV = \frac{1}{2}CV^2. \tag{9.5}$$

9.3 The Electrostatic Energy of Distributed Charges

If dealing with a system of distributed charges, the charge in an element of volume of space dv is ρdv and its potential energy, by analogy with (9.4), is

$$U_e = \frac{1}{2}\int_v \rho V\,dv, \tag{9.6}$$

where V is the potential at the point where the volume charge density is ρ and v is the volume of the region where the charge distribution exits.

For electrostatic systems containing dielectrics, a small change in the energy dU_e can be expressed in terms of field quantities **E** and the variation of **D**

$$dU_e = \int_v \mathbf{E}.d\mathbf{D}\,dv. \tag{9.7}$$

The total electrostatic energy U_e can be calculated by allowing **D** to be brought from an initial value null to its final value **D**.

In a linear isotropic dielectric medium, the electrostatic energy is expressed in terms of the fields as

$$U_e = \frac{1}{2}\int_v \mathbf{D}.\mathbf{E}\,dv. \tag{9.8}$$

This expression can be expressed by saying: where an electrostatic field exists there is a stored energy, or, the electrostatic field has the energy per unit volume

$$u_e = \frac{1}{2}\mathbf{D}.\mathbf{E}, \tag{9.9}$$

known as the electrostatic energy density u_e.

9.4 Relationship Between Force and Electrostatic Energy

If an axis OX is drawn in Fig. 9.1 with its origin in the charge q_1 and directed towards the charge q_2, this has a coordinate x, and (9.2) gives

$$U_e = \frac{q_1 q_2}{4\pi\varepsilon_0 r_{12}} = \frac{q_1 q_2}{4\pi\varepsilon_0 x} \quad \Rightarrow \quad \frac{\partial U_e}{\partial x} = -\frac{q_1 q_2}{4\pi\varepsilon_0 x^2} = -F_x. \quad (9.10)$$

This equation indicates the relationship between the x-component of the electrostatic force acting on the charge q_2 and the electrostatic energy. By generalizing this result, it is possible to arrive at the following theorem: The electrostatic force that is exerted on a charged body with constant charges is equal, except for the sign, to the gradient of the electrostatic energy, that is,

$$\mathbf{F} = -\nabla U_e. \quad (9.11)$$

This result allows the calculation of forces on objects if the expression of the electrostatic energy in terms of the coordinates is known. It is assumed that there is no contribution of energy from other systems and the charges are kept fixed.

9.5 Magnetostatic Energy of Quasi-stationary Currents

Consider the simple circuit shown in the Fig. 9.2. At the instant $t = 0$ the circuit is closed. The application of Ohm's law at instant t gives

$$\mathcal{E} - \frac{d\Phi}{dt} = RI. \quad (9.12)$$

Multiplying all the terms of this equation by I gives

$$\mathcal{E}I - I\frac{d\Phi}{dt} = RI^2 \quad \Rightarrow \quad \mathcal{E}I = I\frac{d\Phi}{dt} + RI^2. \quad (9.13)$$

In the second of these equations, the first term, $\mathcal{E}I$, represents the power injected into the circuit by the generator, the third term, RI^2, represents the power dissipated in the resistance, therefore, through the principle of conservation of energy, the second term $Id\Phi/dt$, must be the power P stored in the circuit, that is, the speed of change of a certain energy U_m:

$$P = I\frac{d\Phi}{dt} \quad \Rightarrow \quad \frac{dU_m}{dt} = I\frac{d\Phi}{dt} \quad \Rightarrow \quad dU_m = Id\Phi. \quad (9.14)$$

Therefore, by the fact of creating a current, an energy U_m is stored in this device. On the other hand, the coil creates a magnetic field, which is why there must be a relationship between the magnetic field and the energy.

The energy stored by the coil when current I flows through it and flux Φ is proportional to I, is

9.5 Magnetostatic Energy of Quasi-stationary Currents

Fig. 9.2 Circuit with resistance and autoinductance

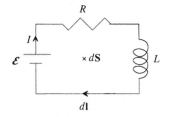

$$\int_0^{U_m} dU_m = \int_0^I I d\Phi = \int_0^I IL dI \quad \Rightarrow \quad U_m = \frac{1}{2}LI^2, \quad (9.15)$$

where L is the self-inductance.

9.6 Generalization

A toroidal solenoid is a circular ring on which a large number of turns of a wire are wound (Fig. 9.3); the magnetic field is confined in its interior. Consider the circumference of mean path length l, cross-sectional area of S, with N windings through which circulates a current of intensity I. Ampère's theorem gives

$$\oint \mathbf{H}.d\mathbf{l} = Hl = NI \quad \Rightarrow \quad H = \frac{NI}{l}. \quad (9.16)$$

Suppose there is a vacuum in the interior of the solenoid, which is magnetically linear ($B = \mu_0 H$), then the magnetic field **B** has the tangential component

$$B = \mu_0 H = \mu_0 \frac{NI}{l}. \quad (9.17)$$

Fig. 9.3 Toroidal solenoid

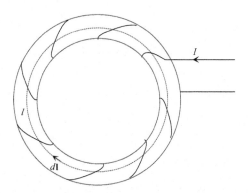

The flux through the whole electrical circuit is

$$\Phi = BNS = \mu_0 \frac{N^2 S}{l} I. \tag{9.18}$$

From (9.14), (9.16) and (9.18), the energy variation of the solenoid due to the variation of field B

$$dU_m = I d\Phi = INSdB = \frac{Hl}{N} NSdB = vHdB, \tag{9.19}$$

where v is the volume of the solenoid. Therefore the variation of energy stored per unit volume is

$$du_m = HdB. \tag{9.20}$$

Although the demonstration of (9.20) is for a very simple particular case, it can be demonstrated rigorously that it is valid for any case. Moreover the magnetic energy density variation caused by an infinitesimal variation of field **B** is

$$du_m = \mathbf{H}.d\mathbf{B}. \tag{9.21}$$

For a magnetically linear and isotropic material $\mathbf{B} = \mu_r \mu_0 \mathbf{H}$, the total magnetic energy density is given by

$$u_m = \int_0^B \mathbf{H}.d\mathbf{B} = \int_0^B HdB = \frac{1}{\mu_r \mu_0} \int_0^B BdB = \frac{1}{\mu_r \mu_0} \frac{1}{2} B^2 = \frac{1}{2} BH = \frac{1}{2} \mathbf{B}.\mathbf{H}. \tag{9.22}$$

Therefore, if there is an electric field and a magnetic field in a linear and isotropic material, then the energy density can be written, in agreement with (9.9) and (9.22), as

$$u = \frac{1}{2} \mathbf{D}.\mathbf{E} + \frac{1}{2} \mathbf{B}.\mathbf{H}. \tag{9.23}$$

9.7 Magnetic Energy in a Hysteresis Loop

Integrating by parts in the (9.21) between state 1 and state 2 gives

$$u_{m2} - u_{m1} = \int_{B_1}^{B_2} \mathbf{H}.d\mathbf{B} = \mathbf{H}.\mathbf{B}\big|_{B_1}^{B_2} - \int_{H_1}^{H_2} \mathbf{B}.d\mathbf{H}. \tag{9.24}$$

If a ferromagnetic material traces a hysteresis loop, where the initial and final states are equal, then the energy introduced per unit volume is

9.7 Magnetic Energy in a Hysteresis Loop

$$u_{mcicle} = -\oint \mathbf{B}.d\mathbf{H}. \qquad (9.25)$$

That is, this energy is introduced for each loop traced but, as the initial and final states are identical, the principle of conservation of energy indicates that when a ferromagnetic material is taken around a complete hysteresis cycle an energy per unit volume equal to the area of the loop is dissipated as heat.

Solved Problems

Problems A

9.1. A hydrogen atom is formed by a proton of charge 1.602×10^{-19} C and an electron of charge -1.602×10^{-19} C. Supposing that the separation between these two charges is 0.5×10^{-10} m, calculate the electrostatic energy of the atom.

Solution

As the charges and distances are known, Eq. (9.2) is applicable:

$$U_e = \frac{q_1 q_2}{4\pi\varepsilon_0 r_{12}} = -\frac{1.602 \times 10^{-19} \times 1.602 \times 10^{-19}}{4\pi 8.8542 \times 10^{-12} \times 0.5 \times 10^{-10}} \text{J} = -4.61 \times 10^{-18} \text{ J}.$$

9.2. Calculate the energy of a system of three particles, each with charge q and located on the vertices of an equilateral triangle of side L.

Solution

Suppose that the three charges are initially located on the vertices of the triangle. The work that the charge q_3 can give to exterior charges in its displacement $d\mathbf{r}$, with q_1 and q_2 remaining fixed, is, by application of the principle of superposition of forces and of (9.1),

$$dW = \mathbf{F}.d\mathbf{r} = (\mathbf{F}_1 + \mathbf{F}_2).d\mathbf{r} = \mathbf{F}_1.d\mathbf{r} + \mathbf{F}_2.d\mathbf{r} = \frac{q_1 q_3}{4\pi\varepsilon_0 r_{13}^2} dr_{13} + \frac{q_2 q_3}{4\pi\varepsilon_0 r_{23}^2} dr_{23},$$

where dr_{12} and dr_{13} are the components of the displacement $d\mathbf{r}$ in the directions of the respective forces. Applying (9.2) to the displacement of q_3 to infinity gives

$$U_{e3} = \frac{q_1 q_3}{4\pi\varepsilon_0 r_{13}} + \frac{q_2 q_3}{4\pi\varepsilon_0 r_{23}}.$$

Charge q_2 then moves to infinity, with q_1 remaining fixed. The energy provided by this charge is, in agreement with (9.2),

$$U_{e2} = \frac{q_1 q_2}{4\pi\varepsilon_0 r_{12}},$$

then the energy of the system of the three charges is

$$U_{e3} = \frac{q_1 q_2}{4\pi\varepsilon_0 r_{12}} + \frac{q_1 q_3}{4\pi\varepsilon_0 r_{13}} + \frac{q_2 q_3}{4\pi\varepsilon_0 r_{23}} = \frac{q_1 q_2 + q_1 q_3 + q_2 q_3}{4\pi\varepsilon_0 L},$$

where it has been taken into account that $r_{12} = r_{13} = r_{23} = L$.

Equation (9.4) could have been applied directly and would give

$$U_e = \sum_{i=1}^{3-1} \sum_{j=i+1}^{3} \frac{q_i q_j}{4\pi\varepsilon_0 r_{ij}} = \sum_{j=1+1}^{3} \frac{q_1 q_j}{4\pi\varepsilon_0 r_{1j}} + \sum_{j=2+1}^{3} \frac{q_2 q_j}{4\pi\varepsilon_0 r_{2j}}$$

$$= \frac{q_1 q_2}{4\pi\varepsilon_0 r_{12}} + \frac{q_1 q_3}{4\pi\varepsilon_0 r_{13}} + \frac{q_2 q_3}{4\pi\varepsilon_0 r_{23}},$$

which can be also expressed as

$$U_e = \frac{1}{2}\sum_{i=1}^{N} q_i V_i = \frac{1}{2}\left[q_1\left(\frac{q_2}{4\pi\varepsilon_0 r_{12}} + \frac{q_3}{4\pi\varepsilon_0 r_{13}}\right) + q_2\left(\frac{q_1}{4\pi\varepsilon_0 r_{12}} + \frac{q_3}{4\pi\varepsilon_0 r_{23}}\right) + q_3\left(\frac{q_1}{4\pi\varepsilon_0 r_{13}} + \frac{q_2}{4\pi\varepsilon_0 r_{23}}\right)\right]$$

$$= \frac{q_1 q_2}{4\pi\varepsilon_0 r_{12}} + \frac{q_1 q_3}{4\pi\varepsilon_0 r_{13}} + \frac{q_2 q_3}{4\pi\varepsilon_0 r_{23}}.$$

Note how this result agrees with the previous results.

9.3. A parallel-plate capacitor has a separation between the plates of $z = 5$ mm, the area of each plate is $200\,\text{cm}^2$, the gap is filled with a dielectric of relative permittivity 5 and the potential between the plates is 300 V. Calculate the energy that it accumulates.

Solution

Applying (9.5) and taking into account the formula of the capacitance of a parallel-plate capacitor gives

$$U_e = \frac{1}{2}CV^2 = \frac{1}{2}\frac{\varepsilon_r \varepsilon_0 S}{z}V^2$$

$$= \frac{1}{2}\frac{5 \times 8.854 \times 10^{-12} \times 200 \times 10^{-4}}{5 \times 10^{-3}} 300^2 \text{ J} = 7.969 \times 10^{-12}\text{ J}.$$

9.4. A system is formed by two flat, opposed plates each of area S, and separated by a small distance d. This system is introduced into a dielectric liquid of relative permittivity ε_r. The plates are connected to a battery of electromotive force V, and they slowly move apart due to the application of an exterior force $\mathbf{F}_e$ to one of plates until their separation is d'. Calculate: (a) the increase of energy of the system; (b) the work of the exterior force; (c) relate the force $\mathbf{F}$ from one plate on the other and the electrostatic energy of the capacitor.

Solution

(a) According to (9.5), the initial energy of the capacitor is given by

$$U_e = \frac{1}{2}CV^2 = \frac{1}{2}\frac{\varepsilon_r\varepsilon_0 S}{d}V^2.$$

The energy that the capacitor has when the distance between the plates is d' can be expressed as

$$U_e' = \frac{1}{2}\frac{\varepsilon_r\varepsilon_0 S}{d'}V^2,$$

therefore the increase of energy is

$$\Delta U_e = U_e' - U_e = \frac{1}{2}\frac{\varepsilon_r\varepsilon_0 S}{d'}V^2 - \frac{1}{2}\frac{\varepsilon_r\varepsilon_0 S}{d}V^2 = -\frac{1}{2}\varepsilon_r\varepsilon_0 SV^2\frac{d'-d}{dd'}.$$

Note that the energy of the capacitor has diminished, because $d' > d$.

(b) Capacitance decreases with the separation of the plates, but the electric voltage remains constant, therefore the charge, $q = CV$, decreases. The charge variation is

$$\Delta q = q' - q = C'V - CV = \frac{\varepsilon_r\varepsilon_0 S}{d'}V - \frac{\varepsilon_r\varepsilon_0 S}{d}V = -\varepsilon_r\varepsilon_0 S\frac{d'-d}{dd'}V.$$

Therefore, the energy that the battery provides is

$$W_{bat} = \Delta q.V = -\varepsilon_r\varepsilon_0 S\frac{d'-d}{dd'}V^2.$$

If W_{ex} is the work provided by the external force, then the principle of conservation of energy can be written thus:

$$W_{bat} + W_{ex} = \Delta U_e \Rightarrow -\varepsilon_r\varepsilon_0 S\frac{d'-d}{dd'}V^2 + W_{ex} = -\frac{1}{2}\varepsilon_r\varepsilon_0 SV^2\frac{d'-d}{dd'}$$
$$\Rightarrow W_{ex} = \frac{1}{2}\varepsilon_r\varepsilon_0 SV^2\frac{d'-d}{dd'}.$$

Note that both the work of the external force and the reduction of the energy of the capacitor have gone into increasing the energy of the battery, that is, in recharging the battery. Furthermore

$$W_{bat} = 2\Delta U_e.$$

(c) For a infinitesimal displacement dx under forces exterior F_e and interior F the principle of conservation of energy gives

$$dW_{bat} + dW_{ex} = dU_e \Rightarrow 2dU_e + F_e dx = dU_e \Rightarrow F_e dx = -dU_e = -\frac{\partial U_e}{\partial x}dx$$

$$F_e = -\frac{\partial U_e}{\partial x}.$$

In equilibrium, the electrostatic force is opposite to the external force, $F = -F_e$, therefore, the electric force between the two plates is

$$F = +\frac{\partial U_e}{\partial x}.$$

Note the plus sign in this case of constant potential. In (9.11), at constant charge, the sign was minus.

9.5. In a region of space that contains no matter, there is an electrostatic field $\mathbf{E} = ay\mathbf{u}_x$. Calculate the energy inside a cube of side L, which supports three edges on the positive part of the coordinate axes. Solve for: $a = 20000 \text{ V/m}^2$ and $L = 0.4$ m.

Solution

Application of (9.8) gives

$$U_e = \frac{\varepsilon_0}{2} \int_v \mathbf{E} \cdot \mathbf{E} dv = \frac{\varepsilon_0}{2} \int_{cube} ay\mathbf{i}.ay\mathbf{i} dv = \frac{a^2\varepsilon_0}{2} \int_{x=0}^{x=L} \int_{y=0}^{x=L} \int_{z=0}^{x=L} y^2 dxdydz$$

$$= \frac{a^2\varepsilon_0}{2} L^2 \frac{1}{3} L^3 = \frac{a^2\varepsilon_0}{6} L^5.$$

With the numeric data provided, this becomes

$$U_e = \frac{a^2\varepsilon_0}{6} L^5 = \frac{20000^2 \times 8.8542 \times 10^{-12}}{6} 0.4^5 \text{ J} = 60.44 \times 10^{-7} \text{ J}.$$

9.6. Calculate the energy of a flat capacitor with capacitance C and charge q by means of the application of (9.7).

Solution

The electric field outside the capacitor is dismissed by considering that the separation between the plates is small, and it is null within each plate since it is a conductor, therefore the only region of space to consider is that between the plates. Consider a rectangular parallelepiped, such as that shown in Fig. 9.4, that includes the inner face of the positive plate, of charge q; the flux of the electric field through its surface is

$$\oint_{paral.} \mathbf{E} \cdot d\mathbf{S} = \int_S EdS = ES,$$

Fig. 9.4 Capacitor and the Gaussian surface

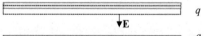

where S is the area of the plate. Applying Gauss's theorem to the parallelepiped gives

$$ES = \frac{q}{\varepsilon_0} \quad \Rightarrow \quad E = \frac{q}{\varepsilon_0 S}.$$

Applying (9.8) to the space between the plates, with a separation of z, gives

$$U_e = \frac{\varepsilon_0}{2} \int_v \mathbf{E}.\mathbf{E} dv = \frac{\varepsilon_0}{2} \frac{q}{\varepsilon_0 S} \frac{q}{\varepsilon_0 S} Sz = \frac{1}{2} \frac{q^2}{\varepsilon_0 S} z.$$

Remembering the formula of the capacity of the flat capacitor, $C = \varepsilon_0 S/z$, finally gives

$$U_e = \frac{1}{2} \frac{q^2}{C},$$

in agreement with (9.5).

9.7. Obtain the expression of the energy of a capacitor by applying (9.6).

Solution

Consider the box of volume v, drawn with dotted lines, in the diagram of the capacitor, shown in Fig. 9.5; all the charges of the capacitor are in its interior. Let q be the charge of the upper plate and V_1 its potential, and $-q$ and V_2 be those corresponding to the lower plate. The integral that appears in (9.6) can be broken down into three integrals, one extended to the space surrounding the interior face of the upper plate, the other extended to the lower plate, and the third to the space between the plates which has no charge:

$$U_e = \frac{1}{2} \int_v \rho V dv = \frac{1}{2} \int_{v1} \rho_1 V_1 dv + \frac{1}{2} \int_{v2} \rho_2 V_2 dv + \frac{1}{2} \int_v 0 V dv = \frac{1}{2} V_1 \int_{v1} \rho_1 dv + \frac{1}{2} V_2 \int_{v2} \rho_2 dv$$

$$= \frac{1}{2} V_1 q + \frac{1}{2} V_2 (-q) = \frac{1}{2} q V,$$

where $V = V_1 - V_2$, i.e. the potential difference between the potential of the upper plate and that of the lower plate.

Note: the exterior field $\mathbf{E}$ has not been considered because the distance between the plates is assumed to be small.

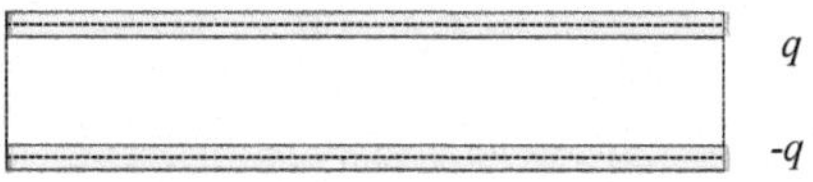

Fig. 9.5 Capacitor and a closed surface (*dotted lines*)

9.8. In the cube described in Problem 9.5, in addition to the indicated electric field, there is a uniform magnetic field of $\mathbf{B} = 0.001\,\mathbf{u}_x$ T. Calculate the magnetic energy stored inside the cube.

Solution

The energy contained in the cube due to the existence of the electric field is, as was seen in Problem 9.5,

$$U_e = \frac{a^2\varepsilon_0}{6}L^5 = 60.44 \times 10^{-7}\,\text{J}.$$

The magnetic energy is, according to (9.22),

$$U_m = \frac{1}{2}\int_v \mathbf{B}\cdot\mathbf{H}\,dv = \frac{1}{2}\int BH\,dv = \frac{1}{2}\int_v \frac{B^2}{\mu_0}\,dv$$

$$= \frac{1}{2}\frac{B^2}{\mu_0}L^3 = \frac{1}{2}\frac{0.001^2}{4\pi\times 10^{-7}}0.4^3\,\text{J} = 25.46\times 10^{-3}\,\text{J}.$$

Therefore, the total energy contained in the cube is

$$U = 60.44\times 10^{-7}\,\text{J} + 25.46\times 10^{-3}\,\text{J} = 25.47\times 10^{-3}\,\text{J}.$$

Problems B

9.9. A spherical region of space of radius R has a uniform charge density of ρ (Fig. 9.6). The charge is made in stages such that in an intermediate stage, when a sphere of radius r has been charged, then a layer of charges of thickness dr is added to it, and so on. Calculate: (a) The electric field E_e outside the sphere of radius r in the intermediate stage. (b) The work necessary to place the charged layer on the spherical surface of radius r. (c) The energy of the charged sphere of radius R. (d) The energy of this sphere if $R = 20\,\text{cm}$ and $\rho = 0.3\,\text{C/m}^3$.

Solution

(a) By applying Gauss's theorem to the spherical surface of radius r_e ($r_e > r$), the outer electric field $\mathbf{E}_e$ is obtained:

$$\oint \mathbf{E}_e\cdot d\mathbf{S} = \oint E_e\,dS = E_e 4\pi r_e^2 = \frac{q_{int}}{\varepsilon_0} = \frac{1}{\varepsilon_0}\rho\frac{4}{3}\pi r^3$$

$$\Rightarrow E_e = \frac{\rho r^3}{3\varepsilon_0}\frac{1}{r_e^2},$$

where it has been taken into account that the charge is located inside the sphere of radius r.

Fig. 9.6 Spherical region of charge density ρ

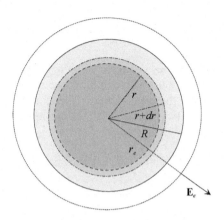

(b) As the charge of the layer is $dq = \rho dv = \rho 4\pi r^2 dr$, the work necessary to approach it from infinity is

$$dW = \int_{r_e=\infty}^{r_e=r} d\mathbf{F}_{ex}.d\mathbf{r}_e = -\int_{r_e=\infty}^{r_e=r} dq E_e dr_e = -\int_{r_e=\infty}^{r_e=r} \rho 4\pi r^2 dr \frac{\rho r^3}{3\varepsilon_0}\frac{1}{r_e^2} dr_e$$

$$= -\frac{4\pi \rho^2}{3\varepsilon_0} r^5 dr \int_{r_e=\infty}^{r_e=r} \frac{dr_e}{r_e^2} = \frac{4\pi \rho^2}{3\varepsilon_0} r^5 dr \frac{1}{r} = \frac{4\pi \rho^2}{3\varepsilon_0} r^4 dr.$$

(c) The energy of the charged sphere is the sum of the work needed for the transport of all the layers from the first, $r = 0$, to the last, $r = R$:

$$U_e = \int_{r=0}^{r=R} \frac{4\pi \rho^2}{3\varepsilon_0} r^4 dr = \frac{4\pi \rho^2}{3\varepsilon_0}\frac{1}{5} R^5 = \frac{4\pi \rho^2 R^5}{15\varepsilon_0}.$$

(d) The application of this result to the data provided gives

$$U_e = \frac{4\pi \rho^2 R^5}{15\varepsilon_0} = \frac{4\pi 0.3^2 \times 0.20^5}{15 \times 8.854 \times 10^{-12}} \text{ J} = 2.725 \times 10^6 \text{ J}.$$

9.10. Calculate the energy of the charged sphere of the previous problem by applying (9.7).

Solution

The electric field $\mathbf{E}$ inside the sphere of radius r is radial, due to the symmetry.
Applying Gauss's theorem it is deduced:

$$\oint \mathbf{E}.d\mathbf{S} = \oint E dS = E 4\pi r^2 = \frac{q_{int}}{\varepsilon_0} = \frac{1}{\varepsilon_0}\rho \frac{4}{3}\pi r^3 \Rightarrow E = \frac{\rho r}{3\varepsilon_0}.$$

Applying this result in (9.8), gives, for the interior of the sphere of radius R,

$$U_{ein} = \frac{\varepsilon_0}{2}\int_v \mathbf{E}.\mathbf{E}dv = \frac{\varepsilon_0}{2}\int_v E^2 dv = \frac{\varepsilon_0}{2}\int_0^R \frac{\rho^2 r^2}{3^2 \varepsilon_0^2}4\pi r^2 dr = \frac{2\pi\rho^2 R^5}{45\varepsilon_0},$$

which does not agree with the result of the previous problem. Why? Equation (9.7) refers to the whole of the space where there is an electric field, not only to the zone where there are charges. Therefore the energy outside the sphere of radius R must be added. The electric field outside this sphere is

$$\oint \mathbf{E}_{ex}.d\mathbf{S} = \oint E_{ex}dS = E_{ex}4\pi r^2 = \frac{q_{int}}{\varepsilon_0} = \frac{1}{\varepsilon_0}\rho\frac{4}{3}\pi R^3 \Rightarrow E_{ex} = \frac{\rho R^3}{3\varepsilon_0 r_{ex}^2},$$

and the energy outside is

$$U_{eex} = \frac{\varepsilon_0}{2}\int_{vex} \mathbf{E}_{ex}.\mathbf{E}_{ex}dv = \frac{\varepsilon_0}{2}\int_{vex} E_{ex}^2 dv$$
$$= \frac{\varepsilon_0}{2}\frac{\rho^2 R^6}{3^2 \varepsilon_0^2}\int_{r_{ex}=R}^{r_{ex}=\infty} \frac{1}{r_{ex}^4}4\pi r_{ex}^2 dr_{ex} = \frac{2\pi\rho^2 R^5}{9\varepsilon_0}.$$

The total energy is

$$U_e = U_{ein} + U_{eex} = \frac{2\pi\rho^2 R^5}{45\varepsilon_0} + \frac{2\pi\rho^2 R^5}{9\varepsilon_0} = \frac{4\pi\rho^2 R^5}{15\varepsilon_0},$$

which coincides, as would be expected, with that of Problem 9.9.

9.11. The capacitor described in Problem 9.3 becomes disconnected from the generator that has charged it. It is wished to calculate the electrostatic force exerted by the plates. Take into account that if, once charged, an external force F is applied to move a plate a distance dz from the other, then the capacitance and the energy vary.

Solution

As the plates of the capacitor have charges of opposite signs, they are attracted to each other. In order to quantify the electrostatic force, we make use of energy methods.

The capacity of a flat capacitor is $C = \varepsilon S/z$, therefore its energy is, according to (9.5),

$$E_p = \frac{1}{2}\frac{q^2}{C} = \frac{1}{2}\frac{q^2}{\varepsilon S}z.$$

When the distance is increased by dz by the application of external force F, which is just the minimum force necessary to keep the plates of the capacitor in equilibrium, then the charge does not change upon being disconnected from the generator, although the energy increases, according to the equation above, by

$$dE_p = \frac{1}{2}\frac{q^2}{\varepsilon S}dz.$$

The energy is supplied to the capacitor by means of the external force F and is Fdz, which, by application of the principle of conservation of energy, must be equal to the increase of energy (given by the previous expression), that is:

$$Fdz = dE_p = \frac{1}{2}\frac{q^2}{\varepsilon S}dz,$$

and hence

$$F = \frac{1}{2}\frac{q^2}{\varepsilon S} = \frac{1}{2}\frac{q^2}{\varepsilon_r \varepsilon_0 S} = \frac{1}{2}\frac{C^2 V^2}{\varepsilon_r \varepsilon_0 S} = \frac{1}{2}\frac{\varepsilon_r \varepsilon_0 S V^2}{z^2}$$
$$= \frac{1}{2}\frac{5 \times 8.854 \times 10^{-12} \times 200 \times 10^{-4} \times 300^2}{(5 \times 10^{-3})^2} N = 1.59 \times 10^{-3} N.$$

It should be noted that the mechanical and electrostatic forces are of equal magnitude and in opposite direction. The result obtained above is coincident with that given by (9.11).

9.12. Suppose that the capacitor described in Problem 9.3 remains connected to the generator that has charged it. It is wished to calculate the force exerted by the plates.

Solution

In this case, when **F** changes the distance between the plates by dz (Fig. 9.7), then capacity of the capacitor changes by dC, but since the tension remains constant, the charge will change by dq. Therefore:
(a) The exterior supplies the mechanical energy Fdz,
(b) The generator supplies the energy

Fig. 9.7 Capacitor connected to a generator

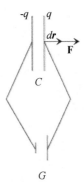

$$dU_g = V\,dq = V^2 dC = -\frac{V^2 \varepsilon S}{z^2}dz,$$

where $C = \varepsilon S/z$, z being the distance between the two plates, and $dC = (-\varepsilon S/z^2)dz$.
(c) The capacitor changes its energy by

$$dU_e = d\left(\frac{1}{2}CV^2\right) = \frac{1}{2}V^2 dC = -\frac{1}{2}\frac{V^2 \varepsilon S}{z^2}dz.$$

The principle of conservation of energy establishes that

$$F\,dz + dU_g = dU_e,$$

which can be written as:

$$F\,dz - \frac{V^2 \varepsilon S}{z^2}dz = -\frac{1}{2}\frac{V^2 \varepsilon S}{z^2}dz \quad \Rightarrow \quad F = \frac{1}{2}\frac{V^2 \varepsilon S}{z^2} = \frac{1}{2}\frac{q^2}{\varepsilon S}.$$

Note that the result is equal to that of Problem 9.11. This equality should be evident, since the distribution of charges is the same in both problems and, therefore, the forces must be equal.

Observe that, if the distance between the plates increases, $dz > 0$, then: (a) the energy supplied to the system from the external force is positive; (b) the energy provided by the generator is negative, that is, the generator is recharged; (c) the capacitor reduces its energy. That is, both the exterior and the capacitor collaborate in recharging the generator.

9.13. The attached figure represents a parallel-plane capacitor with charge q. Each plate has a size $a \times b$ and the separation between them is t, which is small. A dielectric plate, of length a, depth b and thickness t, with dielectric constant ε_r, is introduced a known distance x between the plates, as shown in Fig. 9.8. Calculate the force that the plates exert on the dielectric, discounting friction and neglecting edge effects.

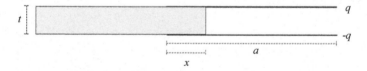

Fig. 9.8 Parallel-plate capacitor and a dielectric plate

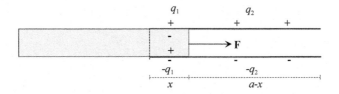

Fig. 9.9 Parallel-plate capacitor. Observe the sense of the force

Solution

On introducing the dielectric, it is polarised as indicated in the figure, and hence the polarisation charges are attracted by those of the plates and, due to the symmetry, the resultant force is expected to be towards the right.

The electrostatic force acting on the dielectric slab is going to be calculated through the energy of the system and application of (9.11), where charges are kept constant.

The system shown in Fig. 9.9 can be considered as two capacitors in parallel. Each of the capacitors has the respective areas, charges, capacities and energies:

$$S_1 = bx, \quad q_1, \quad C_1 = \frac{\varepsilon_r \varepsilon_0 S_1}{t} = \frac{\varepsilon_r \varepsilon_0 b}{t} x, \quad U_{e1} = \frac{1}{2} \frac{q_1^2}{C_1},$$

$$S_2 = b(a-x), \quad q_2, \quad C_2 = \frac{\varepsilon_0 S_2}{t} = \frac{\varepsilon_0 b}{t}(a-x), \quad U_{e2} = \frac{1}{2} \frac{q_2^2}{C_2}.$$

The energy is additive, therefore the energy of the system is the sum of both energies,

$$U_e = U_{e1} + U_{e2}.$$

To calculate the resulting force, it is sufficient to find the component of the force in the x-direction i.e. in the direction towards the right, by applying (9.11). A difficulty remains: to find the values of q_1 and q_2. To this end, use is made of the principle of conservation of electric charge

$$q = q_1 + q_2,$$

and of the fact that, being conducting plates, the potential is the same at all the points of the upper plate, as it is also in the lower plate. For this reason, the difference of potential V is identical in both capacitors, and can be written

$$\frac{q_1}{C_1} = V = \frac{q_2}{C_2} \Rightarrow \frac{q_1}{q_2} = \frac{C_1}{C_2}.$$

These last two equations allow the calculation of q_1 and q_2:

$$q_1 = \frac{qC_1}{C_1 + C_2}, \quad q_2 = \frac{qC_2}{C_1 + C_2}.$$

Replacing both values in the previous equations, the energy is given by

$$U_e = \frac{q^2 t}{2\varepsilon_0 b[(\varepsilon_r - 1)x + a]}.$$

The force is calculated by

$$F = -\frac{\partial U_e}{\partial x} = \frac{q^2 t}{2\varepsilon_0 b[(\varepsilon_r - 1)x + a]^2}(\varepsilon_r - 1).$$

Applying this formula to the particular case $x = a$, which corresponds to the circumstance where the dielectric is totally inserted, gives

$$F = \frac{q^2 t}{2\varepsilon_r^2 \varepsilon_0 b a^2}(\varepsilon_r - 1).$$

That is, when $x = a$, F is not null. However, if the figure is considered when the dielectric is inserted, then the plane perpendicular to that of the drawing which passes through the centre of the plates is symmetric, therefore it is impossible that there is a force towards the right, because the same reasoning would lead to the result that the force is towards the left. As the argument of symmetry is more basic than that used in the solution of the problem, the conclusion is reached that the force is not properly calculated. In the demonstration of the calculation of **F** it is supposed that the electric field is limited to the parallelepiped delimited by the plates, i.e. edge effects have been neglected. However, this assumption is not in complete agreement with the laws of Electromagnetism, therefore, it can be taken as only an approximation of the reality, and therefore the formula obtained for the force is only approximate.

Problems C

9.14. Take a capacitor which is equal to that in the previous problem except it is connected to a battery of electromotive force $\mathcal{E}$ (Fig. 9.10). Calculate the force on the dielectric.

Solution

To establish equilibrium and obtain the solution, the force $\mathbf{F}_{ex} = -\mathbf{F}$ is added to the electric force $\mathbf{F}$ of the system (Fig. 9.11). In a state of equilibrium, the difference of potential between the upper and the lower plates is $V = \mathcal{E}$ and it remains constant

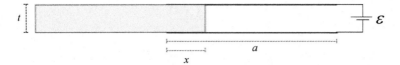

Fig. 9.10 Capacitor connected to an external source

for any value of x because the battery gives, or removes, charges to/from the plates. Let us consider that the dielectric moves the small distance dx towards the right, the energy conservation principle (First Law of Thermodynamics) states that the energy provided by the external force $\mathbf{F}_{ex}$, plus the energy provided by the battery is equal to the increase of energy of the system:

$$dW_{ex} + dW_{bat} = dU_e.$$

Furthermore:
$$dW_{ex} = \mathbf{F}_{ex}.d\mathbf{r} = -\mathbf{F}.d\mathbf{r} = -Fdx,$$

where F is the component on the axis OX of the force $\mathbf{F}$. By definition, the electromotive force is

$$dW_{bat} = \mathcal{E}dq \quad \Rightarrow \quad dW_{bat} = \mathcal{E}^2 dC.$$

The energy is additive, therefore the energy of the system formed by both capacitors is

$$U_e = U_{e1} + U_{e2} = \frac{1}{2}C_1\mathcal{E}^2 + \frac{1}{2}C_2\mathcal{E}^2 = \frac{1}{2}\mathcal{E}^2(C_1 + C_2) = \frac{1}{2}\mathcal{E}^2 C$$
$$\Rightarrow \quad dU_e = \frac{1}{2}\mathcal{E}^2 dC,$$

where $C = C_1 + C_2$ is the capacitance of the system.
Therefore,

$$-Fdx + \mathcal{E}^2 dC = \frac{1}{2}\mathcal{E}^2 dC \quad \Rightarrow \quad Fdx = \frac{1}{2}\mathcal{E}^2 dC.$$

The capacity C of the system is

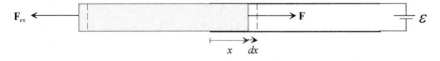

Fig. 9.11 The forces on the dielectric material

$$C = C_1 + C_2 = \frac{\varepsilon_r \varepsilon_0 b x}{t} + \frac{\varepsilon_0 b (a-x)}{t}$$
$$\Rightarrow dC = \frac{\varepsilon_r \varepsilon_0 b}{t} dx - \frac{\varepsilon_0 b}{t} dx = \frac{\varepsilon_0 b}{t}(\varepsilon_r - 1) dx.$$

From the two last equations it is deduced that

$$F dx = \frac{1}{2} \mathcal{E}^2 \frac{\varepsilon_0 b}{t}(\varepsilon_r - 1) dx$$
$$\Rightarrow F = \frac{1}{2} \mathcal{E}^2 \frac{\varepsilon_0 b}{t}(\varepsilon_r - 1) = \frac{q^2 t}{2\varepsilon_0 b [(\varepsilon_r - 1)x + a]^2}(\varepsilon_r - 1),$$

where it has been taken into account that $C = q/V = q/\mathcal{E}$. This result agrees with that of the previous problem.

It should be noted that in this case of constant voltage $dW_{bat} = 2dU_e$ and, therefore, $F dx = dU_e$. Hence, the electrostatic force can be expresses as

$$F = +\frac{\partial U_e}{\partial x},$$

where it is assumed that the voltage between the two plates is kept constant. This equation also leads to the result obtained above for F.

9.15. The coil represented in Fig. 9.12 is of N tightened turns, of small cross-sectional area S, and great mean length l. The circulating current intensity is I. The core material is magnetically linear and of relative permeability μ_r. Calculate the magnetic energy that it contains. Apply: $N = 2000$, $S = 1 \text{cm}^2$, $l = 30 \text{cm}$, $I = 4$ A, and $\mu_r = 10$.

Solution

Ampère's law applied to the inner circumference of length l gives

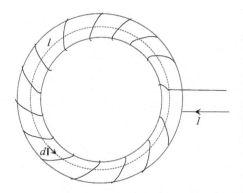

Fig. 9.12 Coil of N turns

$$\oint \mathbf{H}.d\mathbf{l} = Hl = NI \quad \Rightarrow \quad H = \frac{NI}{l}.$$

Although H depends on l, if the solenoid is very thin, then the length of all the inner circumferences are almost equal, and hence H can be assumed independent of the inner point considered. Of course the circulation of H in outer circumferences is null and, therefore, outside the solenoid the magnetic field is null.

Therefore the magnetic field B in the interior is

$$B = \mu_r \mu_0 H = \mu_r \mu_0 \frac{NI}{l}.$$

Hence, the energy per unit volume (9.22) is

$$u_m = \frac{1}{2}\mathbf{B}.\mathbf{H} = \frac{1}{2}BH = \frac{1}{2}\mu_r\mu_0 \frac{NI}{l}\frac{NI}{l} = \frac{1}{2}\mu_r\mu_0 \frac{N^2 I^2}{l^2}.$$

As the volume is $v = Sl$, then the magnetic energy is

$$U_m = u_m Sl = \frac{1}{2}\mu_r\mu_0 \frac{N^2 I^2}{l^2} Sl = \frac{1}{2}\mu_r\mu_0 \frac{N^2 I^2 S}{l}.$$

Using the numerical data provided yields:

$$u_m = \frac{1}{2}\mu_r\mu_0 \frac{N^2 I^2}{l^2} = \frac{1}{2} 10 \times 4\pi \times 10^{-7} \frac{2000^2 \times 4^2}{0.30^2} \frac{\text{J}}{\text{m}^3} = 4468 \frac{\text{J}}{\text{m}^3},$$

$$U_m = \frac{1}{2}\mu_r\mu_0 \frac{N^2 I^2}{l^2} Sl = \frac{1}{2} 10 \times 4\pi \times 10^{-7} \frac{2000^2 \times 4^2}{0.30^2} 10^{-4} \times 0.30 \text{ J} = 0.134 \text{ J}.$$

9.16. The capacitor in Fig. 9.13 has a charge q_0. The switch is closed and it is wished to know the energy of the capacitor and the coil, of null resistance, at instant t.

Solution

The charge that the capacitor contains at any instant must be known in order to calculate the energy stored in it, and the current intensity that circulates around the coil must be determined to obtain its stored energy.

When closing the circuit, a certain current of intensity I will be established at t, and the electromotive force, evaluated around the circuit following the direction

Fig. 9.13 Electric circuit

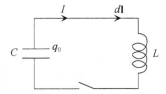

attributed to $d\mathbf{l}$, will be $-LdI/dt$, then the difference of potential between the upper and lower terminal of the coil is LdI/dt. This value agrees with the difference of potential between the upper plate of charge q and lower of charge $-q$, that is,

$$L\frac{dI}{dt} = V = \frac{q}{C}.$$

As the current is equal to the charge that crosses a section of the conductor per unit of time, it must agree, by the principle of conservation of charge, with the reduction of charge q per unit of time: $I = -dq/dt$. Therefore:

$$L\frac{d}{dt}\left(-\frac{dq}{dt}\right) = \frac{1}{C}q \quad \Rightarrow \quad \frac{d^2q}{dt^2} + \frac{1}{LC}q = 0.$$

As this equation is mathematically equal to that which regulates the harmonic movement of a block attached to a spring, the solution must be equal, that is:

$$q = A\cos(\omega t + \varphi) \quad \Rightarrow \quad \frac{dq}{dt} = -A\omega\sin(\omega t + \varphi) \quad \Rightarrow \quad \frac{d^2q}{dt^2} = -A\omega^2\cos(\omega t + \varphi).$$

Since, in $t = 0$, the charge is $q = q_0$, the first of these equalities gives

$$q_0 = A\cos\varphi.$$

From the second equality it is deduced that

$$I = -\frac{dq}{dt} = A\omega\sin(\omega t + \varphi).$$

As at the initial moment, $t = 0$, $I = 0$, therefore

$$0 = A\omega\sin(0 + \varphi) = A\omega\sin\varphi \quad \Rightarrow \quad \varphi = 0.$$

Therefore $q_0 = A\cos 0 = A$, which implies:

$$q = q_0\cos(\omega t) \quad \Rightarrow \quad I = q_0\omega\sin(\omega t).$$

The angular frequency ω can be obtained by substituting q and I in the expression $LdI/dt = q/C$, giving

$$\omega = \frac{1}{\sqrt{LC}} \quad \Rightarrow \quad f = \frac{1}{2\pi\sqrt{LC}},$$

where f is the frequency in Hertz.

Fig. 9.14 Different functions depending on time

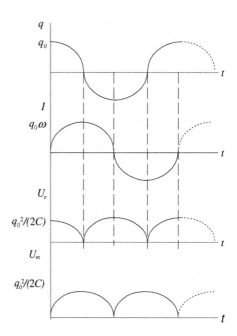

With these results it is possible to deduce the energies stored by the capacitor and the coil:

$$U_e = \frac{1}{2}\frac{q^2}{C} = \frac{q_0^2}{2C}\cos^2(\omega t),$$

$$U_m = \frac{1}{2}LI^2 = \frac{1}{2}Lq_0^2\omega^2\sin^2(\omega t) = \frac{q_0^2}{2C}\sin^2(\omega t).$$

Figure 9.14 represents the charge of the capacitor, the current intensity and the energies of the capacitor and the coil as functions over time. Note how the intensity varies with time (oscillating circuit) and how the capacitor and the coil transfer the energy.

9.17. A ring of linear magnetic material, of relative permeability μ_r, has mean length l and its cross section is a circle of area S (Fig. 9.15). There are N windings of a conductor wire on the ring. It is wished to know the magnetic energy in the device when the current intensity through the conductor is I.

Solution

Ampère's law around the circumference of mean length l gives

$$\oint \mathbf{H}.d\mathbf{l} = Hl = NI \quad \Rightarrow \quad H = \frac{NI}{l}.$$

Therefore, magnetic field **B** is

Fig. 9.15 Ring of a magnetic material

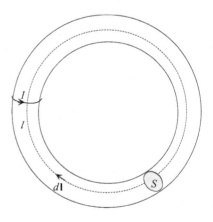

$$B = \mu_r \mu_0 H = \mu_r \mu_0 \frac{NI}{l}.$$

The flux through the whole circuit is

$$\Phi = BNS = \mu_r \mu_0 \frac{NI}{l} NS \quad \Rightarrow \quad d\Phi = \mu_r \mu_0 \frac{N^2 S}{l} dI.$$

Applying (9.14) gives

$$dU_m = Id\Phi = \mu_r \mu_0 \frac{N^2 S}{l} I dI \quad \Rightarrow \quad U_m = \frac{\mu_r \mu_0 N^2 S}{2l} I^2.$$

9.18. A toroidal solenoid has an outer radius $R_e = 40$ cm, an inner radius $R_i = 30$ cm, and $N = 1200$ windings, through which circulates a current of intensity $I = 20$ A. (a) Calculate the energy that it contains. (b) Calculate the energy that it would contain if it were filled successively with a diamagnetic, paramagnetic and ferromagnetic material, of relative permeability 0.9, 1.1, and 50,000 respectively.

Solution

The stored energy is calculated from the density of magnetic energy and the volume of the solenoid.

To calculate the field **H** inside the solenoid, Ampère's law is applied to the mean circumference, giving:

$$\oint \mathbf{H} \cdot d\mathbf{l} = I_{total} \quad \Rightarrow \quad HL = NI \quad \Rightarrow \quad H = \frac{NI}{L}.$$

The length of the solenoid can be estimated with the average radius

$$R = \frac{R_e + R_i}{2} \Rightarrow L = 2\pi R = \pi(R_e + R_i).$$

Therefore

$$H = \frac{NI}{\pi(R_e + R_i)}.$$

The magnetic field $B = \mu_r \mu_0 H$, therefore the density of energy is

$$u_m = \frac{1}{2}\mathbf{B}.\mathbf{H} = \frac{1}{2}BH = \frac{\mu_r\mu_0}{2}H^2 = \frac{\mu_r\mu_0}{2}\left(\frac{NI}{\pi(R_e+R_i)}\right)^2 = \frac{\mu_r\mu_0}{2}\frac{N^2 I^2}{\pi^2(R_e+R_i)^2}.$$

The diameter of the circle that produces a cross section is $R_e - R_i$, then the area of the circle is

$$S = \pi\frac{(R_e - R_i)^2}{4},$$

and hence the volume of the solenoid can be estimated thus

$$v = SL = \pi\frac{(R_e - R_i)^2}{4}\pi(R_e + R_i),$$

and therefore the energy stored with any of the materials is

$$U_m = u_m v = \frac{\mu_r\mu_0}{2}\frac{N^2 I^2}{\pi^2(R_e+R_i)^2}\pi\frac{(R_e-R_i)^2}{4}\pi(R_e+R_i) = \mu_r\mu_0\frac{N^2 I^2}{8}\frac{(R_e-R_i)^2}{R_e+R_i}.$$

For the diamagnetic material this is

$$U_m = \mu_r\mu_0\frac{N^2 I^2}{8}\frac{(R_e - R_i)^2}{R_e + R_i} = 0.9 \times 4\pi \times 10^{-7}\frac{1200^2 20^2}{8}\frac{(0.4 - 0.3)^2}{0.4 + 0.3_i} J = 1.16\,J.$$

For the paramagnetic material with $\mu_r = 1.1$ it is $U_m = 1.42$ J and for the ferromagnetic material with $\mu_r = 50{,}000$ it is $U_m = 64.4 \times 10^3$ J.

Note the large amount of energy accumulated when the introduced material is ferromagnetic.

9.19. At instant $t = 0$, terminal 1 is connected to terminal 2 by means of the switch drawn in Fig. 9.16. After a long time, 1 is disconnected from 2 and immediately connected with 3. (a) During the first operation, calculate the current for the coil, the energy it stores and the power released as heat. (b) Calculate these magnitudes in the second operation.

Solution

(a) In the first stage, Kirchhoff's second law (Fig. 9.17) gives

Fig. 9.16 Electric circuit

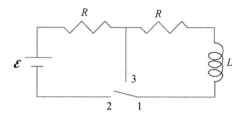

$$\mathcal{E} - L\frac{dI}{dt} = 2RI \quad \Rightarrow \quad \frac{dI}{\mathcal{E} - 2RI} = \frac{dt}{L} \quad \Rightarrow \quad -\frac{1}{2R}\ln(\mathcal{E} - 2RI) = \frac{t}{L} + k.$$

As at the initial instant, $t = 0$, $I = 0$ therefore:

$$I = \frac{\mathcal{E}}{2R}\left(1 - e^{-\frac{2R}{L}t}\right).$$

Note that the current is null when beginning the first stage, and when finishing the first at $t \approx \infty$, it becomes

$$I = \frac{\mathcal{E}}{2R}.$$

The energy stored in the coil, (9.15), is

$$U_m = \frac{1}{2}LI^2 = \frac{1}{2}L\frac{\mathcal{E}^2}{2^2R^2}\left(1 - e^{-\frac{2R}{L}t}\right)^2 = \frac{L\mathcal{E}^2}{8R^2}\left(1 - e^{-\frac{2R}{L}t}\right)^2.$$

The energy when finishing the first stage ($t \approx \infty$) is

$$U_m = \frac{L\mathcal{E}^2}{8R^2}.$$

Note how the stored energy grows until reaching this final value.

The calorific power released is

$$\frac{dQ}{dt} = 2RI^2 = 2R\frac{\mathcal{E}^2}{2^2R^2}\left(1 - e^{-\frac{2R}{L}t}\right)^2 = \frac{\mathcal{E}^2}{2R}\left(1 - e^{-\frac{2R}{L}t}\right)^2.$$

Fig. 9.17 Electric circuit when connecting 1 and 2

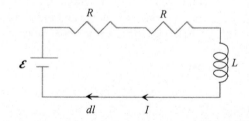

Fig. 9.18 Circuit composed by a resistance R and an autoinductance L

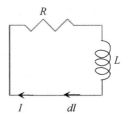

Note that this power begins as null and finishes as $\mathcal{E}^2/(2R)$.

(b) In the second stage, the clock restarting from 0 s at the start of this stage, terminal 1 is connected to terminal 3, see Fig. 9.18. From Ohm's Law, we have for the voltages

$$-L\frac{dI}{dt'} = RI \quad\Rightarrow\quad \frac{dI}{I} = -\frac{R}{L}dt' \quad\Rightarrow\quad \ln I = -\frac{R}{L}t' + k',$$

where time is denoted by t'.

At $t' = 0$, then $I = \mathcal{E}/(2R)$, as demonstrated in section (a), and hence $\ln(\mathcal{E}/(2R)) = k'$ and therefore

$$\ln I = -\frac{R}{L}t' + \ln\frac{\mathcal{E}}{2R} \quad\Rightarrow\quad \ln\frac{2RI}{\mathcal{E}} = -\frac{R}{L}t' \quad\Rightarrow\quad I = \frac{\mathcal{E}}{2R}e^{-\frac{R}{L}t'}.$$

It can be seen how the intensity diminishes over time t' from the final value of the first operation to zero.

The energy stored in the coil at this stage is

$$U_m = \frac{1}{2}LI^2 = \frac{1}{2}L\frac{\mathcal{E}^2}{2^2R^2}e^{-\frac{2R}{L}t'} = \frac{L\mathcal{E}^2}{8R^2}e^{-\frac{2R}{L}t'}.$$

Which indicates that the initial energy in the second stage agrees with the energy at the end of the first, $L\mathcal{E}^2/(8R^2)$, and the energy at the end of the whole process is null.

The power released as heat is

$$\frac{dQ}{dt} = RI^2 = R\frac{\mathcal{E}^2}{2^2R^2}e^{-\frac{2R}{L}t'} = \frac{\mathcal{E}^2}{4R}e^{-\frac{2R}{L}t'},$$

beginning with $\mathcal{E}^2/(4R)$ and becoming null at the end of the experiment.

To sum up, the energy provided by the generator in the first stage is partly stored in the coil and the rest is transmitted to the outside in heat form by the resistance. In the second stage, the electrical energy stored in the coil is transformed into thermal energy of the resistance of the circuit of Fig. 9.18 and released as heat.

Fig. 9.19 Hysteresis loop

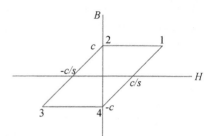

9.20. A ferromagnetic material has such a simple behaviour that its hysteresis loop is in the form of a rhomboid, two sides of which are: segment 1–2 on the line $B = c$ and segment 2–3 on the line $B = sH + c$, where c and s are known positive constants (Fig. 9.19). Calculate the magnetic energy dissipated by each cycle described and per unit volume.

Solution

With the data provided, the complete hysteresis loop 1–2–3–4–1 can be drawn in diagram $B - H$. For an intermediate state, for example on side 4–1 of the loop, a certain H and a certain B exist in the material. If an infinitesimal change of B occurs, dB, the magnetic energy provided to the system per unit volume is $du_m = HdB$.

Applying this change to the four sections of the loop gives:

Section 1–2. This is described by the line $B = c$, therefore

$$u_{21} = \int_{B_1}^{B_2} HdB = 0,$$

since $dB = 0$.

Section 2–3. Its points belong to the line $B = sH + c$. The point 3 is at $B = -c$, therefore $H_3 = -2c/s$, and $H_2 = 0$. Hence,

$$u_{32} = \int_{B_2}^{B_3} HdB = \int_0^{-2c/s} H(sdH + 0) = 2c^2/s.$$

Section 3–4. This section is contained in the line $B = -c$, therefore

$$u_{43} = \int_{H_3}^{H_4} HdB = 0.$$

Section 4–1. This is defined by $B = sH - c$. Therefore

$$u_{14} = \int_{H_4}^{H_1} HdB = \int_0^{2c/s} H(sdH + 0) = 2c^2/s.$$

The total energy supplied in one cycle and per unit volume is the sum of the variations in each section. Therefore

$$u_{\text{mcicle}} = 0 + 2c^2/s + 0 + 2c^2/s = 4c^2/s.$$

However, as the system has recovered its initial magnetic state and therefore its initial magnetic energy, this energy appears as thermal internal energy, which is transmitted to the exterior in the form of heat. That is, the system has "lost" the supplied energy $4c^2/s$ per unit volume by the fact of following a hysteresis loop.

It should be noted that the area of the loop is $4c^2/s$.

9.21. The magnetic circuit shown in Fig. 9.20 has N windings. It is connected to an alternating current generator such that an alternating current of intensity $I = I_0 \sin(\omega t)$ circulates. The material is ferromagnetic with a hysteresis loop for the said current as drawn (Fig. 9.21). (a) Estimate the energy dissipated per cycle. (b) Estimate the dissipated power. (c) Calculate the dissipated power,

Fig. 9.20 Experimental set-up

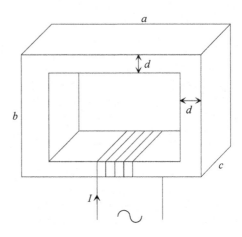

Fig. 9.21 Hysteresis loop

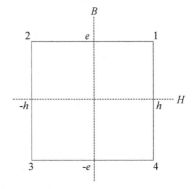

using: $a = 40$ cm, $b = 30$ cm, $c = 20$ cm, $d = 10$ cm, $e = 0.005$ T, $N = 1000$, $I_0 = 6$ A, and current frequency of 50 Hz.

Solution

(a) Ampère's law establishes that

$$\oint \mathbf{H}.d\mathbf{l} = I_{total} = NI \Rightarrow HL = NI \Rightarrow H = \frac{NI}{L}.$$

The mean length can be estimated, from the figure of the ferromagnetic core, as

$$L = 2(a-d) + 2(b-d) = 2a + 2b - 4d,$$

therefore

$$H = \frac{NI}{L} = \frac{NI}{2a+2b-4d} = \frac{NI_0}{2a+2b-4d}\sin(\omega t).$$

The maximum value of H is given when $\sin(\omega t)$ is one and must coincide with the value $H = h$, that is,

$$h = \frac{NI_0}{2a+2b-4d}.$$

Given the data of the hysteresis loop, a change of **B** gives rise to an injection of energy per unit volume of **H.dB**, and therefore for one cycle of the loop, the energy per unit of volume introduced is

$$u_{mcicle} = \oint HdB = \int_1^2 HdB + \int_2^3 HdB + \int_3^4 HdB + \int_4^1 HdB$$
$$= 0 - h(-e-e) + 0 + h(e+e) = 4eh.$$

Therefore the power dissipated per unit volume of material is $4eh$, which coincides with the "area" of the loop, as expected. Substituting the value calculated for h gives

$$u_{mcicle} = 4e\frac{NI_0}{2a+2b-4d} = \frac{2eNI_0}{a+b-2d}.$$

As the volume of the material is

$$v = 2(acd) + 2((b-2d)cd) = 2acd + 2bcd - 4cd^2,$$

the energy given off per cycle in heat form is

$$Q = 4eNI_0\frac{acd+bcd-2cd^2}{a+b-2d}.$$

(b) As frequency f represents the number of cycles per unit time, the dissipated power is
$$P = 4eNI_0 f \frac{acd + bcd - 2cd^2}{a+b-2d}.$$

(c) Application

$P = 4 \times 0.005 \times 1000 \times 6 \times 50 \frac{0.40 \times 0.20 \times 0.10 + 0.30 \times 0.20 \times 0.10 - 2 \times 0.20 \times 0.10^2}{0.40 + 0.30 - 2 \times 0.10}$ W
$= 120$ W.

This is the dissipated power.

Chapter 10
Maxwell's Equations

Abstract In this chapter we will study the Maxwell equations. These equations, together with the material relationships (constitutive) of the system to be analyzed, contain all classical macroscopic information we can obtain from an electromagnetic viewpoint. In fact, the electric and magnetic fields studied in the previous chapters are particular cases of a more general establishment of the problem when the Maxwell equations are considered. Taking into account the high quantity of important developments that have been made in all fields of science and technology by applying these equations, it can be said that the daily life of millions of people has been changed by James Clerk Maxwell and, of course, by the contribution of quantum mechanics.

10.1 Generalization of Ampère's Law

Ampère's law states that the magnetic field **B** and the electric current density **j** obey the equation

$$\frac{1}{\mu_0} \nabla \times \mathbf{B} = \mathbf{j}. \tag{10.1}$$

However, when the currents are neither stationary nor quasi-stationary, experience shows that this law is not correct.

Indeed, a fundamental principle of nature is that of the conservation of charge, which establishes that if a certain electrical charge leaves the boundary of a domain of space per unit of time, then the total charge contained in the domain decreases by the same amount, that is

$$\oint_S \mathbf{j} \cdot d\mathbf{S} = -\frac{\partial}{\partial t} \int_V \rho \, dV. \tag{10.2}$$

Application of the divergence theorem and Gauss's theorem gives

$$\nabla \cdot \mathbf{j} = -\frac{\partial \rho}{\partial t} \quad \Rightarrow \quad \nabla \cdot \mathbf{j} + \frac{\partial \rho}{\partial t} = 0 \quad \Rightarrow \quad \nabla \cdot \left(\mathbf{j} + \varepsilon_0 \frac{\partial \mathbf{E}}{\partial t} \right) = 0. \tag{10.3}$$

© Springer-Verlag Berlin Heidelberg 2017
F. Salazar Bloise et al., *Solved Problems in Electromagnetics*,
Undergraduate Lecture Notes in Physics, DOI 10.1007/978-3-662-48368-8_10

On the other hand, if the identity $\frac{1}{\mu_0}\nabla \cdot \nabla \times \mathbf{B} = 0$ is considered and it is equalized to the above, then the following simple solution is obtained

$$\frac{1}{\mu_0}\nabla \times \mathbf{B} = \mathbf{j} + \varepsilon_0 \frac{\partial \mathbf{E}}{\partial t}. \tag{10.4}$$

This equation is known as the Ampère–Maxwell Law in honour of its discoverers and establishes that when a magnetic field, an electric current and an electric field coexist at a point in space, they are always related by (10.4). This is sometimes interpreted by stating that the magnetic field can be supposed to be generated by the electric current and by the time-varying electric field. When the product of ε_0 times the rate of change of the time-varying electric field is much smaller than the current density, then this law becomes Ampère's law.

10.2 Maxwell's Equations for a Point

The electromagnetic magnitudes: Electric charge density ρ, electric current density $\mathbf{j}$, electric field $\mathbf{E}$, and magnetic field $\mathbf{B}$ are, in general, functions of the coordinates of the point where they are measured and of time. Maxwell grouped all the laws of electromagnetism into a set of four fundamental equations which are called Maxwell's equations, and these are valid for all points in space and for all time instants:

$$\nabla \cdot \mathbf{E} = \frac{\rho}{\varepsilon_0}, \tag{10.5}$$

$$\nabla \times \mathbf{E} = -\frac{\partial \mathbf{B}}{\partial t}, \tag{10.6}$$

$$\nabla \cdot \mathbf{B} = 0, \tag{10.7}$$

$$\frac{1}{\mu_0}\nabla \times \mathbf{B} = \mathbf{j} + \varepsilon_0 \frac{\partial \mathbf{E}}{\partial t}. \tag{10.8}$$

10.3 Maxwell's Equations for a Domain

When considering a domain of space bounded by a closed surface S for the above first or third equation or an open surface S with its boundary closed line Γ for the second or fourth equation, Maxwell's equations acquire the following forms, respectively:

$$\oint_S \mathbf{E} \cdot d\mathbf{S} = \frac{q_{\text{int}}}{\varepsilon_0}, \tag{10.9}$$

10.3 Maxwell's Equations for a Domain

$$\oint_\Gamma \mathbf{E} \cdot d\mathbf{l} = -\frac{\partial}{\partial t} \int_S \mathbf{B} \cdot d\mathbf{S}, \quad (10.10)$$

$$\oint_S \mathbf{B} \cdot d\mathbf{S} = 0, \quad (10.11)$$

$$\frac{1}{\mu_0} \oint_\Gamma \mathbf{B} \cdot d\mathbf{l} = I + \varepsilon_0 \frac{\partial}{\partial t} \int_S \mathbf{E} \cdot d\mathbf{S}. \quad (10.12)$$

The first three equations have already been deduced in previous chapters of this book. The last equation can be obtained by calculating the circulation of **B** along a closed line Γ, by applying the circulation theorem and the Ampère–Maxwell law (10.8).

The first and last equations of this set can be written in terms of the free charge ρ_{np} and of the conduction current intensity $\mathbf{j}_f$, respectively

$$\nabla \cdot \mathbf{D} = \rho_{np} \Leftrightarrow \oint_S \mathbf{D} \cdot d\mathbf{S} = q_{\text{int},np}, \quad (10.13)$$

$$\nabla \times \mathbf{H} = \mathbf{j}_f + \frac{\partial \mathbf{D}}{\partial t} \Leftrightarrow \oint_\Gamma \mathbf{H} \cdot d\mathbf{l} = I_f + \frac{\partial}{\partial t} \int_S \mathbf{D} \cdot d\mathbf{S}. \quad (10.14)$$

10.4 Scalar Potential

In electrostatics (see Chap. 2), the difference in potential can be defined because the curl of the electrostatic field is null. However, Maxwell's second law or law of induction, (10.6), affirms that, in general, this curl is not null, but depends on the rate of change of the magnetic field **B**.

Recalling the definition of vector potential shown in Chap. 5 we can write:

$$\nabla \times \mathbf{A} \equiv \mathbf{B}, \quad (10.15)$$

and (10.6), gives:

$$0 = \nabla \times \mathbf{E} + \frac{\partial \mathbf{B}}{\partial t} = \nabla \times \left(\mathbf{E} + \frac{\partial \mathbf{A}}{\partial t} \right). \quad (10.16)$$

Now, the scalar potential difference can be defined as:

$$V_2 - V_1 \equiv -\int_1^2 \left(\mathbf{E} + \frac{\partial \mathbf{A}}{\partial t} \right) \cdot d\mathbf{l}. \quad (10.17)$$

It is evident that this definition is appropriate for electrostatics, where nothing depends on time, and therefore $\partial \mathbf{A}/\partial t = 0$.

From (10.17) it can be deduced that

$$\nabla V = -\mathbf{E} - \frac{\partial \mathbf{A}}{\partial t}. \tag{10.18}$$

10.5 Surface of Discontinuity

In this field of study it frequently occurs that there are two different materials in contact through a common surface S, for example, air and water in a swimming pool, or glass and a metal in an ordinary mirror. In these cases, Maxwell's laws still hold, and important consequences, which are discussed below, are produced.

Consider medium 1 and medium 2, Fig. 10.1, where P is a point on the separation surface S and $\mathbf{n}$ is the perpendicular unit vector directed from medium 1 to medium 2.

The material located at a point of medium 1, infinitely close to P, has at instant t the electromagnetic properties: charge density ρ_1, surface charge density σ_1, surface current density $\mathbf{j}_{s1}$, electric field $\mathbf{E}_1$ and magnetic field $\mathbf{B}_1$. At another point infinitely close to P, but within medium 2, the electromagnetic properties are ρ_2, σ_2, $\mathbf{j}_{s2}$, $\mathbf{E}_2$, and $\mathbf{B}_2$.

The boundary conditions for the electric and magnetic fields at the interface between two media can be determined as follows:

1°. Consider a small box surrounding point P, with two faces parallel to surface S of area $dl \times dl'$ and a height dh that will tend towards zero. The application of (10.9) to the box gives

$$\oint_S \mathbf{E} \cdot d\mathbf{S} = E_{n2} dl dl' - E_{n1} dl dl' + \mathbf{E}_2 \cdot d\mathbf{S}_{lat2} + \mathbf{E}_1 \cdot d\mathbf{S}_{lat1}$$
$$= \frac{q_{int}}{\varepsilon_0} = \frac{\sigma_2 + \sigma_1}{\varepsilon_0} dl dl' + \rho_2 dV_2 + \rho_1 dV_1. \tag{10.19}$$

When dh tends towards zero, the lateral area and the volume also tend towards zero, and hence

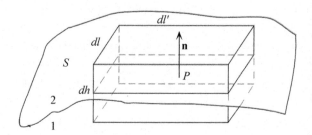

Fig. 10.1 Surface of discontinuity

10.5 Surface of Discontinuity

$$E_{n2}dldl' - E_{n1}dldl' = \frac{\sigma_2 + \sigma_1}{\varepsilon_0}dldl' \Rightarrow E_{n2} - E_{n1} = \frac{\sigma_2 + \sigma_1}{\varepsilon_0}. \quad (10.20)$$

This result indicates that the components of the electric field following the perpendicular to the surface are, in general, different, that is, they undergo a discontinuity.

2°. When considering the rectangular boundary of the anterior face of the box, applying (10.10) and having $dh \to 0$ it gives

$$\oint \mathbf{E} \cdot d\mathbf{l} = E_{t2}dl' - E_{t1}dl' + 0 = -\frac{\partial}{\partial t}\int_S \mathbf{B} \cdot d\mathbf{S} = -\frac{\partial}{\partial t}\int_S \mathbf{B} \cdot d\mathbf{S}_{front} = 0$$
$$\Rightarrow E_{t2} - E_{t1} = 0. \quad (10.21)$$

therefore the tangential component of the electric field is identical on both sides of S.

3°. Following the method applied in section 1°, but with (10.11), the following is obtained

$$B_{n2} - B_{n1} = 0. \quad (10.22)$$

4°. Following the method used in 2°, but with (10.12), the following is obtained

$$B_{t2} - B_{t1} = \mu_0(j_{s2} - j_{s1}). \quad (10.23)$$

Figure 10.2 schematically represents the obtained result, showing the continuity of the tangential component of the electric field and the perpendicular of the magnetic field.

If, instead of (10.9) and (10.13) is applied to the box, then the following equation is derived

$$D_{n2} - D_{n1} = \sigma_{np2} + \sigma_{np1}. \quad (10.24)$$

If (10.14) is applied to a small rectangle that covers a section of boundary and with two sides parallel to this boundary, the following equation results

$$H_{t2} - H_{t1} = \mu_0(j_{fs2} - j_{fs1}). \quad (10.25)$$

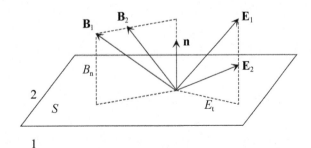

Fig. 10.2 Projection of the fields

Solved Problems

Problems A

10.1. A dielectrically linear material is characterized by the following properties, which do not depend on time: uniform free-charge density ρ_{np}, null-total current density, and electric field **E** variable with x and of a direction parallel to the OX axis. The magnetic field also has the direction of the OX axis and is variable with x. At the point of the material located at the origin of coordinates, **E** and **B** are null. Mathematically express the fields **E** and **B**.

Solution

As the electrical properties do not depend on time, Maxwell's laws are simplified and are expressed by the equations:

$$\nabla \cdot \mathbf{E} = \frac{\rho}{\varepsilon_0}, \ \nabla \times \mathbf{E} = 0, \ \nabla \cdot \mathbf{B} = 0, \ \frac{1}{\mu_0}\nabla \times \mathbf{B} = \mathbf{j}.$$

From the data, it follows that $\mathbf{E} = E(x)\mathbf{u}_x$ and $\mathbf{B} = B(x)\mathbf{u}_x$.

As the charge density data only refers to the free charges and not all the charges, it is convenient to use field **D** instead of **E**. To this end, the relationship (10.13) can be taken into account,

$$\nabla \cdot \mathbf{D} = \rho_{np},$$

and, in addition

$$\mathbf{D} = \varepsilon_0 \varepsilon_r \mathbf{E} \ \Rightarrow \ \mathbf{E} = \frac{1}{\varepsilon_0 \varepsilon_r}\mathbf{D} \ \text{and} \ \mathbf{D} = \varepsilon_0 \varepsilon_r E(x)\mathbf{u}_x.$$

Therefore

$$\nabla \cdot \mathbf{D} = \nabla \cdot [\varepsilon_0 \varepsilon_r E(x)\mathbf{u}_x] = 0 + 0 + \varepsilon_0 \varepsilon_r \frac{\partial E(x)}{\partial x} = \rho_{np} \ \Rightarrow \ E(x) = \frac{\rho_{np}}{\varepsilon_0 \varepsilon_r}x + C$$

$$\Rightarrow \mathbf{E} = \left(\frac{\rho_{np}}{\varepsilon_0 \varepsilon_r}x + C\right)\mathbf{u}_x,$$

where C is a constant of integration. Since for $x = 0$, then $\mathbf{E} = 0$, it follows that $C = 0$, therefore

$$\mathbf{E} = \frac{\rho_{np}}{\varepsilon_0 \varepsilon_r}x\mathbf{u}_x.$$

Similarly, the third equation gives

$$\nabla \cdot \mathbf{B} = \frac{\partial B}{\partial x} + 0 + 0 = 0 \ \Rightarrow \ B = C' \ \Rightarrow \ \mathbf{B} = C'\mathbf{u}_x,$$

where C' is a constant of integration.

The fourth equation leads to

$$\frac{1}{\mu_0} \begin{vmatrix} \mathbf{u}_x & \mathbf{u}_y & \mathbf{u}_z \\ \frac{\partial}{\partial x} & \frac{\partial}{\partial y} & \frac{\partial}{\partial z} \\ C' & 0 & 0 \end{vmatrix} = 0 = 0,$$

which is an identity without utility.
As for $x = 0$, $\mathbf{B} = 0$, it is deduced that

$$\mathbf{B} = 0.$$

Note how the calculations of the expressions of fields $\mathbf{E}$ and $\mathbf{B}$ have been made independently because everything is independent of time.

10.2. The circular hoop of Fig. 10.3 has a radius R and is immersed in a uniform magnetic field, perpendicular to its diameter DD'. (a) Calculate the electromotive force induced in the hoop if the magnetic field is constant and the hoop is immovable. (b) Calculate the electromotive force induced in the hoop if the field varies with time such that $B = B_0 \sin(\omega t)$ and the hoop is motionless. (c) Calculate the electromotive force induced in the hoop if the field is constant and the hoop turns around diameter DD' with the angular velocity ω.

Solution

The application of (10.10) gives, respectively:
(a)

$$\oint \mathbf{E} \cdot d\mathbf{l} = -\frac{\partial}{\partial t} \int_S \mathbf{B} \cdot d\mathbf{S} = -\frac{\partial}{\partial t} \int_S B \cdot dS \cdot \cos\theta = -\frac{\partial}{\partial t} (BS \cos\theta) = 0,$$

Fig. 10.3 Circular loop

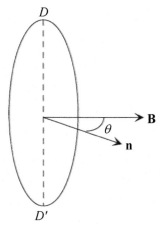

because neither B, nor S, nor θ depends on time. That is, as the magnetic field flux through the hoop is constant, the electromotive force is zero.

(b)
$$\oint \mathbf{E} \cdot d\mathbf{l} = -\frac{\partial}{\partial t}\int_S \mathbf{B} \cdot d\mathbf{S} = -\frac{\partial}{\partial t}\int_S B_0 \sin(\omega t) \cdot dS \cdot \cos\theta$$
$$= -\frac{\partial}{\partial t}(B_0 S \cos\theta \sin(\omega t)) = -B_0 S \omega \cos\theta \cos(\omega t).$$

In this case there is variation in the magnetic field flux because the magnetic field is variable with time at the points of the hoop.

(c)
$$\oint \mathbf{E} \cdot d\mathbf{l} = -\frac{\partial}{\partial t}\int_S \mathbf{B} \cdot d\mathbf{S} = -\frac{\partial}{\partial t}\int_S B \cdot dS \cdot \cos\theta = -\frac{\partial}{\partial t}\int_S B \cdot dS \cdot \cos(\omega t + \theta_0)$$
$$= -\frac{\partial}{\partial t}(BS\cos(\omega t + \theta_0)) = BS\omega \sin(\omega t + \theta_0).$$

In this last case the electromotive force is due to the movement of the points of the hoop in the magnetic field.

10.3. A flat capacitor is formed by two opposed, circular plates of large radii R, with a small separation between them (Fig. 10.4). There is a vacuum between the plates. At the initial instant it is connected to a current generator that provides constant current I. It is required to know the charge at the upper plate and the electric and magnetic fields in the space between the plates at moment t.

Fig. 10.4 Capacitor

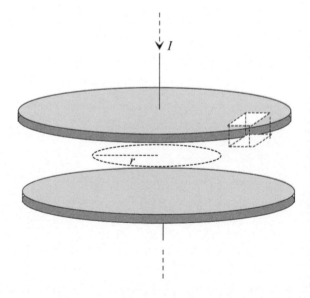

Solution

The device has symmetry of revolution. Considering a circumference of radius r, concentric with the axis of symmetry and located in a plane parallel to the plates and equidistant from them, the tangential component of the magnetic field B_ϕ is of the same value at the points of the circumference. The axial component of the electric field must also have a value E_z equal at all the points on the said circumference.

The total charge stored in the positive plate at moment t is that which has passed through a section of the upper conductor, and this is obtained from the current definition:

$$I = \frac{dq}{dt} \quad \Rightarrow \quad q_{\text{int}} = \int_0^t I.dt = It.$$

If we suppose that the charge is uniformly and rapidly distributed through the plates, then the application of (10.9) to a box that surrounds an interior portion of a plate and does not lie close to the borders gives

$$\oint_S \mathbf{E} \cdot d\mathbf{S} = \oint_S E_z \cdot dS = E_z dS = \frac{dq_{\text{int}}}{\varepsilon_0} \quad \Rightarrow \quad E_z = \frac{1}{\varepsilon_0}\frac{dq_{\text{int}}}{dS} = \frac{1}{\varepsilon_0}\frac{q_{\text{int}}}{S}.$$

Therefore,

$$E_z = \frac{1}{\varepsilon_0}\frac{q_{\text{int}}}{S} = \frac{It}{\varepsilon_0 S} = \frac{It}{\pi \varepsilon_0 R^2} \quad \Rightarrow \quad \frac{\partial E_z}{\partial t} = \frac{I}{\pi \varepsilon_0 R^2}.$$

The application of the Ampère–Maxwell law (10.12) to the circumference, considered as the boundary of the circle of radius r, gives

$$\frac{1}{\mu_0}\oint \mathbf{B} \cdot d\mathbf{l} = \frac{1}{\mu_0}\oint B_\phi \cdot dl = \frac{1}{\mu_0}B_\phi 2\pi r = I + \varepsilon_0 \frac{\partial}{\partial t}\int_S \mathbf{E} \cdot d\mathbf{S} = 0 + \varepsilon_0 \frac{\partial}{\partial t}\int_S E_z \cdot dS$$
$$= \varepsilon_0 \frac{\partial}{\partial t}\int_0^r E_z \cdot 2\pi r dr.$$

Therefore

$$B_\phi = \frac{\varepsilon_0 \mu_0}{r}\int_0^r \frac{\partial E_z}{\partial t} r dr.$$

Note how the magnetic field appears to be produced, not by electrical currents but by a time-variable electric field.

Finally

$$B_\phi = \frac{\varepsilon_0 \mu_0}{r}\int_0^r \frac{I}{\pi \varepsilon_0 R^2} r dr = \frac{\varepsilon_0 \mu_0}{r}\frac{I}{\pi \varepsilon_0 R^2}\frac{r^2}{2} = \frac{\mu_0 I}{\pi R^2}r.$$

10.4. Let us suppose that $n = 1000$ electrons per mm^2 are placed on the upper face of a thin, linear dielectric sheet, of relative dielectric permittivity $\varepsilon_r = 4$. Calculate the electric fields **D** and **E**.

Fig. 10.5 Dielectric sheet

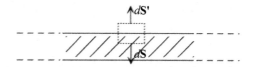

Solution

As the immediately calculable data is the free electrical charge σ_{np}, the (10.13) is applicable. In effect,

$$\sigma_{np} = n(-e) = -ne,$$

where n is known and e is obtained from a table of physical constants.

The sheet appears infinite in length and equal in all directions if it is observed from a point next to the sheet and remote from its borders, and therefore the horizontal component of **D** is null (except close to the borders).

Only the free charges are involved in the calculation of **D**, and therefore the upper surface is a plane of symmetry for **D**.

Application of the (10.13) to the drawn box (Fig. 10.5) gives

$$\oint_S \mathbf{D} \cdot d\mathbf{S} = \int_{S-up} \mathbf{D}' \cdot d\mathbf{S}' + \int_{S'-down} \mathbf{D} \cdot d\mathbf{S} + 0 = D'dS' + DdS = 2DdS = \sigma_{np}dS$$
$$\Rightarrow D = \sigma_{np}/2 = -ne/2.$$

The electric field **E** above the sheet, in vacuum, is obtained from

$$\mathbf{D} = \varepsilon_0 \varepsilon_r \mathbf{E} = \varepsilon_0 \mathbf{E} \Rightarrow E = \frac{1}{\varepsilon_0} D = -\frac{1}{2\varepsilon_0} ne$$
$$= -\frac{1}{2 \times 8.8542 \times 10^{-12}} 1000 \times 10^6 \times 1.602 \times 10^{-19} \text{ V/m} = -9.048 \text{ V/m},$$

whereas on the interior of the sheet $D' = D$, and therefore

$$E' = \frac{1}{\varepsilon_0 \varepsilon_r} D = -\frac{1}{2\varepsilon_0 \varepsilon_r} ne$$
$$= -\frac{1}{2 \times 8.8542 \times 10^{-12} \times 4} 1000 \times 10^6 \times 1.602 \times 10^{-19} \text{V/m} = -2.262 \text{ V/m}.$$

10.5. The adjacent figure represents an ideal parallel-plane capacitor, whose plates are circles of radius $R = 5$ cm, filled in with a dielectric of relative permeability $\mu_r = 10$ and thickness $d = 2$ mm. A current of $I = 1$ mA runs through the wires. Calculate the tangential component of field **B** at the points of a concentric circumference of radius r, located between the plates, as shown in Fig. 10.6. Apply this to $r = 1$ cm.

Fig. 10.6 Plane capacitor

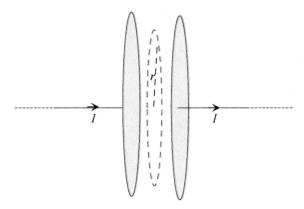

Solution

Given the symmetry of revolution of the problem, (10.14) can be applied to the circumference described, considered as the boundary of the circle. Since the intensity of current through this circle is null, the following is given

$$\oint \mathbf{H} \cdot d\mathbf{l} = \frac{\partial}{\partial t} \int_S \mathbf{D} \cdot d\mathbf{S}.$$

Consider the relationship between **D** and the charge density σ, which is assumed to be homogeneous, by supposing the charge of the capacitor varies slowly with time since the current is small. Thus

$$H_\phi 2\pi r = \frac{\pi r^2}{\pi R^2} \frac{dq}{dt} = \frac{r^2}{R^2} I \quad \Rightarrow \quad H_\phi = \frac{rI}{2\pi R^2} \quad \Rightarrow \quad B_\phi = \mu_r \mu_0 \frac{rI}{2\pi R^2}.$$

Application:

$$B_\phi = \mu_r \mu_0 \frac{rI}{2\pi R^2} = 10 \times 4\pi \times 10^{-7} \frac{0.01 \times 0.001}{2 \times \pi \times 0.05^2} \mathrm{T} = 8 \times 10^{-9} \mathrm{T}.$$

10.6. Consider a very long ideal solenoid of radius R, with n turns per unit length, of constant current intensity I, and which is in a vacuum.
   (a) Obtain the value of **B** at the interior and exterior points, using Maxwell's equations.
   (b) Determine the vector potential at these points.
   (c) Graphically represent the values obtained as functions of the distance from the axis of the solenoid.

Fig. 10.7 Long solenoid

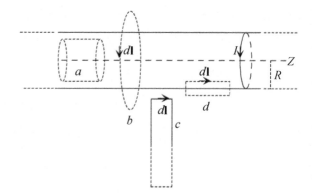

Solution

(a) Application of Maxwell's third (10.11) to the closed cylinder a in Fig. 10.7 gives:

$$0 = \oint_S \mathbf{B} \cdot d\mathbf{S} = \int_{lat} B_r dS_{lat} + \int_{ri} B_z dS_{baseri} + \int_{le} -B_z dS_{basele}$$

$$= \int_{lat} B_r dS_{lat} + 0 = \int_{lat} B_r dS_{lat} \Rightarrow B_r = 0,$$

where it is considered that B_z does not depend on z, except at the proximities of the borders. The radial component B_r is null in both the interior and the exterior.

Applying the fourth law to circumference b, while remembering that everything is independent of time, and that the coil can be replaced by a set of conducting rings, gives:

$$\frac{1}{\mu_0} \oint \mathbf{B} \cdot d\mathbf{l} = \frac{1}{\mu_0} \oint B_\phi \cdot dl = \frac{1}{\mu_0} B_\phi 2\pi r = I + \varepsilon_0 \frac{\partial}{\partial t} \int_S \mathbf{E} \cdot d\mathbf{S} = 0 + 0 = 0.$$

$$\Rightarrow B_\phi = 0$$

The tangential component is therefore null in the interior and the exterior.

Applying the fourth law to the very long exterior rectangle c, gives:

$$\frac{1}{\mu_0} \oint \mathbf{B} \cdot d\mathbf{l} = \frac{1}{\mu_0} B_{zex} \cdot dl = I + \varepsilon_0 \frac{\partial}{\partial t} \int_S \mathbf{E} \cdot d\mathbf{S} = 0 + 0 = 0 \Rightarrow B_{zex} = 0.$$

Applying the fourth law to the rectangle d that includes part of the electrical conductor, gives:

$$\frac{1}{\mu_0} \oint \mathbf{B} \cdot d\mathbf{l} = \frac{1}{\mu_0} (B_{zin} \cdot dl - B_{zex} \cdot dl) = \frac{1}{\mu_0} B_{zin} \cdot dl = I + \varepsilon_0 \frac{\partial}{\partial t} \int_S \mathbf{E} \cdot d\mathbf{S} = nIdl$$

$$\Rightarrow B_{zin} = \mu_0 nI.$$

(b) From the definition of vector potential, the following is deduced:

$$\mathbf{B} = \nabla \times \mathbf{A} \implies \oint \mathbf{A} \cdot d\mathbf{l} = \int_S (\nabla \times \mathbf{A}) \cdot d\mathbf{S} = \int_S \mathbf{B} \cdot d\mathbf{S},$$

which, applied to the exterior circumference, gives:

$$\oint \mathbf{A} \cdot d\mathbf{l} = \oint A_\phi \cdot dl = A_\phi 2\pi r = \int_S \mathbf{B} \cdot d\mathbf{S} = B_{zin} \pi R^2 = \mu_0 n I \pi R^2,$$

$$\implies A_\phi = \frac{\mu_0 n I R^2}{2r},$$

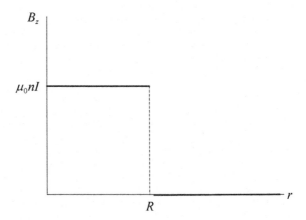

Fig. 10.8 Magnetic field versus distance

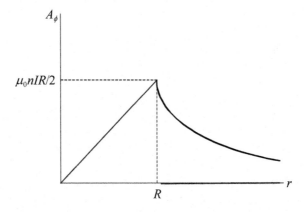

Fig. 10.9 Potential vector **A** against r

and when applied to the interior circumference gives:

$$\oint \mathbf{A} \cdot d\mathbf{l} = \oint A_\phi \cdot dl = A_\phi 2\pi r = \int_S \mathbf{B} \cdot d\mathbf{S} = B_{zin}\pi r^2 = \mu_0 n I \pi r^2$$

$$\Rightarrow A_\phi = \frac{\mu_0 n I r}{2}.$$

(c) Figures 10.8 and 10.9 show the variations of B_z and A_ϕ with r, respectively.

10.7. In a region of space, the vector potential is $\mathbf{A} = C_1 t(z^2 \mathbf{u}_x + x^2 \mathbf{u}_y + y^2 \mathbf{u}_z)$ and the scalar potential is $V = C_2(x^2 y + z^2 x + y^2 z)$. The vectors $\mathbf{u}_x$, $\mathbf{u}_y$, $\mathbf{u}_z$ are the unit vectors along the coordinate axes. Calculate the charge ρ and current $\mathbf{j}$ densities in this region.

Solution

From the data of the problem it follows that:

$$\nabla V = \frac{\partial V}{\partial x}\mathbf{u}_x + \frac{\partial V}{\partial y}\mathbf{u}_y + \frac{\partial V}{\partial z}\mathbf{u}_z = C_2(2xy + z^2)\mathbf{u}_x + C_2(2yz + x^2)\mathbf{u}_y + C_2(2xz + y^2)\mathbf{u}_z$$

$$\frac{\partial \mathbf{A}}{\partial t} = C_1(z^2 \mathbf{u}_x + x^2 \mathbf{u}_y + y^2 \mathbf{u}_z).$$

Applying (10.18) gives

$$\mathbf{E} = -\nabla V - \frac{\partial \mathbf{A}}{\partial t}$$

$$= -\left[C_2(2xy + z^2) + C_1 z^2\right]\mathbf{u}_x - \left[C_2(2yz + x^2) + C_1 x^2\right]\mathbf{u}_y - \left[C_2(2xz + y^2) + C_1 y^2\right]\mathbf{u}_z.$$

Substituting this result in Maxwell's first equation gives

$$\frac{\rho}{\varepsilon_0} = \nabla \cdot \mathbf{E} = -2C_2(x + y + z) \quad \Rightarrow \quad \rho = -2\varepsilon_0 C_2(x + y + z).$$

The expression of $\mathbf{B}$ is obtained from the definition of $\mathbf{A}$:

$$\mathbf{B} = \nabla \times \mathbf{A} = 2C_1 t(y\mathbf{u}_x + z\mathbf{u}_y + x\mathbf{u}_z).$$

Maxwell's fourth equation with the calculated values of $\mathbf{B}$ and $\mathbf{E}$ finally gives:

$$\frac{1}{\mu_0}\nabla \times \mathbf{B} = \mathbf{j} + \varepsilon_0 \frac{\partial \mathbf{E}}{\partial t} \quad \Rightarrow \quad \mathbf{j} = \frac{1}{\mu_0}\nabla \times \mathbf{B} - \varepsilon_0 \frac{\partial \mathbf{E}}{\partial t} = -\frac{2C_1 t}{\mu_0}(\mathbf{u}_x + \mathbf{u}_y + \mathbf{u}_z).$$

10.8. A dielectric sheet of electrical susceptibility $\chi_e = 3$ is polarized with the polarization vector $\mathbf{P}$ perpendicular to the sheet and of modulus $P = 0.3 \, \text{C/m}^2$. Calculate the electric field in the vicinity of the sheet.

Solution

The electric field inside the sheet is obtained directly from the relationship

$$\mathbf{P} = \varepsilon_0 \chi_e \mathbf{E} \quad \Rightarrow \quad \mathbf{E} = \frac{1}{\varepsilon_0 \chi_e} \mathbf{P}$$

$$\Rightarrow \quad E = \frac{1}{\varepsilon_0 \chi_e} P = \frac{1}{8.8542 \times 10^{-12} \times 3} 0.3 \text{ V/m} = 1.129 \times 10^{-14} \text{ V/m}.$$

and is in the same direction as **P**.

In order to calculate the electric field on the exterior, expression (10.20) can be applied with the following considerations. The exterior is separated from the interior by a plane that is the boundary between the dielectric and the vacuum, and there are polarization charges on the surface of the dielectric, but no free charges. The total charge next to this plane is the polarization charge on the surface of the dielectric, which is obtained from

$$\sigma_p = \mathbf{P} \cdot \mathbf{n} = P \cos\theta = P \cos 0 = P = 0.3 \text{ C/m}^2.$$

Equation (10.20) gives

$$E_{n2} - E_{n1} = E' - E = \frac{\sigma_2 + \sigma_1}{\varepsilon_0} = \frac{\sigma_p}{\varepsilon_0}$$

$$\Rightarrow \quad E' = E + \frac{\sigma_p}{\varepsilon_0} = 1.129 \times 10^{-14} + \frac{0.3}{8.8542 \times 10^{-12}} \text{ V/m} = 4.516 \times 10^{-14} \text{ V/m}.$$

This problem can also be solved by calculating field D inside the dielectric and applying the condition of continuity to obtain the value of D outside, and from this point by calculating E outside the dielectric.

10.9. A large, flat, perfectly conducting sheet has a total electric charge q, an area S on each face, and is surrounded by air of relative dielectric permittivity ε_r. Calculate the electric field in the air next to the sheet.

Solution

As the sheet is a perfect conductor, the electric field in its interior is zero; otherwise there would be an infinite current density within it, which makes no sense. Since the tangential component of the electric field **E** is continuous, the tangential electric field in the air must be zero.

By symmetry, the charge is equally distributed on both faces of the sheet.

Drawing a box (see Fig. 10.10) that contains a small part of the sheet, of area dS, the charge within it is

$$dq = \frac{q}{2S} dS.$$

Applying (10.13) gives

Fig. 10.10 Conducting sheet

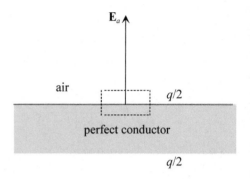

$$\oint_S \mathbf{D} \cdot d\mathbf{S} = \oint_S \mathbf{D} \cdot d\mathbf{S} = 2D_n \cdot dS = q_{\text{int.np}} = \frac{q}{2S} dS \Rightarrow D_n = \frac{q}{4S}.$$

The perpendicular component can be calculated from

$$\mathbf{D} = \varepsilon_r \varepsilon_0 \mathbf{E} \Rightarrow E_n = \frac{1}{\varepsilon_r \varepsilon_0} D_n = \frac{1}{\varepsilon_r \varepsilon_0} \frac{q}{4S}.$$

Problems B

10.10. In the points of a region there is a total charge density ρ_0, null total current density, the electric field $\mathbf{E}$ parallel to the OZ axis and the magnetic field $\mathbf{B} = ax\mathbf{u}_x + bx\mathbf{u}_y$. Obtain the expression of the electric field.

Solution

From the statement it is deduced that $\mathbf{E} = E(x, y, z, t)\mathbf{u}_z$.

Maxwell's first equation gives

$$\frac{\partial E_x}{\partial x} + \frac{\partial E_y}{\partial y} + \frac{\partial E_z}{\partial z} = 0 + 0 + \frac{\partial E}{\partial z} = \frac{\rho_0}{\varepsilon_0} \Rightarrow E = \frac{\rho_0}{\varepsilon_0} z + f_1(x, y, t),$$

where $f_1(x, y, t)$ is an arbitrary function of x, of y, and of time.

From Maxwell's second equation it is deduced that

$$\begin{vmatrix} \mathbf{u}_x & \mathbf{u}_y & \mathbf{u}_z \\ \frac{\partial}{\partial x} & \frac{\partial}{\partial y} & \frac{\partial}{\partial z} \\ 0 & 0 & E \end{vmatrix} = -\frac{\partial \mathbf{B}}{\partial t} = 0 \Rightarrow \frac{\partial E}{\partial y} = 0, \quad \frac{\partial E}{\partial x} = 0, \quad 0 = 0,$$

therefore

$$E = \frac{\rho_0}{\varepsilon_0} z + f_2(t).$$

Maxwell's third equation leads to

$$\frac{\partial B_x}{\partial x} + \frac{\partial B_y}{\partial y} + \frac{\partial B_z}{\partial z} = 0 \Rightarrow a + 0 + 0 = 0 \Rightarrow a = 0 \Rightarrow \mathbf{B} = bx\mathbf{u}_y.$$

This is the only possibility for the magnetic field.

Maxwell's fourth equation provides equality

$$\frac{1}{\mu_0}\begin{vmatrix} \mathbf{u}_x & \mathbf{u}_y & \mathbf{u}_z \\ \frac{\partial}{\partial x} & \frac{\partial}{\partial y} & \frac{\partial}{\partial z} \\ 0 & bx & 0 \end{vmatrix} = \frac{1}{\mu_0} b\mathbf{u}_z = 0 + \varepsilon_0 \frac{\partial f_2(t)}{\partial t} \mathbf{u}_z \Rightarrow \frac{\partial f_2(t)}{\partial t} = \frac{1}{\varepsilon_0 \mu_0} b$$

$$\Rightarrow f_2(t) = \frac{1}{\varepsilon_0 \mu_0} bt + C \Rightarrow E = \frac{\rho_0}{\varepsilon_0} z + \frac{b}{\varepsilon_0 \mu_0} t + C \Rightarrow \mathbf{E} = \left(\frac{\rho_0}{\varepsilon_0} z + \frac{b}{\varepsilon_0 \mu_0} t + C\right) \mathbf{u}_z,$$

where C is a constant of integration.

10.11. In a region of empty space, the electric field is measured and it is concluded that its value is $\mathbf{E} = ayt\mathbf{u}_z$ where a is a constant, y the coordinate corresponding to a Cartesian reference system, and t the time. Find a possible value for the magnetic field in that region.

Solution

In the vacuum there are no charges ($\rho = 0$), nor electric current ($\mathbf{j} = 0$), and hence Maxwell's equations are reduced to

$$\nabla \cdot \mathbf{E} = 0, \quad \nabla \times \mathbf{E} = -\frac{\partial \mathbf{B}}{\partial t}, \quad \nabla \cdot \mathbf{B} = 0, \quad \frac{1}{\mu_0} \nabla \times \mathbf{B} = \varepsilon_0 \frac{\partial \mathbf{E}}{\partial t}.$$

Both the known electric field and the magnetic field yet to be calculated must verify the four equations. Substituting the data $\mathbf{E} = ayt\mathbf{u}_z$ in the first equation gives

$$0 = \nabla \cdot \mathbf{E} = \frac{\partial 0}{\partial x} + \frac{\partial 0}{\partial y} + \frac{\partial (ayt)}{\partial z} = 0.$$

Which demonstrates that $\mathbf{E}$ verifies Maxwell's first equation.

Substituting $\mathbf{E}$ in the second gives

$$\frac{\partial \mathbf{B}}{\partial t} = -\nabla \times \mathbf{E} = -\begin{vmatrix} \mathbf{u}_x & \mathbf{u}_y & \mathbf{u}_z \\ \frac{\partial}{\partial x} & \frac{\partial}{\partial y} & \frac{\partial}{\partial z} \\ 0 & 0 & ayt \end{vmatrix} = -at\mathbf{u}_x \Rightarrow \mathbf{B} = -\frac{at^2}{2}\mathbf{u}_x + \mathbf{F}(x, y, z),$$

where $\mathbf{F}(x, y, z)$ is an arbitrary vector function of the coordinates of the point under consideration. For simplicity, we are going to test an easy solution, $\mathbf{F}(x, y, z) = 0$, with which

$$\mathbf{B} = -\frac{1}{2} at^2 \mathbf{u}_x.$$

This solution verifies the first two basic equations. Substituting it in the third gives

$$0 = \nabla \cdot \mathbf{B} = \frac{\partial B_x}{\partial x} + \frac{\partial B_y}{\partial y} + \frac{\partial B_z}{\partial z} = 0 + 0 + 0 = 0.$$

This identity demonstrates that the adopted solution is correct for the third equation. Substituting it in the fourth gives:

$$\nabla \times \mathbf{B} = \begin{vmatrix} \mathbf{u}_x & \mathbf{u}_y & \mathbf{u}_z \\ \frac{\partial}{\partial x} & \frac{\partial}{\partial y} & \frac{\partial}{\partial z} \\ \frac{1}{2}at^2 & 0 & 0 \end{vmatrix} = 0 = \varepsilon_0 \mu_0 \frac{\partial \mathbf{E}}{\partial t} = \varepsilon_0 \mu_0 a y \mathbf{u}_z.$$

To verify this equation at all points of the space, $a = 0$ must be a condition, but the electric field would then be null. Therefore it is necessary to try another solution, which can be intuited from observing the last equation. Let us try with

$$\mathbf{B} = \left(-\frac{1}{2}at^2 - \frac{1}{2}\varepsilon_0 \mu_0 a y^2 \right) \mathbf{u}_x,$$

which verifies the third equation. Substituting this into the fourth equation gives

$$\nabla \times \mathbf{B} = \begin{vmatrix} \mathbf{u}_x & \mathbf{u}_y & \mathbf{u}_z \\ \frac{\partial}{\partial x} & \frac{\partial}{\partial y} & \frac{\partial}{\partial z} \\ -\frac{1}{2}at^2 - \frac{1}{2}\varepsilon_0 \mu_0 a y^2 & 0 & 0 \end{vmatrix} = 0 + 0 + \varepsilon_0 \mu_0 a y \mathbf{u}_z = \varepsilon_0 \mu_0 a y \mathbf{u}_z$$

$$= \varepsilon_0 \mu_0 \frac{\partial \mathbf{E}}{\partial t} = \varepsilon_0 \mu_0 a y \mathbf{u}_z.$$

Therefore all Maxwell's equations are verified and the second proposal is a correct solution.

Note that the magnetic field obtained is perpendicular to the measured electric field. What creates what? **E** creates **B**, or **B** creates **E**? They simply coexist.

10.12. The attached figure shows a flat capacitor, of circular plates of radius R with a small separation between them (in order to be able to ignore the edge effect). The capacitor is charged by means of a current of small intensity I that runs through the wires.

(a) Determine the flux of **E** and **B** through the cylindrical surface shown in Fig. 10.11, whose bases S_1 and S_2 are of radius $2R$. Base S_2 is between the plates.

(b) Compare the values of the tangential component of field **B** at the points of the two circumferences of radius $2R$.

(c) Obtain the expression for the tangential field inside the capacitor in terms of the distance to the axis.

Fig. 10.11 Plane-parallel capacitor

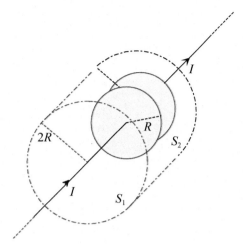

Solution

If the current is connected at instant $t = 0$, the electric charge in the anterior plate is

$$q = \int_0^t I\,dt = It.$$

From the statement of the problem, it is inferred that the electric field is very close to zero, except in the interior of the capacitor where it is uniform and parallel to the axis of the system.

(a) Applying Maxwell's first equation to the drawn cylinder gives

$$\oint_{cil} \mathbf{E} \cdot d\mathbf{S} = \frac{q_{int}}{\varepsilon_0} = \frac{q}{\varepsilon_0} = \frac{I}{\varepsilon_0}t.$$

Applying Maxwell's third equation to the drawn cylinder gives

$$\oint_S \mathbf{B} \cdot d\mathbf{S} = 0.$$

(b) Maxwell's fourth equation applied to the circumference which is the boundary of the circle S_1 gives

$$\frac{1}{\mu_0}\oint \mathbf{B} \cdot d\mathbf{l} = \frac{1}{\mu_0}\oint B_{1\phi} \cdot dl = \frac{1}{\mu_0}B_{1\phi}2\pi 2R = I + \varepsilon_0\frac{\partial}{\partial t}\int_S \mathbf{E} \cdot d\mathbf{S}$$

$$= I + \varepsilon_0\frac{\partial}{\partial t}\int_S 0 \cdot dS = I \Rightarrow B_{1\phi} = \frac{\mu_0 I}{4\pi R}.$$

The same equation applied to the circumference which is the boundary of the circle S_2 gives

$$\frac{1}{\mu_0}\oint \mathbf{B}\cdot d\mathbf{l} = \frac{1}{\mu_0}\oint B_{2\phi}\cdot dl = \frac{1}{\mu_0}B_{2\phi}2\pi 2R = I + \varepsilon_0\frac{\partial}{\partial t}\int_{S_2}\mathbf{E}\cdot d\mathbf{S}$$

$$= 0 + \varepsilon_0\frac{\partial}{\partial t}\int_{S_2}E_n dS = \varepsilon_0\frac{\partial}{\partial t}\int_R E_n dS.$$

The electric field between the plates is calculated from Maxwell's first equation applied to a cylinder with an end between the plates and the other inside the conducting plate:

$$\oint_{cil}\mathbf{E}\cdot d\mathbf{S} = \frac{I}{\varepsilon_0}t \Rightarrow \oint_{cil}E_n \cdot dS = E_n\cdot S = E_n\pi R^2 = \frac{I}{\varepsilon_0}t \Rightarrow E_n = \frac{I}{\pi\varepsilon_0 R^2}t.$$

Therefore

$$\frac{1}{\mu_0}B_{2\phi}2\pi R = \varepsilon_0\frac{\partial}{\partial t}\int_R \frac{I}{\pi\varepsilon_0 R^2}t\, dS = \varepsilon_0\frac{I}{\pi\varepsilon_0 R^2}\pi R^2 \Rightarrow B_{2\phi} = \frac{\mu_0 I}{2\pi R}.$$

This result is identical to the previous one, although it can be said that in S_1 the field $\mathbf{B}$ is due to the current and in S_2 it is due to the magnetic field.

(c) Maxwell's fourth equation applied to a circumference of radius $r < R$ and located between the plates gives

$$\frac{1}{\mu_0}\oint \mathbf{B}\cdot d\mathbf{l} = \frac{1}{\mu_0}B_\phi 2\pi r = I + \varepsilon_0\frac{\partial}{\partial t}\int_S \mathbf{E}\cdot d\mathbf{S} = 0 + \varepsilon_0\frac{\partial}{\partial t}\int_r E_n dS$$

$$= \varepsilon_0\pi r^2 \frac{\partial}{\partial t}\left(\frac{I}{\pi\varepsilon_0 R^2}t\right) = r^2\frac{I}{R^2} \Rightarrow B_{2\phi} = \frac{\mu_0 I}{2\pi R^2}r.$$

Problems C

10.13. The capacitor in Fig. 10.12 is formed by two circular, plane-parallel, metallic plates, of radius R, filled with a dielectric of relative permittivity ε_r. A current flows through the wire such that the charge density on plate a is of the form $\sigma = \sigma_0 \sin(\omega t)$. Obtain the value of the magnetic field $\mathbf{B}$ at a point in a plane between the plates, located at a distance r from the axis of revolution: (a) for $r < R$, (b) for $r > R$.

Solution

(a) For $r < R$, the charge density σ is known, therefore the charge Q of plate a is $Q = \sigma S$, S being the area of the plate and the field $D = \sigma$. Note that between the plates $I = 0$.

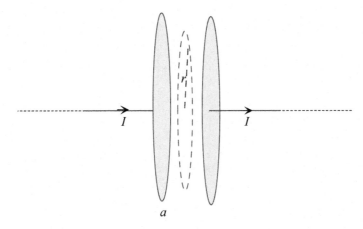

Fig. 10.12 Capacitor formed by to circular parallel plates

For the circle whose boundary is the circumference of radius r, (10.14) gives

$$\oint H_\phi dl = 0 + \frac{\partial}{\partial t} \int_S D_n dS \Rightarrow H_\phi 2\pi r = \frac{\partial}{\partial t} \int_S \sigma dS = \pi r^2 \frac{\partial}{\partial t} \sigma_0 \sin(\omega t)$$

$$= \pi r^2 \sigma_0 \omega \cos(\omega t) \Rightarrow H_\phi = \frac{r\sigma_0 \omega}{2} \cos(\omega t) \Rightarrow B_\phi = \frac{\mu_0 \mu_r r \sigma_0 \omega}{2} \cos(\omega t).$$

Note how the tangential field increases with the distance to the axis.

The null solutions for the radial and axial components are appropriate because, when considering a concentric cylinder that is interior to the dielectric, a zero flux is given for field **B** in agreement with (10.11).

(b) For $r > R$, the calculation is repeated but it must be borne in mind that the electric field is practically limited to the region between the plates:

$$\oint H_\phi dl = 0 + \frac{\partial}{\partial t} \int_S D_n dS \Rightarrow H_\phi 2\pi r = \frac{\partial}{\partial t} \int_S \sigma dS = \pi R^2 \frac{\partial}{\partial t} \sigma_0 \sin(\omega t)$$

$$= \pi R^2 \sigma_0 \omega \cos(\omega t) \Rightarrow H_\phi = \frac{R^2 \sigma_0 \omega}{2r} \cos(\omega t) \Rightarrow B_\phi = \frac{\mu_0 \mu_r R^2 \sigma_0 \omega}{2r} \cos(\omega t)$$

Note how the field diminishes with the distance to the axis, which seems reasonable.

10.14. In a region of space, there is a magnetic field **B**, parallel to an axis, and of modulus

$$B = 0.2r \cos(\pi t) \quad \forall \quad 0 \leq r \leq R$$
$$B = 0 \quad \forall \quad r > R$$

where r is the distance from a point to the axis.

Fig. 10.13 The tangential component of **A** in terms of r

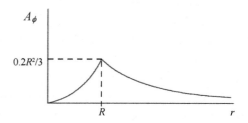

Calculate: (a) the tangential component of the vector potential for $r < R$, (b) the tangential component of the vector potential for $r > R$. (c) Represent this component as a function of r at the instants $t = 0$ and $t = 0.5$ s.

Solution

(a) The definition of **A** is $\nabla \times \mathbf{A} \equiv \mathbf{B}$. Given the symmetry of revolution of the problem, for a concentric circle with the axis and the circumference that is its boundary, of radius $r < R$, the application of the circulation theorem gives

$$\oint \mathbf{A} \cdot d\mathbf{l} = \int_S \nabla \times \mathbf{A} \cdot d\mathbf{S} = \int_S \mathbf{B} \cdot d\mathbf{S} \Rightarrow A_\phi 2\pi r = \int_0^r 0.2r\cos(\pi t) 2\pi r\, dr$$

$$\Rightarrow A_\phi = \frac{0.2}{3} r^2 \cos(\pi t).$$

(b) Similarly, if $r > R$ the following is given

$$\oint \mathbf{A} \cdot d\mathbf{l} = \int_S \nabla \times \mathbf{A} \cdot d\mathbf{S} = \int_S \mathbf{B} \cdot d\mathbf{S} \Rightarrow A_\phi 2\pi r = \int_0^r 0.2r\cos(\pi t) 2\pi r\, dr$$

$$= \int_0^R 0.2r\cos(\pi t) 2\pi r\, dr + \int_R^r 0 \times 2\pi r\, dr \Rightarrow A_\phi = \frac{0.2}{3} \frac{R^3}{r} \cos(\pi t).$$

(c) At $t = 0$, these are, respectively:

$$A_\phi = \frac{0.2}{3} r^2; \quad A_\phi = \frac{0.2}{3} \frac{R^3}{r},$$

the graphical representation of which is shown in Fig. 10.13.
 In $t = 0.5$ s, they are, respectively:

$$A_\phi = \frac{0.2}{3} r^2 \cos(0.5\pi) = 0;$$

$$A_\phi = \frac{0.2}{3} \frac{R^3}{r} \cos(0.5\pi) = 0.$$

Solved Problems

10.15. In a region without electrical currents, there is an electromagnetic field characterised by: (a) its vector potential $\mathbf{A} = \mathbf{u}_x A_0 \sin(kz - \omega t)$, where $\mathbf{u}_x$ is the unit vector in the direction of the OX axis, and A_0 and ω are known constants; (b) its electric field $\mathbf{E} = \mathbf{u}_x E_0 \cos(kz - \omega t)$. Find the values of $\mathbf{B}$, E_0, k and the difference in scalar potential between the points $(x, 0, 0)$ and the origin of the coordinates.

Solution

The relationship between $\mathbf{B}$ and $\mathbf{A}$ is obtained from the definition of $\mathbf{A}$:

$$\mathbf{B} = \nabla \times \mathbf{A} = \begin{vmatrix} \mathbf{u}_x & \mathbf{u}_y & \mathbf{u}_z \\ \frac{\partial}{\partial x} & \frac{\partial}{\partial y} & \frac{\partial}{\partial z} \\ A_0 \sin(kz - \omega t) & 0 & 0 \end{vmatrix} = \mathbf{u}_y A_0 k \cos(kz - \omega t).$$

The data must verify Maxwell's equations. Therefore:

$$\frac{\rho}{\varepsilon_0} = \nabla \cdot \mathbf{E} = \nabla \cdot (\mathbf{u}_x E_0 \cos(kz - \omega t)) = 0 \quad \Rightarrow \quad \rho = 0,$$

$$\nabla \times \mathbf{E} = -E_0 k \sin(kz - \omega t) = -\frac{\partial \mathbf{B}}{\partial t} = -A_0 k \omega \sin(kz - \omega t) \quad \Rightarrow \quad E_0 = A_0 \omega,$$

$$\nabla \cdot \mathbf{B} = 0.$$

$$\frac{1}{\mu_0} \nabla \times \mathbf{B} = \frac{1}{\mu_0} A_0 k^2 \sin(kz - \omega t) \mathbf{u}_x = \mathbf{j} + \varepsilon_0 \frac{\partial \mathbf{E}}{\partial t} = 0 + \varepsilon_0 \mathbf{u}_x E_0 \omega \sin(kz - \omega t)$$
$$\Rightarrow \quad A_0 k^2 = \varepsilon_0 \mu_0 E_0 \omega.$$

Substituting the value of E_0 in the last equation gives:

$$A_0 k^2 = \varepsilon_0 \mu_0 E_0 \omega = \varepsilon_0 \mu_0 A_0 \omega \omega$$
$$\Rightarrow \quad k = \sqrt{\varepsilon_0 \mu_0} \omega \quad \Rightarrow \quad \mathbf{B} = \mathbf{u}_y A_0 \sqrt{\varepsilon_0 \mu_0} \omega \cos(kz - \omega t).$$

The difference in scalar potential is calculated by integrating equation (10.17) along a simple path, the axis OX, whose points are at $y = 0$, $z = 0$ and, therefore, $\mathbf{E} = \mathbf{u}_x E_0 \cos(\omega t)$ and $\mathbf{A} = \mathbf{u}_x A_0 \sin(-\omega t)$:

$$V_{x,0,0} - V_{0,0,0} \equiv -\int_{0,0,0}^{x,0,0} \left(\mathbf{E} + \frac{\partial \mathbf{A}}{\partial t} \right) \cdot d\mathbf{l}$$

$$= -\int_{0,0,0}^{x,0,0} (\mathbf{u}_x E_0 \cos(\omega t) + \mathbf{u}_x A_0 \omega \cos(\omega t)) \cdot (\mathbf{u}_x dx + \mathbf{u}_y dy + \mathbf{u}_z dz)$$

$$= -\int_0^x (E_0 + A_0 \omega) \cos(\omega t) dx = -2 E_0 x \cos(\omega t).$$

10.16. Considering the relationship $\nabla \times \nabla \times \mathbf{A} = \nabla \cdot (\nabla \cdot \mathbf{A}) - \nabla^2 \mathbf{A}$ and forcing vector potential $\mathbf{A}$ and scalar potential V to verify the equation $\varepsilon_0 \mu_0 \partial V / \partial t + \nabla \cdot \mathbf{A} = 0$, calculate $\nabla^2 \mathbf{A}$ and $\nabla^2 V$ from Maxwell's equations.

Solution

From the two given relationships, from the definition of $\mathbf{A}$, and from Maxwell's fourth equation, the following is deduced:

$$\nabla^2 \mathbf{A} = \nabla \cdot (\nabla \cdot \mathbf{A}) - \nabla \times \nabla \times \mathbf{A} = -\nabla \cdot \left(\varepsilon_0 \mu_0 \frac{\partial V}{\partial t} \right) - \nabla \times \mathbf{B}$$

$$= -\varepsilon_0 \mu_0 \nabla \cdot \left(\frac{\partial V}{\partial t} \right) - \mu_0 \mathbf{j} - \varepsilon_0 \mu_0 \frac{\partial \mathbf{E}}{\partial t}.$$

Taking the derivative of (10.18) with respect to time, and substituting the result in the last equation, gives:

$$\nabla^2 \mathbf{A} = -\varepsilon_0 \mu_0 \nabla \cdot \left(\frac{\partial V}{\partial t} \right) - \mu_0 \mathbf{j} - \varepsilon_0 \mu_0 \left(\frac{\partial^2 \mathbf{A}}{\partial t^2} - \nabla \cdot \frac{\partial V}{\partial t} \right) = -\mu_0 \mathbf{j} - \varepsilon_0 \mu_0 \frac{\partial^2 \mathbf{A}}{\partial t^2}.$$

From (10.18) it is deduced:

$$\nabla^2 V = \nabla \cdot (\nabla V) = \nabla \cdot \left(-\mathbf{E} - \frac{\partial \mathbf{A}}{\partial t} \right) = -\nabla \cdot \mathbf{E} - \frac{\partial}{\partial t}(\nabla \cdot \mathbf{A}) = -\frac{\rho}{\varepsilon_0} + \varepsilon_0 \mu_0 \frac{\partial^2 V}{\partial t^2}.$$

10.17. The surface of a swimming pool is flat and separates the water, of relative dielectric permittivity $\varepsilon_{rw} = 1.8$, from the air, of relative dielectric permittivity $\varepsilon_{ra} = 1.0$. The electric field in the air is $\mathbf{E}_a = E_0 \cos\theta \mathbf{u}_y + E_0 \sin\theta \mathbf{u}_z$, where θ is the angle that the electric field forms with the separation plane (Fig. 10.14). Calculate the electric field in the water next to the air.

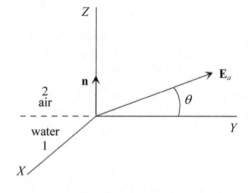

Fig. 10.14 Surface of separation

Solution

As the tangential component of the electric field is conserved, (10.21), it will have the same value in the water

$$E_{tw} = E_{ta} = E_0 \cos\theta.$$

As there are no free charges, (10.24), gives

$$D_{na} - D_{nw} = \sigma_{np} = 0 \;\Rightarrow\; D_{nw} = D_{na} = \varepsilon_{ra}\varepsilon_0 E_0 \sin\theta.$$

Fields **E** and **D** are related by the equation

$$\mathbf{D} = \varepsilon_{rw}\varepsilon_0 \mathbf{E} \;\Rightarrow\; D_{nw} = \varepsilon_{rw}\varepsilon_0 E_{nw} \;\Rightarrow\; E_{nw} = \frac{1}{\varepsilon_{rw}\varepsilon_0} D_{nw}.$$

From these last two equations, the following is derived

$$E_{nw} = \frac{\varepsilon_{ra}}{\varepsilon_{rw}} E_0 \sin\theta = \frac{1.0}{1.8} E_0 \sin\theta = 0.56 E_0 \sin\theta.$$

10.18. The surface of a swimming pool separates the water, of relative dielectric permittivity $\varepsilon_{rw} = 1.8$, from the air, of relative dielectric permittivity $\varepsilon_{ra} = 1.0$. The electric field in the water is $\mathbf{E}_w = E_0 \cos\theta \mathbf{u}_y + E_0 \sin\theta \mathbf{u}_z$, where θ is the angle that the electric field forms with the separation plane. Calculate the electric field in the air next to the water. What happens if the angle θ is large?
Note: The statement of the problem is the same as the previous problem, except here the known electric field is that of the water and the unknown quantity corresponds to the air. The way to solve this problem must be analogous to that of the previous problem.

Solution

As the tangential component of the electric field is conserved, (10.21), it will have the same value in the water

$$E_{ta} = E_{tw} = E_0 \cos\theta.$$

As there are no free charges, (10.24) gives

$$D_{na} - D_{nw} = \sigma_{np} = 0 \;\Rightarrow\; D_{na} = D_{nw} = \varepsilon_{rw}\varepsilon_0 E_0 \sin\theta.$$

Fields **E** and **D** are related by the equation

$$\mathbf{D} = \varepsilon_{ra}\varepsilon_0 \mathbf{E} \;\Rightarrow\; D_{na} = \varepsilon_{ra}\varepsilon_0 E_{na} \;\Rightarrow\; E_{na} = \frac{1}{\varepsilon_{ra}\varepsilon_0} D_{na}.$$

Fig. 10.15 Two media and the electric fields

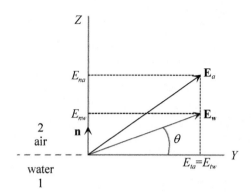

From the two last equations, the following is given

$$E_{na} = \frac{\varepsilon_{rw}}{\varepsilon_{ra}} E_0 \sin\theta = \frac{1.8}{1.0} E_0 \sin\theta = 1.8 E_0 \sin\theta.$$

It can be seen in Fig. 10.15 how the angle, which the electric field vector in the air forms with the separation plane, is greater than that formed by the electric field in the water with the separation plane.

The perpendicular component of $\mathbf{E}_a$ can be written thus: $E_a \sin\theta_a = E_{na} = E_0(1.8\sin\theta)$, where θ_a is the angle that $\mathbf{E}_a$ forms with the separation plane. The maximum value of $\sin\theta_a$ is one, and hence the maximum value of θ is such that $\sin\theta = 1/1.8$ which implies that $\theta_{max} = 33.7°$. Values greater than 33.7° cannot exist.

10.19. A flat boundary separates a paramagnetic material from the vacuum. The relative magnetic permeability of the material is $\mu_r = 4$. The magnetic field $\mathbf{B}$ in the vacuum has a modulus B_v, and forms an angle θ_v with the boundary, and is incoming into the vacuum. There are no surface conduction currents. (a) Calculate the field $\mathbf{B}_m$ in the material; (b) Relate the tangents of the angles formed by $\mathbf{B}$ in both media.

Solution

(a) Applying (10.22) and considering the material as medium 1 and the vacuum as medium 2 gives

$$B_{n2} - B_{n1} = B_{nv} - B_{nm} = 0 \quad \Rightarrow \quad B_{nm} = B_{nv} = B_v \sin\theta_v.$$

Equation (10.25) with $j_{fs} = 0$ gives:

$$H_{t2} - H_{t1} = H_{tv} - H_{tm} = 0 \quad \Rightarrow \quad H_{tm} = H_{tv} = H_v \cos\theta_v.$$

From the relationship between **B** and **H**, we have

$$\mathbf{B} = \mu_r \mu_0 \mathbf{H} \quad \Rightarrow \quad B_{tm} = \mu_r \mu_0 H_{tm}.$$

Therefore

$$B_{tm} = \mu_r \mu_0 H_{tm} = \mu_r \mu_0 H_v \cos\theta_v = \mu_r B_v \cos\theta_v.$$

(b) The angle in the vacuum is such that

$$\tan\theta_v = \frac{B_{nv}}{B_{tv}},$$

whereas in the material it is

$$\tan\theta_m = \frac{B_{nm}}{B_{tm}} = \frac{B_v \sin\theta_m}{\mu_r B_r \cos\theta_m} = \frac{\tan\theta_v}{\mu_r} \quad \Rightarrow \quad \frac{\tan\theta_v}{\tan\theta_m} = \mu_r = 4.$$

10.20. The air gap of a magnet is narrow compared with the diameter of the polar pieces, which are circular. The magnetic field measured in the air gap is $B = 0.1$ T. It is known that the material with which the magnet is constructed has a very complicated relationship between B and H. Calculate B at a point within the iron but next to the air gap.

Solution

Due to the small distance from the surface of the iron to the point at which magnetic field **B** is measured, field **B** is uniform, perpendicular to the polar surfaces (except in the proximities of the borders) and whose direction points from the north pole to the south pole (by definition of north pole). By studying the flux through a box that encloses a part of the north polar surface (see Fig. 10.16), and applying (10.11), it is deduced that

$$0 = \int_S \mathbf{B}\cdot d\mathbf{S} = \int_{S-down} \mathbf{B}.d\mathbf{S} + \int_{S-up} \mathbf{B}'\cdot d\mathbf{S} + \int_{S-lat} \mathbf{B}\cdot d\mathbf{S} = B\cdot d\mathbf{S} + B'\cdot d\mathbf{S} + 0$$

$$\Rightarrow \quad B' = -B,$$

Fig. 10.16 The air gap of a magnet

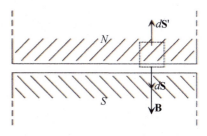

where B and B' are the downward projections of **B** in the air gap and the iron respectively.

Therefore the magnetic field in the interior of the north magnetic pole is downwards, in the same direction as in the air gap. The properties of the material do not influence the result.

10.21. A boundary separates a magnetized ferromagnetic material from the air. Magnetic field **B** in the material has a modulus B_m, which forms an angle θ_m with the boundary, and is incoming into the air. There are no surface conduction currents. Estimate the field $\mathbf{B}_v$ in the air.

Solution

The relative magnetic permeability of the air is approximately $\mu_{ra} = 1$. The relative magnetic permeability of the ferromagnetic material μ_{rm} is not a constant but it is always much larger than one. Applying the result of the problem 10.19 gives approximately

$$\frac{\tan \theta_v}{\tan \theta_m} = \mu_{rm} \gg 1 \quad \Rightarrow \quad \tan \theta_v \gg 1 \quad \Rightarrow \quad \theta_v \approx 90°.$$

Therefore, the magnetic field **B** in the air, next to the surface of the material, always forms an angle close to 90°. The lines of the magnetic field **B** in air are thus practically perpendicular to surfaces of ferromagnetic materials. This important fact is illustrated in Fig. 10.17.

Fig. 10.17 Magnetized ferromagnetic material

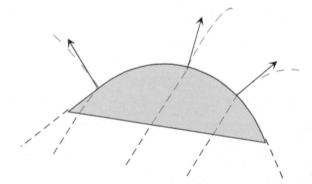

Chapter 11
Motion of Charged Particles in Electromagnetic Fields

Abstract One of the most important applications of the electric and magnetic fields deals with the motion of charged particles. For instance, in experimental nuclear fusion reactors the study of the plasma requires the analysis of the motion, radiation, and interaction, among others, of the particles that forms the system. In biomedicine the use of accelerators, like the cyclotron, allows the preparations of compounds to be employed in diagnostics, such as the FDG (see Problem 11.5), which is used as a tracer in the detection of some kinds of cancer and body diseases by means of the PET technique. Or in a closer case, the Earth, where its magnetic field, acting like a particle mirror, protects us against many of the cosmic rays.

11.1 Lorentz Force

From the definition of the electric field **E**, it follows that the force that an electric field exerts on a particle with charge q is

$$\mathbf{F} = q\mathbf{E}. \tag{11.1}$$

From the definition of magnetic field **B**, it is deduced that the force that a magnetic field applies on a particle is

$$\mathbf{F} = q\mathbf{v} \times \mathbf{B}, \tag{11.2}$$

where **v** is the velocity of the charge.

If the particle is simultaneously submitted to an electric field and a magnetic field, the resultant of the forces is the sum of the two aforesaid forces, that is,

$$\mathbf{F} = q\mathbf{E} + q\mathbf{v} \times \mathbf{B}, \tag{11.3}$$

which is called the Lorentz force.

Furthermore, it is known that all force applied on a particle causes its acceleration **a** in accordance with the fundamental equation of dynamics $\mathbf{F} = m\mathbf{a}$. The particle follows a trajectory and the force **F** generally has a component throughout the trajectory that produces the tangential acceleration, whose value, studied in kinematics, is dv/dt and measures the rate by which the modulus of the velocity varies. The force perpendicular to the trajectory causes an acceleration perpendicular to the trajectory; this component of the acceleration is directed towards the centre of curvature, and measures the rate by which the direction of the velocity varies; its modulus is v^2/R, where R is the radius of curvature of the trajectory.

11.2 Trajectory of a Charge in a Homogeneous Electric Field

Consider a region of space where only a homogeneous and constant electric field **E** exists. At the initial moment, $t = 0$, a particle of mass m and electrical charge q is impelled with an initial velocity $\mathbf{v}_0$. In general, this velocity will form an angle θ with **E**. Three coordinate axes are drawn such that their origin O coincides with the initial position of the particle, such that axis OZ has the same direction as the electric field, and such that axis OX is contained in the plane defined by the vectors **E** and $\mathbf{v}_0$ (see Fig. 11.1).

The electric field can be written in the form of its components thus

$$\mathbf{E} = 0\mathbf{u}_x + 0\mathbf{u}_y + E\mathbf{u}_z = E\mathbf{u}_z. \tag{11.4}$$

Applying the formula of the Lorentz force gives

$$\mathbf{F} = q\mathbf{E} = qE\mathbf{u}_z. \tag{11.5}$$

The fundamental equation of dynamics leads to

$$\begin{aligned} \mathbf{a} &= \mathbf{F}/m = q\mathbf{E}/m = qE/m\,\mathbf{u}_z \\ \Rightarrow\quad a_x &= a_y = 0; \quad a_z = qE/m. \end{aligned} \tag{11.6}$$

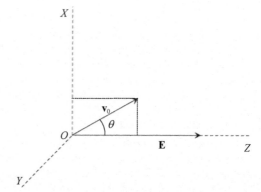

Fig. 11.1 Homogeneous electric field **E**

11.2 Trajectory of a Charge in a Homogeneous Electric Field

That is, the components of the acceleration on axes OX and OY are null and consequently the velocities along the axes OX and OY are constants, that is

$$\frac{dv_x}{dt} = 0 \Rightarrow v_x = C_1 = v_{0x} \Leftrightarrow \frac{dx}{dt} = v_{0x} \Rightarrow x = v_{0x}t$$
$$\frac{dv_y}{dt} = 0 \Rightarrow v_y = C_2 = v_{0y} = 0 \Leftrightarrow \frac{dy}{dt} = 0 \Rightarrow y = 0. \tag{11.7}$$

The third component of the acceleration gives

$$\frac{dv_z}{dt} = qE/m \Rightarrow v_z = qEt/m + C_3 \Rightarrow \tag{11.8}$$

$$v_z = qEt/m + v_{0z} \Rightarrow z = \frac{1}{2m}qEt^2 + v_{0z}t. \tag{11.9}$$

Therefore the particle moves on the plane formed by the vectors $\mathbf{E}$ and $\mathbf{v}_0$, with a constant velocity in the direction perpendicular to $\mathbf{E}$ and with a velocity increasing linearly with time in the direction of the electric field. The trajectory is ascertained by finding the time value through (11.7) and substituting it into the (11.9):

$$z = \frac{1}{2m}qEt^2 + v_{0z}t = \frac{1}{2m}qE\left(\frac{x}{v_{0x}}\right)^2 + v_{0z}\frac{x}{v_{0x}} = \frac{qE}{2mv_{0x}^2}x^2 + \frac{v_{0z}}{v_{0x}}x. \tag{11.10}$$

This equation corresponds to a parabola such as the one drawn in Fig. 11.2.

11.3 Trajectory of a Charge in a Homogeneous Magnetic Field

If there is only one homogeneous and constant magnetic field in a region of space, and a particle with electrical charge q moves within it, then the force that is applied on it is (Lorentz force)

Fig. 11.2 Trajectory of the charged particle

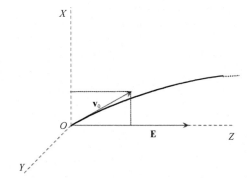

Fig. 11.3 Trajectory in presence of a homogeneous magnetic field

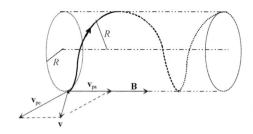

$$\mathbf{F} = q\mathbf{v} \times \mathbf{B}. \qquad (11.11)$$

If the particle has a mass m, then the acceleration to which it is subjected is

$$\mathbf{a} = \frac{q}{m}\mathbf{v} \times \mathbf{B}. \qquad (11.12)$$

Breaking down the velocity vector into a component $\mathbf{v}_{pa}$ parallel to $\mathbf{B}$ and another $\mathbf{v}_{pe}$ perpendicular to $\mathbf{B}$, as shown in Fig. 11.3, gives

$$\mathbf{a} = \frac{q}{m}\left(\mathbf{v}_{pa} + \mathbf{v}_{pe}\right) \times \mathbf{B} = \frac{q}{m}\mathbf{v}_{pa} \times \mathbf{B} + \frac{q}{m}\mathbf{v}_{pe} \times \mathbf{B} = 0 + \frac{q}{m}\mathbf{v}_{pe} \times \mathbf{B} = \frac{q}{m}\mathbf{v}_{pe} \times \mathbf{B}, \qquad (11.13)$$

where it has been taken into account that $\frac{q}{m}\mathbf{v}_{pa} \times \mathbf{B} = 0$ since they are parallel vectors.

Furthermore, $\frac{q}{m}\mathbf{v}_{pe} \times \mathbf{B}$ is perpendicular to both vectors and, as $\mathbf{v}$ is tangent to the trajectory, it means that the acceleration is perpendicular to the trajectory. Therefore, the tangential acceleration is null and consequently the modulus of the velocity is constant and the component $\mathbf{v}_{pa}$ is constant. The acceleration is centripetal, its modulus is v_{pe}^2/R, and therefore

$$\frac{q}{m}v_{pe}B = \frac{v_{pe}^2}{R} \quad \Rightarrow \quad \frac{q}{m}B = \frac{v_{pe}}{R} \quad \Rightarrow \quad R = \frac{mv_{pe}}{qB}. \qquad (11.14)$$

This expression provides the value of the radius of curvature of the trajectory of the charged particle.

From this equation, the following are deduced for the period of revolution and angular frequency, respectively:

$$T = \frac{2\pi R}{v_{pe}} = \frac{2\pi m}{qB}; \quad \omega = \frac{2\pi}{T} = \frac{qB}{m}. \qquad (11.15)$$

While the particle advances with constant velocity $\mathbf{v}_{pa}$ along the direction of $\mathbf{B}$, the projection of the trajectory follows a circumference of radius R, thus tracing a helix.

11.4 Hall Effect

(a) Consider a strip of a material that conducts electric current, of small thickness and of a length much greater than the width w. The strip is connected to an electric battery through a resistor, as indicated in Fig. 11.4. A magnetic field **B** is applied perpendicular to the strip and directed towards the material. A permanent regime is acquired upon the establishment of the current. If the charges that the strip transports are positive and of value q, then they are deflected downwards by the magnetic field force $q\mathbf{v} \times \mathbf{B}$, and hence the lower part acquires a positive charge. This charge, together with the negative of the side above, creates an upward electric field **E**. In the permanent regime the charges go from right to left due to the electric field applied by the battery $\mathbf{E}_b$. Therefore the component of the resultant of the forces in the direction perpendicular to **v** is null, that is

$$q\mathbf{E} + q\mathbf{v} \times \mathbf{B} = 0 \quad \Rightarrow \quad qE = qvB \quad \Rightarrow \quad E = vB. \tag{11.16}$$

Moreover, the electric field is related to the difference of potential and the width of the strip by the formula

$$E = \frac{V_H}{w} \quad \Rightarrow \quad \frac{V_H}{w} = vB. \tag{11.17}$$

The magnitude V_H is called the Hall voltage.

In an experiment, if the strip is metallic, then when measuring the V_H with a voltmeter, the upper part is positive with respect to the lower, and hence the metals conduct by means of negative charges. This effect can therefore serve to ascertain the sign of the charge of the charge carriers.

The Hall effect also serves to measure magnetic field **B** according to (11.17). A Hall probe is applied as a transducer, where a magnetic field **B** produces a measurable voltage V_H.

(b) Certain natural physical phenomena are not in agreement with the classical theory of physics. For example, according to the classical theory of physics, if a body emits a certain amount of energy, this energy can be of any value; nevertheless, according to quantum theory this energy is necessarily an integer multiple of the product hf of a universal constant h, named Plank's constant, times the frequency f of the radiation.

Fig. 11.4 Hall effect

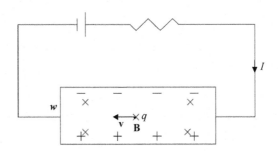

By studying the Hall effect in a conductive sheet through which a current of intensity I circulates at a temperature close to 0 K, and to which an intense magnetic field is submitted, the conclusion is reached that the Hall voltage V_H depends on the magnetic field in a staggered manner, that is, when increasing the field, the voltage remains unchanged until it changes abruptly to a certain higher value and so on. This is called the quantum Hall effect. The Hall resistance defined by the quotient $R_H = V_H/I$ yields the value

$$R_H = \frac{h}{ne^2}, \tag{11.18}$$

where n is an integer, and e is the electrical charge of the electron which is also a universal constant.

As h/e^2 is a universal constant, therefore its value for $n=1$ depends on nothing, and the electrical resistance of the sheet has a fixed value. The sheet therefore serves as a resistance standard and has been assigned the value 25812.807 Ω.

11.5 Trajectory of a Charge in Simultaneous, Homogeneous and Constant, Magnetic and Electric Fields

If a particle of charge q and mass m moves within a region where an electric field and a magnetic field exist simultaneously, then it is subjected to the Lorentz force given by (11.3), and its acceleration is obtained by applying the fundamental equation of dynamics. From this acceleration and from the initial conditions it is possible to obtain the velocity, the displacement, and the trajectory. In general, the solution can be very complicated, and for this reason some simple cases will be studied.

If the fields have the same direction, the calculation is simplified by forming a system of coordinating axes with their origin in the initial position of the particle and axis OZ in the common direction, as indicated in Fig. 11.5. The force on the particle is given by the Lorentz formula (11.3), and hence the acceleration is

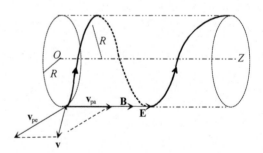

Fig. 11.5 Trajectory in presence of **E** and **B**

11.5 Trajectory of a Charge in Simultaneous, ...

$$\mathbf{a} = \frac{q}{m}\mathbf{E} + \frac{q}{m}\left(\mathbf{v}_{pa} + \mathbf{v}_{pe}\right) \times \mathbf{B} = \frac{q}{m}\mathbf{E} + \frac{q}{m}\mathbf{v}_{pe} \times \mathbf{B}, \tag{11.19}$$

whose component on the *OZ* axis is

$$a_z = \frac{q}{m}E. \tag{11.20}$$

Therefore the projection of the particle position on the *OZ* axis moves with a uniformly accelerated movement caused by the electric field.

The component of the acceleration perpendicular to the OZ axis is

$$\mathbf{a}_{pe} = \frac{q}{m}\mathbf{v}_{pe} \times \mathbf{B}, \tag{11.21}$$

which is identical to (11.13). Therefore, this acceleration causes an equal movement to that studied in Sect. 11.3, that is, the projection of the trajectory on the plane perpendicular to the *OZ* axis is a circumference whose radius is that given by (11.14).

In short, the trajectory is a helix whose pitch increases with time.

11.6 The Mass Spectrometer

Figure 11.6 represents a type of device called a Bainbridge mass spectrometer. A beam of diverse charged particles is accelerated by an electric field and enters an electric field **E** and a magnetic field $\mathbf{B}_1$, which are perpendicular. These fields are such that the Lorentz force is null for certain particles with velocity **v**, that is,

$$\mathbf{F} = q\mathbf{E} + q\mathbf{v} \times \mathbf{B}_1 = 0 \quad \Rightarrow \quad E = vB_1 \quad \Rightarrow \quad v = E/B_1. \tag{11.22}$$

Therefore only the particles that have that velocity are subjected to a null force and move with a constant velocity **v**, and hence their trajectory is straight and they arrive at orifice *H*. This part of the device is called a velocity selector. The particles of selected velocity **v** then enter by orifice *H* into a region with only a magnetic field

Fig. 11.6 Mass spectrometer

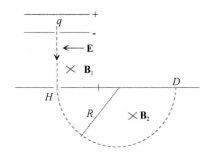

$\mathbf{B}_2$, where the trajectory is circular, and after tracing a semi-circumference they are detected at D. From (11.22) and (11.14), the following is given

$$R = \frac{mv}{qB_1} = \frac{mE}{qB_1B_2}. \tag{11.23}$$

Therefore, if the charge of the all particles is the same, those that have a mass m are detected at the distance $2R$ from the orifice. The other particles of equal electric charge but with different mass can be detected at other distances.

11.7 The Cyclotron

The cyclotron is an apparatus that serves to accelerate charged particles. It was invented by the american Ernest Lawrence in 1929. This is formed by two metallic cases, each in the form of the capital letter "D", open on their rectangular faces.

The two D's are put under a vacuum, subjected to a magnetic field $\mathbf{B}$, and a difference of alternating potential $V = V_0\sin(\omega t)$ between the D's, of angular frequency ω. Figure 11.7 shows a diagram of the device. The trajectory of a charge q is drawn with dashed lines and it can be observed that it is formed by a semi-circumference of radius $R_1 = \frac{mv_1}{qB}$ inside the D_1 case, where v_1 is the velocity in its interior. When the charge arrives at the gap between the D's it meets a difference of potential that impels it, increasing its velocity up to v_2, and therefore its trajectory is a circumference of radius $R_2 = \frac{mv_2}{qB}$, which is greater than R_1. Hence, with a suitable ω, each half cycle increases the velocity of the charge and its energy until it reaches a maximum radius and is extracted from the system.

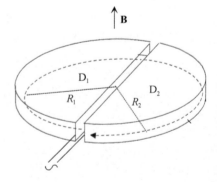

Fig. 11.7 The cyclotron

11.8 The Betatron

The betatron accelerates charged particles by an electric field induced by a changing magnetic field. A charged particle, initially at rest, is required to follow a circular orbit of radius R. To this end, an axially symmetrical magnetic field **B** is produced (as in Fig. 11.8), which is initially null, of modulus variable with the distance to the axis of symmetry, and increasing with time. The field is B_R within the zone of the orbit. The induced electric field is tangent to the orbit due to the symmetry and, at a certain moment, its modulus is obtained by applying the law of induction along the circumference that the particle is going to orbit:

$$E 2\pi R = \frac{d}{dt}\int_S \mathbf{B} \cdot d\mathbf{S} = \frac{d}{dt}\int_S B\, dS \quad \Rightarrow \quad E = \frac{1}{2\pi R}\frac{d}{dt}\int_S B\, dS. \quad (11.24)$$

The tangential field E causes a tangential force $F_t = ma_t = mdv/dt$, therefore, the fundamental equation of dynamics gives

$$m\frac{dv}{dt} = \frac{q}{2\pi R}\frac{d}{dt}\int_S B\, dS. \quad (11.25)$$

Solving (11.14) for v, its derivative with respect to time gives

$$\frac{dv}{dt} = \frac{qR}{m}\frac{dB_R}{dt}. \quad (11.26)$$

Substitution of this last equation into (11.25) yields

$$m\frac{qR}{m}\frac{dB_R}{dt} = \frac{q}{2\pi R}\frac{d}{dt}\int_S B\, dS \quad \Rightarrow \quad \frac{dB_R}{dt} = \frac{1}{2}\frac{d}{dt}\left(\frac{1}{\pi R^2}\int_S B\, dS\right). \quad (11.27)$$

This condition is satisfied if the magnetic field in the orbit is half of the average magnetic field in the circle.

Fig. 11.8 The betatron

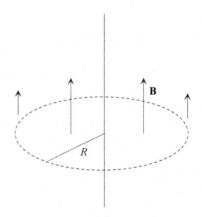

11.9 Relativistic Correction

In some problem in Physics, the velocity of charged particles inside electromagnetic fields is often close to the speed of light in a vacuum, henceforth simply called the speed of light, and it is known to be represented by c. The value of c is $c = 299792458$ m/s (a value of approximately 3×10^8 m/s). The formulation given in classical physics is not adequate in these circumstances and it is necessary to use the formulation given in the theory of special relativity so that the results of applying the equations of physics are consistent with the experimental facts.

When the speed of light is measured, it is observed that it is independent of the state of rest or of movement of the observer. Hence a series of consequences may be deduced, of which two are of interest here:

(a) The force that is applied on a particle is related to its mass and its velocity $\mathbf{v}$ by means of the formula

$$\mathbf{F} = \frac{d}{dt}\left(\frac{m\mathbf{v}}{\sqrt{1 - v^2/c^2}}\right), \tag{11.28}$$

where m is always the mass of the particle, measured when the particle is at rest with respect to the observer. Equation (11.28) is the fundamental equation of relativistic mechanics. When the velocity v of the particle is small (compared with c), the radicand of (11.28) is approximately 1 and (11.28) is reduced to approximately $\mathbf{F} = d(m\mathbf{v})/dt = m\mathbf{a}$, which is the fundamental equation of classical mechanics.

(b) The energy of a particle is expressed by means of the formula

$$E_r = \frac{mc^2}{\sqrt{1 - v^2/c^2}}. \tag{11.29}$$

In order to understand the physical significance of this energy, it is sufficient to make a series expansion for small values of v, that is, for values of v^2/c^2 much smaller than 1. This yields

$$E_r = \frac{mc^2}{\sqrt{1 - v^2/c^2}} = mc^2\left(1 - v^2/c^2\right)^{-1/2} = mc^2\left(1 + \frac{1}{2}v^2/c^2 + \cdots\right)$$
$$\approx mc^2\left(1 + \frac{1}{2}v^2/c^2\right) = mc^2 + \frac{1}{2}mv^2. \tag{11.30}$$

Therefore, the E_r is, for small velocities, the sum of the kinetic energy and the addend mc^2. The addend mc^2 is the energy that the particle holds at rest and is due solely to the fact of having mass. The E_r can be termed relativistic energy.

Solving (11.29) for v, the velocity can be calculated from the relativistic energy, giving

$$v = c\sqrt{1 - (mc^2/E_r)^2}. \tag{11.31}$$

11.9 Relativistic Correction

If the particle has an electrical charge q and is located at a point where the scalar potential is V, then the total energy of the particle is the sum of the relativistic and the potential energy, that is,

$$E_t = \frac{mc^2}{\sqrt{1 - v^2/c^2}} + qV. \qquad (11.32)$$

11.10 A Relativistic Particle in an Electromagnetic Field

As an application of the theory of relativity, the movement of a charged particle in an electromagnetic field will be analysed. Considering the Lorentz force together with (11.28) gives

$$q\mathbf{E} + q\mathbf{v} \times \mathbf{B} = \frac{d}{dt}\left(\frac{m\mathbf{v}}{\sqrt{1 - v^2/c^2}}\right). \qquad (11.33)$$

The movement of charges in stationary electromagnetic fields will be studied, that is, those independent of time.

The elemental work that the electromagnetic field applies on a charged particle that moves by $d\mathbf{r}$ is

$$dW = \mathbf{F}.d\mathbf{r} = (q\mathbf{E} + q\mathbf{v} \times \mathbf{B}).d\mathbf{r} = q\mathbf{E}.d\mathbf{r}, \qquad (11.34)$$

since $(q\mathbf{v} \times \mathbf{B}).d\mathbf{r}$ is null because $\mathbf{v}$ and $d\mathbf{r}$ are parallel. Therefore the work is due solely to the electric field. The magnetic field cannot vary the energy of a particle, however it could vary the velocity direction.

11.11 Charge in a Homogeneous Electric Field

Let $\mathbf{B} = 0$. The coordinate axes are located such that the axis OZ has the direction of the electric field $\mathbf{E}$. It is assumed that, at the initial moment, the particle is impelled from the origin of the coordinates, with the velocity v_0 and in the direction of the electric field.

The component of (11.33) on axis OX is

$$0 = \frac{d}{dt}\left(\frac{mv_x}{\sqrt{1 - v^2/c^2}}\right). \qquad (11.35)$$

Integration of (11.35) gives

$$\frac{mv_x}{\sqrt{1-v^2/c^2}} = k_1 = \frac{mv_{x0}}{\sqrt{1-v_0^2/c^2}} = 0, \qquad (11.36)$$

where k_1 is a constant and it yields $v_x = 0$. Therefore

$$\frac{dx}{dt} = 0, \qquad (11.37)$$

and hence

$$x = k_2 = x_0 = 0. \qquad (11.38)$$

Repeating the calculation for the component of (11.33) on OY axis yields

$$y = 0. \qquad (11.39)$$

The trajectory of the charged particle is thus the OZ axis.

The component of (11.33) on the OZ axis is

$$qE = \frac{d}{dt}\left(\frac{mv_z}{\sqrt{1-v^2/c^2}}\right). \qquad (11.40)$$

Integrating gives

$$\frac{mv_z}{\sqrt{1-v^2/c^2}} = qEt + k_3. \qquad (11.41)$$

Since at $t = 0$, then $v_x = 0$, $v_y = 0$, and $v_z = v_0$, therefore

$$\frac{mv_0}{\sqrt{1-v_0^2/c^2}} = k_3. \qquad (11.42)$$

Hence

$$\frac{mv_z}{\sqrt{1-v_z^2/c^2}} = qEt + \frac{mv_0}{\sqrt{1-v_0^2/c^2}}. \qquad (11.43)$$

This formula gives the velocity of the particle $v = v_z$ at any moment.
Calculating the derivative that appears in (11.40) yields

$$qE = \frac{d}{dt}\left(\frac{mv}{\sqrt{1-v^2/c^2}}\right) = m\left(1-\frac{v^2}{c^2}\right)^{-3/2}\frac{dv}{dt}, \qquad (11.44)$$

11.11 Charge in a Homogeneous Electric Field

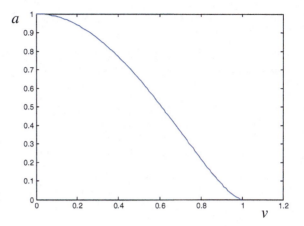

Fig. 11.9 Acceleration versus velocity

and therefore the component of the acceleration along axis OZ is

$$\frac{dv}{dt} = \frac{qE}{m}\left(1 - \frac{v^2}{c^2}\right)^{3/2}. \tag{11.45}$$

Note that if velocity v is null, then the acceleration is qE/m, in agreement with classical mechanics; however if the velocity is equal to that of light, the acceleration is zero. Figure 11.9 represents the acceleration of a charge starting from rest as a function of its velocity according to relativistic theory. The conclusion is drawn that as the particle reaches high velocities, its acceleration becomes smaller, even if the force exerted by the electric field remains constant, that is, the increase of velocity becomes more difficult the faster the particle moves. Therefore, even if the accelerating electric field is large, the particle will not be able to reach the speed of light. According to classical mechanics, a constant force causes a constant acceleration and therefore an increase of velocity that will become infinite. The classical theory is not adequate for the high velocities that are considered here.

Squaring (11.43) yields

$$m^2 v_z^2 = \left(qEt + \frac{mv_0}{\sqrt{1 - v_0^2/c^2}}\right)^2 - v_z^2/c^2 \left(qEt + \frac{mv_0}{\sqrt{1 - v_0^2/c^2}}\right)^2, \tag{11.46}$$

from which the square of the velocity is obtained

$$v_z^2 = \frac{\left(qEt + \frac{mv_0}{\sqrt{1 - v_0^2/c^2}}\right)^2}{m^2 + 1/c^2 \left(qEt + \frac{mv_0}{\sqrt{1 - v_0^2/c^2}}\right)^2}. \tag{11.47}$$

11.12 Charge in a Homogeneous Magnetic Field

Let $\mathbf{E} = 0$. A system of coordinate axes is drawn such that the axis OZ is parallel to magnetic field $\mathbf{B}$. Equation (11.33) is reduced to

$$q\mathbf{v} \times \mathbf{B} = \frac{d}{dt}\left(\frac{m\mathbf{v}}{\sqrt{1 - v^2/c^2}}\right). \tag{11.48}$$

Substituting in this equation that of (11.29) gives

$$q\mathbf{v} \times \mathbf{B} = \frac{d}{dt}\left(\frac{m\mathbf{v}}{\sqrt{1 - v^2/c^2}}\right) = \frac{d}{dt}\left(\frac{E_r \mathbf{v}}{c^2}\right). \tag{11.49}$$

Since E_r is constant in a magnetic field, then

$$q\mathbf{v} \times \mathbf{B} = \frac{E_r}{c^2}\frac{d\mathbf{v}}{dt} \quad \Rightarrow \quad \frac{qc^2}{E_r}\mathbf{v} \times \mathbf{B} = \frac{d\mathbf{v}}{dt}. \tag{11.50}$$

Since the product $\mathbf{v} \times \mathbf{B}$ is perpendicular to $\mathbf{B}$, it is also perpendicular to axis OZ, and therefore the acceleration $d\mathbf{v}/dt$ has a null component along axis OZ. Therefore, the component of the velocity v_z is constant. If the charge is impelled with an initial velocity $\mathbf{v}$ perpendicular to the magnetic field, then $v_{z0} = 0$, and the particle moves in a plane perpendicular to the magnetic field.

Since the product $\mathbf{v} \times \mathbf{B}$ is perpendicular to $\mathbf{v}$, the acceleration $d\mathbf{v}/dt$ is perpendicular to the velocity, that is, to the trajectory, and therefore it is a centripetal acceleration. Equalizing the components of (11.50) along the perpendicular to the trajectory gives

$$\frac{qc^2}{E_r}vB = a_n = \frac{v^2}{R} = \omega^2 R, \tag{11.51}$$

where R is the radius of curvature of the trajectory. Since the first member of (11.51) is constant, so is R and therefore the trajectory is a circumference of radius

$$R = \frac{vE_r}{qc^2 B}. \tag{11.52}$$

The quantity ω represents the angular velocity, or angular frequency, with which the particle travels the circumference and has the value

$$\omega = \frac{qc^2 B}{E_r}. \tag{11.53}$$

Note that, as the centripetal acceleration is towards the centre of curvature, the centre is indicated by the vector $\mathbf{v} \times \mathbf{B}$ (if q is positive).

11.12 Charge in a Homogeneous Magnetic Field

If the velocity of the particle is small ($v \ll c$), then (11.29) gives $E_r = mc^2$ and these last two equations are converted into those formulated with classical mechanics at the start of this chapter.

Solved Problems

Problems A

11.1. A proton of mass $m = 1.672 \times 10^{-27}$ kg and electrical charge $q = e = 1.602 \times 10^{-19}$ C is left without an initial velocity in a homogeneous electric field $E = 20$ V/m. The velocity that the proton acquires and the distance travelled when the elapsed time is 0.08 s are required.

Solution

As the charge and the electric field are known, the force that is exerted on the particle can be calculated. With the force calculated and the mass known, the fundamental equation of dynamics allows the acceleration to be calculated. From the value of the acceleration it is possible to obtain the velocity and the displacement.

The equations (11.7)–(11.9) can be applied directly, in which, the initial velocity is $v_{0x} = v_{0y} = v_{0z} = 0$. If the origin of coordinates is taken as the starting point of the proton, and the axis OZ coincides with the direction of the electric field, then:

$$v_x = 0 \Rightarrow x = 0,$$
$$v_y = 0 \Rightarrow y = 0,$$

$$v_z = \frac{1.602 \times 10^{-19} \times 20 \times 0.08}{1.672 \times 10^{-27}} = 1.5330 \times 10^8 \text{ m/s},$$

$$z = \frac{qE}{2m}t^2 = \frac{1.602 \times 10^{-19} \times 20}{2 \times 1.672 \times 10^{-27}} 0.08^2 \text{m} = 6.132 \times 10^6 \text{ m}.$$

As the obtained velocity is about half the speed of light, these results can only be taken as approximate.

11.2. At a point in space there is an electric field **E** in the direction of the coordinate axis OX, a magnetic field **B** in the direction of the axis OY, and a particle of charge q moving with velocity **v** in the direction of the axis OZ (Fig. 11.10). Calculate the components of the force along the three coordinate axes.

Solution

The force acting on the particle is the Lorentz force

$$\mathbf{F} = q\mathbf{E} + q\mathbf{v} \times \mathbf{B}.$$

Fig. 11.10 Particle with velocity **v**

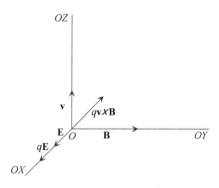

Since **E** has the direction of OX axis, then the components of $q\mathbf{E}$ are, respectively: qE, 0, and 0.

The vectorial product $\mathbf{v} \times \mathbf{B}$ has the opposite direction to OX axis, and its respective components are: $-qvB$, 0, and 0.

Therefore, the components of the force are, respectively: $(qE-qvB)$, 0, and 0 and can be written

$$\mathbf{F} = q(E - vB)\mathbf{u}_x.$$

11.3. An ion of charge $q = e = 1.602 \times 10^{-19}$ C and mass $m = 1.50 \times 10^{-25}$ kg is impelled with a velocity $v = 100000$ m/s perpendicular to a homogeneous electric field $E = 3$ V/m. Calculate the velocity acquired by the ion during the first 0.4 s and draw the trajectory.

Solution

The force on the particle can be calculated from data q and E. The acceleration is calculated from the force and the mass, and hence the velocity and the trajectory.

Drawing coordinate axes with their origin at the initial position of the particle, with the OZ axis in the direction of the electric field and with the OX axis in the direction of the initial velocity, as shown in Fig. 11.11. With respect to these axes, the data can be written thus:

$$v_{0x} = v = 100000 \text{ m/s}, \; v_{0y} = 0, \; v_{0z} = 0.$$

The distance travelled along axis OX in 0.4 s is, according to (11.7),

$$x_{0.4} = v_{0x}t = 100000 \times 0.4 \text{ m} = 40000 \text{ m}.$$

The velocity along axis OZ is, according to (11.9),

$$v_z = \frac{qE}{m}t + v_{0z} = \frac{qE}{m}t = \frac{1.602 \times 10^{-19} \times 3}{1.50 \times 10^{-25}} 0.03 \text{ m/s} = 9.612 \times 10^4 \text{ m/s}.$$

Fig. 11.11 Reference of frame

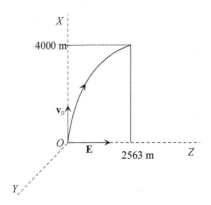

Directly applying (11.10) gives the trajectory

$$z = \frac{qE}{2mv^2}x^2 = \frac{1.602 \times 10^{-19} \times 3}{2 \times 1.50 \times 10^{-25} \times 100000^2}x^2 = 1.602 \times 10^{-4}x^2.$$

The graphical representation is given as a continuous line in Fig. 11.11.

11.4. An electric field has the direction of axis OY and its modulus varies with the point of space in the form $E = E_0 + kz$, where k is a constant. At the initial instant, a particle of charge q and mass m is impelled from the origin of the coordinates, with a velocity v_0 parallel to axis OZ (Fig. 11.12). Find the distance of the particle to the origin of the coordinates at instant t.

Solution

Since the electric field, the charge, and the mass are known, it is possible to calculate the acceleration, the velocity and the components of the displacement along the axes. Given this displacement, the distance can be determined.

The field has the components $E_x = 0$, $E_y = E_0 + kz$, and $E_z = 0$, which cause the respective accelerations $a_x = 0$, $a_y = q(E_0 + kz)/m$, and $a_z = 0$. Therefore:

$$\frac{dv_x}{dt} = 0 \Rightarrow v_x = C_1 = 0 \Leftrightarrow \frac{dx}{dt} = 0 \Rightarrow x = C_2 = 0,$$

$$\frac{dv_y}{dt} = \frac{q}{m}(E_0 + kz),$$

$$\frac{dv_z}{dt} = 0 \Rightarrow v_z = C_3 = v_0 \Leftrightarrow \frac{dz}{dt} = v_0 \Rightarrow z = v_0 t + C_3 = v_0 t.$$

By substituting the third result into the second, this is transformed into

Fig. 11.12 Electric field

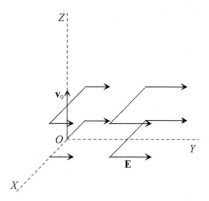

$$\frac{dv_y}{dt} = \frac{q}{m}(E_0 + kv_0 t) \implies v_y = \frac{q}{m}\left(E_0 t + \frac{kv_0}{2}t^2\right) + C_4 = \frac{q}{m}\left(E_0 t + \frac{kv_0}{2}t^2\right)$$

$$\implies y = \frac{q}{m}\left(\frac{E_0 t^2}{2} + \frac{kv_0}{2 \times 3}t^3\right) + C_5 = \frac{q}{2m}\left(E_0 t^2 + \frac{kv_0}{3}t^3\right).$$

The distance d to the centre is therefore

$$d = \sqrt{x^2 + y^2 + z^2} = \sqrt{\frac{q^2}{4m^2}\left(E_0 t^2 + \frac{kv_0}{3}t^3\right)^2 + (v_0 t)^2}.$$

11.5. Electrons, $q = -e$, are used as a test charge to determine a field **B**. This field can be considered homogeneous, stationary, perpendicular to the plane of Fig. 11.13, and confined to the hatched area. The electrons are accelerated starting from rest when passing through plates between which there is a difference of potential $V' = V_2 - V_1$.

(a) Determine the modulus and direction of **B** if, after a certain route through the interior of the magnetic field, the point of impact of the electrons on the screen (plane $x = 0$) is at $(0, a, 0)$. The mass and the charge of the electron are assumed to be known.

(b) Later, at a point P of the previous field **B**, two charged particles are impelled with the same velocity perpendicular to the field (Fig. 11.14). These particles have equal mass. The figure shows the trajectory followed by each particle, recorded by means of a Wilson cloud chamber (basically it contains gas and water vapour). Give reasons for the difference between these particles and explain what may cause the progressive reduction of the radius of curvature.

Fig. 11.13 Experimental set-up with magnetic field

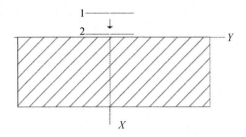

Fig. 11.14 Trajectory of the two particles

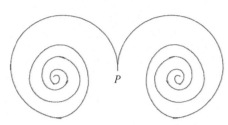

Solution

(a) The principle of conservation of energy establishes that the sum of the kinetic and potential energies in 1 and 2 is the same, that is

$$0 + qV_1 = \frac{1}{2}mv^2 + qV_2 \quad \Rightarrow \quad -eV_1 = 1/2mv^2 - eV_2 \quad \Rightarrow \quad v = \sqrt{\frac{2eV'}{m}}.$$

If $q > 0$, then the impact is to the right, and therefore the half circumference travelled within the magnetic field has its centre at $(0, a/2, 0)$. Since the centripetal acceleration, caused by **B**, points towards the centre of the circumference, then so does the force $q\mathbf{v} \times \mathbf{B} = -e\mathbf{v} \times \mathbf{B}$, and therefore $\mathbf{v} \times \mathbf{B}$ points away from the centre. Hence **B** points out of the paper. Put another way, at the moment when the electron enters the field,

$$\mathbf{F} = -e\mathbf{v} \times \mathbf{B} \quad \Rightarrow \quad F_y \mathbf{u}_y = -ev_x \mathbf{u}_x \times B_z \mathbf{u}_z = ev_x B_z \mathbf{u}_y \quad \Rightarrow \quad F_y = ev_x B_z$$

$$\Rightarrow \quad B_z = \frac{F_y}{ev_x} > 0.$$

Moreover, the fundamental equation of dynamics gives

$$\frac{mv^2}{R} = ev_x B_z = evB \quad \Rightarrow \quad B = \frac{mv}{eR} = \frac{m}{ea/2}\sqrt{\frac{2eV'}{m}} = \frac{2}{a}\sqrt{\frac{2mV'}{e}}.$$

(b) The main difference is that the initial centres of curvature are one to each side, therefore the acceleration is equal except for the direction, and hence the force is equal except for the direction, and therefore the charges must be of opposite signs. The particle to the left must have a negative charge and the particle to the right a positive charge. This allowed experimentally discovery of a new particle: the positron $_{+1}^{0}e$. The reduction of the radius of curvature is due, with high probability, to the reduction of velocity, caused by the collision of the particles with the gas molecules in the cloud chamber.

The existence of the positron was predicted by P. Dirac in 1931, and discovered experimentally by C. Anderson in 1932 while studying cloud chamber photographs of cosmic rays. The positron has the same mass and magnitude of charge, but opposite in sign, as the electron. Rigorously speaking, it constitutes the antiparticle of the electron and it is antimatter. The behaviour of the antimatter is not as usual as thought. In fact, when an electron (matter) coincides with a positron (antimatter) in a region of the space they annihilate each other. As a result two photons (gamma rays) appear moving in opposite directions with energies of 511 keV (momentum conservation law). At first sight it may be thought that the study of the antimatter is only important for the physicists that investigate in quantum field theory, but that is completely wrong. The study of the elementary particles performs a fundamental role in great variety of subjects, such as biology, medicine, chemistry, and of course Physics. By way of illustration, suffice it to say, that the application of the positrons for the diagnostic of some diseases has been used since 1969 (the first time in USA) by means of the PET technique (Positron Emission Tomography). More specifically, by this procedure it is possible to diagnose cancer, degenerative anomalies such as Alzheimer and Parkinson, metabolic disorders, and epilepsy, among others.

PET is a non-invasive method which employs chemical compounds labelled with radioisotopes of short haf-life time, like ^{11}C, ^{13}N, ^{15}O and ^{18}F. These compounds are called tracers and are injected into the body in order to measure where its activity is greatest. The election of the tracer depends on the disease to be investigated. However, one of the most employed tracer is FDG (Fluorodeoxyglucose). As the tumours consume more energy than normal cells, the FDG accumulates more in the regions where the body needs more energy. Due to the fact that the ^{18}F is introduced in the molecule it disintegrates (beta plus decay (β^+), $^{1}_{1}p \to {}^{1}_{0}n + {}^{0}_{+1}e + {}^{0}_{0}\nu$) emitting one positron (${}^{0}_{+1}e$) which annihilates with one of the electrons of the surrounding matter leading to two photons. By using photomultiplier-scintillator detectors located on opposite sides, and computerized tomographic reconstruction based on correlation direction and time coincidence of the photons emitted, it is possible to obtain an image of the regions of more activity. From a medical viewpoint this technique is very sensitive for detecting the activity zones but not the morphology of the tumours. For this reason the new PET-machines bring an incorporated CT[1] scanner, which allows a good reconstruction of the region. As a result by means of the fusion of both data, a precise location of the tumour and its possible malignity are obtained.

11.6. A conductive strip (Hall probe) is located in a region of space where there is a known magnetic field **B**, and voltage V_H is measured when the plane of the tape is perpendicular to the magnetic field. The probe is turned an angle θ around the axis of symmetry parallel to the longest edge. Calculate the Hall voltage that will be measured after the turn.

---

[1]Computerized Tomography (CT) is also a very important technique, but very different from PET. It is based on X rays, and permits physicians to obtain images of plane sections through the body. It is very good for visualizing anatomic structures.

Fig. 11.15 Specimen

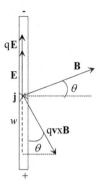

Solution

In the first position of the probe, the voltage is given by (11.17)

$$V_H = wvB \quad \Rightarrow \quad wv = V_H/B.$$

In the second position (see Fig. 11.15), in permanent regime, the carrying charges, if positive, move in the direction of the vector density of current **j** and are subjected to the magnetic field and the electric field caused by the charges that have been deposited at the edges. The resultant of the forces throughout the width of the tape must be null. Therefore

$$qE = qvB\cos\theta \quad \Rightarrow \quad E = vB\cos\theta.$$

The resultant voltage is obtained from

$$E = \frac{V_H'}{w} \quad \Rightarrow \quad V_H' = wE = wvB\cos\theta \quad \Rightarrow \quad V_H' = \frac{V_H}{B}B\cos\theta = V_H\cos\theta.$$

11.7. In a region of space, an electric field **E** and a magnetic field **B** are parallel and homogeneous. A particle with charge q and mass m is impelled with velocity v_0 perpendicular to the fields. Calculate the advance made in the first turn and in the second turn.

Solution

If the OZ axis is drawn in the common direction of the fields and the origin of the coordinates is located at the point of release of the particle, then Fig. 11.16 shows the results. With the data given, the acceleration, velocity and displacement can be calculated. As demonstrated in the theoretical introduction (11.20), the electric field causes an acceleration along the axis OZ of value

$$a_z = \frac{qE}{m} \quad \Rightarrow \quad v_z = \frac{qE}{m}t \quad \Rightarrow \quad z = \frac{qE}{2m}t^2.$$

Fig. 11.16 Electric and magnetic field

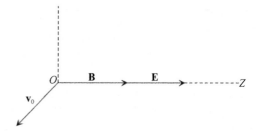

Moreover, the rotation period caused by the magnetic field, (11.15), is

$$T = \frac{2\pi m}{qB}.$$

Substituting this value of time into the previous equation gives the distance travelled in the direction of OZ axis in the first turn thus

$$L_1 = \frac{qE}{2m}T^2 = \frac{qE}{2m}\frac{2^2\pi^2 m^2}{q^2 B^2} = \frac{2\pi^2 mE}{qB^2}.$$

In the time spent in the two turns, $2T$, the distance travelled in the common direction is

$$L_2 = \frac{qE}{2m}(2T)^2 = \frac{qE}{2m}\frac{2^2 \times 2^2\pi^2 m^2}{q^2 B^2} = 4\frac{2\pi^2 mE}{qB^2}.$$

Note how the distance L_2 is not double but four times the distance L_1 in only double the time. The trajectory is not a simple helix but a kind of helix whose pitch increases with time.

11.8. Two isotopes of electrical charge $q = e = 1.602 \times 10^{-19}$ C and masses $m_1 = 1.673 \times 10^{-26}$ kg and $m_2 = 1.743 \times 10^{-26}$ kg, respectively, enter the mass spectrometer described in Sect. 11.6. The electric field applied in the velocity selector is $E = 1000$ V/m and the magnetic fields are equal in the whole device, $B = 0.02$ T. Calculate the velocity of the isotopes on their arrival at the detector and the point where they can be detected.

Solution

By observing the figure of Sect. 11.6, and applying (11.22), the velocity of the isotopes that cross the exit orifice of the velocity selector is calculated

$$v = E/B_1 = 1000/0.02 \text{ m/s} = 50000 \text{ m/s}.$$

Since there is only one magnetic field inside the mass selector, and the modulus of velocity does not vary, then the arrival at the detector is with the velocity of 50000 m/s.

The radius of curvature is obtained by applying (11.23):

$$R = \frac{mE}{qB_1B_2} = \frac{mE}{qB^2}.$$

Therefore, for the isotopes of mass m_1 and m_2, the respective radii are:

$$R_1 = \frac{m_1 E}{qB^2} = \frac{1.673 \times 10^{-26} \times 1000}{1.602 \times 10^{-19}\text{C} \times 0.02^2} = 0.2611 \text{ m},$$

and

$$R_2 = \frac{m_2 E}{qB^2} = \frac{1.743 \times 10^{-26} \times 1000}{1.602 \times 10^{-19}\text{C} \times 0.02^2} = 0.2720 \text{ m}.$$

The points where they can be detected are at the respective distances $2R_1$ and $2R_2$.

11.9. A cyclotron of radius R has a space L between its D's, such that $L \ll R$. There is a magnetic field **B** perpendicular to the plane of the cyclotron. A difference of potential $V = V_0\cos(\omega_i t)$ is applied between the D's, where ω_i is the suitable angular frequency value for each particle. Two different types of particles, of identical positive charges but of respective masses m_1 and m_2, are impelled sequentially. (a) In a first experiment, the particle of mass m_1 and negligible initial velocity is accelerated. In a second experiment, the particle with mass m_2 is accelerated. Calculate the revolutions given by each particle. (b) Determine, by reasoning, the amount of energy supplied by the magnetic field to each of the particles. (c) Obtain the period of rotation of the particle of mass m_1 when the semicircular trajectory of radius $R/2$ is described and compare it with the period corresponding to the trajectory of the last cycle where the radius is R.

Solution

(a) The fundamental equation of dynamics for the particle travelling the final semi-circumference and projected on a radius and towards the centre:

$$qv_f B = \frac{mv_f^2}{R} \Rightarrow v_f = \frac{qBR}{m}.$$

Energy that the electric field between the D's contributes for each cycle (double pass): $2qV_0$.

Applying the principle of conservation of energy to N revolutions, where they reach the final velocity v_f, we have

$$\frac{1}{2}mv_f^2 - 0 = N2qV_0 \Rightarrow N = \frac{m}{4qV_0}v_f^2 = \frac{m}{4qV_0}\left(\frac{qBR}{m}\right)^2 = \frac{qB^2R^2}{4mV_0}.$$

For each particle, substitute m for m_1 or m_2 accordingly. Since the mass is in the denominator, the greater the mass, the fewer the number of cycles traced by the particle.

(b) The energy contributed by the magnetic field can be calculated by means of the work carried out by the force that the magnetic field exerts on the particle, which is

$$W = \int_1^2 \mathbf{F}.d\mathbf{r} = \int_1^2 q\mathbf{v} \times \mathbf{B}.d\mathbf{r} = 0,$$

since $\mathbf{v} \times \mathbf{B}$ is perpendicular to $\mathbf{v}$, that is, to the trajectory, and $d\mathbf{r}$ is tangent to the trajectory.

(c) The period is calculated by applying (11.15)

$$T = \frac{2\pi m}{qB},$$

and therefore the period depends on the mass of the particle, but is independent of the radius of the semi-circumference that it travels.

11.10. An electron with null velocity is injected into a betatron at distance $R = 0.2$ m from its centre. The magnetic field varies from $B = 0$ to $B = B_{max} = 0.005$ T. Calculate the final energy of the electron.

Solution

The velocity at any instant, and the final velocity reached by the electron are obtained from (11.14):

$$v = \frac{qRB}{m} \Rightarrow v_{max} = \frac{qRB_{max}}{m}.$$

Substituting the data from the statement and from the table of constants yields

$$v_{max} = \frac{1.602 \times 10^{-19} \times 0.2 \times 0.005}{9.109 \times 10^{-31}} \frac{m}{s} = 1.759 \times 10^8 \frac{m}{s}.$$

Since this velocity is close to the speed of light in a vacuum, it is not very reliable.

The kinetic energy acquired is estimated by means of substitution of this maximum value into the expression of the kinetic energy:

$$E_k = \frac{1}{2}mv_{max}^2 = \frac{1}{2} 9.109 \times 10^{-31} \times \left(1.759 \times 10^8\right)^2 \text{J} = 1.409 \times 10^{-14} \text{J}.$$

11.11. An electron is pulled by the photoelectric effect, with negligible velocity, from the inner face of the negative plate of a flat capacitor. The separation between the plates is $D = 2$ cm and the difference of potential between them is such that it is at the point of producing a disruptive discharge. Calculate: (a) the energy of an electron on being pulled; (b) the velocity acquired by an

Solved Problems

electron before colliding against the positive armature supposing that there are no collisions against air molecules. Take the data for m and q from the table of physical constants. The dielectric strength of the air is $E = 30000$ V/cm.

Solution

(a) The potential energy is calculated by the work of the force that the electric field applies to the charge:

$$E_p = \int_1^2 q\mathbf{E}.d\mathbf{r} = \int_1^2 qE.dr = -e(V_1 - V_2) \equiv eV.$$

The maximum allowable difference of potential is

$$V = ED.$$

Therefore the energy that it has is

$$E_p = eED \; E_p = eED = 1.602 \times 10^{-19} \times 30000 \times 0.02 \, \text{J} = 961 \times 10^{-19} \text{J}.$$

(b) According to the principle of conservation of energy we can write

$$eED = \frac{1}{2}mv^2 \Rightarrow v = \sqrt{\frac{2eED}{m}} = \sqrt{\frac{2 \times 1.602 \times 10^{-19} \times 30000 \times 0.02}{9.107 \times 10^{-31}}} \text{m/s}$$

$$= 1.453 \times 10^7 \text{m/s},$$

which is, approximately, 5 % of the speed of light; therefore, the classical solution of the problem can give an approximated result, although not exact.

11.12. A synchrotron is formed by an annular vacuum tube of mean radius R (Fig. 11.17). Electrons are required to be accelerated to high velocities while maintaining the radius of the orbit. There is a magnetic field inside the ring, perpendicular to its plane. (a) Given the values for the energy E_r and field **B** at a certain instant, calculate the period T. (b) An accelerating alternating voltage V of constant period T is applied. Calculate the increase of **B** to compensate an increase of energy ΔE_r in a cycle. Given that it is desired to cause an increase per unit time of value $\Delta E_r / \Delta t$, calculate the rapidity of the increase of **B** with time that is needed.

Solution

(a) Since the velocity to be attained is high, it is necessary to apply the formulae of relativistic mechanics. The period is obtained from (11.53)

Fig. 11.17 The synchrotron

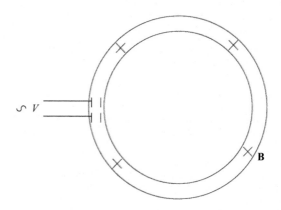

$$T = \frac{2\pi}{\omega} = \frac{2\pi E_r}{ec^2 B}.$$

(b) From this expression B is obtained in terms of E_r

$$B = \frac{2\pi}{ec^2 T} E_r.$$

Therefore, in this problem, B only depends on the variable E_r. The E_r is increased by ΔE_r in each pass by the accelerating electric field due to V, and hence the magnetic field must be increased by

$$\Delta B = \frac{2\pi}{ec^2 T} \Delta E_r,$$

and the increase of B per unit of time is

$$\frac{\Delta B}{\Delta t} = \frac{\Delta B}{T} = \frac{2\pi}{ec^2 T^2} \Delta E_r.$$

Problems B

11.13. A beam of protons, of charge e, homogeneously distributed within a very long cylinder with n protons per unit volume, moves with velocity v along the cylinder axis. Calculate: (a) the electric field existing at distance r from the axis; (b). the magnetic field at this point; (c) the outward radial component of the resultant of the forces on one of the protons.

Solution

(a) As the distribution of charges is known, Gauss's theorem can be applied to any closed surface.

Fig. 11.18 Cylindrical region with protons

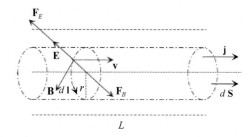

Since there is an axis of symmetry, it is advantageous to consider a cylindrical surface of radius r, concentric with the charge conducting cylinder limited by bases distanced from each other in L (Fig. 11.18).

The flow of the electric field through the specified cylinder is

$$\int E.dS = \int_{lat} E.dS \cos 0° + \int_{ends} E.dS \cos 90° = E \int_{lat} dS = E2\pi rL,$$

where it has been taken into account: (1) the distributive property of the integral, (2) that the electric field is of radial direction due to the symmetry and that E is the component of **E** in the outward radial direction.

The electrical charge density is $\rho \equiv dq/dVol$, therefore the charge within the cylinder is

$$\int \rho dVol = \rho \int dVol = \rho \pi r^2 L = ne\pi r^2 L.$$

Gauss's theorem establishes that

$$E2\pi rL = \frac{ne\pi r^2 L}{\varepsilon_0} \Rightarrow E = \frac{ner}{2\varepsilon_0}.$$

(b) As the system of currents is known and there is cylindrical symmetry, Ampère's law may be applied to a circumference of radius r concentric with the axis of the cylinder.

The circulation of field **B** along the circumference is

$$\oint \mathbf{B}.d\mathbf{l} = \oint B_\phi dl = B_\phi \oint dl = B_\phi 2\pi r = B2\pi r,$$

therefore, through symmetry, the only non-null component of **B** is the tangent to the circumference and B is its projection on $d\mathbf{l}$.

The current density is

$$\mathbf{j} = \rho \mathbf{v} = nev\mathbf{u}_z,$$

and hence the flow of **j**, through the circle whose border is the circumference, is

$$\int \mathbf{j}.d\mathbf{S} = j\pi r^2 = nev\pi r^2.$$

Applying Ampère's law to the circumference gives

$$B 2\pi r = \mu_0 nev\pi r^2 \;\;\Rightarrow\;\; B = \frac{\mu_0 nevr}{2}.$$

(c) The outward radial component of the net force on one of charges q is

$$F = qE - qvB = q\frac{ner}{2\varepsilon_0} - qv\frac{\mu_0 nevr}{2} = \frac{ne^2 r}{2}\left(\frac{1}{\varepsilon_0} - \mu_0 v^2\right).$$

11.14. There are two parallel conducting plates, one of which, called the anode, has an orifice, and a potential $V = 20$ V with respect to the other plate, called the cathode. By illuminating the cathode, electrons of electrical charge $q = -e = -1.602 \times 10^{-19}$ C and of mass $m = 9.107 \times 10^{-31}$ kg, can be extracted from this plate via the photoelectric effect, thereby leaving it with negligible velocity. Calculate the velocity of the electrons that pass through the orifice of the anode.

Solution

Suppose that the distance between the plates is d. With this distance and the difference of potential V, the electric field E can be calculated. With the data q, E and m the acceleration, the velocity and the displacement towards the anode can be calculated successively. In effect

$$E = \frac{V}{d} \;\Rightarrow\; a = \frac{F}{m} = \frac{qE}{m} = \frac{eV}{md} \;\Leftrightarrow\; \frac{dv}{dt} = \frac{eV}{md} \;\Rightarrow\; v = \frac{eV}{md}t,$$

and since the distance travelled from leaving the cathode can be calculated from the velocity, then

$$\frac{dx}{dt} = \frac{eV}{md}t \;\Rightarrow\; d \equiv x = \frac{eV}{2md}t^2 \;\Rightarrow\; t = d\sqrt{\frac{2m}{eV}}.$$

Substitution of this time into the previous equation finally gives

$$v = \frac{eV}{md}t = \frac{eV}{md}d\sqrt{\frac{2m}{eV}} = \sqrt{\frac{2eV}{m}}.$$

In the resolution of this problem, the use of the principle of conservation of energy is advantageous (the sum of the kinetic and potential energies when leaving the cathode is the same as that when arriving at the plane of the anode), in fact,

$$0 + qV_c = \frac{1}{2}mv^2 + qV_a \quad \Rightarrow \quad -eV_c = \frac{1}{2}mv^2 - eV_a \quad \Rightarrow \quad v = \sqrt{\frac{2e(V_a - V_c)}{m}}$$
$$= \sqrt{\frac{2eV}{m}}.$$

11.15. A free particle, of positive charge q and mass m, penetrates into a region of space where a homogenous and stationary magnetic field **B** is present. The velocity v_0 of the particle when entering this region forms an angle α with field **B**. (a) Explain the type of trajectory that the particle will follow in the magnetic field for $\alpha = 0$ and for $\alpha = 90°$. (b) If, in addition to **B** with $\alpha = 90°$, there is also a homogeneous and stationary electric field **E**, what would the direction of **E** have to be to render the movement of the particle rectilinear and uniform?

Solution

(a) The applied force is

$$\mathbf{F} = q\mathbf{v} \times \mathbf{B},$$

which for $\alpha = 0$ gives

$$F = qvB\sin 0 = 0,$$

and since the fundamental equation of dynamics demands that the acceleration is null, therefore the velocity is constant and the trajectory is a straight line and parallel to **B**.

However for $\alpha = 90°$ the modulus of the force is

$$F = qvB\sin 90° = qvB,$$

and the direction is obtained simply by observing Fig. 11.19. A positive charge of initial velocity v_0, undergoes an increase of velocity d**v** in a very short time dt. Since

Fig. 11.19 Initial velocity

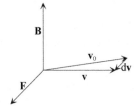

the force is perpendicular to $\mathbf{v}_0$ and $\mathbf{B}$, hence the acceleration $\mathbf{a} = d\mathbf{v}/dt$, and $d\mathbf{v}$ are also perpendicular. Therefore, the new velocity will be in the plane perpendicular to $\mathbf{B}$ and the charge will follow in the plane perpendicular to $\mathbf{B}$.

(b) If the resultant force of both fields must be null, then the following must happen

$$\mathbf{F} = q\mathbf{E} + q\mathbf{v} \times \mathbf{B} = 0.$$

Therefore the vectors $\mathbf{E}$ and $\mathbf{v} \times \mathbf{B}$ must be of equal modulus and opposed directions, that is, $\mathbf{E}$ must be of direction opposite to that of drawn force $\mathbf{F}$ and of modulus $E = vB$.

11.16. A magnetic field $\mathbf{B} = B\mathbf{u}_z$ is homogeneous and stationary. An electron of mass m and charge $q = -e$ is impelled from the origin of coordinates with the velocity $\mathbf{v}_0 = v_{0y}\mathbf{u}_y + v_{0z}\mathbf{u}_z$ (Fig. 11.20). (a) Calculate the radius of the helix that it describes, the pitch OP of the helix, and the coordinates of the point P after the first revolution. (b) If the angle of $\mathbf{v}_0$ with OZ is small, calculate the position of P.

Solution

(a) For this problem the time taken for one turn is obtained from (11.15)

$$T = \frac{2\pi m}{qB}.$$

As the component of the velocity parallel to $\mathbf{B}$ is v_{0z} and remains constant, the distance travelled in the direction of $\mathbf{B}$ in the time T in which the particle completes a turn is

$$OP = v_{0z}T = \frac{2\pi m}{qB}v_{0z}.$$

Therefore, point P has coordinates $(0, 0, \frac{2\pi m}{qB}v_{0z})$.

(b) The component of the velocity on the direction of $\mathbf{B}$ is $v_{0z} = v_0\cos\theta$, where v_0 is the modulus of the velocity and θ is the angle that $\mathbf{v}_0$ forms with $\mathbf{B}$. If the angle is small, a series expansion of $\cos\theta$ allows the following to be written

Fig. 11.20 Trajectory

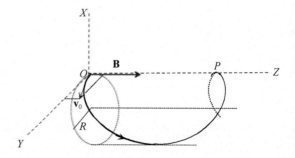

$$v_{0z} = v_0 \cos\theta \approx v_0 \left(1 - \frac{\theta^2}{2} + \cdots\right) \approx v_0.$$

Therefore point P is located in the position $(0, 0, \frac{2\pi m}{qB} v_0)$ for any small value of θ. That is, if a jet of electrons leaves O, each one in a different direction (but directions close to that of **B**), all are focused on the same point. This fact forms the foundation of certain electron microscopes.

11.17. A current **j** circulates through a long conductive sheet of width $w = 15$ mm and thickness $d = 3$ mm. The sheet is introduced into a magnetic field $B = 2$ T perpendicular to the plane of the sheet. The difference of potential between the edge towards which the product $\mathbf{B} \times \mathbf{j}$ points and the opposite edge is $V_H = 3\,\mu$V. The mass density of the sheet is $\rho = 8000$ kg/m³ and consists of atoms of mass $m = 1.7 \times 10^{-25}$ kg. Each atom contributes a charge carrier of unknown sign whose absolute value 1.602×10^{-19} C. Calculate: (a) The velocity of the charge carriers, (b) the intensity of the current in the circuit.

Solution

Suppose that the current density **j** is towards the left. If the transported charges were positive, they would move towards the left and the force that the magnetic field would apply would be downwards; there would be an accumulation of positive charges below and the potential of the lower edge would be positive with respect to the upper edge: this is a contradiction. Therefore, the charge carriers are negative and, as **j** is towards the left, they move towards the right.
(a) By applying (11.17), it is deduced that the modulus of the velocity is

$$v = \frac{V_H}{wB} = \frac{3 \times 10^{-6}}{1.5 \times 10^{-2} \times 2}\,\frac{\text{m}}{\text{s}} = 10^{-4}\,\frac{\text{m}}{\text{s}}.$$

(b) The density of mass can be obtained by multiplying the mass of each atom by the number of atoms in a unit volume, therefore

$$\rho = nm \quad \Rightarrow \quad n = \frac{\rho}{m}.$$

Fig. 11.21 shows that, over time dt, the charges move the distance vdt, and therefore all the charges contained in the parallelepiped of volume $wdv.dt$, which are

Fig. 11.21 Conductive sheet

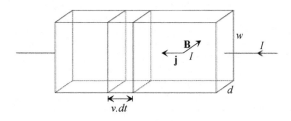

$dq = nqwdv \cdot dt$, cross its face on the right. The current intensity through this face is

$$I = \frac{dq}{dt} = \frac{\rho qwdv}{m} = \frac{8000 \times 1.602 \times 10^{-19} \times 1.5 \times 10^{-2} \times 3 \times 10^{-3} \times 10^{-4}}{1.7 \times 10^{-25}} \text{A}$$
$$= 33.92 \text{ A}.$$

11.18. Figure 11.22 shows a device that corresponds to a curved tube, of average radius R, with two slots S_1 and S_2. Within the device there is a homogeneous field **B** pointing out of the paper. A group of particles are introduced through slot S_1 in the direction drawn, all of mass m and charge q, but which differ in the modulus of velocity. Determine: (a) Which particles pass through the S_2 slot. (b) The time taken for particles to travel from S_1 to S_2. (c) If the particles that leave S_2 continue to be subjected to the same field **B**, determine the modulus and direction of a homogeneous and stationary field **E** outside the tube, such that the movement of particles upon leaving the tube is rectilinear and uniform.

Solution

The trajectory followed within the tube is circular, as drawn, and implies that the acceleration is directed towards its centre of curvature, and since the force $q\mathbf{v} \times \mathbf{B}$ has the same direction as the acceleration, it follows that charge q is positive.
(a) The projection of the fundamental equation of dynamics in the radial direction and towards the centre of curvature gives

$$qvB = m\frac{v^2}{R} \quad \Rightarrow \quad v = \frac{qBR}{m},$$

which provides the dependency of v on the data of the problem and therefore it is the velocity of the particle arriving at slot S2.

Fig. 11.22 Tube

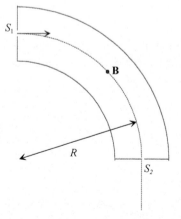

(b) The length of the trajectory is $2\pi R/4$ and the velocity has already been calculated, therefore the time taken for the route is

$$t = \frac{l}{v} = \frac{2\pi R/4}{qBR/m} = \frac{\pi m}{2qB}.$$

(c) For the velocity to be constant on the trajectory once having left the tube, the acceleration must be null, and hence the force of the electromagnetic field must be zero, that is, at any point of the rectilinear trajectory

$$q\mathbf{E} + q\mathbf{v} \times \mathbf{B} = 0.$$

Since $\mathbf{v} \times \mathbf{B}$ is towards the left, $\mathbf{E}$ must be towards the right. Otherwise, the projection of the last equation towards the left is

$$qE + qvB\mathrm{sen}90° = 0 \quad \Rightarrow \quad E = -vB = -\frac{qB^2R}{m}.$$

Problems C

11.19. A beam of particles of charge q enters Aston's mass spectrometer. This apparatus is formed (see the Fig. 11.23) by a section of a cylindrical capacitor of smaller radius R_1 and greater radius R_2 and an aperture S. The particles that pass through S after following the arc of average radius R, perpendicularly penetrate a cylindrical region of radius R', where there is a magnetic field $\mathbf{B}$ inside the cylinder, perpendicular to the plane of the drawing. When leaving the field there is a detector D. The difference of potential between the armature of the capacitor is V. Calculate: (a) The electric field at the points of the arc of radius R inside the capacitor. (b) The velocity of the particles that pass through S. (c) The mass m of those particles which leave towards the detector in a direction perpendicular to that of entry into the magnetic field.

Fig. 11.23 Mass spectrometer

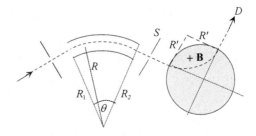

Solution

(a) In order to calculate the electric field, a closed surface is drawn formed by a portion of a cylindrical surface of radius r and length L located between the plates and the rest of the surface enclosing the lower plate. Discounting the effect of the edges, that is, considering that the electric field is null except in the space between the armatures, Gauss's theorem establishes that

$$r\theta L E = \frac{q_i}{\varepsilon_0} \Rightarrow E = \frac{q_i}{\varepsilon_0 r \theta L},$$

where θ is the angle formed by the drawn radii R_1 and R_2, and the electrical charge q_i is on the lower plate.

In order to relate the electric field to the difference of potential between the upper and lower armatures, the following are used

$$V = \int_{R_1}^{R_2} E dr = \int_{R_1}^{R_2} E dr = \int_{R_1}^{R_2} \frac{q_i}{\varepsilon_0 r \theta L} dr = \frac{q_i}{\varepsilon_0 \theta L} \ln \frac{R_2}{R_1}$$

$$\Rightarrow V = ER \ln \frac{R_2}{R_1} \quad \forall r = R$$

$$\Rightarrow E = \frac{V}{R \ln(R_2/R_1)} = \frac{V}{(R_1 + R_2)/2 \times \ln(R_2/R_1)}.$$

(b) The particles that pass through S are those that follow the arc of radius R. Since the magnetic field is perpendicular to this trajectory, the component of that of the fundamental equation of dynamics in the direction perpendicular to the trajectory is

$$qE = m\frac{v^2}{R} \Rightarrow v^2 = \frac{qRE}{m} = \frac{qRV}{m(R_1 + R_2)/2 \times \ln(R_2/R_1)}.$$

Therefore only those particles whose square of their velocity satisfies this equation pass through S, thereby producing a particle selection.

(c) From Fig. 11.23, it is deduced that the radius of the arc travelled inside the magnetic field is R'. The perpendicular component of the fundamental equation of dynamics gives

$$qvB = m\frac{v^2}{R'} \Rightarrow q^2 B^2 = \frac{m^2}{R'^2} v^2.$$

By substituting the square of the velocity obtained in the previous section yields

$$m = \frac{qB^2 R'^2 (R_1 + R_2) \ln(R_2/R_1)}{2RV}.$$

The particles with masses different from this are not detected by D.

Solved Problems

11.20. A magnetic field **B** acts in a cyclotron of radius R. A deuteron (formed by a proton and a neutron) is impelled between its D's with an initial velocity v_0. A difference of alternating potential of type $V = V_0\cos(\omega t)$ is established. Given that the mass and the charge of the deuteron are m and q, respectively, obtain by reasoning: (a) The velocity of the deuteron on completing the first cycle. (b) The energy of the deuteron when exiting the cyclotron. (c) The relationship that must be fulfilled between B, m, q, R and the initial velocity v_0 such that the particle makes at least one cycle. (d) What forces act in the cyclotron? Are they conservative? Explain it.

Solution

(a) The increase of kinetic energy is not due to the magnetic field but to the electric field between the D's. In a cycle of the proton it passes twice between the D's, therefore it absorbs energy $2qV_0$. The increase of kinetic energy in the first cycle is

$$\frac{1}{2}mv^2 - \frac{1}{2}mv_0^2 = 2qV_0 \quad \Rightarrow \quad v = \sqrt{v_0^2 + \frac{4qv_0}{m}}.$$

This is the expression of the velocity on completing the first revolution.

(b) In the last cycle, the radius of curvature is R, and therefore the fundamental equation of dynamics for the final cycle gives

$$qv_f B = m\frac{v_f^2}{R} \quad \Rightarrow \quad v_f = \frac{qBR}{m} \quad \Rightarrow \quad E_{kf} = \frac{1}{2}mv_f^2 = \frac{q^2B^2R^2}{2m}.$$

(c) If N cycles were covered, the variation of the kinetic energy would be

$$\frac{1}{2}mv_f^2 - \frac{1}{2}mv_0^2 = N2qV_0 \quad \Rightarrow \quad N = \frac{mv_f^2 - mv_0^2}{4qV_0}.$$

Since $N \geq 1$ must be true,

$$mv_f^2 - mv_0^2 \geq 4qV_0 \quad \Rightarrow \quad m\left(\frac{qBR}{m}\right)^2 - mv_0^2 \geq 4qV_0 \quad \Rightarrow \quad \frac{q^2B^2R^2}{m} \geq mv_0^2 + 4qV_0.$$

d) The force of magnetic fields acts everywhere. The force of the electric acts in between the D's.
  The work of the forces along a closed line is

$$\oint \mathbf{F}.d\mathbf{l} = \oint q\mathbf{E}.d\mathbf{l} + \oint q(\mathbf{v} \times \mathbf{B}).d\mathbf{l} = qV_0 + qV_0 + 0 = 2qV_0.$$

Magnetic force is not conservative. Electric force is conservative.

11.21. A betatron has an electromagnet whose polar pieces produce a magnetic field $\mathbf{B}_1$ within a circle of radius R_1, and produce a magnetic field $\mathbf{B}_2$ within an annulus concentric with the circle and of inner radius R_1 and outer radius $R_2 = 2R_1$. A proton is introduced so that it reaches energy E_k following the orbit of radius R_2. (a) Calculate the average magnetic field in the circle of radius R_2, assuming B_1 and B_2 are known. (b) Assuming that the magnetic fields $\mathbf{B}_1$ and $\mathbf{B}_2$ are unknown, calculate them.

Solution

(a) The magnetic field averaged over the area of the circle bounded by the orbit is

$$B_m = \frac{1}{S}\int_S B dS = \frac{1}{S}\left(\int_{S_1} B_1 dS_1 + \int_{S_{2-1}} B_2 dS_{2-1}\right) = \frac{B_1 \pi R_2^2/4 + B_2 \pi (R_2^2 - R_2^2/4)}{\pi R_2^2}$$

$$= \frac{1}{4}B_1 + \frac{3}{4}B_2.$$

(b) The charged particle circulates in the orbit of constant radius if the magnetic field on the orbit is half that of the average magnetic field, $B_2 = B_m/2$, and hence, taking into account the value calculated for B_m, gives

$$2B_2 = \frac{1}{4}B_1 + \frac{3}{4}B_2 \Rightarrow B_1 = 5B_2.$$

Furthermore, the fundamental equation of dynamics gives a perpendicular component

$$qvB_2 = m\frac{v^2}{R_2} \Rightarrow v = \frac{qR_2 B_2}{m}.$$

The kinetic energy is related to the velocity by $E_k = mv^2/2$ and, considering the previous equality, gives

$$E_k = \frac{m}{2}\left(\frac{qR_2 B_2}{m}\right)^2 \Rightarrow B_2 = \frac{(2mE_k)^{1/2}}{qR_2} \Rightarrow B_1 = 5\frac{(2mE_k)^{1/2}}{qR_2}.$$

11.22. A force of constant direction and modulus $F = 10000\,\text{N}$ is applied to a stone block of mass $m = 2\,\text{kg}$ at rest. Calculate the velocity of the block after time $t = 20\,\text{h}$. Comment the results.

Solution

Take the OZ axis as the direction of the applied force. With this choice, the components of the force are: $F_x = F_y = 0, F_z = F$.

Application of the fundamental equation of classical dynamics for the components of the acceleration gives: $a_x = a_y = 0, a_z = F_z/m = F/m = 10000/2\,\text{m/s}^2 = 5000\,\text{m/s}^2$.

Therefore, from $\mathbf{v} = \int a dt$, the acquired velocity has the components: $v_x = 0$, $v_y = 0$, $v_z = a_z t = 5000 \times 20 \times 60 \times 60 \, \text{m/s} = 3.6 \times 10^8 \, \text{m/s}$.

As this calculated velocity is greater than the speed of light, this value is impossible. Thus a relativistic approach is necessary.

From the fundamental equation of relativistic dynamics, (11.28), the three components are:

$$0 = \frac{d}{dt}\left(\frac{mv_x}{\sqrt{1 - v^2/c^2}}\right) \Rightarrow \frac{mv_x}{\sqrt{1 - v^2/c^2}} = k_1 = \frac{0m}{\sqrt{1 - 0^2/c^2}} = 0 \Rightarrow v_x = 0.$$

Idem $v_y = 0$

$$F = \frac{d}{dt}\left(\frac{mv_z}{\sqrt{1 - v^2/c^2}}\right) = \frac{d}{dt}\left(\frac{mv_z}{\sqrt{1 - v_z^2/c^2}}\right) \Rightarrow \frac{mv_z}{\sqrt{1 - v_z^2/c^2}} = Ft + k_2 = Ft$$

$$\Rightarrow \frac{m^2 v_z^2}{1 - v_z^2/c^2} = F^2 t^2 \Rightarrow v_z = \left(\frac{1}{\frac{1}{c^2} + \frac{m^2}{F^2 t^2}}\right)^{1/2}$$

$$= \left(\frac{1}{\frac{1}{299792458^2} + \frac{2^2}{10000^2(20\times 60\times 60)^2}}\right)^{1/2} \, \text{m/s} = 2.3037 \times 10^8 \, \text{m/s}.$$

11.23. In a region of space there is an electromagnetic field that, with respect to a reference system, has the components $E_x = 0$, $E_y = 3\,\text{V/m}$, $E_z = 0$, $B_x = 0.2$ T, $B_y = 0$, and $B_z = 0$. Express the equations of motion of a particle of mass m and charge q with a high velocity.

Solution

The force that the field applies on the charge is

$$\mathbf{F} = q(\mathbf{E} + \mathbf{v} \times \mathbf{B}) = q\left(3\mathbf{u}_y + \begin{vmatrix} \mathbf{u}_x & \mathbf{u}_y & \mathbf{u}_z \\ v_x & v_y & v_z \\ 0.2 & 0 & 0 \end{vmatrix}\right) = q\left[(3 + 0.2v_z)\mathbf{u}_y - 0.2v_y\mathbf{u}_z\right].$$

If the velocity is high, the laws of relativistic dynamics (11.33) must be applied, whose three components are:

$$0 = \frac{d}{dt}\left(\frac{mv_x}{\sqrt{1 - (v_x^2 + v_y^2 + v_z^2)/c^2}}\right)$$

$$\Rightarrow \frac{mv_x}{\sqrt{1 - (v_x^2 + v_y^2 + v_z^2)/c^2}} = k_1 = \frac{mv_{x0}}{\sqrt{1 - (v_{x0}^2 + v_{y0}^2 + v_{z0}^2)/c^2}},$$

$$q(3 - 0.2v_z) = \frac{d}{dt}\left(\frac{mv_y}{\sqrt{1 - \left(v_x^2 + v_y^2 + v_z^2\right)/c^2}}\right)$$

$$\Rightarrow \frac{mv_y}{\sqrt{1 - \left(v_x^2 + v_y^2 + v_z^2\right)/c^2}} = q(3 - 0.2v_z)t + k_2 = q(3 - 0.2v_z)t + \frac{mv_{y0}}{\sqrt{1 - \left(v_{x0}^2 + v_{y0}^2 + v_{z0}^2\right)/c^2}}.$$

$$-0.2qv_y = \frac{d}{dt}\left(\frac{mv_z}{\sqrt{1 - (v_x^2 + v_y^2 + v_z^2)/c^2}}\right) \Rightarrow \frac{mv_z}{\sqrt{1 - (v_x^2 + v_y^2 + v_z^2)/c^2}}$$

$$= -0.2qv_y t + k_3 = -0.2qv_y t + \frac{mv_{z0}}{\sqrt{1 - (v_{x0}^2 + v_{y0}^2 + v_{z0}^2)/c^2}}.$$

11.24. A cyclotron has a radius $R = 4$ m and a magnetic field $B = 0.012$ T. (a) Find the velocity of the electrons whose charge is of absolute value $e = 1.602 \times 10^{-19}$ C and whose mass is $m = 9.107 \times 10^{-31}$ kg. (b) For the protons of equal charge to that of the electrons except for the sign and the mass $m_p = 1.673 \times 10^{-27}$ kg, calculate the velocity that they acquire.

Solution

Fundamental equation of classical dynamics

$$\mathbf{F} = m\mathbf{a} \quad \Rightarrow \quad q\mathbf{v} \times \mathbf{B} = m\mathbf{a}.$$

In absolute values

$$qvB = mv^2/R \quad \Rightarrow \quad v = qBR/m.$$

For the electrons, we have

$$v = \frac{1.602 \times 10^{-19} \times 0.012 \times 4}{9.107 \times 10^{-31}} \text{ m/s} = 8.444 \times 10^8 \text{ m/s}.$$

This result is impossible since no object carrying energy can move faster than light in a vacuum. Therefore the fundamental equation of dynamics used is inapplicable in this case.

For the protons

$$v = \frac{1.602 \times 10^{-19} \times 0.012 \times 4}{1.673 \times 10^{-27}} \text{ m/s} = 4.596 \times 10^5 \text{ m/s}.$$

This result is acceptable since protons have a much greater mass than the electrons but have equal energy, and hence their velocity is much smaller.

If the fundamental equation of relativistic dynamics is applied to the electrons, it is sufficient to substitute the value of the energy E_r, (11.29), into the expression of the radius of curvature, (11.52), to obtain

$$v = \left(\frac{R^2 e^2 B^2}{m^2 + R^2 e^2 B^2/c^2}\right)^{1/2}$$

$$= \left(\frac{4^2 \times 1.602^2 \times 10^{-38} \times 0.012^2}{9.107^2 \times 10^{-62} + 4^2 \times 1.602^2 \times 10^{-38} \times 0.012^2/(2.99792458^2 \times 10^{16})}\right)^{1/2} \text{ m/s}$$

$$= 2.7924 \times 10^8 \text{ m/s}.$$

Observe that the result for the electron is very different from that previous and constitutes the only good result.

11.25. Spain's new 3-GeV synchrotron, Alba (Spanish for "dawn light"), appeared on-line in 2010. The first seven beamlines are a mixture of soft X rays, (for applications in material science, solid-state physics, biology, chemistry, and medicine), and of hard X rays for crystallography and absorption studies. The electrons acquire energy of 3 GeV in the accelerator of the machine and enter the storage ring. It is supposed that upon passing through a small zone in this ring where there is a magnetic field of 10 T, the electrons are accelerated by being deflected 3° from their straight trajectory, in order to emit X-rays. Given that the mass of the electron is $m = 9.109 \times 10^{-31}$ kg and its electrical charge is negative and of absolute value $e = 1.602 \times 10^{-9}$ C, and Planck's constant is $h = 6.626 \times 10^{-34}$ Js $= 4.136 \times 10^{-15}$ eVs, calculate: (a) the velocity of the electrons; (b) the time they remain within the magnetic field; (c) the radius of curvature in the magnetic field of 10 T; (d) If the energy of the emitted X-rays is 10 keV, calculate their frequency and wavelength. (1 eV $= 1.602 \times 10^{-19}$ C $\times 1$V $= 1.602 \times 10^{-19}$ J).

Solution

(a) From (11.31), the following is given

$$v = c\sqrt{1 - \left(\frac{9.109 \times 10^{-31} \times 299792458^2}{3 \times 10^9 \times 1.602 \times 10^{-19}}\right)^2} = 0.999999971c = 299792449 \text{ m/s},$$

therefore, the electrons are relativistic.

(b) If the magnetic field of 10 T were unlimited in space, then the period that the protons would remain within the magnetic field would be

$$T = \frac{2\pi r}{v} = \frac{2\pi m}{qB} = \frac{2\pi \times 9.109 \times 10^{-31}}{1.602 \times 10^{-19} \times 10} \text{s} = 357.3 \times 10^{-14} \text{s}.$$

Hence the time spent during the rotation of 3° is

$$t = \frac{3°}{360°} \frac{2\pi m}{qB} = \frac{3}{360} \frac{2\pi 9.109 \times 10^{-31}}{1.602 \times 10^{-19} \times 10} \text{s} = 2.978 \times 10^{-14} \text{s}.$$

(c) The radius is calculated from (11.52), giving

$$R = \frac{vE_r}{qc^2B} \approx \frac{E_r}{qcB} \approx \frac{3 \times 1.602 \times 10^{-19} \times 10^9}{1.602 \times 10^{-19} \times 3 \times 10^8 \times 10} = 1.000 \text{ m}.$$

(d)

$$E_r = h\nu \Rightarrow \nu = \frac{E_r}{h} = \frac{10000 \times 1.602 \times 10^{-19}}{6.626 \times 10^{-34}} \text{Hz} = 2.4177 \times 10^{18} \text{Hz}.$$

Therefore, the wavelength is

$$\lambda = \frac{c}{\nu} = \frac{299792458}{2.4177 \times 10^{18}} \text{ m} = 1.2408 \times 10^{-10} \text{ m} = 0.12408 \text{ nm},$$

which is of the size of the atoms.

Chapter 12
Electromagnetic Waves

Abstract Any varying current or charge distribution varying with time can give rise to radiated electromagnetic fields. Electromagnetic waves, once created, have no connection with the system of charges and currents that produced them. The propagating disturbance travels from one region to another as time passes carrying a certain amount of energy. However, it is only for rapid variations that an appreciable amount of energy is carried away by the wave. In this chapter, we consider the properties of electromagnetic waves derived from the classical electromagnetic theory, Maxwell's equations and Poynting's theorem. The study refers to waves propagating in an uniform, isotropic, and non-conducting medium.

12.1 Electromagnetic Wave Propagation: Wave Equation

Let us assume that there exists time-varying charges or currents within a certain region of space. We will consider here that the electromagnetic waves produced are travelling in an isotropic, homogeneous, and non-conducting linear medium of relative permittivity ε_r and relative permeability μ_r, with no charges, free charge density $\rho_{np} = 0$, and free current density $\mathbf{j}_f = 0$, at any point. Then, propagation may occur not only in dielectric media, where $\varepsilon = \varepsilon_0 \varepsilon_r$ and $\mu = \mu_0 \mu_r$, but also in vacuum, where $\varepsilon = \varepsilon_0$ and $\mu = \mu_0$. Under these assumptions, Maxwell's equations reduce to

$$\nabla \cdot \mathbf{D} = 0 \Rightarrow \nabla \cdot \mathbf{E} = 0, \tag{12.1}$$

$$\nabla \times \mathbf{E} = -\frac{\partial \mathbf{B}}{\partial t}, \tag{12.2}$$

$$\nabla \cdot \mathbf{B} = 0, \tag{12.3}$$

$$\nabla \times \mathbf{H} = \frac{\partial \mathbf{D}}{\partial t} \Rightarrow \nabla \times \mathbf{B} = \mu\varepsilon \frac{\partial \mathbf{E}}{\partial t}. \tag{12.4}$$

By manipulating the above simplified equations, it is found that the fields $\mathbf{E}$ and $\mathbf{B}$ obey the equations

$$\nabla^2 \mathbf{E} = \frac{1}{1/\mu\varepsilon} \frac{\partial^2 \mathbf{E}}{\partial t^2}, \tag{12.5}$$

$$\nabla^2 \mathbf{B} = \frac{1}{1/\mu\varepsilon} \frac{\partial^2 \mathbf{B}}{\partial t^2}. \tag{12.6}$$

These equations are wave equations, with the wave's velocity given by

$$v = \frac{1}{\sqrt{\mu\varepsilon}} = \frac{1}{\sqrt{\mu_0\mu_r\varepsilon_0\varepsilon_r}}. \tag{12.7}$$

It should be noted that in vacuum, the velocity of propagation is $c = 1/\sqrt{\varepsilon_0\mu_0} \simeq 3 \times 10^8$ m/s. Solutions to these equations are electric and magnetic fields that are not independent and together constitute an electromagnetic wave. As (12.5)–(12.6) are linear differential equations, the principle of superposition holds for time-varying electromagnetic fields.

12.2 Plane and Spherical Waves

In Cartesian coordinates, (12.5) for $\mathbf{E}$ is equivalent to three scalar equations, corresponding to the components E_x, E_y, and E_z, respectively. The same can be said for $\mathbf{B}$. Let Ψ be any of the three components of $\mathbf{E}$, or of $\mathbf{B}$, hence the wave equation is simplified to

$$\nabla^2 \Psi = \frac{\partial^2 \Psi}{\partial x^2} + \frac{\partial^2 \Psi}{\partial y^2} + \frac{\partial^2 \Psi}{\partial z^2} = \frac{1}{v^2} \frac{\partial^2 \Psi}{\partial t^2}, \quad v = \frac{1}{\sqrt{\varepsilon\mu}}, \tag{12.8}$$

which is known as the three-dimensional wave equation. For a given orientation of the coordinate system, let us consider that Ψ does not vary in the OY and OZ directions. Then, (12.8) reduces to

$$\frac{\partial^2 \Psi}{\partial x^2} = \frac{1}{v^2} \frac{\partial^2 \Psi}{\partial t^2}, \tag{12.9}$$

which is the one-dimensional wave equation for a non-dispersive medium, in which the velocity is independent of the frequency. The solution of this equation is

$$\Psi(x,t) = f_1(x - vt) + f_2(x + vt). \tag{12.10}$$

These solutions represent waves travelling in opposite directions: $f_1(x - vt)$ corresponds to a waveform moving unchanged in shape with velocity v along the positive OX-axis (Fig. 12.1), whereas $f_2(x + vt)$ represents a waveform travelling in the direction of the negative OX-axis.

12.2 Plane and Spherical Waves

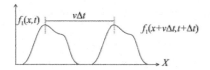

Fig. 12.1 Function $f_1(x - vt)$ at two instants of time. This function represents a travelling wave propagating in the positive OX-direction with velocity v

Let us consider solutions to the wave equation that vary with time according to a harmonic law. Harmonic waves are of special interest because, according to Fourier's theory, any waveform can be obtained by a suitable combination of harmonic waves. A harmonic solution of (12.9) can be written as

$$\Psi(x, t) = \Psi_0 \cos(kx - \omega t + \varphi_0), \tag{12.11}$$

where Ψ_0 is the amplitude, k is the wavenumber, and ω the angular frequency. The phase of the wave is given by $\varphi = kx - \omega t + \varphi_0$, where φ_0 is called the initial phase. This function (12.11) is periodic both in space and time. The distance between equivalent points on successive cycles is equal to the wavelength λ of the wave. The period T is the time required to complete one oscillation. The inverse of the period is known as the frequency $\nu = 1/T$ and is measured in Hz. The wavenumber and the angular frequency are related to the wavelength and the temporal period by $k = 2\pi/\lambda$ and $\omega = 2\pi/T$, respectively. It can be seen that (12.11) is a particular case of (12.10), if the wavenumber and the angular frequency are related to the velocity of propagation of the wave by

$$v = \frac{\omega}{k}, \tag{12.12}$$

which is the rate at which the phase of the wave propagates in space (phase velocity v_p). Note that $v_p = v$ in a non-dispersive medium. Substitution of (12.11) into (12.9) also shows that the former is a solution to the wave equation (12.9).

At a fixed time, the surfaces for which the phase is a constant are called wavefronts, i.e. the surface given by the equation $\varphi = $ constant. If the amplitude of the wave is a constant over the wavefronts, the wave is said to be homogeneous; if not, the wave is nonhomogeneous. Equation (12.11) represents a wave whose wavefront is a plane ($x = $ cte). The wave has the same amplitude Ψ_0 everywhere, and at all points with the same x-coordinate, the wave has the same phase. A wave with a plane phase front is called a plane wave. In general, harmonic solutions of the three-dimensional wave equation, (12.8), representing plane waves can be expressed as

$$\Psi(\mathbf{r}, \mathbf{t}) = \Psi_0 \cos(\mathbf{k} \cdot \mathbf{r} - \omega t + \varphi_0), \tag{12.13}$$

where the vector $\mathbf{r}$ is a position vector of a point in the plane, $\mathbf{k}$ is called the propagation vector and is in the direction of propagation of the wave with magnitude equal to the wave number k, i.e. $\mathbf{k} = (k_x, k_y, k_z) = k\hat{\mathbf{k}}$, $\hat{\mathbf{k}}$ being a unit vector normal to the plane in the propagation direction. In (12.13), the phase is a constant, at a fixed time, in planes $\mathbf{k} \cdot \mathbf{r} = $ cte, that is

$$k_x x + k_y y + k_z z = \text{constant}, \tag{12.14}$$

and this constant is equal to the distance from the origin to the plane. If Ψ_0 does not depend on $\mathbf{r}$, (12.13) represents a homogeneous harmonic plane wave.

If complex notation is used, the harmonic wave is written as

$$\Psi(\mathbf{r}, t) = \text{Re}\{\Psi_0 \exp[i(\mathbf{k} \cdot \mathbf{r} - \omega t + \varphi_0)]\} = \text{Re}\{\Psi_{0c} \exp[-i\omega t]\}, \tag{12.15}$$

where $\Psi_{0c} = \Psi_0 \exp[i(\mathbf{k} \cdot \mathbf{r} + \varphi_0)]$ is called the complex amplitude, which can be considered as a phasor that contains amplitude and phase information but is independent of t. Then, the complex wave can be expressed as

$$\Psi_c = \Psi_{0c} \exp[-i\omega t]. \tag{12.16}$$

The real part of (12.16) actually represents the wave (12.13). If complex notation is used, the subscript "c" is usually omitted for simplicity.

The wave with a spherical phase front is called a spherical wave. It can be shown that the wave equation for spherically symmetric solutions becomes

$$\frac{\partial^2}{\partial r^2}(r\Psi) = \frac{1}{v^2}\frac{\partial^2}{\partial t^2}(r\Psi). \tag{12.17}$$

The general solution to this equation is

$$\Psi(r, t) = \frac{f_1(r - vt)}{r} + \frac{f_2(r + vt)}{r}, \tag{12.18}$$

where $r = |\mathbf{r}|$ is the distance from a point to the source. The harmonic, spherical wave solution is

$$\Psi(r, t) = \frac{A}{r}\cos(kr - \omega t + \varphi_0), \tag{12.19}$$

where A is a constant, and the amplitude A/r is inversely proportional to the distance travelled from the source.

12.3 Harmonic Plane Waves in Unbounded Dielectrics

Any electromagnetic field must satisfy all of Maxwell's equations. Solutions to the wave equations (12.5) and (12.6) that satisfy Maxwell's equations constitute electromagnetic waves. The wave equations have a wide variety of possible solutions. The simplest type of wave that is a solution to (12.5) is a plane wave. Harmonic

12.3 Harmonic Plane Waves in Unbounded Dielectrics

plane waves are important because any three-dimensional wave can be written as a combination of plane waves of different amplitudes, directions, and frequencies. Moreover, far enough away from a source of radiation, the electromagnetic wave can be considered as a plane wave, with considerable approximation. The general expression for a harmonic plane wave solution to (12.5) and (12.6) is of the form

$$\mathbf{E} = \mathbf{E}_0 \cos(\mathbf{k} \cdot \mathbf{r} - \omega t + \varphi_{0E}), \tag{12.20}$$

$$\mathbf{B} = \mathbf{B}_0 \cos(\mathbf{k} \cdot \mathbf{r} - \omega t + \varphi_{0B}). \tag{12.21}$$

This wave is said to be linearly polarized because electric field $\mathbf{E}$ is always parallel to the direction of the amplitude vector $\mathbf{E}_0$. $\mathbf{E}$ points either in the positive or negative direction depending on the instant. Analogously, $\mathbf{B}$ always oscillates in a direction parallel to the amplitude vector $\mathbf{B}_0$.

In order for (12.20) and (12.21) to satisfy Maxwell's equations (12.1)–(12.4), the electric and magnetic fields are not independent, and their amplitudes and phases cannot be specified independently. The following relationships must be satisfied:

$$\nabla \cdot \mathbf{E} = 0, \tag{12.22}$$

$$\nabla \cdot \mathbf{B} = 0, \tag{12.23}$$

$$\mathbf{k} \times \mathbf{E}_0 = \omega \mathbf{B}_0 \Rightarrow \mathbf{k} \times \mathbf{E} = \omega \mathbf{B}. \tag{12.24}$$

From these equations it can be inferred that the electric and magnetic fields are perpendicular to each other, in phase, $\varphi_{0E} = \varphi_{0B} = \varphi_0$, and form a right-handed coordinate system with the propagation vector $\mathbf{k}$, as shown in Fig. 12.2. As $\mathbf{E}$ and $\mathbf{B}$ are both perpendicular to the propagation direction, electromagnetic waves are said to be transverse waves. The magnitude of the magnetic field and that of the electric field are related by $B = E/v$ or $B_0 = E_0/v$. Therefore, the magnitude of the magnetic field in a plane wave is quite small.

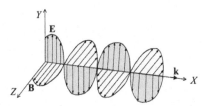

Fig. 12.2 $\mathbf{E}$ and $\mathbf{B}$ fields of a linearly polarized plane wave at an instant of time. The wave propagates in the positive X-direction. The electric field oscillates along the Y-axis and the magnetic field is along the Z-axis and in phase with the electric field

12.4 Polarization

For a plane electromagnetic wave in an isotropic dielectric medium, the electric and magnetic fields are mutually perpendicular and lie in a plane normal to the direction of propagation **k**. The polarization of a plane electromagnetic wave describes the time-varying behaviour of the electric field at a fixed position in space. As the direction of **B** is related to that of **E** by (12.24), a separate study of the behaviour of the magnetic field is not necessary.

The plane of polarization of a wave is defined by the direction of propagation **k** and the direction of oscillation of the electric field **E**. Equation (12.20) describes a plane electromagnetic wave whose plane of polarization is fixed and vector **E** is always in this plane oscillating in direction $\mathbf{E}_0$ (positive or negative). As aforesaid, this wave is said to be linearly polarized. The superposition of two linearly polarized plane waves of the same frequency, propagating in the same direction and with the electric fields oscillating in perpendicular directions, is a plane wave of the same frequency whose state of polarization depends on the phase difference between its components. As a simple example, let us consider the superposition of two plane waves, progressing along $+OX$, which are linearly polarized: one polarized in the Y-direction, and the other polarized in the Z-direction,

$$E_y = E_{0y} \cos(kx - \omega t + \varphi_{0y}), \qquad (12.25)$$

$$E_z = E_{0z} \cos(kx - \omega t + \varphi_{0z}), \qquad (12.26)$$

where E_{0y} and E_{0z} are real numbers denoting the amplitudes, E_{0y}, $E_{0z} > 0$, and φ_{0y} and φ_{0z} are the initial phases (independent on time). The orientation of the resulting electric field (in a plane perpendicular to the direction of propagation) can be fixed or changing with time depending on the phase difference $\delta = \varphi_{0y} - \varphi_{0z}$. The convention used in Optics is followed here. Then, looking into the direction from which the wave is coming, the wave travelling towards us from the source, the electric field **E**, at a fixed position (x = constant), will behave as time progresses as follows:

1. For $\delta = \varphi_{0y} - \varphi_{0z} = 2\pi N$ (N being a whole number), the components are in phase, which results in a vector **E** that oscillates along a line making an angle θ with the Y-axis, where $\tan \theta = E_{0z}/E_{0y}$, which is an constant phase angle. This electromagnetic wave is said to be linearly polarized. It should be noted that if Y and Z components oscillate 180° out of phase, i.e. $\delta = \pi + 2\pi N$, the resulting polarization is also linear.
2. For $\delta = \varphi_{0y} - \varphi_{0z} = \frac{\pi}{2} + N\pi$:

    2.1. If $E_{0y} = E_{0z}$, the resulting vector **E** has a constant magnitude but is continuously changing its direction. In the plane normal to **k**, the tip of the vector **E** describes a circumference with angular frequency ω. If the **E** vector rotates in a clockwise direction, as we view the wave travelling towards us, the wave is said to be right-handed circularly polarized. If the tip of the electric field moves along the circle in counterclockwise direction, the wave is left-handed circularly polarized.

2.2. If $E_{0y} \neq E_{0z}$, the curve described by the tip of the vector **E** is an ellipse in the plane normal to **k**, with its axes aligned with OY and OZ axis. This wave is said to be elliptically polarized, right-handed or left-handed as described for a circularly polarized wave.

3. In the general case, when $E_{0y} \neq E_{0z}$ and the phase difference δ is an arbitrary amount, the vector **E** will be elliptically polarized. The equation of the ellipse described by the tip of **E** is given by

$$\frac{E_y^2}{E_{0y}^2} + \frac{E_z^2}{E_{0z}^2} - 2\frac{E_y E_z}{E_{0y} E_{0z}} \cos \delta = \sin^2 \delta. \tag{12.27}$$

The orientation of the ellipse with respect to the Y-axis is

$$\tan 2\theta = \frac{2E_{0y}E_{0z}}{E_{0y}^2 - E_{0z}^2} \cos \delta, \tag{12.28}$$

as shown in Fig. 12.3.

For the cases considered in 1 and 2, (12.27) becomes a straight line, a circle, or a ellipse with its axes aligned with those of the coordinate system, as shown in Fig. 12.4.

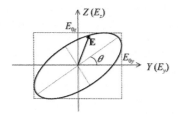

Fig. 12.3 Ellipse described by the tip of vector **E** for the general case given by (12.27)

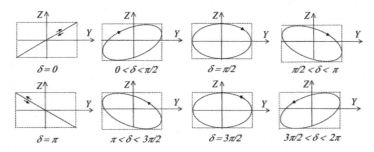

Fig. 12.4 Figures described for the tip of the vector **E** in the YZ-plane perpendicular to the direction of propagation for different phase differences

If the direction of oscillation of the electric field **E** is changing randomly with time, the wave is said to be unpolarized. Only when the varying charges and current distributions producing the waves are suitably controlled do the sources emit radiation of a fixed state of polarization. Plane waves for most visible light sources, except lasers, are unpolarized or randomly polarized.

12.5 Intensity and Poynting Vector

Electromagnetic waves carry with them electromagnetic energy. The energy is transported through space, from one point to another point, by means of waves. Let us consider a finite region of space, with volume V, containing linear isotropic media, and bounded by a surface S. Application of the principle of conservation of energy to the region under consideration leads to the Poynting theorem, which can be expressed as

$$\oint_S (\mathbf{E} \times \mathbf{H}) \cdot d\mathbf{S} = -\frac{d}{dt} \int_V \left[\frac{1}{2}\varepsilon E^2 + \frac{1}{2}\mu H^2 \right] dV - \int_V (\mathbf{j}_f \cdot \mathbf{E}) dV . \quad (12.29)$$

The first term on the right of this equation is the time rate of change of the electromagnetic energy inside the volume V. The last term represents the rate at which the electric field **E** does work on the free charges within the volume V. If there are not sources of e.m.f. in V, and $\mathbf{j}_f$ is a conduction current, the latter is the rate of energy dissipation from Joule's heating, i.e. the ohmic power dissipated in V. The integral on the left represents the amount of energy flowing outwards over the surface S per second. The vector $\mathbf{P}_o = \mathbf{E} \times \mathbf{H}$, is called the Poynting vector, and can be interpreted as the vector giving the direction and rate of electromagnetic energy flow per unit time per unit area. The units of the Poynting vector are W/m^2.

If in the region under consideration there are not free charges, (12.29) gives that the energy of the electromagnetic field in the volume V decreases at a rate equal to the energy flowing out of the surface S per unit time.

The energy density e associated with an electromagnetic field in a linear and isotropic medium, with permittivity ε and permeability μ, is given by (see Chap. 9 (9.23)),

$$e = \frac{1}{2} \mathbf{D} \cdot \mathbf{E} + \frac{1}{2} \mathbf{B} \cdot \mathbf{H} = \frac{1}{2} \varepsilon E^2 + \frac{1}{2} \frac{B^2}{\mu}. \quad (12.30)$$

By using the simple constitutive relations of the medium and the relationship between the magnitude of the magnetic and electric fields in an electromagnetic wave, $B = E/v$, it can be demonstrated that the magnetic energy density is equal to the electric energy density. Hence, the total density energy in an electromagnetic wave can be written as

$$e = \varepsilon E^2 . \quad (12.31)$$

12.5 Intensity and Poynting Vector

Fig. 12.5 The relative directions of electric **E** and magnetic fields **B** and **H**, propagation vector **k**, and Poynting vector **P**$_o$ in a plane wave propagating in a isotropic dielectric medium in the OX-direction

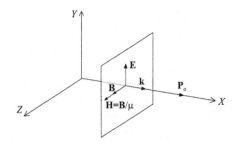

Let us calculate the Poynting vector associated with a plane electromagnetic wave. Using (12.24) and velocity $v = 1/\sqrt{\mu\varepsilon}$, it is found that the instantaneous value of the Poynting vector can be expressed as

$$\mathbf{P}_o = \mathbf{E} \times \mathbf{H} = \mathbf{E} \times \frac{\mathbf{B}}{\mu} = \sqrt{\frac{\varepsilon}{\mu}} E^2 \hat{\mathbf{k}} = v\varepsilon E^2 \hat{\mathbf{k}} = ve\,\hat{\mathbf{k}}. \tag{12.32}$$

Therefore, energy is propagated by the wave in the direction in which the wave propagates at the same velocity as the wave, as shown in Fig. 12.5.

The time average of the Poynting vector over a complete period is given by

$$<\mathbf{P}_o> = v\varepsilon <E^2> \hat{\mathbf{k}} = v<e> \hat{\mathbf{k}}. \tag{12.33}$$

The magnitude of the Poynting vector can be defined as the intensity I_w,

$$I_w \equiv <|\mathbf{P}_o|> = v\varepsilon <E^2> = v<e>. \tag{12.34}$$

This equation gives the average energy flow per unit time across a unit area perpendicular to the direction of propagation (W/m^2).

For a plane harmonic wave, linearly polarized, the fields **E** and **B** are given by (12.20) and (12.21). The instantaneous value of the Poynting vector will be

$$\mathbf{P}_o = \mathbf{E} \times \mathbf{H} = \sqrt{\frac{\varepsilon}{\mu}} E_0^2 \cos^2(\mathbf{k}\cdot\mathbf{r} - \omega t + \varphi_0) \hat{\mathbf{k}} = \varepsilon v E_0^2 \cos^2(\mathbf{k}\cdot\mathbf{r} - \omega t + \varphi_0) \hat{\mathbf{k}}. \tag{12.35}$$

Calculating the average $\mathbf{P}_o$, the average value of $\cos^2(\mathbf{k}\cdot\mathbf{r} - \omega t + \varphi_0)$ over a period is equal to 1/2, thus we have for the intensity I_w of a linearly polarized plane wave

$$I_w = \frac{1}{2}\varepsilon v E_0^2 = v<e>, \tag{12.36}$$

where $<e> = \varepsilon E_0^2/2$ represents the average energy density. Note that the intensity is proportional to the square of the amplitude of the electric field.

12.6 Introduction to Fourier Analysis

In the preceding sections, harmonic solutions to wave equation, given by (12.20) and (12.21), have been considered. In nature, wave disturbances have a finite temporal duration and, therefore, waves propagate as wave trains of finite length. However, by applying Fourier analysis, arbitrary wavefronts can be described in terms of combinations of harmonic plane waves. The Fourier theory states that a periodic function can be expressed as a Fourier series, a sum of sinusoidal functions, while the Fourier transform can be used to describe non-periodic functions.

If a periodic function $f(t)$, with period T_0, has a finite number of points of discontinuity, and has a finite number of maxima and minima in the interval representing the period, then the function can be represented by a Fourier series

$$f(t) = \frac{a_0}{2} + \sum_{n=1}^{\infty} [a_n \cos(2\pi n \nu_0 t) + b_n \sin(2\pi n \nu_0 t)], \quad (12.37)$$

where ν_0 is the fundamental frequency equal to $1/T_0$. The coefficients of the two summations are given by the integrals

$$a_n = \frac{2}{T_0} \int_{-T_0/2}^{T_0/2} f(t) \cos(2\pi n \nu_0 t) dt \quad n = 0, 1, 2 \ldots, \quad (12.38)$$

$$b_n = \frac{2}{T_0} \int_{-T_0/2}^{T_0/2} f(t) \sin(2\pi n \nu_0 t) dt \quad n = 1, 2, 3 \ldots. \quad (12.39)$$

Equation (12.37) shows the expansion of $f(t)$ in terms of sine and cosine functions that are harmonics of the frequency ν_0. The term a_0, associated with zero frequency, represents the average value of $f(t)$ over one period. The plot displaying the coefficients as a function of the frequencies ($n\nu_0$) is called the frequency spectrum (discrete for the series).

We have written (12.37) using time variable, t, the conjugate variable being the frequency ν in Hz. When a spatial variable is used, for instance x, the conjugate variable is called spatial frequency α (in m^{-1}).

Frequency analysis of non-periodic signals and, in general, signals defined in a finite time interval, can be carried out by means of the Fourier transform, and defined by the expression

$$FT\{f(t)\} = F(\nu) = \int_{-\infty}^{\infty} f(t) \exp(-i 2\pi \nu t) dt. \quad (12.40)$$

This integral, which is a function of ν, yields the function $F(\nu)$, and thus we have the Fourier transform of $f(t)$. The non-periodic function $f(t)$ is represented by an infinite number of harmonic functions with frequencies infinitely closely together. $F(\nu)$ gives the contribution of frequency ν to the representation of the function. The absolute value of $F(\nu)$ is called the spectrum of the function $f(t)$. The inverse Fourier transform is defined as

12.6 Introduction to Fourier Analysis

$$FT^{-1}\{F(\nu)\} = f(t) = \int_{-\infty}^{\infty} F(\nu) \exp(i 2\pi \nu t)\, d\nu, \quad (12.41)$$

which allows the determination of a function from its Fourier transform.

There are conditions for the existence of Fourier transforms that are discussed in depth in mathematic treatments, such as that by Bracewell or that by Papoulis. We only point out here that if the integral of $|f(t)|$ from $-\infty$ to ∞ exits and any discontinuities in $f(t)$ are finite, the Fourier transform $F(\nu)$ exits and satisfies the inverse Fourier transform.

Solved Problems

Problems A

12.1 The electric field of an electromagnetic wave is given by the expression:

$$\mathbf{E} = (E_{0x}\,\mathbf{u}_x + E_{0z}\,\mathbf{u}_z)\cos(ky + \omega t).$$

The wave is travelling in a nonmagnetic, dielectric medium with velocity v. Find: (a) The relationship between k and ω so that this field represents a wave propagating without distortion. (b) The wavefront, the direction of propagation, and the state of polarization. (c) The associated magnetic field $\mathbf{B}$. (d) The instantaneous Poynting vector.

Solution

(a) For the electric field to be an electromagnetic wave, components $E_x = E_{0x}\cos(ky + \omega t)$ and $E_z = E_{0z}\cos(ky + \omega t)$ must satisfy the wave equations,

$$\frac{\partial^2 E_x}{\partial y^2} = \frac{1}{v^2}\frac{\partial^2 E_x}{\partial t^2} \quad ; \quad \frac{\partial^2 E_z}{\partial y^2} = \frac{1}{v^2}\frac{\partial^2 E_z}{\partial t^2}.$$

For the X-component, it is found

$$\frac{\partial E_x}{\partial y} = -E_{0x}k\sin(ky + \omega t) \Rightarrow \frac{\partial^2 E_x}{\partial y^2} = -E_{0x}k^2\cos(ky + \omega t),$$

$$\frac{\partial E_x}{\partial t} = -E_{0x}\omega\sin(ky + \omega t) \Rightarrow \frac{\partial^2 E_x}{\partial t^2} = -E_{0x}\omega^2\cos(ky + \omega t)$$

By eliminating the cosine term in the equations above and comparing the equation obtained with (12.9), we find the relationship between phase velocity v, wavenumber k, and angular frequency ω,

$$\frac{\partial^2 E_x}{\partial y^2} = \frac{k^2}{\omega^2}\frac{\partial^2 E_x}{\partial t^2} \Rightarrow v = \frac{\omega}{k}.$$

Analogously, component E_z yields the same relationship.

(b)

- Wavefront: By equating the phase to a constant at a fixed time, it is found that

$$ky + \omega t = \text{constant} \Rightarrow y = \text{constant},$$

therefore, the wavefront is planar.

- The direction of propagation is determined from the condition that **E** does not change as the electric field propagates. Then, the value of **E** at position y and time t is the same as that obtained at $y + \Delta y$ in the time $t + \Delta t$ ($\Delta t > 0$). This requirement is satisfied if the phases are equal,

$$ky + \omega t = k(y + \Delta y) + \omega(t + \Delta t),$$

from which we obtain that $\Delta y < 0$ and $\Delta y = -v\Delta t$. Therefore, **E** represents a cosinusoidal wave travelling in the $-OY$-direction.

- The wave is oscillating in the direction of the vector $(E_{0x}\mathbf{u}_x + E_{0z}\mathbf{u}_z)$, i.e. the direction of **E** remains the same at all times, then the wave is linearly polarized, and thus the equation of the straight line is given by

$$E_z = \frac{E_{0z}}{E_{0x}} E_x \quad \left(z = \frac{E_{0z}}{E_{0x}} x\right).$$

(c) The wave propagates along the $-OY$-axis, the propagation vector is given by $\mathbf{k} = -k\,\mathbf{u}_y$. From (12.24), which relates **B**, **E**, and **k**, the magnetic field is given by

$$\mathbf{k} \times \mathbf{E} = \omega \mathbf{B} \Rightarrow \mathbf{B} = \frac{1}{\omega}\begin{vmatrix} \mathbf{u}_x & \mathbf{u}_y & \mathbf{u}_z \\ 0 & -k & 0 \\ E_{0x}\cos(ky+\omega t) & 0 & E_{0z}\cos(ky+\omega t) \end{vmatrix},$$

$$\mathbf{B} = \frac{1}{\omega}\left[-kE_{0z}\mathbf{u}_x + kE_{0x}\mathbf{u}_z\right]\cos(ky+\omega t) = \frac{-E_{0z}\mathbf{u}_x + E_{0x}\mathbf{u}_z}{v}\cos(ky+\omega t).$$

(d) The instantaneous Poynting vector is calculated from the cross product of **E** and **H**, where $\mathbf{H} = \mathbf{B}/\mu_0$.

$$\mathbf{P}_o = \mathbf{E} \times \mathbf{H} = \left[E_{0x}\cos(ky+\omega t), 0, E_{0z}\cos(ky+\omega t)\right] \times \left[\frac{-E_{0z}}{v\mu_0}\cos(ky+\omega t), 0, \frac{E_{0x}}{v\mu_0}\cos(ky+\omega t)\right]$$

$$= -\frac{\cos^2(ky+\omega t)}{\mu_0 v}(E_{0x}^2 + E_{0z}^2)\mathbf{u}_y,$$

where $v = 1/\sqrt{\mu_0 \varepsilon_0 \varepsilon_r}$ ($\mu_r \simeq 1$). It should be noted that the Poynting vector points in the same direction as the propagation of the wavefronts.

12.2 The electric field of an electromagnetic wave propagating in vacuum is given by

$$\mathbf{E} = \begin{cases} E_x = E_{0x} \cos \frac{2\pi}{3}(z - 3 \times 10^8 t) \\ E_y = E_{0y} \sin \frac{2\pi}{3}(z - 3 \times 10^8 t) \\ E_z = 0 \end{cases}$$

Determine: (a) The wavefront and the wavelength. (b) The state of polarization. (c) The magnetic field.

Solution

(a) By equating the phase to a constant, we have at a fixed time

$$\varphi = \frac{2\pi}{3}(z - 3 \times 10^8 t) = \text{constant} \Rightarrow z = \text{constant (plane wavefront)}.$$

As the wavenumber $k = 2\pi/\lambda = 2\pi/3 \text{ m}^{-1}$, the wavelength is $\lambda = 3$ m.

(b) In order to obtain the curve described by the tip of $\mathbf{E}$ at a fixed location in space, we eliminate the dependence of E_x and E_y on $k(z - ct)$ by calculating the sum of squares of E_x/E_{0x} and E_y/E_{0y},

$$\left(\frac{E_x}{E_{0x}}\right)^2 + \left(\frac{E_y}{E_{0y}}\right)^2 = \cos^2\left[\frac{2\pi}{3}(z - 3 \times 10^8 t)\right] + \sin^2\left[\frac{2\pi}{3}(z - 3 \times 10^8 t)\right] = 1.$$

Therefore, the tip of $\mathbf{E}$ traces an ellipse as time passes and the wave is said to be elliptically polarized.

To determine the direction of rotation of the $\mathbf{E}$ field vector, we examine the locus of $\mathbf{E}$ versus time at $z = 0$, for simplicity. We look at this plane so that we can observe the wave travelling to us, as shown in Fig. 12.6. At $z = 0$, we have

$$\mathbf{E}(0, t) = E_{0x} \cos(-2\pi \times 10^8 t)\mathbf{u}_x + E_{0y} \sin(-2\pi \times 10^8 t)\mathbf{u}_y$$
$$= E_{0x} \cos(-2\pi \times 10^8 t)\mathbf{u}_x + E_{0y} \cos(-2\pi \times 10^8 t - \pi/2)\mathbf{u}_y.$$

The phase difference between φ_{0x} and φ_{0y} is: $\varphi_{0x} - \varphi_{0y} = \pi/2$, which gives right-hand elliptical polarization according to the cases described in the example included in Sect. 12.4 (point 2.2) and in Fig. 12.4, i.e. if a stationary observer faces against the direction of propagation of the wave, the observer sees the vector $\mathbf{E}$ rotating clockwise, and the tip describes an ellipse whenever a period T passes, as shown in Fig. 12.6a.

Let us verify the direction of rotation of the electric field by evaluating the value of $\mathbf{E}$ at different instants of time. Simplifying the above expressions for E_x and E_y at $z = 0$, we have:

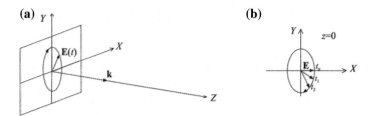

Fig. 12.6 a A stationary observer that faces against the direction of travel sees at $z = 0$ that **E** rotates clockwise around an ellipse. **b** Polarization ellipse at $z = 0$ and electric field **E** at t_0, t_1, and t_2, corresponding to $\omega t = 0$, $\pi/6$, and $\pi/3$, respectively

$$E_x = E_{0x}\cos(-\omega t) = E_{0x}\cos(\omega t),$$
$$E_y = E_{0y}\sin(-\omega t) = -E_{0y}\sin(\omega t),$$

where $\omega = 2\pi \times 10^8$ (rad/s).

At $t = 0$, the **E** field points in the $+X$-direction. As time increases a bit, E_x begins to decrease with time and E_y to increase negatively. Figure 12.6b shows the **E**-field for $\omega t = 0$ (t_0), $\omega t = \pi/6$ (t_1), and $\omega t = \pi/3$ (t_2). The electric field rotates clockwise, as predicted by applying the general results discussed in Sect. 12.4.

(c) The magnetic field **B** is calculated taking into account that $v = \omega/k = c = 3 \times 10^8$ m/s, and $\mathbf{k} = 2\pi/3\,\mathbf{u}_z$ ($\hat{\mathbf{k}} = \mathbf{u}_z$). Then, we have

$$\omega \mathbf{B} = \mathbf{k} \times \mathbf{E} \Rightarrow \mathbf{B} = \frac{1}{c}(\hat{\mathbf{k}} \times \mathbf{E}) = \frac{1}{c}\begin{vmatrix} \mathbf{u}_x & \mathbf{u}_y & \mathbf{u}_z \\ 0 & 0 & 1 \\ E_x & E_y & 0 \end{vmatrix} = \frac{1}{c}\left(-E_y\mathbf{u}_x + E_x\mathbf{u}_y\right),$$

$$\left.\begin{array}{l} B_x = -\frac{E_{0y}}{c}\sin\frac{2\pi}{3}(z - 3 \times 10^8 t) \\ B_y = \frac{E_{0x}}{c}\cos\frac{2\pi}{3}(z - 3 \times 10^8 t) \\ B_z = 0 \end{array}\right\}$$

12.3 Write the expression for the **E** field of a harmonic plane electromagnetic wave with a wavelength of 600 nm and an intensity of 60 W/m². The wave is travelling in the $+OZ$-direction in vacuum and is linearly polarized at an angle of 30° to the OX-axis.

Solution

The amplitude of the electric field of the linearly polarized wave can be calculated from the intensity. Then, from (12.36), for vacuum $\varepsilon = \varepsilon_0$ and $v = c$,

$$I_w = \frac{1}{2}\varepsilon_0 c E_0^2 \Rightarrow E_0 = \sqrt{\frac{2I_w}{\varepsilon_0 c}} = \sqrt{\frac{2 \times 60}{8.85 \times 10^{-12} \times 3 \times 10^8}} = 212.6 \text{ V/m}.$$

Fig. 12.7 Directions of **k** and **E** for the linearly polarized electromagnetic wave

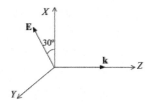

The wave number is determined from the wavelength,

$$k = \frac{2\pi}{\lambda} = \frac{2\pi}{600 \times 10^{-9}} = \frac{\pi}{3} \times 10^7 \text{ m}^{-1}.$$

As the wave is linearly polarized at 30° to OX, the amplitude $E_{0x} = E_0 \cos 30° = 212.6 \cos 30° = 184.1$ V/m, and $E_{0y} = E_0 \sin 30° = 212.6 \sin 30° = 106.3$ V/m. Then, the components of the wave are given by

$$\left.\begin{array}{l} E_x = E_{0x} \cos[k(z-ct)] = 184.1 \cos\left[\frac{\pi}{3} \times 10^7 \left(z - 3 \times 10^8 t\right)\right] \text{ (V/m)} \\ E_y = E_{0y} \cos[k(z-ct)] = 106.3 \cos\left[\frac{\pi}{3} \times 10^7 \left(z - 3 \times 10^8 t\right)\right] \text{ (V/m)} \\ E_z = 0 \end{array}\right\}$$

It should be noted that a cosine-type wave is assumed and both components are in phase, the initial phase assumed being zero. Figure 12.7 shows the **k** and **E** vectors for the linearly polarized wave.

12.4 The electric field of an electromagnetic wave is given by the expression,

$$\mathbf{E} = \left(-\mathbf{u}_x + \sqrt{2}\,\mathbf{u}_y\right) \times 10^2 \cos\left[\frac{5}{3}\pi\left(x + \frac{\sqrt{2}}{2}y\right) 10^5 - 1.9253 \times 10^{14} t\right] \text{ V/m}.$$

Find: (a) Propagation vector and direction of propagation of the wave. (b) Wavelength and phase velocity. (c) State of polarization. (d) Intensity.

Solution

(a) The electric field is of the type:

$$\mathbf{E} = \mathbf{E}_0 \cos(\mathbf{k} \cdot \mathbf{r} - \omega t) = \mathbf{E}_0 \cos(k_x x + k_y y + k_z z - \omega t),$$

where $\mathbf{E}_0 = \left(-\mathbf{u}_x + \sqrt{2}\,\mathbf{u}_y\right) \times 10^2$ V/m, $k_x = (5\pi/3) \times 10^5$ m^{-1}, $k_y = (5\pi/3) \times (\sqrt{2}/2) \times 10^5$ m^{-1}, $k_z = 0$, and $\omega = 1.9253 \times 10^{14}$ rad/s. Then, the propagation vector, $\mathbf{k} = k_x\,\mathbf{u}_x + k_y\,\mathbf{u}_y + k_z\,\mathbf{u}_z$, will be

$$\mathbf{k} = \frac{5}{3}\pi \times 10^5 \left(\mathbf{u}_x + \frac{\sqrt{2}}{2}\mathbf{u}_y\right) \text{ m}^{-1},$$

whose magnitude $|\mathbf{k}|$ is the wave number

$$k = \frac{5}{3}\pi \times 10^5 \times \sqrt{\frac{3}{2}} = 6.41 \times 10^5 \text{ m}^{-1}.$$

An unit vector $\hat{\mathbf{k}}$ in the direction of propagation of the wave is given by

$$\hat{\mathbf{k}} = \frac{\mathbf{k}}{k} = \sqrt{\frac{2}{3}}\mathbf{u}_x + \frac{1}{\sqrt{3}}\mathbf{u}_y = 0.82\,\mathbf{u}_x + 0.58\,\mathbf{u}_y.$$

Note that $\hat{\mathbf{k}} \cdot \mathbf{E} = 0$, as expected.

(b) From the wavenumber k,

$$k = \frac{2\pi}{\lambda} \Rightarrow \lambda = \frac{2\pi}{k} = 9.80 \times 10^{-6} \text{ m (infrared)}.$$

From the angular frequency $\omega = 1.9253 \times 10^{14}$ rad/s, and the value of k calculated in (a), the phase velocity gives

$$v = \frac{\omega}{k} \simeq 3 \times 10^8 \text{ m/s},$$

which shows that the wave is travelling in a material whose velocity is approximately equal to that of vacuum.

(c) The wave is oscillating in the direction of the vector $\mathbf{E}_0 = \left(-\mathbf{u}_x + \sqrt{2}\mathbf{u}_y\right) \times 10^2$, thus the wave is linearly polarized. The equation of the straight line described by the tip of $\mathbf{E}$ is calculated from the quotient,

$$\left.\begin{array}{l} E_x = -10^2 \cos\left[\frac{5}{3}\pi\left(x + \frac{\sqrt{2}}{2}y\right)10^5 - 1.9253 \times 10^{14}t\right] \\ E_y = \sqrt{2} \times 10^2 \cos\left[\frac{5}{3}\pi\left(x + \frac{\sqrt{2}}{2}y\right)10^5 - 1.9253 \times 10^{14}t\right] \end{array}\right\} \Rightarrow \frac{E_y}{E_x} = -\sqrt{2} \Rightarrow y = -\sqrt{2}\,x.$$

Then, $\tan\theta = E_y/E_x = -\sqrt{2}$, $\theta = -54.7°$, as shown in Fig. 12.8. At a fixed position in space, the y and x components oscillate in phase, the tip of the electric field undergoes a simple harmonic motion along the straight line segment of the above calculated line, the amplitude being equal to the magnitude of $\mathbf{E}_0$, $|\mathbf{E}_0| = E_0 = \sqrt{3} \times 10^2$ (V/m).

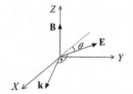

Fig. 12.8 Directions of the fields **E** and **B** and the propagation vector **k** for the linearly polarized wave. Vector **E** points in the direction given by θ

(d) For the linearly polarized wave travelling in a media with electromagnetic properties similar to those of vacuum, (12.36) gives

$$I_w = \frac{1}{2}\varepsilon_0 c E_0^2 = \frac{1}{2} \times 8.85 \times 10^{-12} \times 3 \times 10^8 \times (\sqrt{3} \times 10^2)^2 \text{ W/m}^2 = 39.83 \text{ W/m}^2.$$

12.5 The average intensity of sunlight at the Earth's surface is about 1.36×10^3 W/m². (a) Calculate the amplitude of the electric and magnetic fields at the Earth's surface. (b) Estimate the total power radiated by the Sun. Assume that the radius of Earth's orbit around the Sun is 1.49×10^{11} m. (c) Estimate the power received by the Earth from the Sun. The average radius of the Earth is 6371 km. (d) For a country such as Spain, with a surface of 5×10^{11} m² and an average latitude of 40 degrees, estimate the power received.

Solution

(a) Since the distance between the Sun and the Earth is very large, the wave received on the Earth can be considered to be plane. For simplicity, the wave is assumed to be linearly polarized and the velocity of propagation is $v \simeq c$. The amplitude of the electric field can be calculated from (12.36), which gives the average intensity,

$$I_w = \frac{1}{2}\varepsilon_0 c E_0^2 \Rightarrow E_0 = |\mathbf{E}_0| = \sqrt{\frac{2I_w}{\varepsilon_0 c}} = \sqrt{\frac{2 \times 1.36 \times 10^3}{8.85 \times 10^{-12} \times 3 \times 10^8}} \approx 1012 \text{ V/m}.$$

Then, the amplitude of the magnetic field will be $B_0 = |\mathbf{E}_0|/c = E_0/c = 1012/3 \times 10^8$ T $= 3.38 \times 10^{-6}$ T.

(b) Assuming that the total power from the Sun is radiated isotropically and that there is no loss of energy in travelling through space, it is found that the Sun's total emission power P_s equals the total power spread over a sphere with the Sun at the center. It is assumed that Earth has a "imaginary" circular orbit with a radius of $R_o = 1.49 \times 10^{11}$ m. At the Earth's orbit each square meter of area facing the Sun receives about 1.36×10^3 W ($I_w = 1.36 \times 10^3$ W/m²). Figure 12.9a shows the sphere of radius R_o and area $S = 4\pi R_o^2$. Equating the total power P_s from the Sun to the total power over this sphere,

$$P_s = I_w S = I_w \times 4\pi R_o^2 \approx 1.36 \times 10^3 \times 4\pi \times (1.49 \times 10^{11})^2 \text{ W} \approx 3.79 \times 10^{26} \text{ W}.$$

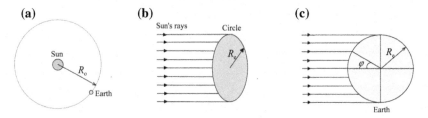

Fig. 12.9 a Earth traveling on an "imaginary" circular orbit around the Sun with radius R_o. b The circle is the projection of Earth's surface on a plane perpendicular to Sun's rays. c The angle at which sunlight strikes the Earth varies with the latitude angle φ

(c) Let us consider a simple model to make an estimate. If the Earth were a flat disk perpendicular to the sunlight rays, with radius R_e, the area of the planet facing the Sun would be the area of the circle, πR_e^2, see Fig. 12.9b, and the power P_e received by the Earth would be given by

$$P_e \approx I_w \times \pi R_e^2 \approx 1.36 \times 10^3 \times \pi \times (6371 \times 10^3)^2 \text{ W} \approx 1.73 \times 10^{17} \text{ W}.$$

(d) The power P_{Sp} received by a country such as Spain with an area $S_{Sp} = 5 \times 10^{11}$ m^2 and a latitude of $\varphi = 40°$ (see Fig. 12.9c) can be estimated by

$$P_{Sp} \approx I_w S_{Sp} \cos \varphi \approx 1.36 \times 10^3 \times 5 \times 10^{11} \times \cos 40° \text{ W} \approx 5.2 \times 10^{14} \text{ W},$$

where $\cos \varphi$ is included in the formula because the surface is not normal to the Sun's rays. Then, the solar power falling on it will be reduced by the cosine of the angle between the line perpendicular to the surface and a central ray from the Sun, as shown in Fig. 12.9c.

12.6 The electric field detected by a radio receiver has an amplitude of 0.15 V/m. Assuming that the wave can be considered linearly polarized, determine: (a) The amplitude of the magnetic field. (b) The intensity of the wave and the energy density. (c) If the distance of the receiver to the transmitter is 1.5 km, calculate the power with which the radio transmitter emits.

Solution

It is assumed that a harmonic electromagnetic wave propagates in vacuum and that the emitter can be considered a point source. As the energy is emitted isotropically in all directions, the energy of the wavefront is spread out over a spherical surface with area $4\pi r^2$, where r is the distance from the transmitter to the receiver, as shown in Fig. 12.10. At a point on the spherical surface, the electric field can be expressed as

$$E(r, t) = \frac{A}{r} \cos(kr - \omega t) = E_0(r) \cos(kr - \omega t),$$

where ω is the angular frequency of the source, A a constant, and $E_0(r) = A/r$.

(a) At the location of the radio receiver, the magnetic field has an amplitude,

$$B_0 = \frac{E_0}{c} = \frac{0.15}{3 \times 10^8} \text{ T} = 5 \times 10^{-10} \text{ T}.$$

Fig. 12.10 Geometry of wave propagation from a point transmitter Q that emits with a power P. Energy is radiated homogenously in all directions. Wavefront is spherical in shape

(b) In calculating the average intensity, we can use (12.36), E_0 being the amplitude of the magnetic field at the detection point. Hence, (12.36) yields for the intensity

$$I_w = \frac{1}{2}\varepsilon_0 c E_0^2 = \frac{1}{2} \times 8.85 \times 10^{-12} \times 3 \times 10^8 \times 0.15^2 \text{ W/m}^2 \approx 3 \times 10^{-5} \text{ W/m}^2.$$

From (12.36), it is inferred that the average energy density $<e>$ is given by

$$<e> = \frac{I_w}{c} = \frac{1}{2}\varepsilon_0 E_0^2 \approx 10^{-13} \text{ J/m}^3.$$

(c) In the case under consideration, the wave energy is conserved as it propagates through the medium. Thus, the average power P emitted by the point source Q equals the average power crossing the spherical wavefronts. Since the distribution of intensity on the spherical wavefronts can be considered homogeneous, application of the conservation of energy principle leads to

$$P = I_w S = I_w 4\pi r^2 = 3 \times 10^{-5} \times 4\pi \times (1500)^2 \text{ W} \approx 848 \text{ W}.$$

Problems B

12.7 An electromagnetic harmonic plane wave of frequency 1 MHz is travelling through vacuum in the direction of the unit vector $(0, \sqrt{3}/2, 1/2)$. The wave is linearly polarized, along the X-direction, and has an amplitude of 0.05 V/m. Write the expressions for the electric and magnetic fields.

Solution

First of all, let us determine the wavelength λ and the wavenumber k:

$$c = \lambda \nu \Rightarrow \lambda = \frac{3 \times 10^8}{10^6} = 300 \text{ m} \quad \text{and} \quad k = \frac{2\pi}{\lambda} = \frac{2\pi}{300} \text{ m}^{-1}.$$

Then, we have for the propagation vector

$$\mathbf{k} = k\hat{\mathbf{k}} = \frac{2\pi}{300}\left(\frac{\sqrt{3}}{2}\mathbf{u}_y + \frac{1}{2}\mathbf{u}_z\right) \text{ m}^{-1}.$$

As the amplitude of the electric field is known, $E_0 = 0.05$ (V/m), as well as its direction of oscillation, parallel to OX, the electric field can be expressed as (Fig. 12.11)

$$\mathbf{E} = \mathbf{E}_0 \cos(\mathbf{k}\cdot\mathbf{r} - \omega t)$$
$$= (0.05\,\mathbf{u}_x)\cos\left[\frac{2\pi}{300}\left(\frac{\sqrt{3}}{2}y + \frac{1}{2}z\right) - 2\times\pi\times 10^6 t\right] \text{ (V/m)}.$$

Fig. 12.11 The electric and magnetic fields of the electromagnetic wave are perpendicular to the direction of propagation

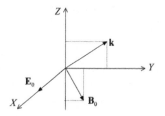

The magnetic field can be easily calculated by (12.24):

$$\mathbf{k} \times \mathbf{E} = \omega \mathbf{B} \Rightarrow \mathbf{B} = \frac{1}{c}\left(\hat{\mathbf{k}} \times \mathbf{E}\right) = \frac{1}{c}\begin{vmatrix} \mathbf{u}_x & \mathbf{u}_y & \mathbf{u}_z \\ 0 & \frac{\sqrt{3}}{2} & \frac{1}{2} \\ E_x & 0 & 0 \end{vmatrix} = \frac{1}{c}\left[\frac{E_x}{2}\mathbf{u}_y - \frac{\sqrt{3}}{2}E_x\mathbf{u}_z\right].$$

$$\mathbf{B} = \frac{1}{3 \times 10^8}\left(\frac{1}{2}\mathbf{u}_y - \frac{\sqrt{3}}{2}\mathbf{u}_z\right) \times 0.05 \times \cos\left[\frac{2\pi}{300}\left(\frac{\sqrt{3}}{2}y + \frac{1}{2}z\right) - 2 \times \pi \times 10^6 t\right]$$

$$= \frac{1}{3 \times 10^8}(0.025\,\mathbf{u}_y - 0.043\,\mathbf{u}_z)\cos\left[\frac{2\pi}{300}\left(\frac{\sqrt{3}}{2}y + \frac{1}{2}z\right) - 2 \times \pi \times 10^6 t\right] \text{ (T)}.$$

12.8 Figure 12.12 shows a sketch of two sources, labelled A and B, which emit synchronously plane electromagnetic waves of frequency 120 MHz propagating in free space (through air). Both waves are linearly polarized, the electric field vibrating along the X-direction, and are directed toward point Q. An intensity of 10^{-6} Wm^{-2} is detected at Q when either A or B is emitting. Write down the expressions for the propagating electric and magnetic fields of the waves emitted by A and B.

Solution

First of all, let us calculate the wavelength, $\lambda = c/\nu = 3 \times 10^8/120 \times 10^6 = 2.5$ m. Then, the wave number is $k = 2\pi/\lambda = 2\pi/2.5 = 2.51$ m^{-1}. The amplitude of the electric field can be calculated from the intensity, (12.36) with $\varepsilon = \varepsilon_0$ and $v = c$,

$$I_w = \frac{1}{2}\varepsilon_0 c E_0^2 \Rightarrow E_0 = \sqrt{\frac{2 I_w}{\varepsilon_0 c}} = \sqrt{\frac{2 \times 10^{-6}}{8.85 \times 10^{-12} \times 3 \times 10^8}} = 2.74 \times 10^{-2} \text{ V/m}.$$

From E_0, the amplitude of the magnetic field results to be, $B_0 = E_0/c = 2.74 \times 10^{-2}/3 \times 10^8$ T $= 9.13 \times 10^{-11}$ T.

Fig. 12.12 Sketch showing two sources labelled A and B and the rays directed toward point Q

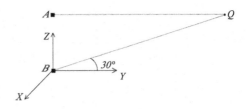

For the wave emitted by A, directed toward Q, the wave vector is $\mathbf{k} = k\,\mathbf{u}_y = 2.51\,\mathbf{u}_y$ m^{-1} and the angular frequency $\omega = 2\pi\nu = 2\pi \times 120 \times 10^6$ rad/s $= 7.54 \times 10^8$ rad/s. Since the wave is linearly polarized, the electric field can be written as (12.20). In the case studied, in which the electric field oscillates along the X-direction, the electric field amplitude vector can be expressed as $\mathbf{E}_0 = 2.74 \times 10^{-2}\,\mathbf{u}_x$ V/m. Then, assuming a cosine type wave with $\varphi_0 = 0$, the electric field is given by

$$\mathbf{E} = \mathbf{E}_0 \cos(\mathbf{k}\cdot\mathbf{r} - \omega t) = 2.74 \times 10^{-2}\,\mathbf{u}_x \cos(2.51\,y - 7.54 \times 10^8 t)\ (\text{V/m}).$$

According to (12.24), we have for $\mathbf{B}$

$$k\,\mathbf{u}_y \times E_x\,\mathbf{u}_x = \omega\,\mathbf{B} \Rightarrow$$

$$\mathbf{B} = \frac{1}{7.54 \times 10^8}[2.51\,\mathbf{u}_y] \times [2.74 \times 10^{-2} \cos(2.51\,y - 7.54 \times 10^8 t)\,\mathbf{u}_x]$$
$$= -9.12 \times 10^{-11}\,\mathbf{u}_z \cos(2.51\,y - 7.54 \times 10^8 t)\ \text{T}.$$

In the same way, for the wave emitted by B, the wave vector is given by

$$\mathbf{k} = 2.51\left(\cos 30°\,\mathbf{u}_y + \sin 30°\,\mathbf{u}_z\right)\ \text{m}^{-1}.$$

In this case (12.20) becomes,

$$\mathbf{E} = \mathbf{E}_0 \cos(\mathbf{k}\cdot\mathbf{r} - \omega t)$$
$$= 2.74 \times 10^{-2}\,\mathbf{u}_x \cos(2.51 \cos 30°\,y + 2.51 \sin 30°\,z - 7.54 \times 10^8 t)$$
$$= 2.74 \times 10^{-2}\,\mathbf{u}_x \cos(2.17\,y + 1.26\,z - 7.54 \times 10^8 t)\ (\text{V/m}).$$

Equation (12.24) yields for $\mathbf{B}$,

$$\mathbf{B} = 9.12 \times 10^{-11}\,(\sin 30°\,\mathbf{u}_y - \cos 30°\,\mathbf{u}_z)\cos(2.17\,y + 1.26\,z - 7.54 \times 10^8 t)$$
$$= (4.56\,\mathbf{u}_y - 7.90\,\mathbf{u}_z) \times 10^{-11} \cos(2.17\,y + 1.26\,z - 7.54 \times 10^8 t)\ (\text{T}).$$

12.9 Determine the result of the superposition of two waves with amplitudes A and B ($A > B$) propagating along OX and that are left and right-handed circularly polarized, respectively. Describe the polarization of the resulting wave.

Solution

Let us consider a left-handed circularly polarized wave $\mathbf{E}_1$ propagating along $+OX$ whose components can be written as

$$\left.\begin{array}{l} E_{1x} = 0 \\ E_{1y} = A \sin(kx - \omega t) \\ E_{1z} = A \cos(kx - \omega t) = A \sin(kx - \omega t + \frac{\pi}{2}) \end{array}\right\}$$

It should be noted that the phase difference is $\varphi_{01y} - \varphi_{01z} = -\pi/2$ and $E_{0y} = E_{0z} = A$, which corresponds to a left-handed circularly polarized wave.

For a right-handed circularly polarized wave, $\mathbf{E}_2$, propagating along $+OX$, E_{2y} leads E_{2z} by $\pi/2$ and $E_{0y} = E_{0z}$. Then, the components of $\mathbf{E}_2$ can be written as

$$\left.\begin{array}{l} E_{2x} = 0 \\ E_{2y} = B \sin(kx - \omega t) \\ E_{2z} = -B \cos(kx - \omega t) = B \sin(kx - \omega t - \frac{\pi}{2}) \end{array}\right\}$$

The components of the resulting field $\mathbf{E} = \mathbf{E}_1 + \mathbf{E}_2$ are:

$E_x = 0$
$E_y = E_{1y} + E_{2y} = A \sin(kx - \omega t) + B \sin(kx - \omega t) = (A + B) \sin(kx - \omega t),$
$E_z = E_{1z} + E_{2z} = A \cos(kx - \omega t) - B \cos(kx - \omega t)$
$\quad = (A - B) \cos(kx - \omega t) = (A - B) \sin\left(kx - \omega t + \frac{\pi}{2}\right)$

The resulting wave propagates along $+OX$, E_y component lags E_z by $\pi/2$, i.e. $\varphi_{0y} - \varphi_{0z} = -\pi/2$, and $E_{0y} \neq E_{0z}$. Then, the wave is left-handed elliptically polarized. The ellipse is inscribed into a rectangle whose sides are parallel to the co-ordinates axes OY and OZ and whose lengths are $2(A + B)$ and $2(A - B)$, respectively.

12.10 A right and a left-handed circularly polarized wave can combine to yield a linearly polarized wave. Prove this statement by expressing a linearly polarized wave of amplitude E_0 travelling along $+OX$ as the sum of two waves with circular polarization and amplitudes A and B for the right and left polarization, respectively. Find A and B in terms of E_0.

Solution

We asume, without loss of generality, that the wave is polarized in the OZ-direction. The wave can be expressed as

$$\mathbf{E} = E_0 \, \mathbf{u}_z \cos(kx - \omega t) = \text{Re}\left\{E_0 \, \mathbf{u}_z \exp[i(kx - \omega t)]\right\},$$

where complex notation is used to simplify the calculation. An expression for a right-handed circularly polarized wave with amplitude A propagating along OX is

$$\mathbf{E}_1 = A\mathbf{u}_y \cos\left(kx - \omega t + \frac{\pi}{2}\right) + A\mathbf{u}_z \cos(kx - \omega t)$$
$$= \text{Re}\left\{A\mathbf{u}_y \exp\left[i\left(kx - \omega t + \frac{\pi}{2}\right)\right] + A\mathbf{u}_z \exp[i(kx - \omega t)]\right\}.$$

In the same way, for the left-handed circular wave we have

$$\mathbf{E}_2 = B\mathbf{u}_y \cos\left(kx - \omega t - \frac{\pi}{2}\right) + B\mathbf{u}_z \cos(kx - \omega t)$$
$$= \text{Re}\left\{B\mathbf{u}_y \exp\left[i\left(kx - \omega t - \frac{\pi}{2}\right)\right] + B\mathbf{u}_z \exp[i(kx - \omega t)]\right\}.$$

Using complex notation, one obtains

$$\mathbf{E} = \mathbf{E}_1 + \mathbf{E}_2,$$

$$E_0 \mathbf{u}_z \exp[i(kx - \omega t)] = A\mathbf{u}_y \exp\left[i\left(kx - \omega t + \frac{\pi}{2}\right)\right] + A\mathbf{u}_z \exp[i(kx - \omega t)]$$
$$+ B\mathbf{u}_y \exp\left[i\left(kx - \omega t - \frac{\pi}{2}\right)\right] + B\mathbf{u}_z \exp[i(kx - \omega t)],$$

$$E_0 \mathbf{u}_z = A\mathbf{u}_y \exp\left[i\frac{\pi}{2}\right] + A\mathbf{u}_z + B\mathbf{u}_y \exp\left[-i\frac{\pi}{2}\right] + B\mathbf{u}_z \Rightarrow E_0 \mathbf{u}_z = iA\mathbf{u}_y + A\mathbf{u}_z - iB\mathbf{u}_y + B\mathbf{u}_z,$$

$$E_0 \mathbf{u}_z = A(\mathbf{u}_z + i\mathbf{u}_y) + B(\mathbf{u}_z - i\mathbf{u}_y).$$

By equating real and imaginary parts,

$$\left.\begin{array}{r}E_0 = A + B \\ 0 = A - B\end{array}\right\} \Rightarrow A = B = \frac{E_0}{2}.$$

Then, a linearly polarized wave with amplitude E_0 can be expressed as the sum of a right-handed circularly polarized wave and a left-handed circularly polarized wave, each having amplitude $E_0/2$.

12.11 Identify the polarization state for the following waves:
(a) $\mathbf{E} = 2\mathbf{u}_x \cos(kz - \omega t) + 3\mathbf{u}_y \cos(kz - \omega t)$,
(b) $\mathbf{E} = \mathbf{u}_x \cos(kz - \omega t) - \sqrt{3}\mathbf{u}_y \cos(kz - \omega t)$,
(c) $\mathbf{E} = 2\mathbf{u}_x \cos(kz + \omega t) + 2\mathbf{u}_y \sin(kz + \omega t)$,
(d) $\mathbf{E} = \mathbf{u}_x \sin(kz - \omega t) + 2\mathbf{u}_y \cos(kz - \omega t)$.

Solution

In this problem, the process followed in determining the polarization state is as follows. Firstly, we obtain the equation resulting from eliminating the dependence of the components on $(kz \pm \omega t)$. Secondly, we plot this equation on a plane perpendicular to the direction of propagation, at a fixed position in space ($z = $ constant). We consider the wave seen by an observer towards whom the wave approaches (optic

convection). Finally, we analyze how the electric field oscillates in the plot of the locus of the tip of the **E** vector versus time.
(a)
$$\left. \begin{array}{l} E_x = 2\cos(kz - \omega t) \\ E_y = 3\cos(kz - \omega t) \end{array} \right\} \Rightarrow \frac{E_y}{E_x} = \frac{3}{2} \ \left(\tan\theta = \frac{3}{2}, \ \theta = 56.3°\right).$$

The propagation vector is $\mathbf{k} = k\,\mathbf{u}_z$ and the two components are in phase. Then, the plane wave is linearly polarized. The electric field oscillates on a straight line that makes 56.3° with the OX-axis, as shown in Fig. 12.13a. The tip of **E** at $z = 0$ will be at the point P when $\omega t = 0$. Its magnitude will decrease toward zero as ωt increases toward $\pi/2$. Then, **E** starts to increase in the opposite direction, toward the point P', where $\omega t = \pi$.
(b)
$$\left. \begin{array}{l} E_x = \cos(kz - \omega t) \\ E_y = -\sqrt{3}\cos(kz - \omega t) = \sqrt{3}\cos(kz - \omega t + \pi) \end{array} \right\} \Rightarrow \frac{E_y}{E_x} = -\sqrt{3} \ (\tan\theta = -\sqrt{3}, \ \theta = -60°).$$

The components are π out of phase; the electric field is linearly polarized oscillating on a straight line that makes $-60°$ with the OX-axis. For $z = 0$, and $\omega t = 0$, $\pi/2$, π, the tip of **E** will be at the points P, O, P', respectively, as shown in Fig. 12.13b.
(c)
$$\left. \begin{array}{l} E_x = 2\cos(kz + \omega t) \\ E_y = 2\sin(kz + \omega t) = 2\cos(kz + \omega t - \pi/2) \end{array} \right\} \Rightarrow E_x^2 + E_y^2 = 4.$$

The components differ in phase by $\pi/2$ and are equal in amplitude. The propagation vector is $\mathbf{k} = -k\,\mathbf{u}_z$. The tip of the electric field traces out a circle of radius 2, as shown in Fig. 12.13c. In examining the direction change of **E** at $z = 0$ as t changes, we set $\omega t = 0$, $\omega t = \pi/2$. For $\omega t = 0$, we get $E_x = 2$ and $E_y = 0$, point P in this figure, whereas for $\omega t = \pi/2$, $E_x = 0$ and $E_y = 2$, point P'. Then, the electric vector

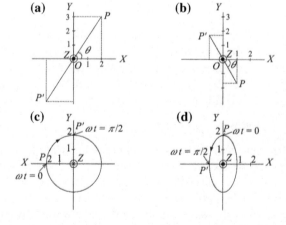

Fig. 12.13 Plot of the locus of **E** for the cases studied in Problem 12.11

moves around the circle in a clockwise direction; this is a right handed circularly polarized wave.

(d)
$$\left.\begin{array}{l} E_x = \sin(kz - \omega t) = \cos(kz - \omega t - \pi/2) \\ E_y = 2\cos(kz - \omega t) \end{array}\right\} \Rightarrow E_x^2 + \frac{(E_y)^2}{(2)^2} = 1.$$

E_x lags E_y by $\pi/2$ and are unequal in magnitude. The propagation vector is $\mathbf{k} = k\,\mathbf{u}_z$. The tip of the electric field traces out an ellipse, as shown in Fig. 12.13d. In this figure the position of $\mathbf{E}$ at $z = 0$ for $\omega t = 0$ and $\omega t = \pi/2$, points P and P', respectively, are also shown; the electric vector rotates counterclockwise as time progresses.

12.12 Determine the polarization state and plot the locus of $\mathbf{E}(0, t)$ for a plane wave with:

(a) $\mathbf{E}(z, t) = 3\,\mathbf{u}_x \cos(kz - \omega t) + 3\,\mathbf{u}_y \cos(kz - \omega t - \pi/4)$,
(b) $\mathbf{E}(z, t) = 3\,\mathbf{u}_x \cos(kz - \omega t + 3\pi/4) + \mathbf{u}_y \cos(kz - \omega t)$.

Solution

(a) In this case, the wave is travelling in the $+OZ$-direction. The x and y components are not in phase; the y component lags the x component by $\pi/4$. Then, according to Sect. 12.4, and for the new orientation of the axes: $OZ \Rightarrow OX$, $OX \Rightarrow OY$, and $OY \Rightarrow OZ$, the resulting phase difference is $\delta = \varphi_{0x} - \varphi_{0y} = \pi/4$. Hence, the wave is elliptically polarized, in accordance with case 3 described in Sect. 12.4. The equation of the ellipse is given by (12.27) with $E_{0x} = E_{0y} = 3$, and the $\mathbf{E}$ vector rotates in the clockwise direction (see Fig. 12.4).

Let us verify the above results by calculating the equation of the ellipse and determining the sense of polarization from the plot of the locus of $\mathbf{E}(0, t)$. The components of the plane wave are described by

$$\left.\begin{array}{l} E_x(z, t) = 3\cos(kz - \omega t) \\ E_y(z, t) = 3\cos(kz - \omega t - \pi/4) \end{array}\right\}$$

At any time and at any position, the phase difference remains constant. Therefore, the electric field will change with time in a similar way at any fixed position in space. For the sake of simplicity, let us study what happens when $z = 0$, the resulting wave is given by:

$$\left.\begin{array}{l} E_x = 3\cos(-\omega t) = 3\cos(\omega t), \\ E_y = 3\cos(-\omega t - \pi/4) = 3\cos(\omega t + \pi/4) \end{array}\right\}$$

The electric vector $\mathbf{E}$ varies with time both in magnitude and in direction. In order to determine the equation described by the tip of $\mathbf{E}$, the term $\cos(\omega t + \pi/4)$, in the second equation, is first expanded. Then, $\cos(\omega t)$ is eliminated between the two equations as follows

$$\left.\begin{array}{l} E_x/3 = \cos(\omega t) \\ E_y/3 = \cos(\omega t + \pi/4) = \cos(\omega t)\cos(\pi/4) - \sin(\omega t)\sin(\pi/4) \end{array}\right\}$$

$$\Rightarrow \frac{E_y}{3} = \frac{E_x}{3} \cos\left(\frac{\pi}{4}\right) - \sin(\omega t) \sin\left(\frac{\pi}{4}\right)$$

Multiplying the first equation by $\sin(\pi/4)$, re-writing the second one, and squaring and adding both equations gives

$$\left.\begin{array}{l} E_x/3 \sin(\pi/4) = \cos(\omega t) \sin(\pi/4) \\ E_x/3 \cos(\pi/4) - E_y/3 = \sin(\omega t) \sin(\pi/4) \end{array}\right\} \Rightarrow \boxed{\frac{E_x^2}{3^2} + \frac{E_y^2}{3^2} - 2\frac{E_x E_y}{9} \cos\left(\frac{\pi}{4}\right) = \sin^2\left(\frac{\pi}{4}\right)}$$

This equation is that of an ellipse in the XY-plane. From (12.28) is obtained that its major axis makes an angle of $45°$ with the OX-axis, as shown in Fig. 12.14. The ellipse is inscribed into a square with sides 3×2 and touches the sides at the points $(\pm 3, \pm 3\cos(\pi/4))$ and $(\pm 3\cos(\pi/4), \pm 3)$. To know the sense in which the end point of the electric vector describes the ellipse, we plot, at $z = 0$, the position of the tip of $\mathbf{E}$ for $\omega t = 0$ and $\pi/2$, the resulting points being P and P', respectively (see Fig. 12.14). Then, the wave is right-handed elliptically polarized since the $\mathbf{E}$ vector rotates clockwise.

The same result is obtained by calculating, for instance at point P (point tangent to the ellipse), the derivative $\left(\partial E_y/\partial t\right)_{t=0} = (-3\omega \sin(\omega t + \pi/4))_{t=0} = -3\omega \sin(\pi/4) < 0$. Hence, E_y decreases with time, and the rotation is clockwise as viewed by an observer that receives the wave.

(b) The same reasoning as in case (a) can be applied (OZ is equivalent to OX, OX to OY, and OY to OZ). The wave is also a plane wave travelling in the $+OZ$-direction. The x component leads the y component by $3\pi/4$ and the amplitudes are different. According to Sect. 12.4, case 3, the wave is elliptically polarized, the equation given by (12.27), with $E_{0x} = 3$, $E_{0y} = 1$, and $\delta = \varphi_{0x} - \varphi_{0y} = 3\pi/4$. Figure 12.4 predicts that the ellipse is described in a clockwise direction.

As in case (a), let us calculate the equation of the ellipse and determine the sense in which the ellipse is described. The components of the wave are

$$\left.\begin{array}{l} E_x(z, t) = 3 \cos(kz - \omega t + 3\pi/4) \\ E_y(z, t) = \cos(kz - \omega t) \end{array}\right\}$$

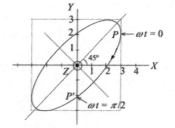

Fig. 12.14 The tip of the electric vector moves along the ellipse. The phase difference between the x and y components is $\pi/4$ and the amplitudes of the two components are equal

Fig. 12.15 Elliptically polarized wave. $E_{0x} = 3$, $E_{0y} = 1$ and the phase difference between the x and the y components is $3\pi/4$

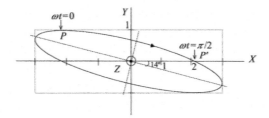

For $z = 0$:
$$\left.\begin{array}{l} E_x = 3\cos(-\omega t + 3\pi/4) = 3\cos(\omega t - 3\pi/4) \\ E_y = \cos(-\omega t) = \cos(\omega t) \end{array}\right\}$$

The equation described by the tip of **E** is calculated as in (a). Firstly, $\cos(\omega t - 3\pi/4)$ is expanded. This gives

$$\left.\begin{array}{l} E_x/3 = \cos(\omega t - 3\pi/4) = \cos(\omega t)\cos(3\pi/4) + \sin(\omega t)\sin(3\pi/4) \\ E_y = \cos(\omega t) \end{array}\right\}$$

Secondly, $\cos(\omega t)$ in the first equation is replaced by E_y and the resulting equation is re-rearranged. Next, the second equation is multiplied by $\sin(3\pi/4)$. Finally, the dependence on t is eliminated by squaring and adding the two equations:

$$\left.\begin{array}{l} E_x/3 - E_y\cos(3\pi/4) = \sin(\omega t)\sin(3\pi/4) \\ E_y\sin(3\pi/4) = \cos(\omega t)\sin(3\pi/4) \end{array}\right\} \Rightarrow \boxed{\frac{E_x^2}{3^2} + E_y^2 - 2\frac{E_x E_y}{3}\cos\left(\frac{3\pi}{4}\right) = \sin^2\left(\frac{3\pi}{4}\right).}$$

This equation describes the ellipse shown in Fig. 12.15, whose major axis makes an angle of about $-14°$ with OX, obtained from (12.28). The ellipse is inscribed into the rectangle with sides 2×3 and 2×1. Figure 12.15 also shows the points P and P' (at $z = 0$), for $\omega t = 0$ and $\pi/2$, respectively. Then, the ellipse is described in the clockwise direction.

12.13 Find the average intensity for an elliptically polarized plane electromagnetic wave propagating in vacuum.

Solution

As an example, let us consider an elliptically polarized plane electromagnetic wave propagating in vacuum in the $+X$-direction, given by the equation

$$\left.\begin{array}{l} E_y(x, t) = E_{0y}\cos(kx - \omega t + \delta) \\ E_z(x, t) = E_{0z}\cos(kx - \omega t) \end{array}\right\}$$

where, in general, $E_{0y} \neq E_{0z}$, and $\delta = \varphi_{0y} - \varphi_{0z}$ is an arbitrary but constant quantity. Equation (12.24) gives for the magnetic field

$$\mathbf{B} = \frac{1}{\omega}\mathbf{k} \times \mathbf{E} = \frac{1}{c}\mathbf{u}_x \times \mathbf{E} = \frac{1}{c}(1, 0, 0) \times (0, E_y, E_z) = \frac{1}{c}\left(-E_z\mathbf{u}_y + E_y\mathbf{u}_z\right),$$

where it has been taken into account that $\mathbf{k} = k\mathbf{u}_x$ and $c = \omega/k$. Then,

$$\mathbf{H} = \frac{\mathbf{B}}{\mu_0} = \frac{1}{\mu_0 c}\left(-E_z\mathbf{u}_y + E_y\mathbf{u}_z\right) = \varepsilon_0 c\left(-E_z\mathbf{u}_y + E_y\mathbf{u}_z\right),$$

and the instantaneous Poynting vector will be

$$\mathbf{P}_o = \mathbf{E} \times \mathbf{H} = (0, E_y, E_z) \times (0, H_y, H_z) = (E_yH_z - E_zH_y)\mathbf{u}_x$$
$$= \varepsilon_0 c\left(E_y^2 + E_z^2\right)\mathbf{u}_x = \varepsilon_0 c\,[E_{0y}^2\cos^2(kz - \omega t + \delta) + E_{0z}^2\cos^2(kz - \omega t)]\mathbf{u}_x.$$

The intensity I_w is equal to the time average of the Poynting vector, which is calculated by integrating the instantaneous Poynting vector over one period and dividing by the period,

$$I_w = <|\mathbf{P}_o|> = \varepsilon_0 c < [E_{0y}^2\cos^2(kx - \omega t + \delta) + E_{0z}^2\cos^2(kx - \omega t)] >$$
$$= \varepsilon_0 c E_{0y}^2 < \cos^2(kx - \omega t + \delta) > + \varepsilon_0 c E_{0z}^2 < \cos^2(kx - \omega t) >,$$

$$<\cos^2(kx - \omega t)> = \frac{1}{2\pi/\omega}\int_0^{2\pi/\omega}\cos^2(kx - \omega t)dt = \frac{1}{2\pi/\omega}\int_0^{2\pi/\omega}\left(\frac{1}{2} + \frac{\cos 2(kx - \omega t)}{2}\right)dt = \frac{1}{2}.$$

In the same way, $<\cos^2(kx - \omega t + \delta)> = 1/2$. Hence, the average intensity results to be

$$I_w = \frac{1}{2}\varepsilon_0 c\,(E_{0y}^2 + E_{0z}^2).$$

The average intensity for an elliptically polarized wave is equal to the sum of the intensities of the components regardless of the phase difference δ. In the particular case, $E_{0x} = E_{0y} = E_0$, the average intensity becomes

$$I_w = \varepsilon_0 c\,E_0^2.$$

It should be noted that this case includes a circularly polarized wave, where the amplitudes are equal and the phase difference is $\delta = \pm\pi/2$.

12.14 Two point radio transmitters, A and B, emit synchronously with a frequency of 75 MHz and a power of 125 and 200 W, respectively. A point Q, on the line joining A and B, is 5 km away from A and 6.950 km from B. (a) Determine the intensity detected at Q when either A or B is emitting. Find also the amplitude of the corresponding electric and magnetic fields. (b) Calculate the intensity detected at Q when both sources are emitting simultaneously. Assume that the electric fields at Q from A and B vibrate in a direction parallel to each other.

Fig. 12.16 Two waves emitted by point transmitters A and B interfere at point Q

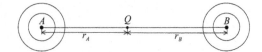

Solution

(a) For a point source emitting in a homogeneous medium such as air, whose electromagnetic properties are approximately the same as those of a vacuum, the resulting wavefront is spherical. As the wave propagates from the source, the power is distributed homogenously over a spherical wavefront of area $4\pi r^2$. When only transmitter A is emitting, the intensity detected at Q and the corresponding amplitudes of the electric and magnetic fields are (Fig. 12.16):

$$P_A = I_{QA}\, 4\pi r_A^2 \Rightarrow I_{QA} = \frac{P_A}{4\pi r_A^2} = \frac{125}{4\pi \times (5\times 10^3)^2}\ \text{W/m}^2 = 3.97\times 10^{-7}\ \text{W/m}^2.$$

$$I_{QA} = \frac{1}{2}\varepsilon_0 c E_{0A}^2 \Rightarrow E_{0A} = \sqrt{\frac{2I_{QA}}{\varepsilon_0 c}} = \sqrt{\frac{2\times 3.97\times 10^{-7}}{8.85\times 10^{-12}\times 3\times 10^8}}\ \text{V/m} = 1.73\times 10^{-2}\ \text{V/m}$$

$$\Rightarrow B_{0A} = \frac{E_{0A}}{c} = \frac{1.73\times 10^{-2}}{3\times 10^8}\ \text{T} = 5.76\times 10^{-11}\ \text{T}.$$

In the same way, when only source B is emitting,

$$P_B = I_{QB}\, 4\pi r_B^2 \Rightarrow I_{QB} = \frac{P_B}{4\pi r_B^2} = \frac{200}{4\pi \times (6.950\times 10^3)^2}\ \text{W/m}^2 = 3.29\times 10^{-7}\ \text{W/m}^2.$$

$$I_{QB} = \frac{1}{2}\varepsilon_0 c E_{0B}^2 \Rightarrow E_{0B} = \sqrt{\frac{2I_{QB}}{\varepsilon_0 c}} = \sqrt{\frac{2\times 3.29\times 10^{-7}}{8.85\times 10^{-12}\times 3\times 10^8}}\ \text{V/m} = 1.57\times 10^{-2}\ \text{V/m}$$

$$\Rightarrow B_{0B} = \frac{E_{0B}}{c} = \frac{1.57\times 10^{-2}}{3\times 10^8}\ \text{T} = 5.24\times 10^{-11}\ \text{T}.$$

(b) When the two sources emit simultaneously, the intensity at Q is that of the wave resulting from the superposition of the wave from A and that from B. As indicated, both waves vibrate in a direction parallel to each other. The two transmitters emit synchronously with the same frequency, therefore the phase difference at Q is only due to the path difference. Path AQ has a length $r_A = 5000$ m, while length of BQ is $r_B = 6950$ m. Hence, the two waves arriving at Q differ in phase by

$$\delta = \varphi_B - \varphi_A = k(r_B - r_A) = \frac{2\pi}{\lambda}(r_B - r_A) = \frac{2\pi}{4}(6950 - 5000) = 975\pi,$$

where $\varphi_A = kr_A - \omega t$ and $\varphi_B = kr_B - \omega t$ are the corresponding phases at Q, and the wavelength $\lambda = c/\nu = 3 \times 10^8 / 75 \times 10^6 = 4$ m. Note that $\cos \delta = \cos(975\pi) = \cos(\pi + 2\pi \times 487) = -1$. As the two electric fields at Q, E_A and E_B, oscillate π out of phase, the resulting electric field at Q can be expressed as

$$E = E_A + E_B = E_{0A} \cos(\varphi_A) + E_{0B} \cos(\varphi_B) = E_{0A} \cos(\varphi_A) - E_{0B} \cos(\varphi_A) = (E_{0A} - E_{0B}) \cos(\varphi_A),$$

whose amplitude is equal to $E_0 = E_{0A} - E_{0B} = (1.73 - 1.57) \times 10^{-2} = 0.16 \times 10^{-2}$ V/m. Finally, the resulting intensity can be calculated by

$$I_w = \frac{1}{2} \varepsilon_0 c E_0^2 = \frac{1}{2} \times 8.85 \times 10^{-12} \times 3 \times 10^8 \times (0.16 \times 10^{-2})^2 \text{ W/m}^2 = 3.40 \times 10^{-9} \text{ W/m}^2.$$

12.15 Two antennas on two satellites, labelled A and B, emit synchronously electromagnetic waves of frequency 10^{10} Hz with a power of 50 kW. The antennas are considered point-like sources emitting isotropically in all directions. Determine the intensity detected at point Q on the Earth's surface. The distance between point Q and antennas A and B is 100 km and 97 km, respectively (Fig. 12.17).

Solution

As both antennas are considered point sources emitting isotropically in all directions, the emitted wavefronts are spherical. Since the distance from the source to the detection point is very large, the wave detected at point Q on the Earth's surface can be considered to be plane. We assume that the electric fields of the two waves at Q oscillate in the same direction. Thus, we can add both waves as if they were scalars. Using complex notation, the superposition of both waves at Q gives an electric field

$$E_c = E_{0A} \exp[i(kr_A - \omega t)] + E_{0B} \exp[i(kr_B - \omega t)] = [E_{0A} \exp i(kr_A) + E_{0B} \exp i(kr_B)] \exp(-i\omega t),$$

where an initial phase is not included because both sources emit synchronously. The complex amplitude E_{c0} of the resulting wave and its conjugate E_{c0}^* are, respectively,

$$E_{c0} = E_{0A} \exp i(kr_A) + E_{0B} \exp i(kr_B) \quad ; \quad E_{c0}^* = E_{0A} \exp i(-kr_A) + E_{0B} \exp i(-kr_B).$$

Fig. 12.17 Sketch showing two rays coming from antennas labelled A and B that interfere at point Q on Earth's surface

Multiplying the complex amplitude by its conjugate, one obtains the amplitude squared, to which the intensity is proportional. Then, the amplitude squared of the electric field resulting from the superposition can be calculated by

$$E_0^2 = E_{c0} E_{c0}^* = [E_{0A} \exp i(kr_A) + E_{0B} \exp i(kr_B)][E_{0A} \exp i(-kr_A) + E_{0B} \exp i(-kr_B)]$$
$$= E_{0A}^2 + E_{0B}^2 + 2E_{0A}E_{0B} \cos[k(r_A - r_B)].$$

Multiplying the above expression by $(1/2)\varepsilon_0 c$, it is found

$$\frac{1}{2}\varepsilon_0 c E_0^2 = \frac{1}{2}\varepsilon_0 c E_{0A}^2 + \frac{1}{2}\varepsilon_0 c E_{0B}^2 + \frac{1}{2}\varepsilon_0 c \, 2E_{0A}E_{0B} \cos[k(r_A - r_B)]$$

$$\Rightarrow I = I_A + I_B + 2\sqrt{I_A}\sqrt{I_B} \cos[k(r_A - r_B)],$$

where I is the intensity detected at Q when both sources are emitting; I_A and I_B are the intensities when either source A or B emits, respectively. The interference term, $2\sqrt{I_A}\sqrt{I_B} \cos[k(r_A - r_B)]$, depends on the phase difference $\delta = k(r_A - r_B)$.

Let us calculate I from I_A, I_B and the phase difference. From the emission power, and the distance from the source to the detection point Q, one obtains for I_A and I_B,

$$P_A = I_A 4\pi r_A^2 \Rightarrow I_A = \frac{P_A}{4\pi r_A^2} = \frac{50 \times 10^3}{4\pi \times (10^5)^2} \, W/m^2 = 3.98 \times 10^{-7} \, W/m^2,$$

$$P_B = I_B 4\pi r_B^2 \Rightarrow I_B = \frac{P_B}{4\pi r_B^2} = \frac{50 \times 10^3}{4\pi \times (97 \times 10^3)^2} \, W/m^2 = 4.23 \times 10^{-7} \, W/m^2.$$

The wavelength results to be $\lambda = c/\nu = 3 \times 10^8/10^{10}$ m $= 0.03$ m. Then, the phase difference resulting from the different paths, r_A and r_B, is equal to

$$\delta = k(r_A - r_B) = \frac{2\pi}{0.03}(100 \times 10^3 - 97 \times 10^3) \, \text{rad} \approx 2\pi \, 10^5 \, \text{rad} \Rightarrow \cos \delta = 1.$$

Finally, the resulting intensity gives

$$I = I_A + I_B + 2\sqrt{I_A}\sqrt{I_B} \cos \delta$$
$$= 3.98 \times 10^{-7} + 4.23 \times 10^{-7} + 2\sqrt{3.98 \, 10^{-7}} \times \sqrt{4.23 \, 10^{-7}} \times 1 \, W/m^2 = 1.64 \times 10^{-6} \, W/m^2.$$

12.16 Two linearly polarized plane waves with equal amplitude E_0 and slightly different angular frequencies ω_1 and ω_2 travel in the same direction at the same velocity v. Find the wave that results from the superposition of the two waves.

Solution

We assume that both waves travel along OZ and the electric fields oscillate in the OX-direction. Then, considering sine functions, the electric fields can be expressed as

$$\mathbf{E}_1 = E_1 \, \mathbf{u}_x = E_0 \, \mathbf{u}_x \, \sin(k_1 z - \omega_1 t),$$
$$\mathbf{E}_2 = E_2 \, \mathbf{u}_x = E_0 \, \mathbf{u}_x \, \sin(k_2 z - \omega_2 t),$$

where $k_1 = \omega_1/v$ and $k_2 = \omega_2/v$. The wave resulting from the superposition will be $\mathbf{E} = \mathbf{E}_1 + \mathbf{E}_2 = (E_1 + E_2) \, \mathbf{u}_x$, oscillating in the same direction as their components. Then, as a result of the superposition we have

$$E = E_1 + E_2 = E_0 \, \sin(k_1 z - \omega_1 t) + E_0 \, \sin(k_2 z - \omega_2 t)$$
$$= 2 E_0 \, \sin\left[\left(\frac{k_1 + k_2}{2}\right) z - \left(\frac{\omega_1 + \omega_2}{2}\right) t\right] \times \cos\left[\left(\frac{k_1 - k_2}{2}\right) z - \left(\frac{\omega_1 - \omega_2}{2}\right) t\right],$$

where the following trigonometric identity has been taken into account:

$$\sin A + \sin B = 2 \sin\left(\frac{A + B}{2}\right) \cos\left(\frac{A - B}{2}\right).$$

We use the following notation: $k_0 = (k_1 + k_2)/2$, $\omega_0 = (\omega_1 + \omega_2)/2$, $k_m = (k_1 - k_2)/2$, $\omega_m = (\omega_1 - \omega_2)/2$.

Since both waves have almost the same frequency ($\omega_1 \approx \omega_2$), we have:

$$\omega_m = \frac{\omega_1 - \omega_2}{2} = \frac{\Delta \omega}{2} \quad ; \quad \omega_0 \approx \omega_1 \approx \omega_2 \quad ; \quad \omega_m \ll \omega_0;$$
$$k_m = \frac{k_1 - k_2}{2} \quad ; \quad k_0 \approx k_1 \approx k_2 \quad ; \quad k_m \ll k_0;$$
$$\lambda_m = \frac{2\lambda_1 \lambda_2}{|\lambda_2 - \lambda_1|} \quad ; \quad \lambda_0 \approx \lambda_1 \approx \lambda_2 \quad ; \quad T_m = \frac{2 T_1 T_2}{|T_2 - T_1|} \quad ; \quad T_0 \approx T_1 = \frac{2\pi}{\omega_1} \approx T_2 = \frac{2\pi}{\omega_2}.$$

Then, the resulting expression obtained for the superposition of two waves with nearly identical frequencies and wavelengths can be expressed as

$$E = 2 E_0 \, \cos(k_m z - \omega_m t) \, \sin(k_0 z - \omega_0 t) = E_0(z, t) \, \sin(k_0 z - \omega_0 t),$$

where the amplitude varies with position z and time t, and can be written as

$$E_0(z, t) = 2 E_0 \, \cos(k_m z - \omega_m t).$$

The above equation obtained for the sum of the two waves represents a rapidly oscillating wave with an angular frequency ω_0, whose amplitude varies slowly with

Fig. 12.18 Sum of two waves of equal amplitudes and slightly different frequencies at a given time

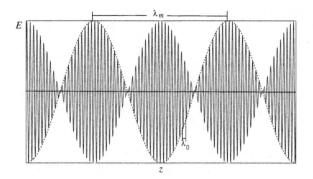

angular frequency ω_m ($\omega_m \ll \omega_0$). Figure 12.18 shows the sum of the two waves at a given time. The wave inside the envelope (carrier wave) has a wave length λ_0, about the same wavelength as the two initial waves, while the amplitude of the envelope (modulating signal) is λ_m ($\lambda_m \ll \lambda_0$). The intensity is proportional to the square of the amplitude

$$E_0^2(z,t) = 4 E_0^2 \cos^2(k_m z - \omega_m t) = 2E_0^2[1 + \cos 2(k_m z - \omega_m t)].$$

At a given position z, $E_0^2(z,t)$ oscillates with an angular frequency equal to $2\omega_m = \omega_1 - \omega_2 = \Delta\omega$, known as "beating frequency". This frequency is double that of the envelope.

The wave inside the envelope propagates with a phase velocity $v_p = \omega_0/k_0$. The velocity of the envelope (the group velocity) can be expressed as $v_g = (\omega_1 - \omega_2)/(k_1 - k_2) = \Delta\omega/\Delta k = d\omega/dk$. In the particular case studied, in which both waves propagate with the same velocity, we have $v_p = v_g = v$.

Problems C

12.17 Two waves with the same amplitude A that propagate in the same direction with velocity v are right and left-handed circularly polarized, respectively. The frequencies ω_1 and ω_2 of the waves are slightly different. Determine the polarization state of the superposition of the two waves.

Solution

Let us consider the superposition of two waves $\mathbf{E}_1$ and $\mathbf{E}_2$ propagating along OZ whose angular frequencies are ω_1 and ω_2, respectively, ($|\omega_1 - \omega_2| \ll \omega_1 \approx \omega_2$). We take $\mathbf{E}_1$ and $\mathbf{E}_2$ to be right and left-handed circularly polarized, respectively. Then, both waves can be expressed as

$$\mathbf{E}_1 = \begin{cases} E_{1x} = A\sin(k_1 z - \omega_1 t + \pi/2) \\ E_{1y} = A\sin(k_1 z - \omega_1 t) \\ E_{1z} = 0 \end{cases} \quad \text{and} \quad \mathbf{E}_2 = \begin{cases} E_{2x} = A\sin(k_2 z - \omega_2 t - \pi/2) \\ E_{2y} = A\sin(k_2 z - \omega_2 t) \\ E_{2z} = 0 \end{cases}$$

The resulting wave from the superposition, $\mathbf{E} = \mathbf{E}_1 + \mathbf{E}_2$, has the following components:

$$E_x = E_{1x} + E_{2x} = A \sin(k_1 z - \omega_1 t + \pi/2) + A \sin(k_2 z - \omega_2 t - \pi/2)$$
$$= 2A \sin\left[\frac{k_1 + k_2}{2} z - \frac{\omega_1 + \omega_2}{2} t\right] \cos\left[\frac{k_1 - k_2}{2} z - \frac{\omega_1 - \omega_2}{2} t + \frac{\pi}{2}\right]$$
$$= -2A \sin\left[\frac{k_1 - k_2}{2} z - \frac{\omega_1 - \omega_2}{2} t\right] \sin\left[\frac{k_1 + k_2}{2} z - \frac{\omega_1 + \omega_2}{2} t\right],$$
$$E_y = E_{1y} + E_{2y} = A \sin(k_1 z - \omega_1 t) + A \sin(k_2 z - \omega_2 t)$$
$$= 2A \sin\left[\frac{k_1 + k_2}{2} z - \frac{\omega_1 + \omega_2}{2} t\right] \cos\left[\frac{k_1 - k_2}{2} z - \frac{\omega_1 - \omega_2}{2} t\right]$$
$$= 2A \cos\left[\frac{k_1 - k_2}{2} z - \frac{\omega_1 - \omega_2}{2} t\right] \sin\left[\frac{k_1 + k_2}{2} z - \frac{\omega_1 + \omega_2}{2} t\right].$$

The results obtained for E_x and E_y represent amplitude-modulated waves. The sine wave oscillates rapidly with angular frequency $(\omega_1 + \omega_2)/2$ ("carrier frequency") and its amplitude is modulated by a slowly varying sine or cosine wave, for E_x and E_y, respectively, whose angular frequency is $(\omega_1 - \omega_2)/2$. Since $\omega_1 \approx \omega_2$, we get $\omega_1 - \omega_2 \ll \omega_1 + \omega_2$, the envelope contains many oscillations of the fast propagating wave, as shown in Fig. 12.18. At a given position z, the amplitude can be considered constant for a short interval of time $\Delta t \ll 2\pi/(\omega_1 - \omega_2)$. However, during this interval many high frequency oscillations for the sinusoidal wave (with the "carrier frequency") can occur.

By denoting,

$$\alpha = \frac{k_1 - k_2}{2} z - \frac{\omega_1 - \omega_2}{2} t \quad \text{and} \quad \beta = \frac{k_1 + k_2}{2} z - \frac{\omega_1 + \omega_2}{2} t,$$

the sum of the two waves can be written as

$$\mathbf{E} = 2A \left[-\sin \alpha \, \mathbf{u}_x + \cos \alpha \, \mathbf{u}_y\right] \sin \beta.$$

At a given position z, α varies slowly with time but the variation of β is very rapid. For a given α, the locus of $\mathbf{E}$ describes a segment of amplitude $2A$ that makes an angle α with the Y-axis, as shown in Fig. 12.19, the oscillation period being $4\pi/(\omega_1 + \omega_2)$.

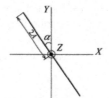

Fig. 12.19 The segment described by the tip of **E** rotates slowly around the OZ-axis

Then, the polarization is lineal but the segment described rotates around the direction of propagation OZ with angular frequency $(\omega_1 - \omega_2)/2$.

12.18 Determine the state of polarization of the superposition of two waves linearly polarized: $\mathbf{E}_1 = A(\sqrt{2}/2\,\mathbf{u}_x - \sqrt{2}/2\,\mathbf{u}_y)\sin(k_1 z - \omega_1 t)$ and $\mathbf{E}_2 = A(\sqrt{2}/2\,\mathbf{u}_x + \sqrt{2}/2\,\mathbf{u}_y)\sin(k_2 z - \omega_2 t)$, where the polarization directions are perpendicular to one another, the propagation vectors are parallel, and their amplitudes are equal. Both waves propagate with the same velocity v and the angular frequencies are slightly different.

Solution

Figure 12.20 shows both waves and the vector $\mathbf{E}$, resulting from the superposition of $\mathbf{E}_1$ and $\mathbf{E}_2$, $\mathbf{E} = \mathbf{E}_1 + \mathbf{E}_2$, which has components:

$$E_x = A\frac{\sqrt{2}}{2}[\sin(k_1 z - \omega_1 t) + \sin(k_2 z - \omega_2 t)]$$
$$= A\sqrt{2}\cos\left[\frac{k_1 - k_2}{2}z - \frac{\omega_1 - \omega_2}{2}t\right]\sin\left[\frac{k_1 + k_2}{2}z - \frac{\omega_1 + \omega_2}{2}t\right],$$

$$E_y = A\frac{\sqrt{2}}{2}[-\sin(k_1 z - \omega_1 t) + \sin(k_2 z - \omega_2 t)]$$
$$= -A\sqrt{2}\sin\left[\frac{k_1 - k_2}{2}z - \frac{\omega_1 - \omega_2}{2}t\right]\cos\left[\frac{k_1 + k_2}{2}z - \frac{\omega_1 + \omega_2}{2}t\right],$$

where the following identities are taken into account:

$$\sin\alpha + \sin\beta = 2\sin\frac{\alpha+\beta}{2}\cos\frac{\alpha-\beta}{2}$$
$$\sin\alpha - \sin\beta = 2\cos\frac{\alpha+\beta}{2}\sin\frac{\alpha-\beta}{2}.$$

By denoting: $\Theta = \frac{k_1-k_2}{2}z - \frac{\omega_1-\omega_2}{2}t$ and $\Phi = \frac{k_1+k_2}{2}z - \frac{\omega_1+\omega_2}{2}t$, the components of $\mathbf{E}$ yield,

$$E_x = A\sqrt{2}\cos\Theta\sin\Phi,$$
$$E_y = -A\sqrt{2}\sin\Theta\cos\Phi,$$
$$E_z = 0.$$

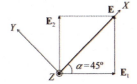

Fig. 12.20 Vector $\mathbf{E}$ is calculated as the sum of the perpendicular vectors $\mathbf{E}_1$ and $\mathbf{E}_2$. X-axis is aligned with $\mathbf{E}$

The angular frequencies associated with Θ and Φ are $(\omega_1 - \omega_2)/2$ and $(\omega_1 + \omega_2)/2$, respectively. As $|\omega_1 - \omega_2| \ll \omega_1 + \omega_2$, Θ varies slowly with time, while for Φ the variation is very rapid. Hence, at a given position z, during short time intervals $\Delta t \ll 2\pi/|\omega_1 - \omega_2|$, the tip of **E** traces out the following curves as time progresses:

- For $\Theta = 0$, since $E_y = 0$, the electric vector undergoes simple harmonic motion along the X-axis with amplitude $A\sqrt{2}$ (linear polarization).
- For $0 < \Theta < \frac{\pi}{2}$, we have:

$$E_x = A\sqrt{2}\cos\Theta\sin\Phi,$$
$$E_y = -A\sqrt{2}\sin\Theta\cos\Phi = A\sqrt{2}\sin\Theta\sin\left(\Phi - \frac{\pi}{2}\right).$$

Eliminating the dependence of E_x and E_y on Φ, it is found that **E** is elliptically polarized, and the vector **E** rotates with angular frequency $(\omega_1 + \omega_2)/2$. The lengths of the two semi-axes of the polarization ellipse are $A\sqrt{2}\cos\Theta$ and $A\sqrt{2}\sin\Theta$, whose magnitudes change with time. Note that the principal axes of the ellipse are aligned with the coordinate axes. As $\varphi_{0x} - \varphi_{0y} = \pi/2$, the **E** vector rotates in a clockwise direction. In the particular case that $\Theta = \pi/4$, the ellipse becomes a circle, i.e. the wave is circularly polarized.
- For $\Theta = \pi/2$, $E_x = 0$ and the ellipse collapses into a straight line along the y direction. Hence, the wave is linearly polarized.
- For $\pi/2 < \Theta < \pi$, $\varphi_{0x} - \varphi_{0y} = -\pi/2$. Thus, **E** is elliptically polarized, rotating in a counterclockwise direction. This is a left-handed elliptically polarized wave. The ellipse becomes a circle for $\Theta = 3\pi/4$.
- For $\Theta = \pi$, the wave is linearly polarized in the X-direction.
- For $\pi < \Theta < 3\pi/2$, the wave is right-handed elliptically polarized since $\varphi_{0x} - \varphi_{0y} = \pi/2$. As in the above cases, for $\Theta = 5\pi/4$ the wave is circularly polarized.
- For $\Theta = 3\pi/2$, the wave is linearly polarized in the Y-direction.
- For $3\pi/2 < \Theta < 2\pi$, **E** is left-handed elliptically polarized. For $\Theta = 7\pi/4$, polarization is circular.

12.19 Find the coefficients of the Fourier series for the periodic rectangular wave shown in Fig. 12.21,

$$f(t) = \begin{cases} 0, & -T_0/2 < t < -\tau/2 \\ H, & -\tau/2 < t < \tau/2 \\ 0, & \tau/2 < t < T_0/2 \end{cases} \quad (12.42)$$

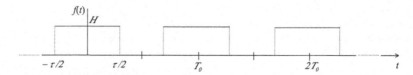

Fig. 12.21 Rectangular wave with period T_0

If $H = 1$ and $T_0 = 2\tau$, write the Fourier series for $f(t)$ and plot the frequency spectrum for the coefficients.

Solution

The Fourier series in terms of the sine and cosine functions are given by (12.37). For a periodic even function, $g(t) = g(-t)$, its Fourier series expansion is expressed as the sum of cosine functions only, since the sine terms make contributions of opposite sign at t and $-t$. Thus, for even functions $b_n = 0$. On the other hand, for odd functions, $g(t) = -g(-t)$, we have coefficients $a_n = 0$. As the periodic rectangular wave given by (12.42) is an even function, coefficients $b_n = 0$. According to (12.38), and taking into account that $\nu_0 = 1/T_0$, coefficients a_n are given by

$$\begin{aligned} a_n &= \frac{2}{T_0} \int_{-T_0/2}^{T_0/2} f(t) \cos(2\pi n\nu_0 t) dt = \frac{2}{T_0} \int_{-\tau/2}^{\tau/2} H \cos(2\pi n\nu_0 t) dt \\ &= \frac{2H}{T_0} \frac{\sin(2\pi n\nu_0 t)}{2\pi n\nu_0} \bigg]_{-\tau/2}^{\tau/2} = \frac{H}{\pi n} 2 \sin\left(2\pi n\nu_0 \frac{\tau}{2}\right) \\ &= \frac{2H}{n\pi} \sin\left(\frac{n\pi\tau}{T_0}\right). \end{aligned} \qquad (12.43)$$

Then, the coefficients of the Fourier series are:

$$a_0 = \frac{2H\tau}{T_0} \; ; \quad a_1 = \frac{2H}{\pi} \sin\left(\frac{\pi\tau}{T_0}\right) \; ; \quad a_2 = \frac{H}{\pi} \sin\left(\frac{2\pi\tau}{T_0}\right) \; ; \quad a_3 = \frac{2H}{3\pi} \sin\left(\frac{3\pi\tau}{T_0}\right) \; ; \quad \ldots$$

For $n = 0$, we have the indetermination form $0/0$, which has been evaluated by applying L'Hôpital's rule.

In the case of a square wave such that $T_0 = 2\tau$ ($\nu_0 = 1/T_0 = 1/2\tau$) and $H = 1$, we have for $n = 0$, $a_0 = 1$, and for $n \neq 0$,

$$a_n = \frac{2}{n\pi} \sin\left(n\frac{\pi}{2}\right).$$

Then, even coefficients are equal to zero as a result. The expansion of the function in harmonic terms can be written as,

$$f(t) = \frac{1}{2} + \frac{2}{\pi} \left(\cos(2\pi\nu_0 t) - \frac{1}{3} \cos(6\pi\nu_0 t) + \frac{1}{5} \cos(10\pi\nu_0 t) \right) + \cdots$$

Figure 12.22 shows the frequency spectrum, that is, the size of the coefficients a_n in terms of $n\nu_0 = n/T_0$. The coefficients represent the amplitudes of each of the harmonic waves in the Fourier series for the square wave. Given this plot, we can simply reconstruct the original square wave by summing the series it represents. As $a_n/a_1 = (1/n) \times \sin(n\pi/2)$, for $n = 9$ we have $a_9 \approx 10\% \, a_1$. Then, with 9 terms, a good approximation of the square wave is obtained. Adding more terms to the series does not significantly modify the result obtained for the square wave.

Fig. 12.22 Coefficients of the Fourier series of a square wave with $T_0 = 2\tau$

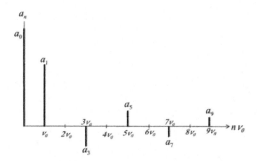

As τ becomes smaller, the width of the pulse decreases. In the case under study, where $T_0 = 2\tau$, the first occurrence of the zero coefficient occurs for $n = 2$ ($a_2 = 0$), the corresponding frequency being $2/T_0$. For $T_0 = 4\tau$; the first zero is obtained for $n = 4$ ($a_4 = 0$) and a frequency of $4/T_0$, while for $T_0 = 8\tau$ the zero amplitude occurs for $n = 8$ and a frequency of $8/T_0$. Then, the frequency n/T_0 at which the first zero coefficient occurs increases as the width of the pulse decreases. The number of coefficients contained in the interval between zero frequency and the frequency for the first zero coefficient also increases as the pulse becomes narrower.

12.20 (a) Find the Fourier transform of a rectangular pulse centered at the origin given by the equation

$$f(t) = \begin{cases} 0, & -\infty < t < -\tau/2 \\ H, & -\tau/2 < t < \tau/2 \\ 0, & \tau/2 < t < \infty \end{cases} \quad (12.44)$$

(b) What is the Fourier transform of the pulse centered at t_0?

Solution

(a) Figure 12.23a shows the pulse defined in (12.44) of height H and width τ and centered at the origin, which can be expressed as $H\Pi(t/\tau)$. To calculate the Fourier transform, we use (12.40) that yields

$$F(\nu) = \int_{-\infty}^{\infty} f(t) \exp(-i2\pi\nu t)\, dt = \int_{-\tau/2}^{\tau/2} H \exp(-i2\pi\nu t)\, dt$$

$$= \frac{H}{-i2\pi\nu} \left[\exp\left(-\frac{i2\pi\nu\tau}{2}\right) - \exp\left(\frac{i2\pi\nu\tau}{2}\right) \right]$$

$$= \frac{H}{\pi\nu} \sin(\pi\nu\tau) = H\tau \frac{\sin(\pi\nu\tau)}{\pi\nu\tau} = H\tau\, \text{sinc}(\nu\tau), \quad (12.45)$$

where the function $\sin(\pi\nu\tau)/\pi\nu\tau$ is called $\text{sinc}(\nu\tau)$. Therefore, the Fourier transform of a rectangular pulse of height H and width τ is equal to

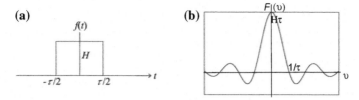

Fig. 12.23 a Rectangular pulse centered at the origin. b The frequency spectrum of the rectangular pulse

$$FT\left\{H\,\Pi\left(\frac{t}{\tau}\right)\right\} = H\tau\,\text{sinc}(\nu\tau).$$

The transform is illustrated in Fig. 12.23b. The function is symmetric with respect to the Y-axis. At frequency $\nu = 0$, the sinc function takes on the indetermination form 0/0. Applying L'Hôpital's rule to determine the value of the function

$$\lim_{\nu \to 0} H\tau\,\text{sinc}(\nu\tau) = \lim_{\nu \to 0} H\tau\,\frac{\pi\tau\cos(\pi\nu\tau)}{\pi\tau} = H\tau.$$

Therefore, at $\nu = 0$, the transform has a value $H\tau$, which is equal to the area under the pulse. The function decreases as ν increases from $\nu = 0$, reaching the first zero when $\pi\nu\tau = \pi$, i.e. $\nu\tau = 1 \Rightarrow \nu = 1/\tau$. Then, the transform alternates between positive and negative values, being zero at $\pi\nu\tau = n\pi$ ($n \neq 0$).

As τ increases, the value of the frequency at which the transform becomes zero decreases, $\nu = 1/\tau$. The wider the pulse, the narrower the sinc function. Conversely, as τ decreases the transform spreads out.

Comparing Fig. 12.23b with the envelope of the coefficients in the Fourier series in Fig. 12.22, we find that there is an equivalence between both figures. It is as if (12.43) were a sampled version of (12.45).

(b) If the rectangular pulse is shifted in time by t_0, as shown in Fig. 12.24, the transform of the function $f_s = f(t - t_0)$ becomes

$$F_s(\nu) = \int_{t_0-\tau/2}^{t_0+\tau/2} H\,\exp(-i2\pi\nu t)\,dt$$

$$= \frac{H}{-i2\pi\nu}\left\{\exp\left[-i2\pi\nu\left(t_0+\frac{\tau}{2}\right)\right] - \exp\left[-i2\pi\nu\left(t_0-\frac{\tau}{2}\right)\right]\right\}$$

Fig. 12.24 Rectangular pulse centered at t_0

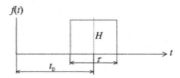

$$= H\tau \exp(-i2\pi\nu t_0) \frac{\sin(\pi\nu\tau)}{\pi\nu\tau}$$
$$= H\tau \exp(-i2\pi\nu t_0) \operatorname{sinc}(\nu\tau).$$

The Fourier transform for the shifted rectangular pulse differs from that for the rectangular pulse centered at the origin by the phase factor $\exp(-i2\pi\nu t_0)$. However, the amplitudes $|F_s(\nu)| = |F(\nu)|$ are equal.

12.21 Figure 12.25a shows a sinusoidal wave train of limited extent given by the equation

$$f(t) = \begin{cases} \Psi_0 \cos(\omega_0 t), & -\frac{\tau}{2} < t < \frac{\tau}{2} \\ 0, & \text{all other } t \end{cases} \quad (12.46)$$

(a) Find the Fourier transform of the finite wave. (b) Estimate the width of the frequency spectrum from the width of the central peak. (c) Evaluate how the pulse width affects the frequency spectrum.

Solution

(a) Equation (12.46) represents the time variation of a wave of finite duration (a pulse) at a given location, for instance at $x = 0$. The time origin is chosen in such a way that the wave is defined on the interval $(-\tau/2, \tau/2)$, thus the duration of the pulse is τ.

The Fourier transform of the cosine wave train is

$$F(\nu) = \int_{-\infty}^{\infty} f(t) \exp(-i2\pi\nu t) dt = \int_{-\tau/2}^{\tau/2} \Psi_0 \cos(\omega_0 t) \exp(-i2\pi\nu t) dt$$
$$= \Psi_0 \int_{-\tau/2}^{\tau/2} \left[\frac{\exp(i2\pi\nu_0 t) + \exp(-i2\pi\nu_0 t)}{2} \right] \exp(-i2\pi\nu t) dt$$
$$= \frac{\Psi_0}{2} \int_{-\tau/2}^{\tau/2} [\exp(-i2\pi(\nu - \nu_0)t) + \exp(-i2\pi(\nu + \nu_0)t)] dt$$

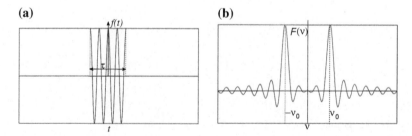

Fig. 12.25 a A wave of frequency ν_0 whose amplitude is modulated by a rectangular pulse. b The Fourier transform for a pulse with width τ

$$= \frac{\Psi_0}{2} \left[\frac{\exp[-i2\pi(\nu-\nu_0)t]}{-i2\pi(\nu-\nu_0)} + \frac{\exp[-i2\pi(\nu+\nu_0)t]}{-i2\pi(\nu+\nu_0)} \right]_{-\tau/2}^{\tau/2}$$

$$= \frac{\Psi_0}{2} \left\{ \frac{\sin[\pi(\nu-\nu_0)\tau]}{\pi(\nu-\nu_0)} + \frac{\sin[\pi(\nu+\nu_0)\tau]}{\pi(\nu+\nu_0)} \right\}$$

$$= \frac{\Psi_0 \tau}{2} \left\{ \frac{\sin[\pi(\nu-\nu_0)\tau]}{\pi(\nu-\nu_0)\tau} + \frac{\sin[\pi(\nu+\nu_0)\tau]}{\pi(\nu+\nu_0)\tau} \right\}$$

$$= \frac{\Psi_0 \tau}{2} \{\text{sinc}[(\nu-\nu_0)\tau] + \text{sinc}[(\nu+\nu_0)\tau]\}.$$

The Fourier spectrum consists of two symmetric terms (sinc functions), one centered at ν_0 and the other at $-\nu_0$, as shown in Fig. 12.25b. The latter is associated with the negative frequency distribution and contains the same information in frequencies as the spectrum for the positive frequencies. Although negative frequencies seem to contain redundant information, they should be retained to recover the original signal from the whole spectrum.

(b) The major contribution to $F(\nu)$ comes from the tallest peak, centered at ν_0, see Fig. 12.26. The frequency that contributes most to the resulting oscillation is ν_0. At ν_0, the numerator and denominator of the function, $\text{sinc}[(\nu-\nu_0)\tau] = \sin[\pi(\nu-\nu_0)\tau]/\pi(\nu-\nu_0)\tau$, give the value of zero, but the limit of the quotient is equal to unity. The spectrum can be evaluated without excessive error by considering only the central peak. Then, the range of "important" frequencies are those that lie within the maximum peak. The width of the peak can be estimated by twice the distance from ν_0 to the first frequency, where $F(\nu) = 0$. The zero occurs when

$$\sin[\pi(\nu-\nu_0)\tau] = 0 \Rightarrow \pi(\nu-\nu_0)\tau = n\pi \Rightarrow \nu = \nu_0 \pm \frac{n}{\tau} \qquad (n=1,2\ldots).$$

For $n = 1$, we have

$$\pi(\nu_1-\nu_0)\tau = \pi \Rightarrow (\nu_1-\nu_0) = \frac{1}{\tau} \Rightarrow \Delta\nu\,\tau = 1, \qquad (12.47)$$

where $\Delta\nu = \nu_1 - \nu_0$. The important frequencies are those in the range

$$\nu_{1'} = \nu_0 - \frac{1}{\tau} < \nu < \nu_1 = \nu_0 + \frac{1}{\tau},$$

Fig. 12.26 Tallest peak of the spectrum of Fig. 12.25b

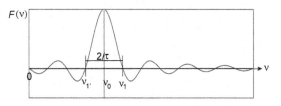

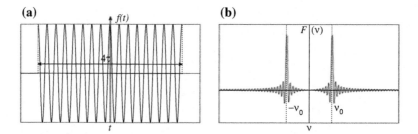

Fig. 12.27 a A wave train with width four times that of Fig. 12.25a. **b** The Fourier transform of the pulse of width 4τ. A wider pulse results in a narrower frequency spectrum

and the width of the range $\Delta\gamma = \nu_1 - \nu_{1'} = 2/\tau = 2\Delta\nu$, which provides an estimate of the spectral width of the pulse.

(c) Equation (12.47) shows that $\Delta\nu$ is inversely proportional to the pulse width τ. There is an inverse relation between spectral width $\Delta\nu$ and the pulse temporal width τ. Then, a narrow pulse has a broad frequency spectrum, whereas a wide pulse has a narrow spectral distribution. The wider the pulse, the narrower the frequency spectrum. For very wide widths, with a large number of cycles, the wave train is nearly a pure cosine wave and the sinc curve becomes a sharp and narrow spike (with its symmetric one).

Figure 12.27a shows a pulse with a width $\tau' = 4\tau$. Its Fourier transform, Fig. 12.27b, has a four times narrower spectrum width than that of Fig. 12.25b.

The same considerations hold for the relationship between the spatial width and the corresponding spectral width. Narrow confinement in space implies a wide spectral distribution. Conversely, a narrow spatial frequency spectrum implies a broad spatial distribution.

12.22 At a given position in space, the waveform can be expressed as

$$E(t) = \begin{cases} A_0 \exp(-at) \cos(2\pi\nu_0 t), & t > 0 \\ 0, & \text{all other } t \end{cases} \quad (12.48)$$

where $a \geq 0$. Find the Fourier transform of this function.

Solution

For $a = 0$, (12.48) becomes a cosine wave with a frequency ν_0. For $a > 0$, the cosine wave has an initial amplitude of A_0, but decreases exponentially in time with the damping constant a, as shown in Fig. 12.28a. At a time $t = 1/a$, the amplitude falls to $1/e$ of its original value A_0. Then, if the period $T_0 = 1/\nu_0 \ll 1/a$, the wave can be considered to be weakly damped. Thus, for small damping $\nu_0 \gg a$.

Equation (12.48) can be expressed as

$$E(t) = A_0 \exp(-at) \cos(2\pi\nu_0 t) = A_0 \exp(-at) \left[\frac{\exp(i2\pi\nu_0 t) + \exp(-i2\pi\nu_0 t)}{2}\right].$$

Then, applying (12.40) to (12.48), the Fourier transform of the damped harmonic wave will result

$$F(\nu) = \int_{-\infty}^{\infty} f(t) \exp(-i2\pi\nu t) dt$$

$$= \int_{0}^{\infty} A_0 \exp(-at) \left[\frac{\exp(i2\pi\nu_0 t) + \exp(-i2\pi\nu_0 t)}{2} \right] \exp(-i2\pi\nu t) dt$$

$$= \frac{A_0}{2} \int_{0}^{\infty} \{\exp[(-i2\pi(\nu - \nu_0) - a)t] + \exp[(-i2\pi(\nu + \nu_0) - a)t]\} dt$$

$$= \frac{A_0}{2} \left[\frac{\exp[(-i2\pi(\nu - \nu_0) - a)t]}{-i2\pi(\nu - \nu_0) - a} \right]_0^{\infty} + \frac{A_0}{2} \left[\frac{\exp[(-i2\pi(\nu + \nu_0) - a)t]}{-i2\pi(\nu + \nu_0) - a} \right]_0^{\infty}$$

$$= \frac{A_0}{2} \left[\frac{1}{i2\pi(\nu - \nu_0) + a} + \frac{1}{i2\pi(\nu + \nu_0) + a} \right].$$

This function is symmetrical and consists of two bell-shaped curves centered at ν_0 and $-\nu_0$, as shown in Fig. 12.28b.

Figure 12.29 shows $|F(\nu)|$ for a damped wave with a frequency of $\nu_0 = 10$ Hz and damping constant $a = 0.5$ s^{-1}. The spectrum for a damping constant of $2a = 1$ s^{-1} is also shown. Both functions are normalized such that the maximum is 1. The difference between the frequencies ν_2 and ν_1, corresponding to the peak width at a height equal to the peak amplitude divided by $\sqrt{2}$, is $\Delta\nu = \nu_2 - \nu_1 = a/\pi$, hence

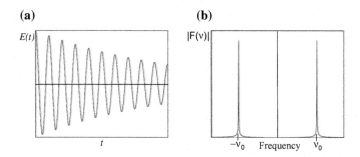

Fig. 12.28 a Damped wave in time domain and b its frequency spectrum

Fig. 12.29 Normalized $|F(\nu)|$ for a frequency $\nu_0 = 10$ Hz and damping constant $a = 0.5$ s^{-1} (*continuous line*) and $2a$ (*dashed line*). As damping increases, spectral width becomes broader

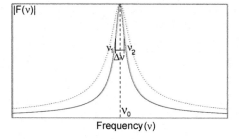

$\Delta\omega = 2a$. Then, spectral width increases with damping. The larger the damping, the more the frequency spectrum is spread over a wider range of frequencies beyond the fundamental ν_0. As the damping goes to zero, the width of the frequency spectrum also vanishes, only the fundamental frequency remains and the Fourier transform becomes a delta function.

12.23 An amplitude modulated wave has a carrier frequency of $\nu_0 = 100$ MHz. Its amplitude is modulated by the function sinc(Ax), where $A = 0.5$ m^{-1}. The wave passes through a low-pass filter, which passes low-frequency signals and attenuates signals with frequencies higher than the cutoff frequency. Find the cutoff frequency so that the bandwidth of the output signal will be four times that corresponding to the incident wave signal.

Solution

At a given time, for instance $t = 0$, using complex notation, the wave can be expressed as,

$$E(x, 0) = E_0 \text{ sinc}(Ax) \exp(i2\pi\alpha_0 x),$$

where $\alpha_0 = \nu_0/c$. In the same way, at a given location, for instance $x = 0$, the signal is given by

$$E(0, t) = E_0 \text{ sinc}(Bt) \exp(i2\pi\nu_0 t),$$

where $B = A c$.

Let us calculate the Fourier transform of the signal in the time domain,

$$FT\{E(t)\} = F(\nu) = \int_{-\infty}^{\infty} E_0 \text{ sinc}(Bt) \exp(i2\pi\nu_0 t) \exp(-i2\pi\nu t)\, dt$$

$$= E_0 \int_{-\infty}^{\infty} \text{sinc}(Bt) \exp[-i2\pi(\nu - \nu_0)t]\, dt$$

If $\nu' \equiv \nu - \nu_0$, the above equation can be written as

$$FT\{E(t)\} = F(\nu') = E_0 \int_{-\infty}^{\infty} \text{sinc}(Bt) \exp(-i2\pi\nu' t)\, dt.$$

The Fourier transform and the inverse Fourier transform can be related by

$$FT\{f(t)\} = FT^{-1}\{f(-t)\},$$

which for even functions, $f_e(t)$, and odd functions, $f_o(t)$, becomes:

$$FT\{f_e(t)\} = FT^{-1}\{f_e(t)\} \quad ; \quad FT\{f_o(t)\} = -FT^{-1}\{f_o(t)\}.$$

Fig. 12.30 The Fourier transform of the amplitude modulated wave by a sinc function

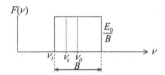

Applying these properties to $F(\nu')$ and taking into account that sinc(Bt) is an even function, we have

$$F(\nu') = E_0\, FT\{\text{sinc}(Bt)\} = E_0\, FT^{-1}\{\text{sinc}(Bt)\} = \frac{E_0}{B}\Pi\left(\frac{\nu'}{B}\right).$$

Finally, the spectrum obtained for the amplitude modulated wave can be expressed as

$$F(\nu') = \frac{E_0}{B}\Pi\left(\frac{\nu'}{B}\right) = \frac{E_0}{B}\Pi\left(\frac{\nu - \nu_0}{B}\right),$$

which is a rectangular pulse centered at ν_0 with height E_0/B and width B, as shown in Fig. 12.30.

In order for the bandwidth of the signal coming out of the filter to be four times that of the input signal, the width of the spectrum must be a quarter of B. As a low-pass filter allows the low frequencies to pass through, the cutoff frequency ν_c will be

$$\nu_c = \nu_1 + \frac{B}{4} = \nu_0 - \frac{B}{2} + \frac{B}{4}$$

$$= \nu_0 - \frac{B}{4} = 100 \times 10^6 - \frac{0.5 \times 3 \times 10^8}{4} = 62.5\text{ MHz}.$$

12.24 The wave arriving at a detector is given by the function: $E(t) = 400\,\text{sinc}[400(t - t_0)]$. The received signal passes through a filter that allows frequencies above 100 Hz to pass ($|\nu| \geq 100$ Hz). Find the output waveform.

Solution

The Fourier transform of the input signal, which is centered at t_0, is

$$FT\{E(t)\} = F(\nu) = FT\{400\,\text{sinc}[400(t - t_0)]\} = \exp(-i2\pi\nu t_0)\,FT\{400\,\text{sinc}(400t)\}$$
$$= \exp(-i2\pi\nu t_0) \times 400 \times FT\{\text{sinc}(400t)\}.$$

When comparing the spectrum of the input signal with that of the signal centered at the origin, there is a phase shift, given by $\exp(-i2\pi\nu t_0)$, but there is no change in amplitude. For the sinc function centered at the origin, the spectrum is given by

$$FT\{\text{sinc}(400t)\} = FT^{-1}\{\text{sinc}(-400t)\} = FT^{-1}\{\text{sinc}(400t)\} = \frac{1}{400}\Pi\left(\frac{\nu}{400}\right).$$

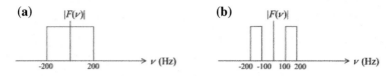

Fig. 12.31 a Fourier transform of the input signal. **b** Spectrum of the output signal

Therefore, the Fourier transform of the input signal and its modulus are, respectively,

$$F(\nu) = \exp(-i2\pi\nu t_0)\, \Pi\left(\frac{\nu}{400}\right) \quad ; \quad |F(\nu)| = \Pi\left(\frac{\nu}{400}\right).$$

Figure 12.31a shows $|F(\nu)|$, a rectangular pulse whose width is 400 Hz.

Frequencies such that $|\nu| \geq 100$ Hz are allowed to pass through the filter. Figure 12.31b represents the intervals of frequencies that pass through the filter. Then, the frequency spectrum of the output signal consists of two rectangular pulses with a width of 100 Hz, centered at 150 and -150 Hz, respectively. Moreover, the term $\exp(-i2\pi\nu t_0)$ produces a phase change. The function $F_{\text{out}}(\nu)$ corresponding to the frequency spectrum of the output signal is given by

$$F_{\text{out}}(\nu) = \exp(-i2\pi\nu t_0)\left[\Pi\left(\frac{\nu+150}{100}\right) + \Pi\left(\frac{\nu-150}{100}\right)\right].$$

The output signal in the time domain $E_{\text{out}}(t)$ can be calculated from the inverse transform,

$$E_{\text{out}}(t) = FT^{-1}\{F_{\text{out}}(\nu)\} = \int_{-\infty}^{\infty} \exp(-i2\pi\nu t_0)\left[\Pi\left(\frac{\nu+150}{100}\right) + \Pi\left(\frac{\nu-150}{100}\right)\right] \exp(i2\pi\nu t)\, d\nu$$

$$= \int_{-\infty}^{\infty} \Pi\left(\frac{\nu+150}{100}\right) \exp[i2\pi(t-t_0)\nu]\, d\nu + \int_{-\infty}^{\infty} \Pi\left(\frac{\nu-150}{100}\right) \exp[i2\pi(t-t_0)\nu]\, d\nu.$$

If we make $t' = t - t_0$, the above equation can be written as

$$E_{\text{out}}(t') = \int_{-\infty}^{\infty} \Pi\left(\frac{\nu+150}{100}\right) \exp(i2\pi t'\nu)\, d\nu + \int_{-\infty}^{\infty} \Pi\left(\frac{\nu-150}{100}\right) \exp(i2\pi t'\nu)\, d\nu$$

$$= FT^{-1}\left\{\Pi\left(\frac{\nu+150}{100}\right)\right\} + FT^{-1}\left\{\Pi\left(\frac{\nu-150}{100}\right)\right\}.$$

When the origin is shifted, we have for the inverse Fourier transform,

$$FT^{-1}\{F(\nu-\nu_0)\} = \exp(i2\pi\nu_0 t) FT^{-1}\{F(\nu)\}.$$

From the above equation, one obtains the output signal in terms of t',

$$E_{\text{out}}(t') = \exp(-i2\pi \times 150t') FT^{-1}\left\{\Pi\left(\frac{\nu}{100}\right)\right\} + \exp(i2\pi \times 150t') FT^{-1}\left\{\Pi\left(\frac{\nu}{100}\right)\right\}$$
$$= \exp(-i300\pi t')\,100\,\text{sinc}(100t') + \exp(i300\pi t')\,100\,\text{sinc}(100t').$$

Finally, the output waveform as a function of t,

$$E_{\text{out}}(t) = \exp[-i300\pi(t-t_0)]100\,\text{sinc}[100(t-t_0)] + \exp[i300\pi(t-t_0)]100\,\text{sinc}[100(t-t_0)]$$
$$= 100\,\text{sinc}[100(t-t_0)]2\left(\frac{\exp[-i300\pi(t-t_0)] + \exp[i300\pi(t-t_0)]}{2}\right)$$
$$= 200\,\text{sinc}[100(t-t_0)]\,\cos[300\pi(t-t_0)],$$

which could be considered as a cosine-type waveform modulated in amplitude by a sinc function.

Chapter 13
Reflection and Refraction

Abstract In the previous chapter, a plane wave propagating in a homogeneous non-conducting isotropic medium was considered. In this chapter, we will examine what happens to an electromagnetic wave at a plane boundary between two non-conducting media with different electromagnetic properties. The laws of refraction and reflection will be applied to determine the directions of propagation of the reflected and transmitted waves. Their amplitudes will be obtained from Fresnel's coefficients. We will also evaluate what fraction of the energy in a plane wave incident on a dielectric boundary is reflected, and what fraction is transmitted. As electromagnetic theory of light states that light is an electromagnetic wave, all included in this chapter is applicable in optics, provided that the size of the object is large compared with the wavelength.

13.1 Laws of Reflection and Refraction

Figure 13.1 shows a plane electromagnetic wave travelling in the direction $\mathbf{k}_i$ that impinges on a plane boundary between two media with different electromagnetic properties. The wave is partially reflected and partially transmitted. The wave velocities of medium 1 and 2 are v_1 and v_2, respectively. The refractive index $n_1 \equiv c/v_1$ characterizes medium 1, which contains the incident and reflected wave, whereas $n_2 \equiv c/v_2$ characterizes medium 2, containing the transmitted wave. The plane defined by the incident wave vector $\mathbf{k}_i$ and the normal to the boundary surface, at the point where $\mathbf{k}_i$ intersects the boundary, is called the plane of incidence. The angle of incidence θ_i, the angle of reflection θ_r, and the angle of refraction θ_t represent, respectively, the angles that the vector of propagation of the incident, reflected, and transmitted waves make with the normal to the boundary. Lines perpendicular to the wavefronts, and, therefore, collinear with the propagation vector $\mathbf{k}$, are referred to as rays. The frequency of the wave is not altered by reflection or transmission. At the boundary between two media, there are relationships which must be obeyed between the fields on the two sides. By requiring that the phase of the wave be continuous across the boundary between two media with different wave propagation velocities, the laws of reflection and refraction can be derived, which can be expressed as:

Fig. 13.1 Directions of propagation for incident plane wave (vector of propagation $\mathbf{k}_i$), reflected wave ($\mathbf{k}_r$), and transmitted wave ($\mathbf{k}_t$). All the vectors are in the plane of incidence

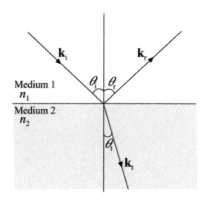

1. The propagation vector of the reflected wave, $\mathbf{k}_r$, and that of the transmitted wave, $\mathbf{k}_t$, lie in the plane of incidence.
2. The angle of reflection is equal to the angle of incidence,

$$\theta_r = \theta_i. \tag{13.1}$$

3. The angle between the propagation vector of the transmitted wave and the normal to the boundary is given by

$$\sin \theta_t = \frac{n_1}{n_2} \sin \theta_i \Rightarrow n_1 \sin \theta_i = n_2 \sin \theta_t \quad \text{(Snell's law)}. \tag{13.2}$$

When $n_2 > n_1$, $\sin \theta_t < \sin \theta_i$, and there is a real angle θ_t of refraction for every angle of incidence. However, if $n_2 > n_1$, $\sin \theta_t > \sin \theta_i$. Since θ_t increases with θ_i, we can find a critical angle of incidence,

$$\theta_c = \sin^{-1} \frac{n_2}{n_1}, \tag{13.3}$$

at which $\theta_r = \pi/2$ and the refracted wave will glaze along the interface. When $\theta_i \geq \theta_c$, there is no refracted wave, and the incident wave is then said to be totally reflected.

13.2 The Fresnel Coefficients

The laws of reflection and refraction give no information about the relations between the magnitudes of the field vectors in the reflected, transmitted, and incident waves. These relations are obtained from Maxwell's equations and the boundary conditions for $\mathbf{E}$ and $\mathbf{B}$, studied in Chap. 10. Figure 13.2 shows the components of the fields. The plane of the figure is the plane of incidence. The electric field in each wave

13.2 The Fresnel Coefficients

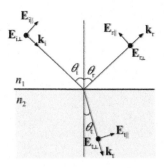

Fig. 13.2 Incident, reflected, and transmitted plane wave fields at a material interface. Vectors are decomposed into components: parallel ($\parallel$) and perpendicular ($\perp$) to the plane of incidence. The latter is assumed to be directed out of the plane of the page and is represented by *dots*

is represented as the sum of two components: $E_\parallel$ denotes the component in the plane of incidence and $E_\perp$ is the component perpendicular to this plane. The ratio of the reflected and transmitted field components to the incident field components are determined by the following coefficients, called the Fresnel coefficients:

$$r_\parallel \equiv \frac{E_{r\parallel}}{E_{i\parallel}} = \frac{n_2 \cos\theta_i - n_1 \cos\theta_t}{n_1 \cos\theta_t + n_2 \cos\theta_i} \quad ; \quad r_\perp \equiv \frac{E_{r\perp}}{E_{i\perp}} = \frac{n_1 \cos\theta_i - n_2 \cos\theta_t}{n_1 \cos\theta_i + n_2 \cos\theta_t}, \quad (13.4)$$

$$t_\parallel \equiv \frac{E_{t\parallel}}{E_{i\parallel}} = \frac{2n_1 \cos\theta_i}{n_1 \cos\theta_t + n_2 \cos\theta_i} \quad ; \quad t_\perp \equiv \frac{E_{t\perp}}{E_{i\perp}} = \frac{2n_1 \cos\theta_i}{n_1 \cos\theta_i + n_2 \cos\theta_t}. \quad (13.5)$$

These coefficients allow us to obtain the amplitude of the components of the electric field for the reflected and transmitted waves in terms of the components of the incident wave. The components for the reflected wave are either in phase with those of the incident wave or shifted by π whereas the phase of the transmitted wave is equal to that of the incident wave.

In Fig. 13.3, the Fresnel coefficients are plotted in the case of air-glass interface. It should be noted that a sign change occurs for $r_\parallel$, which corresponds to a phase shift of π upon reflection. Note in Fig. 13.3b that for $\theta_i = \theta_c$, $|r_\parallel| = |r_\perp| = 1$.

From (13.4), it is obtained that $r_\parallel$ goes to zero when

$$\theta_i + \theta_t = \frac{\pi}{2}, \quad (13.6)$$

which occurs when the reflected and transmitted vectors, $\mathbf{k}_r$ and $\mathbf{k}_t$, are perpendicular to each other. The incident angle that satisfies this equation, in terms of the refractive indices, is found to be

$$\theta_B = \tan^{-1}\frac{n_2}{n_1}, \quad (13.7)$$

which is called Brewster's angle. At Brewster's angle, the electric vector of the reflected wave has no component in the plane of incidence; the reflected wave is said to be linearly polarized, with the electric vector normal to the plane of incidence.

Fig. 13.3 a The Fresnel coefficients in terms of the incident angle θ_i for $n_1 = 1$ and $n_2 = 1.5$. b The reflection coefficients $r_\parallel$ and $r_\perp$ for $n_1 = 1.5$ and $n_2 = 1$

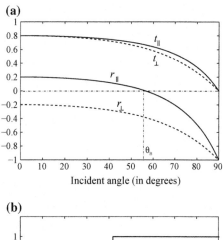

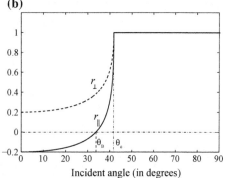

13.3 Reflected and Transmitted Energy

Application of the principle of conservation of energy to the flow across the boundary yields the fraction of the incident energy that is reflected and the fraction that is transmitted. Let us denote by I_i, I_r, and I_t the average intensity for the incident, reflected, and transmitted wave, respectively. The reflection coefficient R (reflectance) is defined as the reflected energy divided by the incident energy,

$$R \equiv \frac{I_r \cos \theta_r}{I_i \cos \theta_i} = \frac{I_r}{I_i}, \tag{13.8}$$

where $I_i \cos \theta_i$ represents the amount of energy that is incident on a unit area of the boundary per second and $I_r \cos \theta_r$ the energy of the reflected wave leaving a unit area of the boundary per second. In the same way, the transmission coefficient T (transmittance) is defined as the energy transmitted divided by the energy incident,

$$T \equiv \frac{I_t \cos \theta_t}{I_i \cos \theta_i}, \tag{13.9}$$

13.3 Reflected and Transmitted Energy

Fig. 13.4 Reflection and transmission of an incident wave front. The energy balance is made across surface A on the interface

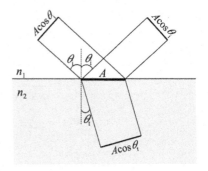

where $I_t \cos \theta_t$ corresponds to the energy of the transmitted wave. Conservation of energy flowing through the boundary surface, see Fig. 13.4, leads to

$$R + T = 1, \tag{13.10}$$

where it has been assumed that absorption is negligible. It is usually convenient to express the reflection and transmission coefficients for the parallel and perpendicular directions. Conservation of the energy, in such directions, leads to the following coefficients for the reflected wave:

$$R_\| = r_\|^2 \quad \text{and} \quad R_\perp = r_\perp^2, \tag{13.11}$$

and for the transmitted wave,

$$T_\| = \frac{n_2 \cos \theta_t}{n_1 \cos \theta_i} t_\|^2 \quad \text{and} \quad T_\perp = \frac{n_2 \cos \theta_t}{n_1 \cos \theta_i} t_\perp^2. \tag{13.12}$$

For each polarization, it can be verified that,

$$R_\| + T_\| = 1 \quad \text{and} \quad R_\perp + T_\perp = 1. \tag{13.13}$$

It should be noted that (13.12) does not hold when the incident angle exceeds the critical angle for total internal reflection.

Solved Problems

Problems A

13.1 Calculate the angle between the refracted ray and the normal to the surface between the two media in the cases shown in Fig. 13.5.

Fig. 13.5 Ray of light incident at 45° on: **a** air-water interface and **b** water-air interface

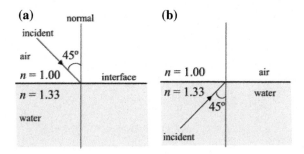

Solution

(a) From Snell's law, $n_1 \sin \theta_i = n_2 \sin \theta_t$, we obtain

$$\sin \theta_t = \frac{n_1}{n_2} \sin \theta_i.$$

Note that the angle of incidence θ_i and that of refraction θ_t are measured with respect to the surface normal. The incident ray, normal, and refracted ray, all lie in the same plane (plane of incidence). n_1 represents the refractive index of the medium in which the incident wave travels, while n_2 is the refractive index of the medium in which the refracted wave moves. In the case under study, we have: $n_1 = 1.00$, $\theta_i = 45°$, and $n_2 = 1.33$. Thus, it follows that

$$\sin \theta_t = \frac{n_1}{n_2} \sin \theta_i = \frac{1.00}{1.33} \sin 45° = 0.532 \Rightarrow \theta_t = \sin^{-1} 0.532 = 32.1°.$$

The angle of refraction is 32.1°, less than 45°, as shown in Fig. 13.6a. In this case that light passes from air to water, i.e. from a medium of lower index to a higher index, the light ray is bent toward the normal.
(b) In the second case, light passes from water, $n_1 = 1.33$, to air, $n_2 = 1.00$. Snell's law yields

Fig. 13.6 As refractive index for water is greater than that for air, in **a** bending is toward the normal while in **b** bending is away from the normal

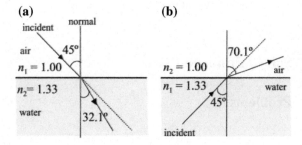

$$\sin\theta_t = \frac{n_1}{n_2}\sin\theta_i = \frac{1.33}{1.00}\sin 45° = 0.940 \Rightarrow \theta_t = \sin^{-1} 0.940 = 70.1°.$$

As light passes from a higher index to lower index, the ray is bent away from the normal, as shown in Fig. 13.6b.

13.2 Figure 13.7 shows a medium (glass) with refractive index $n = 3$, limited by two spherical surfaces whose radii are R and $2R$, respectively. At the center of the sphere of radius R, there is a point light source. Determine the values of α that delimit the external surface through which light does not emerge.

Solution

For any ray emerging from the point source, the angle of incidence at the air-to-glass boundary is zero since the rays propagate in the radial direction. Therefore, the refracted ray does not change its direction at the first interface (air-glass). Figure 13.7 shows a ray emerging from the light source that makes an angle α with the horizontal, which does not suffer a change in direction at point A on the air-glass interface. At the interface between the material and air, the angle between the incident ray and the normal to the interface, the radial direction CB, is denoted by θ_i. Applying Snell's law to the glass-air interface, the critical angle obtained is

$$n\sin\theta_c = 1\times\sin\frac{\pi}{2} \Rightarrow \sin\theta_c = \frac{1}{n},$$

where it is assumed that the refractive index of air is approximately 1. Therefore, the incident ray on the second interface will be totally reflected, if the following condition is satisfied

$$\sin\theta_i \geq \frac{1}{n}.$$

From the figure and the sine theorem, it follows that

$$\frac{\sin\theta_i}{R} = \frac{\sin\alpha}{2R} \Rightarrow \sin\theta_i = \frac{\sin\alpha}{2}.$$

Fig. 13.7 Sphere of radius $2R$ and refractive index $n = 3$. A spherical hole of radius R has been cut in the large sphere. A point light source is located at the center O

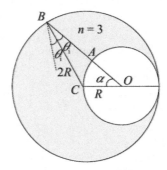

Thus, the values of α for which rays do not emerge from the spherical surface are given by

$$\frac{\sin\alpha}{2} \geq \frac{1}{n} \Rightarrow \sin\alpha \geq \frac{2}{n} = \frac{2}{3} \Rightarrow 41.8° \leq \alpha \leq 138.2°.$$

Due to the symmetry of the problem, this condition is also satisfied in the lower hemisphere.

13.3 A light ray is incident from air on a plane-parallel glass plate at an angle θ_i, as shown in Fig. 13.8. The plate has a refractive index n and a thickness t. Derive the expression for the lateral displacement ℓ of the emerging ray.

Solution

Figure 13.9 shows the path followed by a ray through the plate. Snell's law at the point of incidence A, on the air-glass interface, gives

$$1 \times \sin\theta_i = n\sin\theta_t \Rightarrow \sin\theta_t = \frac{\sin\theta_i}{n} \Rightarrow \cos\theta_t = \sqrt{1 - \frac{\sin^2\theta_i}{n^2}}. \quad (13.14)$$

In the same way, at the interface between air and glass, we obtain

$$n\sin\theta'_i = 1 \times \sin\theta'_t.$$

From the geometry of the ray path shown in Fig. 13.9, we have $\theta'_i = \theta_t$. Then, the above two equations yield the following relationship

$$n\sin\theta'_i = n\sin\theta_t = \sin\theta_i = \sin\theta'_t,$$

which implies that $\theta'_t = \theta_i$, i.e. the direction of the ray emerging from the plate is the same as that of the incident ray impinging on the plate. However, there is a lateral displacement ℓ between the two rays.

From the geometry shown in Fig. 13.9, the lateral displacement ℓ can be calculated as follows

$$\left.\begin{array}{l} CD = AC\sin(\theta_i - \theta_t) \\ AC = \frac{AB}{\cos\theta_t} = \frac{t}{\cos\theta_t} \end{array}\right\} \Rightarrow \ell = CD = \frac{t}{\cos\theta_t}\sin(\theta_i - \theta_t) \Rightarrow$$

Fig. 13.8 Ray impinging obliquely on a glass plate of refractive index n

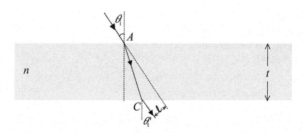

Fig. 13.9 Lateral displacement of a ray after passing through a glass plate

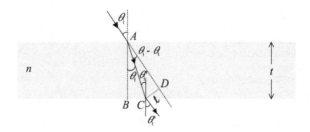

$$\ell = \frac{t}{\cos\theta_t}(\sin\theta_i \cos\theta_t - \cos\theta_i \sin\theta_t) = t(\sin\theta_i - \cos\theta_i \tan\theta_t).$$

The value of $\tan\theta_t$ can be obtained from (13.14), $\tan\theta_t = \sin\theta_i/\sqrt{n^2 - \sin^2\theta_i}$. Substituting $\tan\theta_t$ into the equation obtained for ℓ, it follows that

$$\ell = t\sin\theta_i \left[1 - \frac{\cos\theta_i}{\sqrt{n^2 - \sin^2\theta_i}}\right].$$

This equation gives the lateral displacement ℓ in terms of the angle of incidence θ_i, the refractive index n, and the thickness of the plate t.

13.4 Consider a simple lens formed by a glass half sphere, with radius R and refractive index n_1, surrounded by air, whose refractive index is denoted by n_2. A ray of light, parallel to the axis of rotational symmetry, and at a distance H from the axis, is incident on the flat surface, as shown in Fig. 13.10. If $H \ll R$, find the location of point F where the emerging ray from the glass intersects the axis.

Solution

Figure 13.10 shows a ray of light propagating in a direction parallel to the axis, which impinges from air on the flat surface of a lens at normal incidence. Therefore, the direction of the ray remains unchanged according to Snell's law. Then, the ray passes through the glass and hits the glass-air interface making an angle θ_i with the normal (coincident with the radial line). Finally, the ray emerges from the lens at angle θ_t (the angle of refraction). From the geometry (see Fig. 13.10) and the condition that $H \ll R$, it is found that

$$\sin\theta_i = \frac{H}{R} \approx \tan\theta_i \approx \theta_i,$$

where the small-angle approximation (angle θ approaches 0) has been used, i.e. $\sin\theta \approx \tan\theta \approx \theta$ and $\cos\theta \approx 1$.

By applying Snell's law to the glass-air interface, it follows that the angle of refraction is equal to

$$\sin\theta_t = \frac{n_1}{n_2}\sin\theta_i \Rightarrow \theta_t \approx \frac{n_1}{n_2}\frac{H}{R}.$$

Fig. 13.10 Path followed by a single ray through the lens. Rays travelling parallel to the axis and close to it converge to point F, the focal point of the lens

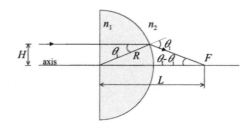

Then, for small angles, the following relation can be obtained

$$\tan(\theta_t - \theta_i) \approx \theta_t - \theta_i \approx \frac{n_1}{n_2}\frac{H}{R} - \frac{H}{R} = \frac{n_1 - n_2}{n_2}\frac{H}{R}.$$

The distance from F, the intersection point of the ray emerging from the glass with the axis, to the flat surface is denoted by L in Fig. 13.10. According to the geometry, point F is located a distance

$$L = R\cos\theta_i + \frac{H}{\tan(\theta_t - \theta_i)} \approx R + H\frac{n_2}{n_1 - n_2}\frac{R}{H} \approx \frac{n_1}{n_1 - n_2}R,$$

from the front of the lens. Therefore, all parallel rays close to the axis converge to point F after passing through the lens. This point is called the focal point of the lens.

13.5 A point light source at an unknown distance H under water yields an illuminated circular area with diameter 12 m, seen from the air side of the interface. Find H. Assume that the refractive index of water is 1.33.

Solution

Figure 13.11 shows a light source at a distance H under water. Rays emerging from O in all directions are incident on the water-air interface. On the water surface, directly above the source, the area illuminated is a circle of diameter D. The maximum radius of this circle is determined by the critical angle for refraction. Refraction can only occur if the ray incident on the interface makes an angle with the normal line smaller than the critical angle.

For ray OO', at normal incidence, there is no change in the direction that the wave is travelling, whereas for ray OP, the incident angle being θ, the transmitted ray into air changes its direction relative to the normal to the surface. As $n_1 = 1.33 > n_2 = 1$, it follows that $\theta_t > \theta$. It should be noted in Fig. 13.11 that the greater the angle that the ray emitted by the source makes with OO', which is equal to the angle of incidence at the water-air interface, the greater the angle of refraction. At angles of incidence greater or equal to the critical angle, light rays will experience total reflection and, hence, no light is emerging from the water to the air. The critical angle at the water-air interface is given by

Fig. 13.11 A point light source under water emitting rays that impinge on the water-air interface

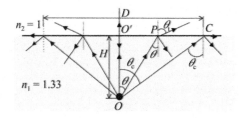

$$n_1 \sin \theta_c = n_2 \sin 90° \Rightarrow \theta_c = \sin^{-1}\left(\frac{1}{n_1}\right) = \sin^{-1}\left(\frac{1}{1.33}\right) = 48.75°.$$

Note in Fig. 13.11 that ray OC hits the interface at the critical angle. Then, the point light source is submerged below surface at a distance,

$$H = \frac{O'C}{\tan \theta_c} = \frac{6}{\tan 48.75°} \text{ m} = 5.26 \text{ m},$$

where $O'C$ is the radius of the illuminated circle.

13.6 A prism is made of a glass whose refractive index varies with wavelength: $n_p = 1.60 - 0.10\,\lambda$, where λ represents the wavelength in vacuum in µm. A beam of white light is incident on a face of the prism at an angle of 45°, as shown in Fig. 13.12. If the wavelength of the red light is 0.750 µm and that of violet light 0.390 µm, find the angular dispersion.

Solution

A ray of white light incident obliquely on a prism is twice refracted as it passes through it. As the refractive index is dependent on the wavelength, the angle of refraction varies with the wavelength and light is hence dispersed into all the colors of the visible spectrum. The visible spectrum ranges in wavelength from approximately 0.390 µm, for violet light, to 0.750 µm, for red light. Longer wavelengths have smaller refractive indexes and are refracted less than shorter wavelengths.

We have for violet light, the lower end of the visible spectrum, a refractive index, n_v, and an angle of refraction, θ_{tv}:

$$n_v = 1.60 - 0.10 \times 0.390 = 1.561, \quad \sin \theta_i = n_v \sin \theta_{tv} \Rightarrow \sin \theta_{tv} = \frac{\sin 45°}{1.561} \Rightarrow \theta_{tv} = 26.94°.$$

From the triangle ABC (see Fig. 13.12) the angle of incidence at B, θ'_{iv}, can be obtained,

$$180° = (90° - \theta_{tv}) + 60° + (90° - \theta'_{iv}) \Rightarrow \theta'_{iv} = 33.06°.$$

Therefore, for violet light, the angle of refraction for the refracted ray emerging from the second face of the prism, θ'_{tv}, will be

Fig. 13.12 Dispersion of light by a prism. $\theta_i = 45°$

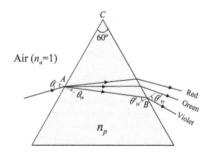

$$n_v \sin \theta'_{iv} = \sin \theta'_{tv} \Rightarrow 1.561 \sin 33.06° = \sin \theta'_{t_v} \Rightarrow \theta'_{tv} = 58.38°.$$

In the same way, for red light, the longer wavelength of the visible spectrum, the refractive index being n_r, we have

$$n_r = 1.60 - 0.10 \times 0.750 = 1.525, \quad \sin \theta_{tr} = \frac{\sin 45°}{1.525} \Rightarrow \theta_{tr} = 27.62°.$$

The angle of incidence at the second face will be

$$\theta'_{ir} = 60° - 27.62° = 32.38°,$$

and the angle of refraction at the second face

$$1.525 \sin 32.38° = \sin \theta'_{tr} \Rightarrow \theta'_{tr} = 54.75°.$$

Therefore, the angular dispersion is given by

$$\theta'_{tv} - \theta'_{tr} = 58.38° - 54.75° = 3.63°.$$

13.7 (a) Determine the Fresnel coefficients for normal incidence. (b) A monochromatic plane wave with an amplitude of 10 V/m is incident normally on the plane surface of an air-glass interface. Find the amplitude and phase for the reflected and transmitted waves in the following two cases: (1) When the wave is incident from the air side; (2) when the wave is incident from the glass side. The index of refraction for glass is 1.5. Take the index of air to be one.

Solution

(a) At normal incidence $\theta_i = 0$, Snell's law (13.2) gives for the angle of refraction: $\sin \theta_t = 0$, $\theta_t = 0$, since $\theta_t \leq \pi/2$. Therefore, the direction of propagation of the incident wave remains unchanged in the process of reflection and refraction.

For $\theta_i = 0$, the meaning of plane of incidence is lost, since the two vectors defining this plane are parallel, and therefore, the distinction between the parallel and perpendicular components is of no interest. The Fresnel coefficients (13.4) and (13.5) for $\theta_i = \theta_t = 0$ reduce to

$$r_\| \equiv \frac{E_{r\|}}{E_{i\|}} = \frac{n_2 - n_1}{n_1 + n_2} \quad ; \quad r_\perp \equiv \frac{E_{r\perp}}{E_{i\perp}} = \frac{n_1 - n_2}{n_1 + n_2},$$

$$t_\| \equiv \frac{E_{t\|}}{E_{i\|}} = \frac{2n_1}{n_1 + n_2} \quad ; \quad t_\perp \equiv \frac{E_{t\perp}}{E_{i\perp}} = \frac{2n_1}{n_1 + n_2}.$$

The $r_\|$ and $r_\perp$ coefficients lead to the same direction of oscillation for the electric field of the reflected wave and, therefore, at normal incidence, the reflection coefficient of a plane wave is independent on the wave's polarization. We can conclude that for normal incidence the ratio of amplitudes of the reflected and incident waves and the ratio of the amplitude of the transmitted wave to that of the incident wave can be expressed, respectively, as,

$$r = \frac{E_r}{E_i} = \frac{n_1 - n_2}{n_1 + n_2} \quad \text{and} \quad t = \frac{E_t}{E_i} = \frac{2n_1}{n_1 + n_2}, \tag{13.15}$$

where a positive r means that reflected wave oscillates in phase with the incident wave. On the other hand, a negative coefficient means that the reflected wave oscillates π radians out of phase with the incident wave.

It follows from (13.15) that if $n_1 < n_2$, the electric vector in the reflected wave is in the opposite direction to that of the incident electric field. As coefficient t always has a positive sign, the transmitted wave oscillates in phase with the incident wave. In this case, the amplitude of the transmitted wave is smaller than that of the incident wave. On the other hand, for $n_1 > n_2$, the reflected wave oscillates in phase with the incident wave, and the amplitude of the transmitted wave is larger than that of the incident wave.

(b) For the air-glass interface ($n_1 = 1$ and $n_2 = 1.5$), (13.15) gives: $r = -0.2$ and $t = 0.8$. Then, for an incident wave with amplitude 10 V/m, the amplitudes of the reflected and transmitted waves will be:

$$E_{0r} = r \times E_{0i} = -0.2 \times 10 = -2 \text{ V/m} \quad ; \quad E_{0t} = t \times E_{0i} = 0.8 \times 10 = 8 \text{ V/m}.$$

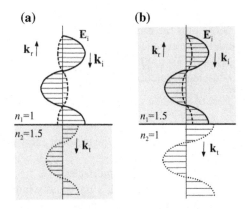

Fig. 13.13 The amplitude and phase relations for harmonic plane waves incident normally on an interface: **a** air-glass and **b** glass-air. The wave is assumed to be linearly polarized

Figure 13.13a shows the amplitude and phase relations for this case; the phase of the reflected wave is shifted by π with respect to the incident wave.

For the glass-air interface ($n_1 = 1.5$ and $n_2 = 1$), the results are shown in Fig. 13.13b, which corresponds to $r = 0.2$ and $t = 1.2$, the resulting amplitudes being::

$$E_{0r} = r \times E_{0i} = 0.2 \times 10 = 2 \text{ V/m} \quad ; \quad E_{0t} = t \times E_{0i} = 1.2 \times 10 = 12 \text{ V/m}.$$

The reflected wave does not change phase, as expected. Figure 13.13 shows the continuity of the tangential component of the electric field across the interface.

13.8 (a) Use the equations for reflectance and transmittance for normal incidence to prove that energy is conserved. (b) For the cases considered in the previous problem for an air-glass interface, determine what fraction of the incident energy is reflected and what fraction is transmitted.

Solution

(a) At $\theta_i = 0$, (13.11) and (13.15) give for the reflectance:

$$R = R_\parallel = R_\perp = r^2 = \left(\frac{n_1 - n_2}{n_1 + n_2}\right)^2. \tag{13.16}$$

Analogously, for the transmittance, (13.12) and (13.15) lead to:

$$T = T_\parallel = T_\perp = \frac{n_2}{n_1} t^2 = \frac{4 n_1 n_2}{(n_1 + n_2)^2}. \tag{13.17}$$

By adding R and T,

$$R + T = \left(\frac{n_1 - n_2}{n_1 + n_2}\right)^2 + \frac{4 n_1 n_2}{(n_1 + n_2)^2} = \frac{n_1^2 + n_2^2 - 2 n_1 n_2 + 4 n_1 n_2}{(n_1 + n_2)^2} = \frac{n_1^2 + n_2^2 + 2 n_1 n_2}{(n_1 + n_2)^2} = 1.$$

As R and T represent, respectively, the reflected and transmitted energy divided by the incident energy (across unit area per unit time), this result means that the reflected energy plus the transmitted energy is equal to the energy incident on the interface, and, therefore, energy (per unit surface area per unit time) is conserved.

(b) For the air-to-glass interface, $n_1 = 1$ and $n_2 = 1.5$, the reflection and transmission coefficients, calculated from (13.16) and (13.17), are:

$$R = 0.04 \quad \text{and} \quad T = 0.96,$$

which means that 4% of the incident energy (per unit area and unit time) is reflected and 96% is transmitted.

For the glass-to-air interface, $n_1 = 1.5$ and $n_2 = 1$, we find that $R = 0.04$ and $T = 0.96$. Then, through a flat-glass window panel, due to the multiple reflection and refraction in the two interfaces, the total power reflected is approximately 8%.

13.9 Plot R and T for normal incidence in terms of the quotient n_1/n_2.

Solution

Let the quotient n_1/n_2 be denoted by x. Then, (13.16) and (13.17) can be written as

$$R = \left(\frac{x-1}{x+1}\right)^2 \quad ; \quad T = \frac{4x}{(x+1)^2}.$$

Figure 13.14 shows reflectance R and transmittance T in terms of $x = n_1/n_2$. For $x \approx 1$, i.e. $n_1 \approx n_2$, all the energy is transmitted. The smaller the difference of the indices of the two media, the less energy is carried by the reflected wave. When R equals T, we have

$$R = T \Rightarrow \left(\frac{x-1}{x+1}\right)^2 = \frac{4x}{(x+1)^2} \Rightarrow x^2 - 6x + 1 = 0 \Rightarrow x = 0.17 \text{ and } 5.83.$$

For $x = 0.17$ and 5.83, it follows that $R \approx T \approx 0.5$. Figure 13.14 shows that for $x > 1$, R and T increases and decreases, respectively, with x. Analogously, for $x < 1$, R increases and T decreases with the same proportion ($T = 1 - R$) as x becomes smaller. Therefore, the greater the difference between the properties of the two media, the greater the energy reflected and the smaller the energy transmitted. Note that R yields the same result for a given value of x and its reciprocal, $1/x$. The same holds for T.

13.10 For an air-to-water interface, plot the reflectance, $R_\parallel$ and $R_\perp$, as a function of incident angle θ_i. Also plot the water-to-air reflectance. Assume that water has an index of refraction of 1.33. Take the index of air to be one.

Solution

It follows from (13.11) and (13.4) that reflectance for the parallel and perpendicular polarization is given, respectively, by

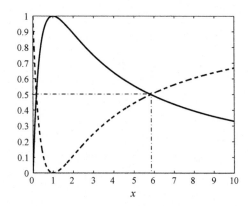

Fig. 13.14 Reflectance (*dashed line*) and transmittance (*continuous line*) in terms of $x = n_1/n_2$ for normal incidence

Fig. 13.15 Reflectance $R_\parallel$ (*continuous line*) and $R_\perp$ (*dashed line*) plotted versus θ_i for the air-water interface, $n_1 = 1$ and $n_2 = 1.33$. Brewster's angle is $\theta_B = 53.06°$

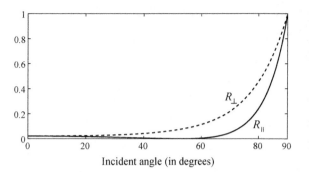

$$R_\parallel = \left(\frac{n_2 \cos\theta_i - n_1 \cos\theta_t}{n_1 \cos\theta_t + n_2 \cos\theta_i}\right)^2 \quad ; \quad R_\perp = \left(\frac{n_1 \cos\theta_i - n_2 \cos\theta_t}{n_1 \cos\theta_i + n_2 \cos\theta_t}\right)^2.$$

We have to represent $R_\parallel$ and $R_\perp$ in terms of θ_i for the air-water interface first and then for the water-air interface.
(a) The plot of the reflectance for the case of $n_1 = 1$ and $n_2 = 1.33$ is show in Fig. 13.15. We find that for normal incidence the fraction of power reflected is 2% ($R_\parallel = R_\perp = 0.02$). For the perpendicular polarization, the fraction of energy reflected increases with the angle of incidence and when θ_i approaches 90°, 100% of the power is reflected. For the component in the plane of incidence, the reflectance decreases to zero at Brewster's angle, $\theta_B = \tan^{-1} = n_2/n_1 = 1.33 = 53.06°$, and then exhibits the same behaviour as the reflectance for the perpendicular polarization.
(b) For the interface water-air, $n_1 > n_2$, we can find a critical angle θ_c at which there is total reflection and then $R_\parallel = 1$ and $R_\perp = 1$. From (13.3), the critical angle $\theta_c = \sin^{-1}(1/1.33) = 48.75°$. Beyond the critical angle, the wave is said to undergo total reflection. Figure 13.16 shows the plot of the reflectance versus θ_i for both components. For angles of incidence greater than the critical angle, the reflectance (modulus) is 1, hence the incident energy is reflected and no energy is transmitted to the second medium. At normal incidence, we find the same values for the reflectance as those for the air-water interface. The reflectance for the perpendicular polarization increases with θ_i until the critical angle is reached. For the component in the plane of incidence, the reflectance becomes zero at Brewster angle $\theta_B = 36.94°$, as expected. For $\theta_i > \theta_B$, $R_\parallel$ increases sharply with θ_i until the incident angle equals the critical angle.

Problems B

13.11 Figure 13.17 shows an optical fiber surrounded by a material of lower refractive index, known as cladding. Find the maximum angle of incidence θ_i for rays incident on the core's end face to be trapped inside the core. Consider that air has a refraction index of 1.

Fig. 13.16 Reflectance $R_\parallel$ (*continuous line*) and $R_\perp$ (*dashed line*) plotted versus θ_i for a water-air interface $n_1 = 1.33$ and $n_2 = 1$. Brewster's angle is $\theta_B = 36.94°$ and critical angle $\theta_c = 48.75°$

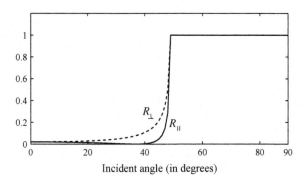

Solution

Figure 13.17 shows a ray meeting the air-core boundary at an angle θ_i, measured relative to a line normal to the boundary. The angle of refraction at P_1 is θ_t. At P_2 on the interface between the core and the cladding, Snell's law gives

$$n_1 \sin \theta'_i = n_2 \sin \theta'_t,$$

where n_1 and n_2 are the indices of refraction of the core and cladding ($n_1 > n_2$), respectively. Note that θ_t and θ'_i are complementary angles, i.e. $\theta_t = 90° - \theta'_i$. Then, we can infer the following relationship, between the angles of incidence at P_1 and P_2, by applying Snell's law at P_1 on the interface between the air and the core:

$$n_a \sin \theta_i = n_1 \sin \theta_t = n_1 \sin(90° - \theta'_i) = n_1 \cos \theta'_i \Rightarrow \cos \theta'_i = \frac{n_a \sin \theta_i}{n_1}, \quad (13.18)$$

where n_a is the refractive index of air ($n_a \approx 1$). The critical angle at P_2 is determined by

$$\sin \theta'_c = \frac{n_2}{n_1}.$$

For internal reflection to take place, the angle of incidence at P_2 must satisfy

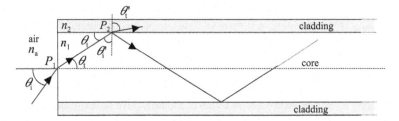

Fig. 13.17 Sketch of an optical fiber showing the core, the cladding, and the path followed by a ray incident at the core's end face with an angle θ_i

$$\theta'_i \geq \theta'_c \Rightarrow \cos\theta'_i \leq \cos\theta'_c = \sqrt{1 - \sin^2\theta'_c} = \sqrt{1 - \left(\frac{n_2}{n_1}\right)^2}. \quad (13.19)$$

Rays that meet the core-cladding boundary at an angle greater than the critical angle are completely reflected. For this condition to be satisfied, it follows from (13.18) and (13.19) that the following equation must hold

$$\cos\theta'_i = \frac{n_a \sin\theta_i}{n_1} \leq \sqrt{1 - \left(\frac{n_2}{n_1}\right)^2} \Rightarrow \sin\theta_i \leq \frac{1}{n_a}\sqrt{n_1^2 - n_2^2}.$$

Then, the maximum angle of incidence is given by

$$\sin\theta_{i,\max} = \frac{1}{n_a}\sqrt{n_1^2 - n_2^2}. \quad (13.20)$$

Using this equation, we can calculate the maximum angle of incidence (acceptance angle) at which the rays incident on the core's end face are trapped inside the core by total internal reflection.

13.12 The refractive index of mammalian tissues can be measured by using a fiber optic cladding method [105] based on substituting the usual cladding by the tissue under study and utilizing the principle of total internal reflection. If a He-Ne laser, with wavelength 632 nm, is used as a light source, the core made from fused quartz with refractive index $n_q = 1.457$ at 632 nm, and the half-angle of the emergent cone of light from the output of the optical fiber is 23.8°, find the refractive index of the tissue.

Solution

In reference [105], a method for measuring the refractive index of mammalian tissues is described, which is based on the principle of internal reflection at the core-cladding interface. If the refractive indices of air and quartz are known, and the angle of the emergent cone of light from the output of the fiber is measured, the refractive index of the tissue can be calculated from (13.20).

Figure 13.18 shows a sketch of a typical optical fiber. The core is made from fused quartz with refractive index $n_q = 1.457$, at the wavelength of He-Ne laser light, and the cladding is a tissue with a refractive index $n_t < n_q$. The incident beam comes from air and enters the fiber at the acceptance angle, θ_a, which is the maximum angle of a ray hitting the fiber core that is kept within the core. Then, total reflection takes place at the quartz-tissue interface. It should be noted that the half-angle of the cone at the exit of the fiber is equal to θ_a. It follows from (13.20) that the refractive index of the tissue, n_t, can be expressed in terms of the refractive index of the core, n_q, the refractive index of air, n_a, and the aperture angle, θ_a,

$$\sin\theta_a = \frac{1}{n_a}\sqrt{n_q^2 - n_t^2} \Rightarrow n_t = \sqrt{n_q^2 - (n_a \sin\theta_a)^2}.$$

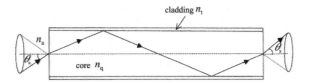

Fig. 13.18 Sketch showing an optical fiber. An incident light ray is first refracted and then undergoes total internal reflection at the core-cladding interface. Cladding is substituted by mammalian tissue for which refractive index is to be measured. Light acceptance cone is shown

By substituting the numerical values into the above equation, the refractive index of the tissue is

$$n_t = \sqrt{1.457^2 - (1 \times \sin 23.8°)^2} = 1.40.$$

13.13 (a) Determine the phase velocity for a harmonic plane wave of frequency f propagating in a homogeneous ionized gas with N electrons per unit volume. (b) Calculate the lowest frequency of the wave that can propagate through the ionized gas. (c) For a wave perpendicularly incident on the interface between vacuum and a layer of an ionized gas with $N = 10^{10}$ m^{-3}, determine the frequencies of the waves that can penetrate into the layer. What are such frequencies for $N = 10^{12}$ m^{-3}? (d) For oblique incidence on an interface between vacuum and a layer of an ionized gas with $N = 10^{12}$ m^{-3}, calculate the lowest frequency of the wave that can penetrate into the layer if the angle of incidence is $\theta_i = 30°$. Mass of electron $m_e = 9.1 \times 10^{-31}$ kg and electronic charge $-e = -1.6 \times 10^{-19}$ C.

Solution

(a) Electrons are much lighter than positive ions and, therefore, they are accelerated more by the electric field of electromagnetic waves passing through the ionized gas. At a given point in the medium, the electric field associated with a plane electromagnetic wave of angular frequency $\omega = 2\pi f$ can be expressed as $\mathbf{E} = \mathbf{E}_0 \sin(\omega t)$. If there are N electrons per unit volume, with mass m_e and charge e, the equation of motion of each electron and the velocity $\mathbf{v}$ due to the electric field of the wave are given by

$$m_e \frac{d\mathbf{v}}{dt} = -e\mathbf{E}_0 \sin(\omega t) \Rightarrow \mathbf{v} = \frac{e}{m_e \omega} \mathbf{E}_0 \cos(\omega t).$$

Note the integration constant can be disregarded by choosing an appropriate origin in time. The current density $\mathbf{j}$ is

$$\mathbf{j} = -Ne\mathbf{v} = -\frac{Ne^2}{m_e \omega} \mathbf{E}_0 \cos(\omega t).$$

Maxwell's equation for $\mathbf{B}$ becomes

$$\nabla \times \mathbf{B} = \mu_0 \mathbf{j} + \mu_0 \varepsilon_0 \frac{\partial \mathbf{E}}{\partial t} = -\frac{\mu_0 N e^2}{m_e \omega} \mathbf{E}_0 \cos(\omega t) + \mu_0 \varepsilon_0 \omega \mathbf{E}_0 \cos(\omega t)$$

$$= \mu_0 \varepsilon_0 \left[1 - \frac{N e^2}{m_e \varepsilon_0 \omega^2} \right] \omega \mathbf{E}_0 \cos(\omega t)$$

$$= \mu_0 \varepsilon_0 \left[1 - \frac{N e^2}{m_e \varepsilon_0 4\pi^2 f^2} \right] 2\pi f \mathbf{E}_0 \cos(2\pi f t).$$

The result obtained shows that the propagation of electromagnetic waves in an ionized gas with N electrons per unit volume can be analyzed as if the wave propagates in a dielectric with "an effective permittivity" ε,

$$\varepsilon = \varepsilon_0 \left(1 - \frac{N e^2}{m_e \varepsilon_0 4\pi^2 f^2} \right) = \varepsilon_0 \left(1 - \frac{f_c^2}{f^2} \right),$$

where $f_c^2 = N e^2 / m_e \varepsilon_0 4\pi^2$. Hence, in the ionized gas, the phase velocity and the refractive index n can be expressed as

$$v = \frac{1}{\sqrt{\mu_0 \varepsilon}} = \frac{1}{\sqrt{\mu_0 \varepsilon_0 (1 - f_c^2/f^2)}} = \frac{c}{\sqrt{1 - f_c^2/f^2}} \Rightarrow n = \frac{c}{v} = \sqrt{1 - \frac{f_c^2}{f^2}}. \quad (13.21)$$

(b) When $f < f_c$, (13.21) shows that the phase velocity is imaginary, which does not make physical sense. Waves with frequencies less than f_c cannot propagate through the ionized gas. On the other hand, if $f > f_c$, electromagnetic waves will propagate in the ionized gas. The frequency f_c is referred to as the critical frequency (or cutoff frequency). For electrons, with $e = 1.6 \times 10^{-19}$ C, $m_e = 9.1 \times 10^{-31}$ kg,

$$f_c = \sqrt{\frac{N e^2}{m_e \varepsilon_0 4\pi^2}} = \sqrt{\frac{N \times (1.6 \times 10^{-19})^2}{9.1 \times 10^{-31} \times 8.854 \times 10^{-12} \times 4\pi^2}} = \sqrt{80.5 N} \approx 9\sqrt{N}. \quad (13.22)$$

(c) For perpendicular incidence, $\theta_i = \theta_t = 0$, i.e. the direction of the refracted ray is the same as that of the incident. The velocity of propagation in the ionized layer is given by (13.21). For $N = 10^{10}$ m^{-3}, (13.22) gives $f_c = 0.9$ MHz, while for $N = 10^{12}$ m^{-3} we have $f_c = 9$ MHz. Therefore, for $N = 10^{10}$ m^{-3}, if a wave with frequency $f < 0.9$ MHz impinges perpendicularly on the ionized gas layer, the wave will be totally reflected. On the other hand, the wave will penetrate through the gas if $f > 0.9$ MHz. For perpendicular incidence and $N = 10^{12}$ m^{-3}, waves with frequencies greater than 9 MHz can penetrate through the gas.

(d) For critical incidence, $\theta_t = \pi/2$, with $n_1 = 1$ and the index of refraction of the the ionized gas given by (13.21), Snell's law gives

$$n_1 \sin \theta_i = n_2 \sin \theta_t \Rightarrow 1 \times \sin \theta_i = \sqrt{1 - \frac{f_c^2}{f^2}} \sin\left(\frac{\pi}{2}\right)$$

$$\Rightarrow \frac{f_c^2}{f^2} = 1 - \sin^2\theta_i = \cos^2\theta_i \Rightarrow f = \frac{f_c}{\cos\theta_i}.$$

Then, for $\theta_i = 30°$, the lowest frequency of the waves that can propagate through the gas is

$$f = \frac{f_c}{\cos\theta_i} = \frac{9}{\cos 30°} \text{ MHz} \approx 10.4 \text{ MHz},$$

where $f_c = 9$ MHz for $N = 10^{12}$ m^{-3}, as seen in (c). Frequencies lower than 10.4 MHz make $\sin\theta_t$ in Snell's law be greater than 1 and, hence, such frequencies cannot penetrate into the gas. On the other hand, waves with $f > 10.4$ MHz will propagate through the gas.

13.14 The ionosphere is a layer of ionized gas around the earth. If the refractive index of the ionosphere can be expressed as $n_i = \sqrt{1 - \frac{\lambda^2}{C}}$, where λ is the wavelength of the wave and C is a constant, (a) find the wavelength of the shortest radio wave that can be totally reflected by the ionosphere. It is assumed that the ionosphere has a sharp boundary at an altitude H above the surface of the Earth. The wave is emitted, at a given angle θ_e, by an emitter E on the surface of the earth, as shown in Fig. 13.19. (b) For $\theta_e = 0$ and $\pi/2$, find the wavelength of the shortest totally reflected wave. (c) For $H \approx 300$ km, $R \approx 6371$ km, and $C \approx 10^3$ m^2, find the wavelength of the shortest radio wave that can be totally reflected from the ionosphere.

Solution

In a region extending from a height of about 50 km to over 500 km, molecules of the atmosphere are ionized by radiation from the Sun. This region is called the ionosphere. The altitude and character of the ionized layers depend on the nature of the solar radiation and on the composition of the atmosphere. Ionization of the ionosphere varies greatly with the time of day, the season, and other factors. An important feature of the ionosphere is that it makes it possible for the reflection of radio waves. However, only those waves within a certain wavelength range (or frequency range) will be reflected. Critical wavelengths (or frequencies) change with

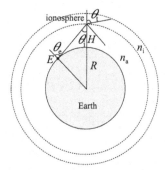

Fig. 13.19 Emitter E on the surface of Earth emits a plane wave with angle θ_e that is reflected when entering the ionosphere. θ_i and θ_t denote the angles of incidence and that of refraction, respectively, at the boundary "atmosphere-ionosphere". R represents the radius of Earth

time of day, atmospheric conditions, and the emission angle. The greater the density of electrons, the shorter the wavelength (or the higher the frequencies) that can be totally reflected. The electron density of the ionosphere ranges from about 10^{10} m^{-3}, in the lowest layer, to 10^{12} m^{-3} in the highest layer. At night, the lower regions become very much depleted of free electrons, and only radio waves with the longest wavelengths can be totally reflected.

In this problem, a simple model of the ionosphere is considered, consisting of a layer at an altitude H above the surface of the Earth. It is assumed that ionization varies sharply at the boundary between the atmosphere and the ionosphere.

(a) Figure 13.19 shows a wave leaving the Earth at an angle θ_e. From Fig. 13.19, it follows,

$$\frac{\sin\theta_i}{R} = \frac{\sin(\pi - \theta_e)}{H+R} = \frac{\sin\theta_e}{H+R} \Rightarrow \sin\theta_i = \frac{R}{R+H}\sin\theta_e = \frac{\sin\theta_e}{1+H/R}.$$

Snell's law at the interface "atmosphere-ionosphere" gives

$$n_a \sin\theta_i = n_i \sin\theta_t \Rightarrow 1 \times \frac{\sin\theta_e}{1+H/R} = \sqrt{1 - \frac{\lambda^2}{C}} \times \sin\theta_t, \qquad (13.23)$$

where the refractive index for the atmosphere n_a is assumed to be approximately equal to 1.

Total reflection begins for $\theta_t = \pi/2$, which gives the smallest wavelength in (13.23) for total reflection to occur. Then, the smallest wavelength, denoted by λ_c, which can be totally reflected for a given θ_e is given by

$$1 \times \frac{\sin\theta_e}{1+H/R} = \sqrt{1 - \frac{\lambda_c^2}{C}} \times \sin\left(\frac{\pi}{2}\right) \Rightarrow \lambda_c = \left\{C\left[1 - \left(\frac{\sin\theta_e}{1+H/R}\right)^2\right]\right\}^{1/2}. \qquad (13.24)$$

Thus, waves with $\lambda < \lambda_c$ are transmitted into the ionosphere, whereas waves with $\lambda > \lambda_c$ are totally reflected.

(b) At vertical incidence, $\theta_e = 0$, (13.24) gives

$$\lambda_c = \sqrt{C}, \qquad (13.25)$$

which corresponds to the largest λ_c. For $\theta_e = \pi/2$, then the wave is sent off in the direction of the horizon, thus it follows from (13.24) that

$$\lambda_c = \left\{C\left[1 - \left(\frac{1}{1+H/R}\right)^2\right]\right\}^{1/2} \approx \left\{C\left[1 - \left(1 - \frac{2H}{R}\right)\right]\right\}^{1/2} = \left[C \times \frac{2H}{R}\right]^{1/2}, \qquad (13.26)$$

which corresponds to the smallest wavelength obtained with the model proposed that can be totally reflected from the ionosphere.

Fig. 13.20 Communications between stations on the Earth using the reflection of waves from the ionosphere

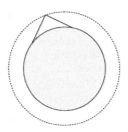

(c) Equation (13.26) provides the smallest λ_c,

$$\lambda_c \approx \left[C \times \frac{2H}{R} \right]^{1/2} = \left[10^3 \times \frac{2 \times 300}{6371} \right]^{1/2} = 9.7 \text{ (m)}.$$

Therefore, if we wish to use the ionosphere as a reflector of radio waves for communicating between stations on the Earth, as shown in Fig. 13.20, waves shorter than approximately 10 m cannot be used. On the other hand, if we wish to communicate with a satellite beyond the ionosphere, we must use shorter wavelengths to ensure wave penetration through the ionosphere.

13.15 Linearly polarized light is incident along the normal of face AB of a glass prism of refractive index $n = 1.5$, as shown in Fig. 13.21. Calculate the percentage of the intensity of incident light reflected back by the prism when light emerges from the glass into the air in the opposite direction of the incident beam.

Solution

When the incident light impinges on face AB at normal incidence, from Snell's law we have: $\theta_i = \theta_t = 0$. At the air-glass interface, (13.5) gives for the transmission coefficient, denoted by t_1,

$$t_1 = t_\parallel = t_\perp = \frac{2n_a}{n_a + n},$$

where n_a represents the refractive index of air, $n_a \approx 1$. Then, the amplitude of the transmitted wave at face AB, E_{0t1}, in terms of the amplitude of the incident wave, E_{0i}, can be expressed as $E_{0t1} = t_1 E_{0i}$.

Intensity I_w (the average Poynting vector) can be expressed in terms of n as

$$I_w = \frac{1}{2} \varepsilon v E_0^2 = \frac{1}{2} \varepsilon_0 n c E_0^2, \qquad (13.27)$$

where it has been taken into account that $n = c/v = \sqrt{\varepsilon_r \mu_r}$, and that for materials with negligible magnetic properties $\mu_r \simeq 1$ and, hence, $n \simeq \sqrt{\varepsilon_r}$. Therefore, at face AB, the intensities of the incident beam I_i and that of transmitted beam I_{t1} can be written, respectively, as

Fig. 13.21 Light enters the prism along the normal of face *AB*, undergoes total internal reflection twice from the sloped faces, and exits again through face *AB*

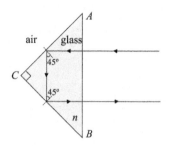

$$I_i = \frac{1}{2}\varepsilon_0 n_a c E_{0i}^2 \quad \text{and} \quad I_{t1} = \frac{1}{2}\varepsilon_0 n c E_{0t1}^2 = \frac{1}{2}\varepsilon_0 n c t_1^2 E_{0i}^2.$$

Then, the ratio of the intensity of the transmitted wave to that of the incident wave, results in

$$\frac{I_{t1}}{I_i} = \frac{n}{n_a} t_1^2 = \frac{n}{n_a}\left(\frac{2n_a}{n_a+n}\right)^2 = \frac{4nn_a}{(n+n_a)^2}. \tag{13.28}$$

The light continues straight on until it hits the back face *AC*. Total internal reflection occurs at *AC* when

$$n \sin\theta_c = n_a \sin\frac{\pi}{2} \Rightarrow \sin\theta_c = \frac{n_a}{n} = \frac{1}{1.5} \Rightarrow \theta_c = \sin^{-1}\frac{1}{1.5} = 41.8°.$$

The light strikes the surface *AC* at 45°, which is greater than the critical angle. After that, the totally reflected ray falls on face *CB* at 45° and it is again totally reflected. Note that no light is refracted out of the prism at faces *AC* and *BC*. Following the ray path shown in Fig. 13.21, the ray then hits surface *AB* along its normal and exits again through the glass-air interface. At this interface, (13.5) gives for the transmission coefficient

$$t_2 = \frac{2n}{n+n_a},$$

where it has been taken into account that $\theta_i = \theta_t = 0$. Denoting the intensity of the beam transmitted to air by I_{t2}, at glass-air interface *AB*, the quotient of the intensities of the transmitted and incident beams can be expressed as

$$\frac{I_{t2}}{I_{t1}} = \frac{n_a}{n} t_2^2 = \frac{n_a}{n}\left(\frac{2n}{n_a+n}\right)^2 = \frac{4n_a n}{(n+n_a)^2},$$

where I_{t1} corresponds to the incident intensity. As the light undergoes total reflection twice, at *BC* and *AC* interfaces, the intensity incident on the glass-air interface (for the exiting beam) is the same as that transmitted through the air-glass interface (for the entering beam).

Substituting I_{t1} given by (13.28) into the above equation, it follows that

$$\frac{I_{t2}}{I_i} = \frac{4nn_a}{(n+n_a)^2} \times \frac{4n_a n}{(n+n_a)^2} = \frac{(4nn_a)^2}{(n+n_a)^4} = \frac{(4 \times 1.5 \times 1)^2}{(1.5+1)^4} = 0.92.$$

Then, the intensity of the beam exiting from face AB is 92% of the intensity of the entering beam.

13.16 A plane harmonic wave linearly polarized is incident at the Brewster angle on an interface between two dielectric media with $n_1 = 1.2$ and $n_2 = 1.5$ (Fig. 13.22). The electric field of the incident wave makes an angle of $60°$ with the normal to the plane of incidence. If the intensity of the incident wave is $2\ \text{Wm}^{-2}$, determine the intensities for the reflected and transmitted waves.

Solution

Equation (13.7) for the Brewster angle gives an angle of incidence of

$$\tan \theta_B = \frac{n_2}{n_1} = \frac{1.5}{1.2} = 1.25 \Rightarrow \theta_B = 51.34° = \theta_i.$$

Then, according to (13.6) the angle of refraction is given by (Fig. 13.22)

$$\theta_t + \theta_B = 90° \Rightarrow \theta_t = 90° - 51.34° = 38.66°.$$

The incident linearly polarized wave can be expressed as

$$\mathbf{E}_i = \mathbf{E}_{0i} \cos(\mathbf{k}_i \cdot \mathbf{r} - \omega t) = [E_{0i\|}\mathbf{u}_\| + E_{0i\perp}\mathbf{u}_\perp]\cos(\mathbf{k}_i \cdot \mathbf{r} - \omega t),$$

$$\left. \begin{array}{l} E_{0i\|} = E_{0i}\sin 60°, \\ E_{0i\perp} = E_{0i}\cos 60°, \end{array} \right\}$$

where $E_{0i} = |\mathbf{E}_{0i}|$ is the amplitude of the incident wave and $E_{0i\|}$ and $E_{0i\perp}$ are the amplitudes of the components parallel and normal to the plane of incidence, respectively. $\mathbf{u}_\|$ and $\mathbf{u}_\perp$ represent the corresponding unit vectors in these directions. The components of the incident wave, parallel and perpendicular to the incidence plane, can then be expressed as

$$\left. \begin{array}{l} E_{i\|} = E_{0i\|}\cos(\mathbf{k}_i \cdot \mathbf{r} - \omega t), \\ E_{i\perp} = E_{0i\perp}\cos(\mathbf{k}_i \cdot \mathbf{r} - \omega t). \end{array} \right\}$$

Fig. 13.22 Plane wave incident obliquely on a plane boundary at the Brewster angle. The reflected wave is linearly polarized along a direction normal to the plane of incidence

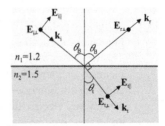

From the intensity of the incident wave I_i, it is found from (13.27) that the amplitude of the incident wave is given by

$$I_i = \frac{1}{2}\varepsilon_0 n_1 c E_{0i}^2 \Rightarrow 2 = \frac{1}{2} \times 8.85 \times 10^{-12} \times 1.2 \times 3 \times 10^8 E_{0i}^2 \Rightarrow E_{0i} = 35.43 \text{ N/C}.$$

Then, the amplitudes of the parallel and perpendicular components are, respectively,

$$\left.\begin{array}{l} E_{0i\parallel} = E_{0i}\sin 60° = 35.43 \sin 60° \text{ N/C} = 30.68 \text{ N/C}, \\ E_{0i\perp} = E_{0i}\cos 60° = 35.43 \cos 60° \text{ N/C} = 17.72 \text{ N/C}. \end{array}\right\}$$

From (13.4), and $n_1 = 1.2$, $n_2 = 1.5$, $\theta_i = 51.34°$, and $\theta_t = 38.66°$, the coefficients for the reflected wave are: $r_\perp = -0.22$ and $r_\parallel = 0$, as expected for Brewster incidence. Then, the amplitudes of the components of the reflected wave are, respectively,

$$\left.\begin{array}{l} E_{0r\parallel} = 0, \\ E_{0r\perp} = r_\perp E_{0i\perp} = -0.22 \times 17.72 \simeq -3.90 \text{ N/C}. \end{array}\right\}$$

Finally, the intensity of the reflected wave will be

$$I_r = \frac{1}{2}\varepsilon_0 n_1 c E_{0r}^2 = \frac{1}{2} \times 8.85 \times 10^{-12} \times 1.2 \times 3 \times 10^8 \times (0 + 3.90^2) \text{ W/m}^2 = 0.024 \text{ W/m}^2.$$

In the same way, for $n_1 = 1.2$, $n_2 = 1.5$, $\theta_i = 51.34°$, and $\theta_t = 38.66°$, (13.5) yields: $t_\parallel = 0.80$ and $t_\perp = 0.78$. Then, the amplitudes of the components of the transmitted wave are found to be

$$\left.\begin{array}{l} E_{0t\parallel} = t_\parallel E_{0i\parallel} = 0.80 \times 30.68 \text{ N/C} = 24.54 \text{ N/C}, \\ E_{0t\perp} = t_\perp E_{0i\perp} = 0.78 \times 17.72 \text{ N/C} = 13.82 \text{ N/C}. \end{array}\right\}$$

Thus, the intensity of the transmitted wave will be

$$I_t = \frac{1}{2}\varepsilon_0 n_2 c E_{0t}^2 = \frac{1}{2}\varepsilon_0 n_2 c \left(E_{0t\parallel}^2 + E_{0t\perp}^2\right)$$
$$= \frac{1}{2} \times 8.85 \times 10^{-12} \times 1.5 \times 3 \times 10^8 \left(24.54^2 + 13.82^2\right) \text{ W/m}^2 = 1.58 \text{ W/m}^2.$$

Problems C

13.17 A linearly polarized plane wave of frequency 100 MHz is incident on an air-glass interface at an angle of 30°. The refractive index of the glass is $n = 1.60$. The electric vector of the incident beam makes an angle of 45° with the plane of incidence and has an amplitude of 25 V/m. (a) Find the reflection and transmission coefficients. (b) Write the expressions for the reflected and transmitted electric fields. (c) Describe the polarization state of the reflected and transmitted beams.

Fig. 13.23 Linearly polarized plane wave incident obliquely on an air-glass boundary

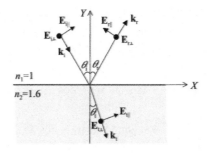

Solution

(a) Application of Snell's law to the air-glass interface with $\theta_i = 30°$, $n_1 \approx 1$, $n_2 = 1.6$ gives $\theta_t = 18.2°$. From (13.4) and (13.5), we have for the Fresnel coefficients

$$r_\| = \frac{1.60 \times \cos 30° - 1 \times \cos 18.2°}{1 \times \cos 18.2° + 1.60 \times \cos 30°} = 0.19 \quad t_\| = \frac{2 \times 1 \times \cos 30°}{1 \times \cos 18.2° + 1.60 \times \cos 30°} = 0.74$$

$$r_\perp = \frac{1 \times \cos 30° - 1.60 \times \cos 18.2°}{1 \times \cos 30° + 1.60 \times \cos 18.2°} = -0.27 \quad t_\perp = \frac{2 \times 1 \times \cos 30°}{1 \times \cos 30° + 1.60 \times \cos 18.2°} = 0.73$$

(b) Let the XY plane be the plane of incidence, as shown in Fig. 13.23. The propagation vector for the incident wave can be then expressed as

$$\mathbf{k}_i = \frac{2\pi}{\lambda} \left(\sin 30° \, \mathbf{u}_x - \cos 30° \, \mathbf{u}_y \right) = \frac{2\pi}{3} \left(\frac{1}{2} \mathbf{u}_x - \frac{\sqrt{3}}{2} \mathbf{u}_y \right) \, \text{m}^{-1}, \quad (13.29)$$

where λ denotes the wavelength in air, equal to $\lambda \approx c/f = 3 \times 10^8 / 100 \times 10^6 \, \text{m} = 3 \, \text{m}$.

The incident wave is linearly polarized along a direction that makes 45° with the plane of incidence. Then, the amplitude of the components parallel and perpendicular to the plane of incidence are $E_{0i\|} = E_{0i} \cos 45°$ and $E_{0i\perp} = E_{0i} \sin 45°$, respectively, where E_{0i} is the amplitude of the incident wave. The components parallel and perpendicular of the incident electric field are then given by

$$E_{i\|} = E_{0i\|} \cos (\mathbf{k}_i \cdot \mathbf{r} - \omega t) = E_{0i} \cos 45° \cos (\mathbf{k}_i \cdot \mathbf{r} - \omega t)$$

$$= 25 \times \frac{\sqrt{2}}{2} \cos \left[\frac{2\pi}{3} \left(\frac{1}{2} x - \frac{\sqrt{3}}{2} y \right) - 2\pi \times 10^8 t \right] \text{V/m}$$

$$= 17.68 \cos \left(1.05 x - 1.81 y - 6.28 \times 10^8 t \right) \text{V/m},$$

$$E_{i\perp} = E_{0i\perp} \cos (\mathbf{k}_i \cdot \mathbf{r} - \omega t) = E_{0i} \sin 45° \cos (\mathbf{k}_i \cdot \mathbf{r} - \omega t)$$

$$= 25 \times \frac{\sqrt{2}}{2} \cos \left[\frac{2\pi}{3} \left(\frac{1}{2} x - \frac{\sqrt{3}}{2} y \right) - 2\pi \times 10^8 t \right] \text{V/m}$$

$$= 17.68 \cos \left(1.05 x - 1.81 y - 6.28 \times 10^8 t \right) \text{V/m},$$

where it has been taken into account that the position vector is $\mathbf{r} = (x, y, z)$, $\mathbf{k}_i$ is given by (13.29), and $\omega = 2\pi f = 2\pi \times 10^8$ s^{-1}.

The propagation vectors for the reflected and transmitted waves are, respectively,

$$\mathbf{k}_r = \frac{2\pi}{\lambda} \left(\sin 30° \, \mathbf{u}_x + \cos 30° \, \mathbf{u}_y \right) = \frac{2\pi}{3} \left(\frac{1}{2} \mathbf{u}_x + \frac{\sqrt{3}}{2} \mathbf{u}_y \right) \text{ m}^{-1},$$

$$\mathbf{k}_t = \frac{2\pi}{\lambda} n \left(\sin 18.2° \, \mathbf{u}_x - \cos 18.2° \, \mathbf{u}_y \right) = \frac{2\pi}{3} \times 1.60 \left(0.31 \, \mathbf{u}_x - 0.95 \, \mathbf{u}_y \right) \text{ m}^{-1}.$$

Then, the reflected wave has the following components

$$E_{r\|} = E_{0i\|} \, r_\| \cos(\mathbf{k}_r \cdot \mathbf{r} - \omega t)$$

$$= 25 \times \frac{\sqrt{2}}{2} \times 0.19 \cos \left[\frac{2\pi}{3} \left(\frac{1}{2} x + \frac{\sqrt{3}}{2} y \right) - 2\pi \times 10^8 t \right] \text{ V/m},$$

$$= 3.36 \cos \left(1.05 x + 1.81 y - 6.28 \times 10^8 t \right) \text{ V/m},$$

$$E_{r\perp} = E_{0i\perp} \, r_\perp \cos(\mathbf{k}_r \cdot \mathbf{r} - \omega t)$$

$$= 25 \times \frac{\sqrt{2}}{2} \times (-0.27) \cos \left[\frac{2\pi}{3} \left(\frac{1}{2} x + \frac{\sqrt{3}}{2} y \right) - 2\pi \times 10^8 t \right] \text{ V/m},$$

$$= -4.77 \cos \left(1.05 x + 1.81 y - 6.28 \times 10^8 t \right) \text{ V/m}$$

$$= 4.77 \cos \left(1.05 x + 1.81 y - 6.28 \times 10^8 t + \pi \right) \text{ V/m},$$

whereas for the components of the transmitted wave, we have

$$E_{t\|} = E_{0i\|} \, t_\| \cos(\mathbf{k}_t \cdot \mathbf{r} - \omega t)$$

$$= 25 \times \frac{\sqrt{2}}{2} \times 0.74 \cos \left[\frac{2\pi}{3} \times 1.60 \left(0.31 x - 0.95 y \right) - 2\pi \times 10^8 t \right] \text{ V/m}$$

$$= 13.08 \cos \left(1.04 x - 3.18 y - 6.28 \times 10^8 t \right) \text{ V/m},$$

$$E_{t\perp} = E_{0i\perp} \, t_\perp \cos(\mathbf{k}_t \cdot \mathbf{r} - \omega t)$$

$$= 25 \times \frac{\sqrt{2}}{2} \times 0.73 \cos \left[\frac{2\pi}{3} \times 1.60 \left(0.31 x - 0.95 y \right) - 2\pi \times 10^8 t \right] \text{ V/m}$$

$$= 12.90 \cos \left(1.04 x - 3.18, y - 6.28 \times 10^8 t \right) \text{ V/m}.$$

(c) The phase difference between the perpendicular and parallel components of the reflected wave is $(\varphi_{0\perp} - \varphi_{0\|})_r = \pi$. Therefore, the wave is linearly polarized, the electric field of the reflected wave makes an angle with the plane of incidence of

$$\tan \theta_r = \frac{E_{r\perp}}{E_{r\|}} = \frac{-4.77}{3.36} \Rightarrow \theta_r = -54.84°.$$

For the transmitted wave, $(\varphi_{0\perp} - \varphi_{0\|})_t = 0$, and, hence, the transmitted wave is linearly polarized, its electric field making an angle with the plane of incidence of

$$\tan\theta_t = \frac{E_{t\perp}}{E_{t\|}} = \frac{12.90}{13.08} \Rightarrow \theta_t = 44.60°.$$

13.18 A right-handed circularly polarized electromagnetic wave is incident on a glass-air interface, as shown in Fig. 13.24. The incident wave has an intensity of 20×10^{-4} Wm^{-2}. When the reflected wave is linearly polarized along a direction perpendicular to the plane of incidence, the angle of incidence is 33.69°. (a) Find the intensity of the reflected wave. (b) Describe the state of polarization of the transmitted wave. (c) Determine the intensity of the transmitted wave.

Solution

Since the incident wave is circularly polarized and the reflected wave is linearly polarized, the wave is incident at the Brewster angle, $\theta_i = \theta_B = 33.69°$. The angle of refraction can be obtained from (13.6), $\theta_t = 90° - 33.69° = 56.31°$. At the Brewster incidence, (13.7) gives for the refractive index of the glass through which the electromagnetic wave is propagating

$$\tan\theta_B = \frac{n_2}{n_1} \Rightarrow \tan 33.69° = \frac{n_2}{n_1} \Rightarrow n_1 = \frac{n_2}{\tan 33.69°} = \frac{1}{\tan 33.69°} = 1.50,$$

where the refractive index of air is assumed to be $n_2 \approx 1$.

The amplitude of the components of the incident wave can be calculated from the intensity of the incident wave I_i, which for a circularly polarized wave is given by

$$I_i = \varepsilon_0 n_1 c E_{0i}^2 \Rightarrow 20 \times 10^{-4} = 8.85 \times 10^{-12} \times 1.5 \times 3 \times 10^8 E_{0i}^2 \Rightarrow E_{0i} = 0.71 \text{ N/C}.$$

Then, the parallel and perpendicular components can be expressed as

$$\left.\begin{array}{l} E_{i\|} = E_{0i}\sin(\mathbf{k}_i \cdot \mathbf{r} - \omega t + \pi/2) = E_{0i}\cos(\mathbf{k}_i \cdot \mathbf{r} - \omega t) = 0.71\cos(\mathbf{k}_i \cdot \mathbf{r} - \omega t) \text{ N/C,} \\ E_{i\perp} = E_{0i}\sin(\mathbf{k}_i \cdot \mathbf{r} - \omega t) = 0.71\sin(\mathbf{k}_i \cdot \mathbf{r} - \omega t) \text{ N/C,} \end{array}\right\}$$

where it has been taken into account that for a right-handed circularly polarized wave $(\varphi_{0\|} - \varphi_{0\perp})_i = \pi/2$.

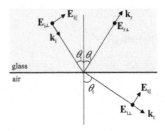

Fig. 13.24 Circularly polarized plane wave incident on a plane glass-air interface

(a) For $n_1 = 1.5$, $n_2 = 1$, $\theta_i = 33.69°$, and $\theta_t = 56.31°$, (13.4) for the reflection coefficients gives $r_\parallel = 0$, as expected, and $r_\perp = 0.38$. Then, the amplitudes of the components of the reflected wave are, respectively,

$$\left. \begin{array}{l} E_{0r\parallel} = 0, \\ E_{0r\perp} = r_\perp E_{0i} = 0.38 \times 0.71 = 0.27 \text{ N/C}. \end{array} \right\}$$

Thus, the intensity of the reflected linearly polarized wave is equal to

$$I_r = \frac{1}{2} \varepsilon_0 n_1 c E_{0r}^2 = \frac{1}{2} \times 8.85 \times 10^{-12} \times 1.5 \times 3 \times 10^8 \times 0.27^2 \text{ W/m}^2 = 1.45 \times 10^{-4} \text{ W/m}^2.$$

(b) Equation (13.5) yields for the transmission coefficients: $t_\parallel = 1.50$ and $t_\perp = 1.38$. The amplitudes of the components of the transmitted wave will be

$$\left. \begin{array}{l} E_{0t\parallel} = t_\parallel E_{0i} = 1.50 \times 0.71 = 1.07 \text{ N/C}, \\ E_{0t\perp} = t_\perp E_{0i} = 1.38 \times 0.71 = 0.98 \text{ N/C}, \end{array} \right\}$$

and the components of the transmitted wave are then given by

$$\left. \begin{array}{l} E_{t\parallel} = 1.07 \cos(\mathbf{k}_t \cdot \mathbf{r} - \omega t) = 1.07 \sin(\mathbf{k}_t \cdot \mathbf{r} - \omega t + \pi/2) \text{ N/C}, \\ E_{t\perp} = 0.98 \sin(\mathbf{k}_t \cdot \mathbf{r} - \omega t) \text{ N/C}, \end{array} \right\}$$

which corresponds to a right-handed elliptically polarized wave since $(\varphi_{0\parallel} - \varphi_{0\perp})_t = \pi/2$ and $E_{0t\parallel} \neq E_{0t\perp}$.

(c) Finally, the intensity of the transmitted wave can be expressed as

$$I_t = \frac{1}{2} \varepsilon_0 n_2 c E_{0t}^2 = \frac{1}{2} \varepsilon_0 n_2 c \left(E_{0t\parallel}^2 + E_{0t\perp}^2 \right)$$
$$= \frac{1}{2} \times 8.85 \times 10^{-12} \times 1 \times 3 \times 10^8 \times \left(1.07^2 + 0.98^2 \right) \text{ W/m}^2 = 2.79 \times 10^{-3} \text{ W/m}^2.$$

13.19 A linearly polarized plane harmonic wave is incident normally on a plane boundary between two dielectric media, denoted 1 and 2, with permeabilities $\mu_1 = \mu_2 = \mu_0$ and refractive indices n_1 and n_2, respectively, see Fig. 13.25a. In order to eliminate the reflected wave, a plane parallel dielectric layer is inserted between the two media, as shown in Fig. 13.25b. Determine the thickness of the layer L and its refractive index n' so that the condition of no reflection is fulfilled.

Solution

Let OY direction be along the polarization direction of the incident electromagnetic plane wave, as shown in Fig. 13.26. Then, for simplicity, using complex notation, the electric field vector of the incident wave can be written as

Fig. 13.25 a A plane wave is incident normally on the interface between two dielectrics. **b** A dielectric layer is inserted between the two media in order to eliminate the reflected wave

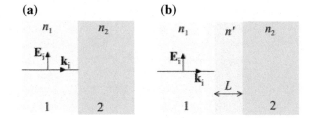

$$\mathbf{E}_i = (E_{0i}\,\mathbf{u}_y)\,e^{j\omega\left(t-\frac{x}{v_1}\right)} = (E_{0i}\,\mathbf{u}_y)\,e^{j(\omega t - k_1 x)},$$

where v_1 is the velocity in medium 1 and the wavenumber $k_1 = \omega/v_1$. The corresponding magnetic field **H** is given by

$$\mathbf{H}_i = \left(\frac{E_{0i}}{\mu_0 v_1}\mathbf{u}_z\right) e^{j(\omega t - k_1 x)} = \left(\frac{n_1 E_{0i}}{\mu_0 c}\mathbf{u}_z\right) e^{j(\omega t - k_1 x)}.$$

Transmitted and reflected waves are generated by the incident wave impinging on the dielectric layer at the interface $x = 0$. In the dielectric layer, the transmitted wave in turn will give rise to reflected and refracted waves at $x = L$. The wave returning back into the layer will be reflected and transmitted, in turn, at $x = 0$. As a result of the multiple reflections at the two boundaries, two set of waves will be propagating in opposite directions through the dielectric layer, denoted by $\mathbf{E}'_t$ and $\mathbf{E}'_r$ in Fig. 13.26. Note that these waves can be considered to be the result of the interference of the multiple reflected and transmitted waves at the two boundaries. The resulting wave from the superposition of the transmitted waves into medium 2 is denoted by $\mathbf{E}_t$ and the resulting reflected wave travelling in medium 1 is represented by $\mathbf{E}_r$. These waves can be expressed as

$$\begin{aligned}
\mathbf{E}_r &= (E_{0r}\,\mathbf{u}_y)\,e^{j\omega\left(t+\frac{x}{v_1}\right)} = (E_{0r}\,\mathbf{u}_y)\,e^{j(\omega t + k_1 x)}, & x < 0, \\
\mathbf{E}'_t &= (E'_{0t}\,\mathbf{u}_y)\,e^{j\omega\left(t-\frac{x}{v'}\right)} = (E'_{0t}\,\mathbf{u}_y)\,e^{j(\omega t - k' x)}, & 0 < x < L, \\
\mathbf{E}'_r &= (E'_{0r}\,\mathbf{u}_y)\,e^{j\omega\left(t+\frac{x}{v'}\right)} = (E'_{0r}\,\mathbf{u}_y)\,e^{j(\omega t + k' x)}, & 0 < x < L, \\
\mathbf{E}_t &= (E_{0t}\,\mathbf{u}_y)\,e^{j\omega\left[t-\frac{(x-L)}{v_2}\right]} = (E_{0t}\,\mathbf{u}_y)\,e^{j[\omega t - k_2(x-L)]}, & x > L,
\end{aligned}$$

where E_{0r}, E'_{0t}, E'_{0r}, and E_{0t} represent complex amplitudes. The boundary conditions must be satisfied simultaneously at the two interfaces.

The field **H** associated with the above expressions for the electric fields can be written as

$$\begin{aligned}
\mathbf{H}_r &= \tfrac{n_1 E_{0r}}{\mu_0 c}(-\mathbf{u}_z)e^{j(\omega t + k_1 x)}, & x < 0, \\
\mathbf{H}'_t &= \tfrac{n' E'_{0t}}{\mu_0 c}(\mathbf{u}_z)e^{j(\omega t - k' x)}, & 0 < x < L, \\
\mathbf{H}'_r &= \tfrac{n' E'_{0r}}{\mu_0 c}(-\mathbf{u}_z)e^{j(\omega t + k' x)}, & 0 < x < L, \\
\mathbf{H}_t &= \tfrac{n_2 E_{0t}}{\mu_0 c}(\mathbf{u}_z)e^{j[\omega t - k_2(x-L)]}, & x > L.
\end{aligned}$$

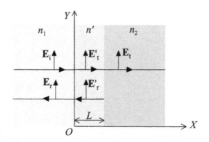

Fig. 13.26 Coordinate system in a plane parallel dielectric layer. E_i is the incident wave, E_r is the reflected wave, and E_t represents the resulting wave transmitted out of the layer. As a result of the multiple reflections at the boundaries, two waves, E'_r and E'_t, are propagating in the layer in opposite directions

The continuity of E_y requires that at $x = 0$

$$E_{0i} + E_{0r} = E'_{0t} + E'_{0r},$$

where the common factor $e^{j\omega t}$ is omitted. At $x = L$, the boundary conditions gives

$$E'_{0t} e^{-jk'L} + E'_{0r} e^{jk'L} = E_{0t}.$$

The continuity of H_z requires that at $x = 0$

$$n_1 E_{0i} - n_1 E_{0r} = n' E'_{0t} - n' E'_{0r}.$$

For $x = L$, the continuity of H_z gives

$$n' E'_{0t} e^{-jk'L} - n' E'_{0r} e^{jk'L} = n_2 E_{0t}.$$

Rearranging the above set of equations, the following system of linear equations is obtained

$$-E_{0r} + E'_{0t} + E'_{0r} + 0 = E_{0i},$$
$$n_1 E_{0r} + n' E'_{0t} - n' E'_{0r} + 0 = n_1 E_{0i},$$
$$0 + E'_{0t} e^{-jk'L} + E'_{0r} e^{jk'L} - E_{0t} = 0,$$
$$0 + n' E'_{0t} e^{-jk'L} - n' E'_{0r} e^{jk'L} - n_2 E_{0t} = 0,$$

where E_{0i} is assumed to be known. From the condition of no reflection, E_{0r} must be zero, and, therefore, by applying Crammer's rule, it follows that

$$\begin{vmatrix} E_{0i} & 1 & 1 & 0 \\ n_1 E_{0i} & n' & -n' & 0 \\ 0 & e^{-jk'L} & e^{jk'L} & -1 \\ 0 & n'e^{-jk'L} & -n'e^{jk'L} & -n_2 \end{vmatrix} = 0.$$

Manipulation of the determinant, dividing the first column by E_{0i}, and subtracting the first row multiplied by n_1 from the second row, gives

$$\begin{vmatrix} 0 & n_1 - n' & n_1 + n' & 0 \\ n_1 & n' & -n' & 0 \\ 0 & e^{-jk'L} & e^{jk'L} & -1 \\ 0 & n'e^{-jk'L} & -n'e^{jk'L} & -n_2 \end{vmatrix} = 0.$$

Multiplying the third row by n' and addition and substraction of the third and fourth rows leads to

$$\begin{vmatrix} 0 & n_1 - n' & n_1 + n' & 0 \\ n_1 & n' & -n' & 0 \\ 0 & 2n'e^{-jk'L} & 0 & -n' - n_2 \\ 0 & 0 & 2n'e^{jk'L} & -n' + n_2 \end{vmatrix} = 0.$$

Evaluating this determinant, we have

$$(n_1 - n')\left(2n'e^{jk'L}\right)(n' + n_2) - (n_1 + n')\left(2n'e^{-jk'L}\right)(-n' + n_2) = 0 \Rightarrow$$

$$(2n_1 n' - 2n' n_2)\cos(k'L) + j\sin(k'L)(-2n'^2 + 2n_1 n_2) = 0.$$

The real part is zero if

$$(2n_1 n' - 2n' n_2)\cos(k'L) = 0 \Rightarrow n_1 = n_2 \quad \text{or} \quad \cos(k'L) = 0 \Rightarrow k'L = \frac{\pi}{2} + N\pi.$$

Only the second solution is valid since the first one would imply that there is no change in medium properties. For the imaginary part to be zero

$$\sin(k'L)(-2n'^2 + 2n_1 n_2) = 0 \Rightarrow \sin(k'L) = 0 \quad \text{or} \quad n' = \sqrt{n_1 n_2},$$

where only the second solution is valid. The first condition cannot be satisfied because the solution chosen for the real part to be zero is $\cos(k'L) = 0$ and, therefore, $\sin(k'L) = \pm 1$.

Therefore, the conditions for no reflection are

$$k'L = \frac{\pi}{2} + N\pi \quad \text{and} \quad n' = \sqrt{n_1 n_2}.$$

Note that the first condition determines the thickness of the layer, whereas the second one the value of the refractive index. The first condition can be expressed as

$$k'L = \frac{\pi}{2} + N\pi \Rightarrow L = \frac{\lambda'}{4}(1+2N) \Rightarrow n'L = \frac{\lambda_0}{4}(1+2N),$$

where λ' is the wavelength in the dielectric layer and λ_0 the wavelength in vacuum. The dielectric layer thickness must be an odd multiple of $\lambda'/4$ for no light to be reflected from the surface and the dielectric layer will then be an antireflection coating.

Chapter 14
Wave Propagation in Anisotropic Media

Abstract In this chapter we will develop the basic electromagnetic characteristics that account when plane electromagnetic radiation propagates throughout an anisotropic linear material. With this aim it is necessary to explain the concept of anisotropy. In order to be clear we will begin with some easy examples that will allow us to understand, not only what does it mean, but also the importance of such a concept in physics.

14.1 Concept of Anisotropy

Let us suppose a rubber balloon is inflated with air. If the balloon is pressed with the same force at two symmetrical points of its surface in opposite directions, it deforms and change its shape. If we repeat the same experiment but from other directions, we will observe the same result, i.e. the system deforms in a similar way. It means that the deformation of the balloon does not depend on the direction of the force applied. In this case we say that the system is isotropic. The reason for this may be found in the specific characteristics of the gases. As it is well known, the interaction between the particles constituting the gas (atoms or molecules) is low enough, then permitting them to move easily inside the volume. As a result, when a force is applied particles change their positions in space adapting themselves to the new form of the boundary (rubber). An identical result is observed if the balloon is filled with a liquid. A typical example in thermodynamics occurs when fluid inside of a cylinder, enclosed with a piston, is compressed. In this case the compressibility coefficient remains constant independently of the direction in which the piston is located over the fluid. When analyzing solids we have many examples too. In fact, let us suppose an elastic body as shown in Fig. 14.1. Let us cut a slender bar in an arbitrary direction (rod 1). If we are interested in determining elastic mechanical properties, we will measure its Young's modulus and Poisson's ratio. Now, choose another direction in space and take another sample (rod 2). If the result obtained is the same we will say that the mechanical behavior of the solid is isotropic with respect to an external load. The same idea applies when the external excitation is the temperature, electric field or

© Springer-Verlag Berlin Heidelberg 2017
F. Salazar Bloise et al., *Solved Problems in Electromagnetics*,
Undergraduate Lecture Notes in Physics, DOI 10.1007/978-3-662-48368-8_14

Fig. 14.1 Samples 1 and 2 in two directions

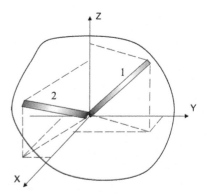

magnetic strength. However, not all systems in nature behave in the same manner, as in the case of crystals.

Unlike the former examples, crystals are composed by an infinite repetition of identical structural units in space. This unit may be one atom or a more complex structure formed by many atoms of molecules. In general all the possibilities we have to form a crystal can be faced by mean points separated periodically from one another at which we have placed a set of atoms. This array of points are called the lattice of the crystal and the group of atoms, ions or molecules is said to be the basis. In this sense we can rigourously define a crystal like a lattice plus a basis.[1] Due to the different dispositions of the units in the space mentioned, the characteristics of crystals are basically anisotropic. With this word we mean that, in general, the physical properties depend on the direction chosen. For instance, it is from well known daily experience that, that when the temperature of a material is modified it deforms. The same occurs in crystals, but the change of the shape depends on the direction examined (anisotropic), therefore this shows that depending on the direction, the thermal expansion behaves in a different way.

In what follows in this chapter, we will focus the analysis on the behavior of the radiation when interacting with matter which is anisotropic to electric excitations. Specifically, we will find that, owing to the non-isotropic polarization response of a system to an external electric field, propagation of a light beam throughout a body differers from the usual behavior in isotropic matter.

An intermediate case is represented by the liquid crystal, which combine properties of liquids and crystalline solids. Liquid crystals share with solids some kinds of partial orientational order (symmetries) and with liquids its possibility to flow like a fluid, thus having properties of both states. Depending on the symmetries and orientation of its molecules we find different families of liquid crystals. In principle, the

---

[1] There are 14 lattices in three dimensions which are known as Bravais lattices, and 7 crystal systems namely, cubic, tetragonal, orthorhombic, monoclinic, triclinic, trigonal, and hexagonal. Besides, from the viewpoint of the crystallography we have 32 crystallographic point groups that a crystal structure may have. They can be generated by symmetry operations which, leaving one point of the crystal fixed, take the structure into itself. Additionally, if we allow translational symmetries in three dimensions, we obtain that a lattice with a basis can have 320 space groups.

basic form of the molecules that compose the liquid crystals are slender and disk-like in shape, which leads to geometrical anisotropy. As a consequence, molecular and atomic interactions are anisotropic too, giving rise to different internal states known as mesophases. In this regard we can divide liquid crystals in some categories as nematics, smetics, cholesterics[2] and discotics. For optical applications nematic and cholesteric liquid crystals are the most used because of their high anisotropic optical properties. They are usually employed in many types of displays and optical system, for instance.

14.2 Susceptibility and Permittivity Tensors Definition

When we studied dielectric materials in Chaps. 3 and 13 we supposed that their properties, from an electrical point of view, were isotropic. It meant the body was described by a dielectric constant only depending on the frequency $\epsilon = \epsilon(\omega)$. As a consequence electromagnetic plane waves with a definite phase velocity and without change in amplitude and polarization were possible. However, when the system is not isotropic, the propagation of waves throughout is very different.

When an isotropic dielectric is submitted into an external electric field, we have seen that its response is to polarize itself. Consequently, a dipolar moment per volume unit **P** appears. In the same way explained in the introduction there are many solids that, due to the arrangement in space and properties of the individual atoms or molecules, their physical response in the presence of an electric field **E** is to polarize too, but differently for each direction. We can understand it by imagining that the atoms regularly spaced (depending on the crystal system) do not have, necessarily, the same electronic distribution for all directions. Due to this fact, it seem to be logical that the polarization, which is very sensitive to molecular geometry and electronic properties, depends on the direction in space.

The polarization **P** that comes up from applying an electric field **E** to an anisotropic medium may be represented by the following expression

$$\begin{bmatrix} P_x \\ P_y \\ P_z \end{bmatrix} = \epsilon_0 \begin{bmatrix} \chi_{11} & \chi_{12} & \chi_{13} \\ \chi_{21} & \chi_{22} & \chi_{23} \\ \chi_{31} & \chi_{32} & \chi_{33} \end{bmatrix} \begin{bmatrix} E_x \\ E_y \\ E_z \end{bmatrix}, \quad P_i = \epsilon_0 \chi_{ij} E_j \quad (14.1)$$

This can be also written as $\mathbf{P} = \epsilon_0 \bar{\bar{\chi}}_e \mathbf{E}$, where $\bar{\bar{\chi}}_e$ is called the susceptibility tensor. Thus, when studying anisotropic media the relation between **P** and **E** is no longer a scalar, as it occurs in isotropic systems. On the other hand, taking into account that $\mathbf{D} = \epsilon_0 \mathbf{E} + \mathbf{P}$, we can write

$$\mathbf{D} = \epsilon_0 \mathbf{E} + \mathbf{P} = \epsilon_0 \mathbf{E} + \epsilon_0 \bar{\bar{\chi}}_e \mathbf{E} = \epsilon_0 (\bar{\bar{I}} + \bar{\bar{\chi}}_e) \mathbf{E} = \epsilon_0 \bar{\bar{\epsilon}}_r \mathbf{E} = \bar{\bar{\epsilon}} \mathbf{E} \quad (14.2)$$

---

[2]Cholesteric liquid crystals are in the literature also known as chiral nematic liquid crystals.

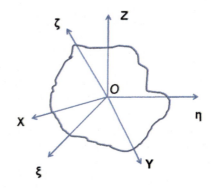

Fig. 14.2 In this figure $OXYZ$ represents an arbitrary coordinate frame in the body, and $\xi\eta\zeta$ are the principal axes of the system

where $\bar{\bar{I}}$ and $\bar{\bar{\epsilon}}_r$ are the identity matrix and the relative permittivity tensor, respectively. Equation (14.2) shows that the displacement vector **D** won't be either, in general, parallel to **E**. Expanding (14.2) yields

$$\begin{bmatrix} D_x \\ D_y \\ D_z \end{bmatrix} = \begin{bmatrix} \epsilon_{11} & \epsilon_{12} & \epsilon_{13} \\ \epsilon_{21} & \epsilon_{22} & \epsilon_{23} \\ \epsilon_{31} & \epsilon_{32} & \epsilon_{33} \end{bmatrix} \begin{bmatrix} E_x \\ E_y \\ E_z \end{bmatrix} \qquad (14.3)$$

It can be demonstrated that susceptibility and permittivity tensors are symmetric. This is to say that $\chi_{ij} = \chi_{ji}$ and $\epsilon_{ij} = \epsilon_{ji}$.[3] This property of symmetry for the tensor does not depend on the reference system chosen for its representation. Due to this symmetry, both tensor are diagonalizable. It means that we can always find a basis where the representation of ϵ_{ij} is diagonal, i.e. we only have terms in the diagonal of the tensor

$$\bar{\bar{\epsilon}} = \begin{bmatrix} \epsilon_1 & 0 & 0 \\ 0 & \epsilon_2 & 0 \\ 0 & 0 & \epsilon_3 \end{bmatrix}. \qquad (14.4)$$

As a consequence it may be demonstrated, that all its eigenvalues are real and the corresponding eigenvectors are perpendicular to each other. The associated directions to the eigenvectors are called principal directions, and its respective eigenvalues ϵ_i ($i = 1, 2, 3$) are known as principal dielectric constants. In these directions the corresponding components of **E**, **D** and **P** are parallel, then it follows (Fig. 14.2)

$$\mathbf{D}_\epsilon = \epsilon_0 \epsilon_{r1} \mathbf{E}_\epsilon = \epsilon_1 \mathbf{E}_\epsilon, \qquad (14.5)$$

$$\mathbf{D}_\eta = \epsilon_0 \epsilon_{r2} \mathbf{E}_\eta = \epsilon_2 \mathbf{E}_\eta, \qquad (14.6)$$

$$\mathbf{D}_\zeta = \epsilon_0 \epsilon_{r3} \mathbf{E}_\zeta = \epsilon_3 \mathbf{E}_\zeta. \qquad (14.7)$$

---

[3] In the development of this chapter we will work with non-active media, then all ϵ_{ij} are real numbers. This situation is different for an active material for which some components of ϵ_{ij} may be complex. In this case holds $\epsilon_{ij} = \epsilon_{ji}^*$, i.e. the tensor is hermitian.

14.2 Susceptibility and Permittivity Tensors Definition

An important fact is that the tensor ϵ_{ij} (χ_{ij}) represents a physical property, which is totally independent from the choice of the reference frame. Therefore, if the axes are modified, the physical property does not change, but only its representation.[4]

14.3 Maxwell's Equations in an Anisotropic Linear Medium Free of Charges and Currents

Let us imagine a magnetically isotropic medium, but with anisotropic dielectric constants. Under these assumptions, we are interested in studying the propagation of plane waves throughout the material, i.e. fields **E**, **D**, and **H** of the form

$$\mathbf{E} = \mathbf{E}_0 e^{j\mathbf{kr}} e^{j\omega t}, \tag{14.8}$$

$$\mathbf{D} = \mathbf{D}_0 e^{j\mathbf{kr}} e^{j\omega t}, \tag{14.9}$$

$$\mathbf{H} = \mathbf{H}_0 e^{j\mathbf{kr}} e^{j\omega t}. \tag{14.10}$$

For this propagation to occur, these waves must verify the Maxwell equations

$$\nabla \cdot \mathbf{D} = 0, \tag{14.11}$$

$$\nabla \times \mathbf{E} = -\frac{\partial \mathbf{B}}{\partial t}, \tag{14.12}$$

$$\nabla \times \mathbf{H} = \frac{\partial \mathbf{D}}{\partial t}, \tag{14.13}$$

$$\nabla \cdot \mathbf{B} = 0, \tag{14.14}$$

and the material equations $\mathbf{B} = \mu_0 \mathbf{H}$ and $\mathbf{D} = \epsilon_0 \bar{\bar{\epsilon}}_r \mathbf{E}$. Then, introducing (14.8)–(14.10) into (14.11)–(14.14) and considering $\omega \in [10^{14}, 10^{16}]\,\mathrm{s}^{-1}$

$$\mathbf{D} \cdot \mathbf{k} = 0 \Rightarrow \mathbf{D} \perp \mathbf{k}, \tag{14.15}$$

$$\mathbf{k} \times \mathbf{E} = \mu_0 \omega \mathbf{H}, \tag{14.16}$$

---

[4]To describe a symmetric tensor of second rank a quadric can be used. More specifically, a quadric can be employed to describe a physical property that is represented by a tensor. In general we can write $(S_{ij}x_ix_j = 1)$, where S_{ij} are its coefficients. They transform in the same manner like the components of a second order symmetric tensor. If the quadric is referred to its principal axes it takes the simplest form (it corresponds to the diagonal tensor), i.e. $(S_1 x^2 + S_2 y^2 + S_3 z^2 = 1)$, S_1, S_1, and S_3 being the principal components of the tensor S_{ij}. Depending on the values and signs of S_i, ($i = 1, 2, 3$), the corresponding quadric surface may be a sphere, an ellipsoid, a hyperboloid of one sheet or a hyperboloid of two sheets (there is also the possibility of imaginary ellipsoids).

Fig. 14.3 a Vector **D** is perpendicular to **k**. b As **D** is perpendicular to **H**, taking into consideration the figure (a), the only possibility is that **D**, **k** and **H** are perpendicular to each other

Fig. 14.4 a Vectors **B** and **H** have the same direction. b Observe that the direction of propagation of the phase vibrations does not coincide with the direction of the electromagnetic beam (Poynting's vector)

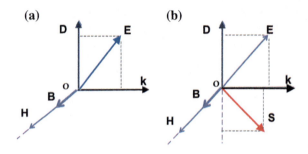

$$\mathbf{k} \times \mathbf{H} = -\omega \mathbf{D}, \tag{14.17}$$

and

$$\mathbf{k} \cdot \mathbf{H} = 0 \Rightarrow \mathbf{k} \perp \mathbf{H}. \tag{14.18}$$

These last equations give the relations among **E**, **D**, **B** and **H** for a given propagation direction **k**. As a result, we can obtain some consequences. In fact, from (14.15) we see that **D** is perpendicular to **k**. Equation (14.17) yields that **D** is perpendicular to **H** (observe the sign (14.17)). In addition, **k** is perpendicular to **H**, then the only possibility is that **D**, **k** and **H** are mutually orthogonal as represented in Fig. 14.3b. Furthermore, **E** is perpendicular to **H** (14.16), but due to the anisotropy, in general, its direction is not the same as **D** (Fig. 14.4a). This means that **E** lies in the same plane defined by O, **D**, and **k** as shown in Fig. 14.4a. Therefore, we can conclude that for anisotropic media the vector **k** and **E** are not perpendicular to one another. This is an important result differing for those of the isotropic bodies.

With respect to **B**, as supposed that the medium is magnetically isotropic, then $\mathbf{B} = \mu_0 \mathbf{H}$. As a consequence, **B** and **H** have the same direction (Fig. 14.4b). Finally, Poynting's vector is perpendicular to **E** and **H**, thus its direction and the direction of **k** are different (Fig. 14.4b). Physically it shows that, in general, the direction of propagation of the wavefront does not coincide with the direction of the energy propagation (Poynting's vector $\mathbf{S} = \mathbf{E} \times \mathbf{H}$). However, **D**, **E**, **k** and **S** are in the same plane (Fig. 14.4b).

14.3 Maxwell's Equations in an Anisotropic Linear Medium ...

Once we know the basic structure of the fields inside of the anisotropic material, we will deduce the equation relating the possible values of **k** with the electric field **E**. With this aim, let us first multiply vectorially both members of (14.16) by **k**

$$\mathbf{k} \times \mathbf{E} = \mu_0 \omega \mathbf{H} \rightarrow \quad \mathbf{k} \times (\mathbf{k} \times \mathbf{E}) = \mu_0 \omega \mathbf{k} \times \mathbf{E}. \tag{14.19}$$

Now, using (14.17) and (14.2) we obtain

$$(\mathbf{k} \cdot \mathbf{E}) \cdot \mathbf{k} - k^2 \mathbf{E} + \mu_0 \omega^2 \epsilon_0 \bar{\bar{\epsilon}}_r \mathbf{E} = 0 \tag{14.20}$$

This result is a system of equations that relates the characteristics of the field **E** for a given direction of **k**.

14.4 Electromagnetic Waves in Uniaxial Dielectrics

In this section we will examine the relations for the electromagnetic fields and the vector **k** when the crystal has two of the three principal dielectric constants equal. Let us suppose that $\epsilon_2 = \epsilon_3 \neq \epsilon_1$. In this case the system to be studied has symmetry of revolution around $O\xi$. When this occurs the crystal is said to be uniaxial and this direction of symmetry is called the *optic axis*. Due to this invariance the pose of the problem may be simplified by restricting **k** to the plane $O\xi\zeta$. In fact, choosing $\mathbf{k} = (k_\xi, 0, k_\zeta)$ and the dielectric tensor referring to its principal axes, and introducing **k** and **E** into (14.20) it leads to (Fig. 14.5) and the following set of equations

$$(-k_\zeta^2 + \mu_0 \omega^2 \epsilon_1) E_\xi + k_\xi k_\zeta E_\zeta = 0, \tag{14.21}$$

$$(-k_\xi^2 - k_\zeta^2 + \mu_0 \omega^2 \epsilon_2) E_\eta = 0, \tag{14.22}$$

$$k_\xi k_\zeta E_\xi + (-k_\xi^2 + \mu_0 \omega^2 \epsilon_2) E_\zeta = 0, \tag{14.23}$$

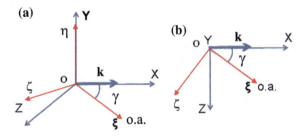

Fig. 14.5 **a** The vector **k** is contained in the plane formed by the principal axes $O\xi$ and $O\zeta$. In order to simplify, in this figure we have chosen **k** in the direction of OX. **b** Plane view of the reference frames

This system has two different solutions. In fact, examining (14.21)–(14.23) we see that this system is verified if $E_\eta \neq 0$ and $E_\xi = E_\zeta = 0$. In this case, from (14.22) we deduce that, for this possibility to occur the bracket in (14.22) must be zero, i.e.

$$-k_\xi^2 - k_\zeta^2 + \mu_0 \omega^2 \epsilon_1 = 0 \longrightarrow k_\xi^2 + k_\zeta^2 = \mu_0 \omega^2 \epsilon_2 = \frac{\omega^2}{c^2} \epsilon_{r2}. \tag{14.24}$$

If we label $\frac{\omega^2}{c^2} = k_0^2$ and take into account that $\epsilon_{r2} = n_2^2$, (14.24) may be rewritten as

$$k_\xi^2 + k_\zeta^2 = \frac{\omega^2}{c^2} \epsilon_{r2} = \frac{\omega^2}{c^2} n_2^2 = k_0^2 n_2^2. \tag{14.25}$$

This last equation shows that the value of the modulus of **k** does not depend on the direction. To see this in another way, let us consider the projections of the vector **k** over the optic axis, i.e. $k_\xi = k \cos \gamma$ and $k_\zeta = -k \sin \gamma$, where γ is the angle between **k** and ξ (see Fig. 14.6). Introducing them into (14.25) we have

$$k^2 = k_0^2 n_2^2 \longrightarrow k = k_0 n_2 = \omega \sqrt{\epsilon_0 \mu_0 \epsilon_{r2}}. \tag{14.26}$$

The propagating waves inside of the crystal for this case are called *ordinary waves* and the principal index n_2 the ordinary index, which we will label n_o. Consequently, the phase velocity of these waves is also constant, with value

$$v_o = \frac{\omega}{k_o} = \frac{1}{\sqrt{\epsilon_0 \mu_0 \epsilon_{r2}}} = \frac{c}{\sqrt{\epsilon_{r2}}}, \tag{14.27}$$

and the refractive index

$$n_o = \frac{c}{v_o} = \sqrt{\epsilon_{r2}}. \tag{14.28}$$

On the other hand, the solution obtained for the electric field means that these waves are linearly polarized.

With this information and applying (14.15)–(14.18), a scheme of the corresponding fields is represented in Fig. 14.6. The second possibility is the opposite to that last one, which means that $E_\eta = 0$, and $E_\xi \neq 0$ and $E_\zeta \neq 0$. If it does occur, the

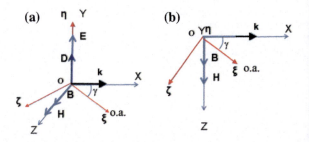

Fig. 14.6 a Schema of the vector fields for the ordinary wave. Observe that **E** and **D** oscillate in the same direction. b Plane view

14.4 Electromagnetic Waves in Uniaxial Dielectrics

only way to have a non-trivial solution is that the determinant of the coefficients in (14.21) and (14.23) be zero, i.e.

$$det \begin{bmatrix} (-k_\zeta^2 + \mu_0\omega^2\epsilon_1) & k_\xi k_\zeta \\ k_\xi k_\zeta & (-k_\xi^2 + \mu_0\omega^2\epsilon_2) \end{bmatrix} = 0. \quad (14.29)$$

The solution of this determinant is

$$(k_\xi^2 n_1^2 + k_\zeta^2 n_2^2)k_0^2 = n_1^2 n_2^2, \quad (14.30)$$

which may be expressed in the following form

$$\frac{k_\xi^2}{n_2^2} + \frac{k_\zeta^2}{n_1^2} = n_0^2. \quad (14.31)$$

In this case the modulus of **k** does depend on the direction of propagation. As we have seen in the case of the ordinary wave, the introduction of $k_\xi = k\cos\gamma$ and $k_\zeta = -k\sin\gamma$ into this former equation leads to

$$k^2 \left(\frac{\cos^2\gamma}{n_2^2} + \frac{\sin^2\gamma}{n_1^2} \right) = k_0^2 \longrightarrow k(\gamma) = \frac{k_0}{\sqrt{\frac{\cos^2\gamma}{n_2^2} + \frac{\sin^2\gamma}{n_1^2}}} = \frac{\omega\sqrt{\epsilon_0\mu_0}}{\sqrt{\frac{\cos^2\gamma}{\epsilon_{r2}} + \frac{\sin^2\gamma}{\epsilon_{r1}}}}. \quad (14.32)$$

This expression for **k** shows that its value depends on the angle γ between the propagation direction and the optic axis ξ.

To determine the polarization of the wave we use (14.21) together with (14.31), obtaining

$$\frac{E_\zeta}{E_\xi} = -\frac{n_1^2 k_\xi}{n_2^2 k_\zeta}, \quad (14.33)$$

which means that the wave is linearly polarized, because the quotient between $\frac{E_\zeta}{E_\xi}$ is a constant for the corresponding values of k_ξ and k_ζ. Besides, as we can see from the second solution the electric field is contained in the plane $O\xi\zeta$. This new wave whose electric field vibrates along the straight line defined by (14.33) is said to be an *extraordinary wave*, and its corresponding principal index n_1 is called the extraordinary index, which we will denote by n_e. As a consequence (14.32), the velocity v_e and refractive index n_e must be functions of γ. Using the definitions (14.27) and (14.28) for n and v, and identifying $\epsilon_{r2} = \epsilon_o$ and $\epsilon_{r1} = \epsilon_e$ with the ordinary and extraordinary relative dielectric permittivities, respectively, we have

$$v_e = \frac{\omega}{k_e(\gamma)} = c\sqrt{\frac{\cos^2\gamma}{\epsilon_{r2}} + \frac{\sin^2\gamma}{\epsilon_{r1}}} = c\sqrt{\frac{\cos^2\gamma}{\epsilon_o} + \frac{\sin^2\gamma}{\epsilon_e}}, \quad (14.34)$$

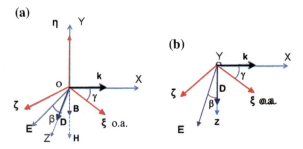

Fig. 14.7 **a** Location of the vector fields in the reference frames $OXYZ$ and $O\xi\eta\zeta$. Observe that for the extraordinary wave **E** and **D** do not have the same direction. Vector **D** is orthogonal to **k** and **H** ((14.15) and (14.17)). **b** Plane view

and the refractive index

$$n_e = \frac{c}{v_e(\gamma)} = \frac{1}{\sqrt{\dfrac{\cos^2\gamma}{\epsilon_{r2}} + \dfrac{\sin^2\gamma}{\epsilon_{r1}}}} = \frac{1}{\sqrt{\dfrac{\cos^2\gamma}{\epsilon_o} + \dfrac{\sin^2\gamma}{\epsilon_e}}}. \quad (14.35)$$

Proceeding in the same way as before for the first solution, by virtue of (14.15)–(14.18), we can draw the electromagnetic fields (Fig. 14.7). Observe that, contrary to what occurs for the ordinary wave, the electric field **E** does not have the same direction as **D**. However, **D**, **E**, **k**, and the optic axis are contained in the same plane. To sum up these results we can say that, when a wave propagates throughout an uniaxial medium it splits into two orthogonal linearly polarized waves, each one travelling with different phase velocities (note that $n_o = \frac{c}{v_o}$ and $n_e = \frac{c}{v_e}$). In this context it is important to understand that the velocity of propagation of a wave (in our case the ordinary and the extraordinary wave) depends on the direction of vibration of the field and not on the direction of the wave propagation. This dependence on the field amplitude oscillation can be understood as follows.

Let us suppose that an anisotropic material is built by non-spherical atoms or molecules as shown in Fig. 14.8. If due to its constitution and geometrical shape the electrons are displaced from the center of charge, the molecule polarizes. The resulting polarization is, in turn, proportional to the electric dipole moment per unit volume and to the local field. This local field is the result of the superposition of external electric field **E** and the dipole fields produced by the set of molecules. Thus, if an external electric field interacts with the atoms in such a way that its direction of oscillation coincides approximately with the polarization direction (Fig. 14.8a), it would increase the polarization of the material. As a result, an increasing of the refractive index will occur, leading to a slow velocity. On the contrary, if **E** acts more or less perpendicular to the dipole fields (Fig. 14.8b), it will not enhance the dipole moments resulting in a increasing of the velocity, and therefore a smaller refraction index. Finally, from this reasoning we conclude that, linear polarized wave fields

14.4 Electromagnetic Waves in Uniaxial Dielectrics

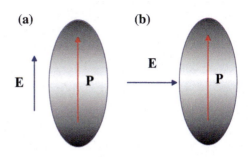

Fig. 14.8 **a** Molecule and electric field along its easy direction. **b** Here the electric field is perpendicular to its symmetry axis

parallel to the polarizable axis of the molecules (usually its symmetry axis) travel slower than linear polarized waves whose vibration directions are perpendicular to the aforementioned axis.

14.5 Propagation of the Energy

As we have seen in Sect. 14.4, due to the fact that in anisotropic media the electric field **E** and the vector **k** are not, in general, perpendicular to each other (see (14.15)–(14.18)) and Fig. 14.4b, the energy does not propagate in the same direction as **k** does. By using the definition of Poynting's vector $\mathbf{S} = \mathbf{E} \times \mathbf{H}$, the structure of the electromagnetic vectors in a plane perpendicular to **H** has been drawn in Fig. 14.9. From this figure we can observe that, the planes of constant phase are perpendicular to the vector **k**, but they do not move in this direction but in the direction of the propagation of the energy given by **S**. This fact did not occur in the case of isotropic materials for which **k** and **S** had the same direction. Due to this fact, the velocity with which energy propagates is different as the phase velocity along the direction of **k**. The relation between both velocities is given by

$$v_S = \frac{v_k}{\cos \beta}, \tag{14.36}$$

where v_k is the velocity with respect to **k** (phase velocity), and v_S is the velocity along **S**, also known as ray velocity. Observe that both velocities coincide if the propagation is along a principal axis of the crystal. In the case of uniaxial crystals we can also deduce the relations among the electromagnetic vectors for the ordinary and extraordinary waves. In effect, by again using (14.15)–(14.18), we obtain a scheme as shown in Fig. 14.10 for the fields corresponding to both waves. Observe that, owing to the electric field **E** having the same direction as **D**, Poynting's vector and **k** have the same direction for the ordinary wave too (Fig. 14.10). On the contrary, for the extraordinary wave **E** and **D** are not collinear, then **k** and **S** propagate through different directions (Fig. 14.11). As we can deduce from Fig. 14.11, the angle between **D** and **E** is the same as the angle formed by **k** and **S**. Skipping the demonstration, this angle may be calculated by means of the following relation

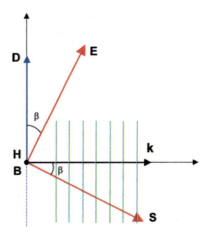

Fig. 14.9 Disposition of the electromagnetic vectors in a general case for an anisotropic media. Observe that the direction of propagation of **k** forms an angle β with the direction of the energy propagation (Poynting's vector $\mathbf{S} = \mathbf{E} \times \mathbf{H}$)

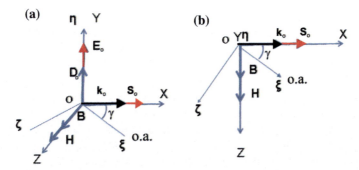

Fig. 14.10 Ordinary wave. **a** Vector **k** is perpendicular to **D** and **H**, and **S** is orthogonal to **E** and **H**. In this case, due to **E** and **H** are collinear, **k** and **S** have the same direction. **b** Plane view

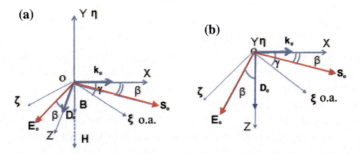

Fig. 14.11 Extraordinary wave. **a** Here the direction of **E** and **D** differ and as a result **k** and **S** form an angle β. **b** Plane view

$$\tan\beta = \frac{(n_1^2 - n_2^2)\tan\gamma}{n_1^2 + n_2^2 \tan^2\gamma} = \tan\beta = \frac{(n_e^2 - n_o^2)\tan\gamma}{n_e^2 + n_o^2 \tan^2\gamma} = \frac{(\epsilon_e - \epsilon_o)\tan\gamma}{\epsilon_e + \epsilon_o \tan^2\gamma}. \quad (14.37)$$

14.6 Geometrical Interpretation

From a geometrical point of view, (14.25) and (14.31) have an easy interpretation. If we look at both equations considering the projections of **k** over the principal axes $O\xi$ and $O\zeta$, we see that they represent two curves in the space of the **k** vectors. In fact, (14.25) corresponds to a circumference and (14.31) to an ellipse. Actually these curves in two dimensions are the intersections of a surface (quadric) with the plane $O\eta$. In general the vector **k** describes a surface which is known as the *wave vector surface*. Therefore, the wave vector surface expressed by (14.25) is a sphere and (14.31) an ellipsoid of revolution, because in uniaxial crystals two of the three principal dielectric constants are equal. Taking into account the characteristics found for the ordinary and extraordinary waves we have demonstrated in the former sections, that it is possible to construct a section of the surface wave vector for uniaxial crystals. In both cases the surfaces are a sphere and an ellipsoid of revolution, whose symmetry axis corresponds to the optic axis of the body. Figure 14.12 represents the two possibilities we can have. The first one (Fig. 14.12a) depicts a negative material

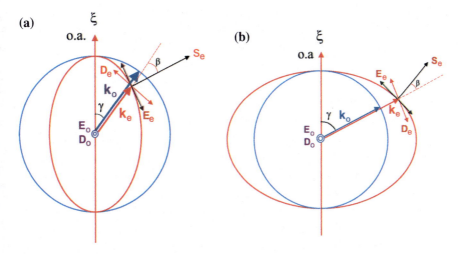

Fig. 14.12 **a** Uniaxial negative crystal ($n_e < n_o$, ($n_1 < n_2$)). **b** Uniaxial positive crystal ($n_e > n_o$). Observe that in the direction of the optic axis $\mathbf{k}_o$ and $\mathbf{k}_e$ are equal. Besides, in this last case, the direction of propagation of the energy is the same for both waves and coincide with of the wave vectors

($n_e < n_o$) and in (b) a positive crystal is shown ($n_e > n_o$).[5] By these drawings we can again examine the relations among the vectors of the waves. In general, for a given direction of propagation we have two different wave vectors labelled as $\mathbf{k}_o$ and $\mathbf{k}_e$, corresponding to the ordinary and extraordinary waves, respectively. From this construction it is obvious that the magnitude of $\mathbf{k}_e$ depends on the angle γ formed by this vector with the optic axis ξ. On the contrary, the modulus of $\mathbf{k}_o$ is always the same. Only if the propagation direction is along ξ vectors $\mathbf{k}_o$ and $\mathbf{k}_e$ are equal, which means that the refractive index for both waves is the same (and their velocities). As a consequence, when a wave travels throughout a crystal in the direction of the optic axis it does not suffer the phenomenon of the double refraction. Physically, in this case the crystal behaves like an isotropic material.

In how the energy propagation is concerned, in the ordinary wave Poynting's vector goes in the same direction as $\mathbf{k}_o$.[6] For the extraordinary wave the direction of $\mathbf{S}_e$ differs from of $\mathbf{k}_e$, except in the direction of the optic axis ($\gamma = 0$).

14.7 Electromagnetic Waves in Biaxial Crystals

This section deals with the most general case of optical anisotropic crystals. In the previous section we studied materials for which two of the three principal dielectric constants were equal. In the present case, we will consider crystals with all principal indices different, i.e. $\epsilon_1 \neq \epsilon_2 \neq \epsilon_3$. Because of these unequal dielectric constants these specimens do not have any axis of symmetry. This kind of material is known as biaxial, and the analysis of the wave propagation is more difficult.

The general pose of the problem can be solved by employing (14.20). Starting from a reference frame $\xi\eta\zeta$ (where the dielectric tensor is diagonal), and setting into (14.20) $\mathbf{k} = (k_\xi, k_\eta, k_\zeta)$, we obtain

$$(-k_\eta^2 - k_\zeta^2 + \mu_0\omega^2\epsilon_1)E_\xi + k_\eta k_\xi E_\eta + k_\xi k_\zeta E_\zeta = 0 \tag{14.38}$$

$$k_\xi k_\eta E_\xi + (-k_\xi^2 - k_\zeta^2 + \mu_0\omega^2\epsilon_2)E_\eta + k_\eta k_\zeta E_\zeta = 0 \tag{14.39}$$

$$k_\xi k_\zeta E_\xi + k_\eta k_\zeta E_\eta + (-k_\xi^2 - k_\eta^2 + \mu_0\omega^2\epsilon_3)E_\zeta = 0, \tag{14.40}$$

---

[5]For example, calcite ($n_e = 1.5534, n_o = 1.6776$), tourmaline ($n_e = 1.638, n_o = 1.669$), sapphire ($n_e = 1.760, n_o = 1.768$) and ruby ($n_e = 1.762, n_o = 1.770$) are negative, and quartz ($n_e = 1.553, n_o = 1.544$), rutile ($n_e = 2.903, n_o = 2.616$), and magnesium fluoride ($n_e = 1.385, n_o = 1.380$) are positive.

[6]In Fig. 14.12 we have not represented the Poynting vector for the ordinary wave in order to not draw many vectors at the same figure. For the ordinary wave $\mathbf{S}_o$ is always orthogonal to the spherical surface, thus following the same direction as $\mathbf{k}_o$.

14.7 Electromagnetic Waves in Biaxial Crystals

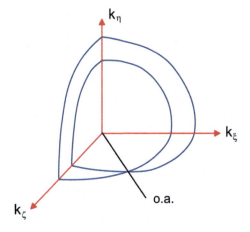

Fig. 14.13 Wave vector surfaces. The intersect at a point for which the k vectors are equal. This direction is called the optic axis of the crystal

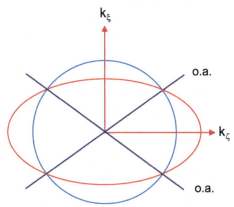

Fig. 14.14 Intersection of the k-surfaces with the plane $k_\eta = 0$. Observe that two optic axes appear

which may be expressed in a matrix form as follows

$$\begin{bmatrix} (-k_\eta^2 - k_\zeta^2 + \mu_0\omega^2\epsilon_1) & k_\eta k_\xi & k_\xi k_\zeta \\ k_\xi k_\eta & (-k_\xi^2 - k_\zeta^2 + \mu_0\omega^2\epsilon_2) & k_\eta k_\zeta \\ k_\xi k_\zeta & k_\eta k_\zeta & (-k_\xi^2 - k_\eta^2 + \mu_0\omega^2\epsilon_3) \end{bmatrix} \begin{bmatrix} E_\xi \\ E_\eta \\ E_\zeta \end{bmatrix} = \begin{bmatrix} 0 \\ 0 \\ 0 \end{bmatrix}. \quad (14.41)$$

This system has a non-trivial solution if the determinant of the matrix coefficients is zero. The solution is composed of two surfaces in **k**-space. However in the case of uniaxial crystals, because of the lack of symmetry, there are more geometrical possibilities for the cross-sections of the wave vector surfaces. Figure 14.13 represents an octant of the double surface. As we can see by inspection, for a chosen direction there exist two distinct wave vectors **k**. However, there is a direction for which the two wavenumbers are the same, which corresponds to the point where the two surfaces intercept and it is defined as the optic axis (see Fig. 14.12).[7] Actually, when representing the whole surfaces in three dimensions, or their intersection with the plane $Ok_\xi k_\eta$ we may note that biaxial crystals have two optic axis (Fig. 14.14).

14.8 Crystal Classification

In the previous sections we have presented the basic characteristics of the uniaxial and biaxial media. The results obtained show that the anisotropy of the materials leads to a tensorial relationship between the displacement vector **D** and **E**. As a consequence the wave propagation differs from that corresponding to isotropic bodies.

To classify optical properties we can use Neumann's principle. This principle says that the symmetry group of any physical property must include the point symmetry group of the crystal. According to this, any physical property (wave normal, velocity surface, etc.) have all symmetry elements corresponding to the crystal. Applying this principle to the wave vector surface (or velocity surface), it follows that crystals belonging to systems triclinic, monoclinic, and orthorhombic must be biaxial. In fact, the symmetry of the surface is *mmm* then it includes the point groups $1, \bar{1}, 2, m, 2/m, 222, mm2$, and *mmm*.

In the case of uniaxial crystals the surface has a higher symmetry, then other crystals verify the conditions. So, following the same reasoning we conclude that the crystalline systems trigonal, tetragonal and hexagonal are uniaxial. For this case the symmetries included in the group *mmm* are $3, \bar{3}, 32, \bar{3}m, 3m, \bar{4}, 4/m, 422, \bar{4}2m, 4mm, 4/mmm, 6, \bar{6}, 6/m\, 622, 6mmm, \bar{6}m2, 6/mmm, \infty, \infty 2, \infty/m, \infty m, \infty/mm$.

The maximal symmetry corresponds to the optical isotropic materials. In this case the wave surface is a sphere ($n_1 = n_2 = n_3$), then only the symmetry groups 23, $m3, 43m, 432, m3m, \infty\infty$, and $\infty\infty m$ are possible, which corresponds to the cubic system.

In the above discussion we did not take into consideration the effects of wave propagation across the crystal when other external perturbations act simultaneously. A more general study of any physical property of a crystal structure should also include the investigation of the symmetry of the external factor. To do this we can apply the well known Curie's principle, together with Neumann's law. The basic idea is the following.

When a physical perturbation acts on a crystal (variation of temperature, electric field, magnetic field, etc.) it can modify its structure. As a result, from the viewpoint of the symmetry, the body will have the symmetry elements which are common for the crystal before applying the external action, and for the field itself. In this regard, Curies's law may be envisaged as a symmetry superposition principle. To illustrate the significance of this principle, we will give an example.

---

[7] Observe that for uniaxial crystals we defined the optic axis as that of the symmetry of the system. Due to the absence of high symmetry in biaxial materials, we must define the optic axis in another way.

14.8 Crystal Classification

Let us suppose a crystal of halite (NaCl). This mineral has symmetry $m3m$ and its refractive indices are $n_1 = n_2 = n_3 = 1.54$, which means that it is optically isotropic. Therefore it will not exhibit optical birefringence. However if we apply an electric field **E** in the direction [001], the index ellipsoid (indicatrix) of the crystal will be deformed as a result of the action of such a field. The state of the crystal is transformed from the spherical symmetry of the initial state (halite) into an ellipsoid of revolution after the external perturbation, and then it becomes optically uniaxial. If the electric field were directed along any direction [$hh0$] which corresponds to a plane of symmetry the ellipsoid would have three different principal axes, converting the crystal to biaxial.

The important consequence that we can get from the above results is that, under some circumstances we can modify optical properties of materials by using external actions. For instance, we can induce optical anisotropies by using electric or magnetic fields, and mechanical stress as well. For instance, the application of an of an electric field to a isotropic substance leads to birefringence. The electrooptical effects that result are called Kerr and Pockels effects.[8] By employing a magnetic field, induced double refraction is also possible. The resulting magnetooptical effect is known as the Cotton-Mouton effect and like the Kerr effect it is proportional to the square of the magnetic field applied. In the same way, by producing deformations in a substance we can induce optical anisotropies which are known as photoelastic effects. In fact, a strain in a body produces a change of the index ellipsoid, which leads to a variation of the polarization constants, and therefore to the apparition of double refraction. In this context it is important to note that there are other new mechanisms for inducing optical anisotropy.

In Chap. 7 (Problems 4 and 5) we studied the behavior of the electric field inside and outside of a dielectric sphere (and metallic) in presence of an external **E**. One of the applications shown was that corresponding to the inclusion of microspheres (dielectric or metallic) into a material. We commented that it is possible to induce surface anisotropy which leads to randomly anisotropic optical properties. Recently it has been demonstrated that inhomogeneous layers have different refractive indices for s and p polarized light. Ultimately, these kind of new materials (layers) behave as having uniaxial properties, which depend on the layer thickness, on the inclusion concentration, and on the *incidence angle*. Besides, in some cases, this surface induced anisotropy leads to large spin Hall effects.[9]

---

[8] In general when an electric field is applied to a material, it can produce linear and non-linear electrooptical effects. In the Kerr effect the difference of indices is proportional to the square of the electric field (non-linear), $\Delta n = K \lambda E^2$, K being the Kerr constant. In the case of the Pockels effect Δn is proportional to the field strength (linear). In substances without center of symmetry the most important effect is the Pockels effect. On the contrary, in materials with center of symmetry the Pockels effect does not exist.

[9] See [56].

14.9 Retarders

One of the most important applications of the optical anisotropic materials we have studied is the construction of slabs for modifying the polarization state of light beams. This anisotropic sheet is constructed in such a way that two principal axes of the material are disposed parallel to its faces. One of them corresponds to the optic axis ξ, then the other one is perpendicular. The idea is based on the decomposition of any radiation that impinges the anisotropic plate into two projections, one over ξ, and the other along OY. As we have previously seen, when a light beam is directed perpendicularly to the optic axis of an anisotropic uniaxial material, inside of the sample two independent linear polarized beams appear, each of them perpendicular to each other. The wave whose electric field vibrates perpendicularly to the symmetry axis direction (optic axis) was labelled as the ordinary wave, and the other perpendicularly to it the extraordinary wave. The beams at the exit of the plate (for this disposition) are superimposed but they differ in their respective propagation velocities, and therefore when light crosses the slab, a phase difference between both components holds. As a result, the emerging light may change its polarization state. Depending on the magnitude of the velocities, the material is said to be positive or negative uniaxial material. If the material is positive ($n_e > n_o$), then the velocity of the ordinary wave is faster than of the extraordinary perturbation (perpendicular direction). The contrary happens when the material is negative ($n_e < n_o$).

In these type of retarders a slow and a fast axis are usually identified. The first one corresponds to the vibration direction of the electric field of the longer optical path, and it is labelled with the word S. The perpendicular, or fast direction, is called the F axis.

For a slab of thickness d the phase difference of two perpendicularly linear polarized waves is

$$\delta = \pm \frac{2\pi}{\lambda}(n_e - n_o)\, d, \qquad (14.42)$$

λ being the light wavelength in vacuum. This equation allows us to calculate the change of the polarization state of a monochromatic radiation that impinges orthogonally onto its optic axis (Fig. 14.15). Rearranging (14.42) we can write the optical path difference as

$$\Delta = r\lambda = \pm(n_e - n_o)\, d, \qquad (14.43)$$

where r is called the retardation parameter. This last expression holds for labelling the different kinds of retarders. In general, a slab which verifies (14.43) is called a r-wave plate. In this way, two of the most used retarders in the laboratories are the quarter and half-wave plates. For them $r = \frac{1}{4}$ and $r = \frac{1}{2}$, respectively. However, there are other types of wave plates which allow the manipulation of a polarized beam in a more extended way. For some application birefringent nanostructured glass plates

14.9 Retarders

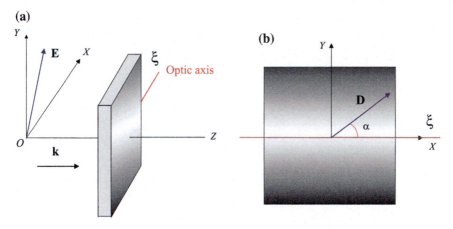

Fig. 14.15 **a** Plane wave directed onto a uniaxial plate whose optic axis is perpendicular to **k** (lateral view). **b** Front view of the displacement vector **D** inside of the plate

are used.[10] These kind of plates are characterized by a spatial distribution of the optic axis that depends on the point over the plate plane.

Solved Problems

Problems A

14.1 Through an optically uniaxial material of refractive indices $n_1 = n_e = 1.7$ and $n_2 = n_o = 1.4$ a wave whose wavefront is perpendicular to the OX axis is propagated. If the optic axis forms an angle of 45° with OX, as shown in Fig. 14.16, obtain: (a) The velocity of the extraordinary ray. (b) Draw the propagation direction of the extraordinary ray.

Solution

(a) As we know from the theory (14.34), $n_1 = \sqrt{\epsilon_{r1}}$ and $n_2 = \sqrt{\epsilon_{r2}}$. There we obtained for the velocity of the extraordinary wave the following relation

$$v_e = c\sqrt{\frac{\sin^2 \gamma}{\epsilon_1} + \frac{\cos^2 \gamma}{\epsilon_2}}, \qquad (14.44)$$

c being the velocity of light in vacuum. Introducing the values of the permittivities and of the angle $\gamma = 45°$ in the latter equation we obtain

$$v_{ext} = 1.96 \times 10^8 \text{ ms}^{-1} \qquad (14.45)$$

---

[10] See, for instance, [40, 71].

Fig. 14.16 Direction of the extraordinary ray in the material

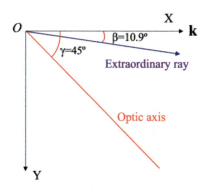

(b) Due to the fact that the impinging beam does not have a direction perpendicular or parallel to the optic axis, it will form an angle β with the direction of **k**. Using (14.37) we obtain

$$\tan \beta = \frac{(n_1^2 - n_2^2) \tan \gamma}{n_1^2 + n_2^2 \tan^2 \gamma} = \frac{(n_e^2 - n_o^2) \tan \gamma}{n_e^2 + n_o^2 \tan^2 \gamma} = 0.19, \tag{14.46}$$

then

$$\beta = 10.9°. \tag{14.47}$$

A picture of the result is shown in Fig. 14.16.

14.2 An uniaxial dielectric material is cut in form of a cube of side 3 cm. The optic axis is located at 45° with respect to the sides as shown in Fig. 14.17. The values of the permittivities are $\epsilon_o = 2.25$ and $\epsilon_e = 4$. A non-polarized plane wave is directed perpendicularly onto the left side (see Fig. 14.18). If the light beam is very thin, obtain the separation between the ordinary and extraordinary beams at the exit of the material.

Solution

To calculate the separation between the ordinary and extraordinary beams when light leaves the material, we must first obtain the angle β that forms both rays inside of the cube. Taking into consideration that $n_o \approx \sqrt{\epsilon_o}$ and $n_e \approx \sqrt{\epsilon_e}$, this angle may be calculated by means of (14.37) as follows

$$\tan \beta = \frac{(n_o^2 - n_e^2) \tan \gamma}{n_o^2 + n_e^2 \tan^2 \gamma} = \frac{(\epsilon_o - \epsilon_e) \tan \gamma}{\epsilon_o + \epsilon_e \tan^2 \gamma} = -0.28. \tag{14.48}$$

Once β is known, the distance between the beams is determined by a simple calculus (see Fig. 14.18)

$$\tan \beta = \frac{\Delta}{L} \implies \Delta = -0.84 \text{ cm}, \tag{14.49}$$

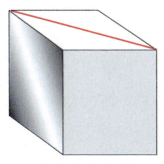

Fig. 14.17 Cube made of an anisotropic material

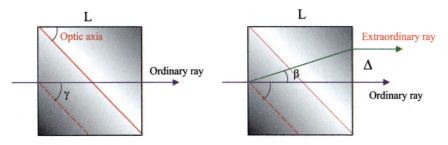

Fig. 14.18 The ordinary ray *o.r.* does not suffer any modification in its trajectory. The extraordinary ray *e.r.* forms an angle β to the *o.r*

Δ being the separation.

14.3 The Fig. 14.19 represents a system of two plane-parallel plates of quartz of indices n_o and n_e. Both plates are placed in such a way that their optic axes are perpendicular to each other. Obtain the phase difference introduced by the system to a plane monochromatic wave that falls perpendicularly upon the first sheet and passes through the second one.

Solution

The light beam that impinges the first uniaxial plate may have any polarization. We have seen in the theory that, in general, when a plane monochromatic wave reaches the face of a uniaxial specimen, two different ray vectors propagating inside the anisotropic material appear. Each one is linearly polarized but the oscillation directions of their respective electric field are perpendicular to each other. These two beams were denoted as ordinary and extraordinary waves. In the present problem the direction of the vector **k** at the entrance of the sample is perpendicular to the optic axis of the material, then we will have two superimposed beams with the same direction of propagation. However, as each wave has a different velocity (their refractive indices are different) a phase difference between them will occur. As a consequence, depending on the thickness of the plate and on the difference $(n_e - n_o)$, the light at the exit of the slab can change its polarization. In other words, the light

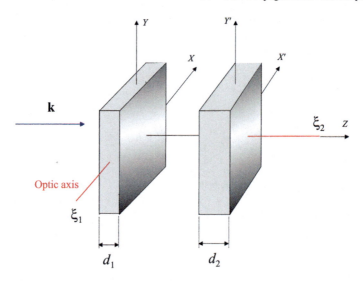

Fig. 14.19 Direction of the extraordinary ray in the material

beam will modify its phase in a quantity of (Fig. 14.19)

$$\varphi = \frac{2\pi}{\lambda_0}(n_e - n_o)d_1, \tag{14.50}$$

where d_1 is its thickness and λ_0 is the wavelength in vacuum of the monochromatic radiation employed. Depending on the specific characteristics of the beam when impinging the first sheet, the value of the phase difference φ introduced by the plate will produce emerging light that could be linear, circular or elliptical. In fact, let us suppose that the incident light is linear polarized. If $\varphi = 0, \pi, 2\pi, \ldots$ the exiting light beam is also linear, but if $\varphi = (2n+1)\frac{\pi}{2}$ it will be in general elliptical. Only in the case that the two components of the incident electric field are equal, the emerging polarization will be circular. All other possibilities, for instance when incident light is elliptical polarized, must be analyzed for each specific case.

The result obtained above holds for the beam leaving the first plate, but after that the emerging light reaches another sheet of thickness d_2, whose optic axis is located parallel to the vector **k** corresponding to the wave leaving the first plate. As this beam goes inside of this second slab in the same direction of the optic axis ξ_2, it does not suffer any modification related with its polarization state. In other words, the light beam at the exit of the second plate is the same as those emerging from the first sheet (neglecting the losses in intensity due to reflections on the boundaries-faces).

Ultimately, when the light beam passes through the first plate there is birefringence but no double refraction, because the propagation is perpendicular to the optic axis. However, when light goes through the second sheet, the beam does not suffer any change, thus there is neither double refraction nor birefringence.

Problems B

14.4 Two thin identical plates of refractive indices $n_1 = 2$ (n_e) and $n_2 = 1.5$ (n_o), and thickness 250 nm, are cut of the same uniaxial material. A linearly polarized plane wavefront of wavelength $\lambda = 500$ nm propagates along the OZ axis and impinges the first plate perpendicularly (see Fig. 14.20). The electric field of the incident field forms an angle of 45° with respect the optical axis of the first sheet. If the light beam at the exit of the first plate strikes perpendicularly the second one, obtain: (a) The polarization state of light at the exit of the system when the two optical axes are parallel to each other. (b) Idem when the two optical axes are perpendicular.

Solution

(a) The vector **k** of the light beam is perpendicular as shown in Fig. 14.20 to the optic axis of the first plate and, the impinging electric field **E** has projections along OY and OX, respectively. As a consequence inside of the slab we will have two electric fields, one vibrating parallel to ξ_1 and another perpendicular to it, but propagating in the same direction with different velocities. This difference of velocities leads to a difference between the phases of the ordinary and extraordinary waves, and then the polarization state of the emerging light may change. The phase difference introduced in the beam by the first anisotropic slab is

$$\delta_1 = \frac{2\pi}{\lambda}(n_e - n_o)d = \frac{2\pi}{500 \cdot 10^{-9}}(2 - 1)250 \cdot 10^{-9} = \frac{\pi}{2}. \quad (14.51)$$

As the two beams go out superimposed, at the exit of this first sample light has modified its polarization from linear to circular. However, before detecting this beam, the light must travel throughout the second plate whose optic axis ξ_2 is parallel to ξ_1. The situation is similar to the first one, but now the impinging light is circularly polarized. As we can imagine the result is the same, i.e. inside of the second plate

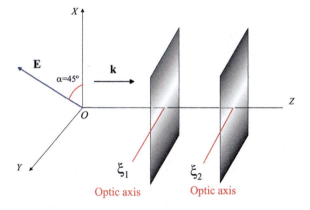

Fig. 14.20 In this set-up the optic axes are parallel to each other

each of the two electric fields sees a different refraction index, thus the change in the phase becomes

$$\delta_2 = \frac{2\pi}{\lambda}(n_e - n_o)d = \frac{\pi}{2}, \qquad (14.52)$$

and hence, the total phase difference introduced by the system of slabs is

$$\delta = \delta_1 + \delta_2 = \frac{\pi}{2} + \frac{\pi}{2} = \pi. \qquad (14.53)$$

This result shows that light travelling throughout this system suffers a change in the polarization. The beam at the beginning is linear polarized but due to the phase difference of π radians introduced by the plates, the light becomes linear polarization too, but modifying its plane of oscillation. In fact, the resulting emerging electric field forms an angle of $\alpha_e = 45°$ with OY but on the right (see Fig. 14.21).

(b) In this case the second plate is rotated in such a manner that its optic axis ξ_2 is perpendicular to ξ_1. To understand what happens we choose the same procedure as in section (a). The first stage is the same, because the first slab does not suffer any modification. This means that after light passes throughout it a change of the phase in $\frac{\pi}{2}$ radians accounts. After that the beam strikes the second plate perpendicularly to the optic axis ξ_2, but in this case the location of it is rotated $\frac{\pi}{2}$ with respect to the first set-up. This fact leads to a phase modification different from before. In effect, the change in the phase when the optic axis is placed as drawn in Fig. 14.22 is

$$\delta_2 = \frac{2\pi}{\lambda}(n_o - n_e)d. \qquad (14.54)$$

Observe that the difference of indices is changed, i.e. $(n_o - n_e)$ instead of $(n_e - n_o)$ (see (14.52)). Combining (14.51) and (14.54) we have

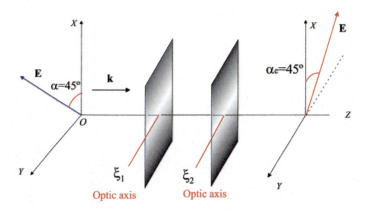

Fig. 14.21 The electric field at the exit forms an angle of $\alpha_e = 45°$ on the right of the OX axis

Fig. 14.22 This figure represents two uniaxial plates with their respective optic axes perpendicular to each other

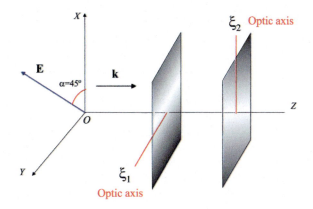

$$\delta = \delta_1 + \delta_2 = \frac{2\pi}{\lambda}(n_e - n_o)d + \frac{2\pi}{\lambda}(n_o - n_e)d = 0, \qquad (14.55)$$

hence we do not have any change of the polarization of light when the beam travels through the system.

14.5 A beam of linearly polarized white light is incident perpendicularly onto a quartz plate of thickness 0.865 nm. Such a plate is cut so that its optic axis lies parallel to its plane. The electric field vibrates along a direction parallel to the plane sheet forming 45° with the OX coordinate axis. The refraction indices of the material are $n_1 = 1.5533$ (n_e) and $n_2 = 1.5442$ (n_o). The emerging beam passes through a polarizer whose transmission axis is perpendicular to the vibration direction of the electric field at the input of all the system. By neglecting the variation of the refractive index with wavelength, what wavelengths between 600 and 650 nm are absent in the transmitted light beam?

Solution

The only possibility for obtaining extinction of light when the radiation crosses the polarizer is that the light beam emerging from the uniaxial plate does not changes its state of polarization. In effect, if that happens the radiation remains linearly polarized in the same plane as it impinges at the entrance of the slab, and therefore, as the electric field on the analyzer would be perpendicular to its transmission axis no light will be detected after the polarizer (Fig. 14.23).

The above reasoning may be expressed as a condition on the phase of the travelling beam by imposing that the phase difference be

$$\delta = \frac{2\pi}{\lambda}(n_e - n_o)d = 2\pi N, \qquad (14.56)$$

N being an integer. Rearranging (14.56) for N, we have

$$N = \frac{(n_e - n_o)d}{\lambda}. \qquad (14.57)$$

Fig. 14.23 Anisotropic plate and polarizer

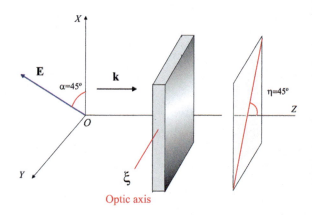

Introducing the first wavelength into (14.57) we can write

$$N_1 = \frac{(n_e - n_o)d}{\lambda} = 13.1. \tag{14.58}$$

As we must seek radiations extincted in the interval $600 < \lambda < 650$, the value $N_1 = 13.1$ represents the limit, then the next valid number must fulfill $N_1 < 13.1$ (observe that the smaller the wavelength the bigger the number N). In the same wave, we impose the same condition to $\lambda = 650$ obtaining

$$N_2 = \frac{(n_e - n_o)d}{\lambda} = 12.1. \tag{14.59}$$

This number is again the limit for $\lambda = 650$, thus the next (for a smaller λ) must verify $N_2 > 12.1$. Combining the results (14.58) and (14.59), the interval of wavelengths we are not able to detect after crossing the polarizer correspond to the interval between N_2 and N_1, then $N = 13$. To determine the for what λ it corresponds we employ (14.56) for $N = 13$, obtaining

$$\lambda = \frac{(n_e - n_o)d}{N} = 605 \text{ nm}. \tag{14.60}$$

14.6 A plate of a negative uniaxial material is located as shown in Fig. 14.24, ξ being the optic axis. The refraction indices of this material are $n_1 = 2$ and $n_2 = 3$. A monochromatic linear polarized radiation beam is directed perpendicularly to the slab, whose electric field forms an angle α with the OY axis. Calculate: (a) The angle α in order that the intensity of the ordinary ray in the material be the half of the corresponding to the extraordinary. (b) The minimum thickness of the plate for which, if $\alpha = 45°$, light exits the system circular polarized.

Fig. 14.24 Direction of the extraordinary ray in the material

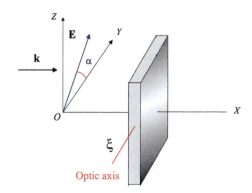

Solution

The anisotropic material is negative which means that $n_e < n_o$, and therefore we have $n_o = 3$ $n_e = 2$. As we have already commented before, when the beam propagates inside of the plate two linearly polarized wave fields appear, one of them called ordinary wave and the other one labelled as extraordinary wave. Inside the slab each wave travels at different velocities because they see unlike refraction index. The projections of the electric field of the impinging wave over the OY axis and along the optic axis ξ (see Fig. 14.24) depend on the angle α. It means that changing this angle we can modify the projections and, as a result, the intensities associated with the ordinary and extraordinary waves will also be changed.

Let us denote the field amplitudes of the ordinary and the extraordinary waves E_o and E_e, respectively. Taking into consideration that both are linearly polarized, their respective intensities inside of the plate are

$$I_o = \frac{1}{2} c n_o \epsilon_0 E_o^2 = \frac{1}{2} c n_2 \epsilon_0 E_o^2 \qquad (14.61)$$

for the ordinary, and

$$I_e = \frac{1}{2} c n_e \epsilon_0 E_e^2 = \frac{1}{2} c n_1 \epsilon_0 E_e^2 \qquad (14.62)$$

for the extraordinary wave. On the other hand, the electric field projections may be related with the angle α as follows (see Fig. 14.25)

$$\tan \alpha = \frac{E_o}{E_e}. \qquad (14.63)$$

The relation between the intensities must be $I_o = \frac{1}{2} I_e$, thus dividing (14.61) by (14.62) we have (Fig. 14.26)

$$\frac{I_o}{I_e} = \frac{\frac{1}{2} c n_2 \epsilon_0 E_o^2}{\frac{1}{2} c n_1 \epsilon_0 E_e^2} = \frac{n_2 E_o^2}{n_1 E_e^2}, \qquad (14.64)$$

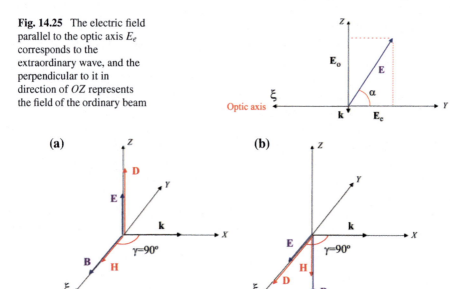

Fig. 14.25 The electric field parallel to the optic axis E_e corresponds to the extraordinary wave, and the perpendicular to it in direction of OZ represents the field of the ordinary beam

Fig. 14.26 a Fields corresponding to the ordinary wave. b Idem for the extraordinary way

hence

$$\frac{I_o}{I_e} = \frac{n_2}{n_1} \tan^2 \alpha = \frac{1}{2}. \tag{14.65}$$

Rearranging this last equation we obtain

$$\frac{3}{2} \tan^2 \alpha = \frac{1}{2} \longrightarrow \alpha = \tan^{-1} \frac{1}{\sqrt{3}} = 30°. \tag{14.66}$$

14.7 A linearly polarized monochromatic radiation of wavelength $\lambda = 400$ nm, and intensity I_0, impinges onto a plate of a uniaxial material whose optic axis is parallel to OX. The thickness of the material is $d = 7$ mm, and the indices $n_1 = 2$ and $n_2 = 1$, respectively. After the plate, a polarizer whose transmission axis as shown in the attached figure is placed. If the polarization plane of the incident light forms 15° with the OY axis, obtain the light intensity registered on a detector placed on the right side of the polarizer (Fig. 14.27).

Solution

To know what happens after the analyzer we must first investigate the effect of the anisotropic slab when light passes through it. With this aim let us employ (14.42) to determine the phase difference between the extraordinary and the ordinary wave

$$\delta = \frac{2\pi}{\lambda}(n_e - n_o)d = \frac{2\pi}{\lambda}(n_1 - n_2)d = 17{,}500\, 2\pi = 2\pi N, \tag{14.67}$$

thus $N = 17{,}500$, which is an even number. As a result, light does not change its state of polarization, i.e. it remains linear polarized when leaving the anisotropic plate. After that, the beam reaches the polarizer whose transmission axis forms $\eta = 45°$ with OY'. Therefore, the electric field when the light passes through it is the corresponding to the projection of E_0 over the direction coinciding with the polarized axis,

$$E_p = E_0 \cos(45 - 15) = E_0 \cos(30) = \frac{\sqrt{3}}{2} E_0. \qquad (14.68)$$

But in the problem the question is about the intensity of the resulting beam, then we must relate (14.68) with the intensity I_0 of the light before reaching the plate. To calculate it let us write this intensity as a function of the electric field E_0

$$I_0 = \frac{1}{2} c n \epsilon_0 E_0^2, \qquad (14.69)$$

which is known. The intensity after the analyzer has a similar expression, but it depends on E_p (14.68)

$$I_p = \frac{1}{2} c n \epsilon_0 E_p^2. \qquad (14.70)$$

Introducing (14.68) into (14.69) yields

$$I_p = \frac{1}{2} c n \epsilon_0 \left(\frac{\sqrt{3}}{2} E_0 \right)^2. \qquad (14.71)$$

and dividing this equation by (14.68) we obtain

$$\frac{I_p}{I_0} = \frac{3}{4} \implies I_p = \frac{3}{4} I_0, \qquad (14.72)$$

which, obviously, is smaller than the intensity I_0.

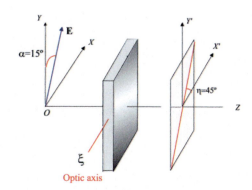

Fig. 14.27 Direction of the extraordinary ray in the material

Problems C

14.8 In the Fig. 14.28 P_1 represents a polarizer whose transmission axis forms an angle of 30° with respect to the vertical axis OX. L is a plate of a uniaxial crystal with thickness d unknown. The optic axis is parallel to its faces. P_2 is an analyzer with a transmission axis that makes an angle φ with respect to the vertical axis. The system is illuminated perpendicularly with a radiation composed by two wavelengths of $\lambda_1 = 400\,\text{nm}$ and $\lambda_1 = 600\,\text{nm}$, respectively. Calculate the minimum thickness d of the plate up to 1 mm so that the analyzer P_2 is able to completely avoid that one of the wavelengths crosses P_2. Explain the value of φ for all the solutions found. Assume that $n_1 - n_2 = 0.01$ for both wavelengths.

Solution

The only possibility for extinguishing with one polarizer any radiation, after light crosses the anisotropic material, is that the beam at the exit of the plate must be linearly polarized. For this reason, in order to ensure that the polarization of light that impinges the analyzer is linear, we will first impose the condition

$$\delta = \frac{2\pi}{\lambda}(n_1 - n_2)d = N\pi. \tag{14.73}$$

Equation (14.73) guaranties a linear state of polarization, but depending on N being either even or odd, the plane of polarization may change.

To face this problem it is important to note that, what we are going to seek is the minimum thickness up to 1 mm for detecting only one of the two radiations (400 or 600 nm) that combined form the light beam. In this way, we will first use (14.73) setting $d = 1\,\text{mm}$, which is the limit. This will give us the value of N that fulfills (14.73) for $d = 1\,\text{mm}$. Introducing the corresponding data we have

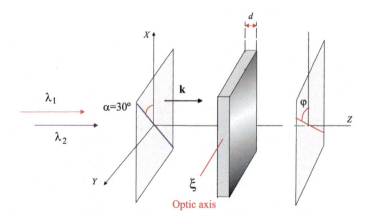

Fig. 14.28 System composed by an anisotropic plate of unknown thickness, one polarizer and one analyzer

$$\delta = \frac{2\pi}{400 \cdot 10^{-9}}(0.01) \, 1 \cdot 10^{-3} = 2\pi \, 25 = N\pi \implies N = 50. \quad (14.74)$$

This result means that there is a change of the phase of light at the exit of the material. However, as N is an even number it is equivalent to say, that no modification of the state of polarization accounts (observe that $\delta = 2\pi. 25$) if the sample would have a thickness of 1 mm. But the specimen must be thicker than $d = 1$ mm, therefore if we substitute into (14.73) the next integer number, i.e. $N = 51$, leaving d as unknown, we can find out the value of d up to 1 mm that verifies the condition required. Note that with $N = 51$ we conserve a linear polarization state of light, which is necessary for guaranteeing the extinction of a wavelength when light crossing the analyzer for some location of it (until now unknown). However, in viewing polarization only, for $N = 51$ it is equivalent to say that $\delta = \pi$, light modifies its plane of vibration. In other words, light impinges on the sheet perpendicularly with its electric field forming an angle $\alpha = 30°$ with the vertical axis, and when the beam exits the material $\alpha = -30°$ (with respect to OX). Therefore, setting $N = 51$ into (14.73) leads to

$$\delta = 51\pi = \frac{2\pi}{\lambda}(n_1 - n_2)d \implies d = \frac{51\pi\lambda}{2\pi(n_1 - n_2)} = 1.02 \text{ mm}. \quad (14.75)$$

Now, let us make the same calculation for the radiation of $\lambda = 600$ nm. To do this, we employ (14.73) again, but for this wavelength obtaining

$$\delta = \frac{2\pi}{600 \cdot 10^{-9}}(0.01) \, 1 \cdot 10^{-3} = 2\pi \, 16.7 = N\pi \implies N = 33.33. \quad (14.76)$$

This result shows that for 1 mm and $\lambda = 600$ light changes the state of polarization, because the phase difference is not an integer number of π. As far as the polarization is concerned, (14.76) is equivalent to say that the beam suffers a phase change of $\delta = 2\pi \, 0.7 \approx 1.33\pi$, which leads to an elliptically polarized light. However it does not matter with our problem. In fact, as we need that the light after crossing the plate be linear polarized, we can impose this condition by introducing $N = 34$ into (14.73), and considering d unknown. Thus it holds

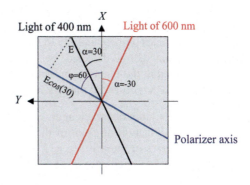

Fig. 14.29 Electric fields for both lights

Fig. 14.30 Electric fields for both lights

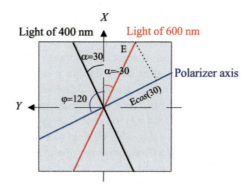

$$d = \frac{34\pi\lambda}{2\pi(n_1 - n_2)} = 1.02 \text{ mm}, \tag{14.77}$$

which gives a value for d identical that obtained for $\lambda = 400$. Nevertheless, as we have used $N = 34$ (even number) in this case, neither the state of polarization of the beam nor its plane of oscillation change.

Once we know the characteristics of the light for the two radiations when passing through the anisotropic plate, we can examine how the polarizer must be located in order to avoid one of the wavelengths crossing throughout. In fact, for $d = 1.02$ mm light comes linearly polarized in both cases, but their respective electric fields do not vibrate on the same plane. The beam of $\lambda = 600$ nm oscillates forming an angle of $\alpha = -30°$ with respect to axis OX, then if we place the analyzer at $\varphi = 60°$ (see Fig. 14.28), this radiation will be eliminated when seeing the phenomenon after the polarizer. In this case the polarization direction and the analyzer axis are perpendicular. The light of $\lambda = 400$ nm will be observed, but not in its totality. As the transmission axis of the analyzer forms $\varphi = 60°$ with OX, the light intensity of 400 nm we will detect corresponds to those of the projection of the electric filed **E** over the polarization axis. Thus, the intensity of the wave leaving the analyzer will be $I_p \sim I_1 \cos^2 30$ (see Fig. 14.29), I_1 being the intensity corresponding to $\lambda = 400$ nm once light has crossed the anisotropic sheet. In the same way, we can investigate what happens with the radiation of $\lambda = 400$ nm. This light goes out from the plate linearly polarized but its vibration axis corresponds to $\alpha = 30°$, therefore if we locate the analyzer at $\varphi = 120°$ (see Fig. 14.30) we note that the electric field of this light oscillates perpendicularly to the analyzer axis. By this procedure no beam of $\lambda = 400$ nm will be detected, but radiation of $\lambda = 600$ nm does. In fact, for this configuration, the intensity measured is again $I_p \sim I_2 \cos^2 30$, where I_2 is the intensity of 600 nm observed directly behind the uniaxial plate.

14.9 A sodium lamp of wavelength λ radiates spherical waves with circular right-hand polarization. In front of the lamp, a lens is placed in order to generate plane waves of intensity $I_0 = 2$ Wm^{-2}. The beam after the lens impinges a uniaxial plate of quartz perpendicularly. The sample is cut parallel to the optic axis and its indices and thickness are $n_1 = 1.5533$ and $n_2 = 1.5442$, and $d = 5.663$ mm, respectively. At the output of this plate, the light crosses a polarizer whose transmission axis is parallel to the aforementioned optic axis. (a) Obtain the polarization state of light after crossing the plate. (b) Calculate the intensity that may be registered at the exit of the polarizer. (c) If the radiation that emerges from the polarizer reaches a sheet of the figure attached under an angle of $\theta = 63.4°$, find the refractive index that the plate has to be in order that no reflected light beam appears (Fig. 14.31).

Solution

(a) When the light passes through the uniaxial plate it suffers a change in phase which may be calculated by using (14.42)

$$\delta = \frac{2\pi}{\lambda}(n_1 - n_2)d = 2\pi\, 87.5 = 2\pi N. \tag{14.78}$$

From the point of view of the polarization, this value obtained means that the change in the phase of the perturbation is equivalent to $\delta = \pi$, and therefore, the light at the exit of the anisotropic plate becomes circular left-hand polarized.

(b) The intensity of the radiation after the slab is that corresponding to circularly polarized light. If we neglects the reflections on the surface boundaries of the sample, the expression for the intensity in the most general case may be expressed as follows (see Problem 12.13)

$$I_0 = \frac{1}{2}cn\epsilon_0(E_x^2 + E_y^2), \tag{14.79}$$

where E_x and E_y are the components of the electric field over the coordinate axes OX and OY, respectively. In the present problem as the light is circular polarized, $E_x = E_y$, and then

$$I_0 = \frac{1}{2}cn\epsilon_0(2E_x^2) = cn\epsilon_0 E_x^2. \tag{14.80}$$

Now, this radiation impinges the analyzer, which only allows passing the projection of the electric field over its axis. Thus, the intensity is

$$I_p = \frac{1}{2}cn\epsilon_0 E_x^2, \tag{14.81}$$

and the relation between this last equation and (14.80) leads to

$$\frac{I_s}{I_0} = \frac{1}{2} \Longrightarrow I_s = \frac{1}{2}I_0 = 1\,\text{Wm}^{-2}. \tag{14.82}$$

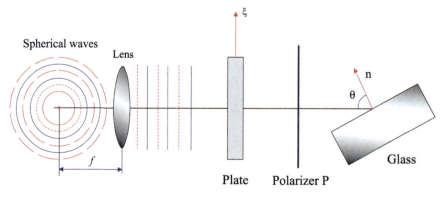

Fig. 14.31 Set-up

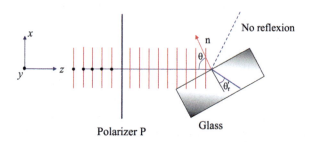

Fig. 14.32 Observe that the electric field of the light before impinging the slab has two components, one along OX and the other over OY. After the pass through the analyzer only the component E_x remains, but when light reaches the glass under conditions of Brewster's angle it losses this component, thus no light is reflected by the surface (we only have a refracted beam at θ'_r.)

(c) In the former Chapter we studied the Brewster angle.[11] There we defined it as the angle for which the reflected light on a surface was only polarized perpendicularly to the plane of incidence (direction OY, see Fig. 14.32). In the present case, because of the polarizer, we only have the component over OX, then when reaching the glass no light is reflect but it does refracted. Applying (14.7), we have

---

[11] It is interesting to note that the definition of Brewster's angle given in Chap. 13 and applied in this problem is only valid for dielectric materials. For planar metallic surfaces, due to the nature of the refraction index ($n(\omega) = n_1(\omega) + in_2(\omega)$), the reflectance at these surfaces for p-waves (parallel to the incidence plane) is nonzero at the Brewster angle. However, in order to address some important features of the reflected light, two different definitions account. So when the modulus of the ratio $\frac{r_p}{r_s} = \frac{r_\parallel}{r_\perp}$ reaches a minimum, the corresponding angle at which it occurs is called the *second Brewster's angle*. The second possibility refers to the incidence angle at which $r_\parallel$ is a minimum. Such an angle is known as the *pseudo-Brewster's angle*. For more details the reader may consult the following references: [3, 5, 47]. A special case of Brewster's angle is referred to rough surfaces. In this case for dielectric interfaces, there is a dip in the p-waves of the scattered light at a specific angle which depends on the dielectric properties of the medium. When studying metallic random surfaces there exist no real solution to the equation which gives Brewster's scattering angle (see, for instance, [57]).

$$\tan\theta_B = \frac{n_g}{n_0} \Longrightarrow n_g = n_0 \tan\theta_B = 1 \cdot \tan(63.4) \approx 2. \quad (14.83)$$

14.10 The system represented in the Fig. 14.33 consists of a polarizer whose transmission axis is parallel to OZ, a crystal of quartz with dielectric constants ϵ_1 and ϵ_2, and optic axis ξ forming an angle γ with OX, and a screen P. A thin light beam impinges perpendicularly to the polarizer. (a) How many light points we will see on the screen P? (b) Obtain the angle that Poynting's vector of the extraordinary ray will form with the optic axis ξ as a function of ϵ_1 and ϵ_2.

Solution

(a) When a plane electromagnetic radiation impinges an uniaxial slab, in principle, two independent waves orthogonally polarized to each other appear, whose phase velocities are different. In other words it means that, for any direction of the vector **k** we have two possible values of the wave-number k. Taking this fact into account, for understanding what happens when a beam strikes an anisotropic plate, we can first decompose the electric field at the entrance of the material into two components, one parallel to the optic axis and the another one perpendicular to it. In the case of this problem, before the beam reaches the slab it passes through a polarizer, then at the exit of the polarizer we only have one component of the electric field, which corresponds to the projection over the OZ axis (see Fig. 14.33), which in turn lies on

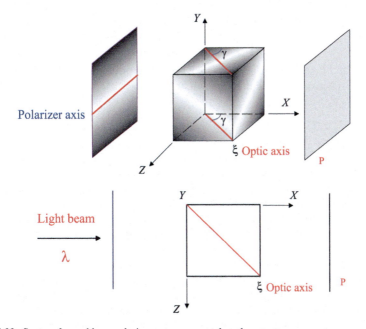

Fig. 14.33 System formed by a polarizer, a quarz crystal, and a screen

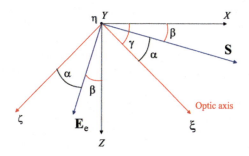

Fig. 14.34 Direction of the extraordinary ray in the material

the plane formed by $O\xi$ and $\mathbf{k}$. As a result, only one wave propagates throughout the material, and therefore, only one point will be shown on the screen P.

(b) As we have seen in the theory, in anisotropic media $\mathbf{k}$ and $\mathbf{E}$ are, in general, not perpendicular to each other. This means that the direction of propagation of the wavefront $\mathbf{k}$ does not coincide with the propagation direction of the energy, which is represented by the Poynting vector $\mathbf{S}$. For obtaining a solution to the question, we will use the principal axes of the material $O\xi\eta\zeta$ with $OXYZ$, simultaneously (see Fig. 14.34). As we can observe, the Poynting vector of the extraordinary ray forms an angle β with OX, and α with $O\xi$. The electric field is perpendicular to $\mathbf{S}$, but not to $\mathbf{k}$, therefore the angle between $\mathbf{E}_e$ and OZ is the same as that between $\mathbf{S}$ and $\mathbf{k}$. Focusing first our attention to the electric field, we can obtain a relation between its components and the angle α as follows (see Fig. 14.35)

$$\tan\alpha = \frac{E_\xi}{E_\zeta}, \tag{14.84}$$

where E_ξ and E_ζ are the projections of $\mathbf{E}_e$ over the principal axes of the crystal $O\xi$ and $O\zeta$, respectively. Along these axes it holds that

$$D_\xi = \epsilon_1 E_\xi \tag{14.85}$$

and

$$D_\zeta = \epsilon_2 E_\zeta, \tag{14.86}$$

then introduction of these values into (14.84) leads to

$$\tan\alpha = \frac{E_\xi}{E_\zeta} = \frac{\epsilon_2 D_\xi}{\epsilon_1 D_\zeta} = \frac{n_2^2 D_\xi}{n_1^2 D_\zeta}. \tag{14.87}$$

Now, in order to eliminate D_ξ and D_ζ from (14.86) we will use the diagram shown in Fig. 14.35. This picture represents the vectors $\mathbf{E}_e$, $\mathbf{D}$, and $\mathbf{S}$ for the extraordinary wave together with the corresponding ellipse (in two dimensions). As we can observe, it is possible to project the components of $\mathbf{D}$ along $O\xi$ and $O\zeta$ (D_ξ and D_ζ) over the

Fig. 14.35 Scheme for the vectors **E**, **D**, and **S**

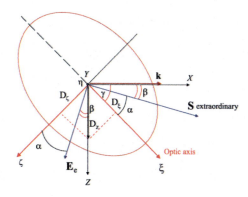

OZ axis of the $OXYZ$ coordinate frame. Then, the new components of **D** are[12] (see Fig. 14.35)

$$D_\xi = D_z \sin\gamma, \tag{14.88}$$

and

$$D_\zeta = D_z \cos\gamma. \tag{14.89}$$

Substituting these equations into (14.86) we obtain

$$\tan\alpha = \frac{\epsilon_2 D_z \sin\gamma}{\epsilon_1 D_z \cos\gamma} = \frac{\epsilon_2}{\epsilon_1}\tan\gamma = \frac{n_2^2}{n_1^2}\tan\gamma. \tag{14.90}$$

This last equation allows us to relate the angle γ formed by the optic axis with OX (direction of **k**-see Fig. 14.35), and the angle α, but it does not give information about β. However, from the Fig. 14.35 we see that $\alpha + \beta = \gamma$, thus we can write

$$\tan(\gamma - \beta) = \frac{\epsilon_2}{\epsilon_2}\tan\gamma, \tag{14.91}$$

and then

$$\beta = \gamma - \tan^{-1}\left(\frac{\epsilon_2}{\epsilon_1}\tan\gamma\right). \tag{14.92}$$

---

[12]The vectors appearing in this figure are not correctly scaled. This is only a scheme for understanding the problem. On the other hand, the semi-axes of the ellipse could be the opposite depending on the optical characteristics of the material-positive or negative.

Appendix A
Matlab Programs

Matlab Program to Calculate Surfaces, Isolines and Gradients

```
figure
[x,y] = meshgrid(-2:.1:2,-2:.1:2);
z = x .* exp(-x.^2 - y.^2); %Function
[px,py] = gradient(z,.1,.1);
surfc(x,y,z)
figure
contour(x,y,z), hold on
quiver(x,y,px,py), hold off, axis image
```

Matlab Program to Calculate Resistance Using Slices

```
rho=1 % Resistivity
%Width of trapezoid
L=6
%Heights of trapezoid
H1=3
H2=5
slop=(H2-H1)/L
%Number of slices
m=10
%Width of each slice
li=L/m
%Heights
h(1:m)=0;
for i=1:m
h(i)=H1+slop*[(i*li)+(i-1)*li]/2;
end
h
%Resistances
R=0;
```

```
for i=1:m
R=R+li/h(i);
end
R=R*rho
```

Matlab Program to Calculate Resistance Using Tubes

```
rho=1 % Resistivity
%Width of trapezoid
L=6
%Heights of trapezoid
H1=3
H2=5
%Slope
sup=(H2-H1)/L;
%Number of tubes
n=8
%
h(1:2,1:n)=0;
l(1:2,1:n)=0;
%Height 1
h(1,1:n)=H1/n
%Length 1
lados(1:2,1:n+1)=0;
lados(1,1)=L;
for i=1:n
lados(1,i+1)=L-i*L/n;
l(1,i)=(lados(1,i+1)+lados(1,i))/2;
end
%Length 2
lados(2,1)=0;
for i=1:n
vert=i*(H2-H1)/n;
horiz=L-lados(1,i+1);
lados(2,i+1)=sqrt(vert^2+horiz^2);
l(2,i)=(lados(2,i+1)+lados(2,i))/2;
end
%Height 2
hdiag=sqrt(H1^2+L^2)/n %Diagonal
hvert=H2/n %Right
a=lados(2,n+1)
b=hvert
c=lados(2,n)
d=hdiag
h(2,1:n)=sqrt(4*(a-c)^2*d^2-(d^2+(a-c)^2-b^2)^2)/(2*(a-c)) %Height of a trapezoid
```

Appendix A: Matlab Programs

```
%Resistances
l
h
Re=0;
Reinv=0;
R(1:n)=0;
for i=1:n
R(i)=l(1,i)/h(1,i)+l(2,i)/h(2,i); %Series
Reinv=Reinv+1/R(i); %Parallel
end
Re=rho/Reinv
```

Appendix B
Electric and Magnetic Properties of Several Materials

See Tables B.1, B.2 and B.3.

Table B.1 Resistivity values and temperature coefficient of various common materials

Material	η (Ωm)	α (K^{-1})
Metals		
Iron	9.71×10^{-8}	0.0065
Aluminum	2.65×10^{-8}	0.0043
Copper	1.67×10^{-8}	0.0039
Silver	1.59×10^{-8}	0.0041
Gold	2.35×10^{-8}	0.004
Nickel	6.84×10^{-8}	0.0069
Mercury	95.8×10^{-8}	0.0009
Tungsten	5.51×10^{-8}	0.0045
Alloys		
Nichrome	100.0×10^{-8}	0.0004
Constantan	49.0×10^{-8}	0.000008
Manganin	48.2×10^{-8}	0.000002

(continued)

Table B.1 (continued)

Material	η (Ωm)	α (K^{-1})
Semiconductors		
Germanium	0.46	−0.048
Silicon	4300	−0.075
Graphite	1.4×10^{-5}	−0.0005
Insulators		
Sulfur	2×10^{15}	
Quartz (SiO$_2$)	1×10^{14}	
Wood	10^8–10^{11}	
Glass	10^{10}–10^{14}	
Lucite	$>10^{13}$	
Mica	10^{11}–10^{15}	
Diamond	10^{13}	

Table B.2 Dielectric constant and dielectric strength of various common materials

Material	ε_r	E_{max} (MV/m)
Air (standard conditions T, P)	$1.00058986 \pm 0.00000050$	3.0
Alumina	10	13.4
Benzene	2.3	163
Distilled Water (20 °C)	80.1	65–70
Glass	4–10	9.8–13.8
Mica	4.5–8	118
Neoprene rubber	6.7	15.7–26.7
Paper	3.7	16
Polyethylene	2.2	18.9–21.7
Polystyrene	2.56	19.7
PTFE (Teflon, Extruded)	2.1	19.7
PTFE (Teflon, Insulating Film)	2.1	60–173
Quartz	3.3	8
Silicone oil	2.5	10–15

Appendix B: Electric and Magnetic Properties of Several Materials

Table B.3 Relative magnetic permeability of various common materials

Sustancia	μ_r
Bismuth	0.99983 (diamagnetic)
Copper	0.9999906 (diamagnetic)
Silver	0.9999736 (diamagnetic)
Lead	0.9999831 (diamagnetic)
Water	0.99999 (diamagnetic)
Air	1.00000036 (paramagnetic)
Aluminium	1.000021 (paramagnetic)
Platinum	1.000265 (paramagnetic)
Palladium	1.0008 (ferromagnetic)
Cobalt	250 (ferromagnetic)
Níckel	600 (ferromagnetic)
Mild Steel (0.2 C)	2000 (ferromagnetic)
Iron (0.2 impurity)	5000 (ferromagnetic)
Permalloy 78 (78.5 Ni)	100000 (ferromagnetic)
Mu-metal (75 Ni, 5 Cu, 2 Cr)	100000 (ferromagnetic)

Bibliography

1. Alexeev, A.I.: Recueil de problèmes d'electrodynamique classique. Mir, Moscou (1980)
2. Alonso, M.y, Finn, E.J.: Física. Vol. II: Campos y ondas. Addison-Wesley Iberoamericana, Argentina (1986)
3. Alsamman, A., Azzam, R.M.A.: Difference between the second-Brewster and pseudo-Brewster angles when polarized light is reflected at a dielectric-conductor interface. J. Opt. Soc. Am. **27**, 1156–1161 (2010)
4. Artsimovich, L.A., Lukyanov, S.Yu.: Motion of Charged Particles in Electric and Magnetic Fields. Mir, Moscow (1980)
5. Azzam, R.M.A.: Explicit equations for the second Brewster angle of an interface between a transparent and an absorbing medium. J. Opt. Soc. Am. **73**, 1211–1212 (1983)
6. Birss, R.R.: Surface stress in the non-uniform effect in magnetostriction. Proc. Phys. Soc. **90**, 453–457 (1967)
7. Boozer, A.: Physiscs of magnetically confined plasmas. Rev. Mod. Phys. **76**, 1071–1141 (2005)
8. Born, M., Wolf, E.: Principles of Optics. Cambridge University Press, UK (1999)
9. Bozorth, R.M.: Ferromagnetism. Van Nostrand, Toronto (1951)
10. Bracewell, R.N.: The Fourier Transform and its Applications. MacGraw-Hill, Tokyo (1978)
11. Brenner, E., Javid, M.: Análisis de Circuitos Eléctricos. Ediciones del Castillo S.A, Madrid (1966)
12. Bronstein, J.N., Semendjajew, K.A.: Taschenbuch der Mathematik (Teil I). B.G. Teubner Verlagsgesellschaft, Stuttgart (1991)
13. Cabrera, J.M., López, F.J., Agulló, F.: Óptica electromagnética. Addison-Wedsley Iberoamericana, Wilmington (1993)
14. Carter, R.G.: Electromagnetism for Electronic Engineers. Chapman & Hall, London (1992)
15. Chapman, S.J.: Electric Machinery and Power System Fundamentals. McGraw-Hill, Boston (2002)
16. Cheng, D.K.: Fundamentals of Engineering Electromagnetics. Addison-Wesley, Massachsetts (1993)
17. Chikazumi, S.: Physics of Magnetism. Krieger, Florida (1978)
18. Conway, J.T.: Exact solutions for the magnetic fields of axisymmetric solenoids and current distributions. IEEE Trans. Magn. **37**, 2977–2988 (2001)
19. Cullity, B.D.: Introduction to Magnetic Materials. Addison-Wesley, Massachusetts (1972)
20. Dalven, R.: Introduction to Applied Solid State Physics. Plenum Press, New York (1981)
21. do Carmo, M.P.: Differential Geometry of Curves and Surfaces. Prentice-Hall, Inc., USA (1976)

22. Dubroff, R.E., Marshall, S.V., Skitek, G.G.: Electromagnetic Concepts and Applications. Prentice-Hall, London (1996)
23. Durand, E.: Électrostatique. Tome I. Masson, Paris (1966)
24. Durand, E.: Électrostatique. Tome III. Masson, Paris (1966)
25. Durand, E.: Magnétostatique. Masson, Paris (1968)
26. Du-Xing, C., Pardo, E., Sánchez, A.: Demagnetizing factors of rectangular prisms and ellipsoids. IEEE Trans. Magn. **38**, 1742–1752 (2002)
27. Edminister, J.A., Nahvi, M.: Circuitos Eléctricos. McGraw-Hill, Schaum, Madrid (1997)
28. Feynman, R., Leighton, R., Sands, M.: Física, vol. 12. Addison-Wesley Iberoamericana, Argentina (1987)
29. Flemming, W.: Functions of Several Variables. Springer, New York (1977)
30. Fouillé, A.: Electrotecnia para Ingenieros. Aguilar S.A, Madrid (1977)
31. Galindo, A., Abellanas, L.: Métodos Matemáticos de la Física. Análisis complejo. Publicaciones de la Universidad de Zaragoza, Zaragoza (1973)
32. Gascón, F.: Electromagnetismo. Servicio de Publicaciones de la Fundación Gómez Pardo, Madrid (1995)
33. Gascón, F., Bayón, A., Medina, R., Porras, M.A., Salazar, F.: Problemas de Electricidad y Magnetismo. Pearson Prentice-Hall, Madrid (2004)
34. Giancoli, D.C.: Physics: Principles with Applications. Prentice Hall, London (1995)
35. Gil, S., Saleta, M.E., Tobia, D.: Experimental study of the Neumann and Dirichlet boundary conditions in two-dimensional electrostatic problems. Am. J. Phys. **70**, 1208–1213 (2002)
36. Gradshteyn, I.S., Ryzhik, I.M.: Table of Integrals, Series, and Products. Academic Press Inc, Boston (1994)
37. Grant, I.S., Phillips, W.R.: Electromagnetism. Wiley, Chichester (1998)
38. Guenther, R.: Modern Optics. Wiley, New York (1990)
39. Haberman, R.: Elementary Applied Partial Differential Equations. Prentice-Hall, London (1998)
40. Hakobyan, D., Brasselet, E.: Left-handed optical radiation torque. Nature Photon. **8**, 610 (2014)
41. Hammond, P., Sykulski, J.K.: Engineering Electromagnetics: Physical Processes and Computation. Oxford Science Publications, Oxford (1994)
42. Harrington, R.F.: Introduction to Electromagnetic Engineering. Dover, New York (2003)
43. Harris, J.W., Stocker, H.: Handbook of Mathematics and Computational Science. Springer, New York (1988)
44. Hecht, E., Zajac, A.: Optics. Addison-Wesley, Reading (1974)
45. Hernando, A., Rojo, J.M.: Física de los materiales magnéticos. Síntesis, Madrid (2001)
46. Hook, J.R., Hall, H.E.: Solid State Physics. Wiley, Chichester (1999)
47. Hüttner, B.: On Brewster's angle of metals. J. Appl. Phys. **78**, 4799–4801 (1995)
48. Imry, J.D., Webb, R.A.: Interferencias cuánticas y efecto Aharonov-Bohm, Investigación y Ciencia, pp. 28–35. Scientific American (1989)
49. Irwin, J.D.: Análisis Básico de Circuitos en Ingeniería. Prentice Hall, México (1997)
50. Jackson, J.D.: Classical Electrodynamics. Wiley, New York (1999)
51. Jackson, J.D.: A curious and useful theorem in two-dimensional electrostatics. Am. J. Phys. **67**, 107–115 (1999)
52. Jackson, R.H.: Off-axis expansion solution of Laplace's equation: application to accurate and rapid calculation of coil magnetic fields. IEEE Trans. Electron. Devices. **46**, 1050–1062 (1999)
53. Jefimenko, O.D.: Electricity and Magnetism. Electret Scientific Company Star City, West Virginia (1989)
54. Joseph, R.I., Schlömann, E.: Demagnetizing field in non-ellipsoidal bodies. J. Appl. Phys. **36**, 1579–1584 (1965)
55. Joss, G.: Theoretical Physics. Dover, New York (1986)
56. Kajorndejnukul, V., Sukhov, S., Haefner, D., Dogariu, A., Agarwal, G.: Surface induced anisotropy of metal-dielectric composites and anomalous spin Hall effect. Opt. Lett. **37**, 3036–3038 (2012)

57. Kawanishi, T.: Brewster's scattering angle in scattered waves from slightly rough metallic surfaces. Phys. Rev. Lett. **84**, 2845–2848 (2000)
58. Kevorkian, J.: Partial Differential Equations. Springer, New York (2000)
59. Kraus, J.D., Fleisch, D.A.: Electromagnetismo con Aplicaciones. McGraw Hill, México (2000)
60. Lampard, D.G.: A new theorem in electrostatics with applications to calculable standards of capacitance. Proc. IEEE **104C**, 271–280 (1957)
61. Lampard, D.G., Cutkosky, R.D.: Some results on the cross-capacitances per unit length of cylindrical three-terminal capacitors with thindielectric films on their electrodes. Proc. IEEE **107C**, 112–119 (1960)
62. Landau, L.D., Lifshitz, E.M.: Classical Theory of Fields. Butterworth-Heinemann, Amsterdam (1990)
63. Landau, L.D., Lifshitz, E.M.: Electrodynamique des milieux continus. Éditions Mir, Moscou (1990)
64. Lavrentiev, M., Chabat, B.: Méthodes de la théorie des functions d'une variable complexe. Éditions Mir, Moscou (1977)
65. Lee, E.W.: Magnetostriction and magnetomechanical effects. Repts. Progr. Phys. **18**, 184–229 (1955)
66. Lide, D.R.: Handbook of Chemistry and Physics. CRC Press Inc., Cleveland (2009)
67. López Pérez, E.y, Núñez Cubero, F.: 100 Problemas de Electromagnetismo. Alianza Editorial, Madrid (1997)
68. Lorrain, P., Corson, D.R.: Campos y Ondas Electromagnéticas. Selecciones Científicas, Madrid (1994)
69. Marsden, J.E., Tromba, A.J.: Cálculo Vectorial. Addison-Wesley Iberoamericana, Argentina (1987)
70. Marshall, S.V., DuBroff, R.E., Skitek, G.G.: Electromagnetismo, Conceptos y Aplicaciones. Prentice Hall Iberoamericana, México (1997)
71. Marruci, L., Manzo, C., Paparo, D.: Optical spin-to-orbital angular momentum conversion in inhomogeneous anisotropic media. Phys. Rev. Lett. **96**, 163905-1-5 (2006)
72. Medina, R., Porras, M.A.: Teoría Elemental de Electrostática y Corriente Continua. Servicio de Publicaciones de la Fundación Gómez Pardo, Madrid (2000)
73. Meyberg, K., Vachenauer, P.: Höhere Mathematik 2. Springer, Berlin (2003)
74. Mishchenko, A., Fomenko, A.: A Course of Differential Geometry and Topology. Mir Publishers, Moscow (1988)
75. Monn, P., Spencer, D.E.: Field Theory Handbook. Springer, Berlin (1961)
76. Muniz, S.R., Bhattacharya, M., Bagnato, V.S.: Simple analysis of off-axis solenoid fields using the scalar magnetostatic potential: application to a Zeeman-slower for cold atoms. Am. J. Phys. **1**, 1–6 (2012). arXiv:1003.3720v1
77. Nelson Jr., R.D., Lide Jr., D.R., Maryott, A.: Selected Values of Electric Dipole Moments for Molecules in the Gas Phase. NSRDS-NBS 10, Washington (1967)
78. Nye, J.F.: Physical Properties of Crystals. Oxford University Press, Oxford (1985)
79. Oliveira, M.H., Miranda, J.A.: Biot-Savart-like law in electrostatics. Eur. J. Phys. **22**, 31–38 (2001)
80. Osborn, J.A.: Demagnetizing factors of the general ellipsoid. Phys. Rev. **67**, 351–357 (1945)
81. Palacios, J.: Electricidad y Magnetismo. Espasa-Calpe S.A, Madrid (1959)
82. Panofsky, W., Phillips, M.: Clasical Electricity and Magnetism. Addison-Wesley, Massachusetts (1978)
83. Papadichev, V.A.: Plate-electrode electrostatic undulators: field calculation using conformal mapping. Nucl. Instr. Meth. **393**, 409–413 (1997)
84. Papoulis, A.: The Fourier Integral and Its Applications. MacGraw-Hill, New York (1962)
85. Pearson, C.E.: Handbook of Applied Mathematics. Van Nostrand Reinhold, New York (1990)
86. Popovic, B.D.: Introductory Engineering Electromagnetics. Addison-Wesley, Massachusetts (1971)
87. Post, R.F.: Un nuevo maglev, Investigación y Ciencia, pp. 61–65, marzo (2000)

88. Puente León, F., Kiencke, U.: Messtechnik. Springer, Heidelberg (2011)
89. Bayerer, J., Puente León, F., Frese, C.: Automatische Sichtprüfung. Springer, Berlin (2012)
90. Purcell, E.M.: Electricidad y Magnetismo, Berkeley Physics Course, vol. 2. Reverté, Barcelona (1992)
91. Råde, L., Westergren, B.: Mathematische Formeln. Sprnger, Berlin (1995)
92. Ras, E.: Transformadores de potencia, de medida y de protección. Marcombo Boixareu Editores, Barcelona (1991)
93. Reitz, J.R., Mildford, F.J., Christy, R.W.: Fundamentos de la teoría electromagnética. Addison-Wesley Iberoamericana, Argentina (1996)
94. Rivera Arreba, I.: Evaluation of Preisach Model of Hysteresis to Model Superconductors. Proyecto de Fin de Grado, Madrid (2014)
95. Rodríguez, M., Bellver, C., González, A.: Campos Electromagnéticos. Publicaciones de la Universidad de Sevilla (1995)
96. Sadiku, M.N.O.: Elements of Electromagnetics. Oxford University Press, New York (2010)
97. Santos, E., Gonzalo, I.: Microscopic theory of the Aharonov-Bohm effect. Europhys. Lett. **45**, 418–423 (1999)
98. Schrüfer, E.: Elektrische Messtechnik. Hanser 9., aktualisierte Auflage, München (1992)
99. Schrüfer, E.: Signalverarbeitung. Hanser 2. Auflage, München (2007)
100. Schwartz, L.: Course d'analyse, vol. I, II. Hermann, Paris (1967)
101. Scott, E.D.: An Introduction to Circuit Analysis: A Systems Approach. McGraw-Hill Inc, New York (1987)
102. Sears, F.W., Zemansky, M., Young, H., Freedman, R.: Física, vol. II. Pearson Educación, México (1999)
103. Serway, R.A.: Electricidad y Magnetismo. McGraw Hill, México (1997)
104. Shadowitz, A.: The Electromagnetic Field. Dover Publications Inc., New York (1988)
105. Singh, S.: Refractive index measurement and its applications. Physica Scripta **65**, 167–180 (2002)
106. Sneddon, I.N.: Elements of Partial Differential Equations. Dover, Mineola (1957)
107. Spiegel, M.R.: Vector Analysis. Schaum's Outline Series. McGraw-Hill, New York (1959)
108. Stone, J.M.: Radiation and Optics. McGraw-Hill, New York (1963)
109. Stoner, E.C.: The demagnetizing factors for ellipsoids. Phil. Mag. **36**, 803–821 (1945)
110. Stratton, J.A.: Electromagnetic Theory. McGraw-Hill Book Company, New York (1941)
111. Sukhov, S., Haefner, D., Kajorndejnukul, V., Dogariu, A.: Surface-induced optical anisotropy of inhomogeneous media. Photonics Nanostructures **11**, 65–72 (2013)
112. Sukhov, S., Kajorndejnukul, V., Brocky, J., Dogariu, A.: Forces in Aharonov-Bohm optical sensing. Optica **1**, 383–387 (2014)
113. Sveshnikov, A., Tikhonov, A.: The Theory of Functions of a Complex Variable. Mir Publishers, Moscow (1973)
114. Thomson, A.M., Lampard, D.G.: A new theorem in electrostatics and its application to calculable standards of capacitance. Nature **177**, 888 (1956)
115. Tipler, P.A., Mosca, G.P.: Physics for Scientists and Engineers (2003)
116. Vallette, A., Indelicato, P.: Analytical study of electrostatic ion beam traps. Phys. Rev. Spec. Top.-Accel. Beams **13**, 114001.1–6 (2010)
117. van der Pauw, L.J.: A method of measuring the resistivity and Hall coefficient on lamellae of arbitrary shape. Phylips Techn. Rev. **20**, 220–224 (1958)
118. Wangsness, R.K.: Electromagnetic Fields. Wiley, New York (1986)
119. Zheng, G., Pardavi-Horath, M., Huang, X., Keszei, B.: Experimental determination of an effective demagnetization factor for non-ellipsoidal geometries Vandlik J. J. Appl. Phys. **79**, 5742–5744 (1996)
120. Zimmerman, N.M.: A primer on electrical units in the Système International. Am. J. Phys. **66**, 324–331 (1998)

Index

A
Acceleration, 639, 642
Aharonov-Bohm effect, 228
Air gap, 324–326
Alzheimer, 646
Ampere, 168
Ampère–Maxwell Law, 600
Ampère's law, 235, 440, 468, 599
    for H, 323
Analytic continuation, 432
Analytic function, 433, 436
Analyzer, 782
Anderson, C., 646
Angular frequency, 640
Antiferromagnetism, 318
Antimatter, 646
Aston's mass spectrometer, 659
Atomic currents, 314
Axisymmetric geometry, 431

B
Bainbridge mass spectrometer, 633
Beta plus decay, 646
Betatron, 635, 650, 662
Biaxial, 764, 765
Biology, 646, 665
Biot–Savart, 254, 257
    law, 76, 229
Birefringence, 765
Bloch wall, 319
Bohr's model, 313
Boundary conditions, 317
    Dirichlet, 420
    mixed, 420
    Neumann, 420
    Robin, 420

Branch, 179
Bravais's lattice, 750
Breakdown voltage, 125
Brewster's angle, 717, 782
Brewster's scattering angle, 782
Bruggeman formula, 468

C
Calcite, 762
Capacitance, 129, 153
    coefficients of, 129
    equivalent, 130–132
    of cylindrical capacitor, 139
    of parallel-plate capacitor, 135
    of spherical capacitor, 137
Capacitor, 129, 575, 577, 587, 589, 609
    cylindrical, 137
    in parallel, 130, 155
    in series, 131, 155
    non-parallel plates, 449
    parallel-plate, 130, 153, 448
    spherical, 136
Cauchy–Riemann conditions, 434, 437, 438, 484
CERN, 225
Charge, **67**
    density
        line, 68
        polarization surface, 123, 132, 140, 147, 149, 151, 157, 161
        polarization volume, 123, 132, 140, 141, 147, 149, 151, 157, 161
        surface, 68
        total, 124
        volume, 67
    distribution, 430

polarization, 122
the principle of conservation of, 67, 150
Chebishev polynomials, 447
Chemistry, 646, 665
Circular metallic ring, 284
Circular right-hand polarization, 780
Circulation, 10, 13, 16, 51
Cloud chamber, 646
Coefficients
  reflection, 717
  refraction, 717, *see also* Fresnel coefficients
Complex
  electric field, 438, 441
Complex analysis
  method, 431
Complex potential, 436–438, 440, 483, 485, 487, 488
  magnetic field, 440
Computerized tomographic, 646
Computerized tomography (CT), 646
Conducting polygon, 289
Conducting sheet, 613
Conductive sheet, 657
Conductivity, 75, 171, 421
Conductor, 75, 421
  in electrostatic conditions, 76
Cone poles, 324
Conformal mapping, 433
Conformal transformations, 434
Conical solenoid, 287
Conservation of energy, 570, 573, 645
Conservative field, 21, 23
The constitutive equation, 125
  for linear dielectrics, 125
    isotropic, 125
Contour line, *see* isoline
Coordinates
  Cartesian, 1
  cylindrical, 1
  spherical, 2
Core arms, 326
Cosmic rays, 646
Cotton-Mouton's effect, 765
Coulomb's law, 68
Counter-electromotive force, 176
Critical angle, 716
Cross section, 326
Crystal, 764, 765
  anisotropic, 762
  biaxial, 762
  uniaxial, 763, 764, 778
Crystal systems, 750

Crystallographic point groups, 750
Curie's principle, 764
Curie temperature, 318
Curl, 13, 38, 39, 41
Current, 167
  density, 167
  direct, 169
  intensity, 168
  mesh, 181
Current density
  surface, 314
  tangential, 260
  volumetric, 314
Current of electrons, 278
Current of protons, 272
Currents
  bounded, 315
  free, 315
Curve, 6, 30, 34
  closed, 7, 225
Curved tube, 658
Curves of constant potential, 484
Cyclotron, 634, 649, 661, 664
Cylindrical capacitor, 435
Cylindrical conductor, 242

D

Del operator, 12
Demagnetized state, 327
Demagnetizing factor, 331, 338
Demagnetizing field H_d, 327
Demagnetizing tensor, 330
Deuteron, 661
D field, 124
Dielectric, 75, 121
  breakdown, 125, 139
  constant, 126, 751, *see also* permittivity, relative
  sphere, 460
  strength, 125, 148
Dielectric sheet, 612
Differential
  length, 4, 9
  surface, 4, 16, 17
  volume, 4
Dipole, 758
  electric, 74
  induced, 122
  moment, 75, 121, 122, 758
    permanent, 122
  point, 75
Dirac, P., 646

Index

Directional derivative, 11, 13
Dirichlet, 420
   boundary conditions, 426, 430, 432, 492, 493
Discotics, 751
Distributed charges, 569
Divergence, 14, 39, 45, 79, 439
   of M, 327
Divergence theorem, 15, 22, 58
Domain
   transformed, 432
Double refraction, 765

E

Effective field, 327
Electric charge density ρ, 600
Electric displacement D, 317, *see also* D field
Electric field, 439, 455, 457, 576, 580, 584, 613–615, 627, 628, 632, 635, 641, 643, 647, 648, 654
   complex potential, 436
   dielectric sphere, 465
   efective, 511, 512
   finite wire, 81
   in plane waves, 671
   induced, 513
   infinite cylinder, 90
   infinite plate, 93
   infinite wire, 80, 484
   metallc sphere, 468
   metallic spherical crown, 89
   spherical crown, 85
Electric force, 661
Electric potential, 455
   properties, 420
Electric susceptibility, 125
Electromagnet, 325, 328
   cylindrical, 398
   variable cross-section, 393, 394
Electromagnet arms, 323
Electromagnetic induction
   examples of application, 514, 515, 517, 519
   general law, 518
   in moving contours, 515
   in stationary contours, 512
Electromagnetic waves, 667
   in dielectric media, 667
   in free space, 667
Electromotive force, 584
Electromotive force (e.m.f.), 174, 175, 192, 511

Electrooptical effect, 765
Electrostatic, 425
Electrostatic energy, 568, 569
Electrostatic field, 70, 576
Electrostatic force, 570, 580, 583
Eletromagnet, 323
Energy, 582
   in electromagnetic waves, 674
   of plane waves, 675
   reflected, 718
   transmitted, 718
Energy density, 572
   in electromagnetic fields, 674
The equation of continuity, 169
Equipotential surfaces, 5, 428
Equivalent solenoid, 381
Ergodicity, 224
Exchange force, 319
Extraordinary wave, 757

F

Farad, 128, 130
Faraday cage, 102
Faraday's law
   general form, 518
   in moving contours, 515, 516
   in stationary loops, 512, 513
Ferrimagnetism, 318
Ferromagnetic
   material, 319, 590, 626
Ferromagnetism, 318
Fictitious charge, 430
Field, 5
   Escalar, 5
   vector, 5
Finite difference method (FDM), 441, 443, 491, 497, 500
Finite element method (FEM), 445, 447
Finite solenoid, 275
   force, 304
   off-axis magnetic field, 308
First magnetization curve, 319, 326
Flat capacitor, 606, 616
Flux, 11, 14, 16–18, 32, 48–50, 57, 79, 147, 157
   electric field, 439
Flux leakage, 326
Force, 574, 575, 580, 584, 635, 645, 656, 662
Forces on currents, 230
Fourier
   series, 676
   transform, 676

Free current
  non-homogeneous, 369
Fresnel coefficients, 716, 717

G
Galerkin method (GM), 446
Gauss
  divergence theorem, 439
  integral theorem, 427
Gauss theorem, 441, *see also* Gauss' law
Gauss' law, 73, 79, 80, 85, 90, 94, 108, 124, 156
  differential form of, 74
  for D, 124, 126, 127, 134, 137–139, 143, 144, 150, 152, 153, 158, 162
Gauss's theorem, 599
Gaussian surface, 73
Generator, 582
Gradient, 12, 29, 31, 34, 36, 45
  in Cartesian coordinates, 12
  in cylindrical coordinates, 30
  in spherical coordinates, 30
Green
  function, 427
  function method, 425
  properties, 425
Green's theorem, 20

H
Hall effect, 631, 632
Hall voltage, 646
Harmonic function, 427
Harmonic plane wave, 670
  electromagnetic fields, 670
  general expression, 670
  linearly polarized, 670
Harmonic wave, 669
  angular frequency, 669
  frequency, 669
  period, 669
  phase, 669
  wavelength, 669
  wavenumber, 669
Helicity, 224
Helix, 656
Helmholtz's theorem, 23
Heterogeneous current density, 369
Heterogeneous magnetization, 362, 385
Hollow cylinder, 286, 366
Hollow magnetized bar, 350, 352
Hollow metallic cylinder, 294
  vector potential, 296

Hysteresis curve, 327, 374
Hysteresis loop, 321, 322, 326, 594, 596
  family, 322

I
Inclusions, 466
Index of refraction, *see* refractive index
Indicatrix, 765
Induced anisotropy, 765
Inductance
  mutual, 520
  self, 519
Induction, 511
Infinite metallic wire, 275
Infinite solenoid, 282
Infinite wire
  charge density, 438
Influence
  coefficients of, 129
  total, 129
Intensity, 674
  of electromagnetic wave, 675
Intersection surfaces, 226
Inverse Fourier transform, 676
Irrotational field, *see* conservative field
Isoline, 5
Isotropic, 749
ITER, 225

J
JET, 225
Joule's
  effect, 173, 175, 177, 193, 194, 196, 201, 203
  law, 173

K
Kantorovich method (KM), 446
Kennelly's theorem, 183, 207, 212, 217, 218
Kerr's effect, 765
Kinetic energy, 636, 661
Kirchhoff, 591
Kirchhoff's circuit laws, 180, 204, 206, 208, 217, 221

L
Laplace
  equation, 74, 76, 128, 419, 420, 422, 427, 434, 443, 491
Laplacian, 20, 42, 47
Lattice, 750

Lawrence, Ernest, 634
Laws of reflection and refraction, 715
Least squares method (LSM), 446
Legendre polynomials, 447
Lenz's law, 513
Level curves, 437
Level surfaces, *see* equipotential surfaces
Linearly polarized light, 773
Liquid crystal, 750
   cholesteric, 751
   nematics, 751
   smetics, 751
Lorentz force, 627, 629, 632, 637

M

Magnesium fluoride, 762
Magnet, 625
Magnetic
   resonance imaging, 224
   circuit, 323, 328
   dipole, 231
   domain, 321
   field, 223, 226
      surfaces, 224
   field lines, 224
   material, 326
   moment, 314
   permeability, 318
   scalar potential, 232
   substances, 318
   susceptibility, 318
Magnetic energy, 572, 578, 586
Magnetic field, 605, 619, 629, 635, 640, 641, 647, 662, 665
   circular ring, 250
   coil, 251
   conical solenoid, 287
   earth, 224
   finite solenoid, 258
   finite wire, 254
   infinite solenoid, 262–264
   infinite wire, 56, 234, 257, 486, 497
   metallic infinite wire, 234
   metallic polygon, 289
   off axis, 231
   rotating disk, 306
   spherical solenoid, 302
   toroidal solenoid, 244
   variable cross-section solenoid, 300
   wire, 254
Magnetic field B, 315
   remanent, 321, 329

Magnetic field H, 315, 316, 318
   circulation, 317
   coercitive, 321
   divergence, 327, 336
Magnetic field lines, 268, 501
Magnetic force, 661
Magnetic material, 589
Magnetic potential, 421, 499
Magnetism
   intense, 318
   weak, 318
Magnetization, 313, 330
   hollow cylinder, 350
   homogeneous, 332
   rotation, 320
   saturation, 320
Magnetization current
   surface, 315
   volumetric, 315
Magnetization M
   divergence, 337
Magnetization rotation, 321
Magnetized
   ball, 381
   bar, 335
   cone, 377
   matter, 440
Magnetohydrodynamics, 224
Magnetometer, 253
Magnetomotive force, 325
Magnetooptical effect, 765
Magnetostatic, 420, 425
   energy, 319
   field H, 327
Magnetostatic field, 223
Magnetostatic potential, 497, 498
Magnetostriction, 319
Mapping, 435
Material
   equations, 326
   negative, 761
   positive, 762
Material equation, 326
Maxwell's equations, 599, 600, 622
   dielectric media, 667
Maxwell's laws, 602, 604
Maxwell-Wagner formula, 466
Mechanical stress, 765
Medicine, 665
Mesh, 179
Mesh current method, 181, 213
Metallic plates, 618
Metallic sphere, 467

Metamagnetism, 318
Method of images, 428, 430, 471, 473, 475, 478, 480
Microspheres, 466
Möbius mapping, 488
Monochromatic radiation, 776
Motional e.m.f., 515
Multimeter, 261
Multiply-valued function, 432
Multivalued function, 499
Mutual inductance, 520
    coefficient, 521, 522
    example of application, 522

N

NaCl, 765
Negative uniaxial material, 774
Network, 179
    active, 186
    passive, 182
Neumann, 420
    boundary conditions, 426, 430, 432, 433, 444
    function, 425–427, 482, 483
Neumann's principle, 764
Node, 179
Non-homogeneous finite solenoid, 298
Normal vector, 9, 11, 14, 16–18, 32
Normalizing condition, 427
Norton
    equivalent, 186, 215, 219
    theorem, 186
Numerical techniques, 441

O

Ohm, 171, 175, 176
Ohm's law, 172, 198, 201, 223, 570, 593
    anisotropic lineal conductors, 171
    for a branch, 179, 201, 202, 205, 213, 218
    for a circuit, 178, 193, 195, 196, 202
    isotropic lineal conductors, 171
    non linear conductors, 171
Open surface, 225
Operating straight line, 326, 329, 345
    slope, 329
Optic axis, 755, 767, 768
Ordinary waves, 756
Ostrogradski-Gauss theorem, *see* divergence theorem

P

Parallel conducting plates, 654
Parallel-plane capacitor, 608
Parallel-plate capacitor, 574
Paramagnetic bar, 371
Paramagnetic material, 624
Paramagnetism, 318
Parkinson, 646
Permanent magnet, 328, 391
    construction, 391
Permittivity
    absolut, 126
    of free space, 69
    relative, 126
Phase velocity, 669, 751, 759
Photoelastic effect, 765
Photoelectric effect, 650
Photomultiplier-scintillator, 646
Plane-parallel, 618
Plasma, 627
Pockels's effect, 765
Poisson
    equation, 74, 229, 419, 426
Polarization, 673, 751
    ellipse, 673
Polarization of plane waves
    circular, 672
    elliptic, 672
    linear, 672
Polarization vector, 122
Polarizer, 777
Pólya field, 438–441
Positive charge, 645
Positron, 646
Positron Emission Tomography (PET), 646
Potential
    coefficients of, 128
    difference, 72, 142, 150, 152, 172, 202, 204, 218, 220, 221
    electrostatic, 71
        of a finite wire, 84
        of a point charge, 72
        of a spherical crown, 87
    field, 21
    newtonian, 45, 48
    scalar, 21, 23–25, 51, 58, 227, 428, 437
Potential energy, 567, 637
Power
    consumed, 177, 201, 203, 216
    converted, 177, 201, 203, 207, 223
    density, 173
    electric, 173
    generated, 175, 203

Index 805

supplied, 175, 193, 203
Poynting
    theorem, 674
    vector, 674, 675, 754, 759, 762, 783
Propagation vector, 669
Proton, 573, 641, 652, 661

Q

Quadric, 753
Quantum Hall effect, 632
Quartz, 762, 769

R

Rayleigh formula, 466
Reflectance, 718
Reflection of plane waves, 715
Refraction
    law of, 715, *see also* Snell's law
Refraction of plane waves, 715
Refractive index, 466, 715
Region
    multiply connected, 6
    simply connected, 6
    star-shaped, 6
Relativistic dynamics, 663
Resistance, 172, 208
    equivalent, 182, 183, 185, 207, 210, 216, 218, 222
    in paralell, 182
    in series, 182
    in star, 183
    in triangle, 183
    internal, 175, 176
Resistivity, 171
Retarders, 766
Riemann
    mapping theorem, 436
    surface, 432
Ritz–Rayleigh method (RRM), 446, 447
Robin, 420
Rotating hollow cylinder, 286
Rotating ring, 284
Rotational, *see* curl
Rotational symmetry, 235, 245
Ruby, 762
Rutile, 762

S

Sapphire, 762
Saturation magnetization, 320
Scalar potential, 622

Second magnetization curve, 319, 321, 327
Self-inductance, 519
    coefficient, 520
    example of application, 520
Semi-analitical methods, 448
Semiconductor, 76
Semi-infinite conducting wire, 292
Semi-infinite solenoid, 282, 283
Semi-infinte wire, 292
Separation of variables method (MSV), 422, 449, 452, 460
Simple domain, 436
Single-valued function, 432
Singularity, 486
Sink point, 15
Snell's law, 716
Solenoid, 587, 609
    equivalent, 334
    finite, 248
    infinite, 262
    toroidal, 245
Solenoidal field, 22, 23
Source point, 15
Space groups, 750
Special relativity, 636
Spherical surface, 430, 578
Spin Hall effect, 765
Stokes's theorem, 15, 21
Superposition principle
    for electric field, 71, 77, 98, 108
    for electrostatic forces, 69
    for electrostatic potential, 72, 78
Surface, 7, 36
Susceptibility, 752
Symmetry
    broken, 435
Synchrotron, 651, 665

T

Temperature coefficient, 171
Tesla, 224
Theory of relativity, 637
Thermal internal energy, 595
Thévenin
    equivalent, 186, 215, 219, 223
    theorem, 186
Third magnetization curve, 319, 321
Thompson-Lampard theorem, 507
Toroidal core, 328
Toroidal solenoid, 270, 571, 590
Torus, 244, 245
Tourmaline, 762

Translational symmetry, 234, 497
Transmitance, 719
Trefftz method (TM), 446

U
Uniaxial, 764
   crystals, 761
   material, 767
   plate, 769, 773
   slab, 783
Uniqueness theorem, 428
Univalued, 434

V
Variational model, 446
Vector lines, 5
Vector potential, 23, 26–28, 60, 62, 227, 253, 266, 267, 269, 314, 371, 468, 513, 621
   circular ring, 252
   infinite wire, 269
Voltage, 523

W
Wave
   plane, 670
   spherical, 670
Wave equation, 667
   electromagnetic fields, 667
   one-dimensional, 668
   three-dimensional, 668
Wave plate, 766
Wave vector
   surface, 763
Wave velocity, 668
Wavefront, 670
White light, 773
W plane, 435

Y
Yoke, 326

Z
Zeeman slowing technique, 310
Z plane, 435

Printed by Printforce, the Netherlands